ILLUSTRANS
COMMODA
VITÆ

THE ARCHIVES
OF THE
ROYAL INSTITUTION OF GREAT BRITAIN

Published in Facsimile

Under the Editorial Supervision of
FRANK GREENAWAY, M.A., Ph.D., F.R.I.C.
Honorary Reader in the History of Science

THE ARCHIVES
OF
THE ROYAL INSTITUTION
OF GREAT BRITAIN

in facsimile

Minutes of Managers' Meetings
1799–1900

Volume V

With an introduction by

Morris Berman

Assistant Professor of History, Rutgers University

Published in association with

THE ROYAL INSTITUTION OF GREAT BRITAIN

by

THE SCOLAR PRESS LIMITED

ISBN
0 85967 225 5

Printed and published in Great Britain by
The Scolar Press Limited, Ilkley, Yorkshire,
and 39 Great Russell Street, London WC1

INTRODUCTION

In terms of actual content, Volume V would appear to be the least exciting section of the Managers' Minutes. If the first three volumes reveal the agricultural orientation of the Institution, and the fourth records the crisis of management that resulted from the R.I.'s exclusive policies, the present volume gives the minute details involved in the transition from a private to a public institution. Adjustment of Memberships, compensation of Proprietors for loss of their hereditary shares in the Institution, discussion of the proper form for nominating Life Subscribers (pp. 155–56), arrangement of new committees (e.g. p. 110) and committee rooms (p. 179), preparation of a new set of bylaws (see especially pp. 83–85 and 101–12)—the details of procedure and administration fill the first half of this volume so completely that the Institution itself almost disappears from sight. The density of the volume—449 pages for little more than four years—reflects the numerous difficulties encountered in the implementation of the new policies. The Minutes do not really return to "normal" until about the middle of the volume, when matters of content, such as the choice of lectures to be held, once again come to the fore.

The necessity for this transition is epitomized by the following excerpt from a Committee Report, quoted on page 15: "it is impossible to go on with the least hope of success under the present Charter & Constitution". But it was one thing to alter the R.I. formally, by Act of Parliament, and another to change its economic and administrative basis. Embedded in the details, then, is an issue of the utmost importance: the attempt to make a scientific institution financially self-sustaining. If the R.I. had attempted to break with the Royal Society's tradition of science-as-polite-learning by adopting an entrepreneurial ideology more attuned to the new industrial age, it now found it necessary to break with the traditional method of scientific support. In neither case, of course, did it succeed as completely as it would have wished. Patronage was becoming increasingly anachronistic in a capitalist framework, but it was only very slowly discarded as a method of financial support. The changes of 1810–14 represent the beginnings of an intermediate stage, that of widening the base of patronage and diluting the exclusiveness of control. Economic independence took nearly three more decades to achieve, and in the interim the Institution's solvency was more often than not the product of generous but random endowments.

Davy was asked by the Managers to discuss the proposed changes with the Proprietors and Subscribers in the lecture theatre on 3 March 1810 (p. 37). In order that "the income of the Institution may be made permanent and sufficient", he told his audience, it "is proposed that the Government [of the R.I.] shall be open and liberal; and that Laws and Regulations may flow from the Members." If the Institution was previously exclusive it would now make the effort, as far as the diffusion of science was concerned, "to offer every facility of admission to the public . . .". "Our doors", he continued, "are to be open to all who wish to profit by knowledge . . . ." It was far from being so straightforward as that. Davy's speech also praised the selective policy of the R.I., and the vagueness and equivocation of the address reflected the cross-purposes of the Institution which the Managers were struggling to resolve. The seemingly uninspired details of the first half of this volume are in fact a record of this struggle: reactions to, and enactment of, the new policies.

The Act of Parliament formalizing these changes (50 Geo. III, c. 51), which is referred to at various points in the early pages of the volume, embodied three measures designed to open the doors of the R.I., as well as its governorship, to a large clientele. The first was the abolition of the Hereditary Proprietorships which, the text of the bill stated, "have operated as Impediments to the general Interest and Co-operation which appear to be essential to such a Publick National and Permanent Establishment . . .". A Proprietor would now be a Member for life only, and would receive as compensation for the loss of the hereditary status of his Membership the privilege of designating his wife, child, or any blood relation a Life Subscriber. If, on the other hand, he did not wish to be made a Member, he would be remunerated for his Proprietorship, which an actuary employed by the R.I. valued at roughly £42 (pp. 59–60).

The second change was the opening up of the election system. During the first decade, governors served for one, two, or three-year terms, and were immediately eligible for re-election directly their term of office expired. In this way, virtually the same men remained in power for the entire period. On the new system, as of 1 May 1811, and every first of May thereafter (unless it fell on a Sunday), the Members would elect from amongst themselves fifteen or more Managers and fifteen or more Visitors, the major part of each group not having held either office the previous year (pp. 22–23). Finally, the third measure abolished the power of the Managers to elect new Proprietors (i.e. Members), and gave this privilege to the Members alone.

If the first measure promised to end the hereditary nature of the R.I., and the third opened the Membership up to popular rather than exclusive selection, it was the second measure in particular that held out the greatest hope for change. For if more than fifty per cent of each set of annually elected governors had not served the previous year, the control of the Institution by the "old guard" would necessarily be attenuated. In fact, the new election system did have an immediate impact. Between 1799 and 1811, more than two-thirds of the governors could be classed as agricultural improvers; between 1811 and 1816, their number dropped to slightly more than forty per cent, and declined steadily thereafter. Professional groups—lawyers, physicians, and men who might be called, even at this early date, "scientific administrators"—comprised nearly one-fifth of the governorship 1811–16, and emerged as the largest single category after 1826. There was, however, a great deal of overlap. If we consult the house-lists, or election slates, in this volume (pp. 208–9, 292–93, 369–70, and 438–39), we find many names surviving from the first decade; to be precise, more than forty per cent of those who held office at some time during the period 1811–16 had done so during 1799–1811 as well. There are other signs of continuity, such as the apparent control of the Institution by the landed interest down to 1825; but the change is more striking than the preservation of the "old guard". The R.I., at this stage, was already experiencing the winds of reform.

With the relative decline of landed influence at the R.I. came a corresponding decrease of agrarian concerns. It is true that Davy attended meetings of the Board of Agriculture through 1817, that Professor John Millington, an active member of the Board, delivered a course of lectures at the R.I. in 1819 on agricultural machinery, and that Sir John Cox Hippisley, a governor from the earliest days, offered the Board, when threatened with dissolution in 1820, a home with the R.I. But one cannot help thinking that events of this sort were now more symbolic than real. The deliberate use of science for specifically agricultural projects, which had characterized the first decade, was absent during the second; and as early as 1810 the Managers found it necessary to cancel the subscription to *The Agricultural Magazine* because it was no longer being read (p. 12). The shift in policy was also accelerated by one of England's worst agricultural depressions, which began in January 1814, and in the face of which the landed interest chose to push through the Corn Law rather than rely on chemistry. Finally, if some of the governors did retain hopes of continuing to use Davy for their private agricultural concerns, they would find that he was no longer strictly an Institution employee. His practical work was turning into national service: ventilating the House of Lords (pp. 28–30), inventing the miners' safety lamp (1815), and working on the protection of copper sheathing of ships for the Navy (1824), to name Davy's most famous efforts in this direction. The very success he had achieved as a technical problem-solver and as a "salesman of science" at the R.I. made him increasingly unavailable for Institutional service. In 1812 he ceased lecturing there and continued on as Professor of Chemistry only in an honorary capacity (p. 299). His function at the R.I. was now almost entirely administrative, and he began to serve as a Manager, chairing the meeting for the first time during this period (p. 388). For various reasons, then, the agricultural orientation of the R.I. was fast disappearing.

The Institution's policy of greater accessibility and diversification was most noticeable in the lecture programme, which reveals, for the first time, an attempt to cater to middle-class needs. In January 1811 the Managers made it possible for anyone who wished to attend the lectures of one professor for a single season to do so at a cost of two guineas (p. 175). Even more significant was the decision to service the medical community. During this time the Society of Apothecaries was waging a campaign for higher professional standards, which included demands for additional scientific education for apprentices and medical students. Informal courses and private schools began to multiply, and contemporary medical writers tended to stress the value of science for medicine. The R.I. sought to take advantage of these events by deciding that "Medical Students alone be admitted to the Lectures without becoming Annual Subscribers and to those courses only which are connected with their Professional Studies on the payment of Two guineas for any single Course" (pp. 250–51). "Medical Students to Professor Davy's Lectures" are listed on pages 265, 275, and 278. It was Brande, however, himself an apothecary and a professor at the Apothecaries' Society, who took the crucial step, requesting permission to transfer the Chemical School in Windmill Street, where he also lectured, to the laboratory of the Royal Institution. The Managers responded with alacrity, predicting that the move would "be highly beneficial to the interests of Science and of the Institution" (p. 394). Lectures on physiology were also made available by the incredibly versatile Peter Mark Roget, physician, inventor, scientific consultant, and compiler of the *Thesaurus* (pp. 269, 305, 380, 406, and 409).

Other aspects of the lecture programme also reveal a new experimental attitude, an attempt to grapple with popular as well as middle-class concerns. The Managers asked the founder of England's working men's colleges, George Birkbeck, to lecture on natural philosophy (p. 254). They reached out to other institutions, such as the East India Company's Haileybury College (see introduction to Volume IV) for a lecturer on the mechanical arts, and to the London Institution for a model of the steam engine (pp. 306–7, 321, and 423). One of the most colourful lectures was provided by Gregor Feinaigle, the German mnemonist, on the art of memory. At first, the Managers declined to host the lecture, since Feinaigle, as was his custom, refused to disclose his secret; but popular pressure from the Members resulted in this decision being reversed (pp. 224–29). According to the report of *The Gentleman's Magazine* for September 1811, the demonstration, which was held on 22 June before an audience of several hundred, was a stunning success. With only a few hours instruction prior to the lecture, one child was taught to multiply eight-digit numbers without pencil or paper, another to repeat the geographical location of fifty cities in degrees and minutes, on request. The Institution's Clerk of Accounts, Joseph Fincher, recited Haüy's mineralogical tables by heart. The pseudo-scientific flavour of the subject, and Feinaigle's highly controversial reputation on the Continent, made the Managers somewhat anxious, and they made a point of publicly disclaiming any endorsement of his system. Nevertheless, the lecture undoubtedly served to give the Institution some of the popularity it was seeking during its transitional phase.

Because of detailed attention to more pressing concerns, items like gifts tend to receive less emphasis in this volume than in the previous four. At one point, for example, a donation is tersely recorded as "Sundry Books" (p. 125). Some of the gifts listed, however, strongly evoke contemporary debates and events, such as David Ricardo's pamphlet on the price of gold (p. 134), Lord Elgin's memorandum on his pursuits in Greece (p. 155), and the presentation of the memoirs of the Reverend William Paley (p. 236), whose tremendously influential *Natural Theology* appeared only a few years before. The names of Subscribers also seem less illustrious, although we can record at least two prominent figures in addition to the future Lady Davy (p. 178): Francis Horner, reformer, Whig spokesman, and founder of *The Edinburgh Review* (p. 335), and the astronomer James South (p. 344).

Rutgers University
February, 1973

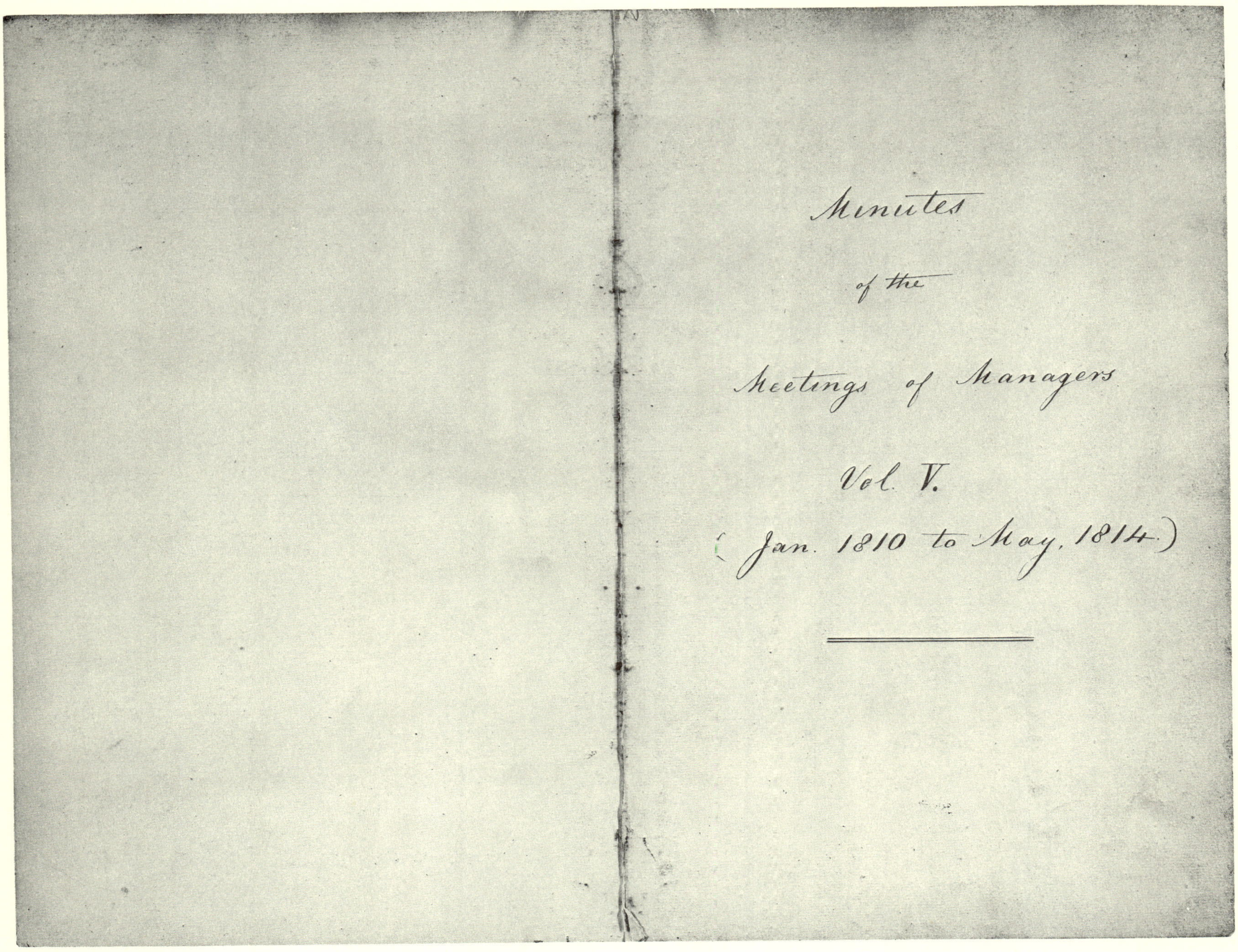

Minutes

of the

Meetings of Managers

Vol. V.

(Jan. 1810 to May, 1814)

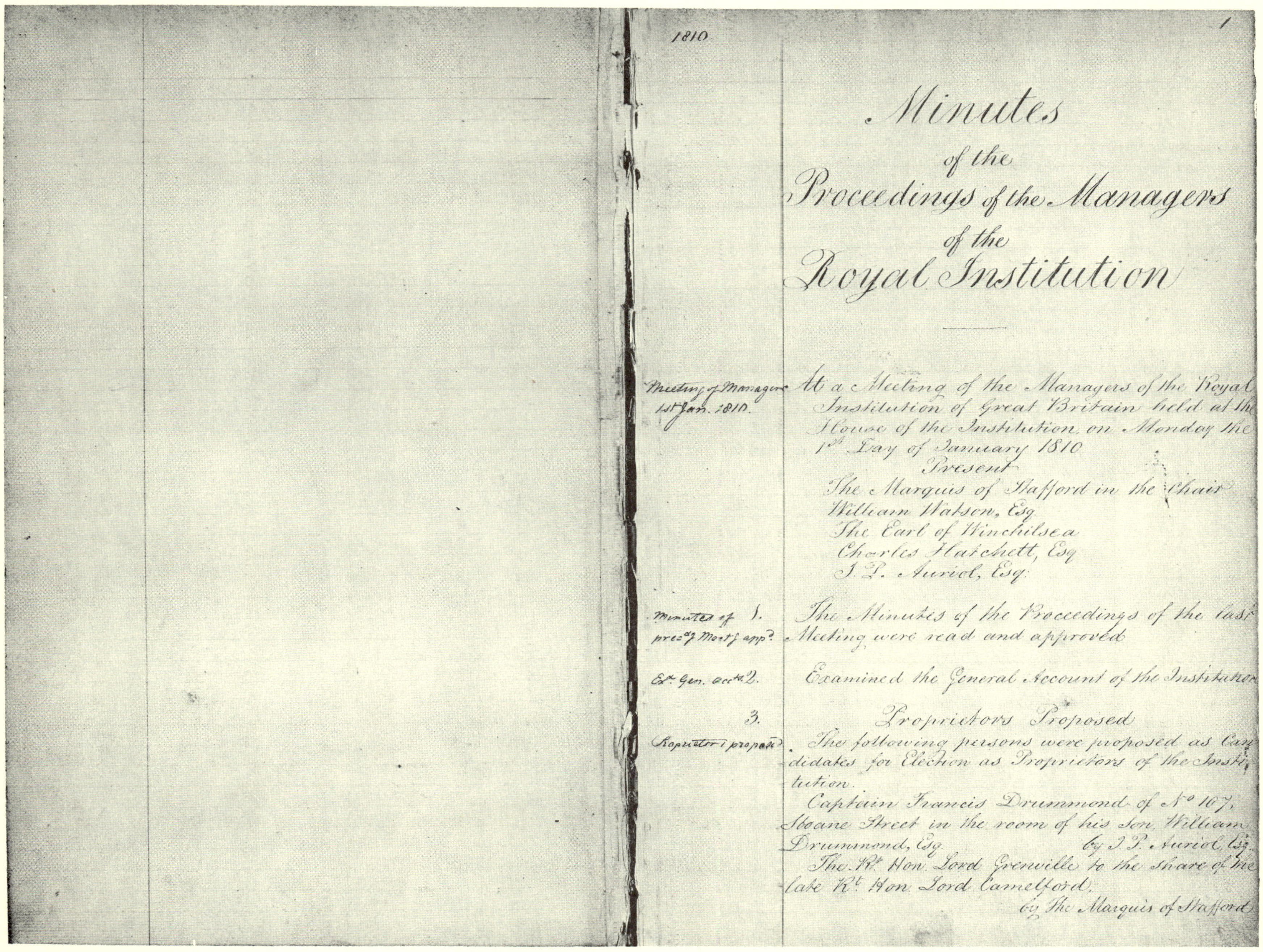

1810 1

Minutes of the Proceedings of the Managers of the Royal Institution

Meeting of Managers 1st Jan. 1810. — At a Meeting of the Managers of the Royal Institution of Great Britain held at the House of the Institution, on Monday the 1st Day of January 1810

Present

The Marquis of Stafford in the Chair
William Watson, Esq
The Earl of Winchilsea
Charles Hatchett, Esq
J. P. Auriol, Esq

Minutes of prec.g Meet.g app.d 1. The Minutes of the Proceedings of the last Meeting were read and approved

Ex.d Gen. acc.ts 2. Examined the General Account of the Institution

3. Proprietors Proposed

Proprietors propos.d — The following persons were proposed as Candidates for Election as Proprietors of the Institution.

Captain Francis Drummond of No 167, Sloane Street in the room of his Son, William Drummond, Esq — by J. P. Auriol, Esq

The Rt. Hon. Lord Grenville to the Share of the late Rt. Hon. Lord Camelford. — by The Marquis of Stafford

John Vere, Esq of Grosvenor Place in the ~~Place~~ room of Richard Debaufre Esq.

4. Ann^l. Subscribers, proposed last Meeting – elected

Annual Subscribers Elected

The following persons proposed as Candidates for Election as Annual Subscribers to the Institution at the last Meeting of the Managers were duly Elected

Richard Cooke, Esq. of Clifton, Somerset
Admiral Nugent, 14 Wigmore Street
N. A. Vigors, Esq. 33 Conduit Street
Sir John Sebright, Bar^t. M.P. Conduit Street
Adam Lymburner, Esq. 47 Bernard Street
Captain Thomas Smyth R.N. 28 Bruton Street
Nugent Kirkland, Esq 8 Bennett Street

5. Ann^l. Subscribers proposed

Annual Subscribers Proposed

The following persons were proposed as Candidates for Election as Annual Subscribers to the Institution

F. A. Moelle, Esq. 29 Suffolk Street, Charing Cross
The Rev B. W. Fletcher 10 Great Cumberland Place
M^r. _ Phibbs, New Bond Street
M^r. _ Webbe, Park Lane
J. Jones Esq. 63 Harley Street
Roger Brograve, Esq. 21 Sackville Street
Bodychan Sparrow Esq. 2 Delahay Street, West^r.
The Rev. William Beckett, Southgate, Middlesex
George Green, Esq. of York and 14 Conduit Street
Charles Madryll Cheere, Esq. 18 Charles Street Berkley Sq
Nathaniel Bishop, Esq 56 Gloster Place New Road
John Ferrie, Esq 34 Westmoreland Place City Road.
D^r. Macmichel 13 Charlotte Street Bedford Squ.
M^r. Bonnell George Thornton Hinde Street
M^r. John Major, Hinde Street
Joseph Sabine, Esq. 4 Duke S^t. Portland Place
Samuel Tolfrey, Esq. Dutchess Street

Withdrawal of member from List of Arrears on explanation

Read a Letter from Joseph Sabine, Esq. and the Managers being satisfied with his explanation,

Ordered That his name be withdrawn from the List of Arrears.

It appearing that Samuel Tolfrey Esq.

quitted this Country for the East Indies at the end of his first year's Subscription with the idea that his engagement then terminated

Ordered that his name be withdrawn from the list of Arrears.

6. Subscribers to Lectures elected.

Subscribers to the Lectures on Natural Philosophy and Chemistry Elected

The following persons ~~were~~ proposed as Candidates for Election as Subscribers to the Lectures on Natural Philosophy and Chemistry at the last Meeting of the Managers were duly elected

M^r. Moore jun^r. Lincolns Inn Fields
Lieu^t. Pemberton 2nd Life Guards

7. Subscribers to Lectures proposed.

Subscribers to the Lectures on Natural Philosophy and Chemistry Proposed

The following persons were proposed as Candidates for Election as Subscribers to the Lectures on Natural Philosophy and Chemistry.

M^r. J. W. Dobbs, 27 Hatton Garden
M^r. Edgar Dobbs 27 Hatton Garden
M^r. Thomas Babington 23 Golden Square
Justin Mac Carthy, Esq. 22 Duke St. Manchester Sq
Jonathan Morgan, Esq. 14 Upper Wimpole Street
J. R. Durham, Esq. 2 Hunter Street
J. W. Unwin, Esq. 17 Queen Street, May Fair
M^r. George Babington, 23 Golden Square
William Cadell, Esq. 66 Berwick Street
M^r. Albert, 13 Lower Grosvenor Place
D^r. Price, York Hospital, Chelsea
M^r. Alexander Gordon, 11 Old Compton Street
Henry Higginson, Esq. N°. 2 Finsbury Square

8. Minutes of Com. of Acct^s approved

Read and approved the Minutes of the Committee of Accounts of this day

Signed an Order for an advance to the Steward of Fifty Pounds.

Letter of Mr Dutens respecting present^g Catalogue of Library to Societies

Read the following Letter from The Rev^d. M^r. Dutens

M^r. Dutens presents his respects to the

Managers of the Royal Institution, and has the honour of informing them that he has directed M^r. Harris to present in their names the Catalogue of the Library of Reference of the Royal Institution to the Societies pointed out by them.

London, 19th December, 1809

10. Ordered, That a Copy of the opinion delivered by M^r. Watson on the queries submitted by M^r. Fletcher on the 24th of April last, be transmitted to M^r. Fletcher.

11. M^r. Ackermann presented N°. 13 and the Supplement of the Repository of the Fine Arts

Ordered that the Thanks of the Managers be returned for this present.

12. Declining Mr Stevens offer of dedicating a work to the Managers of R.I.

Read a Letter from M^r. Stevens.

Resolved, That the Secretary be requested to write to M^r. Stevens to thank him for the honor of his offer to dedicate his work on Music to the Managers of the Royal Institution but they are under the necessity of declining it, in consequence of a Regulation formerly made and which has been uniformly adhered to with the exception of D^r. Garnett's Lectures.

13. Ordered, That the Steward do lay in Fifteen Chaldrons of Coals for the use of the Institution.

Adjourned.

T. Bernard

Meeting of Managers Visitors & Proprietors Jan. 1st. 1810

At an adjourned joint Meeting of the Managers Visitors, and Committee of Proprietors of the Royal Institution of Great Britain, held at the House of the Institution on Monday the 1st Day of January, 1810

Present

The Earl of Winchilsea President The Marquis of Stafford William Watson, Esq. Charles Hatchett, Esq. Thomas Bernard, Esq. J. P. Auriol, Esq.	Managers
Thomas Hammersley, Esq. Scrope Bernard, Esq.	Visitors.
The Rev. Edward Forster W. H. Pepys Henry Warburton Esq James Stodart, Esq. Edward Stracey, Esq.	Proprietors

1. Minutes of prec^g Meet^g approved

The Minutes of the Proceedings of the last Meeting were read and approved.

2. R^d. Plan for improv^g R.I.

Read the Sketch of a Plan for improving the Royal Institution, and erecting it on a Permanent Foundation

Ordered to be printed & distrib^d. to Proprietors.

Resolved, That printed Copies of it be sent round to the Proprietors with the names of such Proprietors as concur in a general Approbation thereof

3. Resolved, That the Proprietors be requested by Letter, should the inclosed Plan meet with their Approbation, to signify their assent to the same at the Royal Institution

4. Committee for prep^g a Draft of Act of Parliament.

Resolved That a Special Committee be appointed to prepare a Draft of an Act of Parliament proper for carrying the said Plan into Execution; and that M^r. Bernard,

6

Mr Watson, and Mr Stracey be requested to form this Committee

Adjourned to Monday the 29th of January 1810, at Two o'Clock precisely.

1810. Jan. 8.

7

Managers' Meeting Jan. 8.

At a Meeting of the Managers of the Royal Institution of Great Britain, held at the House of the Institution, on Monday the 8 Day of January, 1810.

Present

Thomas Bernard, Esq. in the Chair
Charles Hatchett, Esq.
J. P. Auriol Esq. Secretary

Minutes of last Meeting approved. 1.—

The Minutes of the Proceedings of the last Meeting were read and approved

Exd Accounts 2.—

Examined the General Account of the Institution.

Annual Life Subscribers proposed. 3.—

Annual Subscribers for Life Proposed

Colonel Richard Scott, of No 24 Kensington Square was proposed as a Candidate for Election, as an Annual Subscriber to the Institution for Life.

Annual Subscribers elected. 4.—

Annual Subscribers Elected

The following persons, proposed as Candidates for Election as Annual Subscribers to the Institution at the last Meeting of the Managers were duly Elected

F. A. Moelle Esq. 29 Suffolk Street, Charing Cross
Rev: B. W. Fletcher 10, Great Cumberland Place
Mr. Phibbs New Bond Street
Mr. Webbe, Park Lane
J. Jones, Esq. 63 Harley, Street
Roger Brograve, Esq. 21 Sackville Street
Bodychan Sparrow, Esq. 2, Delahoy Street, Westr.
The Revd. William Beckett, Southgate Middlesex
George Green, Esq. of York, and 14 Conduit Street
Charles Madryll Cheere, Esq. 18 Chas. St. Berkley Sq.
Nathaniel Bishop, Esq. 36 Glostr Pla. New Road
John Ferrie, Esq. 34, Westmoreland Place, City Road
Dr. Macmichael, 13 Charlotte Street Bedford Squ.
Mr. Bonnell George Thornton Hinde Str. Manr. Squ.
Mr. John Major Hinde Street Manchester Sq.
Joseph Sabine, Esq. 4 Duke Street Portland Place
Samuel Tolfrey, Esq. Duchess Str. Portland Place

Annual Subscribers proposed 5.—

Annual Subscribers Proposed

The following persons were proposed as Candidates

8

1810

for Election, as Annual Subscribers to the Institution:

Mr Isaac Golledge 41, Jermyn Street
Doctor Müller, 41, Haymarket
John Knowles, Esq. Navy Office
John Frederick Archbold, Esq., 10, Titchfield St
R. Dean, Esq. 2 Bolton Row
George Pearson, Esq. George Street, Hanover Sq.
Richard Neave, Esq. Lincolns Inn
Edward D. Davenport, Esq. 2, James Street Grosv. Sq.
Lt Col. Holland, 21 Upper Cadogan Place

Subscribers to Lectrs on Nat. Phil & Chemy Elected 6.—

Subscribers to the Letures on Natural Philosophy and Chemistry Elected.

The following persons ~~were~~ proposed as Candidates for Election as Subscribers to the Lectures on Natural Philosophy and Chemistry, at the last Meeting of the Managers were duly Elected

Mr. J. W. Dobbs, 27 Hatton Garden
Mr Edgar Dobbs, 27 Hatton Garden
Mr. Thomas Babington, 23 Golden Square
Justin Mac Carthy, Esq 22 Duke Street Manchester Sq.
Jonathan Morgan, Esq. 14 Upper Wimpole Street
J. R. Durham, Esq. 2 Hunter Street
J. W. Unwin Esq. 17 Queen Street, May Fair
Mr. George Babington, 23 Golden Square
William Cadell, Esq. 66 Berwick Street
Mr. Albert, 13 Lower Grosvenor Place
Dr. Price, York Hospital, Chelsea
Mr Alexander Gordon, 11 Old Compton Street
Henry Higginson, Esq. 2, Finsbury Square

Subscribers to Lectures on Nat. Phil. proposed. 7.—

Subscribers to the Lectures on Natural Philosophy and Chemistry Proposed

The following persons were proposed as Candidates for Election as Subscribers to the Lectures on Natural Philosophy and Chemistry

Mr W. Carter, Oriel College, Oxford
Mr Henry Earle, Hanover Square
Mr John Collet junr. Highbury Grove
The Rev. George Stracey, 28 Davies Street

Jan. 8

9

Appd. Minutes of Accts 8.—

Read and approved the Minutes of the Committee of Accounts of this Day

£80 to be paid to Mr Pond for Apparatus for his Lectures 9.—

Ordered that Forty Pounds be paid to Mr. Pond for the Expense of Apparatus purchased for the use of his Lectures for the last year, and a further sum of Forty Pounds for the charge of the present year:

Signed an Order accordingly

Signed an Order for the payment of Mr. Sturt's Salary for the quarter due at Michaelmas Last

Query conc. Mr J. Vere. 10.—

Resolved That the Assistant Secretary inquire whether Mr. John Vere, mentioned as a proposed Transferee of a Share in the Royal Institution, has attained the age of 21 years?

Adjourned.

WWatson

Meeting of Managers. 22nd. Jan. 1810.

At a Meeting of the Managers of the Royal Institution of Great Britain, held at the House of the Institution on Monday the 22 Day of January 1810

Present

William Watson, Esq. in the Chair
John Symmons, Esq.
Sir John Cox Hippisley, Esq.
J. P. Auriol, Esq. Secretary

Minutes of last Meetg approved. 1.— The Minutes of the Proceedings of the last Meeting were read and approved

Exd. Gen. Accts. 2.— Examined the General Account of the Institution

Annl. subscriber for life elected. — Annual Subscriber for Life Elected
Col. Richard Scott, 24 Kensington Square, proposed as a Candidate for Election as an Annual Subscriber for Life, at the last Meeting of the Managers was duly Elected

Annl. Subscribers elected. 3.— Annual Subscribers Elected
The following persons, proposed as Candidates for Election as Annual Subscribers to the Institution at the last Meeting of the Managers, were duly elected
Mr. Isaac Golledge 41, Jermyn Street
Dr. Müller 41 Haymarket
John Knowles, Esq. Navy Office
J. Frederick Archbold, Esq. 10 Titchfield Street
R. Dean, Esq. 2 Bolton Row
George Pearson Esq. George Street, Hanover Square
Richard Neave, Esq. Lincolns Inn
Edward D. Davenport, Esq. 2, James Street Gros. Sq.
Lieut. Col. Holland 21 Upper Cadogan Place

Annl. Subscribers proposed. 4.— Annual Subscribers Proposed
The following persons were proposed as Candidates for Election as Annual Subscribers to the Institution
William Sawyer, Esq. 32 Craven Street Strand
Robert Allen, Esq. Blackheath, & Fentons Hotel
Inigo F. Thomas, Esq. Lower Brook Street
William Seymour, Esq. Margaret Str. Cavendish Sq.

Subscribers to Lectr. on Nat. Ph. & Chem. Elected. 5.— Subscribers to the Lectures on Natural Philosophy and Chemistry Elected
The following persons, proposed as Candidates for Election as Subscribers to the Lectures on Natural Philosophy and Chemistry at the last Meeting of the Managers were duly Elected
Mr. W. Carter, Oriel College, Oxford
Mr. Henry Earle, Hanover Square
Mr. John Collet, jun. Highbury Grove
The Rev. George Straey, Davies Street

Subscribers to Lectr. on Nat. Ph. & Chem. proposed. 6.— Subscribers to the Lectures on Natural Philosophy and Chemistry Proposed.
The following persons were proposed as Candidates for Election as Subscribers to the Lectures on Natural Philosophy and Chemistry.
Mr. Edward Seymour, Margaret Str. Cavendh. Sq.
John Revere, Esq. 8 Dean Street Canterbury Sq.
Col. Brisbane, Ibbotsons Hotel

Appd. minutes of Com. of Accts. 7.— Read and approved the Minutes of the Committee of Accounts of this Day.
~~Day~~
Signed an Order for an advance to the Steward of £50, and for payment of one Quarters Salary to Midsumer, 1809 due to the following persons

Humphry Davy Esq.	100 - 0 - 0
Mr. Edmund Davy	17 - 10 - 0
Mr. William Savage	12 - 10 - 0
Mr. Matthew Casingwoot	25 - 0 - 0
Mr. William Harris	45 - 0 - 0
Mr. David Brooks	13 - 2 - 6
£	213 2 - 6

Books presented. 8.— The Society of Antiquaries presented Archaeologia Vol. 16 4to and
An Index to the first 15 Volumes of Archaeologia, 4to
Dr. Buxton presented his Essay on the use of a Regulated Temperature in Winter-Cough, and Consumption, 12mo
John Disney, Esq. presented The Laws of Gaming, Wagers, Horse-racing, and Gaming-Houses, 8vo

Ordered

Ordered, that the Thanks of the Managers be returned for these Presents

9.— Mr. Davy laid the following Report before the Managers.

Purchase of some odd vols. of Journal de Physique to complete Series.

Mr. Davy has the honour of informing the Managers, that during the time when the Meetings were not attended; an opportunity occurred of completing the "Journal de Physique", by purchasing some volumes which otherwise could not have been procured, and which will very much increase the value of this set of Books; he therefore ventured to commission Mr. Harris to purchase them

Eclectic Review & Quarterly Review ordered. Agricult. Mag. to be discontinued.

Mr. Davy has the honour of informing the Managers that frequent inquiries have been made for the Eclectic Review, and the Quarterly Review: and that the Agricultural Magazine is very little used, (and if he may be permitted to give an opinion) of small merit. He therefore respectfully submits to them whether it may be expedient to order the two Reviews for the Reading Room, and to discontinue the Magazine.

Ordered.

John Symmons

Adjourned

Managers' Meeting. 29 Jan. 1810.

At a Meeting of the Managers of the Royal Institution of Great Britain, held at the House of the Institution on Monday, the 29th Day of January, 1810

Present.

John Symmons, Esq. in the Chair
William Watson, Esq.
J. P. Auriol, Esq.
Sir John Cox Hippisley, Bart.
Sir Thomas Bernard, Bart.

Minutes of last Meetg. appd. 1.— The Minutes of the Proceedings of the last Meeting were read and approved.

Exd. Gen. Accts. 2.— Examined the General Account of the Institution.

Annual Subscribers Elected. 3.— Annual Subscribers Elected.
The following persons proposed as Candidates for Election as Annual Subscribers to the Institution, at the last Meeting of the Managers were duly elected
William Sawyer, Esq. 32, Craven Street Strand
Robert Allen, Esq. Blackheath, & Fenton's Hotel
Inigo F. Thomas, Esq. Lower Brook Street
William Seymour, Esq. Margaret St. Cav. Sq.

Annl. Subscribers propd. 4.— Annual Subscribers Proposed.
The following persons were proposed as Candidates for Election as Annual Subscribers to the Institution.
Thos. Ashton Smith, Esq. M.P. 5 Portman Square
Leonard Worsley Holmes, Esq. M.P. 11 Albemarle St.
Louis Simond Esq. 45 Bryanstone Street,
William Fettes, Esq. 10 Titchfield Street
Ashurst Majendie Esq. Brick Court Temple

Subscribers to Lectr. on Nat. Ph. & Chem. Elected. 5.— Subscribers to the Lectures on Natural Philosophy & Chemistry Elected
The following persons proposed as Candidates for Election as Subscribers to the Lectures on Natural Philosophy and Chemistry at the last Meeting of the Managers were duly elected.

14 1810

1 Mr. Edward Seymour, Margaret Str Cav. Sq
John Revere, Esq. 8 Dean St. Canterbury Sq
Colonel Brisbane, Ibbotson's Hotel

Subscribers to Lect. on Nat. Ph. & Chem. proposed. 6.— Subscribers to the Lectures on Natural Philosophy and Chemistry Proposed.

The following persons were proposed as Candidates for Election as Subscribers to the Lectures on Natural Philosophy and Chemistry.

The Rev. Stephen Sloane, 8, York Street York Place
Mr. J. Simeons, 36 Albemarle Street

Minutes of Comm. of Acct. approved. 7.— Read and approved the Minutes of the Committee of Accounts of this Day.

Signed an Order for payment of the Journal de Physique, reported at the last Meeting, and for Mr. Dibdin's Ames £17-3-0.

Ord. Bell in Porter's hall, with pull to do. in Liby of Refce 8.— Resolved, That Mr. Davy be requested to give directions for a Bell to be hung in the Porter's Hall, with a pull into the Library of Reference.

Spect. Gen. Meetg of Proprietors 9.— Resolved, That a General Meeting of the Proprietors of the Royal Institution be summoned for Wednesday the 7th Day of Feby next at one o'Clock for receiving and considering the Report of the Managers, Visitors and Committee of Proprietors appointed to consider the General State of the Affairs of the Institution

10.— J. P. Fallofield, Esq requested to transfer his transferable Ticket for one year to his Son Mr. W. J. Fallofeild

Resolved That this request be complied with.

Adjourned

Spencer

Jan. 29. 15

Adjourned Meetg of Managers Visitors & Comm. of Proprietors At an adjourned Meeting of the Managers, Visitors and Committee of Proprietors of the Royal Institution of Great Britain, held at the House of the Institution on Monday the 29th Day of January, 1810.

Present.

Sir John Cox Hippisley, Bart. in the Chair.
John Symmons, Esq.
J. P. Auriol, Esq.
Sir Thomas Bernard, Bart.
William Watson, Esq.
The Earl of Morton — Managers.

Alexander Blair Esq. — Visitors.

Robert Bingley, Esq.
The Rev. Edward Forster
James Stodart, Esq.
Sir George Thomas Staunton, Bart.
Wm. H. Pepys, Esq.
Henry Warburton, Esq. — Proprietors.

Draft of Bill read and approved. 1.— The Draft of a proposed Bill prepared by the Committee for "altering and amending the Charter of the Royal Institution of Great Britain, and for enlarging and more effectually promoting the objects of the said Institution" was read and approved and ordered to lie in the Office for the inspection of the Proprietors.

[Re]port of [Co]mmittee for [inquiring] into the State [of the] Institution. 2.— Resolved That the following Report be approved; and that it be printed and sent round to the Proprietors

Report.

The joint Committee of Managers, Visitors, & Proprietors report to the Proprietors that since their appointment in May 1809 they have been occupied in examining into the State of the Institution, from its commencement to the present time; and in looking forward to its probable future Prospects, they are of opinion; that it is impossible to go on with the least hope of success under the present Charter & Constitution.—

16 1810. Jan. 29.

They have also individually & collectively taken into consideration various Plans, that have been suggested by several Members of the Committe, and are, after the maturest deliberation, most decidedly of opinion, that no effectual and permanent Plan for the advantage of the Institution can be adopted, without a previous application to Parliament, for an Act for altering and amending the Charter of the Royal Institution of Great Britain, and for enlarging and more effectually promoting the objects of the said Institution, founded upon the general principles of the "Sketch of a Plan" already sent to each Proprietor by order of the Committee: by which means the Institution will be enabled to get rid of the ineligible and objectionable parts of the present Charter; and be empowered to form a system of Government, liberal in its principles, and extensive in its operations, conducive to the permanent welfare of the society itself, and interesting to the public, and of enlarged and comprehensive utility.

The joint Committee further reports that they have prepared a Petition for leave to bring in a Bill for the purposes aforesaid, and have also prepared a Draft of such Bill for the consideration of the Proprietors, the heads of which they have directed to be printed, and sent round to the Proprietors, as soon as practicable, previous to the Meeting. And that the Draft of the Petition and Bill are now in the Office of the Institution, for the inspection of the Proprietors

The joint Committee have also to report, that the order recently made for Petitions for Private Bills being brought in this Session, on or before the 9th Day of February, has made it requisite that the Petition and the Draft of the Bill, be prepared sooner than had been otherwise necessary.

Valuation of Each proprietors share 3.— Resolved, That the Select Committee for preparing the Bill be requested to cause a Valuation to be made of each Proprietors Share.

17

Adjourned to Monday the 5th Day of February next, at 2 o'Clock.

Dartmouth

18

Managers' Meeting 5 Feb. 1810.

At a Meeting of the Managers of the Royal Institution of Great Britain held at the House of the Institution on Monday the fifth day of February 1810.

Present

The Earl Spencer in the Chair
William Watson, Esq.
John Symmons Esq.
Rt. Hon Lord Dundas
Charles Hatchett, Esq.
The Earl of Dartmouth
Sir Thomas Bernard, Bart.
J. P. Auriol, Esq.

Minutes of last Meetg app. 1. The Minutes of the proceedings of the last Meeting were read and approved

Exd. Gen. acct. 2. Examined the General Account of the Institution

Proprietors elected. 3. Proprietors Elected

The following persons proposed as Candidates for Election as Proprietors of the Institution at a Meeting of the Managers, held on the 1st day of January, 1810, were balloted for and duly elected.

Captain Francis Drummond of 107 Sloane Street in the room of his son William Drummond Esq

The Rt. Hon. Lord Grenville to the share of the late Lord Camelford

No Manager having proposed Mr. Vere he was not Balloted for

Proprietor proposed 4. Proprietor Proposed

Charles Dalston, Nevinson, M. D was proposed as a Candidate for Election as a Proprietor of the Institution in the room of James Sullivan Martin, Esq

by

1810 Feb 5 — 19

Annl. Sub. elected 5. Annual Subscribers Elected

The following persons proposed as Candidates for Election as Annual Subscribers to the Institution at the last Meeting of the Managers were duly Elected

Thos. Ashton Smith, Esq. M. P. 5 Portman Square
Leonard Horsley Holmes, Esq Mr. Albemarle St.
Louis Simond Esq. 45 Bryanstone Street
William Fettes, Esq 10 Titchfield Street
Ashurst Majendie Esq 1 Brick Court Temple

Annual Subscribers proposed. 6. Annual Subscribers Proposed.

The following persons were proposed as Candidates for Election as Annual Subscribers to the Institution

John Story Penleaze, Esq 9 Albany
J. Machado, Esq. 19 York Buildings New Road

Subscribers to Lectn. on Nat. Ph. & Chem. elected. 7. Subscribers to the Lectures on Natural Philosophy and Chemistry Elected

The following persons proposed as Candidates for Election as Subscribers to the Lectures on Natural Philosophy and Chemistry at the last Meeting of the Managers were duly elected

The Rev Stephen Sloane 8 York Street York Place
Mr. J Simeons 36 Albemarle Street

Subscribers to Lectures proposed 8. Subscribers to the Lectures on Natural Philosophy and Chemistry ~~Elected~~ Proposed

The following persons were proposed as Candidates for Election as Subscribers to the Lectures on Natural Philosophy and Chemistry.

Mr. Thomas Dunne 7 Hanover Street
Edward Hanmer, Esq.
Dr. Davidson, 36 Arundel Street Strand

Minutes of Com. of Accts approved 9. Read and approved the Minutes of the Committee of Accounts of this day.

Signed an order for the payment of the following sums

Mr. Allen for Lectures in 1807	105. 0. 0
Do. for Chemicals to 31 Dec. 1808	83. 2. 2
Mr. Dalton on Acct. of his Lectures	42. 0. 0
Mr. Newman, Sundries for Galvanic Appe.	8. 18. 9
Mr. Easingwood Steward	50. 0. 0

20 — 1810

11. The New Tickets of admission being now ready

Ordered That the Proprietors and Subscribers be informed that the Tickets for the year 1809 will not be admissible after the 26 Inst.

Book presented. 12. Mr. Joseph Fox presented The Natural History of the Human Teeth 4to

The Committee of the Athenaeum, Liverpool presented the Catalogue of the Athenaeum

Dr. Highmore presented Jus Ecclesiasticum Anglicanum or The Government of the Church of England Exemplified and Illustrated 4to

Ordered That the Thanks of the Managers be returned for these presents

Seal of Corporation to be affixed to petition to Parliament. Adjourned to Thursday at Two o'clock and ordered that the Managers be summoned accordingly, with notice to affix the Seal of the Corporation to the Petition to Parliament if it should be agreed to at the General Meeting of the Proprietors on Wednesday

Spencer

1810 Feb. 5 — 21

Adjourned Gen. Meeting of Managers, Visitors & Comee. of Proprietors. Feb 5, 1810.

At an adjourned joint Meeting of the Managers Visitors, and Committee of Proprietors of the Royal Institution of Great Britain, held at the House of the Institution on Monday the 5th Day of February, 1810.

Present

The Earl of Dartmouth in the Chair
The Earl Spencer
William Watson, Esq
John Symmons, Esq
Rt. Hon Lord Dundas
Charles Hatchett, Esq.
Sir Thomas Bernard, Bart.
J. P. Auriol, Esq. — Managers

Lord Viscount Bulkeley
Lord Viscount Valentia
Alexander Blair, Esq
Timothy Brent, Esq.
Thomas Hammersley, Esq — Visitors.

Sir H. C. Englefield, Bart.
The Rev. Edward Forster
Sir George Thomas Staunton, Bart.
W. H. Pepys Esq
Robert Bingley Esq.
Henry Warburton, Esq.
James Stodart, Esq. — Proprietors.

Minutes of last Meetg approved. 1. The Minutes of the Proceedings of the last Meeting were read and approved

Addl. Clauses to proposed Bill 2. The following additional Clauses were submitted to the Committee by the Revd. Edward Forster, for the purpose of being inserted in the proposed Bill for amending & altering the Charter of the Royal Institution.

Clauses to be inserted in the Draft of the Bill in folio 6, between the second and third enacting Clauses

Managers "And be it further enacted, that from and after the passing of this Act the Government of the Institution shall be vested in a Council of Fifteen Managers, ~~and~~ one Treasurer

and one Secretary, and that the Marquis of Stafford – Earl of Aylesford – Earl of Chichester – Earl of Dartmouth – Earl of Egremont – Earl of Morton – Earl Spencer – Rt. Hon. Lord Dundas Sir John Cox Hippisley – Sir Thomas Bernard – Rt. Hon. Isaac Corry – Charles Hatchett, Esq. – Benjamin Hobhouse Esq. John Symmons Esq. and William Watson, Esq. (the present Managers) shall be Managers of the said Royal Institution until the first day of May 1811 that on the said first day of May, 1811 and on every succeeding first of May (unless when the same falls on a Sunday and then on the second) the Office and duties of the said Managers for the preceding year shall cease and determine, and the Members of the Royal Institution present at a General Meeting to be held on the first day of May in each year (unless when the same falls on a Sunday and then on the second) shall chuse and elect by Ballot fifteen new Managers of whom eight in the least shall not have been Managers or Visitors during the preceding year (any thing in the said recited Charter to the contrary thereof in anywise notwithstanding)

Visitors. And be it further enacted that from and after the passing of this Act the following persons shall be visitors of the Royal Institution for the purposes mentioned in the said before recited Charter (except such part or parts thereof as is or are repealed altered or made void by this act) namely – The Duke of Somerset – Earl of Bessborough – Earl Bridgewater – Earl of Buckinghamshire – Earl Grosvenor – Rt. Hon. Lord Bulkeley – Lord Viscount Valentia – Rt. Hon. Reginald Pole Carew. – The Hon. The Bishop of Durham – Alexander Blair, Esq. – Timothy Brent, Esq. Thomas Hammersley, Esq. Thomas Hope, Esq. and Samuel Thornton, Esq. until the first day of May 1811. That on the said first day of May 1811 and on every succeeding first of May (unless when the same falls on a Sunday and then on the second) the Office and duties of the said Visitors for the preceding year shall cease and determine and

the Members of the Royal Institution present at a General Meeting to be held on the first day of May in each year (unless when the same falls on a Sunday and then on the second) shall chuse and elect by ballot fifteen new Visitors of whom eight in the least shall not have been Visitors or Managers during the preceding year (any thing in the said recited Charter to the contrary thereof in anywise notwithstanding.

Members. And be it further enacted that all persons desirous of becoming Members of the Royal Institution except such as are at the time of passing this Act Proprietors thereof shall be elected by the Members of the said Institution present at a General Meeting to be held for that purpose, of whom at least two thirds shall be in favour of such person. The Election of such persons being by Ballot. – and Further that a Certificate or Recommendation of such persons desirous of becoming Members shall be signed by at least four Members and placed in one or more more conspicuous part or parts of the House of the Royal Institution (any thing in the said recited Charter to the contrary thereof in any wise notwithstanding)

And be it further enacted that a General Meeting of the Members of the Royal Institution shall be held on the first of the months of January, February, March, April, May, June, July, November and December in each year at such an hour as shall be fixed by the Bye Laws for the purpose of ballotting for and Electing New Members"

Resolved that these Clauses be approved, and be referred to the Committee appointed to draw the Draft of the Bill for the insertion thereof

3. – The following additional Clause was submitted to the Committee by Sir Henry C. Englefield. Bart.

Bye Laws. "And be it further enacted, that from and after the passing of this act there shall be a power vested in the Members of the Royal

Institution to originate and enact Bye-Laws for the Government of the same such Bye-Laws to be proposed at any General Meeting, previous special Notice thereof being given to the Secretary in writing at least fourteen days before such General Meeting. Such Notice being signed by fifteen Members in the least."

Resolved, That this Clause be approved; and referred to the Committee appointed to draw the Bill for the insertion thereof.

4. Resolved, That the Name of the Revd. Edward Forster be added to the Committee for preparing the Bill.

Adjourned to Wednesday the 7th of Feby. instant, at Twelve o'Clock precisely.

W Watson

Adjourned Gen. Meetg. of Mans. & Vis. & Com. 7th Feb. 1810.

At an adjourned Meeting of the Managers Visitors, and Committee of Proprietors of the Royal Institution of Great Britain, held at the House of the Institution on Wednesday, the 7th Day of February, 1810

Present

William Watson Esq: in the Chair.
Sir Thomas Bernard Bart.
The Earl Spencer
J. P. Auriol, Esq:
Charles Hatchett, Esq.
Alexander Blair, Esq:
Timothy Brent Esq.
Lord Viscount Valentia
Thomas Hammersley, Esq:

The Rev. Edward Forster
Henry Warburton, Esq:
Robert Bingley, Esq:
Sir George Thomas Staunton Bart.
James Stodart, Esq:
W. H. Pepys, Esq.

Minutes of last Meeting approved. 1. The Minutes of the Proceedings of the last Meeting were read and approved.

2. Adjourned to Monday next at Two o'Clock

Winchilsea

Managers' Meet^g 8th Feb. 1810.

At a Meeting of the Managers of the Royal Institution of Great Britain held at the House of the Institution on Thursday, the 8th Day of February 1810

Present.

The Earl Spencer in the Chair
Sir Thomas Bernard, Bar^t
William Watson Esq.
Charles Hatchett, Esq.
J. P. Auriol, Esq:

Minutes of last Meet^g approved. 1.— The Minutes of the Proceedings of the last Meeting were read and approved.

Application to Parliament determined on. 2.— Read The following resolution of the General Court of Proprietors which was held yesterday

Resolved unanimously that an immediate application be made to Parliament for an Act for altering and amending the Charter of the Royal Institution of Great Britain and for enlarging and more effectually promoting the objects of the said Institution and that the Managers be desired to affix the Seal of the Corporation to a Petition to Parliament to be presented on or before the ninth Instant

3.— Read the proposed Petition to Parliament and compared it with the Copy entered in the Minutes of the General Court of yesterday

Seal of Corporation affixed to Petition. 4. The Seal of the Corporation was accordingly affixed to the Petition in the presence of George Bramwell, Esq: and signed by all the Managers present at this Meeting.

Winchilsea

Managers Meeting. Feb. 12, 1810.

At a Meeting of the Managers of the Royal Institution of Great Britain held at the House of the Institution on Monday the 12th Day of February, 1810

Present

The Earl of Winchilsea in the Chair
The Earl Spencer
William Watson, Esq:
John Symmons, Esq.
Charles Hatchett, Esq

Minutes of prec^t Meet^g approved. 1.— The Minutes of the Proceedings of the last Meeting were read and approved

Ex^d Gen. Acc^t. 2.— Examined the General Account of the Institution.

Ann^l Subs^{rs} elected. 3.— Annual Subscribers Elected.

The following persons proposed as Candidates for Election as Annual Subscribers to the Institution at the last Meeting of the Managers were duly elected.

John Story Penleaze, Esq: 9, Albany.
J. Machado, Esq, 19, York Buildings, New Road.

Ann^l Subs^{rs} proposed. 4.— Annual Subscribers Proposed.

The following persons were proposed as Candidates for Election as Annual Subscribers to the Royal Institution

Richard Barker, jun Esq. Tavistock Street
John Wheatley, Esq 2 Bolton Row.

Subs^{rs} to Lect^{rs} Elected 5.— Subscribers to the Lectures on Natural Philosophy & Chemistry Elected.

The following persons proposed as Candidates for Election as Subscribers to the Lectures on Natural Philosophy and Chemistry at the last Meeting of the Managers were duly Elected.

Mr Thomas Dunne, 7 Hanover Street
Edward Hanmer, Esq:
Dr Davidson, 36 Arundel Street Strand.

Subscribers

Subsc^{rs} to Lectures proposed. 6.— Subscribers to the Lectures on Natural Philosophy and Chemistry Proposed

The following persons were proposed as Candidates for Election as Subscribers to the Lectures on Natural Philosophy & Chemistry

John Henderson, Esq. 3 Montagu Street
Charles Halford Esq. 42 Norfolk Street Strand

Request from the House of Lords for assistance from the Inst^n in improving the Warmth and Ventilation of the House. 7.— In consequence of the following Letter addressed to the Secretary of the Royal Institution from W. S. Rose Esq: Reading and Committee Clerk to the House of Lords requesting some advice and assistance from the Royal Institution on the subject of warming and ventilating the House of Lords, Mr Auriol and Mr Davy attended a Committee of the House of Lords, and in consequence of the commands of their Lordships, Mr Davy minutely examined the State of the Atmosphere in the House, and the means existing for warming the Air and for changing it and delivered a Report on the subject to their Lordships, of which a Copy is annexed.

Mr. Davy's suggestions approved by the H. of Lords' Committee. Their Lordships were pleased to order that the improvements suggested by Mr Davy should be adopted under his inspection.

Mr Rose's letter. "Sir,

I am commanded by a Committee of the House of Lords to request that you would attend them at four o'Clock on Thursday next and that you would procure the attendance of any Gentleman of your Institution who you think can suggest any effectual means of ventilating the House"

I am Sir
Your Ob^t Humble Servant
W. S. Rose
Reading & Committee Clerk"

Old Palace Yard
Tues. Morn.

Mr Davy's Report. "To the Lords Committees appointed to consider of the best means of Ventilating the House of Lords.

My Lords.— In obedience to the Commands

of your Lordships I have carefully examined the Methods now employed for warming & ventilating the upper House. In this communication I shall do myself the Honour of giving my opinion on these Methods, and likewise of Stating to your Lordships some considerations on the improvements that may be made in them.

The House of Lords is at present warmed by means of three open Fire-places and two flues passing under the floor, which are supplied with hot air from the combustion of Coal, and there is no mode of Ventilation except by opening of the windows or two large Ventilators in the roof of the House. There are no apertures or tubes in the lower part of the House by which fresh air can be admitted, and the air which is heated by the flues is not fresh air but the stagnant air of the House— Even when the Ventilators are at play, and the windows open, no foul air can pass out unless cold air takes its place, which consequently enters in the most inconvenient situation. Under such circumstances it is evident that the air, when the room is at all crowded, must be very impure

Some important alterations for the better, may I conceive, be very easily made— One is that of admitting by means of proper pipes fresh air into the receptacle under the House, containing the flues, so that the heat shall be communicated to this fresh air, the passage of which into the House may be effected by means of a number of small apertures at regular distances round the floor. A second Alteration is by causing the Ventilators to terminate in a curved Tube, which may be opened to any extent that may be necessary, or closed, and which shall suffer the impure air to flow out without admitting of the introduction of any cold air.

By the adoption of these simple improvements, fresh, and moderately warm air would be constantly and gradually admitted into the room without occasioning any currents

30 1810.

and the impure and more heated air would be as constantly got rid of; and by means of proper Dampers, the quantity of fresh air admitted and of hot air thrown off might be regulated according to the state of the attendance of their Lordships in the House.

There are two flues which at present are not used, but which if necessary may be supplied with fresh air in the same manner as the others, but if these are employed it will be proper to diminishe the open Fire Places which always occasion a considerable Draught of air and irregularity of Temerature.

The circumstance of admitting warm fresh air from below is the great circumstance to be attended to; and this and the change in the ventilation may be effected in a very short time and, I should conceive at a very small expense.

I shall say nothing of a Steam Apparatus, because this would require a great length of time for its construction, and could not easily be used during the present Session, and I am inclined to believe that the plan which I venture to propose to your Lordships will be adequate to the desired effect. I shall be proud to give any further assistance in my power towards the accomplishment of an object connected with the health and comfort of the most august Assembly in the world.

I have the honour to be,
My Lords,
Your Lordships Most Obedient
Humble Servant,
Humphry Davy.

Books presented. 8.— Francis Baily, Esq presented "The Doctrine of Life Annuities and Assurances"

Col. De Charmilly presented The Narrative of his Transactions in Spain with the Rt. Hon. John Hookham Frere, &c.

Ordered That the Thanks of the Managers be returned for these presents

Adjourned.

T. Bernard

Feb. 12. 31

Adjourned Gen. Meetg of Managers, Visitors, &c Feb. 12, 1810

At an adjourned joint Meeting of the Managers, Visitors, and Committee of Proprietors, of the Royal Institution of Great Britain, held at the House of the Institution on Monday the 12th Day of February, 1810.

Present

The Earl of Winchilsea in the Chair
The Earl Spencer
John Symmons, Esq:
Charles Hatchett, Esq:
William Watson, Esq:
J. P. Auriol, Esq:
The Earl of Morton
Sir Thomas Bernard, Bart. — Managers.

Alexander Blair, Esq:
Timothy Brent, Esq: Visitors.

James Stodart, Esq:
The Rev Edward Forster
W. H. Pepys, Esq
Henry Warburton, Esq:
Sir H. C. Englefield Bart. Proprietors

Minutes of last Meetg approved. 1.— The Minutes of the Proceedings of the last Meeting were read and approved.

Minutes read 2.— The Minutes of the Proceedings of the General Court of Proprietors of Wednesday the 7th Inst. was read

3.— The Minute No. 4 of the Managers at their Meeting on Thursday the 8th Instant was read, whereby it appears that the Seal of the Corporation was affixed to the Petition in the presence of George Bramwell, Esq: and signed by the Managers present at that Meeting

Petition presented by Sir J. Sinclair. 4.— Mr. Watson reported that the said Petition was presented to the Hon. the House of Commons, on Thursday the 8th Instant by Sir John Sinclair, Bart.

5.— Read

32 1810.

5.— Read a Letter from Sir John Sinclair

Thanks of Meetg to be ~~sent~~ presented to Sir J. Sinclair. Resolved, That the Thanks of the Meeting be presented to Sir John Sinclair for this communication and that his Letter be referred to Mr. Davy, and that he be requested to have an interview with Sir John Sinclair on the Subject

Outline of plan by J. Perry Esq. 6.— Read the Outline of a Plan for the Royal Institution sent by James Perry, Esq:

Resolved, That the Thanks of the Committee be presented to Mr. Perry for his Plan, and that the Committee will avail itself of Mr. Perry's offer to explain any part of it, if, on the consideration of the same it should be found necessary.

Adjourned to Thursday the 15th of February 1810, at two o'Clock precisely.

Somerset.

Feb. 15. 33

Adjourned Joint Meeting 15th Feb. 1810.

At an adjourned joint Meeting of the Managers, Visitors, and Committee of Proprietors of the Royal Institution of Great Britain, held at the House of the Institution, on Thursday the 15th Day of February, 1810,

Present,

The Duke of Somerset, Visitor, in the Chair.
Sir John Cox Hippisley, Bart.
Sir Thomas Bernard, Bart.
The Earl Spencer
J. P. Auriol, Esq. — Managers.

Timothy Brent, Esq, Visitor.

The Rev. Edward Forster.
Henry Warburton, Esq,
James Stodart, Esq. Proprietors.

Minutes of last Meetg approved. 1. The Minutes of the Proceedings of the last Meeting were read and approved.

2. Read a Letter from Thomas Slack, Esq. and it appearing to contain no suggestions applicable to the proposed Plan,

Resolved, That it lie on the Table.

3. Read also a Letter from John Hinckley, Esq. and it appearing to the Committee that there is no suggestion in it applicable to the proposed Act, or which may not come under the Head of Bye-Laws.

Resolved, That the consideration of it be deffrred.

4. Resolved, That the Thanks of the Committee be returned to Mr. Hinckley for this Communication.

5. Resolved, That a considerable part of Mr. Perry's Letter will be more properly the subject of Bye-Laws, and not inconsistent with the Act as proposed.

Inexpedient to extend power of nominating Life-subsrs. beyond original Members. 6 Resolved. That it is the opinion of the Committee that the extending of the Power of nomination of Life Subscribers beyond the original Members will not be consistent with the

future Interests of the Institution

Inexpedient to limit Number of Members by Act of Parliament

7 Resolved. That it is the opinion of the Committee that any Limitation of the Number of Members by the Act may hereafter be inconvenient, & that it may be better provided for by the Bye Laws, in case that upon full consideration it should be found expedient.

8. Resolved That a Copy of the four preceding Resolutions be sent to Mr. Perry with the Compliments of the united Committee

Money for compensation of Proprietors to be raised by loan.

9. Resolved. That a Sum of Money be raised by Loan for making Compensation to such Proprietors as may require it, & for answering the other purposes of the proposed Act. & that the following Preamble be placed at the head of the Subscription

Form for Subscription List.

"Subscription to provide a Fund for making pecuniary compensations in lieu of their Proprietory Shares to such Proprietors as may require it, & for answering the other purposes of the proposed Act for amending the Charter of the Royal Institution; to be advanced by way of Loan for three years without interest, by instalments of £10 per Cent. the first instalment to be paid within the space of one Calendar Month after the passing of the said Act, & the other instalments (if called for) to be each of them paid within one Calendar Month after the same shall have been respectively called for, or required by the Committee of Managers of the said Institution."

Valuation of the House.

10. Resolved. That Sir Thos. Bernard Bart. be requested to apply to Mr. Atkinson to make a valuation of the House of the Institution & other premises, including the Interest of the Corporation in the under Leases of the two adjoining Houses.

Valuation of Books.

11. Resolved. That Mr. Harris be desired

to make a Valuation of all the Books & Manuscripts belonging to the Institution

Notice of Loan

Ordered That Notice of the said Loan be sent round to the Proprietors with the request of the Committee that those who are disposed to contribute to it, will send to or call at the Secretary's Office of the Institution where the Subscription Papers & lists of the Subscribers are left.

Adjourned to Thursday the 22nd Feb. inst. at 2 o'Clock precisely.

Aylesford.

Managers' Meeting, 19th Feb. 1810.

At a Meeting of the Managers of the Royal Institution of Great Britain, held at the House of the Institution on Monday the 19th Day of February 1810,

Present,

Sir Thomas Bernard, Bart. in the Chair
Charles Hatchett, Esq.
John Symmons, Esq.
James Peter Auriol, Esq.

Minutes of last Meeting approved.

1. The Minutes of the Proceedings of the last Meeting were read and approved.

Annl. Subrs. elected.

2. Annual Subscribers Elected.

The following Persons, proposed as Candidates for Election, as Annual Subscribers to the Institution, at the last Meeting of the Managers, were duly elected,

Richard Barker, Junr. Esq. Tavistock Street
John Wheatley, Esq. No. 2, Bolton Row.

Subscribers to Lectures elected

3 Subscribers to the Lectures on Natural Philosophy and Chemistry Elected.

The following Persons, proposed as Candidates for Election, as Subscribers to the Lectures on Natural Philosophy and Chemistry, at the last Meeting of the Managers, were duly elected.

John Henderson, Esq. No. 3, Montagu Street.
Charles Holford, Esq. No. 42, Norfolk Street, Strand.

Books presented.

W. C. Jackson Esqr. Presented "Memoir of the Public Conduct & Services of Wm. Collins Jackson"

4 The Rev. Stephen Weston presented "Remains of Arabic in the Spanish and Portuguese Languages."

Ordered, That the Thanks of the Managers be returned for their Presents.

Special Meeting of Proprietors to consider Bill to be presented to Parliament.

5. Ordered, That a Special Meeting of the Proprietors be called for Tuesday the 27th instant, at Two o'Clock precisely, and the following Notice sent immediately to the Proprietors.

"Royal Institution.

"A General Meeting of the Proprietors will be held at "the House of the Institution on Tuesday, the 27th instant, "to consider the Bill to be presented to Parliament, 'For "'Altering and Amending the Charter of the Royal In-"'stitution of Great Britain, and for enlarging, and more

"'effectually promoting the objects of the said Institution,' "and to authorize affixing the Seal of the Corporation "thereto.

"The Proprietors are informed that the Heads of the "above Bill will be sent round to them as soon as they "can be printed.

"Albemarle Street, "J. P. Auriol, Sec.
19th Feb. 1810.

Letter from Mr. Easingwood to Mr. Savage and Order thereon

6. Mr. Davy reported, that Mr. Easingwood had left the Institution without any notice given to him.

Mr. Savage reported, that he had received a Note by the Twopenny Post from Mr. Easingwood, saying, "I am "in Agonies. I shall perhaps never see any thing of you "again;" and that he, Mr. Savage was indebted to Mr. Easingwood the sum of One Hundred and Thirty Pounds, which he was ready to acknowledge as received of the Managers in part of his Bill of £186.1.3. for Printing in the preceding Year, the Managers making allowance to him for it in his Account with the Institution.

Resolved, That a Receipt be taken of Mr. Savage accordingly, and that the same be deposited in the Hands of Mr. Auriol

Mr. Davy to take direction of Secretary's Office

7. Resolved, That Mr. Davy be requested to take the Direction of the Secretary's Office, &c. and to employ such Person or Persons to attend there as he shall think proper, and also to inspect the Bankers' Books, and the other Accounts; and that this Meeting do adjourn to Wednesday next at Two o'Clock precisely to receive his Report.

8. Resolved, That Mr. Davy be requested to make a prospective State of the present Objects of the Institution in the opening of his next Course of Lectures.

Adjourned.

Spencer

38

Managers' Meeting. Feb. 21. 1810.

1810.

At a Meeting of the Managers of the Royal Institution of Great Britain, held at the House of the Institution on Wednesday the 21st day of February 1810.

Present.

Earl Spencer in the Chair
Sir Thos. Bernard Bart.
Sir J. C. Hippisley Bart.
Charles Hatchett Esqr.
William Watson Esqr.
John Symmons Esqr.
J. P. Auriol Esqr.

Minutes of last Meeting appd. 1.

The Minutes of the Proceedings of the last Meeting were read & approved.-

Mr. Easingwood to be apprehended. 2.

In consequence of the request of the Managers, Mr. Couant attended the Meeting, & undertook that Measures should be proceeded in for apprehending Mr. Easingwood, & that he would from time to time communicate with the Managers on the Subject. -

£198.9 not paid by Mr. Easingwood. 3.

Upon referring to the General Account of the Institution & to the Bankers' book it appeared that the three last Sums stated in the General Account to be paid to the Bankers viz £68.5. on the 29th of January. — £68.5. on the 3rd of February & £61.19. on the 12th of February amounting together to £198.9s. have not been paid in to the Bankers by Mr. Easingwood.-

Statement of Mr. Easingwood's account. 4.

It also appeared that the sum of £43.6.11 remained in Mr. Easingwood's hands. the Balance of the Money advanced to him on account by order

Feb. 21. 39

of the Managers of the 5th instant. & that the Sum of £50 was due to him for half a Year's Salary to Christmas last. besides what had since occurred for the broken quarter. —

The Books &c. to be delivered to Messrs. Hatchett & Davy.

Resolved. That Mr. Hatchett & Mr. Davy be requested to examine the Books. & see if any further deficit appears. —

Resolved. That Mr. Savage be directed to deliver over to Mr. Hatchett & Mr. Davy all the Accounts & Books of the Institution in his charge; together with the Keys in his possession.-

Mr. Savage attended the Managers & undertook to prepare & send to Mr. Davy by tomorrow morning an account of all the Tickets for the present year which he had delivered to Mr. Easingwood to be issued by him.-

Adjourned to Monday the 26th day of February 1810 at two o'Clock

Spencer

40

Adjourned Joint Meeting. 22nd Feb. 1810

1810.

At an adjourned joint Meeting of the Managers, Visitors, & Committee of Proprietors, of the Royal Institution of Great Britain, held at the House of the Institution, on Thursday the 22nd day of February 1810.

Present.

Sir Thomas Bernard Bart. in the Chair
William Watson Esqr.
J. P. Auriol Esqr.
The Earl Spencer — Managers.

Timothy Brent Esqr.
Alexander Blair Esqr. — Visitors.

Sir Geo. Thos. Staunton Bart.
W. W. Pepys Esqr.
James Stodart Esqr.
The Revd. Edwd. Forster
Robert Bingley Esqr.
Henry Warburton Esqr. — Proprietors.

1. Resolution of H. of C. concerning Priv. Bills.

Mr. Watson laid before the Joint Committee a printed Copy of the Resolutions of the House of Commons, whereby it appears that the time for receiving private Bills is fixed for Monday the 19th of March next: & it appearing to the Joint Committee that there will hardly be sufficient time to have the Bill completely settled against the General Meeting to be held on Tuesday next at 1 o'Clock. -

Resolved.

Notices to be sent of motion

That Notice be sent round to the Proprietors that at that Meeting a Motion will be made on the part of the Joint Committee to adjourn the

Feb. 22. 41

Meeting to Friday the 9th day of March at 2 o'Clock precisely, to consider the Bill to be presented to Parliament for altering & amending the Charter of the Royal Institution of Great Britain & for enlarging & more effectually promoting the Objects of the said Institution; & to authorize affixing the Seal of the Corporation thereto. — —

(Signed) J. P. Auriol Secy.

Resolved.

That a List of the Subscriptions to the proposed Loan be sent round with the Notice

Resolved

That this Committee do adjourn to Tuesday next at 1 o'Clock.

Aylesford.

Managers' Meeting. Feb. 26. 1810

At a Meeting of the Managers of the Royal Institution of Great Britain held at the House of the Institution on Monday the 26th day of February 1810.

Present
The Earl Spencer in the Chair
The Earl of Aylesford
The Earl of Egremont
The Marquis of Stafford
The Right Hon. Lord Dundas
Sir Thomas Bernard Bart.
Wm. Watton Esq.
John Symmons Esq.

Minutes of last Meet.g approved — 1. The minutes of the proceedings of the last Meeting were read & approved.

Life Subscribers proposed — 2. Life Subscribers proposed

The following Persons were proposed as Candidates for Election as Life Subscribers

The Revd. John White, Rector of Hardwick Buckinghamshire, by Chas. Hatchett Esq.

Arthur Tegart Esq. No. 46 Pall Mall, by Mr. Harris

Annl. subscribers proposed — Annual Subscribers proposed

The following persons were proposed as Candidates for Election as Annual Subscribers.

The Earl of Clancarty No. 1 Spring Garden Terrace, by G. Holford Esq.

John Dickinson Esq. 49. Welbeck St., by E. Hanmer Esq.

Captn. Elliot of the 1. Regt. of Guards by the Earl of Buckinghamshire

Captn. Edwd. D. King R.N. Kew. by Captn. Willoughby R.N.

James Halford Esq. Junr. No. 41 Norfolk St. Strand. by C. D. Halford Esq.

Subscribers to Lectures proposed — Subscribers to the Lectures proposed

The Dowager Countess of Pembroke
Lady Diana Herbert
by the Countess Spencer.

An estimate of the value of the Philosophical Apparatus by Mr. Newman was laid on the table

Valuation of Minerals — Resolved. That Mr. Davy be requested to apply to Mr. Knight to desire him to make a valuation of the Collection of Minerals & the Apparatus in the Laboratory.—

Estimate of Proprietors' shares. — Resolved. That Mr. Auriol be requested to apply to Mr. Skelton Actuary to the Westminster Insurance Office & to desire him to make an estimate of the value of the Proprietors Shares in the Royal Institution with reference to the interest of the Life Subscribers & Subscribers to the Library.

Omission in Mr. Easingwood's account. — Mr. Davy reported that on inspecting the accounts it was found that a sum paid by Mr. Maxwell to Mr. Easingwood of £4-4- for his annual subscription for the current year was not entered in his Account.—

Book presented. — Presents received.
A defence of Bank Notes by John Grenfell Esq.

Ordered. That thanks be returned to Mr. Grenfell for this his present.

Adjourned

T. Bernard

Joint Meeting 5th March 1810. — At an adjourned Joint Meeting of the Managers, Visitors & Committee of Proprietors, of the Royal Institution of Great Britain, held at the House of the Institution, on Monday the 5th day of March 1810

Present
The Earl of Aylesford, in the Chair.
The Earl Spencer
Sir Thos. Bernard, Bart.
Charles Hatchett, Esq.
John Symmons, Esq.
Wm. Watton, Esq. — Managers.

Scrope Bernard, Esq.
Thomas Hammersley, Esq.
Timothy Brent, Esq. — Visitors.

James Stodart, Esq.
Sir H. C. Englefield, Bart.
The Revd. Edwd. Forster
Henry Warburton, Esq. — Proprietors.

Minutes approved — 1. The Minutes of the Proceedings of the Meeting of the 15th of February, & of the 22nd of February last, were read & approved.

Approval of proposed act — 2. A Draught of the proposed Act was laid before the Joint Committee, & approved, & a fair Copy ordered to be prepared by Mr. Bramwell for the General Meeting.—

3. A Proof Sheet of the Heads of the Act was laid before the Joint Committee, & approved, & ordered to be immediately printed & sent round to the Proprietors.

Valuation of Proprietors' shares. — 4. That Sir Thos. Bernard & Mr. Auriol be desired to employ Mr. Skelton to make a Valuation of the Proprietory Shares, upon the Basis of the other Valuations made of the different Parts of the Property of the Institution.

Adjourned to Friday the 9th of March, at 1 o'Clock precisely.—

Managers' Meeting 5th March 1810

At a Meeting of the Managers of the Royal Institution of Great Britain held at the House of the Institution on Monday the 5th day of March 1810

Present.

Sir Thomas Bernard Bar.t in the Chair
Charles Hatchett Esq.r
W.m Watson Esq.r
John Symmons Esq.r

Minutes approved 1. The Minutes of the last Meeting were read & approved. –

Life Subscribers Elected

The following persons were elected as Life Subscribers to the Institution
The Rev.d John White
Rector of Hardwick Buckinghamshire
Arthur Tegart Esq.r 46 Pall Mall.

Ann.l Subscribers Elected

The following persons were elected annual Subscribers to the Institution
The Earl of Clancarty
Spring Garden Terrace
John Dickinson Esq.r M.P.
49. Welbeck Street
Capt.n Elliot 1st Reg.t of Guards.
Capt.n C. D. King R.N. Kew.
James Halford Esq.r Norfolk St. Strand.

Ladies Subscribers admitted

The following Ladies were admitted as Subscribers to the Lectures.
The Countess of Pembroke
Lady Diana Herbert

Ann.l Subscribers proposed

The following Persons were proposed as Candidates for Election as Annual Subscribers
Robert Eagle Esq.r 32 Upper Berkley St.
Robert Hunter Esq.r Conduit St. Hanover Square

The Rev.d Tho.s White
High Street Mary le bone
was proposed by Sir Tho.s Bernard Bar.t

for Election as a Proprietor of the Royal Institution in the room of Tho.s Slack Esq.r who is desirous of parting with his Share.

Proprietor Proposed.

The Right Hon. Lord Geo.e Cavendish was proposed as a Proprietor in the room of Henry Cavendish Esq.r deceased. by the Earl Spencer. –

Presents

The following presents were received

Two Pamphlets on Oriental Manuscripts by Sir Sidney Smith. –

The Amateur of fencing by M.r Roland.

Ordered. That thanks be returned to Sir Sidney Smith & M.r Roland for these presents.

A letter from Mess.rs Dickins & C.o on Coals was referred to M.r Davy. –

Minutes of Com.ee of Acc.ts approved

Read the Minutes of the Committee of Accounts & approved. –

Resolved. That the Treasurer be desired to pay M.r Edm.d Davy £20 on account & signed a Draft accordingly. –

That the Treasurer be also desired to pay the Sum of £48.10.8 due for Property Tax & signed a Draft accordingly.

Reports to be made on Regulation of Office

Resolved. That Sir Tho.s Bernard Bar.t M.r Symmons, M.r Auriol, M.r Hatchett & M.r Davy be requested to make a Report on the plans to be adopted for the regulation of the Office.

Report resp.g M.r Easingwood.

Read a Report from M.r Hatchett & M.r Davy on the state in which M. Easingwood has left the Office.

Ordered That it be entered on the Minutes.

Report of M.r Easingwood's accounts

Report.

M.r Hatchett & M.r Davy having investigated the Accounts which concern Matthew Easingwood who has absconded Report, that they find in the Bankers Accounts a deficiency of £198.9.0
To which must be added 43.5.11
being Money in M. Easingwood's hands; – – – – – –
And likewise the produce of three Tickets issued to Mess.rs Maxwell, Cowper & amounting to 12.12.0
making the total deficiency in Money amount to £254.7.11

M.r Hatchett & M.r Davy also Report that out of 146 personal Tickets for Life and Annual Subscribers delivered to M. Easingwood, only 123 receipts have been produced, & that no account can be found to explain how the remaining 23 Tickets have been disposed of.

Also. That 26 Transferrable Tickets for Proprietors were delivered to M. Easingwood, but according to his account 41 have been issued, being 15 more than he had received.

In like manner it appears that M. Easingwood has issued 1 Ladies Ticket more than received. –

They likewise find that M. Easingwood received 29 Tickets for Lectures on Natural Philosophy, but his receipts amount to 37 being in excess of Receipts amounting to 8

It appears therefore to M.r Hatchett & M.r Davy, that great irregularities have taken place respecting the issuing of Tickets, & that nothing can be learned from accounts which differ so much as those of M.r Savage & M. Easingwood. –

Lecture ordered to be printed

Resolved. That M.r Davy be requested to allow his Lecture of Saturday last to be printed & sent round to the Proprietors & Subscribers.

Adjourned to Monday the 12th day of March 1810. –

Dundas

Managers Meeting March 12, 1810

At a Meeting of the Managers of the Royal Institution of Great Britain, held at the House of the Institution on the 12th day of March 1810.

Present.

The Rt. Hon. Lord Dundas in the Chair
Sir Thos. Bernard Bart.
Charles Hatchett Esqr.
Wm. Watson Esqr.
John Symmons Esqr.

Minutes approved. 1. The Minutes of the Proceedings of the last Meeting were read & approved.

Gen. Acct. exd. 2. Examined the General Account of the Institution. — —

Annl. Subsrs. Elected 3. Annual Subscribers Elected.
The following persons proposed as Candidates for Election as Annual Subscribers to the Institution, at the last Meeting of the Managers, were duly elected.
Robt. Eagle Esqr. 32 Upper Berkley Street
Robt. Hunter Esqr. Conduit St. Hanover Square

Annl. Subsrs. Proposed. 4 Annual Subscribers Proposed.
The following persons were proposed as Candidates for Election as Annual Subscribers to the Institution
Edwd. Parkins Esqr. 40 Weymouth St. Portland Place
Patrick Bartlet Esqr. 9. Nottingham Place
Mark Byrne Esqr. 46 Park Street
The Abbé Duval 17. Northumberland St.
John Kirkpatrick Esqr. 70. Pall Mall
J. H. North Esqr. 21. Salisbury St. Strand

Subscribers to Lectures proposed 5. Subscribers to the Lectures Proposed
The following persons were proposed as Candidates for Election as Subscribers to the Lectures of the Institution
Henry Taylor Esqr.
Edward Cooke Esqr.

Edward Wakefield Esqr. 41 Piccadilly
Miss Toulston 19. Margaret Street
Mrs Law 18. Portland Place
The Countess of Oxford 14 Cavendish Square
Lady Jane Harley 14 Cavendish Square

Minutes of Comee. of Accts. approved 6. Read & approved the Minutes of the Committee of Accounts of this day. — —

Management of the proposed Bill in Parliamt. 7. Resolved. That Mr Watson be requested to deliver the Bill to Mr Hobhouse with the Request of the Committee that he will take the Conduct of it & employ Mr Bramwell as the Agent for it at the expence of the Corporation; & also that he be desired to request the attendance of such Members as he shall think proper. — —

Porter's Wages raised 8 Ordered That the under Porter's Wages be increased from 18/ to twenty Shillings per Week including his allowance for carrying out the Cards.

Presents. 9. The Society of Arts presented the "Transactions of the Society for 1809."

Colonel Wilkes presented "Historical Sketches of the South of India"

Ordered. That the Thanks of the Managers be returned for these presents. —

Folding Screen ordered. 10. Ordered. That Mr Davy be desired to provide a folding Screen for the library. —

Catalogue of Liby. presented. 11. Ordered. That a Catalogue of the Library of the Royal Institution be presented to the Athenaeum at Liverpool

Transfer. 12 The Transfer from Mr Slack to the Revd. Thos. White 54 High Street Marylebone was laid before the Managers, & ordered

to be produced preparatory to the Balot on the first Monday in April.

Valuation of Library £7095.7. 13. Mr Harris's valuation of the Library was laid upon the Table whereby it appeared that the Value was Seven Thousand & ninety five Pounds, seven Shillings. —

14. Ordered. That Notice be inserted in the Summons, for the next Meeting of Managers, that it is then intended to affix the Seal of the Corporation to the Bill for altering & amending the Charter of the Royal Institution. —

Adjourned.

Winchilsea

Managers Meeting 19th March, 1810.

At a Meeting of the Managers of the Royal Institution of Great Britain, held at the House of the Institution on Monday the 19th day of March 1810 —

Present.

The Earl of Winchilsea President in the Chair
The Marquis of Stafford
The Earl Spencer
The Earl of Chichester
The Right Hon. Lord Dundas
Sir Thos. Bernard Bart.
Chas. Hatchett Esqr.
John Symmons Esqr.
J. P. Auriol Esqr.

Minutes approved 1. The Minutes of the Proceedings of the last Meeting were read & approved. — —

Exd. Gen. Acct. 2. Examined the General Account of the Institution

Annl. Subscribers Elected 3. Annual Subscribers Elected.
The following Persons, proposed as Candidates for Election, as Annual Subscribers to the Institution at the last Meeting of the Managers, were duly elected. —
Edward Parkins Esqr. 40. Weymouth St. Portland Place
Patrick Bartlet Esqr. 9. Nottingham Place
Mark Byrne Esqr. 46. Park Street
The Abbé Duval 17 Northumberland Street
John Kirkpatrick Esqr. 70. Pall Mall.
J. H. North Esqr. 21. Salisbury Street Strand

Annl. Subscribers to Lectures Elected 4. Annual Subscribers to the Lectures Elected. —
The following Persons, proposed as Candidates for Election as Subscribers to the Lectures of the Institution, at the last Meeting of the Managers, were duly elected. —
Henry Taylor Esqr.
Edward Cooke Esqr.
Edward Wakefield Esqr. 41. Piccadilly
Miss Toulston 19. Margaret Street
Mrs Law 18. Portland Place
The Countess of Oxford 14. Cavendish Square
Lady Jane Harley — 14. Cavendish Square

1810

Ann.l Subscribers Proposed 5. **Annual Subscribers Proposed.**

The following Persons were proposed as Candidates for Election as Annual Subscribers to the Institution.

Sir John Swinburne Bar.t 18 Grosvenor Place
Walter Burrell Esq.r Ibbotson's Hotel
Robert Hawthorn Esq.r 8 Palgrave Place Strand

Subscribers to Lectures Proposed 6. **Subscribers to the Lectures Proposed.**

Charles Littlehales Esq.r 8 Bouverie Street.
Benj.n Hall Esq.r Upper Gower Street
Edward Howard Esq.r
Henry Gipps Esq.r 33 Red Lion Square

Minutes of Com.e of Accounts approved. 7. Read & approved the Minutes of the Committee of Accounts of this day.– Signed an Order of £40 advanced to the Steward.–

Bills to be paid 8. Ordered, payment of the following Bills –

G. Tebbs for Chemicals in 1808 & 1809	£ 1-4-6
J. Cuthbertson for Electrical Instruments &c in 1809	2-14-6
James Mitchell for Coals in 1809	8-6-6

Ordered. That Cha.s Royce's Bill for Carpenter's Work done in 1809 amounting to £35-3-6 be referred to M.r Davy.–

Valuation of Chemicals £140. Minerals &c £400. 9. M.r Knight's Valuation of the Chemical Apparatus & Minerals, Cases &c was laid upon the Table, whereby it appeared that he valued

the Chemical Apparatus at	£ 140
The Minerals, Cases &c at	400
	540

Models & Philosoph.l Instruments £787-14-6 10. – M.r Newman's Valuation of the Models, & Mathematical & Philosophical Instruments was laid on the Table, by which it appeared that the total valuation of the same is £787-14-6

Assistant's Wages raised. 11. M.r Davy submitted to the Managers the propriety of raising the wages of W.m Payne

the Assistant in the Laboratory from 18/ to 20/ per Week. – – Ordered.

Mr. Casingwood's accounts. 12. M.r Hatchett & M.r Davy in continuation of their Report of the 5.th instant inform the Meeting that on carefully examining the Books the following Receipts by M. Casingwood do not appear in the General Account Book. – viz. The amount of five Ladies Tickets to

M.rs Skottowe, M.rs West, Miss Halford, M.rs Davenport, Hon M.rs Windham	£ 9-9-0
2 Transferrable Tickets to Proprietors viz M.r Skottowe & M.r Newton – – – –	6-6-0
3. Tickets to Annual Subscribers viz. M.r De Jonville, M.r J. Wyatt, M.r G. Holford	12-12-0
1 Life Subscription M.r Disney – –	31-10-0
Making a further deficit of	£ 59-17-0

Notice of Catalogue. 13. Ordered. That Notice of the Catalogue of the Library of the Royal Institution be inserted on the Cards of the Lectures. – –

Report on the Renewal of Lease of 23 Albemarle St. 14. M.r Savage laid the following Report before the Managers.– –

" M.r Savage Reports to the Managers, pursuant to the directions which he received from them at their last Meeting, that he has had a conversation with M.r Harrison (M.r Tennant being in the Country) respecting the Renewal of the Lease of N.o 23 in Albemarle Street to the Representatives of the late Edward Gray.– –

M.r Harrison stated to him, that they have received £26,700 from the Sale of Estates; which Money is actually paid into the Court of Chancery. – –

That the Master to whom the concerns of the Estate were referred, has reported in favour of the Renewal, & of the payment of the Fines & Interest, as agreed to by the Managers of the Institution & M.r White.–

That a Petition is drawn, stating the facts, & that they have been approved by the Master, which will be presented to the Court on the first

day for receiving Petitions, praying an Order of the Court in favour of M.r White, the Receiver, for the Sum of £2300, to enable him to pay the several Fines and Interest, on producing in Court the several Leases duly executed.– And that M.r Harrison has no doubt, but that the Money will be paid to the Royal Institution in the Month of May.–

M.r Harrison also stated that the delay of the payment of the Fines & Interest, contrary to their promise had arisen from – the principal Mortgagee having refused to sign any separate Deeds, till the whole of his Mortgage on the Estate was paid off, with which they were unable to comply.–

That within the last ten or twelve days, he has agreed to sign separate Deeds, as fast as they can be prepared, on his receiving the proportionate Mortgage on such Property as may be transferred. – – And that this agreement of his enables the Representatives of M.r Gray to apply to the Court of Chancery for the payment of the respective Renewals as fast as they can be prepared, on their being laid before the Court properly executed."–

19 March 1810.– Signed W.m Savage

15. M.r Auriol is requested to give the directions necessary for preparing the usual annual Accounts

Seal of Corporation affixed to Bill 16. The Seal of the Corporation was affixed to the Bill presented to the House of Commons & signed by the President & all the Managers present at this Meeting, & the Secretary, in the presence of M.r Savage. & the same was sent to M.r Bramwell.

Adjourned

Spencer

Managers' Meeting March 26th. 1810. At a Meeting of the Managers of the Royal Institution of Great Britain held at the House of the Institution on Monday the 26.th day of March 1810.

Present.

The Earl Spencer in the Chair
R.t Hon. Lord Dundas
Sir Tho.s Bernard Bar.t
John Symmons Esq.r
W.m Watson Esq.r
Cha.s Hatchett Esq.r
J. P. Auriol Esq.r

Minutes approved 1. Read & approved the Minutes of the last Meeting. – – – – –

Ex.d Gen. Acct 2. Examined the General Account of the Institution

Ann.l Subscribers Elected. 3. **Annual Subscribers Elected**

The following Persons proposed as Candidates for Election as Annual Subscribers to the Institution were duly Elected. – –

- Sir John Swinburne Bar.t 18 Grosvenor Place.–
- Walter Burrell Esq.r Ibbotson's Hotel.–
- Rob.t Hawthorn Esq.r 8 Palgrave Place Strand.

Subscribers to Lectures Elected 4. **Subscribers to the Lectures Elected.**

The following Persons proposed as Candidates for Election as Subscribers to the Lectures of the Institution were duly elected. – –

- Charles Littlehales Esq.r 8. Bouverie Street
- Benj.n Hall Esq.r Upper Gower Street
- Edward Howard Esq.r
- Henry Gipps Esq.r 33 Red Lion Square.–

Ann.l Subscribers Proposed 5. **Annual Subscribers Proposed.**

The following Persons were proposed as Candidates for Election as Annual Subscribers to the Institution. – –

W.m Dickenson Esq.r 28 Bury S.t S.t James's
James L. Grant Esq.r 8 Hanover S.t Hanover Square

58 | 1810 Mar. 26.

Subscribers to Lectures Proposed

6. Subscribers to the Lectures Proposed.

The following Persons were proposed as Candidates for Election as Subscribers to the Lectures of the Institution. – –

~~Mrs Bryan No 1 Gloucester Place Portman Square~~
~~Mrs D. Delap 6 Queen Ann Street Westminr~~
Holt Leigh Esqr 14. Duke St Westminster
Dr Tuthill 13 Soho Square
Revd John Crofts 11. Geo.e St Portman Square.

Minutes of Com.e of Accts Approved

7. Read & approved the Minutes of the Committee of Accounts of this day. – – –

Bills paid. × See over the next leaf. – –

8. Ordered payment of the following Bills viz.
C. Royce for Carpenter's Work in 1809 £ 35 - 3 - 6
J. Newman for Apparatus &c 1809 & 1810 – 7 - 6 -
Mrs Greenwood Sundries for the House &c 1 - 6 - 8

9. – Mr Davy was requested to order 5 Chaldron of Coals & 6 dozn of Candles for the use of the Institution

10. Ordered. That it be inserted in the Cards for the Lectures, that the days of attendance in the Mineral Room will be on Tuesdays & Fridays instead of Mondays & Wednesdays.

Repairs.

11. C. Royce reported that the repairs of the Slating, Chimneys, & Skylights of the Laboratory would cost £ 16. – Ordered to be executed under the Superintendance of Mr Davy. –

Mr Skelton's letter assigning £42.0.10 as the value of each Life-proprietor's Share, & £14.0.2 as that of each Life Subscriber

12. The following valuation of the Proprietors & Life Subscribers Shares of the Institution was received from Mr Skelton Actuary of the Westminster Life Insurance & Annuity Office & read. –

West.r Life Insce & Annuity Office
24th March 1810

"Sir,

I have read the Abstract of the Charter & the Bye Laws of the Royal Institution as also the statement given to me by Sir Thos Bernard & yourself & having considered the subject upon which my Sentiments are desired & attended to the explanations made at our Meeting on Tuesday, I venture my opinion in the hope of its being received as an equitable disposition of the Assets of the Institution in its peculiar situation, though it may probably not meet an entire approval from all who are interested, as the Circumstances may not appear to every one in the same point of view.

The present worth of the property I take from Representation as follows

House and Buildings	£ 14500 - -
Fine due thereon – –	1535 - 10 -
Furniture – – – –	1252 - 9 - 6
Library – – – – – –	7095 - 7 -
Mechanical Apparatus Models &c	787 - 14 - 6
Mineral Collection – – –	400 - -
Laboratory Apparatus –	140 - -
	£ 25711 - 1 -

It appears by the Bye Laws & the Explanation that independent of the Interest of the Proprietors & Life Subscribers, there are some who have peculiar Rights in the Library of Reference to the number of 42 Hereditary & 5 Life Patrons & that a portion of the Property should be as accountable for to them, & that there are likewise demands upon the Institution which should also be admitted & deducted from the value of the Profits

Therefore say value of all the Property		£ 25711 - 1 -
Debts – –	£ 2000 - -	
Privileges & Rights of the Hereditary & Life Patrons	3000 - -	
	– – – –	5000 - -
Remains		£ 20.711 - 1 -

to be divided by 374 Proprietors who have Interests & Privileges so far exceeding those of the 356 Life Subscribers as to give them individually a just & equitable demand upon the Property in the proportion (with submission) of three parts out of four: admitting this as

60 | 1810 Mar. 26.

fact, this Question only remains. What is the value of each Individuals Interest in the £ 20.711 - 1 - And that appears to be £ 42 - 0 - 10 to each of the 374 Proprietors & £ 14 - 0 - 2 to each of 356 Life Subscribers, fractions excepted. – – –

There is another mode differing but little from the foregoing division & keeping nearly the same Idea. divide £ 20.711 - 1 into 4 parts giving three to the 374 – £ 15533 - 5 - 9; & one to the 356 ~~Life Subscribers~~ £ 5177 - 5 - 3; the individual Interest of the 374 will then be £ 41 - 10 - 8 – and of the 356 – £ 14 - 10 - 10 – fractions excepted.

I remain
With much Respect
Yr obedt hble Servt
(Signed) Robt Skelton"

James P. Auriol Esqr

Ordered. That this Letter be sent to Mr Bramwell & that he be desired to furnish 20 Copies of the Bill before Parliament.

13 Resolved. That the sum of £ 42 - 0 - 10 be the Amount to be inserted in the Blank of the Bill as the value of each Proprietor's Share in the Institution. – –

Charter sent to Mr Bramwell.

14. Mr Auriol reported that in consequence of a Note from Mr Bramwell to Mr Savage dated 14th March 1810, desiring that the Charter of the Institution might be produced before the House of Commons – Mr Auriol accordingly attended with it, on the 15th Inst. & delivered it to Mr Bramwell's Clerk whose receipt for the same is deposited in the Iron Chest. – –

× See back a leaf.

Ordered. That a half year's Salary due to Mr Savage at Christmas last be paid amounting to – – – – – £ 25 - 0 - 0.
Signed an Order for payment of the above & also for an Advance of £ 30 – – to Mr E. Davy.

1810 Mar. 26. | 61

Ordered That Mr Savage's Bill for Printing be referred to Mr Davy.

Dundas

62 1810. April 2.

Managers' Meet.g 2nd April, 1810. Monday

At a Meeting of the Managers of the Royal Institution of Great Britain, held at the House of the Institution, on the 2nd day of April 1810

Present.
The Rt. Hon. Lord Dundas in the Chair.
Wm. Watson Esqr.
Sir Thos. Bernard Bart.
Charles Hatchett Esqr.
J. P. Auriol Esqr.

Minutes app.d 1. The Minutes of the Proceedings of the last Meeting were read & approved.

Ex.d Gen. Acct. 2. Examined the General Account of the Institution.

Proprietors Elected. 3. Proprietors Elected.
The following Persons, proposed as Candidates for Election, as Proprietors of the Institution at a Meeting of the Managers, held on the 5th day of March 1810, were balotted for, & duly elected.
The Revd. Thos. White No. 54 High Street Mary le bone, in the Room of Thomas Slack Esqr. who resigns.
The Right Hon. Lord Geo. Henry Cavendish, in the room of Henry Cavendish Esqr. deceased; the Chairman reporting that he had examined the Will of the late Mr. Cavendish, by which it appeared that Lord George Henry Cavendish is left sole Executor.

Proprietors Proposed. 4. Proprietors Proposed.
The following Persons were proposed as Candidates for Election as Proprietors of the Institution.
George Duckett Esqr. in the room of his Father Sir Geo. Duckett Bart. of

63

Upper Grosvenor Street, who is desirous of parting with his Share in favour of his Son & recommends him to be balotted for. By J. P. Auriol Esqr.

William Finch Palmer Esqr. No. 10 Lower Berkley Street in the room of Thomas Palmer Esqr. of Berkley Square, who is desirous of parting with his Share in his favour, & recommends him to be balotted for. By J. P. Auriol Esqr.

Charles Dalston Nevinson M.D. in the room of George Sulivan Marten Esqr. who begs to resign in his favour By Chas. Hatchett Esqr.

Annl. Subscribers Elected. 5. Annual Subscribers Elected
The following Persons, proposed as Candidates for Election as Annual Subscribers to the Institution, were duly elected.
Wm. Dickenson Esqr. No. 28. Bury St. St. James's
James L. Grant Esqr. 8 Hanover St. Hanover Square

Annl. Subsc. Proposed. 6. Annual Subscriber Proposed.
Mr. John Cole No. 30 Norton Street, was proposed as a Candidate for Election as an Annual Subscriber to the Institution

Subscribers to Lectures Elected. 7. Subscribers to the Lectures Elected.
The following Persons, proposed as Candidates for Election as Subscribers to the Lectures of the Royal Institution, at the last Meeting of the Managers, were duly elected.
Holt Leigh Esqr. No. 14 Duke St. Westminster
Dr. Tuthill — 13 Soho Square.
The Revd. John Crofts 11. Geo. St. Portman Square

Subscriber to Lectures Proposed. 8. Subscriber to the Lectures Proposed.
Dr. Morgan No. 109 Great Russel Street was proposed as a Candidate for Election as a Subscriber to the Lectures of the Institution.

9 Examined the following Bills.

64 1810. April 10.

1 Bills Ex.d

D. Ramsay & Son for Edinburgh Evening Courant from July 22. 1808. to Jan. 22. 1810	6. 10. 0
J. Smethurst for Lighting 2 Lamps 2 years to March 1810	7. 4. 0
H. Hole for Winding up Clocks &c in 1809	3. 10. 6

Ordered to be paid by the Steward.

10. Ordered. That the notice from the Sun Fire Office be sent to the Treasurer & that he be requested to pay the Premium amounting to £30. 17. 6.

11. Mr. Davy is requested to insert in the Papers, a proper Advertisement for Dr. Smith's Lectures to commence on Tuesday the 10th of April.

Minutes of Com.e of Acc.ts approved 12. Read & approved the Minutes of the Committee of Accounts of this day.

Dundas

65

Managers' Meet.g 9th April, 1810.

At a Meeting of the Managers of the Royal Institution of Great Britain held at the House of the Institution on Monday the 9th day of April 1810.

Present.
The Rt. Hon. Lord Dundas in the Chair.
Sir Thos. Bernard Bart.
Chas. Hatchett Esqr.
John Symmons Esqr.
J. P. Auriol Esqr.

Minutes approved. 1. The Minutes of the Proceedings of the last Meeting were read & approved.

Ex.d Gen. Acct. 2. Examined the General Account of the Institution.

Annl. Subscriber Elected. 3. Annual Subscriber Elected.
Mr. John Cole No. 30 Norton Street, proposed as a Candidate for Election as an Annual Subscriber to the Institution, was duly elected.

Annl. Subs. Proposed. 4. Annual Subscriber Proposed.
The Revd. Mr. Rowlatt Charter House Square was proposed as a Candidate for Election as an Annual Subscriber to the Institution.

Subscriber to Lectures Elected. 5. Subscriber to the Lectures Elected.
Dr. Morgan No. 109 Great Russel Street, proposed as a Candidate for Election as a Subscriber to the Lectures of the Institution, was duly elected.

Minutes of Committee of Acct.s approved. 6. Read & approved the Minutes of the Committee of Accounts of this day.

7. C. Royce's Estimate of the repairs wanting in the water Closet was laid upon the Table,

by which it appeared that it would cost £6·10·0.
Ordered That the above be executed under the direction of Mr. Davy, & that he be requested to order the broken windows to be repaired. –

Salaries to be paid — 8. Signed an Order for the payment of a quarter's Salary to the following Persons. also for an advance to the Steward of £30·0·0 – –

H. Davy Esqr. due at Michalmas	1809	100·0·0
Mr. Wm. Harris do. – do.		45·0·0
Mr. E. Davy – do. – do.		17·10·0
Mr. D. Brooks – do. – do.		13·2·6
Mr. Sturt to Christmas	1809	17·10·0

9. Mr. T. Perkins an Annual Subscriber, having lost his Ticket of admission. – Ordered. That Mr. T. Perkins have personal admission to the Lectures this Season. –

10. Ordered. That the Visitors be summoned to meet on Friday next at one Oclock. to examine the Annual Accounts of the Institution. – –

Proprietor Proposed — 11. The Minute No. 4. of April 3. 1809. was read & the Nomination of himself as a Proprietor by William Law the Younger, the sole proving Executor under the Will of the late James Law Esqr. was laid before the Managers. –
Resolved. That the Name of the said William Law be balotted for as a Proprietor in the Room of the late James Law Esqr. at the next Monthly Meeting on Monday the 7th. day of May next. – – – –

12. Resolved. That the Treasurer be desired to apply to the Comptroller of the City of

Application for Renewal of Lease — London, for the renewal of the City Lease being immediately executed, & for the Counterpart thereof to be sent to the Institution in order to be executed by the Corporation; & that he be authorized to pay the Fine of £600 for the Lease, together with the Charges, on account of the Institution

T. Bernard

Managers' Meeting. Apl. 16, 1810 — At a ~~Joint~~ Meeting of the Managers ~~& Visitors~~ of the Royal Institution of Great Britain, held at the House of the Institution on Monday the 16th. day of April 1810.

Present.
Sir Thos. Bernard Bart. in the Chair.
John Symmons Esqr. }
Chas. Hatchett Esqr. } Managers.
J. P. Auriol Esqr. }

~~[illegible]~~ }
~~Saml. Thornton Esqr.~~ } ~~Visitors~~
~~Thos. Hammersley Esqr.~~ }

Minutes approved. — 1. The Minutes of the Proceedings of the last Meeting were read & approved. –

Exd. Genl. Acct. — 2. Examined the General Account of the Institution. – – – – –

Annl. Subscriber Elected — 3. Annual Subscriber Elected.
The Revd. Mr. Rowlatt Charter House Square. proposed as a Candidate for Election as an Annual Subscriber to the Institution at the last Meeting of the Managers, was duly elected. –

Annl. Subscrs. Proposed. — 4. Annual Subscribers Proposed.
The following Persons were proposed as Candidates for Election as Annual Subscribers to the Institution. –
The Hon. Peter. D. Burrell. Piccadilly
Saml. Bates Ferris Esqr. 62 South Moulton Street. –

Subscriber to Lectr. Proposed. — 5. Subscriber to the Lectures Proposed.
Mr. John Primatt Maud 66. Aldersgate Street. was proposed as a Candidate for Election, as a Subscriber to the Lectures of the Institution. – – – –

Present. — 6 The Right Hon. John Trevor presented The late Lord Hampden's Poems.
Ordered. That the thanks of the Managers

be returned for this Present. –

Minutes of Comm. of Accts. approved. — 7. Read & approved the Minutes of the Committee of Accounts of this day. –

Proprietor proposed — 8. Sir Thos. Bernard stated to the Committee, that Mrs. Chaplin had represented that the Proprietor's Share now vested in her, is in Trust for Sir Godfrey Webster, the eldest Son of the late Proprietor, to whom she is desirous of transferring it. before the Bill passes: Sir Thomas therefore proposes the Name of Sir Godfrey Webster Bart. as a Proprietor in the room of Mrs. Chaplin. – – –
Resolved. That in case a proper Nomination as to the said Share, be sent in due time to the Managers signed by Mrs. Chaplin; the Name of Sir Godfrey Webster be balotted for on the first Monday in May.

Treasurer instructed to attd. at Office of Controller of City of London to execute Lease. — 9. Mr. Savage reports that he has seen the Comptroller for the City of London, respecting the Renewal of the Lease of the Premises held by the Royal Institution under the City, & that he had stated it would be necessary to attend at his Office on Tuesday morning with the original Lease & Assignment, & the Amount of the Fine & Expences attending the Renewal which will be about £626 – – when the Lease would be executed without any delay; otherwise the Business will be referred to a Committee, which would occasion considerable delay & expence. – –
Resolved. That the Treasurer be requested to attend at the Office of the Comptroller for the City of London tomorrow morning at ten Oclock, or to send a person with the above Sum, & that Mr. Savage do also attend with the original Lease & assignment. –

Proposed Report of Visitors — 10. The Report of the Visitors proposed to be made at the next Annual General Meeting

70 — 1810.

of the Proprietors, having been read at a Joint Meeting of Managers & Visitors this day.–

Ordered. That it be printed & circulated to the Proprietors with the usual Notice of the General Meeting to be held at the House of the Institution on the first day of May next at one OClock.–

Managers recommended for Election — Resolved. That the following Managers be recommended for Election,– at the Annual General Meeting of Proprietors

The Earl of Winchilsea as President–
The Earl of Dartmouth
The Earl of Egremont
The Earl of Chichester
The Rt. Hon. Isaac Corry [illegible] &
John Symmons Esqr. as Managers
for three years

Secretary. — Resolved. That James Peter Auriol Esqr.– be recommended for Election as Secretary for the ensuing year.––

Dundas

Apl. 16.

Joint Meeting Apl. 15, 1810 — At a Joint Meeting of the Managers & Visitors of the Royal Institution of Great Britain held at the House of the Institution on Monday the 16th. day of April 1810.

Present.
Sir Thos. Bernard Bart. in the Chair.
John Symmons Esqr.
Charles Hatchett Esqr.
J. P. Auriol Esqr. —— Managers.

The Lord Visct. Valentia
Thomas Hammersley Esqr.
Samuel Thornton Esqr. —— Visitors.–

Minutes approved 1. The Minutes of the last Meeting of the Visitors were read & approved.–

Report to be printed 2. Read a Draft of a Report, & ordered the same to be printed.––

Visitors recommended 3. Resolved. That The Earl of Bridgewater The Earl of Buckinghamshire. The Lord Visc. Valentia. The Right Hon. Lord Boringdon & The Right Hon. R. P. Carew be recommended for Election as Visitors for the next three years; & Scrope Bernard Esqr. as Treasurer for the ensuing year

Dundas 71

72 — 1810.

Manager's Meeting. 23 Apl. 1810. — At a Meeting of the Managers of the Royal Institution of Great Britain, held at the House of the Institution, on Monday the 23d. day of April 1810.–

Present
The Rt. Hon. Lord Dundas in the Chair.
The Earl Spencer
Sir Thos. Bernard Bart.
Sir J. C. Hippisley Bart.
J. P. Auriol Esqr.

Minutes approved. 1. The Minutes of the Proceedings of the last Meeting were read and approved.–

Exd. Gen. Acct. 2. Examined the General account of the Institution.–

Annl. Subsrs. Elected 3. Annual Subscribers Elected.
The following Persons proposed as Candidates for Election, as Annual Subscribers to the Institution, at the last Meeting of the Managers, were duly elected.––
The Hon. Peter D. Burrell. Piccadilly.
Saml. Bates Ferris Esqr. 62. South Moulton St.

Subscriber to Lectures elected 4. Subscriber to the Lectures Elected.
Mr. John Primatt Maud 66 Aldersgate St. proposed as a Candidate for Election, as a Subscriber to the Lectures of the Institution, at the last Meeting of the Managers was duly elected.––

Subscriber to Lectures Proposed 5. Subscriber to the Lectures Proposed.
The Revd. Mr. Archdeacon Cambridge. Twickenham was proposed as a Candidate for Election as a Subscriber to the Lectures of the Institution

Minutes of Com. of Accts. approved. 6. Read & approved the Minutes of the Committee of Accounts of this day. & signed an Order for an advance of £30–– to the Steward; also an Order for the advance of £30–– to Mr. Savage in part

Apl. 23.

pt. payment of his Bill for Printing.––

7. Resolved. That Mr. Auriol be requested to order the Return to be made for the Property & Assessed Taxes in the usual manner.

8. Ordered. That a hundred weight of Soap be procured under the direction of Mr. Davy.

Royal Assent to Bill 9. The Secretary read the following Report from Mr. Bramwell

Sir, Temple. 18th. April 1810
I have the Pleasure to inform you, that the Royal Institution Bill, has this day received the Royal Assent
I have the Honour to be
J. P. Auriol Esqr. &c. (Signed) Geo. Bramwell.

Ordered. That the above be reported at the General Meeting on the first of May

Proposition for Code of Bye Laws — Resolved. That it be submitted to the General Meeting, whether it will not be proper to appoint a Committee from their Body to confer with the Managers & Visitors upon the subject of preparing a Code of Bye Laws under the New Act.––

500 Copies of Act to be printed & distributed — Ordered. That 500 Copies of the Act be printed, & that one be sent to each Member of the Institution.–––

10. Resolved. Unanimously. That the Thanks of the Managers be returned to Sir Jno. Sinclair for the attention he has given in Conducting the Bill of the Institution through the House of Commons.

First Report from Mr. Savage about terms of Renewal, Lease having lapsed 11. Mr. Savage Reports to the Managers, that he attended at the Office of the Comptroller for the City of London, with the original Lease & Assignment of the Premises held by the Institution

73

Report concerning Renewal of Lease

Institution, according to their directions; & that Mess.rs Ransom, Morland, & Co. tendered £.626 being the Sum which the Comptroller specified; & that this was not received, the answer being "that a much larger sum was to be paid."—

—In consequence of this answer, Mr. Savage immediately waited on the Comptroller to ascertain the truth of these contradictory statements; who informed him, that in the first instance he had mistaken the time, believing it to have been a Lady Day Renewal, in which case it would have been done as a matter of course; but that on examining the Books he found it ought to have been renewed at Michaelmas, that the Lease had lapsed, & that of necessity it must go before the Committee; & that the Arrears of Ground Rent, for 2¼ years must be paid with the Fine, also the Interest for the Time elapsed, & the Fees of Lease, & that the Committee would expect a Present to the Poor's Box; for which he supposed they would be satisfied with Five Guineas, or ten at the most.——

Mr. Savage thought he was acting right, in signing the Petition for the Renewal, to prevent any further delay; & in promising to attend the Committee on Wednesday the 2nd of May, at One O'Clock, for the purpose of paying the Money, when, the Comptroller told him, he had no doubt the Committee would immediately order the Lease to be executed.—

The Sum that will be required, according to the Comptroller's Account will be.

Fine & Interest about	£ 618 - 0 - 0
Rent	337 - 10 - 0
Fees of Lease	22 - 17 - 6
Poor's Box	5 - 5 - 0
	£ 983 - 12 - 6

Resolved. That a Copy of this Report be sent to the Treasurer, & that he be requested to take Measures for the Discharge of the above & obtaining the Renewal of the Lease.—

12. Ordered. That Mr. Savage Report to the Managers at their next Meeting, what Arrears are due to the Institution, from the Under Tenants for Ground Rent.—

Present to Library 13. Lord Visct. Valentia having presented to the Library of Reference, the large Paper Copy of his Work value Fifty Guineas.

Resolved. That it be submitted to the Committee of Patrons whether Lord Valentia shall not be balotted for as a Patron of the Library for Life.

T. Bernard

Managers Meeting 30 Apl. 1810

At a Meeting of the Managers of the Royal Institution of Great Britain held at the House of the Institution, on Monday the 30th day of April 1810.

Present.

Sir Thomas Bernard Bart. in the Chair.
Sir J. C. Hippisley Bart.
William Watson Esqr.
John Symmons Esqr.
J. P. Auriol Esqr.

Minutes approved 1. The Minutes of the Proceedings of the last Meeting were read & approved.—

Exd. Genl. Acct. 2. Examined the General Account of the Institution

3. Subscriber to the Lectures Elected.

Subscriber to Lectures Elected

The Revd. Archdeacon Cambridge Twickenham, proposed as a Candidate for Election as a Subscriber to the Lectures, at the last Meeting of the Managers, was duly elected.

Minutes of Comm. of Accts approved. 4. Read & approved the Minutes of the Committee of Accounts of this day.—

Annl. Subsr. 5. Mr. Thomas Babington a Subscriber to the Lectures, wishes to become an Annual Subscriber, by paying the difference.

Ordered. That his request be complied with.

Ticket lost 6. Mr. Alexr. Garden a Subscriber to the Lectures, having lost his Ticket of admission.

Ordered. That Mr. Alexr. Garden have personal admission to the Lectures for this Season & that the ticket be stopped, if presented.

Mr. Davy's Lecture. 7. Resolved. That the Lecture of Mr. Davy which has been printed, be advertized for Sale, under the Orders of Mr. Davy; & notice of it inserted on the Weekly Cards.

Ground Rent due to Institution 8. Mr. Savage reported "That at Midsummer next, there will be two years Ground Rent due to the Institution from No. 22, and one Year due from No. 23.

2 Years from No. 22	£96 · 0 · 0
1 year from No. 23	74 · 0 · 0
	£170 · 0 · 0

Committee for regulating the Secretary's Office 9. Resolved: That Sir Thos. Bernard, Sir J. C. Hippisley & Mr. Auriol be a Committee to consider the Regulation of the Secretary's Office, & the proper Establishment for conducting the Business thereof in future, in consequence of the absence of Mr. Easingwood, & to make their Report to the Board.

Payment of Sum due for Renewal of Lease from Exors of Mr Gray. 10. Resolved. That Mr. Davy be requested to call on Mr. Tenant to enquire from him, when the Institution may rely on being paid the Sum due for the renewal of the Lease to the Executors of Mr. Gray, stating to him the inconvenience which has been sustained by the Institution from the delay, particularly in the extra charge they have been put to, for renewing the Lease under the City of London.—

T. Bernard

78

Managers' Meet. May 7. 1810.

At a Meeting of the Managers of the Royal Institution of Great Britain, held at the House of the Institution, on Monday the 7th day of May 1810.

Present.

Sir Thos. Bernard Bart. in the Chair.
Sir J. C. Hippisley Bart.
Wm. Watson Esqr.
Jno. Symmons Esqr.
J. P. Auriol Esqr.

Minutes approved. 1. — The Minutes of the Proceedings of the last Meeting were read & approved.

Ex. Gen. Acct. 2. — Examined the General Account of the Institution.

Annl. Subscriber proposed. 3. Annual Subscriber Proposed.
Mr Josh. Trowd Spencer, Henrietta Street Covent Garden, was proposed as a Candidate for Election as an Annual Subscriber to the Institution.

Minutes of Com. of Accts approved. 4. Read & approved the Minutes of the Committee of Accounts of this day.

Present. 5. Present received.
Bracy Clark Esqr. presented his "Series of Experiments on the Foot of the living Horse."
Ordered. That the Thanks of the Managers be returned for this Present.

Communication from Mr Tennant on Sums due to the Institution by Execrs of Mr Grey. 6. Mr Davy made the following Report.
"Mr Davy has the honour of informing the Managers, that in consequence of their desire, he has had an interview with Mr Tenant, who informed him on Saturday, that he had no doubt but that the Fine due to the Royal Institution, from the Representatives of the late Mr Grey, will be paid within the present Month." —
Monday May 7th 1810

1810. May 7. 79

Report recommending the appointment of 2 Clerks at £100 each per ann. 7. — "The Select Committee report their opinion that it will be proper in the present state of the Institution, to have two regular Clerks employed; one the Clerk and Assistant in the Laboratory, whose business it shall be, under the direction of Mr Professor Davy, to prepare an Account of the Scientific progress of the Institution; attending in the Mineralogical Collection, or Secretary's Office a limited time each day to give any explanation that may be wanted on that subject. For this appointment they beg leave to recommend Mr Edmund Davy. *(Mr Ed. Davy)*

"The other Clerk who will be wanted will be "Clerk of Accounts;" who shall be constantly in the Secretary's Office, & be employed in keeping the Accounts, receiving Subscriptions, issuing Tickets, & doing the ordinary Business of the Office; both of the Clerks attending the Meetings of the Managers when wanted, to give any information or assistance in their respective departments. — For this purpose Sir John C. Hippisley has recommended Mr Joseph Fincher whose character & capacity he is well acquainted with; & for whom he is ready to give a Security in the sum of £1000. — *(Mr J. Fincher)*

"They recommend as the Salary of each of these Clerks, the Sum of £100 a year, & that Mr Edmund Davy have Mr Easingwood's Apartment subject to its being occasionally used, when wanted for Committees. As it may not be convenient at present to accommodate the Clerk of Accounts in the House, they suggest an allowance pro tempore of £20 a year to him for his Apartment, Coals, & Candles out of the House.

"Taking the subject of the Office into consideration, they submit the propriety of providing by the Bye-Laws for the Election

80 1810.

some one of the Members to act as Asst. Secretary

"of some one of the Members of the Institution, to act as Assistant Secretary; & of Mr Davy (in case he should be induced to accept it) & some other Scientific person to be Secretary & Assistant Secretary of the Laboratory. — These three Appointments to be gratuitous & honorary.

"In addition to this, they suggest that no receipt shall in future be given for Subscriptions nor Tickets issued except with the signature of the Secretary or Assistant Secretary, countersigned by the Clerk of Accounts, with exception only of Subscriptions received by one of the Bankers of the Institution. They also suggest that a Book shall be kept in the Secretary's Office, & laid before the Managers at all their Meetings, containing an Account of all Monies due to or from the Institution; so that as to Rents, Monies due in respect of Election of Members, & all other Sums due to the Institution, except Annual Subscriptions, & also as to all Bills & Debts due from the Institution, a statement thereof may be always before the Managers at all their Meetings.

Mr Savage's resignation as Assistant Secretary. 8. The following Letter from Mr Savage addressed to Sir Thos. Bernard, was laid before the Managers.

"Sir,
"In addition to the many Acts of kindness which I have received from you, since I have had the honour of being known to you, I now take the liberty of asking another; which is, that you, Sir, will have the goodness to tender my resignation to the Managers, at their Meeting on Monday, as Assistant Secretary to the Royal Institution. — To the Managers I beg leave to

May 7. 81

"return my most unfeigned thanks, for the numerous favours which I have received from them during a term of nearly ten years, which I have been in their Service; & to request that my Resignation may take place on the 12. Instant, when that term will be completed.

"The increase of my private business has rendered it impossible that I could attend properly both to that & the duties attached to my Situation in the Institution; & I have long determined to resign my Office at the expiration of the ten years; but the late transaction, which unfortunately took place, induced me to postpone the notice to the present time. —
"With my sincere wishes for the prosperity & utility of the Institution.
"I have the Honr. to be &c Willm. Savage.
Sir Thos. Bernard Bart.
& — & — &

9. Resolved. That the Resignation of Mr Savage as under Secretary be accepted; & that he be continued in his Situation of Printer to the Royal Institution, the Managers being fully sensible of the many Services which he has rendered to the Institution, during the preceding ten years.

Propositions respecting Secretaries &c. 10. Resolved. That the following propositions be taken into consideration by the Board on Monday next.

That instead of an Arrangement of an Assistant Secretary (Mr Savage) at £50 a year, a Steward & Clerk of Accounts (Mr Easingwood) at £100 a year, & an Assistant in the Laboratory (Mr Edm. Davy) at £70 a year, the following arrangement be made pro tempore at Salaries amounting to the same Total Sum.
Mr Edmd. Davy to be Clerk & Assistant in the Laboratory at £100 a year to commence

82 — 1810.

Mr. Joseph Fincher to be clerk of Accts at £120 per annum.

from the 19th. of February last, the day of Mr. Easingwoods quitting the Institution:

Mr. Joseph Fincher to be Clerk of Accounts at £100 a year from this day, with an allowance of £20 a year for finding his own Apartment, Coals, & Candles, until he can be accommodated in the House.

That the other Matters recommended by the said Report, be referred to the Secretary to direct the same, (so far as he shall find it expedient) to be carried into Execution.

Dundas

May 11. — 83

Joint Meeting 11th May 1810.

At a Meeting of the Managers, Visitors, & Committee of Members of the Royal Institution of Great Britain, held at the House of the Institution, on Friday the 11th. day of May 1810.

Present.

The Rt. Hon. Lord Dundas. *in the Chair*
Sir Thos. Bernard Bart.
Willm. Watson Esqr.
J. P. Auriol Esqr. — *Managers.*

The Hon. Robt. Clifford
Sir H. C. Englefield Bart.
The Revd. Edwd. Forster
James Stodart Esqr.
W. H. Pepys Esqr.
Henry Warburton Esqr.
Humphry Davy Esqr. — *Committee.*

Arrangement to be adopted in forming Bye Laws.

Resolved Unanimously. That the following mode of Proceeding be adopted.

1. That the Committee do decide on certain heads to form the Chapters of the Bye Laws.
2. That each head be taken up separately, & treated of completely, at the same Meeting, if possible.
3. That at each Meeting, a fair Copy be produced and read of the Proceedings of the former Meeting, when every thing not objected to, shall stand confirmed.
4. That the Chapters so confirmed, be immediately sent to the Printer.
5. That the Committee do adjourn de die in diem after Monday, the 14th, and meet at 12 precisely: and that 3 Members be a quorum.
6. That these Resolutions be printed & sent round to the Members of the Joint Committee, as a General & Final Summons. (Signed) Dundas.

84 — 1810.

Resolved. That the following be the Provisional Arrangement of the chapters of the Bye-Laws.

1. Of the Government, and of the making and repealing of Bye-Laws.
2. Of the Election, Rights, & Privileges of the Members.
3. Of Honorary Members.
4. Of the Election of Officers.
5. Of the Duties of the President, & Vice Presidents.
6. Of the Duties of the Committee of Managers.
7. Of the Duties of the Committee of Visitors.
8. Of the Duties of the Treasurer.
9. Of the Duties of the Secretary, & his Assistants.
10. Of the General Meetings of the Members.
11. Of the Property of the Institution.
12. Of the Common Seal, and of Deeds.
13. Of Life and Annual Subscribers.
14. Of the Causes and Forms of Ejection.
15. Of the Internal Regulation, of the House of the Institution.
16. Of Scientific Committees.
17. Of the Weekly Meetings of the Members.
18. Of the Professors, Lecturers, and Lectures.
19. Of the Publications.

May 11. — 85

20. Of the Libraries, & Mineralogical Collection.
21. Of the Laboratory, Apparatus & Model Room.
22. Of Benefactors.

Resolved. That the Committee do adjourn to Monday the 14th. Instant, at 12 o'Clock precisely, when the attendance of the Members is requested.

(Signed) Dundas. [illegible]

Managers' Meeting, 14 May 1810.

At a Meeting of the Managers of the Royal Institution of Great Britain, held at the House of the Institution, on the 14th day of May 1810.

Present

The Rt. Hon. Lord Dundas, in the Chair
Sir Thos. Bernard Bart.
B. Hobhouse, Esqr.
J. P. Auriol, Esqr.

Minutes approved. 1. The Minutes of the Proceedings of the last Meeting were read and approved. —

Exd. Genl. Acct. 2. Examined the General Account of the Institution. —

Annl. Subr. Elected. 3. Annual Subscriber Elected.
Mr. Joseph Frowd Spencer No. 22 King Street Covent Garden proposed as a Candidate for Election as an Annual Subscriber to the Institution, at the last Meeting of Managers was duly elected.

Minutes of Com. of Accts. approved. 4. Read and approved the Minutes of the Committee of Accounts of this day.

Ticket retd. 5. The Rt. Honble Isaac Corry's Ticket having been transmitted to him in Ireland by mistake and paid for but not having been used. Ordered that upon its being sent back the money be returned.

Report of Select Committee read and adopted. 6. Read the Report of the Select Committee to the Committee of Monday last, together with Mr. Savage's Resignation and the Propositions or Resolutions proposed on Monday last for Consideration on this day.
Resolved. That the same be confirmed and adopted with this variation that no change shall at present be made in the Apartments of the House. That Mr. Fincher's Salary and Allowance commence from this day, Sir John Hippisley executing the proposed Security to the satisfaction of the Secretary.

Commencement of Mr. J. Fincher's Salary and Allowance.

Spencer.

Managers' Meetg. May 21. 1810

At a Meeting of the Managers of the Royal Institution of Great Britain, held at the House of the Institution on the 21st day of May 1810.

Present

The Earl Spencer, in the Chair.
Sir Thos. Bernard, Bart.
Benjn. Hobhouse, Esqr.

Minutes approved. 1. The Minutes of the proceedings of the last Meeting were read and approved.

Exd. Genl. Acct. 2. Examined the General Account of the Institution.

Subsr. to Lectures proposed. 3. Subscriber to the Lectures proposed.
George Burrows Esqr. No. 34 Bloomsbury Square was proposed as a Candidate for Election as a Subscriber to the Lectures of the Institution.

Minutes of Comee. of Accts. approved &c. 4. Read and approved the Minutes of the Committee of Accounts this day and signed an Order for an advance to Mr. E. Davy of £5.14.10 the Balance of his Account and to Mr. Joseph Fincher of Thirty Pounds on account of the Institution.

Proprietor's compensation claimed. 5. A Letter was received from Mr. Benjn. Blake of Hans Place giving Notice that he will receive the compensation of £42.0.10 for the relinquishment of his proprietory share in the Royal Institution and requesting to know when the Money due would be paid.
Ordered. That Mr. Blake be informed that the Money is expected to be paid in the course of two Months, and that as soon as the day of payment is fixed, due Notice shall be given thereof.

Form for keepg. Regular Acct. of Monies due to & from the Institution. 6. Sir Thomas Bernard laid before the Managers a form for keeping a regular Account of Monies due to and from the Institution
Ordered. That the Clerk of Accounts prepare a List of the Debts and Credits of the Institution arranged according to the form.

Bond of fidelity of £1000 executed by Sir J. C. Hippisley & J. Fincher as Clerk of Accounts. A Bond of Security in the penalty of One thousand Pounds dated the 19th May 1810 and executed by Sir J. C. Hippisley and Mr. Joseph Fincher, for the faithful —

discharge of the duty of the said Joseph Fincher as Clerk of Accounts to the Royal Institution was laid upon the Table and Ordered to be delivered by Mr. E. Davy to Mr. Auriol to be deposited with the other Papers in the Royal Institution.

T. Bernard

Managers' Meeting, 28 May 1810

At a Meeting of the Managers of the Royal Institution of Great Britain, held at the House of the Institution on the 28th Day of May 1810.

Present

Sir Thos. Bernard Bart. in the Chair
Sir J. C. Hippisley Bart.
J. P. Auriol Esqr.

Minutes approved. 1. The Minutes of the proceedings of the last Meeting were read and approved.

Exd. Gen. Acct. 2. Examined the General Account of the Institution.

Subsr. to Lectrs. Elected. 3. Subscriber to the Lectures Elected
George Burrows Esqr. No. 34 Bloomsbury Square proposed as a Candidate for Election, as Subscriber to the Lectures of the Institution at the last Meeting of the Managers was duly elected.

Minutes of Com. of Accts. approved. 4. Read and approved the Minutes of the Committee of Accounts of this day.

Presents. 5. Presents received
Mr. Penn presented "Illustrations of Virgils Fourth Eclogue"
Mr. T. Brande presented "Observations on the Effects of Magnesia &c."
Ordered. That the Thanks of the Managers be returned for these presents.

6. Ordered That Mr. Brooks be desired to collect the unpaid Subscriptions due to the Voltaic Battery and that Mr. J. Fincher do furnish him with a List of the same.

7. Read a Letter from Mr. Douglas Guest offering to give a gratuitous course of six Lectures on the Arts during each Season, until such period as change or circumstances may dictate.
Resolved That it be referred to Lord Dundas, Mr. Hatchet and Mr. Davy to consider the same.

Duties of Mr. E. Davy & Mr. J. Fincher. 8. Duties of Mr. E. Davy and Mr. J. Fincher.

That Mr. Edmund Davy be informed that it is his duty

90 1810

to attend as Clerk of the Committee on Bye Laws and Scientific Committees and to assist and attend the Professor of Chemistry in the Laboratory and Lecture Room and upon all Records of Scientific Investigation.

That Mr. Fincher be informed that it is his duty constantly to attend as Clerk and have the care of the Secretary's Office, and to keep the Accounts, receive Subscriptions, issue Tickets, keep the Minutes and Books, belonging to the Managers and Visitors, and the Patrons of the Library and to attend at their Meetings.

Winchilsea

91

Joint Meeting 14th May 1810.

At a Meeting of the Managers, Visitors & Committee of Members of the Royal Institution of Great Britain, held at the House of the Institution the 14th day of May 1810.

Present.

Timothy Brent Esqr. in the Chair
The Right Hon. Lord Dundas.
The Hon. Robt. Clifford
The Revd. E. Forster
Henry Warburton Esqr.
J. P. Auriol Esqr.
Humd. Davy Esqr.

Minutes confirmed. 1. The Minutes of the last Meeting were read & confirmed.

Byelaws Chap I 2. Resolved. That the first Chapter of the Bye Laws be "Of the Government, and of the enacting and repealing of Bye Laws" & stand as follows. Vide. Minutes the 29 May 1810.

" Chap. II 3. Resolved. That the second Chapter of the Bye Laws be "Of the Election, Rights, and Privileges of the Members" & do stand as follows. Vide Minr. the 29. May 1810.

Timy. Brent

92 1810

Joint Meeting 15th May 1810

At a Meeting of the Managers, Visitors, & Committee of Members of the Royal Institution of Great Britain, held at the House of the Institution, the 15th day of May 1810.

Present.

Timothy Brent Esqr. in the Chair.
Sir H. C. Englefield Bart.
Sir Thos. Bernard Bart.
The Hon. Robt. Clifford
The Revd. E. Forster
J. P. Auriol Esqr.
Humphry Davy Esqr.

Minutes confirmed 1. The Minutes of the last Meeting were revised & confirmed.

Byelaws. Chap. 3. 2. Resolved That the third Chapter of the Bye Laws be "Of Honorary Members" & do stand as follows. Vide Minr. the 29 May 1810.

" Chap. 4. 3. Resolved. That the fourth Chapter of the Bye Laws be "Of the Election of the Officers" & do stand as follows. Vide Minr. the 29 May 1810.

" Chap 5 4. Resolved. That the fifth Chapter of the Bye Laws be "Of the Duties of the President, and Vice Presidents." & do stand as follows. Vide. Minr. the 29. May 1810.

Timy. Brent

93

Joint Meetg. 16 May 1810.

At a Meeting of the Managers, Visitors & Committee of Members of the Royal Institution of Great Britain held at the House of the Institution the 16th day of May 1810.

Present.

The Revd. E. Forster in the Chair.
The Hon. Robt. Clifford
J. P. Auriol Esqr.
Timothy Brent Esqr.

Minutes confirmed 1. The Minutes of the last Meeting were revised and confirmed.

Bye Laws Chap 6–10. 2. Resolved. That Chapter six of the Bye Laws be "Of the Duties of the Committee of Managers." That Chap. seven be "Of the Duties of the Visitors." That Chap. 8. be "Of the Duties of the Treasurer". That Chap. nine be "Of the Duties of the Secretary & his Assistants." & do stand as follow. Vide Minutes the 29. May. 1810.

Edward Forster

94 1810.

Joint Meeting 17th May 1810.

At a Meeting of the Managers, Visitors, & Committee of Members of the Royal Institution of Great Britain, held at the House of the Institution, the 17th day of May. 1810.

Present.
Timothy Brent Esqr. in the Chair.
The Hon. Robt. Clifford
The Revd. E. Forster
Jas. Stodart Esqr.

Minutes confirmed 1. The Minutes of the last Meeting were revised and confirmed.

Bye Laws Ch. 10. 2. Resolved. That Chapter ten of the Bye Laws be "Of the general Meetings of the Members." & do stand as follows.
Vide Minr. the 29. May 1810.

Timy. Brent

95

Joint Meetg. 18 May, 1810.

At a Meeting of the Managers, Visitors, & Committee of Members of the Royal Institution of Great Britain, held at the House of the Institution the 18th day of May 1810.

Present.
The Revd. E. Forster in the Chair.
The Hon. Robt. Clifford
Wm. H. Pepys Esqr.
Timothy Brent Esqr.

Minutes Confirmed 1. The Minutes of the last Meeting were revised and confirmed.

Bye Laws. Chap. 11. 2. Resolved. That Chapter eleven of the Bye Laws be "Of the Property of the Institution."
" 12 That Chap. twelve be "Of the common Seal, and of Deeds."
" 13 That Chap. thirteen be "Of Life Subscribers, and Annual Subscribers." & do stand as follow. Vide Minr. 30 May 1810.

Edward Forster

96 1810.

Joint Meeting May 19. 1810

At a Meeting of the Managers, Visitors, & Committee of Members of the Royal Institution of Great Britain, held at the House of the Institution, the 19. day of May 1810.

Present.
Sir J. C. Hippisley Bart. in the Chair.
The Hon. Robt. Clifford
The Revd. E. Forster

Minutes confirmed 1. The Minutes of the last Meeting were revised and confirmed.

97

Joint Meeting May 22nd 1810.

At a Meeting of the Managers, Visitors & Committee of Members of the Royal Institution of Great Britain, held at the House of the Institution the 22nd day of May 1810.

Present.
Sir H. C. Englefield Bart. in the Chair.
The Hon. Robt. Clifford
The Revd. Edwd. Forster
Timothy Brent Esqr.
Wm. Watson Esqr.

Bye Laws. Ch. 14–16 1. Resolved. That Chapter fourteen of the Bye Laws be "Of the Causes and Form of Ejection from the Institution."
That Chap. fifteen be "Of the internal Regulation of the House of the Institution."
That Chap. sixteen be "Of Scientific & Literary Committees." & do stand as follow.
Vide Minr. May 30. 1810.

H Englefield

98 1810.

Joint Meeting 23d May 1810.

At a Meeting of the Managers, Visitors & Committee of Members of the Royal Institution of Great Britain, held at the House of the Institution the 23d. day of May 1810.

Present

Sir H. C. Englefield in the Chair.
Sir Thos. Bernard Bart.
The Hon. Robt. Clifford
The Revd. E. Forster
Henry Warburton Esqr.
J. P. Auriol Esqr.
Timothy Brent Esqr.
Humphry Davy Esqr.

Minutes confirmed with exception — 1. The Minutes of the preceding Meeting were revised and confirmed; with the exception of Articles 20. & 21. to be transferred to the Chapter "of Publications."

Bye Laws. Chap. 17. — 2. Ordered That Chapter seventeen be "Of Publications" & do stand as follows Vide Mins. the 30th. May 1810.

Ch. 18. — 3. Ordered. That Chapter eighteen be "Of Weekly Meetings." & do stand as follows. Vide Mins. the 30th. May. 1810.

H C Englefield

99

Joint Meeting 24. May. 1810.

At a Meeting of the Managers, Visitors, & Committee of Members of the Royal Institution of Great Britain, held at the House of the Institution the 24th day of May 1810.

Present.

Sir H. C. Englefield Bart. in the Chair.
The Hon. Robt. Clifford
The Revd. E. Forster
Timothy Brent Esqr.
Humphry Davy Esqr.

Minutes confirmed — 1. The Minutes of the last Meeting were revised and confirmed.

Bye Laws. Chap. 19 — 2. Resolved. That Chapter nineteen of the Bye Laws be "Of Professors, Lecturers, & Lectures. & do stand as follows. Vide Mins. May 30th. 1810.

H C Englefield

100 1810

Joint Meeting May 25. 1810.

At a Meeting of the Managers, Visitors, & Committee of Members of the Royal Institution of Great Britain held at the House of the Institution the 25th day of May. 1810.

Present.

Timothy Brent Esqr. in the Chair.
The Hon. Robt. Clifford
The Revd. E. Forster.

Minutes confirmed. — 1. The Minutes of the preceding Meeting were revised and confirmed.

Bye Laws. ch. 20–22 — 2. Resolved. That Chapter twenty of the Bye Laws be "Of the Libraries & Mineralogical Collections."
That Chap. twenty one be "Of the Laboratory, Apparatus, and Model Room."
That Chap. twenty two be "Of Benefactors." & do stand as follow. Vide Mins. May 30th. 1810.

Timy Brent

101

Joint Meeting May 29. 1810.

At a Meeting of the Managers, Visitors, & Committee of Members of the Royal Institution of Great Britain, held at the House of the Institution the 29. day of May 1810.

Present.

Timothy Brent Esqr. in the Chair
The Hon. Robt. Clifford
The Revd. E. Forster
James Stodart. Esqr.
Humphry Davy Esqr.

Minutes confirmed. — 1. The Minutes of the last Meeting were read and confirmed.

Bye Laws ch. 1–10 confirmed. — 2. The following first ten Chapters of the Bye Laws were finally revised & confirmed.

CHAPTER I.

Of the Government. And of the enacting and repealing of Bye Laws.

ARTICLE 1. The Corporation shall be governed by
One President,
Fifteen Managers,
Fifteen Visitors,
One Treasurer, and
One Secretary;

All of whom are to be chosen annually from among the Members. But the major part of the said fifteen Managers, and fifteen Visitors, so chosen, shall not have served the office of either Manager or Visitor during any part of the preceding year.

ART. 2. All Bye Laws are to be made, altered, or repealed, and may be suspended by the majority of the Members present at a general monthly Meeting.

ART. 3. Notice of any intended alteration, repeal, enactment or suspension of a Bye Law, shall be given in writing, at a monthly Meeting, which Notice shall be read from the Chair, and if signed, either at the Meeting, or previous thereto, by fifteen Members, it shall be hung up in a conspicuous and convenient place in the House of the Institution for general inspection; and it shall also be sent round to the Members at the expense of the Corporation, at least fourteen days before the next general monthly Meeting, when, if carried in the affirmative and confirmed at the following monthly Meeting, it shall be a law.

ART. 4. Any alteration, repeal, or enactment of a Bye Law, shall be put to the vote Article by Article; and shall be entered on the Minutes as distinct paragraphs.

CHAPTER II.

Of the Election, Rights, and Privileges of the Members.

ART. 1. The Members of the Royal Institution are to be elected by ballot.

ART. 2. Every person, desirous of becoming a Member of the Royal Institution, shall be proposed and recommended at a general monthly Meeting by four Members at least, who shall ~~then deliver to the Secretary~~ [ter], a paper signed by themselves, with their own names, specify~~ing~~ the name, quality, and profession of the Candidate, who desires to be a Member; also his knowledge [or] ~~and~~ love of science, and the usual place of his residence; and that all such particulars be certified from their personal acquaintance with him, or with his writings: which paper, ~~with the date of the day when~~ [shall be] delivered [to] and countersigned by the Secretary of the Meeting, [and] shall be suspended in some conspicuous and convenient part of the house, until the following general monthly Meeting, at which the ballot shall take place. Provided always, that all Peers, Peer's sons, and Privy Counsellors, being recommended by four Members, shall be immediately balloted for.

ART. 3. The ballot for the election of Members shall take place at any of the general monthly Meetings, twenty-one Members at least being present and balloting. A majority of two-thirds must be in favour of each Candidate for his election.

ART. 4. The balloting box shall be placed, whenever there is an election at any of the general monthly Meetings, in some convenient situation, with the name of the person to be balloted for before it (the drawers having been first examined and locked by the President or Chairman), and the Secretary, or persons officiating for him, shall give a ball to each Member, when he comes to vote, and at the same time mark his name in a list of the Members.

Art. 5. When all the Members present, who choose to vote, have balloted, the President or Chairman shall publicly examine the drawers, and having audibly counted the votes for and against the candidate, shall declare the event of the ballot; and shall also insert it with his own hand on the Recommendation delivered according to Article 2; which paper shall be preserved by the Secretary, until the ensuing monthly Meeting; when the President or Chairman, shall compare the Minutes with the above papers.

Art. 6. Every person elected a Member, shall, before his admission, pay into the hands of the Treasurer or persons officiating for him, the sum of *eight guineas* as his admission fee; and enter into a bond for the annual payment in the month of January in each year, of the sum of *four guineas*, or pay at once the sum of *fifty guineas*, including the admission fee in lieu of, or as a compensation for, such annual payment.

Art. 7. Every Member shall on his admission, or within three years after, at his option, pay the sum of five Guineas to the Library Fund. Ch. XX. Art. 4.

Art. 8. Life Subscribers who have compounded, or who shall be named under the Act of Parliament by former Proprietors, shall be eligible as Members (subject to the regulation of the Bye-Laws relating to election), paying an admission fee of *four guineas*, and *two guineas* annually, or a composition of *twenty-five guineas*, including the admission fee.

Art. 9. Every Member, who has given bond, may, at any time, compound for his future annual contribution, by paying at once the sum of *forty-two guineas*, if his annual payment be four Guineas or the sum of twenty one Guineas, if his annual payment be two Guineas, which payment being made, his bond shall be given up.

Art. 10. If any person shall refuse, or fail to pay the admission Fee, and to sign the Bond for annual Payments, or to compound for the same, within six monthly Meetings, from the day of his Election, he shall not be admitted, and his Election shall be void, unless the time be upon the representation of an adequate and sufficient cause extended by a vote at a general monthly Meeting.

Art. 11. A Register shall be kept of all the Members of the Institution, according to the dates of their Admission, bearing the following title:

The Register
of the
MEMBERS
of the
ROYAL INSTITUTION
of
Great Britain,
~~Began~~ Established under a Charter granted
by
His Majesty King George the Third,
in the 40th year of his reign,
Enlarged, and confirmed,
by
Act of Parliament,
in the 50th year of his said Majesty's reign,
and in the year of our Lord,
M DCCC X.

Art. 12. Every person to be admitted a Member shall, before his admission, subscribe an obligation in the words following:

"We, whose names are hereunto subscribed, do hereby promise each for himself, that we will, to the utmost of our power, promote the honour and interest of the Royal Institution of Great Britain; that we do recognize, and will observe, the Statutes, Laws, and Regulations of the said Institution, so long as we shall continue Members thereof."

Which having signed, he shall be presented by a Member to the President or Chairman at some general monthly Meeting, who shall address him in these words.

"A. B. I do, by virtue and authority of my office, admit you a Member of the Royal Institution of Great Britain."

Art. 13. In the Register, previous to the Declaration to be signed by the Members on their admission, the Names of the original Members (formerly Proprietors) shall be inserted, with the respective dates of their admission, leaving an opposite Column for the signature of those Members when convenient.

Art. 14. All Members shall have notice of their election sent them by the Secretary; together with a book containing the Charter, Act of Parliament, Bye Laws, and List of Members.

Art. 15. Every Member after his admission, shall have a certificate of his admission, of the form (A) in the Appendix, ~~sealed with the Corporation Seal,~~ signed by the President, and countersigned by the Secretary.

Art. 16. All Members, who were Proprietors, shall have the power of nominating by writing under their hands or by Will one person, being the wife, child, or other relation in blood, to be admitted to the privileges of a Life Subscriber to the Institution, and until such nomination, such Members shall have the privilege of admitting one person to the Lectures, Collection, Libraries and Reading Rooms, whenever they shall personally attend, but not otherwise.

Art. 17. All Members shall have the right of adding the letters M. R. I. to their names.

Art. 18. All Members have a right of admission to all public parts of the Institution; to all the Lectures and public Experiments, to the Repositories, Libraries, Reading-Rooms, Collections, and Laboratories.

Art. 19. All Members have a right of sending to the Institution any Specimens of Minerals, or other Substances likely to be useful in Arts or Manufactures, with a request that they may be examined, and if necessary analized and reported upon, and their probable applications stated.

Art. 20. It shall be lawful at any General monthly meeting to proceed to the ballot for the election of all persons who have been proposed at any General monthly meeting since the passing of the Act, so soon as the Code of Bye laws shall have been passed.

CHAPTER III.

Of Honorary Members.

Art. 1. Honorary Members shall be elected by Ballot, in the same manner as Peers. See Chap. II, Article 2. Such Honorary Members being British subjects of the Blood Royal, or such Foreigners as are of elevated rank, or distinguished for scientific or literary attainments.

Art. 2. Honorary Members shall have admission to all the public parts of the Institution, and to all Lectures, but are not eligible to any of the Offices of the Institution, nor entitled to vote, at any Meetings of the Members.

CHAPTER IV.

Of the Election of the Officers.

Art. 1. The Election of the President, Managers, Visitors, Treasurer, and Secretary, shall take place annually at the House of the Institution, on the First day of May; or when that day falls on a Sunday, on the next day.

Art. 2. If the President, or any of the Managers or Visitors, the Treasurer, or the Secretary, resign, or die, the remaining Officers shall continue to act until the next general monthly Meeting, when the Vacancy or Vacancies shall be declared from the chair. And at the following general monthly Meeting, such Vacancy or Vacancies shall be filled up, after the same form or manner, as far as circumstances will permit, as at the general annual Meeting, from among those of the Members, who would be eligible to such vacant office or offices at the general annual election on the following first of May, according to the Charter, Act of Parliament, and Bye-Laws.

Art. 3. The President, Managers, Visitors, Treasurer, and Secretary, shall be elected by a majority of the Members present, their Votes being taken by Ballot.

Art. 4. On the day of annual election, the President or Chairman shall take the chair precisely at two o'clock, P. M. and balloting-glasses having been provided and placed before the chair; and a sufficient number of Balloting Lists, according to the form (B) in the Appendix, having been prepared, and previously sent to all the Members, with the Notice of the general annual Meeting, the balloting-glasses shall remain open until three o'clock, P. M. at which time the Ballot shall close, unless it shall be moved and carried, that it shall be continued longer open. Three or more Scrutineers shall, previous to or during the Ballot, be appointed by the Chairman of the Meeting, to examine the Lists, and report the result of the Ballot.

Art. 5. When the Ballot is closed, the President shall deliver to the Scrutineers the balloting glasses, and they shall retire to a room set apart for them, and into which no other person shall be admitted. One of the Scrutineers shall then open and read aloud each List deposited in the balloting glasses, and file it, while the other Scrutineers shall mark the names: and when all the Lists shall have been opened and read, the Scrutineers shall cast up the number of the votes for each person, and report the same to the chair: when the President or Chairman shall declare the names of the persons elected to the respective offices.

Art. 6. When at the election of the President, Managers, Visitors, Treasurer, or Secretary, the suffrages in favour of two or more candidates are equal, the election shall be decided by lots, which lots shall be prepared by the Scrutineers, and drawn by the Chairman of the Meeting.

Art. 7. Printed Notices of all elections of officers, together with balloting Lists, shall be sent to the Members at least five days previous to the General Meeting, at which such election shall take place.

CHAPTER V.

Of the Duties of the President and Vice Presidents.

Art. 1. The President is an ex-official Manager and has a right of being present at all Meetings of Committees, the Committee of Visitors excepted.

Art. 2. The President for the time being, shall have full power to elect, chuse, and nominate, from and among the Managers for the time being, as many Vice-Presidents as to him shall seem meet and needful for the regular holding of the several meetings and other business.

Art. 3. The duty of the President, or of a Vice-President in his absence, is to preside at all Meetings of the Institution, of the Committees of Managers, to state and put questions both in the affirmative and negative, according to the sense and intention of the Meeting: to check irregularities, and keep all persons in order: to direct summonses for all extraordinary and special general and other Meetings: to execute, or see to the execution, of all the laws and rules of the Institution, and to certify the confirmation of all Minutes of the monthly Meetings in these terms—"Confirmed this day of 18 signed ——— President." Or if any part of the Minutes has been objected to, in these terms—"Confirmed, save and except the Paragraph (or Article) beginning with these words * * * * * and ending * * * * *, this day of 18 Signed ——— Pres.

Art. 4. The President shall take place of every Member of the Institution at their ordinary place of Meeting; and also in all other places, where any number of Members meet as a Society or Committee; and being in the Chair, is to be covered, while speaking unto or hearing particular Members, notwithstanding their being uncovered.

CHAPTER VI.

Of the Duties of the Committee of Managers.

Art. 1. All the affairs and concerns of the Institution are under the direction and government of the Committee of Managers, consisting of the President, Fifteen Managers, the Treasurer, and the Secretary.

Art. 2. The President presides at all Meetings of the Committee of Managers; and in case of his absence, one of the Vice-Presidents: and if neither shall be present, one of the Managers shall take the chair.

Art. 3. The Meetings of the Committee of Managers shall be held at the House of the Institution: and no Meeting shall be competent for the transaction of business, unless three of the Members of the Committee be present. And in case, at any time fixed for a Meeting of the Committee of Managers, there shall not be three Members present, then such Member or Members may (if he or they shall think fit) adjourn the Meeting for any time, so as notice of such adjournment be sent to the other Members of the Committee, at least two days previous to such adjourned Meeting.

Art. 4. The Meetings of the Committee of Managers for the dispatch of business, shall be held on the first Monday in every month, precisely at one o'clock, P.M. and oftener if the Managers think proper.

Art. 5. Special Meetings of the Committee of Managers may be called and held as often as the President, or in his absence, a Vice-President, shall deem such Meeting to be necessary; or as often as any three Members of the Committee of Managers, by a writing under their hands, shall express to the President, or, in his absence, to one of the Vice-Presidents, their request, that such Meeting should be called and held.

Art. 6. Printed Notices of all Meetings of the Committee of Managers, shall be sent to each Member, at least two days before such Meetings. And when a Special Meeting of the Committee is held the object, or objects of such Meeting, shall be mentioned in the notice.

Art. 7. When in deciding upon any question by the Committee of Managers at any of their Meetings, the voices for and against the question proposed shall be equal, the President or presiding Member shall have a second or casting vote.

Art. 8. The Committee of Managers may make such regulations respecting the mode of transacting their business, as they shall think proper: and repeal or alter such regulations, as often as they shall find cause so to do: provided such regulations be not repugnant to the Charter, Act of Parliament, or to the Bye-Laws.

Art. 9. The Committee of Managers, for the time being, shall cause fair and just accounts, and registers in writing, to be kept of all receipts and payments and doings by them their officers and agents respectively: and shall take care that they be annually made up to the 31st day of December in each year; and laid, with the vouchers, before the Committee of Visitors, on or before the 25th day of March ensuing; that they may be examined and audited by them.

Art. 10. The Annual Account, Books of Account, and Vouchers, shall be open to the inspection and examination of every Member of the Royal Institution, from twelve to three o'clock, P. M. from the ~~1st~~ 10th to the 30th of April inclusive.

Art. 11. The Annual Accounts of the Receipts and Expenditure shall be laid before the Members for their approbation at the general annual Meeting on the first of May, by the Committee of Visitors. The form of the Account shall be similar to that in the Appendix (C) and shall have the Account of the preceding Year reprinted by the side of it.

Art. 12. The Committee of Managers shall have the superintendance and management of the Repository for the Models, Machinery, and Apparatus, belonging to the Institution, and of any further Articles of the same nature which may be presented to the Institution, by persons desirous of encouraging the same, or which shall be purchased for its use. They shall direct the application of all Funds allotted for that purpose, in purchasing or procuring Models or Drawings, of the most useful Machines and mechanical Inventions, and also of philosophical and other Apparatus, which shall from time to time be made or discovered.

Art. 13. The Committee of Managers may engage fit and suitable persons, as Lecturers: and cause to be given occasional Courses of Lectures in Astronomy, experimental and natural Philosophy, and Chemistry, and in different Branches of Science, Literature, and the Arts.

Art. 14. The Committee of Managers shall be responsible, that no subjects be introduced at the Lectures, but such as are connected with the objects of the Institution.

Art. 15. One Member of the Committee of Managers or of Visitors shall preside at every Lecture: and a place shall be set apart for him.

Art. 16. The Committee of Managers have ~~individually~~ power to admit to the Lectures, to the Repository, Libraries, Mineralogical Collection, and other public parts of the Institution, Ambassadors, Envoys, and Ministers of foreign Princes and States, resident at the Court of Great Britain, and also other foreigners of high rank, or of distinguished scientific or literary acquirements, who may occasionally visit this country, during their temporary abode in the metropolis.

Art. 17. The Committee of Managers shall have individually power to issue, on the application in writing, personal admissions for the day, for the purpose of introducing persons, who are not residents in town, and more particularly for the introduction of persons engaged in public education. Such Admissions shall be reported at the next Committee of Managers by the grantor, and entered on a Register kept for that purpose.

Art. 18. The Committee of Managers shall elect and appoint all Officers and Servants, and remove them when they see cause so to do, except those elected by the Members at large, and those appointed by the Patrons of the Library.

Art. 19. The Committee of Managers have the direction of the House; and may make such standing Orders and Regulations, for the preservation of order and decorum therein, as they from time to time deem to be just and proper, consistent with the Bye Laws.

Art. 20. The Committee of Managers may, if they think proper, appoint a person as Superintendant of the House, to preserve order and decorum therein—to controul the current expenses, and to inspect and direct the inferior Officers and Servants in the discharge of their respective duties.

Art. 21. The Committee of Managers may accommodate with apartments such Officers and Attendants, as may with advantage to the Institution be resident in the house; but persons so accommodated, shall not be allowed to bring children into the house, and lodge them therein.

Art. 22. No Member of the Committee of Managers shall vote or be present at any Meeting of the Committee upon any question that concerns himself personally.

Art. 23. The Committee of Managers shall, and they are hereby ~~requested~~ required to, take Security, with two good and sufficient Sureties, from all Officers and Servants through whose hands any Monies belonging to the Institution shall pass; or under whose Custody any Books, Medals, Models, Minerals, or Scientific Apparatus shall be deposited, for the due and faithful execution of their respective Offices.

CHAPTER VII.

Of the Duties of the Visitors.

Art. 1. The Institution is, from time to time, once at least in six months, exclusive of the annual audit, to be inspected and examined by the Committee of Visitors.

Art. 2. The Meetings of the Committee of Visitors shall be held at the House of the Institution: and no Meeting shall be competent to the transaction of business, unless three or more Members be present: and in case, at any time fixed for a Meeting of the Committee of Visitors, there shall not be three Members present, then such Member or Members may (if he or they think fit) adjourn the Meeting for any time, so as notice of such adjournment be sent to the other Members of the Committee, at least two days previous to such adjourned Meeting.

Art. 3. The Meetings of the Committee of Visitors shall be held on the second Mondays in January, April, June, and October, in every year, at one o'Clock, P. M. precisely, or oftener if they think proper.

Art. 4. There shall also be a Meeting of the Committee of Visitors on the 25th day of March in each year (except that day be a Sunday, and then on the next day), in order to receive the annual accounts from the Committee of Managers; and they shall continue this Meeting by adjournments (if necessary) until they have audited the accounts, which shall be finished by the 31st of March, previous to their being open for the inspection of the Members of the Institution.

Art. 5. The Committee of Visitors shall draw up a Report of the State of the Institution in all its branches annually: which Report shall contain an abstract of the receipts and expenditure of the two formers years according to the Form (C) in the Appendix. The Committee shall have this Report printed and sent to the Members at least five days previous to the general Meeting on the 1st of May.

Art. 6. Special Meetings of the Committee of Visitors may be called and held as often as any three of the Visitors, or the Committee of Managers shall express to the President, or one of the Vice Presidents, their request, by a writing under their hands, that such Meeting should be called and held.

Art. 7. Printed Notices of all Meetings of the Committee of Visitors shall be sent to each Member at least two days before such Meeting: and when a Special Meeting of the Committee is held, the object or objects of such Meeting shall be mentioned in the Notice.

Art. 8. The Committee of Visitors shall elect from their own Members a Secretary, who shall take minutes of their proceedings, and direct the same to be fairly entered: and shall also direct Summonses or Notices to be sent, and do such other business as appertains to the duty of Secretary.

Art. 9. When in deciding upon any question by the Committee of Visitors, at any of their Meetings, the voices for and against the question proposed shall be equal, the presiding Member shall have a second or casting vote.

Art. 10. The Committee of Visitors may make such regulations respecting the mode of transacting their business, as they shall think proper: and repeal or alter such regulations, as often as they shall find cause so to do: provided such regulations be not repugnant to the Charter, Act of Parliament, or to the Bye-Laws.

CHAPTER VIII.

Of the Duties of the Treasurer.

ART. 1. The Treasurer is an ex official Manager. All receipts according to Form (D) in the Appendix for contributions of every kind from Members or Subscribers, shall be signed by the Treasurer, or by a Manager or Visitor at his request; such Manager or Visitor putting the letters Mr. or Vr. after his name. The Treasurer to receive the book of receipts from the Committee of Managers.

ART. 2. The Treasurer is not to keep any money belonging to the Institution in his hands, but the whole of it is to remain in the hands of the Bankers of the Institution appointed by the Committee of Managers, and all subscriptions and payments are to be entered in the Bankers' books under his direction.

ART. 3. The Treasurer is to order the payment of such draughts as shall be made on him by the Committee of Managers, as also of such bills and other disbursements as they shall specially direct to be paid by him.

ART. 4. The Treasurer shall enter into a bond with two sureties, to be approved of by the Committee of Managers in the sum of five thousand pounds, on condition that he duly account for, pay, and apply, all such sums of money or other property and effects, belonging to the Institution, as shall come into his hands as Treasurer.

ART. 5. The Treasurer shall keep a general cash book of all his receipts and payments, which shall be laid upon the table of the Committee of Managers at all their meetings. And he shall make up his accounts to the 31st day of December in every year, and lay them before the Committee of Managers annually at their first meeting in the month of February following, in order to their being audited by them, and prepared for the annual inspection and audit of the Committee of Visitors.

CHAPTER IX.

Of the Duties of the Secretary and his Assistants.

ART. 1. The Secretary is an ex official Manager. He attends the general Meetings of the Members and the Meetings of the Committee of Managers. He is to take down in writing, and enter or cause to be entered in the minute books, the minutes of the proceedings of such Meetings, and to give instructions and directions to the assistant Secretaries and Clerks in every thing relating to the business of his office. He is to see that notices are given in due form of all general Meetings of the Members, and of the Meetings of the Committee of Managers.

ART. 2. The Secretary is to superintend the foreign and domestic correspondence: and to do such other acts and deeds, in his official capacity, as are or shall be specially directed by the Bye Laws and regulations.

ART. 3. The minute books, and in general all written documents and papers belonging to the Institution and relating to the business and correspondence thereof, are under the care and direction of the Secretary.

ART. 4. The Secretary shall open all Letters and Communications addressed to the Royal Institution, and shall report them, either to the Committee of Managers, or to the first general monthly Meeting, as he shall think fit. But if a Meeting of the Committee of Managers intervenes, he shall submit them to that Committee.

CHAPTER X.

Of the general Meetings of the Members.

ART. 1. The annual Meeting of the Members is to be held on the 1st of May, or when that falls on a Sunday, on the second,—the Chair shall be taken at two o'clock precisely.

ART. 2. At the annual Meeting the Report of the Committee of Visitors shall be read by the Secretary, or his Assistant, and submitted to consideration and approval: which Report shall be open to the observation and inquiries of any Members present.

ART. 3. At the annual Meeting No business shall be entered upon or transacted, nor any Resolutions nor Motions brought forward, but what relate to the Election of Officers, and the consideration of the Report of the Committee of Visitors.

ART. 4. General monthly Meetings of the Members shall be held at the House of the Institution, on the first Monday in the months of February, March, April, May, June, July, August, November, and December, in every year, at two o'clock P. M. precisely.

ART. 5. When the annual Meeting in May falls on a Monday, the general monthly Meeting of that Month, shall be held on the second Monday instead of the first.

ART. 6. The President or in case of his absence, a Vice President, shall preside at each general Meeting: but when neither the President nor any of the Vice Presidents are present at such Meeting, one of the Committee of Managers or Visitors shall preside, or, if none of the Members of those Committees are present, the Chairman shall be elected by and from among the Members, who are present.

ART. 7. No business shall be transacted at any general Meeting, unless thirteen or more Members, not being of the Committee of Managers, are present. In that case, the Meeting shall adjourn, at half past two.

ART. 8. As soon as the Chair is taken, the order of business shall be nearly as follows. The reading—of the Minutes of the former Meeting for confirmation—of the names of Members proposed. The considering—of any Matters relating to Bye Laws—of any Proposition or Report from the Committees of Managers or Visitors—of any Propositions or Reports from the Scientific and Literary Committees according to their dates—of all Letters or Communications to be submitted to the Meeting from Members or others—of any new matter Members may wish to propose. The Ballot for Members may be continued during the business of the Meeting.

ART. 9. The Minutes of the Transactions at every general Meeting of the Members shall be entered during the progress of the business, by the Secretary, or in case of his absence, by some other person, being a Member, whom the President or Chairman of the Meeting, shall name, to act in the place of the Secretary, on such an occasion, which Minutes, so entered, shall be signed by the President or Chairman of the Meeting, before the Meeting is dissolved. The said Minutes shall also be fairly copied into the general Minute Book, and signed by the President or Chairman, when read and confirmed at the subsequent Meeting.

ART. 10. Any Motion, or Resolution, may be made, seconded and decided upon at any general monthly Meeting, without prior notice being given, unless otherwise provided for and regulated by the Charter, Act of Parliament, or the Bye Laws: but in all such cases, the same shall be brought forward at the following general monthly Meeting for confirmation. Notice of such Motion and Resolution so brought forward for confirmation, shall be sent round to the Members of the Institution, with the notice of the general monthly Meeting, at which such Motion, or Resolution, is to be proposed for confirmation, at least five days prior to such general monthly Meeting.

ART. 11. Notice of all general monthly Meetings shall be sent to the Members at large by the Secretary, at least five days before such Meeting.

ART. 12. The opinion of the Members at any general Meeting shall be taken by a, shew of hands, and the President or Chairman may, if he pleases, name two Tellers; but the question shall be put again by Ballot, if a Ballot be demanded by five Members in writing within five minutes.

ART. 13. The general monthly Meetings may be continued by adjournment to the day or hour most convenient to transact any business *stated* at the Meeting, but which could not be disposed of on that day, and Notice of the Adjournment, together with the nature of the business, shall be hung up in the Hall of the Institution. And no new matter shall be taken up or discussed at any adjourned monthly Meeting.

ART. 14. Special general Meetings may be held by order of the President,—or of the Committee of Managers,—or in consequence of a requisition in writing, addressed to the President, Managers, or Secretary, signed by *fifteen* Members, and expressing fully the nature of the business intended to be brought forward at such special Meeting; in which case, the President, Managers, or Secretary, shall call such Meeting.

ART. 15. Notice of all special general Meetings shall be immediately sent to the Members, at least *five days* before the Meeting; and in this Notice the objects of the Meeting are to be clearly expressed.

ART. 16. In all Meetings ~~and Committees,~~ the President or presiding Member, shall only vote when the Suffrages are equal.

ART. 17. No Meeting shall be adjourned until the question of Adjournment is put and carried; nor shall the question be put until the names of the Candidates proposed at such Meeting have been read.

Timy Brent

Joint Meeting May 30th 1810.

At a meeting of Managers, Visitors, and Committee of Members of the Royal Institution of Great Britain, held at the house of the Institution the 30th day of May 1810

Present

Timothy Brent Esqr in the Chair
The Honble Robt Clifford
The Revd E: Forster
Henry Warburton Esqr

Minutes confirmed

1 The minutes of the last meeting were read and confirmed.

Bye Laws Chap 11–22 confirmed.

2 The following last twelve Chapters of the Byelaws were finally revised & Confirmed.

Chapt. XI

Of the Property of the Institution.

ART. 1. The property of the Institution, of every kind and sort whatsoever, is solely vested in the Corporation. And no sale, mortgage, incumbrance or other disposition of any lands, tenements, hereditaments, belonging to the Institution, whether freehold or leasehold, or of any stock or money of the Corporation which shall be invested in the public Funds, shall be made, except with the approbation, and by vote of a majority of the Members present at the monthly general Meetings: such vote being passed at one general monthly Meeting, and confirmed at the following. And notice of such intended question being sent to the Members at least fourteen days before such a Meeting, signed by the Committee of Managers, or by fifteen Members.

ART. 2. The Rents, Revenues, and annual Income of the Institution, are to be laid out and employed by the Committee of Managers for the following purposes, namely;

First, In paying the Rents and Taxes, and defraying the expenses of the necessary Repairs and Alterations of the Buildings belonging to the Institution.

Secondly, In the payment of the Salaries and Wages of the persons necessarily employed, and in defraying the incidential Expenses of the House and Establishment.

Thirdly, In the purchase of Books, of Newspapers and Journals, foreign and domestic, and other scientific and literary Publications, for the use of the Libraries and Reading Rooms.

Fourthly, In the purchase of Philosophical Apparatus and Models of Machinery; provided always, that no sum beyond One Hundred Pounds shall be expended on such Apparatus and Models, until the same shall have been proposed by the Committee of Managers to the Members, and approved of by the majority present, at one general monthly Meeting, and confirmed at the next.

ART. 3. The Surplus of all the annual Income and Receipts above the Expenditure in the four departments before enumerated, shall be annually invested by the Committee of Managers, in the public Funds, in the name of the Corporation, in order either to form a permanent Fund for the benefit of the Institution, or to be applied, as required, to any object connected with the prosperity of the Institution.

ART. 4. All Monies not permanently invested, and not wanted for defraying current and probable expenses shall, from time to time, be vested by the Committee of Managers in floating public Securities, to be disposed of, when the produce shall be wanted.

ART. 5. No Monies, nor any other property of value, belonging to the Institution, shall be disposed of, or given, by the Committee of Managers, or by any other person or persons, as *presents* or *rewards*, to any officers or servants, except by vote of a general Meeting duly confirmed.

ART. 6. No Money shall, under any circumstances whatever, be borrowed by the Committee of Managers, nor any debt be incurred by them, further than the floating and ordinary expenses of the Institution, without the consent of the majority of Members present at one general monthly Meeting, and confirmed at the next, except the Loan raised under a Resolution of the joint Committee of Proprietors, Managers, and Visitors, held at the Institution, February 15th, 1810.

CHAPTER XII.

Of the Common Seal, and of Deeds.

ART. 1. The Common Seal of the Institution shall be kept in an iron Chest, having *two* Locks, with *four* Keys; and these Keys shall be kept—one by the President, or, in his absence, by one of the Vice-Presidents,—one by a Manager appointed by the Committee of Managers to that trust,—one by the Treasurer,—and the other by the Secretary.

ART. 2. When any person holding one of the beforementioned Keys, leaves his Residence in the Metropolis, for a longer period than *fourteen days*, he shall deliver his Key into the keeping of some one of the Managers, at his option, who remains in town, and inform the Secretary of his having done so: Provided always that the Manager so named, has not in his custody either of the remaining three keys.

ART. 3. Whenever the Common Seal of the Corporation is to be affixed to any Deed or Writing, notice thereof, stating the nature of such Deed or Writing, and that the common seal is to be affixed, shall be sent to each of the Managers, at least two days previous to the completion of the business; and all such Deeds and Writings shall be passed and sealed at a Meeting of the Managers, and countersigned by the President, or Chairman, the majority of the Managers present, and the Secretary; and an account of the proceeding shall be entered in the Minutes of the Meeting.

CHAPTER XIII.

Of Life Subscribers and Annual Subscribers.

ART. 1. The Committee of Managers, on a Recommendation, signed by four Members of the Institution, according to form (E) in the Appendix, shall have the power of admitting Life Subscribers, and shall report the admission to the next general monthly Meeting.

ART. 2. Every Life Subscriber shall, on his admission, pay into the Funds of the Institution, the sum of four Guineas as his admission Fee, and sign a Bond for the annual payment of three guineas on the 1st day of January in each year, or pay the sum of thirty Guineas in lieu of, or as a compensation for, such annual payment.

ART. 3. Every Life Subscriber shall, on his admission, or within three years after, pay the sum of five Guineas into the Library Fund. See Chapter XX, Art. 4.

ART. 4. Life Subscribers shall have admission to all the public parts of the Institution, and to the Public Lectures, but not to any of the general, or other, Meetings of the Members.

ART. 5. Any person desirous of becoming an Annual Subscriber may be admitted by the Committee of Managers, upon a recommendation of four Members of the Institution, according to Form in the Appendix, (E) and paying, on his admission, the sum of four guineas, for the current year, and continuing to pay the same sum annually upon every succeeding first of January. Such Annual Subscribers being allowed the same privileges as the Life Subscribers, as long as they continue to make their annual Payment. Their admission shall be reported at the next general monthly Meeting.

ART. 6. The Committee of Managers, upon the recommendation of four Members of the Institution, according to the Form in the Appendix, (E) shall have the power of admitting Ladies as Subscribers to the Lectures, with admission to the Collection of Mineralogy and Geology, upon the payment, on admission, of the sum of two guineas for the current year, and the same sum annually on the 1st of January.

ART. 7. All annual Subscribers on their admission shall pay the sum of one Guinea into the Library Fund. See Chapter XX, Art. 4.

ART. 8. Particular Courses of Lectures may be attended upon the payment of a sum, to be fixed by the Committee of Managers, and to be known, as soon as convenient, after the arrangement of each Course, by application at the Clerk's Office. Persons desirous of attending such Course or Courses, to be recommended by four Members of the Institution, according to the Form (F) in the Appendix, to the Committee of Managers who may issue the Admission Ticket.

CHAPTER XIV.

Of the Causes and Form of Ejection from the Institution.

ART. 1. Any Member or Subscriber who shall, contemptuously, or contumaciously, disobey and transgress the Laws and Regulations of the Institution; or shall, by speaking, writing, or printing, or in any other manner, publicly defame the Institution; or advisedly, maliciously, or dishonestly, do any thing to the damage, detriment, or dishonour thereof, shall be ejected therefrom, under the obligation signed on admission, and cease to be a Member thereof.

ART. 2. No Subscriber is permitted, nor in future shall be permitted, to announce to the public, by any inscription on any part of the house which he or she occupies, or to publish in any advertisement, either printed or written, that he or she is a Subscriber to the Institution; and all persons so offending against this regulation shall be proceeded against as transgressors of the Bye Laws, and ejected from the Institution.

ART. 3. When either the Committee of Managers, or the President of the Institution, shall see cause for the ejection of any Member or Subscriber, the question of ejectment shall be moved at one of the Meetings of the Committee of Managers; and, at the next Meeting, the said question shall be decided by ballot; and if a Subscriber, the person shall be ejected. But if a Member, the vote of the Committee of Managers shall be reported at the next general monthly Meeting of the Members, and if then approved, and confirmed at the following monthly Meeting, the ejection shall take place.

ART. 4. When a Member or Subscriber is ejected, his or her name shall be struck out of the List of Members or Subscribers, at a Meeting of the Committee of Managers, and entered on the Minutes of the Meeting, and a copy of such Minute shall be forwarded to the Member or Subscriber ejected.

CHAPTER XV.

Of the internal Regulation of the House of the Institution.

Art. 1. The House of the Institution shall (with exception of Sundays, Christmas Day, Good Friday, and days appointed for Fasts and Thanksgivings,) be open from Nine o'clock in the Morning to Eleven o'clock at Night; after which hour no person shall be admitted, except such as may reside in the House; and all other persons are required to leave the House before Half past Eleven o'clock at Night.

Art. 2. All matters relating to the internal regulation of the Inhabitants, Officers, or Servants of the House, are subject to the Committee of Managers.

Art. 3. All Repairs or Alterations to be made in the house not exceeding an expense of £100 shall be made by order of the Committee of Managers, but when a greater expense is to be incurred, the proposition shall be submitted to the Members at a general monthly Meeting.

CHAPTER XVI.

Of Scientific and Literary Committees.

Art. 1. The better to carry into effect the object of the Royal Institution, there shall be established three Committees, under the following titles:

A Committee of Mathematics, Mechanics, and Mechanical Inventions—of 25 Members.

A Committee of Chemistry, Geology, and Mineralogy—of 25 Members.

A Committee of general Science, Literature, and Arts—of 25 Members.

Art. 2. The Committees shall be chosen by ballot from among the Members, who are not of the Committees of Managers or Visitors, in the week that follows the annual election of Officers. Notice of the day of election, together with Balloting Lists, according to the form in the Appendix, shall be sent with the Notice for the annual election of the 1st of May.

Art. 3. Ten of the Members of each Committee shall be annually chosen from among those Members of the Institution, who have not been on the same Committee the preceding year, the other fifteen may be re-elected.

Art. 4. At the first Meeting of the several Committees, each Committee shall by ballot elect (from among the 25 Members) two Chairmen, and also a Secretary, who may be a Professor of the Institution. The result of the Election shall be immediately notified in writing to the Committee of Managers, and reported by the Chair to the Members at the general monthly Meeting of June, immediately after the reading of the Minutes of the preceding Meeting.

Art. 5. The President, Managers, Visitors, Secretary, and Treasurer, are ex official Members of all Committees, and the President shall, when he attends, take the Chair, if he pleases, and shall be attended by the Chairman on his right hand.

Art. 6. The Professors of the Institution may attend all Committees, and shall vote, if Members of the Institution.

Art. 7. The Members of the other Committees may attend as auditors, but it shall be in the power of any Member present, to move that the room be cleared.

Art. 8. The general monthly Meeting of Members, or the Committee of Managers may refer any matters relative to the Object of the Institution, to single, or joint Committees; and such Committees shall report thereon to a general monthly Meeting, or to the Committee of Managers.

Art. 9. The Committees shall consider all matters referred to them by a general monthly Meeting, and all Communications from the Committee of Managers—They shall examine (each according to its particular province) all foreign and domestic Journals, relating to the Objects of the Institution, and shall point out to the Managers such things as are worthy of investigation. They shall have the power—of entering on investigations connected with the Objects of the Committee, but shall not incur expense without the consent of the Committee of Managers—of recommending Lectures to be given—and of reporting to the Members assembled at their general monthly Meetings by means of their Chairman, or any other Member of the Committee.

Art. 10. Whenever Reports are to be made, that require to be illustrated by experiments, such Reports shall be made at an adjourned monthly Meeting, when a place shall be set apart for the Committee from whence the Report comes.

Art. 11. Four Committee Minute Books shall be kept, one for each Committee, and the 4th for joint or extraordinary Committees.

Art. 12. The Committees shall meet weekly, but may continue or adjourn their Meetings as they think proper; subject to the internal regulations of the Institution, the 1st Committee shall meet on Mondays,
the 2d on Tuesdays,
the 3d on Fridays.

Art. 13. There shall be constantly hung up in a convenient part of the House, a List of Committees, specifying the business they are to consider, and the days they are appointed to meet. Also a List of Reports ready to be made.

Art. 14. There shall be preserved in the Secretary's Office a List of all the References to the several Committees, in order that the subjects of them may be regularly taken up, at least as far as convenient.

Art. 15. The Chairmen shall, according to seniority, preside at all Meetings of their respective Committees, unless the President of the Institution be present.—They shall take care, that regular Meetings of their Committees be appointed—They shall see, that all business be regularly taken up; the Minutes correctly entered, and the Names of Members attending inserted in the Minute Book, and shall sign the same—They shall not allow any question to be put to the Committee unless three Members be present—They shall have a second and casting Vote when the numbers are equal—They shall have the superintendance generally of the affairs of their Committee, and when no particular direction has been given by their Committee, they shall be the means of Communication between it and the Committee of Managers, or general monthly Meetings.

Art. 16. A Fund not exceeding £10 shall be placed at the disposal of the Chairman of each Committee by the Managers for current expenses, which Fund shall be renewed from time to time as the Committee of Managers shall judge expedient. Where larger sums are required, application must be made to the Committee of Managers.

Art. 17. The Committees shall have the power to request the attendance (by a written Note signed by their Chairman) of any person whom they may wish to consult. Such Stranger may deliberate on the day of their attendance, but not vote.

Art. 18. Each Committee shall have the power of declaring an open Committee to be held in an ensuing week, at which all Members of the Institution may attend. The Chairman shall send immediate Notice to the President and Secretary, who shall order Notice of such open Committee to be hung up in the Hall, and to be inserted on the Cards.

Art. 19. The Committee shall have the power, during their sittings, to send for any Books from the Libraries, by a written order from the Chairman, which Books shall be returned at the conclusion of the sitting.

Art. 20. When a Report, made to a general monthly Meeting, or any original Communication, made by a Member at a general monthly Meeting shall be approved of by the Meeting, and a Motion made and carried that such Report or Communication be printed, it shall be sent to the Committee of Managers to be printed in the Records of the Institution. The Managers may, if they judge it necessary, require the aid of a Committee of Papers. In which case, each Scientific Committee shall depute three of its Members, who shall always be presided by a Manager, if not, by some Member, called to the Chair.

CHAPTER XVII.

Of the Publications.

Art. 1. The publication of all Experiments, Researches, and Discoveries, made in the House, and with the Apparatus of the Royal Institution, whether by the Members, or the Professors, or other persons employed by the Institution or by persons not belonging to the Body, who may have obtained permission to use the Apparatus or Laboratory, is the sole right of the Institution, and previous to their own publications, no Accounts of such Experiments or Researches shall be printed, either in separate Publications, or in the Journals or Records of any other Society, without the leave of the Committee of Managers of the Royal Institution, first obtained in writing, which permission shall *shall be specified in every such publication or* be printed at the head of every such Publication in form and manner following:

~~Published by the Permission of~~ *Communicated by* *the Royal Institution of Great Britain.*

And if any one shall neglect or refuse to prefix such Permission to their Publications, future use of the Apparatus of the Royal Institution shall be denied them, except by a Vote of the Members at a general monthly Meeting; and also, if any publisher shall not prefix the said Form to their Publications of such Communication, no further Communication shall be made to them except by a Vote of the Members at a general monthly Meeting.

Art. 2. The Publications of the Royal Institution shall bear the Title of Records of Science, *published by the Royal Institution of Great Britain*, and shall be published from time to time as occasion shall require.

Art. 3. Besides the Reports of Committees, and papers ordered for publication at the general monthly Meetings, the Managers shall be at liberty to print any original papers written by Members or others, and communicated to them, or any other Matters which they may deem worthy of publication; and they may, as they shall judge proper, require the assistance of any of the Scientific Committees, in deciding on the Merits of such Papers or Communications.

Art. 4. The Records of the Royal Institution shall be published in an uniform neat manner, in the 4to. size, and shall be delivered to the Members, either Gratis, or at a reduced Price, or at full Price, according as the Committee of Managers shall judge it to be most conducive to the interests of the Institution. The Members of the Committee making the Report, and of the Committee of Papers, shall always receive the part containing their Report Gratis. And also twelve copies of any original paper communicated by any person—and printed in the Records, shall be sent Gratis to such person.

CHAPTER XVIII.

Of the Weekly Meetings of the Members.

Art. 1. There shall be held in the House of the Institution Weekly Meetings of the Members during the months of January, February, March, April, May, June, July, November, and December in each year.

Art. 2. These Meetings shall be held in the Library of Reference, which shall be open for that purpose on Friday Evenings, at Eight o'clock, and shut at Eleven.

Art. 3. The object of these weekly Meetings is to collect and assemble together scientific and literary Members, for the purpose of general and free conversation upon scientific and literary Subjects, and upon Matters relating to the good and prosperity of the Institution.

CHAPTER XIX.

Of Professors, Lecturers, and Lectures.

Art. 1. For the better carrying into effect the objects and views of the Royal Institution, there shall be Professorships established as occasion shall require. Notice of any Professorship, proposed to be established, shall be given in writing, at a monthly Meeting, which Notice shall be read from the Chair, and if signed, either at the Meeting, or previous thereto, by fifteen Members, it shall be hung up in a conspicuous and convenient place in the House of the Institution for general inspection; and it shall also be sent round to the Members, at the expense of the Corporation, at least fourteen days before the next general monthly Meeting, when, if carried, and confirmed at the following monthly Meeting, such Professorship shall be established.

Art. 2. The Professors of the Royal Institution shall be elected annually by ballot; and on the same days as the Members of the Scientific Committees: and shall be put in nomination at the general monthly Meeting in April, from the Chair.

Art. 3. All persons are eligible to the Professorships, whether Members of the Institution or not, with the exception of the Managers or Visitors. But all Candidates shall have delivered one or more Courses of Public Lectures in the House of the Institution, within two years previous to his Election.

Art. 4. Whenever a Professorship shall be vacant, public notice shall be given thereof; stating the Nature of the Professorships, together with the immediate Salary to be allowed: which notice shall be given at a general monthly Meeting, at least one month previous to the Election.

Art. 5. It is the duty of the Professors to deliver various Courses of Lectures, under the direction of the Committee of Managers, upon the different objects of their several Professorships—to superintend all Experiments, ordered by the Committee of Managers—to carry on Experiments and Investigations connected with the Objects of the Institution within their several Departments,—to point out, and to procure, under the Direction of the Committee of Managers, such Apparatus as may be necessary;—to take care of the Apparatus within their several Departments,—and to attend to such other matters as may be directed by the Committee of Managers.

Art. 6. In case of the neglect of duty of a Professor, the Committee of Managers shall have the power of suspending him immediately from all or any of his functions and privileges; but shall report the same to the next general monthly Meeting. And if it be then moved and seconded, that the Professor be removed, the same formalities shall be observed for his removal, as in his appointment.

Art. 7. The Committee of Managers may appoint Lecturers at their discretion, for the purpose either of giving regular Courses, or for single and occasional Lectures; upon such terms as they may think proper; provided always, that no Manager or Visitor shall receive any remuneration for any Lectures delivered by him, whilst he continues a Manager or Visitor.

Art. 8. The several Courses of Lectures shall commence on or after the first of January, and shall be completed on or before the 31st of December in each year.

Art. 9. The Lectures, whether by Professors or others, delivered at the Royal Institution, shall, as far as possible, be given from a written text, or at least from written notes, which shall, if required by the Committee of Managers, be submitted to their inspection.

CHAPTER XX.

Of the Libraries and Mineralogical Collections.

Art. 1. The Library of Reference and Mineralogical Collection, is vested in the Corporation, and is under the same direction and government as the other parts of the Institution; subject only to such Privileges as are hereinafter mentioned.

Art. 2. Any Member contributing to the Library, or Collection of Minerals, or to both, to the amount of 100l. or upwards, is a Patron for Life, of the said Library and Collection; and such Patron has power by writing to appoint any other person (being a Member of the said Institution) to be a Patron for Life of the said Library and Collection.

Art. 3. Each Patron shall have the power of introducing, or of recommending, in writing, one person every day to the said Library and Collection, when the same shall be open.

Art. 4. There shall be formed a Library Fund, the application of which, together with all Sums subscribed to the Library or Mineralogical Collection, shall be under the direction of the Patrons.

Art. 5. A Sum not exceeding £50 shall be placed at at the disposal of the Patrons of the Library, which Fund shall be renewed from time to time as the Committee of Managers shall judge expedient. When larger Sums are required, application shall be made to the monthly Meetings.

Art. 6. The Patrons may from time to time, with the consent and concurrence of the Committee of Managers, make such Regulations, Rules, and Orders, respecting the Library of Reference and Mineralogical Collection, and the Persons to be admitted or employed therein, as they shall deem expedient, provided the same be not contrary to, or inconsistent with, either the Charter, the Bye Laws, or Regulations of the Institution.

Art. 7. The Library and Mineralogical Collection shall be open for the Members and Subscribers, and persons recommended by the Patrons, on all days on which the Institution is open, from ten to four o'clock.

Art. 8. No person shall take down any of the Books in the Library of Reference, or Specimens in the Collections; but a note or card thereof must be given, with his name, to the Keeper, or one of the Attendants, who shall immediately supply him with the Book required.

Art. 9. No person shall be capable of exercising his right as a Patron, except during such time as he shall be a Member of the Institution.

CHAPTER XXI.

Of the Laboratory, Apparatus, and Model Room.

Art. 1. The Laboratory, Apparatus, and Model Room, are under the direction and government of the Committee of Managers.

Art. 2. There shall be distinct Catalogues or Inventories of all things, appertaining to the Laboratory, Apparatus, and Model Rooms, which shall be submitted to the Visitors at their inspection.

CHAPTER XXII.

Art. 1. All Benefactors to the Royal Institution shall have their Names registered in a Catalogue of Benefactors, with a specific account of the nature of the Gift or Benefaction; Copies of such Lists shall be printed at the end of each volume of the Records.

Vinl Brent

The undersigned present the foregoing Code of Bye Laws drawn up under the Charter and Act of Parliament by the Joint Committee of Managers, Visitors, and Members, nominated for that purpose at a General Monthly Meeting held the 7th of May last.

Winchilsea
Dundas
Wm Watson
Edward Forster
J P Auriol
T. Bernard
Robt Clifford
E Bracey
Henry Warburton
James Stodart
Vinl Brent
S. Bernard
H. C. Englefield
James Stodart
Wm Haseldine Pepys
J. C. Hippisley

Managers' Meeting. 11th June 1810.

At a Meeting of the Managers of the Royal Institution of Great Britain, held at the House of the Institution on the 11th. Day of June 1810.

Present

The Earl of Winchilsea President in the Chair
The Right Honble Lord Dundas
Sir Thomas Bernard Bart.
Sir John C. Hippisley Bart.

Minutes approved — 1. The Minutes of the proceedings of the last Meeting were read and approved.

Exd. Genl. Acct. 2. Examined the General Account of the Institution.

Minutes of Com. of Accts approved — 3 Read and approved the Minutes of the Committee of Accounts of this day.

Resignation of J. Auriol Esq. as Secretary — 4 Read the following Letter from J. P. Auriol Esqr.

"To the Managers of the Royal Institution.
My Lords & Gentlemen

"I have now had the honor of "filling the Office of Secretary to the Royal Institution "by successive Elections from the Year 1800 to the present "time and during that period I have endeavoured to "discharge the duties of the Situation with punctuality "to the best of my ability and to promote the success "of the Establishment; I have also united by my "assent and contributed by my feeble efforts to bring the "Institution into the enlarged and improved state in "which it is now placed by Act of Parliament, but "as I find that I shall not be able, without much "inconvenience to myself, to give that further attention "to the Office which it undoubtedly requires, especially "during the remainder of this Year when my absence "from Town will preclude my attendance both at the "Weekly Meetings of the Managers, and the Monthly "Meetings of the Members; I should be sorry to stand "in the way of a more efficient person, I entreat therefore "that you will do me the favour to accept this intimation "of my intention to resign the Office of Secretary to the Royal "Institution when it can be done with the least possible inconveniency and allow

"me at the same time to testify my undiminished good "wishes for the prosperity and success of this important "Establishment."

Park Street
4th June 1810

I have the honor to be
My Lords & Gentlemen
Your most obedient and faithful
humble Servant

(Signed) J. P. Auriol

Resolved. That the Committee receive with regret Mr. Auriol's intimation of his intention to resign the appointment of Secretary which he has so long filled with ability and attention and with so much advantage to the Institution and hope he will be able without inconvenience to hold the appointment for some time longer until they can propose a Successor.

J. Hinckley to be appointed Asst. Secy to Mr. Auriol — That John Hinckley Esqr. be requested to give his aid to Mr. Auriol as Assistant Secretary and that Summonses be sent him accordingly.

Proprietors desiring Compensation. — Read Letters from
The Honble George Villiers
Thos. Wright Esqr.
Wm. Boulter Esqr.
Thos. Palmer Esqr.
and
Saml. Lanford Esqr.
expressing their desire to accept the proposed Compensation of £42.0.10 for their Shares as Proprietors.

Application from an original Proprietor to resign and appoint a Nominee as his Successor. — Read the following Letter from George Sullivan Marten Esqr.

"Finding that various circumstances have prevented the Royal Institution from accepting my resignation, as an original Proprietor, in favor of Dr. Charles Dalston Nevinson M.D. of Somerset Street Portman Square. I now beg leave to nominate this Gentleman as a Life Proprietor, and I trust you will be empowered to inform Dr. Nevinson of such appointment having taken place as soon as possible.

I have the honor to remain
Your obedt. humble Servant

(Signed) — Geo. Sul. Marten"

Referred to Sir J. C. Hippesly — Resolved – That Sir J. C. Hippisley be requested to communicate with Mr. Marten about it, and report whether the nomination comes within the Provision in the Act.

5 Ordered That the Treasurer be requested to advance the Clerk £30 on his receipt and signed an Order accordingly.

Presents received.

Presents. 6. Mr. Richd. Carmichael presented "An Essay on the nature of the Scrofula."

Mr. Andrew Mathias presented "An enquiry into the Mercurial Disease."

Ordered. That the Thanks of the Managers be returned for these presents.

T. Bernard

Managers' Meetg 18th June, 1810.

At a Meeting of the Managers of the Royal Institution of Great Britain, held at the House of the Institution on the 18th day of June 1810.

Present.

Sir Thomas Bernard Bart. in the Chair
Sir John Cox Hippisley Bart.
William Watson Esqr.

Minutes Confirmed — 1. The Minutes of the proceedings of the last Meeting were read and approved.

Exd. Gen. Accts. 2. Examined the General Account of the Institution.

Minutes of Com. of Accts. 3. Read and approved the Minutes of the Committee of Accounts of this day.

Proprietors desiring to receive Compensation. — 4. Read Letters from

The Rt. Honble Lord Ossulston.
The Honble Admiral Digby
Richard Harford Esqr.
and
Mrs. Susannah Day.

expressing their desire to accept the proposed Compensation of £42.0.10 for their shares as Proprietors. –

Resolved.

That Mr. Auriol be requested to write to The Right Honble Lord Ossulston informing him that his Notice demanding the pecuniary compensation for his Interest in the Institution is sufficient and that he will be informed as soon as the necessary arrangements are complete.

Mr. Scrope Bernard the Treasurer to be appointed Receiver of fine due by Mr. Grey's Exors. — 5 Mr. Savage reported that the Fine due from the Representatives of Mr. Gray will be paid in the course of the present week and that he is desired by Messrs. Green & Tennant to request a Receiver may be appointed by the Managers.

Resolved.

That the Treasurer Scrope Bernard Esqr. be appointed the Receiver of the above mentioned Fine, and that he be informed that by the Calculation which has been made by the Clerk of the Institution,

118 1810.

it appears that there is due for the same Fine and Interest up to Lady day last the sum of £1535. 10. 0.

Resolved.

That a Copy of this Resolution be carried to Mess.rs Green & Tennant by M.r Savage.

That the Treasurer be also informed that strictly speaking the Institution is intitled to Interest calculated up to the day of payment; but that it is of so much importance that the Money should be received while it may be had, that the Managers beg he will consider the matter as left to his discretion.

6. Ordered that M.r Brooks be directed to pay Seventy five Pounds ten shillings received for Voltaic Subscriptions to Mess.rs Ransom & C.o on account of the Institution.

Tender of Coals. 66s.

7. Read the following Letter from Mess.rs Dickins & C.o – Gentlemen

"We beg leave to offer to lay in the best Wallsend Coals for the use of the Royal Institution at 66.s ℔ Chaldron.

At the same time we beg to state as the general opinion of the Trade they will not be cheaper than the price above named being at present worth 70.s ℔ Chaldron, but having bought a considerable quantity previous to the rise enables us to offer the above price: the honor of an early answer will much oblige

Gent.n your most obed.t Serv.ts

To the Managers of the Royal Institution } signed) "Dickins & Fourdrinier"

Ordered. That Mess.rs Dickins & C.o send in 20 Chaldrons of Coals at the above price.

Nomination of a Member by Proprietor.

8. Read a nomination by M.rs Blaauw appointing William Henry Blaauw Esq. to succeed to his late Fathers share which nomination was transmitted to the Secretary about the time of the passing of the Act.

T Bernard

119

Managers' Meeting June 25th 1810.

At a Meeting of the Managers of the Royal Institution of Great Britain, held at the House of the Institution, on the 25.th Day of June 1810.

Present

Sir Tho.s Bernard Bar.t in the Chair

John Symmons Esq.r

J. P. Auriol Esq.r

Minutes confirmed

1. The Minutes of the proceedings of the last Meeting were read and approved.

Ex.d Gen Acc.ts

2. Examined the General Account of the Institution.

Minutes of Com. of Acc.ts approved

3. Read and approved the Minutes of the Committee of Accounts of this day.

Proprietors desiring Compensation

4 Read Letters from

James Gorst Esq.r

expressing his desire to accept the proposed Compensation of £42. 0. 10 for his share as Proprietor.

and – George Sulivan Marten Esq.r

acknowledging that the Act precludes his former request read 2.d April last, from being granted, but hoping that if any relief can be afforded he may not be forgotten.

5. Resolved.

That M.r Hinckley be requested to report to the Managers on Monday next the circumstances of the cases of M.rs Blaauw, M.r Marten and such other Proprietors whose Transfers or Nominations were before the Managers at the time of passing the Act of Parliament.

Donation of £10 from Col. Noel.

6. M.r Davy read the following Extract of a Letter from Daniel Moore Esq.r

"Dining with Col.l Noel on Friday last and the subject of the Institution entering on the topics, ~~when~~ he deputed me to present the enclosed Ten Guineas as a Subscription to the Galvanic Battery and to cause his name to be subscribed to the Loan for £100 – He further expressed his disposition to afford such other pecuniary aid as the principal Members may contribute."

---

120 1810.

Resolved. That M.r Davy be requested to write to M.r Moore and to present the Thanks of the Managers through M.r Moore to Col.l Noel on account of his very liberal offer. –

7. Present received

Present.

The Rev.d Edw.d Forster presented a Section of "The British Gallery of Engravings N.o 6."

Ordered. That the Thanks of the Managers be returned for this present.

Printing

8 Sir Thomas Bernard reported that he had examined M.r Savage's Bill for printing from the 1.st of January last amounting to £172. 13. 3 and that he does not think it overcharged.

Resolved. That it be taken into consideration at the next Meeting.

Adjourned till one o'Clock on Monday 2.d July.

W Watson

121

Managers' Meeting 2nd July 1810

At a Meeting of the Managers of the Royal Institution of Great Britain, held at the House of the Institution, on the 2.nd Day of July 1810.

Present

W.m Watson Esq.r in the Chair.

Charles Hatchett Esq.r

Sir Tho.s Bernard Bar.t

Minutes approved

1. The Minutes of the proceedings of the last Meeting were read and approved.

Ex.d Gen. Acc.t

2. Examined the General Account of the Institution

Minutes of Com. of Acc.ts approved

3. Read and approved the Minutes of the Committee of Accounts of this day.

Proprietor desiring Compensation &c

4. Read Letters from

Mess.rs Pybus & C.o stating that General Charles Morgan will thro' their House receive the proposed Compensation of £42. 0. 10 when the Money is ready.

& – The Rev.d W.m J. G. Phillips respecting his Proprietory Share.

Resolved.

That it be referred to M.r Hinckley.

Presents. 5

Presents received

The American Philosophical Society presented "Transactions of the American Philosophical Society at Philadelphia Vol 6 part 2 with a Letter to the President of the Royal Institution from their Librarian J. Vaughan Esq.r of Philadelphia and a Letter also to the President of this Institution from W.m Vaughan Esq.r of Mincing Lane. –

Ordered. That the Thanks of the Managers be returned to the American Philosophical Society for this present Also to John Vaughan Esq.r × thro' whom it is received and to W.m Vaughan Esq.r –

× Librarian of the American Phi. Soc.y

– Mincing Lane

Dr. Cha.s Burney presented.

Tentamen de metris ab Æschylo in Choricis Cantibus

adhibitis.

Ordered. That the Thanks of the Managers be returned for this present.

Payment of Mr. Savage's bill.

6. **Ordered**—

That as soon as the sum due for the Fine from Mr. Gray's Executors is received Mr. Savage's Bill for £172.13.3 up to the 9th of June last be paid together with £26.1.3 for the Balance of his Account up to the 31st of December last.

Request for leave of Absence

7. Mr. Davy had the honour of laying before the Managers a request from Mrs. Greenwood the Housekeeper that she may be permitted to be absent from the Institution for a Month from the middle of July till the Middle of August; Mr. Davy conceives that at that time of the year a compliance with Mrs. Greenwoods request will not materially interfere with the domestic business of the Institution.

Ordered. That Mrs. Greenwoods request be complied with.

8. **Resolved** – That the sum of £30 be advanced to the Clerk of Accounts.

9. Mr. Hinckley presented his Report.

Ordered That the same be referred to Sir Thos. Bernard Bart.

T. Bernard

Managers' Meeting 9 July 1810

At a Meeting of the Managers of the Royal Institution of Great Britain, held at the House of the Institution, on the 9th Day of July 1810.

Present

Sir Thos. Bernard Bart. in the Chair
Wm. Watson Esqr.
James Peter Auriol Esqr.

Minutes approved

1. The Minutes of the Proceedings of the last Meeting were read and approved.

Exd. Gen. Acct.

2. Examined the General Account of the Institution.

Minutes of Com. of Accts. approved

3. Read and approved the Minutes of the Committee of Accounts of this day.

4. Mr. Hinckley having delivered his Report on the Cases referred to him by the Meeting of the Committee of Managers on the 25th of June last

Ordered. That it be kept among the rough Minutes of this day – and the following Resolutions of the Managers be laid before the next General Monthly Meeting.

Application for admission to Membership of Royal Institution by nomination of Proprietor.

In consequence of applications to them, the Committee of Managers submit to the General Meeting, a few cases in which some Individuals have suffered from the peculiar circumstances attending the passing of the Act respecting the Royal Institution.

The first class consists of those cases, where upon the death of Proprietors, unavoidable delays have taken place in the admission of the Successors without any default on their part; and where by the rapidity with which the Act passed, they were precluded from being admitted.

These cases are five in number, and have occurred on the decease of the following Proprietors.

The Right Honble Lord Kinnaird; the Nominee for whose share is his eldest Son, the present Lord Kinnaird

Sir Richd. Joseph Sullivan Bart. the Nominee for whose share is his eldest Son, now

Sir Henry Sullivan Bart.

Lieut. General John Graves Simcoe, the Nominee for whose share is his eldest Son F. G. Simcoe Esqr.

William Blaauw Esqr. the Nominee for whose share is William Henry Blaauw his eldest Son – and

David Scott Esqr. the Nominee for whose share is his eldest Son David Scott Esqr.

As to these Nominees the Managers beg leave to recommend, as an act of right and justice, that they vizd. Lord Kinnaird, Sir Henry Sullivan, F. G. Simcoe Esqr., William Henry Blaauw Esqr., and David Scott Esqr. be admitted Members of the Institution, free of any expense; and with the same power as is possessed by the other Members, who were Proprietors, of nominating each a Life Subscriber.

The second class consists of only two cases, where Transfers had been prepared, but where the Act unexpectedly passed before the necessary legal steps could be taken by the Managers to give effect to the Transfers.

These are Transfers by the two following Proprietors.

Sir George Duckett, Bart. having executed a Transfer in favor of his Son, George Duckett Esqr. M.P. – and

George Sullivan Marten Esqr. having executed a Transfer of his Share in favor of Dr. Charles Dalston Nevinson M.D.

As to these two cases the Managers only recommend, that upon nominations being made under the Act by Sir George Duckett as to Mr. Duckett, and by Mr. Marten as to Dr. Nevinson, Mr. Duckett and Dr. Nevinson be admitted (instead of Life Subscribers) Members of the Institution.

There is another Case of a Member, The Revd. William Joseph George Philips of Grosvenor Place, applying to the Managers and afterwards to the Monthly Meeting for a Transferrable Ticket, on account of his not being able in future to be much resident in Town. This the Managers beg leave to submit to the consideration of the Members whether any Regulation should be adopted in such

cases conceiving that if it is not adopted in other similar instances, it cannot be admissible in any.

Collection of Arrears suspended

5. Read a Note from Mr. Este.

Ordered that Mr. Brooks be directed not to collect the Arrears of the Years 1809 and 1810 from Mr. Este but that Mr. E. be informed if his name be not withdrawn he must continue to pay his Subscription in future.

Mr. Fincher's Salary to be paid Monthly

6. Mr. Auriol reported that Mr. Fincher had requested him to state to the Managers that it would be an accommodation to him if his Salary could be paid Monthly instead of Quarterly.

Resolved That Mr. Fincher be authorized to charge his own Salary Monthly, delivering regular receipts for the same.

Presents

7. Presents received

John Hinckley Esqr. presented Sundry Books value Six Guineas.

Ordered That the Thanks of the Managers be returned for these presents.

750 Copies of Charter &c ordered.

8. That Mr. Auriol be desired to order 750 Copies of the Charter Act of Parliament and Bye Laws and also forms of Receipts &c according to the Appendix, and likewise such other forms as are requisite for effecting the purposes of the Bye Laws.

Payments of Newly elected Members.

9. **Resolved**. That the newly elected Members be informed that if they will have the goodness to pay in their Qualifications to the Account of the Institution at Messrs. Ransom Morland & Co. regular Receipts, as soon as prepared shall be exchanged for the Bankers receipts.

Part of Mr. Savage's bill paid

10. **Ordered** That the Treasurer be desired to pay Mr. Savage £70 on account of his Bill and to Mr. J. Stuart £57.10 for a Quarters Salary to the 31st of March last and signed an Order accordingly.

11. Read a Letter from

Propr. desiring to receive Compensation

The Revd. Edmund Ferrers.
expressing his desire to accept the proposed Compensation of £42.0.10 for his share as a Proprietor.

Concerning Fine due by Mr. Gray's Exors.

12 **Resolved.** That Mr. Auriol and Mr. Hinckley be requested to take the necessary measures to obtain payment of the Fine due to the Institution from the Executors of the late Mr. Gray.

Apparatus belonging to late Mr. Cavendish.

13. Mr. Davy requests permission of the Managers to place some Chemical Apparatus which belonged to the late Mr. Cavendish in one of the Garrets of the Royal Institution. It is Mr. Davys intention to employ them in the experiments and lectures of the Royal Institution. Agreed to.

T. Bernard

Managers Meeting 30th July 1810

At a Meeting of the Managers of the Royal Institution of Great Britain, held at the House of the Institution, on the 30th day of July 1810.

Present

Sir Thos. Bernard Bart. in the Chair
Scrope Bernard Esqr.
J. P. Auriol Esqr.

Minutes approved

1. The Minutes of the proceedings of the last Meeting were read and approved.

Exd. Gen. Acct.

2. Examined the General Account of the Institution.

Minutes of Comee. of Accts. approvd.

3. Read and approved the Minutes of the Committee of Accounts of this day.

Proprietors desiring to receive compensation

4. Read Letters from
George Grant Esqr. dated 3d. July 1810
Samuel Kenyon Esqr. " 13th. July "
expressing their desire to accept the proposed Compensation of £42.0.10 for their Shares as Proprietors.
and — Matthew Raper Esqr.
with thanks for his Election and ^expressing his^ hopes to attend at the second General Monthly Meeting.

Presents

5. Presents received.

David Pike Watts Esqr. presented Nares's Remarks on a version of the new Testament 8vo.
The Revd. Joseph Harpur presented An Essay on the principles of Criticism 4to.
The Revd. Henry De la Fite presented De Luc's Geological Travels 8vo.
Directors of the Globe Insurance Compy. presented Trial between Wm. Lowerthy and themselves in April 1810. 8vo.
Ordered. That the Thanks of the Managers be returned for these presents.

Proposal from Patrons for raising Mr Sturt's Salary

6. Read the following proposal from the Patrons of the Library.

Mr. Dutens proposed to recommend to the Managers of the Royal Institution the raising the Salary of the Assistant in the Library, Mr. Sturt, whose assiduity and

and diligence deserves to be taken into consideration, especially as the other Servants of the Institution have had some addition made to their Salaries.

Resolved. That this proposal be taken into consideration as soon as the finances of the Institution will permit.

7. Mr. Harris begs permission for leave of absence for a Month.

Resolved. That Mr. Harris's request be granted, he giving Mr. Davy previous notice of his departure.

Balloting Box.

8. That Mr. Hinckley be requested to provide a proper Balloting box.

Reference to Experiments in Laboratory by Mr Davy in Trans. of R. S.

9 Mr. Davy requests permission of the Managers of the Royal Institution to refer to some experiments which he has lately made in the Laboratory of the Institution in a paper which the Council of the Royal Society have ordered to be printed.

Resolved. That Mr. Davy's request be complied with, he conforming to the Regulations in the Bye Laws.

Life Subscr. becomes Member.

10 The Treasurer having reported that Mr. Petit has paid in £31 " 10 0

Resolved. That the General Monthly Meeting be informed that Louis Hayes Petit Esqr. a Life Subscriber to the Institution has paid in 25 Guineas his Qualification as a Member and 5 Guineas his Subscription to the Library and Mineralogical Fund.

Mr. Savage's Bill

11 **Resolved.** That Mr. Savage be desired to charge the Bill for printing ordered by Mr. Davy at the Managers request to the General Account of the Institution.

Forms to be printed.

12. Mr. Auriol reports to the Managers that in consequence of their Resolution at the last Meeting he has ordered the following forms to be printed viz.

Notification of the Election of Members (not included in the Bye Laws)
Bonds for Members.
Certificates of Election.
Receipts
Copies of the Charter, Act of Parliament, Bye Laws and List of Members.

Recommendation of Life or Annual Subscribers.
Recommendation for particular courses of Lectures.
Personal Admissions for the day.

All these according to the forms in the Bye Laws except the first which is submitted for approbation.

Resolved that it be approved with this alteration proposed.

13 **Resolved** That Mr. Davy be requested to employ Mr. Royce to examine into and report the value of the Windows now out of use.

W. Watson

Managers' Meeting, Aug. 6. 1810.

At a Meeting of the Managers of the Royal Institution of Great Britain, held at the House of the Institution on the 6th day of August 1810.

Present

John Symmons Esqr. in the Chair
Wm. Watson Esqr.
J. P. Auriol Esqr.

Minutes approved. 1. The Minutes of the proceedings of the last Meeting were read and approved.

Exa. Gen. Acct. 2. Examined the General Account of the Institution.

Minutes of Com. of Accts. appd. 3. Read and approved the Minutes of the Committee of Accounts of this day.

4. Read Letters from

Davies Giddy Esqr.

in reply to the Notification of his Election stating, that he will attend when he arrives in Town on the Meeting of Parliament.

Life Subscriber. Samuel Foart Simmons M.D.

nominating his Son Richard Simmons of Christ Church College Oxon M.D. to be admitted to the privileges of a Life Subscriber.

& Mr. Rt. Fidler

requesting payment of his small Bill for Apparatus &c. amounting to £8.1.6.–

Presents 5. Present received.

Mr. Ackerman presented.

An Account of the sacrifices and sufferings of the Inhabitants of the Tyrol and Voralberg.

Ordered. That the Thanks of the Managers be returned for this present.

Payment from new Members. 6. J. Fincher reported that he has received in the last Week from new Members the following sums

Thos. Snodgrass Esqr.	52 . 10 . 0
Subscription to the Library	5 . 5 . 0
H. Davy Esqr. Admission fee	8 . 8 . 0
Duty on Bonds	1 . 0 . 0

which he has paid into Messrs. Morland & Co's Bank; he

has also inquired this day of the several other Bankers of the Institution, and finds, the following sums have been paid into Messrs. Hammersley & Co's Bank by

John Sanders a life Subscriber £26 . 5 . 0

and into Messrs. Herries & Co's Bank by

John Disney Esqr. (a life Subr.)
his admission fee of £4 . 4 . 0
& Subscript. to the Liby. 5 . 5 . 0 — 9 . 9 . 0

7. Resolved. That the sum of £50 be advanced to the Clerk.–

Valuation of Windows 8. Mr. Davy reported that he had obtained from Mr. Royce a Valuation of the Glass Windows referred to him, vizt. at the sum of £10.

Resolved. That the same be referred back to Mr. Davy to obtain a counter-valuation

9. Ordered. That in addition to the Resolutions voted on 9th Ult. to be laid before the General Monthly Meeting of this day the following Report be there read

Proprietors claiming Compensation: The Managers also report that the following fifteen persons have claimed the Compensation of £42 . 0 . 10 as mentioned in the Act.

Benjamin Blake Esqr.
The Honble George Villiers
Thos. Wright Esqr.
Wm. Boultbee Esqr.
Thos. Palmer Esqr.
Saml. Lanford Esqr.
The Right Honble Lord Ossulston
The Honble Admiral Digby
Richd. Harford Esqr.
Mrs. Susannah Day
James Gorst Esqr.
General Sir Charles Morgan Bart. M.P.
The Revd. Edmund Ferrers
George Grant Esqr.
Saml. Kenyon Esqr.

Life Subscriber. They further report. That Samuel Foart Simmons M.D. has nominated his Son to enjoy the privileges of a Life

Subscriber under the Act by the following Letter.

My Lords and Gentlemen Augt. 6th 1810

Having expressed my wish to relinquish my Hereditary share in the property of the Royal Institution. I beg leave to nominate my Son Richard Simmons, of Christ Church College, Oxford, Doctor of Physic, as a Life Subscriber to the said Institution; and have the honor to be,

My Lords & Gentlemen
Your faithful humble Servant
S. F. Simmons

It being further stated

That A. F. Haldimand Esqr. having sent in his nomination of his Son Wm. Haldimand Esqr. previous to the passing of the Act, and a late Clerk of the Institution to whom it was delivered having neglected his duty Mr. H. submits, whether his Son should not be admitted as of right.

T. Bernard

Managers' Meetg. 27th Aug. 1810.

At a Meeting of the Managers of the Royal Institution of Great Britain held at the House of the Institution on the 27th day of August, 1810,

Present

Sir Thos. Bernard Bart. in the Chair
Scrope Bernard Esqr.
W. Watson Esqr.

Minutes approved 1. The Minutes of the proceedings of the last Meeting were read and approved.

Exd. Gen. Acct. 2. Examined the General Account of the Institution.

Minutes of Com. of Accts. approved 3. Read and approved the Minutes of the Committee of Accounts of this day.

Payment of Bills &c. 4. Signed an Order for an advance to the Clerk and for payment of the following Bills.

J. Fincher on account of the Royal Institution	100 . 0 . 0
Ditto on account of Voltaic Apparatus	34 . 6 . 4
Dr. Smith for course of Lectures &c.	66 . 4 . 6
Mr. Dalton balance of acct. for Lectures	42 . 0 . 0
Mr. Pond on account of 2 courses	52 . 10 . 0
One Years Parochial Rates due 25th March	55 . 0 . 0
½ " Property tax "	48 . 10 . 8
½ " Land-tax	7 . 5 . 6
½ " Assessed taxes	68 . 16 . 7
½ " Landlords' property tax	25 . 0 . 8
Mr. Savage for Printing &c. to 9th June	163 . 18 . 10
Mr. H. Davy one quarters Saly. to 31st Decr.	100 . 0 . 0
Mr. E. Davy up to 30th June	63 . 8 . 0
Mr. Harris ½ year to 31st March	90 . 0 . 0
Mr. Sturt ¼ Salary up to 30th June	17 . 10 . 0
Mr. Bramwell for Act of Parlt.	233 . 19 . 11
Mr. Knight to 31st Decr. 1808	51 . 13 . 0
Mr. Allen on account	100 . 0 . 0
	£1320 . 4 . 0

5. Ordered That the Clerk pay the following Bills

Mrs. Wilmot for washing	13 . 9 . 9
Dixon for 1000 Cards	1 . 10 . 0
	14 . 19 . 9

Brot forward.	14 .. 19 . 9
Argand & Elgar for use of Lamps	0 . 8 . 6
Bird – Bricklayer	5 .. 11 . 6
Papworth Plaisterer	2 .. 18 . 4
Fores for Pasteboards &c	2 . 0 . 0 pd by Mr Pond out of cash advanced
Jent for Earthenware	20 . 0 .. 11
Cuthbertson for Apparatus	2 . 14 . 6
	£ 48 .. 13 . 6

Bill partly paid 6. **Resolved**. That Mr Allen be informed that the Committee has ordered £100 to be paid him on account and that the Committee hope to be able to settle the remainder of his Account out of two Fines now due and promised to be paid. But, his Account having been lost in consequence of their Clerk having absconded, the Committee request he will favor them with a Statement of the particulars, either on account of Bills or Lectures. –

Honr Member. 7 Read Letters from His Excellency
Prince Castelcicala thanking the Society for electing him an Honorary Member. &
Claim for Compensation Miss Ann Hays
stating her having already intimated some Months ago her Claim of the pecuniary Compensation under the Act of £42. 0. 10
– **Ordered**. That the same be added to the List of Claims

Presents **Presents received.**

8. The Revd Stephen Weston presented his translation of the Emperor Kienlung's Poem on the Conquest of the Miao-tse from the Chinese – &
David Ricardo Esqr presented his Pamphlet on the high price of Bullion.
Ordered. That the Thanks of the Managers be returned for these presents.

Acct of Monies recd & paid. 9. **Ordered**. That a separate Account be kept of all monies received and payments made on account of the Library from 31st March last.

10 **Resolved**. That Mr Hinckley be requested to ascertain the circumstances on which Mr Haldemand's Claim is founded and to report to the Managers.

Fine from Mr. Gray's Exors. paid – £1517. 11. The Treasurer Reported that he received on Wednesday last the 22d August from Messrs Tenant & Co. Solrs to Gray's Executors the Fines due from the late Mr Gray with Interest to Michas last amounting together to £1517 – And it was inserted in the receipt that the remaining Interest from Michaelmas to the 22d Augt should be paid by Mr James White the Receiver for the Estate.

T. Bernard

Managers' Meetg 8th Oct. 1810.

At a Meeting of the Managers of the Royal Institution of Great Britain held at the House of the Institution on the 8th day of October 1810.

Present
Sir Thos Bernard Bart in the Chair
John Symmons Esqr
J. P. Auriol Esqr

Minutes confirmed 1. The Minutes of the proceedings of the last Meeting were read and approved.

Exd Gen. Acct 2. Examined the General Account of the Institution.

Minutes of Comee of Accts approved 3. Read and approved the Minutes of the Committee of Accounts of this day and also of the preceding Meetings of the 30th day of July, 6th, 13th, 20th, & 27th of August, 3d & 17th Septr, & 1st October. –

4. It appeared that Mr Pond had paid the £2 for Pasteboards ordered to be paid by the Clerk. –

5 Read Letters from
Mr George Glover
requesting payment of his Bill. –
Life Subscr proposed. Wm Saunders M.D. of Russell Square
desiring the Managers to appoint his Nephew William Ward Esqr of New Bond Street to the privileges of a Life Subscriber.
Ordered – That the Clerk refer Dr Saunders to the Act which requires him to make the appointment himself
& Mr Wm Haldimand stating his having presented a claim of transfer to him, of his father's proprietory share before the Act past

Form of Bond. 6. Read Mr Hinckley's report of Mr Butlers opinion concerning the form of the Bond.
Ordered. That Mr Hinckley communicate the thanks of the Managers to Chas Butler Esqr for his opinion on the form of the Bond.

Presents received 7 **Presents received.**
Edward Geoghegan Esqr presented his Commentary on the treatment of Ruptures.
John Grenfell Esqr presented two Copies of his defence of Bank Notes and letters to Frans Horner Esqr M.P.
Ordered. That the thanks of the Managers be returned for these presents. –

Acct of Debts 8. **Ordered** That the Clerk prepare an Account of Debts owing to and from the Institution by way of Dr & Cr. –

W Watson

Managers' Meetg 15th Oct. 1810.

At a Meeting of the Managers of the Royal Institution of Great Britain, held at the House of the Institution, on Monday the 15th day of October 1810

Present.
Wm. Watson Esqr. in the Chair
John Symmons Esqr.
Sir Thos. Bernard Bart.
J. P. Auriol Esqr.

Minutes approved. 1. The Minutes of the proceedings of the last Meeting were read and approved.

Exd. Gen. Acct. 2. Examined the General Account of the Institution.

Minutes of Comee of Accts approved. 3. Read and approved the Minutes of the Committee of Accounts of this day, and signed an Order for an Advance to the Clerk of £50.

Life Subscriber appointed 4. Read a Letter from Wm. Saunders M.D. appointing his Nephew Wm. Ward Esqr. of New Broad Street to the privileges of a life Subscriber to the Institution.
Ordered.– That his name be added to the List of Life Subscribers.

5. A Bill being presented for John Leeson Constable for attendance & delivering Cards.
Ordered.– That it be referred to Mr. Hinckley for examination and to order payment if right.

Stamp Duty 6 Ordered. That the duty of 6d. on Mr. Davy's printed Lecture be paid at the Stamp Office.

Report on Question of gratuitous admission of Life subscriber. 7. Mr. J. Hinckley reports that, in consequence of a reference to him of 1st October to ascertain the facts concerning Mr. Haldimand's alleged transfer to his Son previous to the Act, your reporter waited on Messrs. Haldimand, father & son, the next day, when having heard their narrative he requested that the whole might be stated in writing; and on being asked from whom, expressly

requested that each would relate his own transactions. This produced in the evening of the same day the letter from Mr. Haldimand junr. of 2d. Instant to which perhaps no other objection can be made than its being the report of only one of the parties, whereas it might have been confirmed by three or even four others therein mentioned. But as these parties were so early informed by some of their nearest connexions of an expected interference of irristible force by an Act of Parliament, it may be lamented that thro' the same relative (who is said almost habitually to have frequented the Institution and to be intimately acquainted with its most distinguished professor) they did not in the course of so many months once find a moment of relaxation to inquire, whether the forms of Office had been duly satisfied and the transfer of its now so much valued privileges confirmed to the new Candidate. The regret excited by the apparent forgetfulness of the existence of such an interest in the affairs and success of the Institution arises not only from a sincere desire in your reporter to reckon the very respectable candidate in question among its members, but also from an ardent wish to give general satisfaction: whereas should that wish lead to a hasty compliance with the request of Mr. Haldimand, it might excite dissatisfaction in others, whose similar noncompliance with forms established by themselves in their own Bye Laws has disabled them in like manner from divesting themselves of privileges now confirmed to them for life by the power of Parliament. For, notwithstanding any act, whether of the said Mr. Haldimand senior, or of the Managers, or of the whole body of the members of the Royal Institution, still Mr. Haldimand senior would remain a member of the Institutn. as much after the gratuitous admission of one son as a member, and perhaps of another son by appointment of the latter to the gratuitous privileges of a life Subsriber, as he is now because there exists no power except a new act of the whole national Legislature, whereby he can be divested of those privileges. The object of the present application may therefore be said to resolve itself into nothing short of a gratuitous admission of Mr. Haldimand junior as a member of the Royal Institution. It is however very fortunate that there still exists a mode by which a refusal if unavoidable may be softened; since Mr.

Haldimand, who cannot certainly complain of being taken by surprize, may both continue to enjoy the privileges of a Member, and give his Son by his own absolute appointment in writing all the privileges of a life Subscriber.–

Order in consequence. Ordered. That Mr. Hinckley's report be approved and that a copy of it be sent to Mr. Haldimand, with the Managers' regret, that they cannot propose to the General Meeting any thing more to be done, than what has lately been done in the stronger case of Sir George Duckett; where his Nominee has been admitted a Subscriber for Life.
That the thanks of the Managers be given to Mr. Hinckley for his very able and satisfactory report.

Acct. of Debts & Credits 8. An Account of Debts & Credits having been laid on the table.
Ordered. That it be laid on the table at every meeting.–

9 Ordered. That the Clerk provide 1/4 Cwt of Soap & 4 dozen of Candles or 1 Cwt of Soap if cheap.–

T. Bernard

Managers' Meetg Oct. 22, 1810.

At a Meeting of the Managers of the Royal Institution of Great Britain, held at the house of the Institution, on Monday the 22d. day of October 1810.

Present
Sir Thos. Bernard Bart. in the Chair
Wm. Watson Esqr.
John Symmons Esqr.
Scrope Bernard Esqr.

Minutes approved. 1. The Minutes of the proceedings of the last Meeting were read and approved.
2. Examined the General Account of the Institution.

Minutes of Com. of Accts approved 3 Read & approved the Minutes of the Committee of Accounts of this day.

Genl. Monthly Meetgs. 4 Ordered. That the Clerk under the direction of the Secretary send Summonses to the Members of the Institution for the General Monthly Meetings of the ensuing Season including the Members elect who have not yet taken their seats.
Ordered. That the form be printed as follows on a Card.

Form of Summons. Royal Institution 22d. Octr. 1810

The General Monthly meetings for the ensuing Season are to be held on the following days at 2 o'Clock precisely.
Monday 5th November 1810
——— 3d. December "
——— 4th February 1811
——— 4th March "
——— 1st April "
——— 6th May "
——— 3d. June "
——— 1st July "
——— 5th Aug "

New Solicitor appointed 5 Mr. Fletcher who has acted as the Solicitor of the Institution having left England.
Resolved. That Mr. Seymour of Margaret Street Cavendish

142 | 1810. Oct 29.

Square be desired to give his assistance to the Institution as the Solicitor. And that Mr. Watson be requested to communicate the above to Mr. Seymour, and to desire that he will take measures for obtaining the City Lease to the Corporation, and for enabling the Managers to receive the Fines & Rents due for the two adjoining houses.—

6. Ordered. That a List of Members to be balloted for at the next General Monthly Meeting: of those to be proposed whose signatures are compleat; and of those wanting signatures; be laid on the table at every Meeting.

John Symmons

143

Managers' Meet 29th Oct. 1810.

At a Meeting of the Managers of the Royal Institution of Great Britain held at the House of the Institution, on Monday the 29th October 1810.

Present
John Symmons Esqr. in the Chair
Sir Thos. Bernard Bart.
Scrope Bernard Esqr.

Minutes approved — 1. The Minutes of the proceedings of the last Meeting were read and approved.

Exd. Gen. Acct. — 2. Examined the General Account of the Institution.

Minutes of Comee. of Accts. approved — 3. Read and approved the Minutes of the Committee of Accounts of this day.

Letters read — 4 Read Letters from
Mr. Seymour
with his acknowledgments for appointing him Solicitor to the Institution.
& Dr. Marcet
announcing his intention of laying before a General Meeting the particulars concerning Mr. Haldimand's proposed transfer of his late proprietory share.—

Seal of Corpn. to be affxd. to Lease. — 5 Ordered.—That notice be given in the next Summons ~~sen~~ to the Managers, stating that the Corporation Seal is then intended to be affixed to a Counterpart of the renewal Lease from the City of London to the Institution.

6 Ordered—That the Clerk wait upon or write to Mr. Auriol concerning the keys of the Iron Chest in order to the affixing the Seal of the Corporation as above.

7 Ordered—That the Clerk deposit with Mr. Seymour, Solicitor, the Book of Leases belonging to the Royal Institution.

T. Bernard

144 | 1810.

Managers' Meeting, Nov. 5. 1810.

At a Meeting of the Managers of the Royal Institution of Great Britain held at the House of the Institution, on Monday the 5th of November 1810.

Present
Sir Thos. Bernard Bart. in the Chair
Wm. Watson Esqr.
J. P. Auriol Esqr.

Minutes approved — 1. The Minutes of the proceedings of the last Meeting were read and approved.

Exd. Gen. Acct. — 2. Examined the General Account of the Institution

3 Read and approved the Minutes of the Committee of Accounts of this day.

Bills (books & binding) — 4 Read the following Resolution of ~~a~~ the Patrons of the Library.
Resolved. That the Managers be acquainted that there are two Bills presented by Mr. Harris; one commencing November 1807 of £87. 8. 0 due to Mr. Payne for Books and the other due to Mr. Searle for Bookbinding amounting to £15. 0. 9.
Ordered. That the above be taken into consideration on Monday the 26th Instant.

5 Ordered. That the affixing of the Corporation Seal to the City Lease be adjourned till Monday next and that notice be given in the Summons thereof.

T. Bernard

145

Managers' Meet. 12 Nov. 1810.

At a Meeting of the Managers of the Royal Institution of Great Britain held at the house of the Institution on Monday the 12th of November 1810.

Present
Sir Thos. Bernard Bart. in the Chair
Wm. Watson Esqr.
Sir J. C. Hippisley Bart.

Minutes appd. — 1. The Minutes of the proceedings of the last Meeting were read and approved.

Exd. Gen. Acct. — 2. Examined the General Account of the Institution.

Minutes of Com. of Acct. approved — 3. Read and approved the Minutes of the Committee of Accounts of this day.

Presents received — 4
Presents received
Lord Teignmouth presented Wm. Dealty M.A. Vindicatn. of the Bible Society in a reply to the Revd. Dr. Wordsworth's letter to his Lordship.
Ordered. That the thanks of the Managers be returned for this present.
Vol. 1. No. 1. of Dr. Bradley & Shearman's New Medical and Physical Journal was left at the Royal Institution.

Affxg. of Corpn. Seal to Lease adjd. — 5. Ordered. That the affixing of the Corporation Seal to the City Lease be adjourned till Monday next and that notice be given in the Summons thereof.—

Lectures — 6 Ordered. That Mr. Hinckley be requested to ascertain how many Lectures Mr. Allen delivered in consequence of his Agreement in November 1809.

7 Resolved. That it be referred to Mr. Hinckley to get some Cards of Subscription printed.

8 Ordered. That the Clerk pay £3 due for the Water Rate.—

Newspapers. — 9. Ordered. That the Steward provide a second-

Copy of the Times Newspaper and discontinue the second copy of the Morning Post. —

Money ordered for taxes &c. — 10 Signed an Order for an advance to the Steward of £50. 0. 0 and for payment of the following sums.

½ years Kings taxes due Michas 1810.	103. 8. 9
½ years Parochial rates	26. 5. 0
Mr E. Davy ¼ Salary to 30th September 1810	25. 0. 0
	£204. 13. 9

T. Bernard

Managers Meets 19th Nov. 1810

At a Meeting of the Managers of the Royal Institution of Great Britain held at the house of the Institution on Monday the 19th of November 1810 —

Present
Sir Thos Bernard Bart. in the Chair
Sir J. C. Hippisley Bart.
John Symmons Esqr
Scrope Bernard Esqr

Minutes approved 1. The Minutes of the proceedings of the last Meeting were read and approved.

Ext. Gen. Acct 2 Examined the General Account of the Institution

Minutes of Com. of Accts approved 3 Read and approved the Minutes of the Committee of Accounts of this day.

Lecture on Music proposed. 4 Read Letter from Dr Kemp purposing to give Lectures on Music.
Ordered that a copy be sent to Mr Hatchet with a request that he and the rest of the Committee of Science will consider and report thereon.

5. Present received.

Present recd. Mr Wm Savage presented 2d & 3d Vols & No. 19 of the Librarian by James Savage, and No. 1 of Joseph Knight's Horticultural Essays.
Ordered. That the thanks of the Managers be returned for this present. —

Manager's death. 6. Mr J. Hinckley reported a vacancy in the management by the death of the late Lord Dartmouth.
Ordered. That it be reported from the Chair to the next General Monthly Meeting and that the Managers take into their consideration at their next Meeting whom to recommend in his place.

7 Mr J. Hinckley reported that nine Lectures by Mr

Mr Allen's Lectures — Allen were announced in the Cards of the last Season viz.

	Lecture 1.	9th Febry 1810
	2.	16th
The agreement was	3.	23
12 Lectures	4.	16 March
for 60 Guineas and	5 –	23 –
therefore the proportion due	6 –	30
to him would be	7 –	13 April
45 Guineas	8 –	27
	9 –	11. May

Ordered. That the sum of 45 guineas be admitted as due to Mr Allen for the above Lectures.

8. Ordered. That the Visitors be summoned for one o'clock on Wednesday the 28th Instant.

Rooms desired for Mr Harris granted subject to restriction. 9. Mr Symmons reported Mr Harris's wish to be put in possession of Mr Easingwoods rooms.
Ordered. That Mr Harris be put in possession of the rooms subject to the temporary occupation of them by all Committees whatsoever.

Fixing Vane on Mr Colborne's Chapel? 10 Ordered. That Mr Harris be desired to enquire what will be the expense if leave can be obtained of fixing a vane on the cupola of Mr Colborne's Chapel.

11. Ordered That the affixing of the Corporation Seal to the City Lease be adjourned till Monday next and that notice be given in the Summons thereof.

T. Bernard

Managers' Meeting Nov. 26. 1810.

At a Meeting of the Managers of the Royal Institution of Great Britain, held at the house of the Institution, on Monday, the 26th day of Novr 1810.

Present
Sir Thos Bernard Bart. in the Chair
Charles Hatchett Esqr
John Symmons Esqr
Wm Watson Esqr
Sir J. C. Hippisley Bart
Earl of Morton
J. P. Auriol Esqr

Minutes appd 1 The Minutes of the proceedings of the last Meeting were read and approved.

Ext. Gen. Acct. 2 Examined the General Account of the Institution

Minutes of Com. of Acct. approved 3 Read and approved the Minutes of the Committee of Accounts of this day. — and signed an Order for an advance to the Clerk of £30. —

Nomination of a Life Subscriber 4 Read the following Letter from Samuel Kenyon Esqr

Gentn

"I sometime ago wrote to Mr Auriol to state to him that in consequence of my great distance from London and having no intention of again residing in the Metropolis, I had determined on receiving the sum specified as the proportion allotted to each original Proprietor, but since that period, a Nephew of mine (whose health has been exceedingly impaired by his having been engaged as an Officer, in the unfortunate Expedition under Sir John Moore in Spain and since in the Island of Walcheren, where he caught the fever) has fixed his residence in London, and it is my wish to waive the claim I made, and to appoint him to the privileges of a Life Subscriber And I do hereby nominate and appoint Edward Kenyon Esqr of Holborn Court Grays Inn my Nephew, to the privileges of a Life

"Subscriber to the Royal Institution."
I have the honor to be
20th Nov. 1810. Gentn. Your obedt. & humble Servt.
signed, Saml. Kenyon

To the Managers of the Royal Institution.

Resolved.
That the above letter from Samuel Kenyon Esqr. be read at the next General Monthly Meeting of the Members and submitted to their consideration.

Resignation of Mr. Auriol as Secy. & Mr. A proposed as Manager. Mr. Hinckley as Secy.

5. Mr. Auriol having signified his desire of resigning the Office of Secretary to the Royal Institn.
Resolved. That the same be reported to the next General Monthly Meeting and, though the Managers feel very great regret on his resignation, that it be recommended to the Members to accept it. And it be further recommended to the General Meeting to allow Mr. Auriol's name to be proposed for election as a Manager in the vacancy occasioned by the death of the Earl of Dartmouth and the name of John Hinckley Esqr. for election as Secretary of the Institution in the room of Mr. Auriol. —

Music Lecture declined

6. Mr. Hatchett reported on the part of the Committee of Science that having taken into consideration Dr. Kemp's proposal to give a course of Lectures on Music illustrated by assistance of Vocal and Instrumental performers, it appears that this proposal cannot be accepted, it being inadmissible under the present regulations of the Royal Institution.
Resolved. That the Secretary do write to Dr. Kemp to thank him for his offer and to acquaint him that it cannot be accepted under the present regulations of the Royal Institution. —

7. **Resolved.** That it be submitted to the General Meeting that it would be very convenient that Mr. Seymour, who acts gratuitously as the Solicitor of the Institution, should have notice of the Meetings of the Members, in order that he may be present when requisite to give any information to the Meeting that may be wanted on official business

Seal affixed to Counterpart of Lease.

8. Notice having been sent to all the Committee of Managers two days previous. "That at this Meeting the Corporation Seal was to be affixed to a counterpart of the City renewal Lease." — The said Counterpart was passed and sealed at this Meeting and countersigned by Sir Thos. Bernard Bart. Vice President & Chairman of the Meeting, by Earl Morton Vice President and Manager, Sir John Coxe Hippisley Bart., Manager William Watson Esqr. Manager, and John Symmons Esqr. Manager, and Charles Hatchett Esqr. manager being a majority of the Managers present and by James Peter Auriol Esqr. Secretary. —

Privileges granted to Mr. Seymour.

9. **Resolved.** That Mr. Seymour the Solicitor have personal admission to all the reading rooms and libraries of the Institution, with liberty to attend all the public Lectures and that Mr. Auriol be requested to inform him thereof. —

John Symmons

Managers' Meetg. Dec. 3. 1810.

At a Meeting of the Managers of the Royal Institution of Great ~~Britain~~ Britain held at the house of the Institution on Monday the 3rd day of December 1810.

Present
John Symmons Esqr. in the Chair
Charles Hatchett Esqr.
Scrope Bernard Esqr.

Minutes appd.

1. The Minutes of the proceedings of the last Meeting were read and approved.

Exd. Gen. Acct.

2. Examined the General Account of the Institution.

Minutes of Com. of Accts. approved.

3. Read and approved the Minutes of the Committee of Accounts of this day.

4. Read Letters from

Nomination of Life Subscriber.

Dr. Charles Burney of Greenwich nominating his Son Charles Parr Burney B. A. of Merton College Oxford to the privileges of a life Subscriber.
Ordered. That his name be added to the List of Life Subscribers. —

Committee Room.

from The Revd. Edward Forster, as one of the Honorary Librarians reciting Chap: 20. Art: 7 of Bye Laws and recommending the preparation of a Committee room; the consideration of which was postponed till next Meeting. —

Reviving old Bye Law.

from William Thornton Esqr. to the Honble Robert Clifford proposing the revival of one of the former Bye Laws which restrained all Servants of the Institution from receiving any present or reward from Proprietors now Members or Subscribers.

5. **Ordered.** That Mr. E. Davy be put in possession of Mr. Harris's late rooms.

6. Read a proposition and request from the Patrons of the Library, that the double windows be placed in the Library during the Winter.

On Warming Library.

Mr. J. Hinckley reported that the double windows having been already sold and the Patrons being uninformed thereof when the above request was voted, The Reverend Mr. Forster one of the Hony. Librarians had expressed a wish that in Lieu thereof heated air be introduced at the skirting of the Library by a pipe or pipes from the vicinity of the fire place in the Reading Room below — postponed the consideration hereof till next Monday. —

T. Bernard

154 — 1810

Managers' Meeting, 17 Dec. 1810

At a Meeting of the Managers of the Royal Institution of Great Britain held at the house of the Institution on Monday the 17th day of December 1810.

Present
Sir Thos. Bernard Bart. in the Chair
John Symmons Esqr.
Scrope Bernard Esqr.

Minutes approved. 1. The Minutes of the proceedings of the last Meeting were read and approved.

Exd. Gen. Acct. 2. Examined the General Account of the Institution

Minutes of Com. of Accts. appd. 3. Read and approved the Minutes of the Committee of Accounts of this day, and signed an Order for an advance to the Clerk of £60. —

4 Read Letters from
The Revd. Wm. Cockburn

Vane on St. George's Chapel. handsomely consenting to erect a Vane on St. George's Chapel —

George Hayden

Printer to the Royal Instn. soliciting to be Printer to the Institution.

& W. Bulmer to the like effect as follows.

Shakespear Printing Office,
Cleveland Row, St. James's
Dec: 17th 1810.

Sir

Mr. Savage having mentioned to me his intention of resigning his situation as Printer to the Royal Institution, this day, I take the liberty of offering my services to fill the vacancy occasioned by his resignation. — It may live in the recollection of yourself, as well as many other Gentlemen, that I printed all the Preliminary papers during the formation of the Institution, and continued to execute the whole of the printing business till the Managers resolved on keeping a press within the walls of their own establishment, to facilitate the accomplishment of which object I lent them every assistance in forwarding their wishes on that head. With these pretensions, I submit my application to the consideration of the President, Vice President, and Managers of the Royal Institution, and have the honor to remain Sir,

Your obliged humble Servant
(signed) W: Bulmer"

Sir T. Bernard Bart.

Dec. 17 — 155

Presents received.

Presents recd. 5 Edward Hawke Locker Esqr. presented Coxe's Life of Benjamin Stillingfleet.

The Reverend Edward Forster presented British Gallery of Engravings, No. 7.

The Revd. Edmund Davies presented Mythology and Rites of the Druids.

John Disney junr. Esqr. presented Theatrum Insectorum by Wotton & others printed in 1634.

Resolved. That the thanks of the Managers be returned for these presents.

Memorandum on Lord Elgin's pursuits in Greece was left at the Royal Institution.

Letter to Subscribers to Loan. 6 Resolved. That the following letter be written by the Secretary to the Subscribers to the loan for providing a fund for making pecuniary compensation to Proprietors relinquishing their interest in the Institution and living in England;

Royal Institution 17 Decr. 1810.

"I am desired by the Managers to inform you that the sum required on account of the loan for providing a fund for making pecuniary compensation to Proprietors relinquishing their interest in the Institution is five per Cent on the sum subscribed, and that your subscription being for the sum of £ , they request you will order the sum of £ , to be paid to Messrs. Morland Ransom Morland & Co. on account of the Royal Institution."

Mr. Bulmer re-appointed Printer. 7 Resolved. That in case Mr. Savage be unable to continue to act as Printer Mr. Bulmer be restored to his situation of Printer to the Institution.

Form of Nominating Life Subscriber. 8 Resolved, That the form of nominating a life Subscriber under the Act be as follows.

"Sir,

I hereby nominate my ——— of ——— to be admitted to the privileges of a Life Subscriber to the Royal Institution.

156 — 1810

I am Sir,
Your obedient Servant

To the Secretary of the Royal Institution"

Ordered. That 250 Copies be printed.

on warming the Liby. 9. Resumed the consideration of the Patron's request on warming the Library.

Ordered. That the Clerk enquire what effect the introduction of hot air by pipes from below would have on the Insurance.

10. Postponed the consideration of Mr. Forster's letter on Chap: XX. Art: 7 till next Meeting.

11. The Housekeeper having applied for an increase of Salary.

Ordered. That it be considered at the first full meetings of Managers.

12 Ordered. That the Housekeeper procure
10 doz: dips candles @ 11d. per doz.
6 doz: Moulds @ 12d. per doz:

13. Resolved. That the summoning the Scientific Committees be considered this day fortnight.

Arrangement for Lectures 14. Resolved. That Lord Dundas, Mr. Hatchett & Mr. Davy be requested to prepare an arrangement for the lectures of the approaching season.

Life Subscriber 15. Ordered. That the name of Edward Kenyon Esq. of Holborn Court Grays Inn be added to the List of Life Subscribers in consequence of a nomination by Samuel Kenyon Esqr.

16. J. Fincher reported. That from the shortness of time he had left after the General Monthly Meeting to pay the money into the Bankers hands he overpaid the account 16d. which is due to him in change.

Ordered. That he be allowed to charge it in his account.

Dec. 17. — 157

Annl. Subscribers 17 Ordered. That the following persons be admitted Annual Subscribers on paying their contributions.

Chevalier de Broval — of 34 Suffolk Street attached to the Neapolitan embassy under a nomination by the King of Naples introduced by
Prince Castelcicala
& recommended by Sir Thos. Bernard Bart: J. Hinckley E. Forster — John Symmons.

Arthur Mower. of 6. Robert Street Bedford row signed — Jos: Banks — Hump: Davy — T. Bernard. J Hinckley

Postponed the rest for want of signatures and forms.

Dundas

Managers' Meet. 24 Dec. 1810.

At a Meeting of the Managers of the Royal Institution of Great Britain, held at the house of the Institution on the 24th day of December 1810. —

Present
Rt Honble Lord Dundas in the Chair
The Earl of Aylesford
Sir Thos. Bernard Bart.
John Symmons Esqr.
J. P. Auriol Esqr.

Minutes appd. 1. The Minutes of the proceedings of the last Meeting were read and approved.

Ex. Gen. Acct. 2. Examined the General Account of the Institution.

Appd. Mintes of Com. of Accts. 3. Read and approved the Minutes of the Committee of Accounts of this day.

Resignation of J Symmons Esq. as Managers 4. Read Letter from
John Symmons Esqr.
proposing to resign his Office of Manager.
Resolved. That the Managers express great regret at the receipt of this letter and will take it into consideration on the last Monday in February.

Life Subscr. 5 In consequence of a nomination by Sir Thos. Bernard Bart. — James Baker Esqr. Fellow of New College Oxford was admitted to the privileges of a Life Subscriber to the Royal Institution.

Proposition for Committee Room. 6 Read the following letter from The Revd. Edwd. Forster.

"One of the Honorary Librarians begs to draw the attention of the Committee of Managers to the followg. Bye Law, and takes the liberty of submitting that, and the Statement which follows, to their consideration.

Chapter XX. Art: 7.

The Library & Mineralogical Collection shall be open for Members, Subscribers and persons recommended by the Patrons, on all days, on which the Institution

is open from ten till four o'clock."

As by the above Law, passed last July, the Library is obliged to be open every day from ten till four; it is suggested to the Committee of Managers, that any Committee, sitting in the Library during those hours, is liable to interruption from the Members and Subscribers, while at the same time any Committee, holding its meetings there, is a serious inconvenience to all those who frequent that part of the Institution. — It is therefore submitted to the Committee of Managers, whether it would not be proper and consistent with the existing Bye Laws for them to provide another room for all the meetings of the Committees of Managers, of Visitors, and of Patrons; and to order that no meeting of any Committee whatever shall be held in the Library during the hours it is open to the Members and Subscribers at large.

The honorary Librarian should not have taken the liberty of troubling the Committee of Managers on the subject, had he not been in the habit of hearing constant complaints.

Edward Forster
Honry. Librarian

3d Decr. 1810
To the Committee of Managers
of the Royal Institution.

Ordered. That it be laid before the next General Monthly Meeting with a report which Mr. Auriol is requested to prepare on the subject.

On Warming Liby. 5 J. Fincher reported that the introduction of hot air by pipes will increase the rate of premium in proportion to the risk which must be ascertained by the Surveyor of the Fire Office but in all events it will increase the rate of Insurance.

Ordered. That the consideration of it be postponed till further inquiry.

Annl. Subsrs. 6. The following persons having been duly recommended were elected as Annual Subscribers for the year 1811.
The Earl of Tyrconnel.
Major John Harding.

Mr. Samuel Parker
Edward Bilke Esqr.
L'Abbe Ratel
Capt. C. Irvine Royal Navy.
Mr. Christie
Mr. Rodriguez

Cases of Sir G. Duckett, &c. 7. Read the reference from the General Monthly Meeting of 5th November 1810. referring back to this Committee the cases of Sir George Duckett & George Sullivan Marten Esqr.
Resolved. That Mr. Auriol be requested to prepare a report on these cases and on that of A. F. Haldimand Esqr.

Life Subscriber 8 In consequence of a nomination by the Earl of Aylesford — Lord Guernsey was admitted to the privileges of a Life Subscriber to the Royal Institution.

Payment of Bill for Philos. Insts. 9 Mr. Davy presented two bills from John Newman Mathematical Instrument-maker for repairs and additions to the Apparatus of the Institn.
— the first from Decr. 1809 to 20th Decr. 1810 £27 . 8 . 11.
the second from April 1809 to June 1810 . . . 7 . 14 . 6
amounting together to . . . . £35 . 3 . 5

Signed an Order for the payment of the above two bills of £35 . 3 . 5. —

Philos. Instts. lent. 10. Mr. Davy having kindly lent the Apparatus, presented to him by Lord George Cavendish, and formerly the property of Mr. Cavendish deceased, for the Lectures of the Royal Institution and the Carriage of them amounting to £9 . 1 . 6.
Ordered. That this amount be paid by the Clerk

Mr. Davy's Lectures. 11. Mr. Davy proposes (if the plan should meet the approbation of the Managers) to give a course of Lectures on Chemical Philosophy — & another on Electro-chemical Science the first to commence by an introductory Lecture on the third Saturday in January. —

T. Bernard

Managers Meeting Dec. 31. 1810.

At a Meeting of the Managers of the Royal Institution of Great Britain, held at the house of the Institution on the 31st day of December 1810.

Present
Sir Thos. Bernard Bart. in the Chair.
John Symmons Esqr.
J. P. Auriol Esqr.
Earl of Morton
Earl of Aylesford

Minutes appd. 1. The Minutes of the proceedings of the last Meeting were read and approved.

Ex. Gen. Acct. 2. Examined the General Account of the Institution

Mints. of Com. of Accts. appd. 3. Read and approved the Minutes of the Committee of Accounts of this day.

Subject for Inspn. 4. Read Letter from
— William Armstrong of Bexley Kent
wishing to submit a Mechanical subject to the inspection of the Institution.
Ordered. That it be referred to Mr. Davy.

Life Subsr. 5 In pursuance of a nomination by Richard Barker, his Son Richard Barker junior was admitted to the privileges of a Life Subscriber to the Royal Institution.

Annl. Subsrs. 6 The following persons having been duly recommended were elected as Annual Subscribers for the year 1811.
to the Lectures only — Mrs. & Miss Smith — Park Lane Grosvenor Gate
Annual — Samuel Weller Singer Esq. Duke Street St. James
" — Walter Drew Esqr. — Gower Street Bedford Square
" — Andrew Henry Lynch Esqr. — Salisbury Street Strand
" — Patrick Paterson Esqr. — 6 Maddox Street.
" — Robt. Bakewell Esq. — 10 Bury Street.

Resolution of Patrons of Liby. claimd. 7. Read the following resolution of the Patrons of the Library.
Extract from the Bye Laws Chap XX. Art: 4.
"There shall be formed a Library & Mineralogical

acct. to Bye Laws £101. 17; and referring to Bye Laws Chap. 20 Art. 5.

"fund, the application of which, together with all "sums subscribed to the Library or Mineralogical "Collection shall be under the direction of the "Patrons"

It appears by the above Bye Law, that a fund is formed for the purchase of Books and Minerals, which is under the sole direction of the Patrons; which fund arises,–

1st/ By Art: 8 Ch: II/ from a subscription or admission fee from every member on his admission or within three years after of 5 guineas.

2dly/ By Articles 3 & 7 Chap. XIII/ from a subscription or admission fee from Life and Annual Subscribers respectively of five guineas and one guinea, with the exception of Ladies.

3dly/ By Art: 5 Chap XX/ from any sums not exceeding £50 at any one time, which the Committee of Managers are empowered by the said law to place at the disposal of the Patrons of the Library

4thly From sums voted for the above purpose at any General Meeting of Members.

5thly By individual or other donations & bequests.

Now it appears, that there has already been received by the Clerk or Secretary of the Royal Institution the sum of £101. 17. 0 as admission fees on account of the Library & Mineralogical fund, and that the same has been paid into the hands of the Bankers of the Committee of Managers or general Bankers of the Institution. The Committee of Patrons therefore have to request that the Managers will give orders to have the said sum of £101. 17. 0 paid to the Library Account under the direction of T. Hammersley Esqr. the Treasurer appointed by the Patrons; and to order the Clerk or Secretary, or whoever may receive the same, to pay over at once, whatever sums may be received at any time on account of the Library and Mineralogical fund to the account of the Patrons at their Bankers.

And as by Article 5 of Chap: XX the Committee of Managers have the power of aiding the Library and Mineralogical fund by the occasional application of any sum (not exceeding at one time the sum of

£50) from the general income or funds of the Institution, the Patrons request the Committee of Managers will take the said Bye Law into their consideration, as they have several long outstanding bills for Books &c to provide for.

Resolved. That this application be taken into consideration at the next Meeting.

8. Resolved. That the Scientific & Literarary Committee be summoned for Friday Evening 11th January at 8 o'Clock precisely.

Coals 9. Ordered. That it be referred to Mr. Auriol to order any quantity of Coals from 5 to 20 Chaldron according to the price.

Report concerning Committee Room 10. In pursuance of a reference of 24th Ult: of Mr. Forster's letter on a Committee room to Mr. Auriol in order to prepare a report thereon, he presented the following.

The Managers of the Royal Institution report that they at first held their meetings in the room then called the Managers' room, but since allotted for English Newspapers – They were induced to give up this accomodation for the convenience of the Proprietors and Subscribers as being a more spacious Apartment than that which was before appropriated for reading the daily Papers; and the Managers removed thence into a smaller room which they have likewise given up for the collection of Minerals; from the time of relinquishing this last appartment they have held their meetings in the Library of Reference. – In ordinary cases these meetings are only once a week, on a Monday; and seldom continue longer than from half past one to three o'Clock; so that it cannot be any great privation to the members to be debarred the entrance into that room for so short a space of time.

The Managers in giving up their two first rooms consulted only the benefit of the Establishment but they admit they have since experienced personal conveniency by assembling in the Library, from the

advantage of finding a well aired room to receive them instead of a damp and comfortless Apartment, in which no fire was lighted but on the days of meeting.

The only Apartments on the Ground floor in which the Managers could now be accomodated are the News room or the room beyond the Newsroom. – The former they conceive could not be shut up during the time of their meetings without interfering with the ordinary resort for that interval of more Members and Subscribers than usually frequent the Library of reference at the same hour. The latter, that is the room beyond the News-room, is now employed for keeping the Aparatus, and Mr. Davy reports that it cannot be spared, as no other part of the house is so fit for the purpose; and having a stone floor, it could not be converted into a Managers' room without considerable expense in the first instance for flooring, fitting up and furnishing, and a further continued expense in keeping constant fires in it during the Winter, without which it could not be made habitable. The Managers therefore submit to the General Meeting, whether it is not more advisable that no alteration at present should be made with regard to the place of holding their meetings, especially as the short periods in which the room is occupied by them in their sittings cannot prove a material inconvenience to any Member of the Institution.

Resolved. That the foregoing report be presented to the next General Monthly Meeting.

T. Bernard

Managers' Meeting Jan. 7 1811

At a Meeting of the Managers of the Royal Institution of Great Britain, held at the house of the Institution on the 7th day of January 1811.

Present
Sir Thos. Bernard Bart: in the Chair
John Symmons Esqr.
Scrope Bernard Morland Esqr.
J. P. Auriol Esqr.

Minutes appd. 1. The minutes of the proceedings of the last meeting were read and approved.

Ex. Gen. Acct 2. Examined the General Account of the Institution.

Minutes of Com. of Accts approved 3. Read and approved the minutes of the Committee of Accounts of this day.

Coals. 4. Mr. Auriol reported that in consequence of the reference to him at the last meeting he had ordered 20 Chaldrons of Coals to be immediately supplied by Messrs. Dickins & Co. at the house of the Royal Institution at 72s. ℔ Chaldron exclusive of meetage and shooting.

Orders for payment of Bills 5. Signed an order for an advance of £50. 0. 0 to the Clerk and for payment of the following sums viz).

Mr. E. Davy ¼ Saly to Xmas 1810	25.	0.	0
Mr. Allen on account of his bill for Chemicals which is referred to Mr. Davy for his report.	100.	0.	0
Mr. Henry for Coals &c. in 1808	43.	3.	11
Mr. Shepherd for Upholstery in 1808	39.	1.	10
Mr. Nevill for Turnery from 1807 to Decr. 1809	18.	0.	5
Messrs. Wrangham for Stationary from Jany 1808 to December 1809	148.	13.	2
Messrs. Dickins & Co. for Coals from Octr. 1809 to Jany 1810	98.	12.	6
Mr. Harris Librarian ¼ yrs Saly to Midsumr	45.	0.	0
Mr. Sturt assist. do. ½ yrs do. to Xmas 1810	35.	0.	0
which two last sums are to be charged to the Library account till the opinion of a General meeting can be taken thereon.	£602.	11.	10

166 | 1811. Jan 7.

Duke of Bedford elected Member of Lity & Scientific Comees — 6. Read Letter from His Grace the Duke of Bedford acknowledging the notice of his election on one of the Scientific & Literary Committees a meeting of which is appointed for friday evening when if in town he will make a point of attending.

Increase of Housekeeper's Salary — 7. Ordered. That the housekeeper's Salary be increased from £50 to £60 ꝑ Ann: commencing from 1st January 1811.

Form of Certificate — 8. A form for a Certificate of admission of Members was laid before the Managers and being approved was ordered to be engraved on similar paper to the last.

Resignation of Mr Auriol deferred — 9. The Managers took into consideration the minute of the 26th of November 1810 respecting the desire signified by Mr Auriol of resigning his office of Secretary to the Royal Institution and no resolution having passed the General Meeting for the acceptance of that resignation. Resolved. That Mr Auriol be requested to defer his resignation until the annual meeting in May next.

Cases of Sir Geo Duckett &c. — 10. Mr Auriol presented his report on a reference to him of the cases of Sir George Duckett Bart. George Sullivan Marten and A. F. Haldimand Esqrs. Resolved. That it be approved, and that the future report from the Managers to the next General Monthly Meeting be laid before the next meeting of this Committee.

Sir Thos Bernard Bart. having occasion to quit the room, the Chair was taken at his request by his Brother
Scrope Bernard Morland Esqr.

Life subscribers — 11. In pursuance of nominations under the Act, the following persons were admitted to the privileges of Life Subscribers.
William Gill Paxton Esqr.
of Buckingham Street Strand nominated

167

by his father ~~nominated~~ Archibald Paxton Esqr. –
& – The Revd George Stacey
of 28. Davies Street – Berkeley Square
by his brother Edward Stacey junr. Esqr.

W Watson Chairman

168 | 1811. Jan 14.

Managers Meeting Jan 14. 1810

At a Meeting of the Managers of the Royal Institution of Great Britain, held at the house of the Institution on the 14th day of January 1811.

Present
William Watson Esqr. in the Chair
John Symmons Esqr.
Sir Thos. Bernard Bart.
Scrope Bernard Morland Esqr. Treasurer
Charles Hatchett Esqr.
Earl of Winchilsea President
Earl of Aylesford
J. P. Auriol Esqr. Secretary

Minutes approved. — 1. The minutes of the proceedings of the last meeting were read and approved.

Exd Gen. Acct. — 2. Examined the General Account of the Institution

Minutes of Com of Accts approved. — 3. Read and approved the minutes of the Committee of Accounts of this day.

Reports from Scientific Committees — 4. Read reports from the following Scientific Committees who in pursuance of the 4th Article of Chap XVI of the Bye Laws assembled at the house of the Institution on the 11th Instant & made the following elections vizt.

Mathematics, Mechanics, & Mechanical Inventions.
The Earl Stanhope – Chairman
Honble Robt Clifford – Chairman
Thos Day Esqr. – Secretary

Chemistry, Geology, & Mineralogy
Sir Joseph Banks – Chairman
Humphry Davy Esqr. – Chairman
James Laird M.D. – Secretary

General Science, Literature & the Arts
Lord Holland – Chairman
Daniel Moore Esqr. Chairman
John Disney Esqr. Secretary

169

5. Mr Auriol in consequence of the desire expressed by the Managers at their last meeting – that he would defer his resignation of the Secretaryship till the Annual meeting in May next has the honor to signify his acquiescence.

Report on the applications respecting the transfer of life-shares — 6. The following report having been read and agreed to.
Resolved That it be presented at the next General Monthly Meeting of the Members.
The Managers of the Royal Institution taking into their further consideration the cases of Sir George Duckett Bart. George Sullivan Marten Esqr. & A. F. Haldimand Esqr. who had applied to make transfers of their respective shares in the Institution antecedent to the passing of the Act of Parliament of 50 Geo: 3d. which cases were referred back from the General meeting of the Members to the Managers; they think it proper to note that the particular circumstances of each of these cases are as follows. –
The Certificate of George Sullivan Marten Esqr. desiring to make the transfer of his share to Dr. Charles Dalston Nevinson is dated 2d February 1810, and was presented before the Managers at their Monthly Meeting of the 5th February, but wanting the formality of Dr. N's being recommended for ballot by a Manager according to the second Article of the 7th Section of the former Bye Laws, it went over to the next Monthly Meeting of the 5th March, when also on the same account it was passed over a second time to the monthly meeting of the 2d of April; and at this meeting being recommended by Mr Hatchett a manager Dr. Nevinson would have been ballotted for at the following monthly meeting in May had not the Act of Parliament which passed on the 18th of April arrested this and the following transfer before they were completed.
The Certificate of Sir George Duckett Bart. desiring to make a transfer of his share to his son George Duckett Esqr. bears date on the 2d of March 1810 and was presented to the Managers at their Monthly Meeting on the 5th of that month with

a proper recommendation from a Manager, but on account of some supposed informality it was not ordered to be entered on the minutes of that day, in consequence of which it went over with the foregoing to the next monthly meeting held on the 2d. of April, when both were recorded, and the parties would have been balloted for at the monthly meeting in May, had not the passing of the Act in that interval prevented them.

The case of A. F. Haldimand Esqr. desiring to transfer his share to his son Wm. Haldimand Esqr has been very circumstantially enquired into and reported on to the Managers, and the report has been read at the General meeting. – Certain it is that Mr. Haldimand's Certificate never came into the hands of the Secretary nor before the Managers; but the delivery thereof to the late Clerk Mr. Easingwood has been attested, and is now further confirmed by the personal knowledge of Professor Davy; tho' the date of the Certificate and the exact period of its delivery cannot now be ascertained. It is to be remarked that until the absence of the late Solicitor of the Corporation it was ever usual to pass the Certificate thro' his hands to the Managers. This Certificate may therefore have been sent by Mr. Easingwood to him, in which case it may now be among his Papers. The manner of Mr. Easingwoods absconding from the Institution and the confusion in which the Clerk's Office was purposely left by him are well known to the Members, but as his departure took place on Saturday the 17th of February there was full sufficient time for the transfer by Mr. Haldimand to have passed through all the requisite formalities as well as the two foregoing; It must therefore be admitted that these three applications were all made in due time by the parties, that they can form no precedents for other cases, and as no prejudice or inconvenience to the Institution can arise from admitting the transfers in their fullest extent, the Managers upon maturely reconsidering the subject are induced to recommend them to the favorable consideration of the General

Meeting, and if the Members should be inclined to allow them, the Managers would suggest the propriety of balloting for Col. George Duckett, Dr. Nevinson, and Wm. Haldimand Esqr. at the ensuing Monthly Meeting of the Members and in case they should be admitted to succeed to the full privileges of original Proprietors in virtue of these transfers it may be considered whether Sir George Duckett, Mr. Marten & Mr. Haldimand, should be required to sign a formal surrender of all their rights in the Institution as intended by them in offering to make these transfers, tho' it is not to be presumed that they will ever think of claiming or exercising any of their former privileges, the transfers thus confirmed being a ~~complete~~ virtual surrender of them. –

On Payment of Fines – Notice in Consequence

7. Resolved. That Mr. Seymour be desired to give notice to the Lessees of the two houses adjoining to the Royal Institution that if the fine and fees upon the proposed renewals of their respective houses are not paid with interest on or before the 26th Instant the Managers propose to recommend to the General Meeting of the first Monday in February, to order that no further fine shall be received or renewal granted, on account of the Lessees not having conformed to the conditions of the covenant of renewal.

8. Resolved. That Mr. Seymour be also desired to give notice to the said Lessees that in case the arrears of rents due to the Royal Institution in respect of the said houses be not forthwith paid, the Managers will think it their duty to order the said arrears to be levied by distress. –

Annl. Subscrs.

9 The following persons having been duly recommended were elected as Annual Subscribers, and Subscribers ~~to~~ to the Lectures

Lady Cahir of Blakes Hotel
Mrs. & Miss Cullen of Lambs Conduit Place
Ann. Subscribers
Lord Gray of Scotland

Mr. Thomas Kinder junr. of Great Russell St.
John Langley Esqr. " Howland Street
Stephen Winthrop M.D. . New Cavendish Street
Dr. Thos. Smith – of Piccadilly.
Revd. Jeremiah Smith 55 Jermyn Street
Le Commandeur Braancamp – Curzon St. Mayfair
Edwd. S. Hardisty Esqr. junr. of Great Marlbro' Street

10. Read a letter from Alexr. Donovan Esqr. dated Dublin 31st. Decr. 1810.
Ordered. That the name of Mr. Donavan be withdrawn from the List of Annual Subscribers and that no arrears be demanded of him.

11. Ordered. That three blank books be provided for the Scientific Committees

12. Resolved. That the former rule with respect to the Lecturers tickets be continued in their favor

13. Resolved. That each person proposed for election as Member of the Institution upon paying 4 guineas do receive tickets of admission to the Lectures, such sum to go in part payment of Annual Contribution or admission fee upon election.

Porter's dress.

14 Ordered. That a new dress be made for the Porter of the Institution under the direction of Professor Davy.

T. Bernard

Manager's Meeting, 21. Jan. 1811.

At a Meeting of the Managers of the Royal Institution of Great Britain, held at the house of the Institution, on the 21st day of January 1811.

Present
Sir Thos. Bernard Bart in the Chair
John Symmons Esqr.
Charles Hatchett Esqr.
Right Honble. Isaac Corry
J. P. Auriol Esqr. Secretary

Minutes approved 1 The minutes of the proceedings of the last meeting were read and approved.

Exd. Gen. Acct. 2. Examined the General Account of the Institution.

Minutes of Comee. of Accts. apprd. 3. Read and approved the minutes of the Committee of Accounts of this day.

4. Read Letters from
Messrs. Jones & Yarrell Newsmen.
requesting payment of their sundry bills.
Resolved. That £140 be paid them on account and Signed an Order accordingly.

admission to lectures only for 2.2. ?

& from — Thos. Kinder junr. Esqr.
requesting to be admitted to the Lectures only, for 2 guineas and have the other 3 guineas returned to him, others having been admitted the same day at 2 guineas each.
Ordered. – That this be referred to Mr. Hinckley to confer with Mr. Kinder.

Presents. 5. Presents received

John Disney Esqr. presented his Collection of Acts of Parliament concerning County & Borough Elections. with references to reported cases and a copious index.
Saml. Ware Architect – presented his treatise on arches & abutment piers.
John Hinckley Esqr. presented – East India Register
Resolved. That the thanks of the Managers be returned for these presents –

6\. Life Subscribers. In pursuance of nominations under the Act, the following persons were admitted to the privileges of Life Subscribers.

Andrew Henry Thomson Esqr of Albany
by his father John Thomson Esqr
John Rennie Esqr of Stamford S.^t
by his father John Rennie Esqr
Charles Pott Esqr of Albion Place Blackfriars Br.
by his father Robert Pott Esqr
The Honble Douglass Ja.^s Wm Kinnaird
by his Brother The Right Honble Lord Kinnaird
Thomson Bonar jun.^r Esqr New Broad Street
by his father Thomson Bonar Esqr
& — Henry Campbell White Esqr of Devonshire place
by his brother The Revd Thos White. —

7\. Annual Subscribers. The following persons having been duly recommended were elected as Annual Subscribers & Subscribers to the Lectures.

Rachel Stacey 68 Lambs Conduit Street
Mrs Browning - Audley Square
Lady Olivia Sparrow
Miss Sparrow.
Mrs Busk 9. Portland Place
Miss Grant 42 Mortimer Street
Mrs Chambers 43 d.^o
Mrs Burrowes 7 Hereford Street
Mrs Bonar – Broad S.^t
Mrs Iremonger 15 upper Grosv.^r S.^t
Sarah Beaumont 68 Lambs Conduit S.^t

Annual Subscribers

Turner Macan Esqr of Orchard Street
John Surtees Esqr of North End Fulham
Fra.^s Ribr.^s Setaria of 3 Castle St Leicester Sq.^e
The Honble Richd Fitzgibbon of Saville row.
J. R. White Esqr 14. Bury St St James's

8\. Admission to Lectures for £2.2.0. J Fincher reported that the following Gentlemen had been admitted to Mr Davys first Lecture on depositing 2 guineas each till an arrangement was made by the Managers for that purpose.

Alex.^r Stratton Esqr of Ibbetsons Hotel – by H Davy

Thos Musgrave Esqr Ibbetsons Hotel – by H Davy
Master Sparrow ———— d.^o
Thos Thomas Esqr ———— d.^o
Mr Richd Cruickshank 41 upper Baker S.^t – John Pond
Arthur Pott Esqr of Castle St Southwark – Robt Pott
Robert Jas.^s Chambers Esqr 43 Mortimer S.^t H Davy
Alex.^r Nowell Esqr – Wimpole Street d.^o
Lord Visc.^t Jocelyn ——. H. Davy.
R. Johnston Esqr 15 Percy St Bedford Square H Davy
Thos Broadley Esqr Tavistock Hotel – W. Wilberforce
Mr Richd Wilson Brown of Bartholomews Hospital
Mr Josias Baker – of Chapple St Bedford row.
Mr Thos Hare of Charles Street Soho Square
Mr John Dunn of Westminster Hospital
Drummond Delap Esqr — by H Davy
Mr Doumolin of Soho Square
Mr Paterson – Old Broad St R. Debaufre
Mr G. Standish Barry
M.^r Le Mann 74 upper Berkeley Street
Col. Roger Elliott Roberts.
Wm Fitton M D. 3 Great Portland S.^t H Davy
The Right Honble John Forster.

Resolved. That those persons who have already been admitted to the Lectures on Chemistry and Natural Philosophy for the present Season at two guineas each be permitted to attend such Lectures during the present year and that such persons as subscribed on the same terms last year be permitted to renew their subscription for the Current year ending 31.st Decr 1811. – And the above mentioned persons having been duly recommended were accordingly elected and admitted as Subscribers to the Lectures on Chemistry and Natural Philosophy for the present year.

9\. Admission to Lectures of one Professor to be £2.2.0. Resolved. That in future persons desirous of attending the Lectures of any one Professor or Lecturer during a single Season be permitted to attend such Lectures on the payment of Two guineas.

10\. Election of Manager. Resolved. That the following resolution be taken into consideration at the next Meeting to be recommended (in case of approval) to the consideration of the next General Monthly Meeting of the Members. —
That no Member be eligible as a Manager (except by re-election) unless a previous declaration

be made that he consents to being proposed and in case of election will attend as Manager.

11 Communication concerning Committee room &c. In pursuance of Art: 15 – Chap: XVI of the Bye-Laws The Honble Robert Clifford as Chairman of the Committee of Mathematics had a conference with the Managers and communicated the following report.

1.st As to the providing of a proper Committee Room.
2.^d As to the manner of setting apart a place for the Committee at the General Monthly Meeting when a report is made according to Art: 10. Chap: XVI.
3.^d As to a proper frame the better to distinguish the business of the Committee from that of the Institution according to Art: 13.
4.th As to the propriety of inserting notices of Committee Meetings on the back of the Weekly Cards.
5.th As to a box for preserving papers.
6.th As to having a stock of ruled and uniform paper at the Clerks Office, to supply those Members who may choose to purchase it, for the purpose of making communications

Resolved. That it be taken into consideration at the next Meeting.

J. C. Hippisley

Managers' Meet.^g Jan. 28, 1811.

At a Meeting of the Managers of the Royal Institution of Great Britain, held at the house of the Institution, on the 28.th day of January 1811.

Present

Sir John Cox Hippisley Bart. in the Chair.
William Watson Esqr
Charles Hatchett Esqr
Right Honble Isaac Corry
Sir Thos Bernard Bart.
John Symmons Esqr
Scrope Bernard Morland – Treasurer
J. P. Auriol Esqr – Secretary.

1\. Minutes approved. The minutes of the proceedings of the last Meeting were read and approved.

2\. Ex.^d Genl Acc.^t Examined the General Account of the Institution.

3\. Minutes of Com. of Acc.^ts approved. Read and approved the Minutes of the Committee of Accounts of this day.

4\. Present rec.^d

Present received.

J. Rickman Esqr presented A Map of Roman Britain after Ricardus Corinensis. —
Resolved. That the Thanks of the Managers be returned for this present. —

5\. Life Subscribers. In pursuance of nominations under the Act the following persons were admitted to the privileges of Life Subscribers.

Gilbert Gardner Blane Esqr of Cleveland row
by his father Gilbert Blane ~~Esqr~~ M.D
& Mrs Henrietta Balfour – of Curzon S.^t
by her Husband John Balfour Esqr

6\. The following persons having been duly recommended were elected as Life & Annual Subscribers, and Subscribers to the Lectures vizt

178 — 1811.

Subscribers to Lectures

To the Lectures.

Mrs Dr Gillis of 9 upper Seymour Street
Mrs Odell of 10 Queen Ann St West.
Mrs Robley No 1 Russell Square
Mrs Apreece 16. Berkeley Square
Miss White 22 Wimpole Street.
Miss Ware of New Bridge Street

Life Subsr

Life Subscriber.

Allowed to become an Annual Subsr Vide Minut. of 1st April 1811 — William Taylor Esqr of 15 Albemarle Street.

Annl Subsr

Annual Subscribers.

Daniel Leonard Esqr 22 Bell yard Lincolns Inn
Charles Leonard Esqr do
A. S. Henderson Esqr 6 Air Street Piccadilly
Captn Leake 53. Harley Street
Mr Chandler of Stamford St Blackfriars road

Subsrs to Lectures.

The following persons having been duly recommended were admitted as Subscribers to the Lectures on Chemistry & Geology for the Season.

Mr Emlen of 22 Great Maddox St
Mr E. A Brande of Arlington Street.
Mr Sloan of 33 Great Maddox St
Dr Hue of Chancery Lane
Mr W. Hill of Princes St Bedford Row
J. H. James Esqr Featherstone Buildgs Holborn
Mr Lawrence 17. Giltspur Street
Henry Combe Esqr Caroline St Bedford Sq.
Mr John Webb of St Barthomews Hospital
Wm Cother Esqr of Southampton Buildgs
Thos Alcock Little Windmill Street.
Mr Fletcher —
J. Blackford Esqr British Hotel Jermyn Street

Assent of every person nominated as Manager to be obtained

7. Took into consideration the resolution of the last meeting No 10, ordered for consideration this day concerning a previous declaration of assent by Managers to be elected.

<u>Ordered</u>. That it be added to the report from the Managers to the next General Monthly Meeting.

8. Took into consideration the resolution from the Patrons of 31st Decr 1810 concerning the paying over

Jan. 28 — 179

the contributions for the Library to the Treasurer thereof. —

Monies paid on account of Library

<u>Resolved</u>. That an Account be laid before the Managers of all monies paid by the Institution on account of the Library, and also of all monies received under the new Bye Laws on account of the Library, together with the times of the respective payments and receipts. —

Committee Room. &c.

9. Took into consideration the report of the Committee of Mathematics and as to the first article concerning a Committee Room.

<u>Ordered</u>. That Mr Royce be directed to attend Mr Clifford Chairman of the Committee and deliver an estimate of the expense.

As to Article 2. concerning setting a part a place at the General monthly meeting for the Committees

<u>Ordered</u>. That three pieces of Cloth be provided for the purpose under the direction of Mr Clifford.

As to Art: 3 concerning a frame for the business of the Committees.

<u>Ordered</u> That a frame be provided under the direction of Mr Clifford.

As to Art: 4 concerning notices of the meetings of the Scientific Committees.

<u>Ordered</u>. That notices be inserted in the next Weekly Cards. —

As to Art: 5 concerning a box for papers.

<u>Ordered</u> That a box be provided under the direction of Mr Clifford.

As to Art: 6. concerning ruled paper.

<u>Ordered</u>. That it be provided under the direction of Mr Clifford. —

"Day" newspaper ordered.

10. <u>Ordered</u>. That the Newspaper called the Day be taken till the 1st June.

11. Ordered. That Lists of the Committee be sent round with the next Weekly Cards to the Members.

Resignation of Sir T. Bernard tendered

12. Sir Thomas Bernard having retired, delivered a letter to the Chairman which was read tendering the resignation of his situation as Manager and intimating

180

that if the Annual General Meeting should hereafter elect him a Visitor he would with great pleasure continue his services in that situation.

The Committee on this occasion cannot avoid expressing their deep regret at the prospect of losing the assistance of a Gentleman whose zeal abilities & indefatigable industry have so eminently contributed to the prosperity of the Institution, they therefore unanimously express their wish that Sir Thomas Bernard will consent to defer his resignation till the Annual meeting in May next.

Read a letter from The Honble Robert Clifford on the affairs of the Institution. —

Spencer.

181

Managers' Meetg Feb. 4. 1811.

At a Meeting of the Managers of the Royal Institution of Great Britain held at the House of the Institution, on Monday the 4th day of February 1811.

<u>Present</u>

Earl Spencer <u>in the Chair</u>.
Rt Honble Lord Dundas.
Charles Hatchett Esqr.
Rt Honble Isaac Corry.
Earl of Morton.
Earl of Aylesford.
Scrope Bernard Morland Esqr <u>Treasurer</u>.
J. P Auriol Esqr <u>Secretary</u>.

Minutes approved.

1. The minutes of the proceedings of the last meeting were read and approved.

Exd Gen. acct

2. Examined the General Account of the Institution

Minutes of Gen. com. of Accts approved

3. Read and approved the minutes of the Committee of Accounts of this day & Signed an Order for an advance to the Steward of £50.

4. Read Letters from C. B. Impey Esqr on a claim under his fathers Will.

<u>Ordered</u>. — That the Secretary answer the letter. —

& from Sir Thos Bernard Bart confirming his former letter and declining to suspend his resignation as Manager till 1st May. —

Library Acct

5. Read the Account of the Receipts & Disbursements for the Library.

<u>Ordered</u>. That it be added to the Managers' report to the General monthly meeting.

6. The Treasurer laid his Account before the Committee according to Art: 5. Ch: VIII of the Bye Laws. —

Present received

7. <u>Present received</u>.

Robert Bakewell Esqr presented his observations on the influence of soil & climate on Wool — the mode of improving it & the management

182 | 1811. | Feb. 11. | 183

of Sheep particularly the Spanish breed. With notes by Lord Somerville
Ordered. That the thanks of the Managers be returned for this present,

8\. The following persons having been duly recommended were admitted as Annual Subscribers and Subscribers to the Lectures viz:

Subscribers to Lectures.

To the Lectures.
- Mrs Thomas Hope – Duchess Street
- Miss Warre – 30 George St. Hanover Square
- Mr John Bainbridge 40 Harley Street
- Mr Braancamp 12 Curzon Street
- Mrs Castor – do
- Miss Benar – New Broad Street
- Mrs Porden – 59 Berners Street
- Lady Sidney Smith of Cleveland row

Annual Subsc.rs

Annual Subscribers
- Mr Robt. Bakewell junr. of Bury St. St James's
- Mr Thrasea Poetus Bakewell. of do.
- Henry Tighe Esqr.
- Captn. Redwood of 41 Somerset Street

Subsc.rs to Lectures.

9\. The following persons having been duly recommended were admitted as Subscribers to the Lectures on Chemistry & Geology for the Season. –
- Sir George Shee Bart. 5 Cumberland place
- Henry Charles Litchfield of John St. Bedford row
- Mr W. King of upper Harley Street
- Mr Richd. Bright junr. Guys Hospital.
- Dr Berger of Warwick Court Holborn

10\. *Ordered*. That balloting lists be laid on the table of the General Monthly Meeting with the name of William Blake Esqr. for Manager. – And in case the General meeting be of opinion to proceed to the election of a Secretary – that Balloting Lists be laid on the table with the name of J. P. Auriol Esqr. – Secretary.

Spencer

At a Meeting of the Managers of the Royal Institution of Great Britain, held at the house of the Institution, on Monday the 11th day of February 1811. –

Present
Earl Spencer in the Chair
John Symmons Esqr.
Rt. Honble. Lord Dundas.
William Blake Esq.
Earl of Morton.
Wm. Watson Esqr.
J. P. Auriol Esqr. Secretary

1\. The minutes of the proceedings of the last meeting were read and approved.

Minutes of Genl. Meetings to be read at next Meetg. of Managers

2\. *Resolved*. That the minutes of all General Meetings be read at the next following meeting of Managers.
The Minutes of the Monthly Meeting of the 4th Instant were accordingly read.

Wm. Blake Esq. elected Manager.

3\. William Blake Esq. having been elected Manager at the last General Monthly Meeting, and J. P. Auriol Esqr. re-elected Secretary, took their seats at the Board. –

Payments to Library

4\. *Resolved*. That the Contributions to the Library already received, be paid over to the Treasurer of the Patrons of the Library; and that all future subscriptions be continued to be paid in like manner pursuant to the Bye Law Cap: 20 Art. 4.

Exd. Genl. acct.

5\. Examined the General Account of the Institution.

Approved minutes of acct.

6\. Read and approved the Minutes of the Committee of Accounts of this day.

7\. In pursuance of nominations under the Act the following persons were admitted to the privileges of Life Subscribers. –

184 | 1811. | Feb. 11 | 185

Life Subscribers

Honble. William Fredk. Mackenzie of Seaforth by his father The Rt. Honble. Lord Seaforth
& John Ranking Esqr. of 30 St James's Street by his brother George Ranking Esqr.

8\. The following persons having been duly recommended were admitted as Annual Subscribers & subscribers to the Lectures viz:

Subsc.rs to Lectures

To the Lectures.
- Viscountess Powerscourt.
- Miss Clarke 13 Albemarle Street
- Lady Frances Fitzwilliam 6 upper Wimpole St.
- Miss Isabella Dundas 19 Arlington St.

Annl. Subsc.rs

Annual Subscribers
- John Meares Esqr. Park Lane
- James Sutherland Esq. 14 Haymarket
- George Barker Esqr. of Birmingham
- Edward Wakefield Esqr. 41 Piccadilly

Subsc.rs to Lectures

9\. The following persons having been duly recommended were admitted as Subscribers to the Lectures on Chemistry & Geology for the Season.
- Dr Charles Henry Hagen of Berwick Street
- Wm. Loftus Esqr. – Bartholomew Close
- Mr D. Jones – Thanet place Strand.
- George Dangerfield Esqr. 14 Bedford row
- Joseph Adams. M.D. of Hatton Garden
- James Rush M.D.
- F. Van Zeller Esqr. Welbeck Street

Presents.

10\. Presents received. –
Edward Rudge Esqr. F.R.S. presented his
Plantarum Guianæ rariorum Icones et descriptiones hactenus ineditæ, folio, Vol I.
Mr James Sowerby F.L.S. &c. presented his
New elucidation of Colours 4to.
Ordered. That the thanks of the Managers be returned for these presents.

11\. Took into consideration Mr Royce's estimate of the probable expense of preparing the Repository for a Committee room amounting to £60.

Committee Room

Postponed the consideration of this subject. ~~[illegible]~~.

Screen or Frame ordered

12\. Mr J. Hinckley reported that the Lecturer on Astronomy wishes for a frame or skreen to the Lecture Room for exhibiting Diagrams &c.
Ordered. That a frame or skreen be prepared under the direction of Mr Pond. –

Certificates for ballots.

13\. *Ordered*. That the Certificates for ballot be suspended in the Library, being the place where the Weekly Meetings of Members are held; and that Mahogany battens with hooks be prepared under the direction of Mr J. Hinckley. –

John Symmons

Managers Meeting. 18th Feb. 1810

At a Meeting of Managers of the Royal Institution of Great Britain, held at the house of the Institution on Monday the 18th of February 1811 –

Present

John Symmons Esqr. in the Chair.
Earl of Winchilsea, President.
Scrope Bernard Morland Esqr. Treasurer.
William Blake Esqr.
Earl of Morton.
Earl Spencer.
Rt. Honble. Isaac Corry.
J. P. Auriol, Esqr. Secretary.

Minutes approved. 1. The minutes of the proceedings of the last meeting were read and approved.

Exd. Gen. acct. 2. Examined the General Account of the Institution

Minutes of Com. of Accts. approved 3. Read and approved the Minutes of the Committee of Accounts of this day.

Presents.

Presents received

4\. The Revd. John Hey D.D. presented his Lectures in Divinity, delivered in the University of Cambridge. 4 Vol. 8vo.

~~Pr.~~ Le Chevalier De Sade presented his work De la Tydologie, ou la Science des Marées; & Memoire en forme d'Instruction pour l'Astronome de l'Observatoire fondé à Milford Haven, – 8 vo. –

Humphry Davy L.L.D. presented his Experiments on some of the combinations of Oxymuriatic Gas & Oxygene, &c 4to.

Ordered. That the thanks of the Managers be returned for these presents.

Life subscriber nominated. 5 In pursuance of a nomination under the Act by Charles Dalston Nevinson M.D, his Nephew – George Robert Marten of Sandridge Lodge Herts was admitted to the privileges of a Life Subscriber.

Subscribers to Lectures 6 The following persons having been duly recommended were admitted as Subscribers viz.t

To the Lectures.

- √ Miss Trecothick – Argyle Street
- Mrs. Jeffery.
- The Countess of Gosford.
- √ Lady Craufurd.
- √ Miss Craufurd.
- √ Mrs. Sotheby.

Annl. Subscribers. Annual Subscribers.

- √ John Dickinson Esqr. 49 Welbeck Street
- √ Wm. Richardson Esq. 67 Baker Street
- √ George Stratton Esqr. – 46 do. –
- √ Joshua Wm. Brooks Esqr. Boyle Street
- √ Hugh Rose Esqr. 36 Beaumont St.
- √ Captn. Ewing – do.
- √ The Revd. Henry Kett 13 Park St. Westmr.

Subscribers to Lectures. To the Lectures on Chemistry & Geology for the Season.

- √ Lieut. Col. Paterson 3d. Guards Bury St.
- √ Wm. Henry Blodget Esqr. Maddox St.
- √ Thos. Nolan Esqr. Moore place Lambeth
- √ Wm. Henley Esqr. 7 Sols row Hampstead road

Vacancy in Managers. 7. Resolved. That the Earl of Darnley be recommended to succeed to the Vacancy in the management occasioned by the resignation of Sir Thomas Bernard, and that the following form be circulated to the Members, according to Bye Laws Chap: 4 Art: 2 and appendix B. –

A vacancy having occurred by the resignation of Sir Thomas Bernard Bart: Manager of the Royal Institution, you are requested to ballot for a Successor, in pursuance of the 2d. Article of Chap: IV. of the Bye Laws.

Manager	
The Earl of Darnley	

If you wish to substitute any other name, write it opposite, and pass your pen thro' the printed name.

Standing Order 8. Ordered. That the Clerk insert in a Book, and preserve, one copy of every paper ordered to be printed for the Institution

Admission to Lectures 9. The Wives, and Daughters, of Proprietors, having under the former Bye Laws, been admitted to the Lectures on paying 1 guinea each.

See Minutes of the next meeting No. 1.

Resolved. That in future the Wives & Daughters of Members be admitted to the Lectures on the following terms viz.t

(see XIII. 6)

For the first Lady being a Wife or Daughter of a Member .... 2 guineas
For each Lady after the first –– 1 guinea.

Vane 10. Mr. Bunning Architect of St. George's Chapel having desired to know whether he may order the Smith to prepare the vane, for erecting which he has received Mr. Cockburne's authority;
Resolved. That it be referred to Mr Davy.

Payment of compensation to late Proprietors £630. 12. 6. 11. Signed an Order for payment of Compensations to the following persons, late Proprietors, who have claimed them under the Act.

	£. s. d.
Benjn. Blake Esqr.	42. – 10
The Honble. George Villiers	42. – 10
Thos. Wright Esqr.	42. – 10
Wm. Boultbee Esqr.	42. – 10
Thos. Palmer Esqr.	42. – 10
Saml. Lawford Esqr.	42. – 10
The Rt. Honble. Lord Ossulston	42. – 10
The Honble. Admiral Digby	42. – 10
Richd. Harford. Esqr.	42. – 10
Mrs. Susannah Day	42. – 10
James Gorst Esqr.	42. – 10
Genl. Chas. Morgan	42. – 10
The Revd. Edmund Ferrers	42. – 10
George Grant Esqr.	42. – 10
Lady Ann Hayes	42. – 10
	£630. 12 6

Attendance of Officers at Genl. Meetings 12. Ordered. That all the inferior Officers and Servants of the Institution be in attendance on the General meetings at the discretion of the Secretary.

13 Read the following resolution referred to this Committee by the Visitors. –

Suggestion of Visitors concerning Accts. approved. "That it is the opinion of this Committee that, in keeping the Accounts, items not exceeding 10s. each, had better be put together in one quarterly bill; and that this proposal be submitted to the Managers."
Agreed to, and Ordered accordingly.

14. Read the following resolution from the Committee of Chemistry &c of 13th. Inst.t
"That the notice of Mr. Davy's Experiments be transmitted from this Committee to the Managers at their next Meeting with a recommendation that it be laid before the Royal Society."
Agreed to. –

15. Read a letter from
Mr. M Craig offering to read gratuitously between 2d. April & 1st. June "Six Discourses on the principles of painting &c with reference to specimens of great Masters now in this Country and accessible to the public."
Referred to the Committee of Lectures.

Order for paying £142. 16. to the Patrons of Liby. 16. Signed an Order for the payment of £142 . 16 . 0, received for the Library & Mineralogical fund, to the Patrons of the Library.

Candidates for Ballot 17 Read the names of Candidates for ballot at the next General Monthly Meeting as follows.

Thomas Collins Esqr.
Nathl. Conant Esqr.
Wm. Murdock Esqr.
Wm. Brande Esqr.
George Aust Esqr. –
Thomas Harrison Esqr.
Thos. Calverley Esqr.
Revd. Dr. Robt. Thomson LL.D.
Wm. Ayrton Esqr.
Jas. Curry M.D.
A.P. Buchan M.D.
Edward Lee Esqr.
Willm. Salmond Esqr.
Wm. Boultby Sleath M.D.

190 1811.

Admiral Nugent.
Thos. Mather Esqr.
Sir Oswald Mosley ~~Bart~~ & M.P.
Benjn. Hall Esqr. M.P.

Confirmed with the Exception of No. 9

Spencer.

Feb. 25. 191

Managers' Meetg 25th Feb. 1811.

At a Meeting of the Managers of the Royal Institution of Great Britain held at the house of the Institution on Monday the 25th of February 1811.

Present

Earl Spencer. in the Chair
Wm. Watson Esqr.
Wm. Blake Esqr.
Rt. Honble. Isaac Cory.
Charles Hatchett Esqr.
J. P. Auriol Esqr., Secretary.

minutes approved, ext. No. 9. — 1. The minutes of the proceedings of the last meeting were read and approved except No. 9 which was altered as follows.

Admission of Wives and daughters of Proprietors. — The Wives and Daughters of Proprietors having been under the former Bye Laws admitted to the Lectures on paying One guinea each.

Resolved. — That in future the Wives & Daughters of Members who enjoyed that privilege, when the Act passed, be agreeably to the provisions thereof, admitted to the Lectures on the same terms so long as they continue to pay their subscriptions. —

Exd. Gen. acct. — 2. Examined the General Account of the Institution.

Min. of Com. of Accts. approved. — 3. Read and approved the minutes of the Committee of Accounts of this day And signed an Order for payment of the following sums vizt.

	£	s	d
Mr. Feetham for Ironmongery from Jany. to 31st. Decr. 1808	167	19	5
John Gardner for Oil to 31st Decr. 1809	50	16	6
Messrs. Knights Chemicals to do.	47	5	6
Messrs. Dalton & Keys Cards to do.	30	12	0
Messrs. Jones & Yarrell Newspaprs. do.	41	7	11
Mr. J. Hammond Linen-draper to do.	23	12	6
Patent Oil Compy. — to do.	17	6	6
Wedgwood & Co. Sundrs. for Laboraty. do.	7	4	11
Howis & Co. Sundries for the house do.	18	3	0
G. Glover Soap & Candles to May 1810	73	1	8
Carr. forwd. —	£477	9	11

192 1811.

	£	s	d
Bro. forwd.	477	9	11
½ yrs property tax to Sept. 1810	36	0	8
H. Davy Esqr. ½ yrs Saly. to June 1810	200	–	–
David Brooks ¾ yrs do. — to do.	39	7	6
	£752	18	1

4. Mr. Auriol produced a letter from John Symmons Esqr. confirming his wish to resign his situation of Manager the consideration of which was deferred from the 24th. Decr. to this day.

Resignation of Mr. Symmons, as Manager. — Resolved. That the regret of the Managers be repeated to Mr. Symmons on his present determination and that his resignation be reported at the next General Monthly Meeting. —

Read a letter from E. B. Impey Esqr. on his claim to be admitted to the privileges of a life Subscriber under his father's will.

Ordered. That the Clerk write to Mr. Litchfield the Executor referring the Executors to Mr. Seymour the Sol. to the Institution who will confer with them on their producing the will to him.

Life Subrs. — 5. In pursuance of nominations under the Act the following persons were admitted to the privileges of Life Subscribers.

Frances Lady Davis of Albemarle Street by her Husband Sr. John Brewer Davis Knt.
& Elizth. Hatchett of Clarges Street by her Husband Charles Hatchett Esqr. —

Subscribers to Lectures. — 6. The following persons having been duly recommended were admitted as Subscribers vizt. —

To the Lectures.

Miss Cole 23 Bloomsbury Sq.
Miss Snow — do.
Miss Churchill 60 Lower Grosvenor St.

annual Subsrs. — Annual Subscribers.

P. M. Wagner Esqr.
Barney Smith Esqr. Park Lane
Lord Compton. Great George St. Westmr.
Thos. Lennard Esqr. Bruton Street
Cecil Smith Esqr. York Street St. James's

Feb. 25. 193

Vane on Chapel opposite the Royal Instn. — 7. ~~Mr. Davy reported that the Alfred would contribute 3 guineas towards the expense of a Weather Vane on the Chapel opposite the Royal Institution.~~

~~Resolved. That Mr. Davy be requested to direct the Vane to be placed there accordingly, and that Mr.~~ Bunning be referred to Mr. Davy.

astronomical & clock presented by Sir H. Englefield. — Mr. I Hinckley reported that Sir Henry C. Englefield Bart. has kindly offered to present an Astronomical Clock for the Library with a small transit for its regulation, provided the Institution will undertake the regulation thereof, and on condition of reverting to the donor, if the Institution cease to make use of it.

Your reporter further stated, that Mr. Pond has kindly offered to fulfil this condition by undertaking the regulation of the above mentioned Clock.

Ordered. That the thanks of the Managers be returned for this present And that it be placed in the antiroom to the Reading room under the direction of Mr. Pond.

9. Ordered. That 3 doz. wax Candles, and 1 Cwt of Soap, be provided and paid out of the Current expenses.

Spencer.

194 — 1811.

Managers' Meeting 4th March 1811.

At a Meeting of the **Managers** of the Royal Institution of Great Britain, held at the house of the Institution, on Monday, the 4th of March 1811. —

Present
Earl Spencer, in the Chair.
Wm. Watson, Esqr.
Rt. Honble. Isaac Corry.
Wm. Blake Esqr.

Minutes approved 1. The minutes of the proceedings of the last meeting were read and approved.

Exd. Gen. acct. 2. Examined the General Account of the Institution

Minutes of Com. of Accts. approved 3. Read and approved the minutes of the Committee of Accounts of this day, & signed an Order for an Advance to the Clerk of £50.

4. Read a letter from John Symmons Esqr. in reply to the vote of the last meeting of Managers on his resignation.

Presents. **Presents received.**

5. — Saml. Highley Esqr. of Fleet Street presented
A popular treatise on the natural and artificial causes of disease in general, with the means of prevention; & rules for Diet, Regimen &c by John Robertson, 2 vol. 8 vo.
Also a Work
"On the morbid sensibility of the eye, commonly called weakness of sight, by John Stevenson 8 vo.

The Society for bettering the condition of the Poor, presented 31st. & 32d. reports of their Society.
John Robertson Esqr. presented his
"Popular treatise on the causes of disease &c 2 vol. 8vo.
" On diseases of the Generative System, 8 vo.
Ordered. That the thanks of the Managers be returned for these presents. —

Mar. 4. — 195

Life Subscriber 6. In pursuance of a nomination under the Act by Charles Marsh Esqr., his Nephew
The Revd. Thos. Vialls ...
was admitted to the privileges of a Life Subscriber.

7. The following persons having been duly recommended were admitted as Subscribers vizt.

Subsrs. to Lectures — **To the Lectures.**
Mrs. Davies Giddy — 6 Holles Street
Mrs. Barton. 25 Cumberland St.
Miss Towers 45 Queen Ann St. West
Mrs. Fawkes 23 Portman Square.
Mrs. Dickinson 49 Welbeck Street

Annl. Subsrs. — Annual Subscribers.
James Putnam, Esqr. Holles St. Cavendish Sq.
Ward Chipman, Esqr. 22 Bell yard.

8. J. Fincher reported that the following Subscribers to the Lectures are desirous of becoming Annual Subscribers upon paying the difference of their Subscriptions.

Annual Subsrs. —
John Blackford Esqr.
F. van Zeller Esqr.
Wm. Fletcher Esqr.
Wm. King Esqr.
Richd. Crookshank Esqr.

Ordered. That their request be complied with.

Standing Order respects Coachmen 9. Resolved. That the following Standing Order be printed on the Card.
"The Managers request that Coachmen be instructed to set down and take up with the horses toward Grafton Street."

Dundas

196 — 1811.

Managers' Meetg. 11th Mar. 1811.

At a Meeting of the **Managers** of the Royal Institution of Great Britain, held at the house of the Institution, on Monday, the 11th. of March 1811.

Present
Rt. Honble. Lord Dundas, in the Chair.
Wm. Blake Esqr.
Earl of Darnley.
Charles Hatchett Esqr.
Rt. Honble. Isaac Corry.
Benjn. Hobhouse, Esqr.
Earl of Aylesford.
J. P. Auriol Esqr. Secretary.

Minutes appd. 1. The minutes of the proceedings of the last meeting were read and approved.

Exd. Gen. acct. 2. Examined the General Account of the Institution

Minutes of Comee. of Accts. 3. Read and approved the minutes of the Committee of Accounts of this day.

Advance to Patrons of Library 4. In consequence of its appearing to the Managers, that the funds of the Library, are insufficient at the present time to pay the arrears of Salaries due to the Officers of the Library.
Resolved. — That an advance be made to Mr. Harris of £90, being the amount of half a years Salary due to him on 31st. December last.

Payment of Bills. 5. Signed an Order for payment of the following Bills &c. —

Searle — Bookbinder from Jany. 1808 to Decr. 1809	£46 . 10 . 7
De Boffe — Bookseller to Apl. 1810	57 . 7 . –
Feetham — Ironmonger from Jany. to Decr. 1809	69 . 10 . 2
Wm. Harris ½ yrs Salary to Decr. 1810	90 . 0 . 0
David Brooks Collector ½ yrs do. to do.	26 . 5 . –
	£289 . 12 . 9

Mar. 11 — 197

Minutes of Gen. Meetg. read 4. Read the minutes of the last General Monthly Meeting of the 4th Inst.

Presents. — **Presents received.**

5 Ant. Fr. Bertrand de Moleville presented
A Chronological abridgment of the history of Great Britain, from the first invasion of the Romans to the present reign, Vol. 1 & 2, fine paper.
John Disney Esqr. Secy. of the Committee of General Science & Literature presented.
The Life of Sir Michael Foster, Knt. sometime one of the Judges of the Court of King's Bench, by Michael Dodson Esqr.
Ordered. That the thanks of the Managers be returned for these presents.

Life Subscribers 6. In pursuance of a nomination under the Act by James Warre, Esqr. his Daughter Ellen Warre of George Street Hanover Square, was admitted to the privileges of a life Subscriber.

7. The following persons, having been duly recommended were admitted as Subscribers vizt.

Subsrs. to Lectures.
To the Lectures —
Miss Hatchett No. 2 Clarges St.

To the Lectures on Chemistry & Geology
Sydney Shore Esqr. Cecil St. Strand
Edward Lutwyche Esqr. Phillimore place Kensington

8. Ordered. — That a meeting of the Patrons of the Library with the Managers be requested for Monday at o'clock precisely and that the Clerk issue Summonses accordingly.
Ordered. — That Mr. Davy's Lecture on the plan of the Institution be given to such Members or Subscribers as choose to accept it.

Annl. Subsr. 9. J. Fincher reported that Thomas Allcock Esqr. a Subscriber to Professor Davy's Lectures; is desirous of becoming an Annual Subscriber, on paying the difference

on his subscription.
Ordered. That Mr. Allcock's request be complied with. —

10. Ordered. That Mr. Hurt be desired to give such assistance to the Clerk as may not interfere with his duty in the Library or as the Secretary may direct.

Astronomical Clock. 11. Mr. J. Hinckley reported that he had communicated to Sr. Henry C. Englefield the resolution of the Managers No. 8. of the 25th Ult.o and that he had consented to the Clock's being fixed in the Anti-room to the Reading room instead of the Library of reference and requested that the removal to the Institution might be under the direction of his own Clock-maker.
Resolved accordingly; and that Mr. Hinckley be requested to communicate the same to Sir H. Chas. Englefield.

Sir J. Sebright appointed Manager. 12. Resolved. That it is the opinion of this Meeting that Sir John Saunders Sebright Bart. is a proper person to succeed John Symmons Esq. as Manager. —

Confirmed with the Exception of No. 8.

Spencer.

Managers' Meet.g 18th March 1811.

At a Meeting of the Managers of the Royal Institution of Great Britain, held at the house of the Institution, on the 18th day of March 1811.

Present
Earl Spencer, in the Chair.
Charles Hatchett, Esq.r
Wm. Watson, Esq.
Wm. Blake, Esq.r
Earl of Winchilsea.
Benj.n Hobhouse, Esq.r
Earl of Darnley.

Minutes app.ved 1. The minutes of the proceedings of the last meeting were read and approved.

Ex.d Gen. Acct. 2. Examined the General Account of the Institution.

Minutes of Com. of Acct.s approved. 3. Read and approved the minutes of the Committee of Accounts of this day, & Signed an Order for payment of the following Bills

Wm. Feetham Ironmongery to June 1810	24 . 3 . 5
Mess.rs Knights Chemicals to Dec.r 1810	30 . 19 . 9
Hammond Linen-draper to d.o	12 . 5 . 2
Henry, Charcoal to d.o	39 . 15 . –
Howis, Sundries for the house d.o	20 . 14 . 2
Mr. Allen for Lectures in 1810	45 . – . –
Mess.rs Wrangham Stationary to June 1810	38 . 8 . 10
Mr. Nevill – Sundries for the house to d.o	5 . 1 . 11
	£216 . 8 . 3

Signed a request to Scrope Bernard Esq.r Treasurer to receive the following sums and place them to the account of the Institution

Mess.rs Hammersley & Co.	£33 . 11 . 0 Gen.l Acct. / 10 . 0 . 0 Loan Acct.	43 . 11 . 0
—— Herries & Co.		19 . 19 . –
—— Pybus & Co.		5 . 10 . –
—— Hoare & Co.		12 . 14 . –
		£81 . 14 . 0

Presents received.

Presents. 4. ~~Gen.l Sarrazin presented~~
~~Confession du General Buonaparte à l'Abbé Maury 8vo 1811.~~
~~The same in English,~~
1. Reponse du Général Sarrazin au Rapport fait au Général Bonaparte par le General Clarke, 8vo 1810.
The same, in English
Reflexions sur les Notes du Moniteur, au 14 Septembre, 8vo 1810.
Reflexions sur les Notes du Moniteur des 16, 23, 29 et 30. Novembre; suivies des Notes Biographiques des Généraux Junot, Massena, Ney et Regnier, 8vo. 1810.
Capt.n Foskett presented
2 The Rights of the Army vindicated; in an Appeal to the public, on the case of the author 8vo 1811. —
Johnson Grant, M.A. presented
3 A Summary of the history of the English Church, and of the Sects which have departed from its communion; with answers to each dissenting body on its pretended grounds of separation, vol 1. 8vo 1811.
4 The Conductors of the British Review &c. presented
No. 1. of the British Review & London Critical Journal
Thomas Taylor Translator of Plato & Aristotle presented
"Elements of the true Arithmetic of Infinites.
Ordered. That the thanks of the Managers be returned for these presents.

Life Subscriber. 5. In pursuance of a nomination under the Act by Henry Arthur Herbert Esq.r his Son, Charles Herbert Esq. of Muckross in the County of Kerry Ireland. was admitted to the privilege of a life Subscriber.

6. The following persons having been duly recommended were admitted as Subscribers viz!

Subscribers to Lect.res

To the Lectures.
Mrs. Leeds – Croxton Park
Mrs. Davis } Grosvenor Square
Miss Davis }

Miss Birch, Cumberland St. Portman Square
Mrs. Searle 4 Seymour St. d.o

Ann.l Subscribers.

Annual Subscribers.
George Wm. Leeds. Croxton Park Cambridge
Capt.n Maling R.N. 39 South Audley St.
John Searle 4 Seymour St. Portman Square

To the Lectures on Chemistry & Geology.
Frances Welles Esq.r of Cheltenham.

Mr. Pime having applied to continue his Subscription for his unmarried Daughters on the same terms as last year.
Resolved. That his request be granted for the current year he having enjoyed that privilege when the Act passed.

Confirmed with the exceptions of thanks for No. 2 & 3 in Article 4.

Dundas

Managers' Meeting. 25th March. 1811.

At a Meeting of the Managers of the Royal Institution of Great Britain, held at the house of the Institution, on the 25th. of March 1811. —

Present
Rt. Honble Lord Dundas, in the Chair.
Charles Hatchett Esqr.
Wm. Watson Esqr.
Rt. Honble Isaac Corry.
Sir J. C. Hippisley, Bart.
Benjn. Hobhouse Esqr.
J. P. Auriol Esqr. — Secretary.
Scrope Bernard Morland, Esqr. Treasurer.

Minutes approved, with exceptions — 1. The minutes of the proceedings of the last meeting were read and approved, with the exceptn. of thanks for Nos. 2 & 3 in Article 4.

Exd. Gen. Acct. — 2. Examined the General Account of the Institution.

Minutes of Com. of Accts. appd. — 3. Read and approved the minutes of the Committee of Accounts of this day.

Life Subscriber. — 4. In pursuance of a nomination under the Act by Timothy Brent Esqr. his Nephew William Brent Brent of Lincolns Inn, Esqr., was admitted to the privileges of a Life Subscriber. —

5. Thomas Broadley Esqr.
a Subscriber to the Lectures on Chemistry & Geology having been duly recommended, was admitted a Life Subscriber on paying the difference of his subscription.

Present received — 6. Present received.

Christr. Johnson Esqr. M.R.I. presented
Kelham's Domesday Book illustrated, 8vo.
Ordered. That the thanks of the Managers be returned for this present.

Insurance. — 7. Received notice from the Sun Fire Office that the Insurance of £12,000 on the house &c. will expire

in 15 days from this day unless the premium being £15. 17. 6
& the Duty 15. –. –
£30. 17. 6
be paid within that time for the next years Insurance
Resolved. That the Treasurer be requested to pay the above, and to repeat the payment annually and for this purpose that he desire the Office to send the Notices in future to the Bank of Messrs. Morland & Co. instead of delivering them at the Institution. —

Spencer.

Managers' Minutes April 1, 1811.

At a Meeting of the Managers of the Royal Institution of Great Britain held at the house of the Institution, on Monday, the 1st of April 1811

Present
Earl Spencer, in the Chair.
Wm. Blake Esqr.
Chas. Hatchett Esqr.
Rt. Honble Isaac Corry
Benjn. Hobhouse, Esqr.
Scrope Bernard Morland Esqr. Treasurer
Earl of Morton, V.P.
J. P. Auriol Esqr. Secretary.

Minutes approved. — 1 The minutes of the proceedings of the last meeting were read and approved.

Exd. Gen. Acct. — 2. Examined the General Account of the Institution.

Minutes of Com. of Accts. approved — 3. Read and approved the minutes of the Committee of Accounts of this day.

4. Read a letter from Dr. E. Smith on the subject of his Lectures.
Resolved. That it be referred to the Committee of Lectures. —

Presents received. — Presents received.

5. John George Children Esqr. F.R.S. M.R.I. &c. presented
Guill. Bowles, Introduccion á la Historia Natural, y á la Geografia Fisica de España, 4to.
Ordered. That the thanks of the Managers be returned for this present.

6. William Taylor Esqr. having been proposed and elected a Life Subscriber on the 28th. January last, instead of an Annual Subscriber, and being desirous of admission as an Annual Subscriber;
Ordered. That the name of William Taylor Esqr. be struck out of the list of Life Subscribers, and that he

be admitted an Annual Subscriber.

7. Mr. Disney Secretary of the Committee of General Science being admitted to a Sitting;
Read the minutes of the Committee of General Science &c. of the 22d. March last.
Resolved. That this reference be taken into consideration at the next meeting.

Grates, &c. — 8 Mr. Harris reported, That the brick-work of the Grates in both the Reading rooms wants to be restored, and the Chimney in the first room is subject to smoking, which Mr. Feetham undertakes to prevent at an expense not exceeding £7.
Ordered. That it be executed under the direction of Mr. Davy.

Morton V.P.

ORDER NUMBER
PO NYAATP08

BOOK NUMBER
5554113

BAY NUMBER
8001-0

AUTHOR	Berman, M.
TITLE	The Archives of the Royal Institution of Great Britain in Facsimile: Minutes of Managers' Meetings 1799-1900 (Volume 5)

DESCRIPTION

Volume 5. This is an ex-library book and may have the usual library/used-book markings inside.This book has hardback covers. In good all round condition. No dust jacket.

Monday 8th April 1811.

Managers Meetg adjourned.

This day being fixed for a meeting of the Managers, and the Rt. Honble Isaac Corry, & J. P. Auriol Esqr being the only two Managers who attended; adjourned the meeting to Thursday the 11th Instant. —

Morton V.P.

At an adjourned Meeting of the Managers of the Royal Institution of Great Britain, held at the house of the Institution on Thursday the 11th of April 1811. —

Present
Earl of Morton, in the Chair
Lord Dundas
Earl Spencer
Rt. Honble Isaac Corry.
J. P. Auriol Esqr Secretary.

Minutes approved — 1. The minutes of the proceedings of the last meeting were read and approved.

Exd Gen. Acct. — 2. Examined the General Account of the Institution.

Minutes of Gen. Acct. approved. — 3. Read and approved the minutes of the Committee of Accounts of the 8th Instant, & Signed an Order for an advance to the Clerk of £70. —

Letters received.

4. Read a letter from J. Mitchell – House of Commons – requesting to be informed who will pay the usual fees to the Vote Office.
Ordered. That Sir J. C. Hippisley be requested to give an order for the amount due, payable by the Clerk of the Institution, And that it be paid accordingly.
& from Mr. Pond on allowance for Apparatus stating that £30 would balance the Account for this and the last year, and that he should consider the arrangement as very liberal and satisfactory.
Ordered. That £30 be paid to Mr Pond, and signed an Order accordingly.

Life Subscriber — 5. In pursuance of a nomination under the Act by Francis Gregor, Esqr his Wife, Jane Gregor of Trewarthenwick Coy of Cornwall, was admitted to the privileges of a Life Subscriber.

6. The following persons having been duly recommended

were admitted as Subscribers viz.

Annl Subscribers — Annual Subscribers.
E. Hodgkinson Esqr Pall Mall Court
Monsr de Bettancour.
To Lectures on Chemistry & Geology
Charles Thomson Esqr of Russell Sqr

7. Took into consideration the formation of a House List to be laid on the table on the 1st of May next.

8. Ordered. That a letter of the under written tenor be immediately sent to the following Members.

For Managers	For Visitors.
Rt. Honble Lord Barrington	George Aust Esqr.
Robt. Fergusson Esqr.	Duke of Bedford.
Davies Giddy Esqr. M.P.	Rt. Honble Ld Braybrook
Earl of Hardwicke	Wm. Dunlop Esqr.
Earl of Mansfield	George Ellis Esqr M.P.
Wm. Marsden Esqr.	Daniel Giles Esqr.
Wm. Sotheby Esqr.	Richd. Thompson Esqr.
Edwd. Stracey junr Esqr.	Genl. Thornton.

Copy of letter — Royal Institution

Form of Letter for requesting consent to be nominated Manager or Visitor

"The Managers being expected to form a house-list of members to be balloted for, as Managers & Visitors for one year on the 1st of May next, have directed me to acquaint you that they are desirous of inserting your name as a (Manager / Visitor) and, wishing to know whether you are willing to accept of the appointment for the ensuing year; I request the favor of your determination as soon as convenient.

House List. — Selected for the House-list as

Managers	Visitors.
Wm Blake, Esqr.	Earl of Bessborough
Earl of Darnley.	Alexr. Blair, Esqr.
Charles Hatchett, Esqr.	Timy Brent, Esqr.
Earl of Morton.	Rt. Honble Regd. Pole Carew
Sir John Saunders Sebright Bt.	Thos. Hammersley, Esq

Earl Spencer	Duke of Somerset.
Wm Watson Esqr.	Saml Thornton, Esqr MP.

Read the minutes of the proceedings of the General Monthly Meeting of the 1st Inst. and of the adjourned General Monthly Meeting held on the 5th Inst.
Ordered.
That the consideration of printing the report on Edwd. Massey's Sounding Machine be postponed till after the next General Monthly Meeting.

9. Ordered. That the following Articles be procured under the direction of Mr Davy.
4 doz Candles, 28 yards of Sheeting @ about 3/6d pr yd.
& Cloth for Dusters & Pillow-cases.

Confirmed with the Exception of substituting the Name of James Peter Auriol Esqr for that of Charles Hatchett Esqr in No 8.

Spencer.

Managers' Meeting 22nd April, 1811.

At a Meeting of the Managers of the Royal Institution of Great Britain, held at the house of the Institution on Monday the 22d. of April 1811.

Present
Earl Spencer, in the Chair.
Wm. Blake, Esqr.
Rt. Honble. Lord Dundas.
Wm. Watson, Esqr.
Charles Hatchett, Esqr.
Earl of Darnley.
Benjn. Hobhouse, Esqr.
Earl of Morton.
J. P. Auriol Esqr. Secretary.

Minutes appd. except No. 8. 1. The minutes of the proceedings of the last meeting were read and confirmed with the exception of substituting the name of James Peter Auriol, Esqr. for that of Charles Hatchett, Esqr. in No. 8. —

Exd. Gen. acct. 2. Examined the General account of the Institution.

Minutes of Com. of Accts. approved 3. Read and approved the minutes of the Committee of Accounts of this day, and signed an order for payment of the followg. sums vizt.

J. Fincher on acct. of the Royal Institution	£50 . 0 . 0
J. Gardner for Oil from Jany. to Decr. 1810	132 . 11 . 0
Messrs. Jones & Yarrell Newspapers – do.	98 . 8 . 6
Mr. W. Savage balce. of Acct. for printg. do.	56 . 9 . 0
J. Searle Bookbinding from Jany. to Decr. 1810	23 . 9 . 9
Messrs. Dickins & Co. — Coals in 1810	84 . 11 . 8
Mr. W. Allen bal. of accts. for Chemicals to Decr. 1810	179 . 2 . —
Messrs. Dalton & Keys – Cards & blottg. paper to Decr. 1810	66 . 8 —
Mr. Pond for Lectures in 1810	52 . 10 —
H Davy Esqr. ½ yrs Saly. to Xmas 1810	200 . —
½ years Parochial rates to March 1811	26 . 5 . —
	£969 . 14 . 11

Letters received.

4. Answers having been received from Members intended to be proposed for election on the 1st of May – viz.
The Earl of Mansfield.
Robert Fergusson, Esqr.
Willm. Marsden, Esqr.
having declined being proposed as Managers.
Resolved. That the following names be substituted
Rt. Honble. Sir Joseph Banks Bart. K.B. &c &c
Sir Henry C. Englefield Bart.
Wm. Hyde Wollaston M.D. Sec R.S.
and that letters be written to them accordingly.
George Ellis, Esqr.
having declined being proposed as a Visitor.
Resolved. That the name of Thomas Murdoch, Esqr. be substituted for that of Mr. Ellis.

Presents.

Presents received.

5. His Majesty's Commissioners on the public Records presented
Valor Ecclesiasticus temp. Henr VIII. Vol 1.
Rotulorum Originalium in Curiâ Scaccarii, vol 2.
The Society of Arts presented,
28th. Vol of their transactions.
Ordered. That the thanks of the Managers be returned for these presents.

Life Subsr. 6. In pursuance of a nomination under the Act by Emanuel Baruh Lousada Esqr. his niece Abigail Baruh Lousada of Devonshire Square was admitted to the privileges of a Life Subscriber.

Subscribers 7. The following persons, having been duly recommended were admitted as Subscribers vizt.

to Lectures. To the Lectures. —
Honble. Lady Stanley of Portland Place.

Annl. Subsrs. Annual Subscribers.
Lieut. Col. Macauley, St. Albans St.
Sir Egerton Leigh Bart. Portland St.

Respecting Fines. 8. Ordered. That the Clerk write to Mr. Seymour the Solicitor, to know what he has done with respect to the Fines due from the adjoining houses.

Annual Report of Visitors 9. The Committee of Visitors having prepared their annual report which was read.
Ordered. That it be printed and circulated with the balloting-lists. —

Charles Hatchett.

Managers' Meetg. 29th Apl. 1811

At a Meeting of the Managers of the Royal Institution of Great Britain, held at the house of the Institution, on Monday, the 29th. of April 1811.

Present
Charles Hatchett, Esqr. in the Chair.
Wm. Watson, Esqr.
Wm. Blake, Esqr.
Scrope Bernard Morland, Esqr. Treasurer.
J. P. Auriol, Esqr. Secy.

Minutes approved 1. The minutes of the proceedings of the last meeting were read and approved.

Exd. Gen. Acct. 2. Examined the General Account of the Institution.

Minutes of Com. of Accts. 3. Read and approved the minutes of the Committee of Accounts of this day.

Letters received.

Sir J. Banks Manager. Genl. Thornton Visitors 4 Read a letter from the Rt. Honble. Sir Joseph Banks Bart. accepting the nomination for election as Manager. — from General Thornton accepting the nomination for election as Visitor. — from Wm. Hyde Wollaston M.D. declining the nomination for election as Manager. —

Fines & from Mr. Seymour the Solr. stating that the under-lessees of the corner house of Grafton Street had appointed to-morrow the 30th. April to pay all money due from them to the Royal Institution, but that he has not received any answer from the under-lessee of No. 22 concerning the Fine.

Present received

Present received.

5 1st. Report of Commissioners of Bogs in Ireland.

6. Ordered. That the name of Henry Warburton Esqr. be substituted in the house list for that of Dr. Wollaston.

7. Ordered. That the Managers & Visitors be summoned for 1st. May at ½ past 12, o'Clock.

214 1811.

Subscribers to Lectures. 8. The following persons having been duly recommended were admitted as Subscribers viz:

To the Lectures.

Miss Eliza Wilbraham 56 upper Seymour St.
Miss Ann Wilbraham d°.
Miss Story.
Hon.ble Mrs Damer Upper Brook St.
Lady Sitwell.
Miss Sitwell.

Annual. Annual Subscribers.

G. Wilbraham Esqr. junr. Green St. Grosr. Sqr.
Honble Mr Clifford 25 Portland Place.

Sub. to Lectures. To the Lectures on Geology.

Henry Westmacott Esqr. 25 Mount St.
John Abernethey Esqr. 14. Bedford row.
Wm. Clifford Esqr. 74 lower Grosvenor St.
John Cleaver Banks Esqr. Park row Knightsbridge

Ventilation of Lecture Room. 9. Mr J. Hinckley reported that he had for some years heard many complaints of the neglected Ventilation of the Lecture-room which remains from lecture to lecture without any regulation for changing the air and with doors so accurately closing of themselves as to cause on sensible frames, a very material and painful depression of the powers of intellect and attention during the whole of the Lecture.

Your reporter therefore begs to propose that orders be given to the Porters to set the doors open by means of wedges from the conclusion of each lecture till within half an hour of the next and to keep three of the windows open till the same hour during Summer and to use the same precautions during an hour early in the morning in winter.

Also that blinds be prepared, and uses in shading the sun from the lecture room.

Referred to Mr Davy.

10. Ordered. That House Lists be prepared according to the selection of Managers & Visitors and laid on the table of the General Meeting.

Spencer.

215

Managers Meeting. May 1. 1811.

At a Meeting of the Managers of the Royal Institution held by adjournment at the house of the Institution, on Wednesday the 1st May 1811.

Present

Earl of Winchilsea, President. in the Chair.
Earl of Morton V.P.
Earl of Chichester.
Rt. Honble Isaac Corry.
Earl of Darnley
Wm. Blake Esqr.
Wm. Watson Esqr.
Rt. Honble Lord Dundas.
Earl Spencer.
Scrope Bernard Morland, Esqr. Treasurer.
Charles Hatchett Esqr.
J. P. Auriol Esqr. Secretary.

Rough Minutes approved. 1. Read and approved the rough minutes of Monday last their not having been time to enter them fair.

2. Read a letter from Henry Warburton Esqr. accepting the honor intended of placing him upon the house List for Managers.

J. Guillemard proposed as Secretary. 3. It was moved by Wm. Watson Esqr. and seconded by Wm. Blake Esqr. That the blank in the house List for Secretary be filled up by the name of John Guillemard Esqr.

which being agreed to unanimously.

Ordered. That the Lists be filled up accordingly.

Resignation of Mr. Hinckley as Hon. secy. asst. 4. Read the following letter from John Hinckley Esqr.

My Lords & Gentlemen,

Finding that the understood pledge under which I have given my assistance during eleven months at the Board of Managers – that of being recommended in the house list for election as honorary Secretary to the Institution may by dividing interests which ought to be united, interfere with my primary wish and that of the

216 1811.

Managers, an entire union of all hearts in the advancement of the interest of the Royal Institution, I have only to lament that some friendly and confidential communication had not explained this matter earlier – so that I might have had the pleasure of contributing my mite to the common good, by withdrawing as I now do my pretensions, & concurring in the recommendation of my worthy friend Mr Guillemard, than whom I think a more proper selection could not have been adopted.

I only desire to be understood that the offer of recommending me to succeed our present most worthy Secretary was wholly unsought, and unthought of by me when it was proposed, as was also that of filling for an intermediate interval the office which I was merely requested to accept as a preparation for it.

I have the honor to be
My Lords & Gentlemen
Your obed. Serv.t
(Signed) John Hinckley

Royal Institution,
1st May 1811

To the Managers of the Royal Institution.

It was moved by Mr Watson & seconded by Mr Blake that the thanks of the Managers be given to John Hinckley Esqr. for his ready & zealous services in the situation of honorary assistant to the Secretary of the Royal Institution.

Agreed to, unanimously.

Fines recd. £546. 14. 5. 5. The Treasurer reported that he received yesterday in two payments £546. 14. 5 being the amount of Fines Interest & Ground rent from No. 23 in this Street to the 25th March last.

Spencer

217

Managers' Meeting 6 May, 1811.

At a Meeting of the Managers of the Royal Institution of Great Britain held at the house of the Institution on Monday the 6th of May 1811.

Present.

Earl Spencer, in the Chair.
Rt. Honble Lord Carrington
Wm. Watson Esqr.
Wm. Blake Esqr.
Wm. Sotheby, Esqr.
J. P. Auriol Esqr.
Sr Henry C. Englefield Bart.
Henry Warburton Esqr.
Scrope Bernard Morland Esqr. Treasurer.
John Guillemard, Esqr. Secretary.

Minutes appd. 1. The minutes of the proceedings of the 2 last meetings were read and approved.

Exd. Gen. Acct. 2. Examined the General Account of the Institution.

Minutes of Com. of Accts approved 3. Read and approved the minutes of the Committee of Accounts of this day.

Present recd. 4. Present received.

Matthew Yatman presented
Hamstedds Essay on the cause of Gravity, 8vo.

Ordered. That the thanks of the Managers be returned for this present.

Subscribers 5. The following persons having been duly recommended were admitted as Subscribers vizt.

To the Lectures.

Viscountess Milton
Miss Margt. Craufurd. George St. Hanovr. Square

To the Lectures on Geology.

Hugh Leycester Esqr. 3 Lincolns Inn.
Arch. Hamilton Esqr. 2 Cumberland Street
Owen Williams Esqr. 41 Berkeley Sqr.
Fredk. Pigou Esqr. 12 Hill Street.

6. Resolved. That in future the meeting of the Committee

of Accounts be at eleven o'Clock, and that the Meeting of Managers be at half past eleven on every Monday Morning and that the notices be corrected accordingly.

Mr Guillemard Secretary. 8. Mr. Auriol surrendered the key of the Iron Chest, the key of the box containing the Seal of the Royal Institution which is kept in the Iron Chest, and the key of the Port-folio also the seal which was in his charge and they were accordingly delivered to Mr. J. Guillemard the present Secretary.

J P Auriol

Managers' Meeting 13th May 1811

At a Meeting of the Managers of the Royal Institution of Great Britain, held at the house of the Institution, on Monday the 13th of May 1811.

Present

J. P. Auriol, Esq. in the Chair.
Edward Tracey Esq.
Davies Giddy, Esqr.
Wm. Sotheby Esqr.
Wm. Blake Esqr.
Sir John Saunders Sebright Bart.
John Guillemard, Esqr. Secretary.

Minutes appd. 1. The minutes of the proceedings of the last meeting were read & approved.

Exd Gen. Acct 2. Examined the General Account of the Institution.

Minutes of Com. of Accts approved. 3. Read & approved the minutes of the Committee of Accounts of this day And signed an Order for the payment of the following sums.

H. Davy Esqr. 1/4 years Saly to Ladyday	£100. – . –
Mr. E. Davy do.	25. – . –
David Brooks do.	13. 2. 6
1/2 years assessed taxes to do.	71. 2. 7
1/2 do. Land tax do.	7. 5. 6
1/2 do. Landlords property tax do.	25. – . 8
1/2 do. Property tax	36. – . 8
	£277. 11. 11

Present recd. 4. Present received.
H Davy Esqr. LLD. & &c. presented his paper "On a combination of Oxymuriatic Gas & Oxygene Gas.
Ordered. That the thanks of the Managers be returned for this present.

Visitors. 5 It appearing that General Thornton, & Thos. Murdock Esqr. elected of the Committee of Mathematics &c are Visitors Ordered that their names be omitted in the printed List And that the name of Thos. Day Esqr. being returned as the

26th on the List, be inserted.

Meeting of Committees. Ordered – That the following Committees be requested to meet in the ensuing week for the election of two Chairmen, and a Secretary for each Committee.

Mathematics, Mechanics &c. on Monday, one o'Clock
Chemistry, Geology &c. ——— Tuesday do.
General Science &c ——— Wednesday do.

Read the minutes of the proceedings of the last General Monthly Meeting.

The following persons having been duly recommended, were admitted as Subscribers viz.

Annl. Subscrs. Annual Subscribers.
Col Burke of Albany.
Wm Cathcart Esqr. 31 Half moon St. Piccadilly
Lieut Col Thos. Munro Duke St. Manchr. Sqr.
Captn. Burroughs 2d Regt. of Guards.

Resolution relative to Patrons of Liby paying their Officers. Resolved. That as the Patrons of the Library appear now to have sufficient funds it be referred to them to consider whether they will not resume the payments of the Salaries of their Officers attendant in the Library.
Ordered. That a copy of the above resolution be laid before the Patrons at their next Meeting.

Ordered. That the Newspapers be put under the care of Mr. Hurt.

Mr Davy stated that Mr. Brande had applied for leave to exhibit 3 Pictures belonging to the Institution at his Lectures.
Ordered. That this request be complied with.
Confirmed, with the exception of the order relative to the care of the Newspapers —

Wm. Sotheby —

Managers Meeting 20 May 1811.

At a Meeting of the Managers of the Royal Institution of Great Britain, held at the house of the Institution, on Monday, the 20th May 1811.

Present

Wm. Sotheby, Esqr. in the Chair.
Wm. Watson, Esqr.
Davies Giddy, Esqr.
J. P. Auriol, Esqr.
John Guillemard, Esqr.

Minutes appd. 1. The minutes of the proceedings of the last meeting were read and approved, with the exception of the Order relative to the care of the Newspapers.

Exd. Gen. Acct. 2. Examined the General Account of the Institution.

Min. of Com. of Accts approved 3. Read and approved the minutes of the Committee of Accounts of this day, And signed an Order for an advance to the Clerk of £50.

Ground rent 4. Ordered. That the Treasurer be requested to pay the sum of £135 for one years Ground rent due to the City of London on lady-day last.

Ventilation 5 Ordered. That the ventilation of the News-room be under the direction of Mr J. Fincher.

6 Ordered. That the care of the Newspapers be under the direction of the Clerk, And that the Porters be ordered to obey his directions.

Subscribers. 7. The following persons having been duly recommended were admitted as Subscribers viz.

Annual Subscriber
George Wagner Esqr. Pall Mall.
To the Lectures on Geology.
Captn. Wilbraham.

J P Auriol

222 — 1811. — May 27 — 223

Managers' Meeting. May 27. 1811

At a Meeting of the Managers of the Royal Institution of Great Britain, held at the house of the Institution on Monday, the 27th May 1811.

Present
J. P. Auriol Esq.r, in the Chair.
W.m Blake, Esq.r
Edw.d Stacey, Esq.r

Minutes approved. 1. The minutes of the proceedings of the last meeting were read and approved.

Ex.d Gen. Acct. 2. Examined the General Account of the Institution.

Minutes of Com.e of Acct.s approved. 3. Read and approved the minutes of the Committee of Accounts of this day.

Reports from Scientific Committees 4. Read reports from the following Scientific Comm.s who in pursuance of the 4th Article of Chap. XVI of the Bye Laws assembled at the house of the Institut.n on the 21st & 22d Instant and made the following elections viz.t

Chemistry Geology & Mineralogy.

Hump.y Davy Esq.r LL.D. — Chairman.
Charles Hatchett Esq.r — Chairman.
James Laird M.D. — Secretary.

General Science Literature & the Arts.

Daniel Moore Esq.r FRS & LS. Chairman.
John Disney Esq.r — Chairman.
John Hinckley Esq.r — Secretary.

5. M.r Pond having given 24 Lectures in the year 1810, instead of the 20 which according to the minutes of 20th November 1809 he had agreed to deliver.
Resolved. That he be paid the difference at 5 guineas ℔ lecture amounting to £21. and signed an Order accordingly. —

6. Resolved: That the consideration of the reference from the Committee of General Science respecting M.r Feinaigle be postponed till a fuller meeting.

Adjourned.

Davies Giddy

224 — 1811. — June 3 — 225

Managers' Meeting 3rd June 1811.

At a Meeting of the Managers of the Royal Institution of Great Britain, held at the house of the Institution, on Monday the 3rd of June 1811.

Present
Davies Giddy, Esq.r in the Chair.
J. P. Auriol, Esq.r
W.m Watson Esq.r
Edw.d Stacey, Esq.r
W.m Blake, Esq.r
Scrope Bernard Morland, Esq.r Treasurer.
Earl of Darnley.
Henry Warburton, Esq.r
John Guillemard, Esq.r Secretary.

Minutes approv.d 1. The minutes of the proceedings of the last meeting were read and approved.

Ex.d Gen. Acct. 2. Examined the General Account of the Institution.

Minutes of Com. of Acct.s app.d 3. Read and approved the minutes of the Committee of Accounts of this day.

Library Officers paid by Managers. 4. A Statement of the Accounts of the Library was laid before the meeting by The Rev.d Lewis Dutens, The Rev.d Edw.d Forster, and William Dunlop, Esq.r deputed by the Patrons in consequence of the resolution of the Managers of the 13th Ult.o to hold a conference with the Managers
Resolved. That the Salaries to the Officers of the Library be continued for the present to be paid by the Managers on account of the Library; And Signed an Order for the payment of £17.10.0 to M.r Sturt for ½ Sal.y due Lady day last.

5 The Managers not understanding that M.r Feinaigle has any intention of making a public disclosure of his system — are of opinion that consistently with the principles on which the Royal Institution is founded such request of the Committee of General Science cannot be acceded to.

6 ~~Signed an Order for payment to D.r Smith of £63 for Lectures delivered in 1811, and £3. 16. 0 for sundries used in the Lectures.~~

7 Resolved. That for the future no Newspapers be bound except the Times & Courier.

Confirmed with the exception of Article 5.

S. Bernard Morland

Managers Meeting June 10, 1811.

At a Meeting of the Managers of the Royal Institution of Great Britain held at the house of the Institution on Monday, the 10th of June 1811. —

Present.

Scrope Bernard Morland, Esq.r in the Chair.
William Blake, Esq.r
John Guillemard, Esq.r Secretary.

Minutes app.d with exception — 1. The minutes of the proceedings of the last meeting were read and approved, except Art: 5 respecting M.r Feinagle which was reserved for further consideration. —

Ex.d Gen. acc.t — 2. Examined the General Account of the Institution.

Min of Com. of Acc.ts approved — 3. Read and approved the minutes of the Committee of Accounts of this day, And signed an Order for an advance to the Clerk of £50, as also for the payment of Mess.rs Dickinson & Co's bill for Coals amounting to £76 . 13 . 4. —

Coals. — 4. Ordered. That 20 Chaldron of Coals be immediately supplied by Mess.rs Dickinson & Co. at the house of the Royal Institution. —

Present rec.d — 5. Present received.

The Rev.d Edward Forster presented
N.o 8 of the British Gallery of Engravings.
Ordered. That the thanks of the Managers be returned for this present.

Refusal to sanction any system the details of which are not fully disclosed — 6. The Managers having resumed the consideration of the recommendation of the Committee of General Science to permit M.r Feinaigle to repeat his experiments on l'Art Mnemonique before the Institution at large, and of the 5 Art: of the minutes of their meeting of the 3.d of June, came to the following resolution.

The Managers not understanding that M.r Feinaigle has any intention of making a public disclosure of the system on which his experiments are founded are of opinion that the exhibition of such experiments is not consistent with the principles on which the Royal Institution is established; in deference however, to the recommendation of the Committee and of the opinions expressed by many Members at the General Monthly Meetings they will not object to his repeating his experiments before the Institution at large. At the same time they beg it may be distinctly understood that in granting this permission they do not consider themselves as in the slightest degree recommending his system or as giving any opinion on its merits.

Ordered. That the above resolution be transmitted to the Committee of General Science. —

Ordered. That the substance of the above resolution be inserted in the Card announcing M.r Feinaigle's exhibition

7. Read a letter from I Pouget Esq.r an Annual Subscriber, stating that he has resided out of Town since July last and been but once to the Lectures in this year, conceiving himself not liable to be called on for his subscription for the present year.

Ordered. That he be informed that the Managers being bound by General rules do not consider themselves at liberty to consider M.r Pouget as having ceased to be an Annual Subscriber, no notice having been given of his intention so to do. —

8. The Porter having requested that his wages be advanced from 1 guinea to £1. 5. 0 p.r Week.
Resolved. That his request cannot be complied with.

9. M.r Edmund Davy requested leave of absence for six weeks to go into the Country.

Henry Warburton

Confirmed with the exception of N.o 8 which is referred to M.r Davy.

Managers' Minutes 17th June, 1811.

At a Meeting of the Managers of the Royal Institution of Great Britain held at the house of the Institution, on Monday the 17th of June 1811. —

Present

Henry Warburton, Esq.r, in the Chair.
J. P. Auriol, Esq.r
Jn.o Guillemard, Esq.r Secretary.

Minutes app.d with exception of art. 8. — 1 The minutes of the proceedings of the last meeting were read and approved except Art: 8 which is referred to M.r Davy.

Ex.d Gen. Acc.t — 2. Examined the General Account of the Institution

Min.tes of Com. of Acc.ts approved — 3. Read and approved the minutes of the Committee of Accounts of this day, And signed an Order for an advance to the Clerk of £50.

Life Subscriber. — 4. In pursuance of a nomination under the Act by James Peter Fallofield Esq.r his Son The Rev.d William Fallofield of Albemarle Street, was admitted to the privileges of a Life Subscriber.

Presents. — Presents received.

5 Jonathan Williams Esq.r presented
Thermometrical Navigation &c.
Charles Marsh, Esq.r presented
L'Allegro, Miltoni Poema 4.to
Resolved. That the thanks of the Managers be returned for these presents. —
& Observations on Mineralogical systems, by R Chenevix Esq.r from the Translator.

M.r Feinagle experiments repeated — 6 Read a report from the Committee of General Science requesting that notice be given that M.r Feinaigle will repeat his experiments on his Art of Memory on the 22.d Instant.

Ordered. That notice be given to all persons entitled to admission in the Institution.

Admission of Visitors — 7. ~~Having seen a resolution of the Committee of General Science of Friday the 14th Inst.t "That the Chairmen do invite such persons as they may think desirable to M.r Feinaigle's exhibition on Saturday next."~~
~~Resolved. That as the Managers have not~~ power to sanction such general admissions the Comm.ee be requested to refer to Art 17 of Ch: VI of the Bye Laws.

Report from Committee of Visitors — 8. Read the following report from the Committee of Visitors on the state of the Institution.

1st White-washing is generally wanted for the Staircase & passages of the house as well as for the walls of the Laboratory, the ceiling of which likewise wants repair.

2dly Chairs are wanting in the room N.o 1 which they think should be strong & plain.

3dly The green cloth on the tables in the rooms N.os 1 & 2 is in a very dirty state, and they think it would be an improvement if those tables were covered with black leather, particularly the table in the room N.o 1. — that being most conspicuous. —

4thly The lamps in the reading room N.o 2. are on a bad construction, and they are of opinion that 4 lamps (instead of the six at present there) on a good construction would give more light, be less expensive and occasion less heat. —

5thly On the Committee observing to Professor Davy who attended them round the house that the heat in the Theatre was occasionally very oppressive he said that it appeared to him to arise in some measure from the foul air collected under the seats and thought that he could point out a remedy. —

6thly A List of the furniture not having been made out for some years, they are of opinion that a new one should be ordered and that the old one should be preserved.

7thly It was reported to them that the Catalogue of the minerals is making out, but is not completed, and that a Catalogue of the Laboratory, Apparatus & model rooms is wanting. —

8thly It was likewise reported to them that the List of the Managers, Visitors, Members & Subscribers is not yet printed, and they are of opinion that they

"places of abode should be added where they can be obtained.

9thly They recommend that when a new Edition of the Bye Laws is printed it shall be sold to the Subscribers and to the public at a moderate price, And that it shall include a List of the Subscribers as well as the Members with their places of abode.—

Orders in consequence — Ordered. That Mr. Royce make an estimate on Art. 1.—

That Mr. Harris be requested to make an estimate on Art: 2 & 3.—

That Mr Davy be requested to give his advice as to the best method of lighting the reading room and ventilating the Lecture room.— Art: 4 & 5.

That Mr Edmund Davy make a List of the furniture as also the Catalogue of Minerals, the Laboratory & model rooms. Arts 6, & 7.

List of Members to be printed — That the recommendation of the Visitors respecting printing the List of Managers Visitors, Members & Subscribers, be further considered when Orders are given for printing the Bye Laws.— Art: 8, & 9.

9. Ordered. That it be referred to Mr Davy either to continue the present Porter or provide another at his discretion.—

Vane of St George's chapel 10. Ordered. That the Clerk pay the bill for erecting the Vane on St. Georges Chapel on obtaining 1/3d from the Alfred Club.

J P Auriol

Managers' Minutes July 1, 1811.

At a Meeting of the Managers of the Royal Institution of Great Britain, held at the house of the Institution, on Monday, the 1st day of July 1811.—

Present

J. P. Auriol Esqr. in the Chair.

Scrope Bernard Morland Esqr. Treasurer.

Davies Giddy, Esqr.

Earl of Morton.

John Guillemard, Esqr. Secretary.

Minutes appd 1. The minutes of the proceedings of the last meeting were read and approved.

Exd Gen acct. 2. Examined the General Account of the Institution.

Min. of Com. of Accts appd 3. Read and approved the minutes of the Committee of the two last Weeks.

Voluntary Subscriptions proposed. 4. Read a letter from John Hinckley Esqr. suggesting the idea of opening a book entitled "Voluntary Subscriptions for a permanent fund to secure the success and general purposes of the Royal Institution", adding that he should be happy to add his mite.

& also a letter from John Hinckley Esqr. on his late non attendance on the board of Managers.

Resolved.

That John Hinckley Esqr be requested to accept the Office of Honorary foreign Assistant Secretary to the Royal Institution.—

Life Subscriber 5. In pursuance of a nomination under the Act by Roger Wilbraham, Esqr. his Nephew George Wilbraham Esqr of No. 6 Green Street Grosvenor Square was admitted to the privileges of a Life Subscriber to the Royal Institution.

Presents. Presents received.

6. John Sanders Esqr. presented

Lieut Col. Imrie's description & section of the Strata of the Grampian Mountains.

Dr Smith presented Linnaeus's Tour in Lapland, 2 vol.

The Revd. Theophilus Mauzit presented

La Liturgie de l'Eglise Anglicane,

Mr Wm Wadd, presented Observations on strictures in the Uretha,

Mr D. Boileau, presented "An introduction to the study of political economy.

Ordered. That the thanks of the Managers be returned for these presents.

7. Mr Davy reported that John Payne had quitted the service of the Institution and that in compliance with the request of the Managers he had engaged Stephen Statter as Assistant Porter, Thomas Statter for the present acting as Porter.

Resolved. That John Payne receive a months wages on quitting the Institution.

Salary of asst. in Lby. increased. 8. Read a recommendation of the Patrons of the Library of 30th July 1810 for the raising of Mr Sturts Salary.

Resolved. That Mr Sturt's Salary be increased from £70. to £80 pr Ann: to be reckoned from the 1st of January last.

Payment of Bills, &c. 9. Signed an Order for payment of the following sums.

Mr. Harris 1/4 yrs Sal. to March 1811	£45. 0. 0
Sturt 1/4 — Do. — to June	57. 10 —
Do — 1/2 years advance to do	5. — —
Fincher on acct. of the Royal Instn.	50. —
Sewers rate	16. 13. 4
	£134. 3. 4

10. Resolved. That the Meetings till November next be held Monthly viz. The Committee of Accounts at 1/2 past 12. & the meeting of the Managers at One, on the first Monday of every month; and that meetings in the intermediate weeks be suspended unless summoned for Special business.

S Bernard Morland

Managers' Minutes, Aug. 5 1811.

At a Meeting of the Managers of the Royal Institution of Great Britain, held at the house of the Institution, on Monday the 5th of August 1811.—

Present

Scrope Bernard Morland, Esqr. in the Chair.

Henry Warburton, Esqr.

Davies Giddy, Esqr.

John Guillemard, Esqr. Secretary.

Minutes appd. 1. The minutes of the proceedings of the last meeting were read and approved.

Exd Gen. Acct. 2. Examined the General Account of the Institution.

Min. of Com. of Accts approved 3. Read and approved the minutes of the Committee of Accounts of this day, and of the 4 last weeks viz. 8th. 15th. 22d. & 29th. July.

4 Letters received.

Read a letter from the Committee of Mathematics, Mechanics & Mechanical Inventions stating that in pursuance of Art: 4 Chap: 16 of the Bye Laws they had elected

The Earl of Stanhope — Chairman.

Honble Robt. Clifford — Chairman.

Thos. Day Esqr. — Secretary.

& from Mrs Elizth Christie requesting information whether she can transfer the share of her late husband Major Christie to his Sons.

Resolved. That a letter be written to her requesting inform.n as to the time of Major Christies death.

Presents Presents received.

5. His Majesty's Commissioners on the public Records presented The Statutes of the Realm vol 1.

The Conductors of the British Review presented The British Review, No. 2.

Richd. Duppa Esqr. presented An account of the subversion

of the Papal Government 1798.
Dr. Buchan presented
A treatise on Sea Bathing with remarks on the use of the Warm Bath.

Ordered. That the thanks of the Managers be returned for these presents.

Life Subs.r 6. In pursuance of a nomination under the Act by Tho.s Harris Esq.r of Covent Garden Theatre his Son Henry Harris Esq.r was admitted to the privileges of a Life Subscriber to the Royal Institution.

Resig. of Visitor 7. Mr. Davy reported to the Managers that he had authority from Mr. Murdock to give in his resignation of Visitor to the Royal Institution.

Repairs 8 Received Mr. Harris's estimate and ordered the articles mentioned to be procured – and that the Chimney in the Foreign Newspaper Room be repaired according to the estimate and engagement of Mr. Wm. Feetham.

9 Received Mr. Royce's estimate amounting to £43.5.0 and Ordered that the same be carried into effect with the exception of £7 for painting in the Laboratory.

Leave of Absence to J. Fincher 10 Mr. Fincher requested leave of absence for a few days in the course of the next and following week.
Leave was granted.

11 Signed an Order for payment of the following sums viz.t

J. Fincher on account of the Royal Institution			£70 —
J. Newman for Voltaic Apparatus			19 . 3 . 6
H. Davy Esq.r	1/4 yrs salary to June 1811		100 . —
E. Davy	d.o	d.o	25 . —
David Brooks	d.o	d.o	13 . 2 . 6
Wm. Harris	1/4 d.o	d.o	45 . —
			£272 6 .

12 The Hon.ble Rob.t Clifford attended to state on the part of the Committee of Mathematics that it is expedient to fit up a Room for their Meetings and proposed the Model room for that purpose.

Resolved.
That the consideration thereof be deferred. –

J. P. Auriol

Managers' Meet.g 2nd Sept. 1811.

At a Meeting of the Managers of the Royal Institution of Great Britain, held at the house of the Institution, on Monday the 2.d of September 1811. —

Present
J. P. Auriol, Esq.r in the Chair.
Scrope Bernard Morland Esq.r Treasurer.
Thomas Murdock Esq.r

Minutes app.d 1 The minutes of the proceedings of the last meeting were read and approved.

Ex.d Gen. Acct. 2. Examined the General Account of the Institution.

App.d Minutes of Com. of Acc.ts 3. Read and approved the minutes of the Committee of Accounts of this day, and of the 3 last weeks, viz.t 12th 19th & 26th Ult.o And signed an Order for an advance to the Clerk of £30.

4. Read a letter from Mr. J. A. White an Annual Subscriber for life; stating that a severe change in health prevents him residing in Town and wishes to be considered as an Annual Subscriber for the past time, and that he be allowed to discontinue his subscription for the future.

Managers cannot permit Subscriptions to be ~~discontinued~~ returned

Ordered. That a letter be written to Mr. White – informing him that the Managers have not power to give up any subscription, it being the property of the Institution. –

Present 5.

Presents received.
George Wilson Meadley Esq.r presented
Memoirs of William Paley, D.D.
Directors of the Union Fire Office presented
The Trials of The Rev.d Rob.t Bingham,
Society for bettering the condition of the Poor, presented
Cottage Dialogues, & Martin's letter to Lord Pelham on the state of Mendicity.
The Theory of Money; or practical enquiry into the present state of the circulating medium. By the Author.

Ordered. That the thanks of the Managers be returned for these presents.

Read the minutes of the last General and Adjourned Monthly Meeting.

Resolved. That Mr. Harris be requested to make out a Catalogue of the furniture of the Institution it falling within the duty of the assistant superintendent of the house.

Ordered. That duplicates of the Times, Chronicle, & Morning Post, be discontinued till 1st January next

Mr. Sturt requested permission to be absent from the Institution for 3 weeks. – &
The Porter requested permission to be absent for a week. –
Resolved. That the above requests be complied with. –

S. Bernard Morland

Managers' Meeting 7th Oct. 1811.

At a Meeting of the Managers of the Royal Institution of Great Britain, held at the House of the Institution on Monday, the 7th of October 1811. —

Present
Scrope Bernard Morland, Esqr. Treasr. in the Chair.
Thomas Murdoch, Esqr.
John Guillemard Esqr. Secretary.

Minutes appd. 1. The minutes of the proceedings of the last meeting were read and approved.

Exd. Gen. Acct. 2. Examined the General account of the Institution.

App. Minutes of Com. of Acct. 3. Read and approved the minutes of the Committee of Accounts of this day, and of the 4 last weeks vizt. 9th 16th 23d & 30th Ult. And Signed an Order for payment of the following sums vizt.

J Fincher on account of the Royal Institution	£30 .. 0 .. 0
Mr. Sturt 1/4 years Salary to 30th Septr.	20 .. 0 .. 0
Mr. Newman sundries for the Laby. from Jany. to Septr. 1811	36 .. 10 .. 10
	£86 .. 10 .. 10

4 Read the following letter from Wm. Seymour Esqr. Solicitor to the Royal Institution.

October 4th 1811

My Lords & Gentlemen /

I have perused the Probate of the Will of the late Sr. Elijah Impey, dated 30th March 1806, & a Codicil dated 12th June 1809, executed in the presence of 2 witnesses, and find that he thereby gave & bequeathed to his Son, Elijah Barwell Impey, his "Share right title "& interest in the Royal Institution and to his Executors "Administrators & Assigns for ever." I learn that Sr. Elijah died on the 1st October 1810, without having previously signified that he would not relinquish his Hereditary or saleable share in the Institution, and without claiming compensation under the Act of Parliament; whereby I

consider that he will be deemed to have given up his Hereditary and Saleable share, but to have possessed the power of nominating by Will, a Son to be admitted to the privileges of a Life Subscriber, and that, consequently, in the events which have happened, Mr. Elijah Barwell Impey has become entitled to be admitted to the privileges of a Life Subscriber to the Royal Institution.

I have the honor to be,
My Lords & Gentlemen,
Your obedient humble Servt.
(signed) Wm. Seymour.

To the Managers
of the Royal Institution.

Life Subscriber — Resolved. That Elijah Barwell Impey be admitted to the privileges of a Life Subscriber.

Presents. — Presents received.

5. Wm. Thos. Brande Esqr. presented
Experiments upon fermented Liquors, &
Lieut. Genl. Thornton presented Becaria on Crimes and Punishments.
Ordered. That the thanks of the Managers be returned for these Presents.

Renewed City Lease recd. 6. J. Fincher reported that he had received from Wm. Seymour Esqr. the Solicitor the renewal Lease from the City of London to the Royal Institution dated 2d. May 1810.

Thos. Murdoch

Managers' Meetg. Nov. 4. 1811.

At a Meeting of the Managers of the Royal Institution of Great Britain, held at the house of the Institution on Monday, the 4th of November 1811 —

Present
Thomas Murdoch, Esqr. in the Chair.
William Watson, Esqr.
Scrope Bernard Morland, Esqr. Treasurer.
John Guillemard, Esqr. Secretary.

Minutes appd. 1. The minutes of the proceedings of the last meeting were read and approved —

Exd. Gen. Acct. 2. Examined the General Account of the Institution.

Minutes of Com. of Acct. approved. 3. Read and approved the Minutes of the Committee of Accounts of this day, and of the 3 last weeks vizt. 14th 21st & 28th Ult. And Signed an Order for payment of the following sums.

Mr. Harris on account of Salary, due 30th Sept. last.	£20. —
Mr. E. S. Lees for Dublin Journals from Nov. 1809 to Nov. 1811.	16. —
	£36. —

4. Read a letter from Mrs. Elizth. Christie stating that Major Christie died at Paris 4th July 1809.
Ordered,
That the above letter together with her former letter of application be referred to Mr. Seymour Solicitor to this Institution.

Presents — Presents received.

5. His Majesty's Commissioners on the public Records. presented
Indexes to Domesday Book, folio
Placitorum abbreviatio, folio, — &
The Revd. H. De la Fite presented
De Luc's Geological Travels, vol. 2 & 3, 8vo.
Ordered. That the thanks of the Managers be returned

for these presents.

6. Ordered. That Mr. Shepherd, Upholsterer, cover with green Cloth the 3 Library Tables in the Reading Rooms, and 1 in the Clerk's Office, at an expense not exceeding £10 agreeable to his estimate.

Annl. Subs. 7. The following person having been duly recommended was admitted an Annual Subscriber

Col. Henry White — of Albemarle Street

Weekly Meetgs. resumed 8 Ordered. That the weekly meetings of the Managers be now resumed. —

9. Ordered. That application be made to Lord George Cavendish, Sr. H. C. Englefield, Col. Noel, Thos. Wm. Carr Esq. & Wm. H. Pepys Esqr. for their subscription to the Loan Account. And also to the Executors of the Earl of Dartmouth, Sir Francis Baring Bart. and to the Agent of Sir George Staunton Bart.

10. Ordered. That a letter be written to Mr. Hatchett requestg. to know whether a report is about to be made on the subject of the Lectures for the ensuing year.

Wm. Blake

Managers' Meeting 18th Nov. 1811.

At a Meeting of the Managers of the Royal Institution of Great Britain, held at the house of the Institution on the 18th November 1811 —

Present

Wm. Blake, Esqr. in the Chair
Scrope Bernard Morland, Esqr. Treasurer
William Sotheby, Esqr.
John Guillemard, Esqr. Secretary.

Minutes appd. 1. The minutes of the proceedings of the last meeting were read and approved.

Exd. Gen. Acct. 2. Examined the General Account of the Institution.

Min. of Com. of Accts. approved 3. Read and approved the minutes of the Committee of Accounts of this day.

Life Subscriber. 4. In pursuance of a nomination under the Act by Scrope Bernard Morland, Esqr. his Son Francis Bernard Morland Esqr. was admitted to the privileges of a Life Subscriber to the Royal Institution —

Present. Present received.

C. H. Tatham Esqr. presented
Representations of a Greek Vase, 4to.

Ordered.
That the thanks of the Managers be returned for this present.

5. J Fincher reported that Mr. Watson & Mr. Murdoch being the only Managers present at the meeting of the Committee of Accounts of the 11th Instant, and an Advance being wanted for the Current expences of the house they had signed an Order for the sum of £40. and the same was ordered to be paid. —

W Watson

Managers' Meeting 25 Nov. 1811.

At a Meeting of the Managers of the Royal Institution of Great Britain, held at the house of the Institution, on the 25th of November 1811. —

Present

William Watson, Esqr. in the Chair.
J. P. Auriol, Esqr.
Thos. Murdoch, Esqr.
William Blake, Esqr.
John Guillemard, Esqr. Secretary.

Minutes approved. 1. The minutes of the proceedings of the last meeting were read and approved.

Exd. Gen. Acct. 2. Examined the General Account of the Institution.

Minutes of Com. of Accts. approved. 3. Read and approved the minutes of the Committee of Accounts of this day. —

Life Subscriber. 4. In pursuance of a nomination under the Act by William Watson, Esqr. his Daughter Cecilia Watson of Queen Square was admitted to the privileges of a Life Subscriber to the Royal Institution

Report concerning printing. 5. The Secretary having reported that a large quantity of useless printed paper amounting to nearly ½ a Ton weight had been found in different parts of the house and that after having duly inspected the same he had ordered it to be sold for the benefit of the Institution. It was moved and seconded that he be requested to make a report of the expense of printing from the first establishment of the Royal Institution to the present time distinguishing as far as possible the amount incurred in each year.

Lectures 6. Resolved. That Mr. Blake be requested to enquire whether Mr. Farish be willing to give a course of Lectures in the ensuing Season and on what terms

7. Resolved. That Mr. Blake be also requested to enquire whether Dr. Wollaston be willing to give a course of Lectures

in the ensuing Season and on what terms.

8. Resolved. That the Secretary be requested to enquire whether the Astronomer Royal, Dr. Smith & Dr. Crotch be willing to give courses of Lectures in the ensuing Season and on what terms.

Resolved. That Mr. Murdoch be requested to enquire whether Mr. Lawrence would be willing to give a course of Lectures in the ensuing Season and on what terms. —

Thos. Murdoch

Managers' Meeting 2nd Dec. 1811.

At a Meeting of the Managers of the Royal Institution of Great Britain, held at the house of the Institution, on Monday the 2d. day of December 1811. —

Present

Thomas Murdoch, Esqr. in the Chair.
William Blake Esqr.
John Guillemard, Esqr. Secretary.

Minutes approved 1. The minutes of the proceedings of the last meeting were read and approved.

Exd. Gen. Acct. 2. Examined the General Account of the Institution.

Minutes of Com. of Accts. 3. Read and approved the minutes of the Committee of Accounts of this day. —

Lectures. 4. Mr. Blake reported that he had applied through the Dean of Carlisle to Mr. Farish but has not yet received a definitive answer.

5. Mr. Blake likewise reported that he had made a similar application to Dr. Wollaston who wishes to decline giving Lectures at the Institution

6. The Secretary reported that he had made application to the Astronomer Royal and Dr. Smith but has not yet received from them definitive answers.

7. Read a letter from Dr. Crotch in answer to Mr. Guillemard's application stating that he is willing to give two courses of Lectures one in the present and another in the ensuing Season at 8 guineas a Lecture.

8. Mr. Murdoch reported that he had applied through Richard Hart Davis Esqr. to Mr. Lawrence but has not yet received a definitive answer. —

9. The Secretary reported that he had received a letter from Mr. Davy stating that if the Managers approve of it he intends giving two courses of Lectures in the ensuing Season — one

on Electro-Chemical Science and another on General Chemistry as applicable to the phænomina of nature.

Present 10. Present received.

A table to calculate Interest & Salaries from the Calculator.—

11. The Secretary read his report on the expenses of Printing incurred by the Royal Institution.

12. Mr Blake presented plans for the future mode of keeping the Accounts of the Institution which were ordered to be adopted.—

Thos Murdoch

Managers Meetg 9th Dec. 1811.

At a Meeting of the Managers of the Royal Institution of Great Britain, held at the house of the Institution on the 9th day of December 1811.—

Present
Thomas Murdoch, Esqr. in the Chair.
William Blake, Esqr.
Scrope Bernard Morland, Esqr. Treasurer.
William Sotheby, Esqr.
John Guillemard, Esqr. Secretary.

Minutes appd. 1. The minutes of the proceedings of the last meeting were read and approved.—

Exd. Gen. Acct. 2. Examined the General Account of the Institution.

Minutes of Comee of Accts. approved 3. Read and approved the minutes of the Committee of Accounts of this day.

Lectures 4. Mr Blake reported that he had received a letter from Mr Farish declining on account of the present state of his health giving Lectures at the Institution.

do 5. Mr Murdoch read a letter addressed by Mr Lawrence to Richard Hart Davis Esqr declining giving Lectures at the Institution.

do 6. Resolved. That the Secretary write to request Mr Pond will have the goodness to inform him whether it will be convenient to Mr Pond to deliver a course of Lectures on Astronomy & Mechanics in the ensuing season as the Managers wait his answer to enable them to announce the different courses of Lectures in the public papers.

7. Resolved. That in the next weekly Notice the Clerk insert a request that the Managers would make a point of attending as business of a particular and urgent nature will be brought before them.

8 The following persons having been duly recommended

were admitted as Annual Subscribers for the year 1812.—

Annl. Subscribers

The Revd. Peter Gandolphy Spanish Place Manchr. Sq.
I. Russell, Esqr. 22 Bentinck Street.
John Wright Esqr. 75 Welbeck Street
Honble General Broderick 3 Foley Place

Bills, payment of 9. Signed an Order for payment of the following sums vizt.

I. Fincher on account of the Institution		£40.—
I. Reeve for ½ years Prop. Tax to Michas last	25.0.8	
½ years Land tax to do	7.5.6	
½ years Assessed taxes do	71.2.7	103 " 8.9
		£143. 8. 9

WWatson

Managers Meeting 16. Dec. 1811.

At a Meeting of the Managers of the Royal Institution of Great Britain, held at the house of the Institution on the 16th of December 1811.—

Present
William Watson Esqr. in the Chair.
Thomas Murdoch, Esqr.
Henry Warburton, Esqr.
I. P. Auriol, Esqr.
Wm Blake, Esqr.
Wm Sotheby, Esqr.
John Guillemard, Esqr. Secretary.

Minutes appd. 1. The minutes of the proceedings of the last meeting were read and approved.

Exd. Gen. Acct. 2. Examined the General Account of the Institution.

Mins. of Com. of Accts appd. 3. Read and approved the minutes of the Committee of Accounts of this day.

Complaint of Conduct of Officers of house 4. Read a letter from The Revd. Peter Gandolphy complaining of the conduct of one of the Officers of the house.
Resolved.— That Mr Guillemard be requested to apply to Mr Gandolphy to state the particulars of the conduct of which he complains.

Presents 5 Presents received.
The Revd. Edward Forster presented
No. 9, of the British Gallery of Engravings
The Conductors of the British Review presented
The British Review No. 4.
Sir J. C. Hippisley Bart. presented
2 copt. of his speech on the Catholic Question
Ordered.
That the thanks of the Managers be returned for these presents.—

6 Resolved. That the Duke del Infantando be informed that the Committee of Managers have given orders for his admission to the Lectures & other public parts of the Institution.

7. **Resolved**. That a copy of the recommendation in consequence of which Mr. Russell was elected an Annual Subscriber be sent to him.

8. The following persons having been duly recommended were admitted as Annual Subscribers for the year 1812 —

Ann.l Subs.rs — Daniel Mussanden, Esq.r 30, Half Moon St.
Francis Parkman Esq.r 36 Norfolk St. Strand

Lectures 9. The Secretary reported that he had received a letter from Dr. Smith stating that he is willing to give a course of Lectures on the Philosophy of Botany to commence in April next.

do. 10 **Resolved**. That Mr. Davy be requested to confer with Mr. Pond on the subject of Lectures on Astronomy &c Mechanics.

do. 11 **Resolved**. That Mr. Sotheby be requested to write to Mr. Campbell to enquire whether he be willing to give a course of Lectures on Poetry in the ensuing Season and on what terms.

do. 12. **Resolved**. That Mr. Warburton be requested to write to Dr. Clarke to enquire whether he be willing to give a course of Lectures on Mineralogy in the ensuing Season and on what terms.

do. 13. **Resolved**. That Mr. Davy be requested to apply to Mr. Shield to enquire whether he be willing to give a course of Lectures on Music in the ensuing Season and on what terms.

do. 14 **Resolved**. That Mr. Davy be requested to apply to Mr. Dibdin to enquire whether he be willing to give a course of Lectures in the ensuing Season and on what terms.

Admission of Med.l Students. 15. **Resolved** That for the future Medical Students alone be admitted to the Lectures without becoming Annual Subscribers and to those courses only which are connected with their Professional Studies on the payment of Two guineas for any single Course.

Confirmed, with the exception of Number 14 inserted by mistake.
Wm. Sotheby —

Managers' Meeting 23 Dec 1811.

At a Meeting of the Managers of the Royal Institution of Great Britain held at the house of the Institution on the 23d of December 1811. —

Present
Wm. Sotheby, Esq.r in the Chair.
Wm. Watson, Esq.r
Thos. Murdoch, Esq.r
Wm. Blake, Esq.r
Henry Warburton, Esq.r
John Guillemard, Esq.r

Minutes app.d ext. No. 14 — 1. The minutes of the proceedings of the last meeting were read and approved with the exception of No. 14 inserted by mistake.

Ex.d Gen. acct. — 2. Examined the General Account of the Institution.

Mins of Com. of Acct.s approved. — 3. Read and approved the minutes of the Committee of Accounts of this day.

4 Read a letter from Mr. Russell thanking the Managers for having elected him an Annual Subscriber.

Letter from Mr. Campbell conc.g his course of Lectures. — 5 Read the following letter from Mr. Campbell in consequence of the application made to him by Mr. Sotheby viz.t:

Sydenham Sat.y Dec.r 1811.

My dear Sir

I am very sensible of the honor which the Managers of the Royal Institution have conferred upon me in wishing me to deliver a course of Lectures on Poetry and I will with the utmost pleasure set about composing those lectures provided they are not required from me before it is possible for me to execute so important a task — I have no objections to the terms you mentioned when you favoured me with an interview — Having never yet lectured in public I perhaps over-rate the time necessary for preparation and it is possible that if I were allowed till after Easter I might be actually ready for the course a good deal sooner — If so, I should eventually engage myself to begin sooner — But having no ambition to come forward with crude lectures I would rather decline the engagement, honorable and flattering as I feel it, than agree to commence before Easter under a positive promise. —

I shall await your answer with some anxiety but not without hopes that the Managers of the Institution making allowance for the difficulty attending a first appearance of this kind before the public will not consider my unwillingness to open the course immediately as any thing but a mark of respect for the Institution and the Audience &c of proper caution respecting an undertaking in which the little favor I may possess with the public is committed to trial. — I am exceedingly ambitious however to accept this offer and it only remains with them to consider if they can afford me the time I require for preparation. — When I receive your answer and send my reply the affair I suppose will not require any more formal conclusion. — I remain with great gratitude for your kindness in this as in other instances your obliged & respectful friend.

Signed) T. Campbell

Resolved. That Mr. Sotheby be empowered to accept Mr. Campbell's proposal and be requested to signify the same to him. —

Presents 6 — Presents received.

Mr. Benj.n Travers presented
"An Inquiry into the process of nature in repairing injuries of the Intestines, 8vo

Dr. Buchan presented
Bionomia

Resolved. That the thanks of the Managers be returned for these presents —

Lectures 7. Mr. Davy reported that he had written to Mr. Pond but had not yet received any answer.

8. Mr. Warburton reported that he had written to Dr. Clarke but had not yet received any answer.

254 1811.

9. Mr. Davy reported that he had written to Mr. Shield but had not yet received any answer.

10. Resolved. That a letter be written to Dr. Crotch informing him that the Managers decline entering into an engagement with him on the subject of the Lectures on Music.—

Explanation of No. 4. last Minutes 11 Mr. Guillemard reported that he had seen Mr. Gandolphy on the subject of his letter read at the last meeting and that a satisfactory explanation had taken place.—

12. The following person having been duly recommended was admitted as an Annual Subr. for the year 1812—

Annl. Subsr. Geo: Henry Longridge Esqr. 17 Lower Brook St.

13. Resolved. That Mr. Davy be requested to apply to Dr. Birkbeck to enquire whether he be willing to deliver a course of Lectures on Natural Philosophy and on what terms.—

W. Watson

255

Managers Meetg. 30. Dec. 1811.

At a Meeting of the Managers of the Royal Institution of Great Britain, held at the house of the Institution, on the 30th day of December 1811.—

Present
William Watson, Esqr. in the Chair.
Thomas Murdoch, Esqr.
Henry Warburton, Esqr.
John Guillemard, Esqr. Secretary.

Minutes appd. 1 The minutes of the proceedings of the last meeting were read and approved.

Exd. Gen. Acct. 2. Examined the General Account of the Institution

Minutes of Com. of Accts. 3. Read and approved the minutes of the Committee of Accounts of this day and signed an order for an advance to the Clerk of £20, & to Mr. Sturt ¼ years Salary to 31st December 1811.—

Life Subscriber. 4. In pursuance of a nomination under the Act by Peter Nouaille, junr. Esqr. his Nephew Edward John Rudge Esq. of Caius College Cambridge was admitted to the privileges of a Life Subscriber.—

5 The following persons having been duly recommended were admitted as Annual Subscribers for the year 1812.—

Annl. Subscribers.
John William Roberts, Esqr. of Albany.
Thomas Porteus Esqr. 34 Half Moon Street.

6 Finding it impossible to announce the Lectures for the ensuing Season.
Resolved—That a special Summons be issued to each Manager in the following words & signed by the Members present.

Board of Managers
We the undersigned Managers earnestly request your attendance on Monday next the 6th January 1812 upon business which must be decided on that day.—

Wm. Blake

256 1812

Managers Meeting 6th Jan. 1812

At a Meeting of the Managers of the Royal Institution of Great Britain held at the house of the Institution, on the 6th of January 1812.—

Present
William Blake, Esqr. in the Chair.
Wm. Watson, Esqr.
Edwd. Stracey, Esqr.
Wm. Sotheby, Esqr.
Thos. Murdoch, Esqr.
Rt. Honble. Sir Josh. Banks Bart.
Davies Giddy, Esqr.
Scrope Bernard Morland, Esqr. Treasurer.
Earl of Darnley.
Henry Warburton, Esqr.
Rt. Honble. Lord Carrington.
John Guillemard, Esqr. Secretary.

Minutes approved 1. The minutes of the proceedings of the last meeting were read and approved.

Exd. Gen. Acct. 2 Examined the General Account of the Institution

Minutes of Com. of Accts. appd. 3 Read and approved the minutes of the Committee of Accounts of this day.

Presents. Presents received.
4 The Revd. Stephen Weston presented
A collection of Chinese Characters,
The Editors of the Philanthropist,
The Philanthropist No. 5.
The Revd. Peter Gandolphy presented
A defence of the Ancient Faith.
Ordered
That the thanks of the Managers be returned for these presents.—

On Lectures 5 Mr. Warburton reported that Dr. Clarke declines giving Lectures at the Institution.

6 Mr. Davy reported that he had made application to

Jan. 6. 257

Dr. Birkbeck who has intimated that he will give his answer in the course of the present week.—

On Lectures 7 Mr. Davy reported that Mr. Shield being out of Town he has not been able to obtain any answer.

8. Sir Josh. Banks reported that Mr. Pond would complete his course of Lectures on Astronomy in the course of the ensuing Season.—

9. Ordered. That the lectures be announced in the usual manner to begin on the 25th Inst.

10. Mr. Davy stated that Mr. Moore who had assisted in the Chemical lectures delivered by himself in Dublin was willing to give his services in the Laboratory of the Royal Institution gratis, on being allowed an Apartment in the house.
Resolved. That his services be accepted, and that he be accomodated with the room lately occupied by Mr. Sturt.—

Subscribers 11. The following persons having been duly recommended were admitted as Subscribers vizt.
To the Lectures.—
Miss Sturt of upper Seymour Street
Mrs. Guillemard of Gower Street
Miss Neale, 82 Charlotte St. Portland Place
Annl. Annual Subscribers—
Alexr. H. Everett, Great Maddox St.
Thos. Hill Mortimer Esqr. Albany
Fredk. Watson Esqr. Cleveland Court.
Jn. Ewen, Esqr. Piccadilly
Matthw. Montagu Esqr.—Montagu house Portman Sqr.

12. Resolved.
Committee for Publishing Journal. That in order to carry into effect the Bye Law relating to the publication of the Journals of the Institution the following Members be appointed a Committee for that purpose.
Willm. Blake, Esqr.
H. Davy, Esqr.
Davies Giddy, Esqr.

Thomas Murdoch, Esqr
Wm Sotheby, Esqr
Henry Warburton, Esqr
Wm Hyde Wollaston. M.D.
Jno Guillemard, Esqr

J.P. Auriol

Managers' Meet.^g 13 Jan. 1812.

At a Meeting of the Managers of the Royal Institution of Great Britain, held at the house of the Institution, on the 13.th day of January 1812.

Present
J. P. Auriol, Esqr in the Chair.
Davies Giddy, Esqr
Wm Blake, Esqr
Edwd Stracey, Esqr
Henry Warburton, Esqr
Thos Murdoch, Esqr
Jno Guillemard, Esqr Secretary.

Minutes app.^d 1. The minutes of the proceedings of the last meeting were read and approved.—

Ex.^d Gen. Acc.^t 2. Examined the General Account of the Institution.

Minutes of Com. of Acc.^ts approved. 3. Read & approved the minutes of the Committee of Accounts of this day, and Signed an Order for an advance to the Clerk of £40.

Shares in R.I. not disposable 4. Read a letter from Mr Balme enclosing one from the Bishop of Clogher

Resolved. That he be informed that the 4.th Clause of the Act of Parliament precludes Members from now disposing of their present shares in the Institution.

Presentation of Catalogues. 5 Ordered. That copies of the Catalogue of the Library be presented to the Geological Society and to the Royal Academy.

6. Ordered. That the sum of £7.17.6 be paid for these Catalogues and those already presented according to Orders of the 11.th Decr 1809 & 12.th March 1810; to Mr Harris.

Subscribers. 7. The following persons having been duly recommended were admitted as Subscribers vizt

To the Lectures.
Miss Murdoch.
Miss Julia Murdoch.
Miss Louisa Murdoch.

Annual Subscribers.
Scott, Esqr St. Albans Street
Revd Jos.^h Fortune – Moathouse Pembrokeshire

Wm Blake

Managers' Meeting Jan. 20, 1812.

At a Meeting of the Managers of the Royal Institution of Great Britain, held at the house of the Institution, on the 20.th of January 1812.—

Present
Wm Blake, Esqr in the Chair.
J. P. Auriol, Esqr
Thomas Murdoch, Esqr
Edward Stracey, Esqr
Davies Giddy, Esqr
Henry Warburton, Esqr
Jno Guillemard, Esqr Secretary.

Minutes app.^d 1. The minutes of the proceedings of the last meeting were read and approved.

Ex.^d Gen Acc.^t 2. Examined the General Account of the Institution.

Min.^s of Com. of Acc.^ts approved. 3. Read and approved the minutes of the Committee of Accounts of this day.

Life Subscribers 4. Read a letter from Mr Seymour Solicitor to the Institution in which he gives his opinion that the Executors of Major Christie may now nominate a person to be admitted a Member.—

Resolved.

That Langham Christie Esqr eldest Son of Major Christie deceased and nominated by his Executrix, be considered as having succeeded to the rights of his father who was a Proprietor of the Royal Institution and that the said Langham Christie, Esqr in virtue of those rights is now a Member thereof.—

Present. Present received.

Sir Frederick Baker Bart.^t presented
Virgilii Opera, – Didot's splendid Edition in folio, —

Ordered.

That the thanks of the Managers be returned for this present.

262 1812.

Subscribers — The following persons having been duly recommended were admitted as Subscribers vizt.

To the Lectures. —
Miss Parkins – 40 Weymouth Street
Mrs Horner – 48. Bernard S^t.
Mrs Winthrop, 6 New Cavendish S^t.
Miss F. C. Horner } 6 Russell Square
Miss A. W. Horner }
Miss Brown – 59 Berners S^t.
Miss Ferrington, 15 Kings arms Yard
Mrs Mackenzie 49 Albemarle S^t.
Miss Wilson — } 31 Harley S^t.
Miss Harriet Wilson }
Miss Clarke – 3 Kenton S^t.
Mrs Bernard Bosanquet – Montague place

Annual Subscribers. —
John Shaw, Esqr. } 34. Soho Square
John Cross, Esqr. }
John Fuller, Esqr. Great S^t. Helens London
George Vernon Esqr. } 40 Grosvenor Sq^r.
Leveson Vernon Esqr. }
Saml. Chilver, Esqr. New Burlington S^t.
Phineas Bond, Esqr. His Majesty's Consul Genl. of the Middle & Southern States of North America. —
Thos. Lawrence Esqr. Frith S^t.
Daniel L. Maude Esqr. G^t. Portland S^t.

Resolved.
That all Subscribers to particular Lectures of the preceding year be allowed to attend the courses of the same Professor or Lecturer on the same terms as last year.

Wm Watson

263

Managers Meet^g 27th Jan. 1812.

At a Meeting of the Managers of the Royal Institution of Great Britain, held at the house of the Institution on the 27th January 1812. —

Present
W^m. Watson, Esqr. in the Chair.
Thos. Murdoch, Esqr.
Henry Warburton, Esqr.
J. P. Auriol, Esqr.
Davies Giddy, Esqr.
Jno. Guillemard, Esqr. Secretary.

Minutes approved — The minutes of the proceedings of the last meeting were read and approved. —

Ex^d Gen. Acc^t 2. Examined the General Account of the Institution.

Minutes of Com. of Acc^t approved. 3. Read and approved the minutes of the Committee of Accounts of this day, & signed an Order for an Advance to the Clerk of £40.

4 Read a letter from Samuel Bishop claiming compensat^n for painting certain pictures for the Royal Institution. Ordered with reference to the Minute of the 23^d Novr. 1807 that no answer be given to the same. —

Present — Present received.
5 John Beames, Esqr. presented Translation of Glanville.
Ordered.
That the thanks of the Managers be returned for this present. —

6 Arthur Stanhope Esqr. Comptroller of the Foreign Letter Office having applied for a renewal of his ticket of admission
Resolved
That the said application be referred to the next General Monthly Meeting.

7. The following persons having been duly recommended were admitted as Subscribers vizt.

264 1812

Subscribers. Ladies to the Lectures.
Mrs Liston – Pall Mall
Lady Ford – Gloster place
Mrs Hawkins Browne, South Audley S^t.
Mrs Gurney, Portman Square
Mrs Gordon, Hill Street
R^t. Honble Lady Mary Markham – Park place
R^t. Honble Lady Caroline Thynne Grosv^r. S^t.
Mrs Forbes } Berkeley Square
Miss Forbes }

Ann^l. Annual Subscribers.
L^d. Ennismore, Wimpole S^t.
Edwd. Hanmer, Esqr. 42 upper Seymour S^t.
Thos. Slack, Esqr. 16 Gloucester place New road
S. Shaen, Esqr. Princes S^t. Cavendish Square
Henry Ellis, Esqr., 6 Chapple S^t. West.
Lieu^t. Col. White } 29 S^t. Albans S^t.
Captn. Murchison }
Isaac Hawkins Browne Esqr. South Audley S^t.
Hudson Gurney, Esqr. Portman Sq^r.
Robt. Neilson, Esqr. Stedings Hotel
L^d. Calthorpe, Grosvenor Square
Geo: Gillichan M.D. Maze Pond Guys Hospital
George Hayley Esqr. of New York
Jacob Bath Esqr.
D^r. R. M. Paterson, Bentinck S^t. Manchester Sq^r.
Major Genl. J. Bradshaw N° 1. Albany
Jos^h. Robley, Esqr.
D^r. Goldie – 27 South S^t. Grosvenor Square
Ja^s. Boswell Esqr. Temple
His Excell^y Robt. Liston Esqr. Pall Mall.
M^r. Jekeyll – 70 Mary^l. Street
James Ratton Esqr. – Bedford Place.
F. K. Hasler, Esqr. 8 Michaels Grove Brompton
Alex^r. Doorman, Esqr. Bridge S^t. Blackfriars
George Canning Esqr. M.P.
Honble & Rev^d. M^r. Crosbie, 81 Jermyn S^t.
Lieu^t. Col. The Honble Robt. Dundas, Arlington S^t.
Saml. Hinds Esqr. Verulam Build^s.
M^r. Gregson, Charles S^t. Grosvenor Sq^r.
Anth^y Tod Thomson Esqr. Sloane S^t.
Spencer Stner, Esqr. 2 Hanover Square

Jan 27. 265

Medical Students to Professor Davy's Lectures.
John Haviland, Esqr. Garden Court Temple
Henry Earle, Esqr. Hanover Square
Martin Ware, Bridge S^t. Blackfriars
W^m. Hill, Esqr. — Bedford row.

J. P. Auriol

Managers Meetings Feb. 3rd, 1812.

At a Meeting of the Managers of the Royal Institution of Great Britain, held at the house of the Institution, on the 3rd of February 1812. —

Present
I. P. Auriol, Esqr. in the Chair.
Wm. Blake, Esqr.
Thos. Murdoch, Esqr.
Henry Warburton, Esqr.
Scrope Bernard Morland, Esqr. Treasurer.
Edward Stracey, Esqr.
Earl of Winchilsea, President.
Jn.o Guillemard, Esqr. Secretary.

Minutes approved. 1. The minutes of the proceedings of the last meeting were read and approved.

Exd. Gen. acct. 2. Examined the General Account of the Institution

Minutes of Com. of acct approved 3. Read and approved the minutes of the Committee of Accounts of this day.

4. Read a representation from M. L. Este Esqr. on the subject of the Lectures on comparative Physiology.
Ordered. after discussion, that no notice be taken of the said Memorial. —

Subscribers. 5. The following persons having been duly recommended were admitted as Subscribers vizt.

Ladies to the Lectures.
Miss Boddington. } 31. upper Brook St.
Mrs. Harford. }
Mrs. Leckie. — 17 Everett St. Russell Sq.
Miss Dinwiddie. 11 Caroline St.
Mrs. Blackburn. }
Miss Blackburn. } 25 Cavendish Sq.
Miss D. Blackburn. }
Lady Carnegie. —
Miss Carnegie. —
Mrs. Ranking, & 2 Miss Rankings, Cheapside

Annual Subscribers.
Richard Coad, Esqr. Oxford St.
Revd. A. Irvine, Portland Place
Charles Ross, Esqr. do.
Wm. Brant, Esqr. Oriel College, Oxford.
Stephen Payne Galwey Esqr. Montagu St.
Thos. Cotton, Esqr. 18 Basinghall St.
Henry Norton Willis Esqr. – Kensington palace
General Agnew,
Dr. Skey, Russell St.
Wm. Adams Esqr. Albemarle St.

Spencer.

Managers Minutes, 10th Feb. 1812.

At a Meeting of the Managers of the Royal Institution of Great Britain, held at the house of the Institution, on the 10th February 1812. —

Present
Earl Spencer, in the Chair.
Scrope Bernard Morland, Esqr. Treasurer.
I. P. Auriol, Esqr.
Wm. Blake, Esqr.
Thos. Murdoch, Esqr.
Henry Warburton, Esqr.
Jn.o Guillemard, Esqr. Secretary.

Minutes appd. 1. The minutes of the proceedings of the last meeting were read and approved.

Exd. gen. acct. 2. Examined the General Account of the Institution.

Min. of Com. of acct. approved 3. Read and approved the minutes of the Committee of Accounts of this day.

Admittance to Lectures refused 4. Read a letter from M. L. Este, Esqr. stating that he was refused admittance to the Lectures —
Ordered
That Mr. Fincher inform him that Mr. Brooks acted correctly in refusing admittance to Mr. Este, it appearing that he is not a Subscriber. —

Presents

Presents received.

5. The Earl of Carlisle, presented
The Mausoleum at Castle Howard
M. Bertrand De Moleville, presented
A chronological abridgment of the History of Great Britain, vol. 3 & 4.
Ordered.
That the thanks of the Managers be returned for these presents. —

6. Read the Minutes of the last General Monthly Meeting

7. In pursuance of a nomination under the Act by

Life Subscriber. John Mc. Arthur L.L.D., his Son
John Joseph Alphonso Mc. Arthur Esqr. of Lincolns Inn was admitted to the privileges of a Life Subscriber to the Royal Institution. —

Audited acct. 8. Audited the Accounts of the Treasurer up to the 31st. Decr. 1811, and found a balance on that day in favor of the Institution of £3. 9. 11. —

Conc. Lectures 9. Ordered. That the sum of £50 be paid to Mr. Pond on account of his Lectures for the year 1811. —

10. It having been omitted at the time the Lecturers were engaged to record the pecuniary compensations to be made.
Resolved.
That the sums to be paid to the Lecturers of the present year are as follow. —

To Dr. Smith — 60 guineas for his course.
Dr. Roget 75 guineas for his course.
T. Campbell, Esqr. 100 guineas for his course.

11. As it does not appear that any specific sum was agreed upon as a compensation to Mr. Pond for his lectures of last year.
Resolved. That the remuneration to be made to him be referred to the next meeting of Managers.

Subscribers 12. The following persons having been duly recommended were admitted as Subscribers vizt. —

Ladies to the Lectures.
Mrs. Golborn } Welbeck Street
Miss Golborn }
Mrs. Clerk, — }
Miss Clerk, — } upper Seymour St.
Miss Marianne Clerk }
Mrs. Gandolphy, 25 upper Seymour St.
Mrs. Edwd. Rogers Clapham
Rt. Honble. Lady Jane Montagu
Miss Wrightson South Audley Street.
Annual Subscribers. —
Dr. Satterley, Berners St.

Lieut. Col. Jas. Bathurst, Wimpole St.
Captn. Wm. Maughan, 57 Hans place.
Honble. Mr. Stourton, 63 Baker St.
Thos. Lynch Goleborn, Esqr. 27 Welbeck St.
Lieut. Col. Congreve. 13 Cecil St. Strand.
Josh. Alcock jun. Esqr. of the Treasury.
C. Hewett, Esqr.
Genl. Nesbit Balfour, 5 Tenterden St.
Wm. Hy. Whitbread, Esqr. Dover St.
Wm. Murray, Esqr. Garden Ct. Temple.
Jno. English, Esqr. Conduit St.
Robt. Chaloner Esqr. Hanover Square
Henry Bright Esqr. Harcourt Buildgs. Temple.

Henry Warburton

Managers' Meetg. 17 Feb. 1812.

At a Meeting of the Managers of the Royal Institution of Great Britain, held at the House of the Institution, on the 17th day of February 1812. —

Present
Henry Warburton, Esq. in the Chair. —
Wm. Blake, Esqr.
Thos. Murdoch, Esqr.
J. P. Auriol, Esqr.
Earl Spencer.
Wm. Sotheby, Esqr.
Jn. Guillemard, Esqr. Secretary.

Minutes approved 1. The minutes of the proceedings of the last meeting were read and approved. —

Ext. Gen. Acct. 2. Examined the General Account of the Institution.

Minutes of Com. of Accts. approved 3. Read and approved the minutes of the Committee of Accounts of this day.

Presents

Presents received.

4. John Disney, Esqr. presented
A practical abridgment of Election Law, 8vo.
The Geological Society presented
Vol. 1 of their Transactions. —
Mr. Thos. Haynes, presented
An improved system of Nursery gardening. —
Ordered.
That the thanks of the Managers be returned for these presents.

5. Signed an Order for payment of the following sums.

I Fincher on account of the Institution	£.50. —
3/4 year's parochial rates to Xmas 1811 £37..10.. — ; Assessed taxes . — to do. — 105..15..7½	143..5..7½
Dickins & Co. — Coals in 1811	76..2..6
Mr. Parker — Glazier — for the years 1808, 9, 10 & 11	86..11..7.
Thompson — Soap & Candles in 1811	15..17. —
Mrs. Nevill — Sundries for the house in 1811	10..9..3

6 Resolved
That the sum to be paid to Mr. Pond for his course of Lectures commenced last year and to be concluded this Season, be 100 guineas. —

Warming Mineral Museum 7. It having been reported to the Managers that the Rooms containing the Mineralogical Collection are not sufficiently warmed.
Resolved.
That a communication to that effect be made to the Patrons of the Library and Mineralogical Collection with a request that they would be pleased to give the necessary orders for the removal of that inconvenience. —

Subscribers 8. The following persons having been duly recommended were admitted as Subscribers vizt.

Ladies to the Lectures.
Miss Fox. Hertford St.
Mrs. Oom. — } Bedford Square
Miss Hankey. }
Miss Haden, New Burlington St.

Life Subscriber. —
Paul Kneller Smith, Esqr. Cheapside.

Annual Subscribers.
Wm. Battie Wrightson, Esqr. South Audley St.
Scott Murray, Esqr. Cavendish Sq.
Richards Jennings, Esqr. Portland Place
Jn. Shute Duncan, Esqr. Lincolns Inn
Mr. Pozzo di Borgo 5 Manchester St.
Thos. Hoblyn, Esqr. Sloane St.
Geo: Fredk. Albert Esqr. Gt. Ryder St.

J P Auriol

Managers' Meetg. Feb. 24, 1812.

At a Meeting of the Managers of the Royal Institution of Great Britain, held at the house of the Institution, on the 24th of February 1812. —

Present
J. P. Auriol, Esq. in the Chair.
Wm. Blake, Esqr.
Davies Giddy, Esqr.
Sr. John S. Sebright, Bart.
Thos. Murdoch, Esqr.
Henry Warburton, Esqr.
Earl of Darnley.
Jn. Guillemard, Esqr. Secretary.

Minutes appd. 1. The minutes of the proceedings of the last meeting were read and approved.

Ext. Gen. Acct. 2. Examined the General Account of the Institution.

Minutes of Com. of accts. approved 3. Read and approved the minutes of the Committee of Accounts of this day, and signed an Order for payment of the following sums vizt.

Humphry Davy Esqr. 1/2 year's Salary to Decr. 1811.	£ 200. —
Mr. Edmd. Davy — do. — do.	50. —
Mr. Brooks Collr. — do. — do.	26..5. —
Mr. Wm. Harris — 1/4 year's salary do.	45. —
Do. — in balance of Salary due 30th Septr. last	25. —
Mr. Gardner Oil from Jany. to Decr. 1811.	143..3..8
Mr. Henry — Charcoal — do.	10..16. —
Mr. Allen — Chemicals — do.	54..11..8
Messrs. Knight & Co. — do. — do.	3..7..6
Searle — Bookbinding — do.	21..5..9
Harris & Co. Sundries for the house do.	19..9..7.
	£ 598..19..2

Life Subscriber. 4. In pursuance of a nomination under the Act by Thomas Hammersley, Esqr. of Pall Mall, his Son
Hugh Hammersley, Esqr.
was admitted to the privileges of a Life Subscriber to the Royal Institution. —

5 *Ordered.*
That 20 Chaldrons of Coals be supplied by Messrs Dickins & Co at the house of the Institution.

Repairs 6. *Ordered.* That Mr Royce examine the roof of the Lecture room and of the house in general and give in an estimate of whatever he considers as necessary repairs.

7. *Ordered.* That a Liverpool Lamp be purchased & placed in the Hall for the purpose of an experiment on the burning inferior Oil —

8. *Ordered.* That a table be provided for the use of persons wishing to inspect the bound up Newspapers in the passage leading to the News-room. —

Subscribers. 9. The following persons having been duly recommended were admitted as Subscribers vizt

Ladies to the Lectures.

Mrs Irvison, Chapel St Lisson Green
Miss Edgell Wyatt, Grosvenor St
Mrs Baring, 43 Clarges St Piccadilly.
Miss Sebright } Curzon St
Miss Emily Sebright }
Honble Fras Calthorpe, } Grosvenor Square
Honble Hariet Calthorpe }

Annual Subscribers.

Charles Pallmer, Esqr. Manchester St
Geo: Buchan, Esqr. 34 Duke St Manch: Sq.
Alexr. George Mackay, Esqr. 47 Baker St
Captn. Wm Maude, Great George St Westmr.
Geo: Baring, Esqr. Clarges St
Wm Haygarth, Esqr. 27 South Audley St
Richd Oswin, Esqr. 61 upper Morton St
Honble Henry Bennet.
S. Chandos Pole, Esqr. Gloucester place
Mr Wm Blackburn junr. 25 Cavendish Sqe
Jno Caldwell, Esqr. 27 Bury St
Col. Ellice, 43 Conduit St
Jas Malony, Esqr. Blenheim Hotel —

Honble Gilbert Elliot, Royal Hotel
Henry Sheatfield, Esqr. 8 Kings Bench walks Temple.

Medical Student.
Mr Thos Roper. —

W Watson

Managers' Meeting March 2nd, 1812

At a Meeting of the Managers of the Royal Institution of Great Britain, held at the house of the Institution, on the 2d of March 1812 —

Present
William Watson, Esqr. *in the Chair.*
William Blake, Esqr.
J. P. Auriol, Esqr.
Scrope Bernard Morland, Esqr.
Earl of Morton.
Henry Warburton, Esqr.
Thos Murdoch, Esqr.
John Guillemard, Esqr. *Secretary.*

Minutes approved 1. The minutes of the proceedings of the last meeting were read and approved. —

Exd Gen. Acct 2. Examined the General Account of the Institution.

Minutes of Com. of Accts approved 3. Read and approved the minutes of the Committee of Accounts of this day.

Life subscribers 4. In pursuance of nominations under the Act, the following persons were admitted to the privileges of Life Subscribers to the Royal Institution. —

Miss Harriet Boddington, of Clapton
by her Father Thos Boddington Esqr.
George Henry Elliott Esqr. of Sackville Street.
by his Grand-father
The Revd Samuel Glasse, D. D. —

Recommendation of Visitors respecting arrears, and the Managers' answer there to. 5. The Visitors communicated to the Managers the following Resolution taken at an adjourned Meeting of the Visitors on Monday the 24th February last.

The Visitors at several meetings having taken into consideration the large arrears on the books of the Royal Institution and having had the same divided into classes.

Resolved.
That those lists be laid before the Managers

with remarks made against each name in pencil by the Collector as to the probability of recovering the same And that it be recommended to the Managers to appoint three or more Members to form a Committee, to be composed from the Managers only, or from the Managers and Visitors, or from Visitors only, or from the Members in general, for the purpose of inquiring into the large arrears which appear on the books to be due, many of which are of very old standing, and after collecting such information as they may think necessary with regard to each arrear, to give them power and authority to decide on each case in a summary manner, either by insisting that the whole shall be paid or by striking out the whole sum, or accepting part of such sum as to them shall appear most expedient and just, & making from time to time a report of their proceedings to the Managers —

Resolved.
That the thanks of the Managers be returned to the Visitors for the communication of their Resolution respecting the appointment of a Committee to examine into the arrears & for the information conveyed in the Lists transmitted therewith. And that they be informed that the Managers conceiving they do not possess the authority to delegate the power proposed in the resolution of the Visitors they will continue to take such steps as they may think most expedient in each particular case to obtain payment of the arrears —

6. Jos. Fincher reported that he laid the Resolution of the Managers of the 17th Ulto before the Committee of Patrons on which they Resolved That the thanks of the Patrons be returned to the Committee of Managers for that communication. —

7. *Ordered.* That the consideration of the renewal fine for No 22 in Albemarle Street be resumed at the next Meeting.

Subscribers The following persons having been duly recommended were admitted as Subscribers vizt

Ladies to the Lectures.
Mrs Adair. Pall Mall.

278 | 1812.

Mrs. Adams, 15 St James's place
Mrs. Cleaver, Bruton St.
Mrs. Haldimand } Curzon St. May-fair.
Miss Prinsep }
Mrs. Russell } 52 Beaumont St.
Miss Russell }

Annual Subscribers.
Honble. Wm. Howard, 24 Bury St. St James's
Wm. Dacres Adams, Esqr. 15 St. James's Place
Richd. Prime, Esqr. 8 old Buildings Lincolns Inn
Honble David Erskine, 32 Weymouth St.

Medical Student.
Mr. William Hawke 44 Southampton Build gs.
Dr. H. Holland 3 Lisle St. to Professor Davy's Lectures

Spencer.

279

Managers Meeting 9th March 1812

At a Meeting of the Managers of the Royal Institution of Great Britain, held at the house of the Institution on the 9th March 1812.–

Present
Earl Spencer, in the Chair.
Henry Warburton, Esqr.
Wm. Blake, Esqr.
Wm. Watson, Esqr.
J. P. Auriol, Esqr.
Earl of Darnley.
Wm. Sotheby, Esqr.
Jn: Guillemard Esqr. Secretary.–

Minutes approved 1. The minutes of the proceedings of the last meeting were read and approved.–

Exd. Gen. Acct. 2. Examined the General Account of the Institution.–

Minutes of Com. of Accts. 3. Read and approved the minutes of the Committee of Accounts of this day, & Signed an Order for payment of the following bills–

Wm. Bulmer for printing to 31st Decr. 1811 –	£107. 14. 6
Messrs. Dalton & Keys for Lecture cards & tickets	64. 12. –
Mr. Newman cleang apparatus &c	10. 10. –
	£182. 16. 6

Life Subscriber 4. In pursuance of a nomination under the Act by Robert Ward, Esqr. his Son Wm. Ward, Esqr. of Trinity College–Cambridge was admitted to the privileges of a Life subscriber to the Royal Institution.–

5 Read the minutes of the last General Monthly Meeting.

6. Referred the consideration of the renewal fine to the next meeting of the Committee of Accounts.–

Appointment of Committee to examine into arrears. 7. Resolved. That a Committee be appointed to examine into the arrears due to the Institution from Subscribers and that the following Visitors be requested to be of the Committee with such of the Managers as find it convenient to

280 | 1812.

attend–
Lieut: Genl. Thornton
Sir Thos. Bernard Bart.
Daniel Giles Esqr.
any three of whom shall be competent for the transaction of business And that they be requested to report to the Committee of Managers as speedily as possible.

8 Referred the estimate of Mr. Royce as to repairs &c to Mr. Davy–

Subscribers. 9. The following persons having been duly recommended were admitted as Subscribers vizt.

Ladies to the Lectures–
Miss Lewis, Montagu place
Miss Trusler,–
Miss Cleaver } Bruton St.
Miss Cathe. Cleaver }
Mrs. Vivian } Lincolns Inn fields
Miss Vivian }
Miss Whitbread, Dover Street
Countess de Salis.
Miss Pigou, Hill St. Berkeley Square

Annual Subscribers.
Garnet Warburton, Esqr. Ibbetsons Hotel
Captn. Chas. Ogle, of the Royal Navy – Hill St.
Mr. de St. Amand, 25 York Build gs New Road

&
Life Subscriber.
John Forster, Esqr. Bentinck Street.

Thos. Murdoch

281

Managers Meeting 16th March, 1812.

At a Meeting of the Managers of the Royal Institution of Great Britain, held at the house of the Institution on the 16th March 1812.–

Present
Thomas Murdoch, Esqr. in the Chair.
Earl Spencer,
Earl of Morton,
Henry Warburton, Esqr.
Wm. Blake, Esqr.
J. P. Auriol, Esqr.

Minutes approved. 1. The minutes of the proceedings of the last meeting were read and approved.

Minutes of Com. of Accts. approved 2. Read and approved the minutes of the Committee of Accounts of this day, and signed an Order for payment of the following sums.–

J Fincher on acct. of the Royal Institution	£60. – –
Wm. Feetham for Ironmongery to Decr. 1811	37. 6. 3
De Boffe – Books – to Septr. 1811 –	45. 6. 6
Mr. Sheppard. Furniture – to Novr. 1811 –	35. 2. –
Mr. Hammond – Sheeting – to – do. –	17. 4. 10
Jones & Yarrell Newspapers – for the year do. –	98. 6. 6
Wrangham & Co. Stationary from June 1810 to Decr. 1811.	58. 6. 4
Do. – Newspapers from do. to do. –	65. 14. 7
Royce – Carpenter – to – do. –	44. 2. 1
Bird – Bricklayer – do. –	10. 13. 6
Papworth – Plaisterer – do. –	34. 3. 11
Wedgwood – Sundries for the Lectures – do.	14. 6. 4
	£521. 2. 10

Lectures. 3. Earl Morton communicated to the Managers a letter addressed to him from Earl Percy containing a proposal from Professor Coleman of the Veterinary College to be allowed to give a course of Lectures at the Institution, on the structure of the Horse's foot, in the Month of May next:
Resolved
That Ld. Morton be requested to inform Earl Percy that the arrangements for the Lectures at the Institution for

this Season having been settled they cannot comply with his request.—

Presents. **Presents received.**

4 The Society of Antiquaries presented.
Archaeologia, vol: 16. part 2d.
Sir Christopher Hawkins Bart. presented
Observations on the Tin Trade of the Ancients in Cornwall.
Sir Thos. Bernard Bart. presented.
New system of Instruction at the Barrington School.
The Editors of the British Review presented.
The British Review, No. 5.

Ordered.
That the thanks of the Managers be returned for these presents.—

Life subscriber 5. In pursuance of a nomination under the Act by John Skottowe Esqr. his Wife
Mrs. Sarah Honeywood Skottowe of Wimpole St.
was admitted to the privileges of a Life Subscriber to the Royal Institution.

Fine of 22 Albemarle Street 6. Resolved. That Mr. Seymour Solr. of the Royal Institution be requested to repeat his application to the Ground Tenant of the house No. 22 in Albemarle St. for payment of the fine of renewal of Lease due from him at Michas 1809.—

Subscribers 7. The following persons having been duly recommended were admitted as Subscribers vizt.

Life Subscriber.—
Henry Westmacott Esqr. Mount St.

Annual Subscribers.
Wm. Boulger, Esq. – Bradfield house near Reading
Henry Dalton, Esqr. Knaith Lincolnshire.

To the Lectures.
Mrs. Henry Brooke, Gower St.

J. P. Auriol

Managers' Meeting 23d March, 1812.

At a Meeting of the Managers of the Royal Institution of Great Britain, held at the house of the Institution on the 23d of March 1812 –

Present
I. P. Auriol, Esqr. in the Chair.
Wm. Watson, Esqr.
Earl Spencer,
Wm. Sotheby, Esqr.

Minutes appd. 1. The minutes of the proceedings of the last meeting were read and approved.—

Minutes of com. of Accts. appd. 2. Read and approved the minutes of the Committee of Accounts of this day.—

3. Read a letter from Robert John Thornton M.D. respecting his Son's Subscription for the year 1812, and adding that he should feel happy to deliver one or two Lectures gratuitously on Poetic Botany"

Ordered.
That he be informed that the Managers cannot dispense with the obligation to which his Son has subscribed And that the arrangements for the Lectures at the Institution for this Season having been settled they cannot comply with his request.—

Present 4. **Present received.**

The Society for bettering the condition of the Poor presented The "Barrington School."

Ordered. That the thanks of the Managers be returned for this present –

Subscribers. 5. The following persons having been duly recommended were admitted as Subscribers vizt.

Ladies to the Lectures.

Miss Alexander. Lincolns Inn fields
Mrs. Trail, 35 Wimpole St.

Miss Trail.
Mrs. Loftus.
Mrs. Rattray.
Miss Moubray.
} 35 Wimpole St.
Mrs. Houlton – Hartwell. Somerset ~~[illegible]~~ house

Adjourned to Monday the 6th. April.

4.6 Henry Warburton

Managers' Meeting 6th April, 1812.

At a Meeting of the Managers of the Royal Institution of Great Britain, held at the house of the Institution, on the 6th of April 1812 —

Present
Henry Warburton, Esqr. in the Chair.
J. P. Auriol, Esqr.
Edwd. Stracey, Esqr.
Wm. Blake, Esqr.
Scrope Bernard Morland, Esqr. Treasurer.

Minutes appd. 1. The minutes of the proceedings of the last meeting were read and approved.—

Minutes of Com. of Accts. appd. 2. Read and approved the minutes of the Committee of Accounts of this day. And Signed an Order for paymt. of £20 to Mr. Hurt for 1 quarters Saly to 31st. March.

Present, **Present received.**

3 The Right Honble. Nicholas Vansittart presented.
Three Letters on the subject of the British & Foreign Bible Society.

Ordered.
That the thanks of the Managers be returned for this present —

Alteration of Bond. 4. The Managers conceiving the present form of the Bond executed by persons who have been admitted as Members of the Royal Institution to be incorrect.

Resolved
That the said form be thus corrected. – By leaving out the following words "during such time as he shall continue a Member of the said Royal Institution," and by inserting in lieu thereof during the term of his natural life." and also by leaving out "and as long as he shall continue a Member of the Royal Institution".

5 Read the following communication from the Patrons of the Library.
"The Patrons of the Library & Mineralogical

286 — 1812 April 6.

Communication from Patrons of Library respecting the warmth &c. of Mineral Room.

collection having received the communication from the Managers on the 26th February last respecting the proper warming of the room, in which the Minerals are kept, took the earliest opportunity of enquiring into the same by a Committee of their own body, and beg leave to state, that it appears to them, that for every purpose of containing and preserving the collection the present accomodations are sufficient, and the Room itself fully ventilated & warmed.

—But as it has appeared to the Patrons, that the object of the Managers in making such a communication may perhaps be, the making that room more generally useful; and to apply it to the purpose of the meetings of their own Committees or others, arising out of the Institution at large, or for any other general purpose, they beg leave to state to the Managers, that they have examined the same and have taken the advice of experienced persons and are of opinion that sufficient warmth and ventilation may be procured with ease and safety by the following means.

By fixing a light frame of Glass to the lower ceiling of the room so as to prevent the hot air from ascending and being cooled within the present skylight.—

By fixing a stove, or small furnace, in the Laboratory for heating air, to be conveyed in a warm state to several parts of the Mineralogical room, which Stove or furnace might also be made equally useful for Chemical and other purposes, in the Laboratory itself, notwithstanding its supplying the upper room with sufficient hot air.— Or by an alteration of the present fire place, and bringing it further into the room, by which means a current of warm air might be introduced.—

The expense of these alterations the Patrons are led to suppose will amount to about £60, and if the Managers think the same necessary and will give the requisite orders, they will be happy to superintend the alterations.— At the same time the Patrons beg leave to state that they are of opinion, that neither that Room, nor the Library ought to be appropriated to any purpose unconnected with, or detrimental to the

287

objects to which they belong; and that no Committee whatever, except Committees of Books & Minerals, should be allowed to meet in either place: and they have found this particularly inconvenient with respect to the Library, where they are obliged to hold their own meetings, subject to the intrusion of any one coming there; as indeed are the meetings of the Committee of Managers.— Upon this subject they beg leave to refer to a letter, addressed by one of the honorary Librarians to the Committee of Managers dated 3rd Dec.r 1810.

Subscribers

The following persons having been duly recommended were admitted as Subscribers viz.t

Annual Subscribers.

Tho.s Mills, Esq. — Clarges St. Piccadilly
W.m Trotter, Esq. — 30, York Build.gs
Enosh Durant, Esq. — Kings Arms Yard Coleman St.
H. B. Sawbridge, Esq. — Temple

Subscribers to the Lectures.—

Miss Smith, Park St. Westm.r
M.rs Maling }
Miss Darwin } Sloane Street
Miss Hulse.

W.m Blake

288 — April 13, 1812.

Managers Meeting 13th April 1812

At a Meeting of the Managers of the Royal Institution of Great Britain, held at the house of the Institution, on the 13th of April 1812.—

Present

Earl of Darnley, in the Chair.
Henry Warburton, Esq.r
W.m Blake, Esq.r
J. P. Auriol, Esq.r
Jn.o Guillemard, Esq.r Secretary.

Minutes app.d 1. The minutes of the proceedings of the last meeting were read and approved.

Minutes of Com. of Acc.ts approved. 2. Read and approved the minutes of the Committee of Accounts of this day and signed an Order for payment of the following sums.

J. Fincher on acc.t of the Instit.n £60.—
M.r Newman, for Glass for the year 1811 24. 4. 6
£84. 4. 6

Com.n from Patrons of Liby. to be considered. 3. **Resolved.** That the communication from the Patrons of the Library entered on the last Minutes be taken into consideration at the next Meeting together with the Bye Laws relating to the Library & Mineralogical fund.—

Presents. 4. Presents received.—

Persian Recreations, or Oriental stories, 8vo
from the Author.—
Brief Remarks on the proposed Regents Canal, 8vo
from the Author.—
Richard Saumarez Esq.r presented
"The principles of Physiological & Physical Science". 8vo—

Ordered. That the thanks of the Managers be returned to M.r Saumarez for his present.—

Subscribers 5. The following persons having been duly recommended

289

were admitted as Subscribers viz.t

To the Lectures.

Lady Ennismore, 14 Upper Wimpole St.
Miss Anne Smith }
Miss Fanny Smith }

6. Read the minutes of the last General Monthly Meeting.—

J. P. Auriol

Managers' Meeting 20 April, 1812

At a Meeting of the Managers of the Royal Institution of Great Britain, held at the house of the Institution, on the 20th April 1812.—

Present

J. P. Auriol, Esqr. <u>in the Chair</u>.
Earl Spencer,
Henry Warburton, Esqr.
Thomas Murdoch, Esqr.
Wm. Blake, Esqr.
Jn. Guillemard, Esqr. <u>Secretary</u>.

Minutes Appd. 1. The minutes of the proceedings of the last meeting were read and approved.—

Minutes of Com. of Accts. approved. 2. Read & approved the minutes of the Committee of Accounts of this day.—

Present.

<u>Present received.</u>—

The Society of Arts presented the 29th. volume of their Transactions.

3 <u>Ordered</u>. That the thanks of the Managers be returned for this present.—

Communicated from ~~Cor~~ Visitors 4. Read the following communication from the Committee of Visitors—

"The Visitors made their second annual inspection of the house of the Royal Institution, & submit the following deficiencies to the consideration of the Managers—

The east end of the roof of the Laboratory wants repair.—

The great Stair-case & ceiling are very dirty, and require painting & whitewashing

The carpets in the two Libraries, and the Chairs in the lower Library require repair.

The Oil Cloth which covered the wooden stairs of the back Stair-case is worn out and fresh Oil Cloth is wanting.—

Remuneration to Mr Fincher for additional trouble

The Visitors beg leave to recommend to the Managers to make a present of 5 guineas to Mr. Fincher for the additional trouble he has had in preparing the Accounts of the arrears for their information, and for the information of the Managers."—

5. In consequence of the above recommendation of the Visitors to make a present to Mr. Fincher of 5 guins. and from their own knowledge and sense of his activity and diligence in preparing & making up the Accounts of the Institution.—

<u>Ordered</u>.

That Ten guineas be paid to him as a reward for his attention & labours & signed an Order accordingly

6 <u>Resolved</u>. That the communication from the Patrons of the Library be deferred to a future day.—

7. Mr. Auriol having quitted the Chair and Lord Spencer having retired.

Henry Warburton, Esqr. <u>in the Chair</u>.

8 <u>Resolved</u>

That an application be made to the Earl of Winchilsea on the subject of his election as President of the Royal Institution for the ensuing year.—

Subscriber. 9. The following person having been duly recommended was admitted as a Subscriber to the Lectures, vizt.

The Honble. Miss Dundas, Arlington Street.

10. <u>Resolved</u>.

That a special meeting be held on Friday next the 24th. Inst. for the purpose of settling the House List of Members recommended as Officers for the year ensuing.—

Winchilsea P.

Managers' Meeting, Apl. 24, 1812.

At a Special Meeting of the Managers of the Royal Institution of Great Britain, held at the house of the Institution, on Friday the 24th. day of April 1812.—

Present

Earl of Winchilsea, President. <u>in the Chair</u>.
Earl Spencer.—
Earl of Morton.
Earl of Darnley.
Wm. Blake, Esqr.
Edward Stacey, Esqr.
Wm. Watson Esqr.
J. P. Auriol, Esqr.
Thos. Murdoch, Esqr.
John Guillemard, Esqr. <u>Secretary</u>.

House List. 1. The purpose of the meeting having been declared.—

<u>Ordered</u>. That the house Lists be filled up as follows—

President, Earl of Winchilsea & Nottingham K.G. F.

Treasurer.—Scrope Bernard Morland Esqr. M.P. L.L.D.

Secretary.—John Guillemard, Esqr. F.R.S.—

<u>Managers</u>.

J. P. Auriol, Esqr. F.R.S
Wm. Blake, Esqr. F.R.S
Earl of Darnley. F.R.S
Lord Dundas. F.R.S & F.S.A.
Charles Hatchett, Esqr. F.R.S. & F.S.A
Sir John Cox Hippisley Bart. M.P. L.L.D. F.R.S.&
Benjamin Hobhouse, Esqr. M.P. F.R.S & F.S.A
Leonard Horner, Esqr.
Alexr. Marcet M.D. F.R.S
Thomas Murdoch, Esqr. F.R.S & F.S.A
Viscount Palmerston, M.P.
Earl Spencer. K.G. F.R.S & F.S.A

Sir John Thos. Stanley, Bart. F.R.S. & F.S.A
Henry Warburton, Esqr. F.R.S.
Wm. Watson, Esqr. F.R.S.

<u>Visitors</u>.

Earl of Aylesford. F.R.S & F.S.A
Sir Thos. Bernard, Bart. L.L.D.
Lord Boringdon, F.R.S
Timothy Brent, Esqr.
Rt. Hon: Reginald Pole Carew, M.P. F.R.S & F.S.A
Earl of Chichester. F.R.S.
The Hon: Lord Bishop of Durham.
Wm. Dunlop Esqr.
Earl Grosvenor.
Thos. Hope, Esqr. F.R.S & F.S.A
Earl of Morton. F.R.S. & F.S.A.
Richd. Horseman Jolly, Esqr. F.R.S & F.S.A
Duke of Somerset. F.R.S
Lt. Genl. Wm. Thornton.
Lord Viscount Valentia. F.R.S: F.S.A & F.L.S

Winchilsea P.

294

Managers' Meeting Apl. 27, 1812 — April 27 1812

At a Meeting of the Managers of the Royal Institution of Great Britain, held at the house of the Institution, on the 27th April 1812.

Present
Earl of Winchilsea, President, in the Chair.
Earl of Morton.
Earl Spencer.
J. P. Auriol, Esqr.
Wm. Blake, Esqr.
Scrope Bernard Morland, Esqr. Treasurer.
Wm. Watson, Esqr.
Thos. Murdoch, Esqr.
Wm. Sotheby, Esqr.
John Guillemard, Esqr. Secretary.

Minutes approved 1. The minutes of the proceedings of the 2 last meetings were read and approved with the exception of the Resolution No. 8 in the minutes of the 20th April which appears to have been irregular.—

Minutes of Com. of Accts. 2. Read and approved the minutes of the Committee of Accounts of this day.

Present. Present received.

3. Mr. Wyatt presented "A proposed plan for a new Theatre."
Ordered That the thanks of the Managers be returned for this present.—

Subscribers 4. The following persons having been duly recommended were admitted as Subscribers vizt.

Annual Subscribers.
Crackanthorpe, Esqr. 15 Sackville St.
Henry Bertram Ogle, Esqr. 12 Stratton St.

Subscribers to the Lectures.
Mrs. Span of Nottingham place.
Countess of Gosford } 4 Cumberland place
Miss Milbank }

295

Miss Montgomery.— 43 upper Seymour St.
Honble. Miss Clinton, Foley place.
Miss Churchill, Grosvenor place

Expediency of arranging the Lectures before Summer Vacation

Resolved.
That it is highly expedient that the appointment of Lecturers for the next Season be taken into consideration at as early a period as possible, in order that the arrangements for that purpose may be completed before the Summer vacation.—

J P Auriol

296

Managers' Meeting 4th May, 1812 — 1812.

At a Meeting of the Managers of the Royal Institution of Great Britain, held at the house of the Institution, on the 4th of May 1812.—

Present
J. P. Auriol, Esqr. in the Chair.
Charles Hatchett, Esqr.
Lord Dundas.
Sr. Jno. Thos. Stanley, Bart.
Earl of Winchilsea.— President
Alexr. Marcet, M.D.
Earl of Morton.
Wm. Watson, Esqr.
Jno. Guillemard, Esqr. Secretary.

Minutes approved. 1. The minutes of the proceedings of the last meeting were read and approved.

Minutes of Com. of Accts approved 2. Read and approved the minutes of the Committee of Accounts of this day.—

3. Read a letter from Mr. Cullen requesting that his Brother Mr. A. D. Cullen an Annual Subscriber might be permitted to withdraw his Subscription for the present year with liberty on his return to England to renew his Subscription retaining in the mean time his name on the books—
Ordered. That he be informed that the Managers have it not in their power to grant his request.

Lectures declined Read a letter from Mr. Gregson, stiling himself interior Surveyor, purporing to deliver a course of Lectures on what he terms "Practical Philosophy"
Ordered.
That he be informed that the course proposed to be delivered by him is not within the objects of the Institution—

Present Present received.

4. Mr. Hornor presented Specimens of an improved mode of delineating Estates.—
Ordered. That the thanks of the Managers be returned

297

May 4

for this present.—

Repairs 5. Mr. Guillemard reported that he had examined the house with reference to an estimate made by Mr. Royce of necssary repairs, and was of opinion they were requisite and ought to be immediately attended to.
Ordered. That the said repairs be immediately begun and completed under the superintendence of the Secretary.—

Subscribers. 6 The following persons having been duly recommended were admitted as Subscribers vizt.—

Annual Subscribers.
J. R. Matthews, Esqr. Park Street
Jn: Blake Kirby, Esqr. 27 Gt. Portland St.

To the Lectures.
Mrs. Matthews, Park Street.
Miss Joanna Smith.
Mrs. Edwd. Stanley.

Spencer

Managers Meeting 11th May, 1812

At a Meeting of the Managers of the Royal Institution of Great Britain, held at the house of the Institution, on the 11th of May 1812.—

Present
Earl Spencer, in the Chair.
I. P. Auriol, Esqr.
Charles Hatchett, Esqr.
Wm. Blake, Esqr.
Leonard Horner, Esqr.
Alexr. Marcet, M.D.
Sir J. C. Hippisley, Bart: M.P.
Thos. Murdoch, Esqr.
John Guillemard, Esqr.—Secretary.

Minutes approved 1. The minutes of the proceedings of the last meeting were read and approved.

Minutes of Com. of Accts approved 2. Read & approved the minutes of the Committee of Accounts of this day.—

3. Read a letter from Mr. Chas. Pilgrim on the subject of his arrears as annual Subscriber.
Ordered. That he be informed that on his representation that he gave notice 2 years since of his intention to discontinue his subscription, he be considered as having ceased to be an Annual Subscriber from that time.—

4. Read a letter from Mr. Wildman soliciting the attention and approbation of the Managers to a Work entitled "The Scripture Atlas."
Ordered. That the letter and the above work be transmitted to the Patrons of the Library.—

Presents.

Presents received.

5. Mrs. Eliza Rogers presented.
"Lives of the Twelve Cæsars, 5 vol with an Atlas of Maps.
& 4 Copies of Correspondence and proceedings in the negociation for a renewal of the East India Company's Charter—presented by the Publisher.—
Ordered. That the thanks of the Managers be returned

to Mrs. E. Rogers for her present.—

Lectures. 6. Ordered. That letters be written to Dr. Smith, Dr. Roget, and Mr. Campbell, the present Lecturers on Botany, Physiology, & Poetry requesting them to inform the Managers whether they are disposed to deliver Lectures on the same subjects in the next Season.—

Commn. from Sir H. Davy declining Lectures 7. Mr. Hatchett reported that Sir Humphry Davy tho' he cannot pledge himself to deliver Lectures will be willing to accept the offices of Professor of Chemistry, & Director of the Laboratory and Mineralogical Collection, without Salary.
Resolved.
That the Managers hear with great regret the notification they have just received that Sr. H. Davy cannot pledge himself to continue the Lectures which he has been accustomed to deliver with so much honour to the Institution, and advantage to the public, but at the same time congratulate themselves on the liberal offer which Sir Humphry has made to superintend the Chemical Department and to assist and advise any Lecturer the Managers may be pleased to appoint.
Ordered.
That Sir Humphry Davy be appointed Director of the Laboratory & the Mineral Collection.—
Resolved further That the Managers cannot refrain from expressing their high sense of Sir Humphry Davy's past services and returning him their thanks.

Nomination of Sir H. Davy as Professor of Chemistry. 8. Ordered.—That a Special General Meeting of the Members be called for Thursday the 21st. Instant for the purpose of putting Sir Humphry Davy in nomination from the Chair as Professor of Chemistry.

9. Read the Minutes of the last General Monthly Meeting.

Subscribers 10. The following persons having been duly recommended were admitted as Subscribers vizt.

To the Lectures.
Rt. Honble. Lady Mary Bennet.—
Honble. Mrs. Greville Howard, Grosvenor place

Miss Davenport, 28 Lower Brook St.

Annual Subscriber.
Captn. Montagu, 8 upper Wimpole St.

except No. 8 the consideration whereof is adjourned to Thursday next (18th. May 1812)

J P Auriol

Managers' Meetg. 18th May, 1812

At a Meeting of the Managers of the Royal Institution of Great Britain, held at the house of the Institution on the 18th day of May 1812.—

Present
J. P. Auriol, Esqr. in the Chair.
Charles Hatchett, Esqr.
Leonard Horner, Esqr.
Wm. Blake, Esqr.
Lord Dundas.
Thos. Murdoch, Esqr.
Earl of Winchilsea, President.
Sr. John Thos. Stanley, Bart.
Alexr. Marcet. M.D.

Minutes approved except No. 8. 1. The minutes of the proceedings of the last meeting were read and approved with the exception of No. 8 the Confirmation of which is suspended to an adjourned meeting of the Managers to be held on Thursday next.

Minutes of Com. of Accts approved. 2. Read and approved the minutes of the Committee of Accounts of this day.—

2nd & 6th Rules & Standg orders rescinded (at Meetings of Managers. 3. Resolved. That the second & sixth of the standing Rules and Orders to be observed at the meetings of the Managers be rescinded.—

Life Subscriber 4. In pursuance of a nomination under the Act by Thomas Coutts, Esqr. his Daughter Susan, Countess Dowager of Guilford of Piccadilly widow was admitted to the privileges of a Life Subscriber.—

5. Adjourned to Thursday next at One o'Clock.—

Spencer

302 — 1812.

Adj.d Meeting of Managers May 21, 1812.

At an adjourned Meeting of the Managers of the Royal Institution of Great Britain, held at the house of the Institution, on the 21st May 1812 —

Present
Earl Spencer, in the Chair.
Wm. Blake, Esq.r
Charles Hatchett, Esq.r
Alex.r Marcet M.D.
J. P. Auriol, Esq.r
Tho.s Murdoch, Esq.
Sir John Tho.s Stanley, Bart.
John Guillemard, Esq.r Secretary.

1. Read and Confirmed the minute, N.o 8, of the Meeting held on the 11th Inst.t

Th. Murdoch

1812 — 303

Managers Meet.g 25 May, 1812.

At a Meeting of the Managers of the Royal Institution of Great Britain, held at the house of the Institution on the 25th May 1812 —

Present
Tho.s Murdoch, Esq.r in the Chair.
Cha.s Hatchett, Esq.r
Leonard Horner, Esq.r
Wm. Blake, Esq.r
Wm. Watson, Esq.r
Lord Dundas.
J. P. Auriol, Esq.r
Sir John Tho.s Stanley, Bart.
John Guillemard Esq.r Secretary.

Minutes app.ved 1. The minutes of the proceedings of the last meeting were read & approved —

Minutes of Com. of Acc.ts approved. 2. Read and approved the minutes of the Committee of Accounts of this day, & Signed an Order for payment of the following sums viz.t

S.r H. Davy, 1/4 years salary to 31st March	£100. —
M.r E. Davy — d.o — d.o	25. —
M.r Brooks Collector — d.o	13. 2. 6
M.r Harris — Librarian — d.o	45. —
Jo.s Fincher on acc.t of the Institution	60. —
	£ 243. 2. 6

Lectures 3. Read a letter from D.r Smith in answer to an application from the Secretary stating that for the present he begs to postpone the consideration of any additional course of Lectures on account of his health —

Presents

Presents received.

4. Wm. Tho.s Brande, Esq.r presented
Chemical Researches on the Blood & some other Animal Fluids —
Jos.h Palmer, Esq.r presented
A Fortnight's Ramble to the Lakes.
Rev.d James Plumptre — presented English Drama, 3 vol.s

304 — 1812.

Edward Harwood, Esq.r presented
Selecta Numismata Græca, 4to. —
John Hinckley, Esq.r presented.
Emancipation or the Complaint answered. —

Ordered.
That the thanks of the Managers be returned for these presents. —

Offer of Mr. Brand to deliver a course of Lectures 5. M.r Hatchett reported that M.r Brande is willing to deliver a course of Lectures on General Chemistry including Electro-Chemical Science.
Resolved. That his offer be accepted. —

6. Ordered on the proposal of M.r Hatchett that the sum of 100 guineas be paid to M.r Brande as a compensation for the above Lectures. —

Subscribers. 7 The following persons having been duly recommended were admitted as Subscribers viz.t
Major Munt George St. Hanover Square
Hon.ble Capt.n W. Waldegrave —

Charles Hatchett. —

The following Minute having been omitted by mistake is now inserted.

Repetition of Mr. Campbell's Lectures declined. M.r Campbell having communicated to the Board of Managers through M.r Jotheby that he was willing to read the same lectures next year as he had done this and upon the same terms —
Resolved. That M.r Campbell be informed that the Managers are apprehensive that it will not be for the benefit of the Institution that the same lectures should be repeated without variation. —

Charles Hatchett. —

305

Managers' Meet.g 1st June 1812.

At a Meeting of the Managers of the Royal Institution of Great Britain, held at the house of the Institution on the 1st of June 1812 —

Present
Charles Hatchett, Esq.r in the Chair.
J. P. Auriol, Esq.r
Wm. Blake, Esq.r
Earl of Morton.
Earl Spencer.
Sir John Tho.s Stanley, Bart.
Wm. Watson, Esq.r
Tho.s Murdoch, Esq.r
Alex.r Marcet M.D.
Jn.o Guillemard, Esq.r Secretary.

Minutes app.d 1. The minutes of the proceedings of the last meeting were read & approved.

Minutes of Com. of Acc.ts app.ved 2. Read and approved the minutes of the Committee of Accounts of this day. —

Dr Roget's Lectures 3. Read a letter from D.r Roget stating that he is willing to read a new course of Lectures on Comparative Physiology. —
Ordered
That his offer be accepted and that the Sum of One Hundred Guineas be paid to him as a Compensat.n for the above Lectures. —

Present.

Present received.

4. James Fayting Gyles, Esq.r presented
An Essay on the Authenticity of the new Testament 8vo. —
Ordered.
That the thanks of the Managers be returned for this present. —

Wm. Blake

Managers' Meeting. 8th June, 1812.

At a Meeting of the Managers of the Royal Institution of Great Britain, held at the house of the Institution, on the 8th of June 1812 –

Present
Wm. Blake, Esqr. in the Chair.
Scrope Bernard Morland, Esqr. Treasurer.
Thomas Murdoch, Esqr.
Charles Hatchett, Esqr.
Wm. Watson, Esqr.
Sir John Thos. Stanley, Bart.
Earl of Morton.
Earl of Winchilsea, President
J. P. Auriol, Esqr.
John Guillemard Esqr. Secretary.

Minutes approved 1. The minutes of the proceedings of the last meeting were read and approved.

Minutes of Com. of Accts. approved. 2. Read & approved the minutes of the Committee of Accounts of this day, & Signed an Order for payment of 75 guineas to Dr. Roget for Lectures delivered in 1812. –

Lectures 3. Read a letter from Dr. Smith stating that he hopes he shall soon be able to deliver his 8 remaining Lectures, & submitting himself entirely to the decision of the Managers.
Resolved
That Dr. Smith be respectfully informed that the Managers are of opinion that the Lectures on Botany had better be deferred till the next Season.

4. Read a letter /in answer to an application from Mr. Blake/ from Mr. Farish who declines undertaking to read himself Lectures on Mechanics & the Mechanical Arts, but offering every assistance in his power either in procuring a Competent Lecturer or advising the best mode of constructing the necessary Apparatus.

5 Ordered. That Mr. Blake be requested to make a similar

application to Professor Dealtry of the East India Company's College at Haileybury –

Mr. Campbell engaged for a 2nd course of Lectures on Poetry. 6. Mr. Campbell having communicated to the Managers thro' Sir John Stanley, & Mr. Blake that there was a mistake in the report made by Mr. Sotheby on the subject of his Lectures, And that Mr. Campbell is willing to read a new course on Poetry next Season, and desires that it may be understood that he never intended to read the same lectures without variation.
Ordered. That a letter be written to him informing him that the Managers accept his offer with pleasure, and request if it is not inconvenient that he will give a part of the Course before Easter. –

Presents

Presents received.

7 I D'Israeli, Esqr. presented
Calamities of Authors; including some Inquiries respecting their moral & literary Characters, 2 vol., 8vo.
& William Fitton M.D. presented.
Notes on the Mineralogy of part of the Vicinity of Dublin, principally from Papers of the late Revd. Walter Stephens M. A.
Ordered. That the thanks of the Managers be returned for these presents –

– 8. Read the minutes of the last General Monthly Meeting. –

Committees. 9. Ordered – That the following Committees be requested to meet in the ensuing week for the election of two Chairmen, and a Secretary for each Committee –

Mathematics, Mechanics, &c. on Monday at One o'Clock
Chemistry, Geology, &c. —— Tuesday – do.
General Science, &c. —— Wednesday – do.

Lectures 10. Ordered. That the Earl Spencer be requested to make an application to Mr. Shee to inquire whether he is willing to read a course of Lectures on Painting in the next Season. –

11. Ordered. That letters be written to Ld. George Cavendish

the Executors of Sir Francis Baring, & Messrs. Carr, Smith & Pepys to remind them that the Instalments of their Subscriptions still remain unpaid, and that the Managers are thereby prevented from settling the balance of the Loan Account.

Spencer

Managers' Meetg. 15th June, 1812.

At a Meeting of the Managers of the Royal Institution of Great Britain, held at the house of the Institution, on the 15th of June 1812 –

Present
Earl Spencer, in the Chair.
Chas. Hatchett, Esqr.
Leonard Horner, Esqr.
J. P. Auriol, Esqr.
Benjn. Hobhouse, Esqr. – M.P. –
Thos. Murdoch, Esqr.

Minutes appd. 1. The minutes of the proceedings of the last meeting were read and approved –

Minutes of Com. of accts. appd. 2. Read & approved the minutes of the Committee of Accounts of this day. –

3. Read a reference from the Committee of Visitors entered in their proceedings of the 11th Inst.

Appointment of W. H. Pepys, Esq. to superintend apparatus. 4. Wm. Hasledine Pepys, Esqr. a Member of the Royal Institution having offered his services to superintend the Apparatus belonging to the Institution.
Resolved. That William Hasledine Pepys Esqr. be appointed honorary inspector of the Apparatus & Models belonging to the Institution. –
Ordered. That Mr. Newman the person employed for cleaning & repairing the Apparatus be placed under the direction of Mr. Pepys – And that Mr. Pepys be requested to report to the Managers what is necessary to be done for putting in order such part of the Apparatus as may immediately want it with an estimate of the expense. –

Inventory of furniture ordered. Ordered. That Mr. Harris be desired to proceed with & complete the Inventory of Furniture with as little delay as possible. –
Ordered.
That the Floor-Cloths be removed according to the recommendation of the Committee of Visitors.
Ordered. That the Carpet and Blinds in the small Library be repaired under the direction of the Housekeeper.

Ordered. That the Clerk procure six small Chairs for the small Library.

Compensation refused 5 Read the following communication from Sir Robt. Ainslie Bart.

Sir Robert Ainslie Bart. requests that the Managers will permit him to give up his Hereditary interest in the Institution for the compensation of £42.0.10 as allowed by the Act of Parliament.

The application of Sr. Robt. Ainslie having been taken into consideration, Resolved. That it cannot be complied with, as no compensation can be given to any Proprietor, "unless such Proprietor shall within 3 Calendar Months after the passing of the Act by writing under his hand signify that he will not agree to relinquish and give up his hereditary & saleable share or property in the said Institution." which Act passed into a Law on the 18th. of April 1810. See Clause 2 of page 6 of the printed Act.

Ordered. That a Copy of this Minute be transmitted to Sr. Robt. Ainslie.

6 Earl Spencer reported that he had seen Mr. Shee in consequence of the Resolution at the last meeting And that Mr. Shee had desired him to express his thanks for the honor done him but that he was under the necessity of declining to deliver a course of Lectures at the Institution.

Annl. Subsr. 7. The following person having been duly recommended was admitted an Annual Subscriber viz:

Mr. Engstrom 48 Curzon St.

W Watson

Managers' Meetg. 22nd June 1812.

At a Meeting of the Managers of the Royal Institution of Great Britain, held at the house of the Institution on the 22d. of June 1812.

Present

Wm. Watson, Esqr. in the Chair.
Sir J. C. Hippisley, Bart.
J. P. Auriol, Esqr.
Earl Spencer.
Earl of Winchilsea, President.
Thos. Murdoch, Esqr.

Minutes approved. 1. The minutes of the proceedings of the last meeting were read and approved.

Minutes of Com. of Accts. approved. 2. Read and approved the minutes of the Committee of Accounts of this day, & Signed an Order for payment of the following sums, viz:

T. Campbell for Lectures in 1812. £105.
Josh. Fincher on acct. of the Instit.n 60.

Letter from Mr. Horner stating Mr. Pepys's agreement to inspect apparatus &c. 3. Read the following letter from Leonard Horner, Esqr.

Sir,

As it will not be in my power to attend the Board of Managers this day I have to report that in compliance with the Resolution passed at the last Board, I waited on Mr. Pepys, who with great readiness has undertaken to inspect the Apparatus of the Institution & to see it put in order.

He will take an early opportunity of informing the Managers what parts of it require immediate attention, & he will at the same time state the probable expense of repairing them.

I have the honor to be
Sir,
Your very obedt. Servt.
(signed) Leonard Horner.

To, The Chairman
of the Board of Managers.

Present received.

Present 4. His Grace the Archbishop of Canterbury presented A Catalogue of the Manuscripts in the Library of Lambeth Palace.

Ordered. That the thanks of the Managers be returned for this present.

Chairmen &c. of Committees 5 Josh. Fincher reported that the following Committees met on the 16th. & 17th. Inst. and made the followg. elections viz:

Chemistry, Geology, & Mineralogy.

Sr. John Saunders Sebright Bart. Chairman.
Wm. Hyde Wollaston, M.D. Chairman
Wm. Thos. Brande Esqr. Secretary.

General Science, Literature & the Arts.

Sir Frederick Baker Bart. Chairman.
John Disney Esqr. Chairman
John Hinckley, Esqr. Secretary.

Benjamin Hobhouse

Managers' Meeting, 29th June 1812

At a Meeting of the Managers of the Royal Institution of Great Britain, held at the house of the Institution, on the 29th of June 1812.

Present

Benjn. Hobhouse, Esqr. M.P. in the Chair.
Thos. Murdoch, Esqr.
J. P. Auriol, Esqr.
Leonard Horner, Esqr.
Earl Spencer.
Jn. Guillemard, Esqr. Secretary.

Minutes confd. 1. The minutes of the proceedings of the last meeting were read and approved.

Minutes of Com. of Accts. approved. 2. Read and approved the minutes of the Committee of Accounts of this day.

Presents received.

Present. 3 Mr. Harris reported that he had received from Messrs. Burley & Moore the Executors of The Revd. Mr. Dutens 89 Vols. of Miscellaneous Tracts.

The Revd. Edward Forster presented.

The British Gallery of Engravings, No. 10.

Sir Frederick Baker Bart. exchanged the Copy of Virgil presented by him in January last, for one more splendidly bound with proof impressions of the plates.

Ordered. That the thanks of the Managers be returned for these presents.

Mr. Flaxman engaged to lecture. 4. Earl Spencer reported that Mr. Flaxman is willing to deliver one or more Lectures on the Arts connected with Sculpture.

Ordered. That his offer be thankfully accepted.

Annl. Subsr. 5 The following person having been duly recommended was admitted an Annual Subscriber.

Haughton James, Esqr. of Piccadilly.

Thos. Murdoch

Managers Meeting, 6th July, 1812.

At a Meeting of the Managers of the Royal Institution of Great Britain, held at the house of the Institution, on the 6th of July 1812.—

Present
Thos. Murdoch, Esqr. in the Chair.
Alexr. Marcet. M.D.
John Guillemard, Esqr. Secretary.

Minutes appd. 1. The minutes of the proceedings of the last meeting were read and approved.

Minutes of Comee. of Accts. approved. 2. Read and approved the minutes of the Committee of Accounts of this day, & Signed an Order for payment of the following sum, vizt.

Mr. J. Sturt for 1 quars Salary to 30th June £20

Sir H. Englefield's Time-keeper &c. returned. 3. Sir Henry C. Englefield Bart. having requested by letter that his Time-keeper & Transit Instrument may be returned.—
Ordered. That they be delivered to whomsoever Sir H. Englefield may appoint.—

Presents

Presents received.

4 Sir Fredk. Baker, Bart. presented.
Anacreon Gr. from the Vatican M.S. with paintings, folio, superbly bound.
The Conductors of the British Review, presented
The British Review, No. 6.
Sir J. C. Hippisley Bart. presented
His Speech on the Catholic Question.
Ordered.
That the thanks of the Managers be returned for these presents.—

5. On the representation of Sr. H. Davy that the following persons may be usefully employed.
Ordered. That Mr. Edmund Davy complete the Catalogue of the Minerals and Mr. Moore be permitted to make
WW experiments in the Laboratory provided they do not put the Institutn.

to expense And that Wm. Payne be employed in cleaning & repairing the Apparatus in conjunction with Mr. Newman under the superintendence of Mr. Pepys.—

6 Mr. Harris having requested permission to have the Garret No. 26 for his Niece.—
Ordered. That his request be granted on condition that no inconvenience be found to arise therefrom.—

appointment of Mr. Harris as superintendent of the house. 7. Ordered That Mr. Harris be appointed Superintendant of the House till further orders, And that it be his duty to see that every thing is kept in a regular and proper state, and the Servants of the Establishment are hereby directed to attend to and obey his instructions.—

W Watson

Managers' Meeting 13th July. 1812.

At a Meeting of the Managers of the Royal Institution of Great Britain held at the house of the Institution on the 13th of July 1812.—

Present
Wm. Watson, Esqr. in the Chair.
Leonard Horner, Esqr.
Thos. Murdoch, Esqr.

Minutes approved. 1. The minutes of the proceedings of the last meeting were read and approved.

Minutes of Com. of Accts. approved. 2. Read and approved the minutes of the Committee of Accounts And Signed an Order for payment of the following sum.
1/4 years assessed taxes to 25th. March last £69. 14. 11.

3. Read a letter from Mr. Sturt Assistant Librarian on the subject of an increase of his Salary.
Ordered.
That the consideration of it be postponed for the present.—

Morton C. P.

Managers' Meetg. 20th July, 1812.

At a Meeting of the Managers of the Royal Institution of Great Britain held at the house of the Institution on the 20th. of July 1812.

Present
Earl of Morton, in the Chair.
Scrope Bernard Morland, Esqr. M.P. Treasurer.
John Guillemard, Esqr. Secretary.—

Minutes appd. 1. The minutes of the proceedings of the last meeting were read and approved.

Minutes of Com. of Accts. appd. 2. Read and approved the minutes of the Committee of Accounts of this day.

Presents

Presents received.

3. The Secretary, presented
Theorie des Peines & Recompenses.
Benjn. Hobhouse, Esqr. presented
A number of Volumes of Journals of the House of Commons to complete the set now in the Library of the Institution.—
Ordered. That the thanks of the Managers be returned for these presents.

Life Subscriber. 4 In pursuance of a nomination under the Act by The Revd. Anthony Hamilton D.D. his Son William Hamilton Esqr. of Dean St. Park Lane was admitted to the privileges of a Life Subscriber to the Royal Institution

W Watson

Managers' Meeting, 3rd Aug. 1812.

At a Meeting of the Managers of the Royal Institution of Great Britain, held at the house of the Institution, on the 3rd of August 1812.—

Present
Wm. Watson, Esqr. in the Chair.
Wm. Blake, Esqr.
Jas. Peter Auriol, Esqr.
Scrope Bernard Morland, Esqr. Treasurer.
Jno. Guillemard, Esqr. Secretary.

Minutes appd. 1. The minutes of the proceedings of the last meeting were read & approved.

Minutes of Com. of Accts. appd. 2. Read & approved the minutes of the Committee of Accounts of this day, & Signed an Order for an advance to the Clerk of £70.

Present 3. Present received.
G. W. Meadley, Esqr. presented,
Mr. Wyvill's Papers on Toleration.
Ordered. That the thanks of the Managers be returned for this present.—

Papers discontinued 4. Ordered. That the duplicates of the Times, Morning Chronicle, & Morning Post be discontinued until the 1st of January next.—

5. Wm. Payne being in possession of the Garret No. 26. & Mr. Harris having requested to have the use of the Room No. 15 instead of the said Garret.
Ordered. That his request be granted till the 1st of January next, subject however in the mean time to any arrangements the Managers may find it expedient to make respecting Committee Rooms.—

Increase of Clerks' Salary 6. Resolved. That the Salary of the Clerk be increased to the sum of £135 to be reckoned from 30th June last.—

7. Resolved. That Mr. Edmund Davy be directed to state

to the Managers at their next Meeting the progress he has made in the Catalogue of the Mineralogical Collection, what number of classes remain to complete the Catalogue, and when it may be expected to be finished.

10. Resolved further that Mr. E. Davy be directed to prepare an Inventory of the Chemical Utensils preparations and other Articles in the Laboratory And Ordered that the Room be cleansed preparatory to an arrangement of the said Articles in proper order under the direction of Mr. Davy.—

Adjourned to the first Monday in September.—

W. Watson

Managers' Meeting 7th Sept. 1812

At a Meeting of the Managers of the Royal Institution of Great Britain, held at the house of the Institution, on the 7th Septr. 1812.—

Present
Wm. Watson, Esqr. in the Chair.
Wm. Blake, Esq.
Jno. Guillemard, Esqr. Secretary.

Minutes appd. 1. The minutes of the proceedings of the last meeting were read and approved.

Minutes of Com. of Accts. 2. Read and approved the minutes of the Committee of Accounts of the 10th 17th 24th & 31st Ulto. and of this day, and Signed an Order for an advance to the Clerk of £50.—

Presents 3. Presents received.
Sir Humphry Davy, presented,
Elements of Chemical Philosophy, vol. 1.
Wm. Adams, Esqr. presented,
Observations on the Diseases of the Eye.
Francis Maseres, Esqr. Cursitor Baron of the Exchequer presented,
May's History of the Long Parliament, and General Ludlows Letters,—
Ordered. That the thanks of the Managers be returned for these presents.—

4. Read the minutes of the last General Monthly Meeting.—

Report from Mr. Pepys. 5. Read a report from Wm. Hasledine Pepys, Esqr. on the Instruments & Apparatus in the Institution.
Ordered. That the thanks of the Managers be returned to Mr. Pepys for the present report and that his liberal offer to proceed to inspect the Instruments & Apparatus of the Laboratory be gratefully accepted.
Ordered. That the Instruments mentioned in the report of Mr. Pepys as requiring repair be compleated under his direction.—

6. Mr. Blake reported that he had received a letter from Professor Dealtry expressing his high sense of the honour done him in the proposal of the Managers but stating that the duties of his Professorship at the East India College render it impossible for him to deliver any Lectures in London.—

Privileges of Ambassadors. 7. Ordered. That the same privileges be allowed to the Count Fernando Nunez the present Spanish Ambassador that were granted to his predecessor the Duke del Infantado And that His Excellency be acquainted therewith through Mr. Manning.—

Mr. Brande permitted to make experiments in Laboratory previous to his course of Lectures 8. Mr. Wm. Thos. Brande having made application thro' Mr. Blake for permission to make experiments in the Laboratory of the Institution preparatory to his course of Lectures.—
Ordered. That his request be granted.

Report of Mr. E. Davy concerning Minerals 9. Mr. Edmund Davy in consequence of the resolution of the Managers at their last Meeting presented the following report.—

Mr. Edmund Davy has the honor to inform the Managers of the state of the Mineralogical Collection belonging to the Royal Institution.—

He has found it necessary to have all the Minerals in the Museum cleaned, which has accordingly been done.

With the exception of the inflamable & saline compounds he has nearly finished the Catalogue of the Mineralogical arrangement, including all the common metalic combinations to which are added the genera, Barium, Strontium, & Calcium, &c. The Geological part of the Collection is arranged nearly in the order in which it will appear in the Catalogue.— The fossils which contain organic remains including petrifactions are not yet arranged.— The Minerals which are deficient in the Museum have been noticed and a Catalogue of them accompanies the present report.—

Mr. E. Davy begs to state to the Managers, that there are in the Collection many duplicate specimens, and many of comparatively little value; these might be exchanged for such as are wanting, or be sold.— He conceives it will take several Months to render the Catalogue complete.—

The Catalogue would have been in a more advanced state, if Mr. E. Davy's time had not been occupied in repeating the experiments illustrative of the important discoveries made in the Institution. These experiments were made for the information of His Excellency the Count de Funchal at the request of the Professor of Chemistry. —

Resolved. That there being nothing now to impede his progress Mr. Edmd. Davy be directed to dedicate the whole of his time to the completion of the Catalogue.

Leave of absence refused.

Mr. E. Davy having further requested to be excused from the duties of his station for one Month.

Resolved. That this request cannot be acceded to, till the Catalogue of the Minerals is completed. —

Wm. Payne having requested leave of absence for a fortnight.

Ordered. That he be informed that it will be his duty together with Mr. Newman to assist Mr. Pepys in the inspection & arrangement of the Laboratory And that when that is completed his request of absence for a fortnight be acceded to —

Adjourned to Monday the 26th. October

Charles Hatchett.

Managers' Meeting 26th. Oct. 1812.

At a Meeting of the Managers of the Royal Institution of Great Britain held at the house of the Institution, on the 26th. October 1812.

Present
Charles Hatchett, Esqr. in the Chair.
Wm. Blake, Esq.
Alexr. Marcet, M.D.
John Guillemard, Esqr. Secretary.

Minutes appved. 1. The minutes of the proceedings of the last meeting were read and approved. —

Minutes of Com. of Accts. appved. 2. Read and approved the minutes of the Committee of Accounts of the 14th., 21st. & 28th. Ulto. and also of the 5th. 12th. 19th. & 26th. Instant: — and signed an order for payment of the following sums. —

Mr. Sturt 1 quars. Saly. to 30th. Septr. last	£20. – –
Messrs. Dickins & Co. for Coals in 1812	64. 13. 4
	£84. 13. 4

Mr. Campbell apologizes for not fulfilling his engagement. 3. Read a letter transmitted by Mr. Sotheby from Mr. Campbell stating that it is with infinite regret he finds himself precluded by ill health from being able to lecture next Season at the Royal Institution —

Ordered. That a letter be written to Mr. Campbell expressive of the regret of the Managers that such a disappointment should have been occasioned by his ill health and of their hope that he may still be able to resume his engagement.

Presents.

Presents received.

4. Francis Maseres Esqr. Cursitor Baron of the Exchequer presented Sr. John Temple's History of the Irish Rebellion, 4to.
Charles Norris Esqr. presented Etchings of Tenby, 4to.
Ordered.
That the thanks of the Managers be returned for these presents —

5. Ordered. That Twenty Chaldrons of Coals be laid in for the use of the Royal Institution. —

Wm. Blake

Managers' Meeting 2nd Nov. 1812

At a Meeting of the Managers of the Royal Institution of Great Britain, held at the house of the Institution on the 2nd. of November 1812 —

Present
Wm. Blake, Esqr. in the Chair.
Scrope Bernard Morland, Esqr. Treasurer.
John Guillemard, Esqr. Secy.

Minutes appved. 1. The minutes of the proceedings of the last meeting were read and approved.

Minutes of Com. of Accts. appved. 2. Read and approved the minutes of the Committee of Accounts of this day, and Signed an Order for payment of the following sum. —

Mr. Harris for 1 quars. Saly. to 30th June £45. —

Present.

Present received.

3. Gilbert Mathison, Esqr. presented.
Kirby's Perspective of Architecture, folio — &
Cantemir's History of the Othman Empire, folio.
Ordered. That the thanks of the Managers be returned for this present. —

4. Ordered. That Mr. Royce be desired to prepare an estimate of the expense of laying down a temporary floor in the Model Room. —

5. Ordered. That Mr. Feetham be desired to prepare an estimate of the expense of putting the Steam Apparatus in order for the purpose of warming the Lecture Room. —

6. Mr. Blake reported that he had received a letter from the Revd. Mr. Hudson of Trinity College Cambridge requesting to know what remuneration would be given to a Lecturer on Natural Philosophy. —
Ordered. That Mr. Blake be authorized to offer a remuneration of 100 guineas for a course of from 12 to 20 Lectures. —

Benjamin Hobhouse

Managers' Meetg. 9th. Nov. 1812.

At a Meeting of the Managers of the Royal Institution of Great Britain, held at the house of the Institution on the 9th. of November 1812. —

Present
Sir Benjn. Hobhouse, Bart. M.P. in the Chair.
Wm. Blake, Esq.
Scrope Bernard Morland, Esqr. Treasurer.
Alexr. Marcet M.D.
Thos. Murdoch, Esqr.

Minutes appved. 1. The minutes of the proceedings of the last meeting were read and approved. — Signed an Order for an advance to the Clerk of £50.

Minutes of Com. of Accts. appved. 2. Read and approved the minutes of the Committee of Accounts of this day.

Flooring of Model Room — £24 3. Ordered. That Mr. Royce proceed to lay down a deal floor in the Model Room agreeable to his estimate at an expense not exceeding £24. —

4. Read the minutes of the last General Monthly Meeting.

5. Mr. Hinckley the Secy. of the Committee of General Science having expressed a wish that notices be sent round to the Members of the Meetings of the Scientific Committees.
Ordered.
That Mr. Edmund Davy as Clerk of the Scientific Committees be desired to send Notices accordingly.

W. Watson

326

Managers Meeting. 16th Nov. 1812.

1812.

At a Meeting of the Managers of the Royal Institution of Great Britain held at the house of the Institution on the 16th of November 1812. —

Present

Wm Watson, Esqr. in the Chair.
Leonard Horner, Esqr.
Thos. Murdoch, Esqr.
J. P. Auriol, Esqr.
Wm Blake, Esqr.

Minutes appved. 1. The minutes of the proceedings of the last Meeting were read and approved. —

Minutes of Com. of Accts. appved. 2. Read and approved the minutes of the Committee of Accounts of this day.

Letter from Mr. Campbell announcing his recovery. 3. Read the following letter from Mr Thos. Campbell

Sydenham 11th Novr. 1812.

Sir

Six weeks ago I wrote to Mr. Sotheby to apprize the Managers of the Royal Institution that being in exceedingly bad health I dreaded the prospect of lecturing next Winter. — It gave me great pain to give up so honorable an engagement but I thought it to be my duty to provide in time against any chance of the Institution being disappointed by my failure.

Contrary to my expectations I have recovered very speedily and if the Royal Institution be not provided with a lecturer on the same subject I shall still feel proud and happy to resume the engagement.

An answer from the Managers on this subject will greatly oblige your very humble Servt.

Thos. Campbell

To the Secy. of the Royal Institution.

Ordered.

That Mr. Fincher be desired to write the following answer to Mr. Campbell. —

Board of Managers 16th Novr. 1812.

Sir

I am directed by the Managers to express their

327

Nov. 16.

sincere pleasure at the recovery of your health and that the Institution will have the advantage of your course of lectures in the ensuing Season. — The Managers further direct me to say that they will feel obliged by your mentioning at what period it will be most convenient for you to commence your Lectures that they may be enabled to announce them to the Public. —

4 Read a letter from Mr. Edmund Davy on the subject of his attendance as Clerk of the Scientific Committees.

Ordered. —

That this letter be taken into consideration on Monday the 30th Inst. and that special notices be sent round to the Managers accordingly. —

On Nomination of a Life Subscriber 5. Read a letter from The Honble. Robt. Clifford M.R.I. on the subject of the late Lieut. F. G. Simcoe's nomination of a Life Subscriber. —

Ordered. That the same be submitted to the next General Monthly Meeting and that a letter be written to The Honble. Robt. Clifford acquainting him therewith. —

Warming Lecture Room — £7. 6. Ordered. That Mr Feetham be directed forthwith to put the Steam Apparatus for warming the Lecture Room in proper order at an expense not exceeding £7. —

Report on State of Laboratory 7. Read a report from Mr. Pepys on the state of the Laboratory. —

Ordered. That the consideration of it be adjourned until a communication shall have been had with Mr. Brande the Lecturer on Chemistry on the subject. —

Scrope Bernard Morland

328

Managers' Meeting 23rd Nov. 1812

1812.

At a Meeting of the Managers of the Royal Institution of Great Britain held at the house of the Institution on the 23rd Novr. 1812. —

Present

Scrope Bernard Morland, Esqr. in the Chair.
William Blake, Esqr.
Thomas Murdoch, Esqr.
Alexr. Marcet M.D.

Minutes appved. 1. The minutes of the proceedings of the last meeting were read and approved. —

Minutes of Com. of Accts. appved. 2 Read and approved the minutes of the Committee of Accounts of this day. —

Time of Mr. Campbell's Lectures. 3 Read the following letter from Mr. Campbell.

Sir

Sydenham Novr. 21st 1812 —

I beg through you respectfully to inform the honorable Managers of the Institution that for many reasons I should desire to defer my lectures till the beginning of April — for one reason in particular that I may have the honor of a crowded audience — By taking April & May I conceive I shall have the best chance of the west-end of the Town being full. —

I remain Your obliged & obedt. Servt.

Thos. Campbell

To the Secy. of the Royal Institution

Ordered.

That the following answer be written to Mr. Campbell. —

Board of Managers
Novr. 23d. 1812.

The Managers have desired me to acknowledge the receipt of your letter of the 21st Inst. and to state that they shall be happy to accede to the lectures being given at any period most agreeable to yourself they have desired me however to add that they conceive it would be very much for the benefit of the Institution if you could make it

329

Nov. 23.

convenient to give one or two lectures before the Easter recess.

4. Mr Brande having communicated with the Managers respecting Mr. Pepys's report.

Ordered.

That the further consideration of it be postponed to Monday the 30th Inst.

5. Ordered. That the thanks of the Managers be returned to Mr. Pepys for his report on the Instruments and Apparatus. —

Mr Powell engaged as Lecturer. 6. Mr. Blake reported that he had received a letter from Mr. Hudson of Trinity College, Cambridge strongly recommending The Revd. John Powell Fellow of Trinity College Cambridge as a Lecturer on Natural Philosophy.

Resolved. —

That Mr. Blake be authorized to conclude the arrangement with Mr. Powell and to offer him apartments in the Institution. —

Wm. Blake

Managers Meet.g 30 Nov. 1812.

At a Meeting of the Managers of the Royal Institution of Great Britain, held at the house of the Institution on the 30th of November 1812 –

Present
William Blake, Esq.r in the Chair.
Charles Hatchett, Esq.r
Leonard Horner, Esq.r
Thomas Murdoch, Esq.r
Alex.r Marcet, M.D.

Minutes app.ved 1. The minutes of the proceedings of the last meeting were read and approved.

Minutes of Com.e of Acc.ts app.ved 2. Read and approved the minutes of the Committee of Accounts of this day.

Present 3. The Conductors of the British Review presented
N.o 7 of the British Review.
Ordered.
That the thanks of the Managers be returned for this present.

A Communication from Mr. Brande and order thereon 4. Mr. Brande having communicated with the Managers on the subject of the Apparatus mentioned in Mr. Pepys report and shewn them a list of the Articles necessary for the Lectures.
Resolved.
That Mr. Brande be requested to procure the Articles specified in his list in the most economical manner –

Newspapers. 5. Ordered. That duplicates of the Times, Chronicle & Morning Post be continued during the session of Parliament. –

Adjourned to Thursday next.

J.P. Auriol

Managers' Meeting, Dec. 3. 1812

At an adjourned Meeting of the Managers of the Royal Institution of Great Britain, held at the house of the Institution on Thursday the 3rd of December 1812 –

Present
W.m Blake, Esq.r in the Chair.
Tho.s Murdoch, Esq.r
Alex.r Marcet M.D.
Scrope Bernard Morland Esq.r Treasurer.
Sir Benjamin Hobhouse, Bart. M.P.
Leonard Horner, Esq.r

An adjournment having taken place on Monday the 30th of November in consequence of there being only four Members left to transact the business and it being thought that the subject before the Board was of sufficient importance to require a fuller attendance of Managers And two of the Managers of the present meeting having been obliged to retire in consequence of appointment elsewhere so as to leave the Board in precisely the same circumstances as at the former meeting.
Resolved.
That this Meeting be further adjourned to Monday the 7th Instant.

J.P. Auriol

Managers Meeting 7th Dec 1812.

At a Meeting of the Managers of the Royal Institution of Great Britain held at the house of the Institution on the 7th of December 1812 –

Present
James Peter Auriol, Esq.r in the Chair.
Charles Hatchett, Esq.r
W.m Watson, Esq.r
Tho.s Murdoch, Esq.r
Leonard Horner, Esq.r
Alex.r Marcet, M.D.
S. Bernard Morland, Esq.r Treasurer.
W.m Blake, Esq.r

Minutes app.ved 1. The minutes of the proceedings of the 2 last meetings were read and approved.

Minutes of Com.e of Acc.ts app.ved 2. Read and approved the minutes of the Committee of Accounts of this day. –

Lectures. 3. Sir Humphry Davy having communicated a proposal from The Rev.d Edward Forster for a course of Lectures on Oratory.
Resolved. That as the Managers have already engaged six Lectures for the ensuing Season which they conceive to be as many as can be employed with advantage to the Institution, they cannot under the circumstances avail themselves of Mr. Forster's offer And they request Sir Humphry Davy to have the goodness to communicate this answer to Mr. Forster. –

4. Mr. Edmund Davy's letter having been taken into consideration. –
Resolved. That in consequence of the changes that have taken place since the last season Mr. Edmund Davy's services will no longer be principally required in the Laboratory And that his duties in future will be to attend as Clerk on the Scientific Committees to arrange the Mineral Collection and make the Catalogue and give his assistance to the various Lecturers in the preparation of their Lectures and in the Lecture Room. –

Resignation of Mr. E. Davy

Mr. Edmund Davy having been called in and the detail of the above duties having been explained to him by the Chairman he immediately expressed his wish to resign and begged that his declaration might be considered as a formal resignation.
Resolved. That Mr. Edmunds Davy's resignation be accepted and that his salary be paid up to the end of the Current quarter. –

Hours of Lectures 5. Resolved.
That the hour of lectures during the month of January be at 2 o'Clock and during the remainder of the Season at 3 o'Clock. –

Life Subscriber. 6. The following person having been duly recommended was admitted as a Life Subscriber viz.t
Daniel Alban Durtnall, Esq.r
N.o 1. Bishops Place, Queens Elm Brompton.

Confirmed with the exception of No 3. Mr Forster's Offer having been Misunderstood } Leonard Horner

Managers' Meeting 14th Dec 1812

1812.

At a Meeting of the Managers of the Royal Institution of Great Britain held at the house of the Institution on the 14th December 1812 –

Present
Leonard Horner, Esqr. in the Chair.
William Blake, Esqr.
Jas. Peter Auriol, Esqr.
Thos. Murdoch, Esqr.
Alexr. Marcet, M.D.

Minutes appvd. with exception of No. 3. — 1. The minutes of the proceedings of the last meeting were read and approved with the exception of No. 3 –

Resolved.

Mr. Forster's gratuitous Lectures on Oratory acceptd. — That the proposal of The Revd. Edward Forster respecting a course of lectures on ancient oratory having been misunderstood and it having been stated to the Managers that it is his intention to deliver them gratuitously.

Ordered. – That his offer be accepted and that the thanks of the Managers be sent to him for his liberal proposition. –

Minutes of Comee. of Accts. approved — 2. Read and approved the minutes of the Committee of Accounts of this day, and Signed an Order for payment of the following sums. –

	£
Mr. Edmund Davy 3/4 year's salary to 31st Decr. 1812 –	75. – –
1/2 years Kings taxes to Michas 1812	104. 2. 9
Josh. Fincher on acc. of the Royal Institution	50. – –
	£229. 2. 9

3. Read minutes of the last General Monthly Meeting.

4. Mr. Powell having requested thro' Mr. Blake to know whether he could be accomodated with an Apartment in the Institution for his servant.

Resolved. That the accomodation be granted.

Dec 14.

Furniture ordered — 5. Ordered. That a Table, Carpet, & 12 Chairs be purchased by the Clerk for the Committee Room.

Subscribers. — 6. The following persons having been duly recommended were admitted as Subscribers vizt.

To the Lectures.
Mrs. W. Parry, Montagu Square.

Annual Subscribers.
Lieut Col. C. Edwards, 10 Little Stanhope St.
Revd. E. Lipscomb, 19 Michaels Place Brompton
Revd. Dr. Knox. 1. Adam Street Adelphi.
Francis Horner, Esqr. 7 New Square, Lincolns Inn.
Chas. Augustus Tulk Esq. Marble hall Twickenham
Sir James Duberly, Albemarle Street.
Sir Francis Milman Bart. 47 Lower Brook St.
John Horner, Esqr. 6 Russell Square
Captn. Andree, 10 London St. Fitzroy Square
Maxwell Constable, Esqr. Everingham near York.

W Watson

Managers' Meeting 21st Dec 1812.

1812.

At a Meeting of the Managers of the Royal Institution, held at the house of the Institution on the 21st of Decr. 1812.

Present
Wm. Watson, Esqr. in the Chair.
J. P. Auriol, Esqr.
Wm. Blake, Esqr.
Leonard Horner, Esqr.

Minutes approved. 1. The minutes of the proceedings of the last meeting were read and approved.

Minutes of Com. of Accts. appvd. — 2. Read and approved the minutes of the Committee of Accounts of this day.

Subscribers. — 3. The following persons having been duly recommended were admitted as Subscribers vizt.

To the Lectures.
Honble. Mrs. Bruce 8 upper Montagu St.
Miss H. Dashwood 56 Montagu Square

Annual Subscribers.
John Fleming, Esqr. 104 Gloster Place
Thos. Halford Esqr. Piccadilly –
Chas. Kar, M.D. 10 Sloane St.
Alexr. Baillie Esqr. 20 Albany
Josh. Hayes Lyon, Esqr. 14 Paper Build gs. Temple
Jno. Hodgson, Esqr. 35 Soho Square.

W Watson

Managers' Minutes 28th Dec. 1812.

At a Meeting of the Managers of the Royal Institution of Great Britain, held at the house of the Institution on the 28th of December 1812 –

Present
Wm. Watson, Esqr. in the Chair.
Wm. Blake, Esqr.
Leonard Horner, Esqr.
Thos. Murdoch, Esqr.

Minutes appvd. 1. The minutes of the proceedings of the last meeting were read and approved.

Minutes of Com. of Acc. appvd. 2. Read and approved the minutes of the Committee of Accounts of this day.

Present. — Present received.

Richard Duppa Esqr. M.R.I. presented Illustrations of the Lotus of Antiquity. 4to.

Ordered. That the thanks of the Managers be returned for this present.

Leonard Horner

338

Managers' Minutes Jan 4. 1813.

1813.

At a Meeting of the Managers of the Royal Institution of Great Britain held at the house of the Institution on the 4th of January 1813.—

Present
Leonard Horner, Esqr. in the Chair.
Wm. Watson, Esqr.
Alexr. Marcet, M.D.
Thos. Murdoch, Esqr.
Wm. Blake, Esqr.

Minutes appvd. 1. The minutes of the proceedings of the last meeting were read and approved.—

Minutes of Com. of Accts. appved 2. Read and approved the minutes of the Committee of Accounts of this day.— and Signed an Order for payment of the following sums.

Mr. Josh. Fincher on acct. of the Royal Institution	£50 " —
Mr. Sturt ¼ year's salary to 31st. Decr. 1812	20 " —
	£70 " —

Presents. Presents received.—

3. Mr. James Tyson presented,
Historical view of the causes of the decline of the Commerce of Nations, 8vo.
Society for bettering the condition of the Poor presented their 33rd. Report.—
Ordered. That the thanks of the Managers be returned for these presents.—

Subscribers 4. The following persons having been duly recommended were admitted as Annual Subscribers, vizt.

Pelham Hood, Esqr. Beaumont St. Portland place
Wm. Greenwood, Esqr. Cary St. Lincolns Inn
Jno. Allan Esqr.
A. C. Koosen, Esqr. 17 York Place Portman Sqr.
Alexr. Fraser, Esqr. 6 Arlington Street
Dr. Mac Michael, 18 Bedford place.

339

Jan. 4.

5. Mr. Buller, M.R.I having sent a specimen of Ore to ascertain the proportion of Manganese contained in it.— Ordered. That it be referred to Sir Humphry Davy.

6. Resolved. That the hour of 3 o'Clock being found inconvenient for Members of both houses of Parliament the Lectures be continued during the whole of the season to be given as heretofore at Two o'Clock.—

Thos. Murdoch

340

Managers' Meeting Jan. 11. 1813.

1813

At a Meeting of the Managers of the Royal Institution of Great Britain held at the house of the Institution on the 11th of January 1813.—

Present
Thomas Murdoch, Esqr. in the Chair.
Wm. Blake, Esqr.
J. P. Auriol, Esqr.
Alexr. Marcet M.D.

Minutes appd. 1. The minutes of the proceedings of the last meeting were read and approved.

Minutes of Com. of Accts. approved 2. Read and approved the minutes of the Committee of Accounts of this day.

Subscribers 3. The following persons having been duly recommended were admitted as Subscribers vizt.

To the Lectures.
Marchioness of Stafford
Lady Charlotte Leveson Gower.
Mrs. Hall Devonshire St.
Mrs. Murdoch Portland Place
Mrs. Morris } Wandsworth Surry.
Miss Morris }
Miss C. Cubitt Charlotte row

Annual Subscribers.
Sr. Robt. Heron Bart. M.P.
Mr. John Murray Albemarle St.
David Hall, Esqr. Devonshire St.
Chas. Morris Esqr. Wandsworth
Dr. John Bright Cleveland row

Benjamin Hobhouse

341

~~Minutes of~~ Managers' Meeting 18 Jan. 1813

At a Meeting of the Managers of the Royal Institution of Great Britain, held at the house of the Institution on the 18th of January 1813—

Present
Sir Benjn. Hobhouse, Bart. in the Chair.
Chas. Hatchett, Esqr.
Wm. Blake, Esqr.
Alexr. Marcet, M.D.
Scrope Bernard Morland, Esqr. Treasurer.

Minutes appvd 1. The minutes of the proceedings of the last meeting were read and approved.

Minutes of Com. of Accts. appvd 2. Read and approved the minutes of the Committee of Accounts of this day.

Lectures 3. Read a letter from Mr. Thos. Campbell, stating that he flattered himself he should be able to deliver 2 Lectures before the middle of April next.

4 Ordered. That an abstract of Mr. Payne's duties be prepared and submitted to the Managers at their next meeting.

Assistt. in Labortry Salary raised 5. Resolved. That Mr. Wm. Payne's salary be increased from £1 to £1. 5. 0 pr. Week to commence from 1st. Jany. And if at the end of 1 year from that period he shall be found to have given satisfaction in his new capacity the Managers will take into consideration the further advance of his Salary.

Recommendation from Visitors. 6. Read the following recommendation from the Committee of Visitors.

The Visitors recommend that in the small Library the two extreme Lamps over each table be removed. And that two Lamps of the form of the centre Lamp be provided and placed over each table which alteration will save the consumption of the Oil of two Lamps and the light will be more extensively thrown upon the tables.—

Resolved. — That this request be complied with.

Subscribers — The following persons having been duly recommended were admitted as Subscribers vizt.

To the Lectures.

Viscountess Bernard.
Lady Nicholl.
Miss Nicholl.
Miss I. Nicholl.
Mrs. Wigram, Albemarle Street.
Miss Hayes No. 1 Lower Seymour St.
Mrs. Martin 22 Russell Square.

Annual Subscribers.

Earl Gower — Cleveland house.
Col. Adams. South Audley St.
Richd. Parry Welbeck St.

Charles Hatchett —

Managers' Meetg. Jan. 25th. 1813.

At a Meeting of the Managers of the Royal Institution of Great Britain held at the house of the Institution on the 25th. January 1813. —

Present
Chas. Hatchett, Esqr. in the Chair.
Wm. Watson, Esqr.
Wm. Blake, Esqr.
S. Bernard Morland, Esqr. Treasurer.

Minutes appvd. 1. The minutes of the proceedings of the last meeting were read and approved.

Minutes of Com. of Accts. appvd. 2. Read and approved the minutes of the Committee of Accounts of this day. And signed an Order for an advance to the Clerk of £50. —

Mr Payne's duties 3. Mr. Wm. Payne's Duty.
Resolved.
That it is the duty of Mr. Wm. Payne to attend and assist the Lecturers and Professors in preparing for and during their Lectures where any Instruments or Apparatus may be required. — To attend to their careful removal from the Model-room and Laboratory to the Lecture room and to clean and replace them after being used, reporting to the Managers such accidents as shall require repair. A constant diary being kept by him for that purpose. — That on one day in each week he be employed in keeping clean the models in the Repository, And that all the Instruments in the Glass cases be cleaned and dusted at least once within the month. —

Subscribers 4. The following persons having been duly recommended were admitted as Subscribers vizt.

To the Lectures.

Miss Smith, 23 Leicester Square
Miss Bignell, Russell Square
Mrs. Holland. } Upper Brook St.
Miss Hollands }

Mrs. Dickinson } Devonshire place
Miss Dickinson }
Mrs. Chevalier } South Audley St.
Miss Chevalier }
Mrs. Jn. Hankey } Lincolns Inn fields
Miss Hankey }
Mrs. C. Nevinson } Somerset St.
Miss Moody. }
Mrs. Protheroe, 39 Harley St.
Mrs. Barclay — } 56 Conduit St.
Miss E. B. Barclay }
Miss M. B. Barclay }
Miss Legh, — Bedford Square.
Mrs. Blake, Portland Place.
Mrs. Gordon — } 15 Devonshire St.
Miss Dalrymple }
Mrs. Bernard Morland.
Miss Bernard Morland.
Miss Maryanne B. Morland.
Mrs. Benjn. Wyatt, Foley Place
Miss Mary Weston 16 St. Albans St.

Annual Subscribers.

John Baines, Esqr. Bolton Row
James South, Esqr. High St. Southwark.
Mr. Wm. Cornwell, South Audley St.
Samuel Omay, 26 Bury St.
Earl of Leitrim, Gt. Cumberland place
Mr. Savory, New Bond St.
John Legh, Esqr. Bedford Square
Lt. Col. Robt. Barclay Conduit St.
William Hill, Esqr. Seale St.
Wm. Clark, Esqr. Craven St.

J P Auriol

Managers' Meetg. 1st Feb. 1813

At a Meeting of the Managers of the Royal Institution of Great Britain, held at the house of the Institution on the 1st. of February 1813. —

Present
J. P. Auriol, Esqr. in the Chair.
Wm. Blake, Esqr.
Leonard Horner, Esqr.
Alexr. Marcet, M.D.
Thos. Murdoch, Esqr.
Chas. Hatchett, Esqr.
Sir Jno. Thos. Stanley Bart.

Minutes appvd. 1. The minutes of the proceedings of the last meeting were read and approved.

Minutes of Com. of Accts. appvd. 2. Read and approved the minutes of the Committee of Accounts of this day.

Lectures 3. Read a letter from Dr. Roget requesting permission for a Room to put the preparations and drawings in which he shall want for his lectures.
Resolved. That the Room lately occupied by Mr. Edmd. Davy be allotted to him for that purpose. —

Renewal of Lease 22 Albemarle Street 4. Read a letter from Messrs. Mills & Trower requesting to obtain from the Managers of the Royal Institution a renewal of the Lease of No. 22 Albemarle Street in favor of Mr. James Spooner the Nephew of the late Mr. Swinton on payment of a Fine of £240 with Interest from the period of renewal in Sept. 1809. —
Resolved.
That the Board are disposed to grant a renewal of the Lease of No. 22 on payment of the Fine due with interest and arrears. —
Ordered. That a Copy of this Resolution be sent to Mr. Seymour the Solicitor to the Royal Institution for that purpose. —

5. Present received.

346

Presents — Revd. Peter Gandolphy, presented.

Liturgy, or a Book of Common Prayers.

A Letter to the Revd. Herbert Marsh.

Ordered.

That the thanks of the Managers be returned for this present.–

6. Mr. Hatchett having stated a request on the part of the Lecturers.

Lecturers permitted to introduce by tickets 4 persons to his own, and (by personal introduction) one to other Lectures.

Resolved. That each Lecturer be allowed 4 Tickets of admission to each of his own Lectures, and permission to introduce one person to any other Lecture during the Season provided the said Lecturer be present himself to introduce them.–

Subscribers 7. The following persons having been duly recommended were admitted as Subscribers vizt.

To the Lectures.

Mrs. Stodart } Russell Square
Miss Stodart }
Mrs. Flaxman, Buckingham St.
Honble Miss Brodrick Grosvenor Place
Countess of Selkirk Grosvenor St.
Rt. Honble Lady Cathne. Douglas.
Mrs. Nesbitt. Sloane St.
Mrs. Hodge, Craven St. Strand
Mrs. Peters. } 56 Park St.
Miss G. Peters }
Lady Troubridge, Albemarle St.
Mrs. Robertson } Lincolns Inn fields
Miss Robertson }
Miss Margaret Agnew, Russell Sq.
Mrs. Roberts } Upper Grosvenor St.
Miss Wake }
Mrs. Achard } Bloomsbury Square
Miss Achard }

Annual Subscribers.

Mr. Plato Petrides.–
Revd. Chas. Este, St. Georges row Hyde Park.
Robt. Forster, Esqr. Suffolk St.

1813.

347

Feb. 1.

Henry Sutton, Esqr. Southampton row.
Adam Gordon, Esqr.
Robt. Gordon, Esqr. M.P. 23 Bruton St.

Spencer

348

1813.

Managers Minutes 8th Feb. 1813

At a Meeting of the Managers of the Royal Institution of Great Britain, held at the House of the Institution on the 8th of February 1813.

Present

Earl Spencer, in the Chair.
Wm. Watson, Esqr.
J.P. Auriol, Esqr.
Chas. Hatchett, Esqr.
Earl of Winchilsea, President.
Thos. Murdoch, Esqr.
Wm. Blake, Esqr.
Sir Jn. Thos. Stanley, Bart.
Earl of Morton.

Minutes appved. 1. Read and approved the minutes of the proceedings of the last meeting.

Minutes of Com. of Accts. approved. 2. Read and approved the Minutes of the Committee of Accounts of this day, and Signed an Order for payment of the following sums, viz!

Assessed taxes to Xmas 1812 —	£110 . 12 . 7
Mr. Newman Sundrs. for the Laby	28 . 2 . 10
Do. — — Glass Acct. for 1812.	27 . 4 . 4
	£165 . 19 . 9

Request from Mr. E. Davy not acceded to. 3. Read a letter from Mr. Edmund Davy requesting of the Managers the favor of remaining in the House a short time longer, to enable him to prepare courses of Lectures on Chemistry & Geology.

Resolved.

That the Managers having appropriated the Room hitherto occupied by Mr. Edmund Davy to another purp[ose] That Mr. Fincher be directed to inform him that his request cannot be complied with, And that the Managers expect that he will make his arrangements for quitting the House of the Institution on as early a day as may be convenient in the present week.

Presents received.

349

Feb. 8.

Presents 4. The Society for the encouragement of Arts &c. presented,

Their Transactions, vol. 30.

The Proprietors of the New Review, presented.

The New Review, No. 2.–

Ordered.

That the thanks of the Managers be returned for these presents.–

5. Read the Minutes of the last General Monthly Meeting.–

Vice Presidents 6 Lord Winchilsea reported that he had named the following Vice Presidents.

Earl Spencer.
Earl of Morton.
Earl of Darnley.
Lord Dundas.

Subscribers. 7. The following persons having been duly recommended were admitted as Subscribers vizt.

To the Lectures.

Mrs. Jodrel, – Portugal St. Grosr. Square
Rt. Honble. Lady Elizth. Thynne.
Mrs. Anthy. Thomson, 92 Sloane St.
Miss Mills, } Portman Square.
Miss S Miss }
Miss Dinwiddie
Mrs. Shaw Lefevre
Miss E. Hamilton } 19 Orchard St. Portman Sq.
Miss F. Hamilton }

Annual Subscribers.

Peter Boissier Esqr. Middle Temple.
Jas. Agnew, Russell Sq.
Edward Protheroe Esqr. M.P. } Harley St.
Mr. Edwd. Protheroe junr. }
Wm. Porden, Esqr. Berners St. –
The Mudeleer Balthazar de Sarem.
John Walker, junr. Esqr. – Bedford Sq.
Dr. Jams. Emlen, Southampton Builds.
Lewis Way, Esqr. Grosr. place.

Spencer

Managers' Meeting 15th Feb. 1813.

At a Meeting of the Managers of the Royal Institution of Great Britain, held at the house of the Institution on the 15th February 1813.—

Present

Earl Spencer V.P. in the Chair.
Chas. Hatchett, Esqr.
Lord Dundas, V.P.
Wm. Blake, Esqr.—
Alexr. Marcet, M.D.
Sir Jn. Thos. Stanley, Bart
Thos. Murdoch, Esqr.

Minutes appved. 1. The Minutes of the proceedings of the last meeting were read and approved.—

Minutes of Com. of Acct. appved. 2. Read and approved the minutes of the Committee of Accounts of this day.—

3. Sir Humphry Davy having stated to the Managers that £300 remained due to him on account of his salary to the 31st Decr. last.
Ordered.
That it be paid and signed an Order accordingly.

4. Resolved. That the bills of the last year be referred to the Committee of Accounts for the purpose of considering what part of them ought to be discharged.

5. Ordered. That 20 Chaldrons of Coals be laid in for the use of the Royal Institution.—

Present 6. Present received.
W. J. Tilford M.D. presented
Sketches towards a Hortus Botanicus Americanus, coloured plates 4to.—
Ordered
That the thanks of the Managers be returned for this present—

Subscribers 7. The following persons having been duly recommended were admitted as Subscribers viz.

To the Lectures.
Miss Tighe, St. James's St.
The Honble. Miss Somerville } Hill Street
Miss H. Somerville }
Lady Sitwell, New Norfolk St.—

Annual Subscribers
James Wentworth Murray, Esqr. Sloane St.
Stamp Brooksbank, Esqr. Lower Brook St.
Revd. J. H. Banks, Park row, Knightsbridge

8. Signed an Order for payment of the following sums.

Sr. Humphry Davy 3/4 yrs Salary to 31st Decr. 1812	£300. – –
Messrs. Dickins & Co. Coals in 1812	64. 13. 4
Josh. Fincher on acct. of the Royal Institution	50. – –
	£414. 13. 4

Winchilsea P.

Managers' Meeting 22nd Feb. 1813.

At a Meeting of the Managers of the Royal Institution of Great Britain, held at the house of the Institution, on the 22d of February 1813.—

Present

Earl of Winchilsea, President in the Chair.
Lord Dundas V.P.
Leonard Horner, Esqr.
Wm. Watson, Esqr.
Chas. Hatchett, Esqr.
Alexr. Marcet, M.D.
J. P. Auriol, Esqr.
Thos. Murdoch, Esqr.
Earl Spencer, V.P.

Minutes confd. 1. The minutes of the proceedings of the last meeting were read and approved.

Minutes of Com. of Accts. appved. 2. Read and approved the minutes of the Committee of Accounts of this day, & Signed an Order for payment of the following sums, vizt.

Mr. Butler Parochial rates from Jany. to 25th Decr. 1812	£31. 5. 0
Mr. Harris 1/2 years salary to 31st Decr. 1812	90. – –
Mr. Brooks Collr. 3/4 yrs do. to do.	39. 7. 6
	£180. 12. 6

Resignation of Mr Guillemard 3. Read a letter from John Guillemard, Esqr. to Sir H. Davy begging him to state to the Managers that he shall be obliged to resign his Secretaryship at the next Annual Meeting in May.—

Read a letter from Mr. Sturt Assistant Librarian on the subject of an increase of Salary.
Resolved
That Mr. Sturt's letter be referred to the Patrons of the Library.

4. Mr. Hatchett communicated a request from Mr. Powell

Lectures that after Tuesday 23rd Inst. he be permitted to give his lectures on the alternate weeks.
Resolved. That this request be complied with.

5. Josh. Fincher reported that The Revd. Edward Forster would commence his lectures on or after the 14th April.—

Dismissal of Mr Payne 6. Mr. Harris the Superintendant of the House reported that on Friday Evening last hearing a great noise in the Lecture-room he went to see the cause of it when he found Mr. Payne, & Mr. Newman at high words and Newman complained of having been struck by Payne for representing to him his neglect of duty in being about when he should have attended on Mr. Brande.—

Resolved
That Mr. Wm. Payne be dismissed from The Royal Institution and that a gratuity of £10 be paid him in consideration of his long services.

Subscribers. 7. The following persons having been duly recommended were admitted as Subscribers vizt.

To the Lectures.
Miss Elizth. Forster, Bentinck St.
Mrs. James Warre.
Lady Charlotte Scott, Portman Square.
Lady Drake, Nottingham St.
Mrs. Dixon, Portland Place

Annual Subscribers.
Edward Dawson, Esqr. Lancaster
Chas. Mayne Young, Esqr. Duchess St.
Robt. Hawthorn, Esqr. Gower St.
Wm. Rashleigh, Esqr. Sackville St.

Charles Hatchett.—

354 — 1813.

Managers' Meetg March 1. 1813

At a Meeting of the Managers of the Royal Institution of Great Britain held at the house of the Institution on the 1st March 1813 —

Present
Charles Hatchett, Esqr. in the Chair.
Wm. Watson, Esqr.
Thos. Murdoch, Esqr.
Sir Benjn. Hobhouse, Bart. M.P.
J. P. Auriol, Esqr.
Earl Spencer, V.P.
Wm. Blake, Esqr.
Earl of Morton, V.P.
Scrope Bernard Morland, Esqr. M.P. Treasurer

Minutes appoved 1. The minutes of the proceedings of the last meeting were read and approved.

Minutes of Com. of Accts approved 2. Read and approved the minutes of the Committee of Accounts of this day, and Signed an Order for payment of the following sums.

Mr. Howis Sundries for the house	£19 . 9 . 8
Mr. Searle Bookbinding in 1812	9 . 4 . 6
De Boffe for Books in 1812	40 . 11 . 6
Mr. Bulmer for printing in 1812	110 . 16 —
Messrs. Dalton & Keys Cards in 1812	39 . 4 . 3
Mr. Wrangham — Newspapers in 1812	39 . 2 . 8
Do. — Stationary — do.	30 . 15 . 5
Mr. Henry — Charcoal — do.	7 . 19 —
Mr. Thompson for Soap & Candles	48 . 14 —
G. Hammond Linen-drapery	10 . 8 . 2
Mr. Gardner Oil in 1812	117 . 5 . 7
	473 . 10 . 9

Presents. 3. Presents received.
Mr. John Fearn, presented,
An Essay, on Consciousness, or a series of evidences of a distinct mind, 4to. —

1813 Mar. 1 — 355

Mr. John Stevenson presented.
A Practical Treatise on Cataract 8vo.
Papers from the Protestant Union
& Voltaire's Massacre of St. Bartholomew,
By the Translator.
Ordered. That the thanks of the Managers be returned for those presents.

Assistant Librn's Salary increased 4. Mr. Sturt's letter having been referred to the Patrons of the Library, Joseph Fincher the Clerk of the Library reported that the Committee of Patrons had directed him to lay before the Managers the following Resolution "The Patrons recommend to the Committee of Managers that the Salary of Mr. Sturt, the Assistant Librarian be increased from £80 to £90 ℘ Ann: in consideration of his long services"
Ordered. That Mr. Sturt's Salary be increased to £90 ℘ annum to commence from 31st. Decr. last.

Appointment of Michael Faraday 5. Sir Humphry Davy has the honor to inform the Managers that he has found a person who is desirous to occupy the situation in the Institution lately filled by Wm. Payne His name is Michael Faraday He is a youth of 22 years of age; As far as Sir H. Davy has been able to observe or ascertain he appears well fitted for the situation His habits seem good, his disposition active and chearful and his manner intelligent He is willing to engage himself on the same terms as those given to Wm. Payne at the time of his quitting the Institution.

Resolved. That Michael Faraday be engaged to fill the situation lately occupied by Mr. Payne on the same terms, And that the Clerk furnish him with a Copy of the Order relating to his duties accordingly.

6. Mr. Hatchett stated that he had been desired by the Earl of Winchilsea to acquaint the Managers that Ld. Palmerston has accepted the office of a Vice President.

Mr. Brande's Lectures 7. Mr. Hatchett reported that it is the intention of Mr. Brande to extend the number of his Lectures from ten to eleven or twelve —

356 — 1813.

Increase of Clerks' Salary to £180. 8. Resolved.
That the Salary of Joseph Fincher, Clerk to the Institution be increased to £180 ℘ ann: to commence from the 31st Decr. last in consideration of his faithful services. —

Subscribers. The following persons having been duly recommended were admitted as Subscribers viz:

To the Lectures.
Mrs. Aubin, — } 13 Bulstrode St.
Miss Mary Louisa Ross } Manchester Sq.

Annual Subscribers.
George Tritton Esq. of East Hill Wandsworth.
Benjn. Lester Lester, Esqr. M.P. Bury St. —
Henry Weymouth, Esqr. — Wandsworth.
Christr. Spurrier, Esqr. Gordon's Hotel.
Robt. Wm. Newman, Esqr. M.P. Sackville St.
Honble Henry Verney, Hill Street.
Hugh Miles Bunbury, Esqr. Percy Hotel.

Ordered.
That rolling Canvass blinds be purchased for the Committee Room. —

Dundas

357

Managers' Meetg 8th Mar. 1813.

At a Meeting of the Managers of the Royal Institution of Great Britain held at the house of the Institution on the 8th of March 1813. —

Present
Lord Dundas, V.P. in the Chair.
Earl Spencer, V.P.
Wm. Watson, Esqr.
Chas. Hatchett, Esqr.
Wm. Blake, Esqr.
J. P. Auriol, Esqr.
Thos. Murdoch, Esqr.
Alexr. Marcet, M.D.

Minutes appoved 1. The minutes of the proceedings of the last meeting were read and approved.

Minutes of Com. of Accts approved 2. Read and approved the minutes of the Committee of Accounts of this day, and Signed an Order for an advance to the Clerk of £50. —

Ticket Lost. 3. Read a letter from Robt. Hunter, Esqr. an Annual Subscriber requesting that the Managers would grant an Order for admission to Miss E. Barclay to the Lectures she having lost her ticket. —
Ordered.
That Miss E. Barclay be personally admitted during the remainder of the Season, And that the Door-keeper be directed to stop the ticket. —

Mr. Campbell's Lectures Read a letter from Mr. Thos. Campbell, stating that he should deliver his last Lecture on Saturday the 10th. of April. —
Ordered. That a letter be written to Mr. Campbell requesting to know on what day he proposes to deliver his first Lecture. —

Presents 4. Presents received.
The Revd. Robert Hall presented
" An address to the Public on the renewal of the Charter of the East India Company. —

358 1813

2 Copies of Debates at the Courts of East India Proprietors.
Mr. Granville Sharp, presented.
"The Protestant Union No. 6, & 7. &c. &c.

Ordered.

That the thanks of the Managers be returned for these presents.—

5. Read the minutes of the last General Monthly Meeting—

Life Subscriber 6. In pursuance of a nomination under the Act by Henry Goodwyn, Esqr. his Son Thomas Wildman Goodwyn of Blackheath was admitted to the privileges of a Life Subscriber to the Royal Institution.—

Wm. Payne's dismissal. 7. Wm. Harris as Superintendant of the House begged to know if Wm. Payne continued in the House with the concurrence of the Managers.

Ordered.

That in pursuance of the former Order of the Board of Managers Wm. Payne be desired to quit the house of the Institution.

8. The following persons having been duly recommended were admitted as Subscribers, vizt.

To the Lectures.
Miss E. Bignell, Russell Square
Miss Manningham, Queen Ann St.

Life Subscriber.
John Engelberts Liebenrood, Esqr. 7 Holles St.

Annual Subscriber.
Willm. Jope, Esqr. Scotland yard.—

Winchilsea

359

Managers Minutes 15th March, 1813.

At a Meeting of the Managers of the Royal Institution of Great Britain, held at the house of the Institution on the 15th of March 1813.—

Present
Earl of Winchilsea, President. in the Chair.
J. P. Auriol, Esqr.
Chas. Hatchett, Esqr.
Wm. Watson, Esqr.
Earl of Morton, V.P.

Minutes appved. 1. The minutes of the proceedings of the last meeting were read and approved.

Minutes of Com. of Acct. appved. 2. Read and approved the minutes of the Committee of Accounts of this day.—

Mr. Campbell's Lectures 3. Read a letter from Mr. Thos. Campbell stating that he meant first instead of last Lecture.

Ordered.

That Mr. Campbell be informed that it will be perfectly convenient to the Managers that he should give his first Lecture on Saturday the 10th. of April as he proposes.—

Mr. Flaxman engaged to lecture on Sculpture 4. Read a letter from Earl Spencer stating that Mr. Flaxman will be ready to begin his Lectures the middle of next week; and he proposes to read only two Lectures—the one on the Art of Sculpture as practiced in England before the Establishment of the Royal Academy; the other on the Sculpture of the Antients in its best times; both these lectures he will illustrate by drawings. His Lordship took the liberty of proposing to the Managers that Mr. Flaxman's lectures should be fixed as follows—the first on Tuesday March 30th. & the second on Tuesday April 13th. being the weeks when Mr. Powell will have no Lecture.

Resolved.

That Mr. Flaxman's first lecture be given on Tuesday March 30th. and the second Lecture on Tuesday April 13th. and that the Clerk be directed to inform Lord Spencer that the Managers are much indebted to His

360 1813.

Lordship for this communication.—

5. Resolved.

That Mr. Seymour be requested to attend the Board of Managers on Monday next.—

6. Mr. Hatchett reported that the Pipe which conveys water to the Cistern in the Room adjoining the Lecture room and the Pipe conveying the water from the Cistern are both stopped.—

Ordered.

That the Plumber examine and repair the same.—

Life Subscriber. 7. In pursuance of a nomination under the Act by Christopher Idle, Esqr. his Son George Idle, Esqr. of the Adelphi Terrace was admitted to the privileges of a Life Subscriber to the Royal Institution.

Subscribers. 8. The following persons having been duly recommended were admitted as Subscribers, vizt.

To the Lectures.
Miss Susan Mordaunt. Hereford St.
Miss Johnson.
Miss Cure. } 37 Great George St. Westmr.
Miss E. Cure.
Miss Harriet Holmes, Mansfield St.

Annual Subscribers.
J. S. B. Morritt, Esqr. Portland Place.
Richd. Mountney Jepson, Esqr. upper Berkley St.
Wm. Pym, Esqr. Gower St.—

Spencer

361

Managers Meeting 22nd March. 1813

At a Meeting of the Managers of the Royal Institution of Great Britain, held at the house of the Institution on the 22d. March 1813.—

Present
Earl Spencer, V.P. in the Chair.
Sir J. C. Hippisley, Bart.
Wm. Watson Esqr.
Scrope Bernard Morland, Esqr. Treasurer.

Minutes appved. 1. The minutes of the proceedings of the last meeting were read and approved.

Minutes of Com. of Acctt. appved. 2. Read and approved the minutes of the Committee of Accounts of this day.—

Resignation of Mr. Guillemard as Secretary. 3. Read the following letter from John Guillemard, Esqr. communicated by the Earl of Winchilsea.—

Mr. Guillemard has the honor of presenting his respects to the Managers of the Royal Institution and of requesting permission to resign his office of Secretary, from which he is now sensible, he ought to have retired as soon as he found it expedient to pass the winter in Cornwall. With most grateful acknowledgments of the civilities and friendly attentions which he has received at the hands of every Member of the Society and particularly from the late and present Board of Managers and the Professor of Chemistry, he announces with regret that he has no longer any hope to make up by future diligence and zeal for his past deficiencies, and scarcely considers himself as entitled on any account to ask the favour of being permitted to withdraw himself without public censure from a service which he is conscious he has performed very ill.—He does however ask it and relies much on the candour and indulgence of the Society.

Resolved.

That Ld. Winchilsea be requested to return the grateful thanks of the Managers to Mr. Guillemard for his valuable services rendered to the Institution and to express the sincere regret of the Managers at the loss of

so excellent an Officer. –

Present **Present received.**

4. James Smithson, Esq.r M.R.I presented.

A Paper on a substance called Ulmin – 2 Cop.s –

Ordered.

That the thanks of the Managers be returned for this present. –

Open Comm.e of Chemistry 5. The Chairman of the Committee of Chemistry having thro' Sir H. Davy expressed a desire that an open Committee of Chemistry be held in the ensuing Week.

Resolved.

That the following notice be hung up in the Hall of the Institution and be inserted on the Cards.

A Committee of Chemistry open to all Members and Gentlemen personally introduced by Members will be held in the Theatre of the Royal Institution on Wednesday the 31st Instant at 3 o'Clock when the Professor of Chemistry will demonstrate a new series of facts on the fluoric principle.

6. The following persons having been duly recommended were admitted as Subscribers, viz.t

Life Subscriber. Life Subscriber.

Viscount Clive – Berkeley Square.

Annual Subscriber.

Scrope Davies, Esq.r Limmers Hotel.

Spencer.

Managers' Meet.g 29 March 1813

At a Meeting of the Managers of the Royal Institution of Great Britain, held at the house of the Institution on the 29th of March 1813 –

Present

Earl Spencer, V.P. in the Chair.
Cha.s Hatchett, Esq.r
Tho.s Murdoch, Esq.r
W.m Blake, Esq.r
Earl of Winchilsea, President.
Alex.r Marcet, M.D.

Minutes app.ved 1. The minutes of the proceedings of the last meeting were read and approved.

Minutes of Com. of Acc.ts 2. Read and approved of the minutes of the Committee of Accounts of this day, And Signed an Order for payment of the following sums.

1 Years Ground rent to the City of London to 25th March 1813	£135. 1. 8
Jos.h Fincher, on account of the Royal Institution	60. –. –
	£195. 1. 8

3. Ordered. –

That a ticket of admission be sent to His Excellency Prince Koslovsky according to Art:16. Chap VI. of the Bye Laws –

Subscribers. 4. The following persons having been duly recommended were admitted as Subscribers viz.t

To the Lectures.

M.rs Stonar – Green St. Grosvenor Square
Miss Curran, 47 upper Norton St.
M.rs Liebenrood 7. Holles St.

Annual Subscribers.

Francis Morgan, Esq.r Somerset Street.
Lieut. Col. Stonar. Green Street, Grosv.r Square.

George R. Philips, Esq.r Mount Street
Viscount Lowther – Pall Mall.
Tully Higgins, Esq.r Limmers Hotel.
M.r Timothy Bramah – Eaton Street.

Winchilsea

Managers' Meeting 5th Apl. 1813.

At a Meeting of the Managers of the Royal Institution of Great Britain, held at the house of the Institution on the 5th of April 1813 –

Present

Earl of Winchilsea, President, in the Chair
Earl Spencer. V.P.
Lord Dundas, V.P.
Leonard Horner, Esq.r
Charles Hatchett, Esq.r
J.P. Auriol, Esq.r
W.m Blake, Esq.r
Earl of Darnley, V.P.
Scrope Bernard Morland, Esq.r Treasurer.
Tho.s Murdoch, Esq.r
Sir Benj.n Hobhouse, Bart.
Alex.r Marcet, M.D.

Minutes app.ved 1. The minutes of the proceedings of the last meeting were read and approved.

Minutes of Com. of Acc.ts app.ved 2. Read and approved the minutes of the Committee of Accounts of this day And Signed an Order for payment of the following sum. –

M.r Newman for clean.g & repairing Apparatus in 1812 under the direction of M.r Pepys	£39. 11. 0

Mr. Brande's Lectures. 3. M.r Hatchett has stated to M.r Brande the desire of the Managers that he would deliver a short course of 4 or 5 Lectures after Easter, and M.r Hatchett now reports that M.r Brande will at that time deliver four Lectures on Vegetable Chemistry, which if possible he would wish to give on Saturdays as usual, but if this should be found inconvenient on account of any previous arrangement, then M.r Brande will deliver his Lectures at 2 o'Clock on Thursdays. –

Resolved.

That the days for these Lectures shall be taken into consideration at the next meeting.

366

1813.

4 The following persons having been duly recommended were admitted as Subscribers vizt.

To the Lectures.

Miss Baker Hill Street

Mrs Philips 111 Mount Street

Annual Subscribers.

Major Gilbert Cooper Bury St.

Revd. Jn. Cook, 20 Charles St.

Revd George Doyly, Lambeth Palace

Spencer.

367

Managers' Meeting 12th April 1813

At a Meeting of the Managers of the Royal Institution of Great Britain, held at the house of the Institution on the 12th of April 1813.

Present

Earl Spencer, in the Chair.

J. P. Auriol, Esqr.

Wm Blake, Esqr.

Thos. Murdoch, Esqr.

Earl of Morton.

Alexr. Marcet, M.D.

John Guillemard, Esqr. Secretary.

Minutes appved 1. The minutes of the proceedings of the last meeting were read and approved.

Minutes of Com. of Accts appved 2. Read and approved the minutes of the Committee of Accounts of this day.—

Presents 3. Presents received.

Debates at the General Court of Proprietors of East India Stock on the 17th & 23d. of February.

A. Letter to the Earl of Buckinghamshire, 2 cops. of each.

Dr. Young presented.

An introduction to Medical Literature, 8vo

Ordered. That the thanks of the Managers be returned for this present.—

Subscribers. 4. The following persons having been duly recommended were admitted as Subscribers vizt.

To the Lectures.

Honble Miss Lucy Cust. 30 Hill Street.

Mrs Baker, of Hill St.

Mrs Lance — Nottingham Place.

Annual Subscribers.

Sir James Riddell, Bart. Albany.

J. P. Merier, Esqr. Wimpole Street.

Winchilsea

368.

1813.

Managers' Meeting Apl 19, 1813.

At a Meeting of the Managers of the Royal Institution of Great Britain, held at the house of the Institution on the 19th of April 1813.—

Present

Earl of Winchilsea, in the Chair.

Leonard Horner, Esqr.

Alexr. Marcet, M.D.

Wm Blake, Esqr.

J. P. Auriol, Esqr.

Scrope Bernard Morland, Esqr. Treasurer.

Thos. Murdoch, Esqr.

Earl Spencer.

John Guillemard, Esqr. Secretary.

Minutes appved 1. The minutes of the proceedings of the last meeting were read and approved.

Minutes of Com. of Accts appved 2. Read and approved the minutes of the Committee of Accounts of this day, and Signed an Order for an advance to the Clerk of £60.

Present. Present received.

3. William Tighe, Esqr. presented

Statistical Observations relative to the County of Kilkenny.—

Ordered. That the thanks of the Managers be returned for this present.

Mr Flaxman's lectures gratuitous. 4. Mr. Fincher having informed the Board that Mr. Flaxman declines receiving any Compensation for his Lectures on Sculpture.

Resolved. That the thanks of the Managers be returned to him for this mark of his liberality.

Resolved further.

Life Subscriber. That John Flaxman, Esqr. be elected a Life Subscriber of the Royal Institution.—

Ordered. That the sum of 35 guineas be paid to the Treasurer on account of Mr Flaxman in lieu of his Composition of Life Subscriber & Signed an Order accordingly

369

Apl. 19. 5. The following person having been duly recommended was admitted as an Annual Subscriber vizt.—

Earl Nugent of Pall Mall.

House List 6. Ordered. That the House Lists be filled up as follows.

President, Earl Spencer. K.G.

Treasurer. Scrope Bernard Morland, Esqr. M.P. L.L.D.

T. Harrison, Secy. Secretary. Thomas Harrison Esqr. F.R.S.

Managers.

J. P. Auriol, Esqr. F.R.S

Wm Blake, Esqr. F.R.S

Sir H. Davy, LLD FRS, &c.

George Duckett, Esqr. FRS.

John Guillemard, Esqr. F.R.S

Charles Hatchett, Esqr. F.R.S.

Leonard Horner, Esqr.

Marquis of Lansdowne F.R.S.

Alexander Marcet, M.D. F.R.S

Thos. Murdoch, Esqr. F.R.S &c.

Earl of Selkirk, F.R.S.

Wm Sotheby, Esqr. F.R.S.

John Walker, Esqr. F.R.S. &c.

Wm Watson, Esqr. F.R.S.

Earl of Winchilsea & Nottingham K.G. FRS &c.

Visitors.

George Aust, Esqr. F.R.S.

Duke of Bedford F.S.A.

Timothy Brent, Esqr.

Wm Dunlop, Esqr.

Earl of Hardwicke K.G. F.R.S. &c. &c.

Daniel Moore, Esqr. F.R.S. &c.

Earl of Morton, K.T. F.R.S &c.

Claude Scott, Esqr.

Sir Jn. Saunders Sebright, Bart. M.P.

Richd. Horseman Solly, Esqr. F.R.S. &c.

Duke of Somerset F.R.S.

Lord Teignmouth,
Lt. General William Thornton, M.P.
Samuel Thornton, Esq.r F.A.S.
Wm. Hyde Wollaston, M.D. Sec: R.S.

Benj.n Hobhouse

Managers' Meet.g 26 Apl. 1813.

At a Meeting of the Managers of the Royal Institution of Great Britain, held at the house of the Institution on the 26th April 1813.—

Present

Sir Benj.n Hobhouse, Bart. in the Chair.
Wm. Watson, Esq.r
Charles Hatchett, Esq.r
Thos. Murdoch, Esq.r
Alex.r Marcet, M.D.
James Peter Auriol, Esq.r
John Guillemard, Esq.r Secretary.

Minutes app.ved 1. The minutes of the proceedings of the last meeting were read and approved.—

Minutes of Com. of Acc.ts app.ved 2. Read and approved the minutes of the Committee of Accounts of this day.

Presents received.

Present. 3. Abstract of the Minutes of Evidence before the house of Commons on the affairs of the East India Company, 2 copies.
Debate at a Court of Proprietors of East India Stock, on Lord Castlereagh's propositions, 2 copies.
G. Dyer, Esq.r presented four letters on the English Constitution.

Ordered.

That the thanks of the Managers be returned for this present.

Irreg.y in rec.ts Foreign Periodicals 4. It appearing that for a considerable time past there has been great irregularity in the receipt of the foreign periodical publications.—

Resolved.

That Dr. Marcet be requested to inquire into the best means of obtaining these publications with greater regularity and to make a report thereon to the Managers.—

Resignation of Lord Winchilsea 5. The Earl of Winchilsea having stated that his public avocations have for some time past demanded and still continue to demand so considerable a portion of his time as to preclude his giving that attendance which under other circumstances he would have been most ready and desirous to give and which he considers as essential to the prosperity of the Establishment, and having in consequence expressed a wish to resign his Office as President, which resignation was accordingly accepted at the last Meeting.—

Resolved.

That the Managers deeply sensible of the great benefits which the Royal Institution has derived from the services of the Earl of Winchilsea as its President from its commencement to the present time request His Lordship's acceptance of their unanimous and warmest thanks.—

Resolved.

That the Secretary be directed to transmit to the Earl of Winchilsea a Copy of the above Resolution.

6. The following person having been duly recommended was admitted as a Subscriber, viz.t
Edward Dicey, Esq.r Durham place Chelsea.

J P Auriol

Managers' Meet.g May 3, 1813.

At a Meeting of the Managers of the Royal Institution of Great Britain, held at the house of the Institution on the 3rd of May 1813.—

Present

James Peter Auriol, Esq.r in the Chair.
Marquis of Lansdowne,
Wm. Sotheby, Esq.r
Leonard Horner, Esq.r
Jn.o Guillemard, Esq.r
Cha.s Hatchett, Esq.r
Sir Humphry Davy
Wm. Blake, Esq.r
Tho.s Harrison, Esq.r Secretary.

Minutes app.ved 1. The minutes of the proceedings of the last meeting were read and approved.

Minutes of Com. of Acc.ts app.ved 2. Read and approved the minutes of the Committee of Accounts of this day, and signed an Order for payment of the following sum.

Mr. Fhirt 1/4 year's Salary to 30th March £22.10.—

Letters received.

Letter from Ld. Winchilsea. 3. Read the following letter from the Earl of Winchilsea.

Sir,

I have had the honor to receive your letter enclosing the resolution entered into by the Managers of the Royal Institution on Monday last, and request the favor of you to return to them my sincere thanks for this very flattering testimony of their approbation of my conduct in the office of President, and to assure them that nothing but a conviction that I had it not in my power to give as much attendance at the meetings and attention to the business of the Institution as I felt I ought to do, could have induced me to relinquish my pretensions to an office so highly honorable and flattering to me, and which I had held ever since the formation of this Corporation by the unanimous voice of the Proprietors.

I request of you to accept of my sincere thanks for the very obliging manner in which you have made this communication to me.

I have the honor to be
Sir, Your most obedient humble Serv.t
Winchilsea.

To Jn.o Guillemard Esq.r
Sec.y R.I.

Lectures. 4. Read a letter from M.r Landseer to M.r Hatchett proposing to deliver a course of Lectures on the Philosophy of Art.
Resolved. That M.r Hatchett be requested to communicate further on this subject with M.r Landseer. –

Subscribers. 5. The following persons having been duly recommended were admitted as Subscribers viz.t

To the Lectures.
M.rs Adams, 15 S.t James's Place
Life Subscriber.
The Rev.d W.m Winthrop – New Cavendish S.t
Annual Subscribers.
W.m Wilbraham, Esq.r Mount S.t
Sydney Beresford, Esq.r Strand.

J P Auriol

Managers' Meeting. May 10 1813.

At a Meeting of the Managers of the Royal Institution of Great Britain held at the house of the Institution on the 10th May 1813. –

Present
James Peter Auriol, Esq.r in the Chair.
John Guillemard, Esq.r
John Walker, Esq.r
George Duckett, Esq.r
Tho.s Murdoch, Esq.r
Earl Spencer, President.
Sir Humphry Davy.
W.m Watson, Esq.r
Alex.r Marcet, M.D.
W.m Sotheby, Esq.r
W.m Blake, Esq.r
Tho.s Harrison, Esq.r Secretary.

Minutes app.ved 1. The minutes of the proceedings of the last meeting were read and approved.

Minutes of Com. of Acc.ts app.ved 2. Read and approved the minutes of the Committee of Accounts of this day. –

Presents. Presents received.

3. Isaac Hawkins Brown, Esq.r ~~M.P.~~ presented
Theological Disquisitions; or An Enquiry into those principles of Religion which are most influential in directing and regulating the passions and affections of the mind By T. Cogan, M.D. 8vo
John Weyland, jun.r Esq.r presented
"Letter to Sir Hugh Inglis Bart. on the state of Religion in India".
Rev.d Peter Gandolphy presented
"A second Letter to Rev.d Herbert Marsh, D.D. confirming the opinion that the vital principle of the Reformation has been lately conceded by him to the Church of Rome.
Tho.s Chevalier Esq.r F.R.S. presented
A treatise on Gun-Shot wounds.

Ordered. That the thanks of the Managers be returned for these presents.

Life Subscriber 4. In pursuance of a nomination under the Act by Earl Spencer his Son John Charles commonly called Viscount Althorp of S.t James's Place was admitted to the privileges of a Life Subscriber to the Royal Institution.

Earl Spencer president 5. The President reported that he had named the following Vice Presidents – viz.t

Vice Presidents.
The Marquis of Lansdowne.
Earl of Winchilsea.
Earl of Selkirk.
Sir Humphry Davy.

6. Resolved. That the Professor of Chemistry be requested to procure models of the Abbe Haüy's Crystalography. –

7. The following persons having been duly recommended were admitted as Subscribers viz.t
To the Lectures.
Miss Froude 40 Craven S.t Strand.
Annual Subscriber
John Abernethey, Esq.r Bedford row.

Spencer.

Managers' Meet.g 17th May. 1813.

At a Meeting of the Managers of the Royal Institution of Great Britain, held at the house of the Institution on the 17th of May, 1813. –

Present
Earl Spencer, President, in the Chair.
Charles Hatchett, Esq.r
W.m Watson, Esq.r
J. P. Auriol, Esq.r
Jn.o Walker, Esq.r
Leonard Horner, Esq.r
George Duckett, Esq.r
W.m Sotheby, Esq.r
Tho.s Murdoch, Esq.r
W.m Blake, Esq.r
Sir Humphry Davy, V.P.
Alex.r Marcet, M.D.
Tho.s Harrison, Esq.r Secretary.

Minutes app.ved 1. The minutes of the proceedings of the last meeting were read and approved. –

Minutes of Com. of Acc.ts app.d 2. Read and approved the minutes of the Committee of Accounts of this day, and signed an Order for payment of the following sums, viz.t –

M.r Royce for repairs of the house &c	£ 180. 12. –
Jos.h Fincher on account of the Royal Institution	60. – –
	£ 240. 12. –

Lectures. 3. Resolved. That the following persons be forthwith applied to for the purpose of ascertaining whether they will be disposed to give Lectures at the Royal Institution the next Season.
D.r Smith – Botany
D.r Roget – Comparative Physiology.
M.r Campbell. Poetry.
M.r Flaxman. Sculpture.
Rev.d Edw.d Forster.

Mr Landseer. Philosophy of the Arts.

4. The following persons having been duly recommended were ~~as~~ admitted as Subscribers vizt.

Subscribers — To the Lectures.

Miss Cullen, 4 Marlborough place, Chelsea
Miss Rich 3. Great Cumberland place
Mrs Jennings Portland Place.

Annual Subscriber.

Robt Anderson, Esqr 38 York place Portman Square

Spencer.

Managers' Meeting 24 May 1813

At a Meeting of the Managers of the Royal Institution of Great Britain, held at the house of the Institution, on the 24th of May 1813.

Present
Earl Spencer Pt in the Chair.
Scrope Bernard Morland, Esqr Treasurer.
George Duckett, Esqr
Charles Hatchett, Esqr
Sir Humphry Davy, V.P.
Wm Watson, Esqr
Leonard Horner, Esqr
J. P. Auriol, Esqr
Wm Sotheby, Esqr
Alexr Marcet M.D.
Wm Blake, Esqr
Thos Harrison, Esqr Secretary.

Minutes appd 1. The minutes of the proceedings of the last meeting were read and approved.

Minutes of Comm of Accts approved 2. Read and approved the minutes of the Committee of Accounts of this day.

Present 3. Present received.
Sir Humphry Davy, presented Elements of Agricultural Chemistry, in a Course of Lectures for the Board of Agriculture 4to.
Ordered.
That the thanks of the Managers be returned for this present.

Mr Flaxman's Lectures. 4. Earl Spencer reported that he had seen Mr Flaxman and that he will endeavour to write one or two new Lectures adapted to the audience of the Royal Institution.

Lectures. 5. Sir Humphry Davy reported that Dr Smith is willing to deliver a Course of Lectures on Botany but hopes to be allowed to have two successive days in the weeks of his Lectures.

6. Read a letter from Mr Landseer addressed to Charles Hatchett, Esqr stating that he should be prepared to read from 4 to 6 Lectures on the Philosophy of the Arts during the next Season.

7. Mr Horner reported that Dr Roget is disposed to give from 10 to 12 Lectures on Physiology embracing more of the Natural History of Animals in the ensuing Season.

8. Mr Sotheby reported that Mr Campbell is under the necessity of declining to give any Lectures the next Season.

9. Resolved. That Mr Sotheby be requested to apply to Mr Moore and ascertain whether he would be inclined to give a Course of Lectures on Poetry during the ensuing Season.

10. Resolved. That Sir Humphry Davy be requested to apply to the Revd Edward Forster, to know whether he will deliver a Course of Lectures on Oratory the next Season.

11. The following persons having been duly recommended were admitted as Subscribers viz.t

To the Lectures.
Countess of Jersey.
Miss Mortimer, 36 Northumberland St New road.

JP Auriol.

Managers' Meeting 31 May 1813.

At a Meeting of the Managers of the Royal Institution of Great Britain, held at the house of the Institution on the 31st of May 1813.

Present
James Peter Auriol, Esqr in the Chair.
George Duckett, Esqr
Wm Blake, Esqr
Jno Walker, Esqr
Thos Murdoch, Esqr
Thos Harrison, Esqr Secretary.

Minutes appd 1. The minutes of the proceedings of the last meeting were read and approved.

Minutes of Com. of Accts appd 2. Read and approved the minutes of the Committee of Accounts of this day And Signed an Order for payment of the following sums viz.t

Mr Brande for Lectures in 1813	£105
Revd Mr Powell — do	105
	£210

Spencer

1813.

Managers Meeting 7th June. 1813.

At a Meeting of the Managers of the Royal Institution of Great Britain, held at the house of the Institution on the 7th of June 1813. —

Present

Earl Spencer, President in the Chair.
Sir Humphry Davy V.P.
Wm. Sotheby, Esq.
Chas. Hatchett, Esq.
Thos. Murdoch, Esq.
Thos. Harrison, Esq. Secretary.

Minutes appved. 1. The minutes of the proceedings of the last meeting were read and approved. —

Minutes of Com. of Accts. appved. 2. Read and approved the minutes of the Committee of Accounts of this day, and signed an Order for payment of the following sums, viz.

Dr. Roget for Lectures delivd. in 1813	£105 —
Dr. Smith for do. — in 1812 & 1813 —	63 —
Sundries for Dr. Smith's lectures —	4.12.6
	£172.12.6

3. Ordered. That an Advertizement be inserted in the Morning Chronicle & Morning Advertizer for 30 Chaldrons of Walls End Coals Pool measure to be delivered at the Royal Institution free of all expence any time before the 1st of October next at the option of the Seller — Payment within 3 months after delivery. Offers to be made to the Clerk before Monday next. —

Lectures declined by Mr. Moore 4. Read letters from Mr. Moore addressed to Wm. Sotheby Esq. declining giving Lectures at the Royal Institution in the ensuing Season.

5. Resolved. That Mr. Sotheby be requested to inquire of Mr. Southey whether he will be disposed to give a Course of Lectures on Poetry at the Royal Institution during the ensuing Season. —

W Watson

June 14. Managers Meeting

At a Meeting of the Managers of the Royal Institution of Great Britain, held at the house of the Institution on the 14th of June 1813. —

Present

Earl Spencer, President in the Chair.
William Watson, Esq.
J. P. Auriol, Esq.
Leonard Horner, Esq.
Jn. Guillemard, Esq.
Thos. Murdoch, Esq.
Thos. Harrison, Esq. Secretary

Minutes appved. 1. The minutes of the proceedings of the last meeting were read and approved.

Minutes of Com. of Accts. approved. 2. Read and approved the minutes of the Committee of Accounts of this day and signed an Order for payment of the following sums, viz.

Josh. Fincher on account of the Royal Instn.	£70. —
1/4 years Assessed taxes to 25th. March 1813 —	71.11.11
1/4 years Parochial rates to do. —	12.18.4
Mr. Newman Hend. for Mr. Powell's Lectures	12.13 —
	£167.3.3

3. Read a letter from Messrs. Wollett & Co. offering to deliver 30 Chaldrons of Walls-end Coals at 56/9. free of any extra expense. Ordered. That the Clerk mention this offer to Mr. Dickins and enquire whether he will supply them at that price.

Mr. Campbell's Lectures 4. Read a letter from Thos. Campbell, Esq. respectfully requesting that this fifth & next lecture may be allowed to be the last of his Course in consequence of the ill state of his health and the lateness of the Season. — Resolved. That this request be complied with.

Lectures on Oratory by Mr. Forster 5. Read a letter from The Revd. Edward Forster to Sir Humphry Davy, communicated by Mr. Guillemard, offering to give a Course of 12 Lectures on Oratory on an extended scale during the ensuing Season. —

1813.

6. Resolved. That Mr. Guillemard be requested to engage Mr. Forster to give such a Course of Lectures on Oratory to consist of not fewer than 12 Lectures at a sum not exceeding 100 guineas.

Presents received.

Presents. 7. Messrs. Burley & Moore as Executors of the late Revd. Louis Dutens presented

Jo. Temporarii Chronologicarum Demonstrationum Libri III. —

bequeathed by him to the Royal Institution. —

Dr. Buchan, M.R.I. presented

Notice sur la Vie et les Travaux de M. Maskelyne, par M. Le Chevalier Delambre. —

Ordered. That the thanks of the Managers be returned for these presents.

8. Ordered. That the Clerk be authorized to pay the Waterman one guinea for the contribution of the Institution towards watering the Street before the house during Summer. —

Thos Murdoch

Managers Meeting 21 ~~May~~ June 1813.

At a Meeting of the Managers of the Royal Institution of Great Britain, held at the house of the Institution on the 21st. of June, 1813. —

Present

Thos. Murdoch, Esq. in the Chair.
Jn. Guillemard, Esq.
Thos. Harrison, Esq. Secretary.
Sir Humphry Davy, V.P.
Wm. Watson, Esq.
J. P. Auriol, Esq.

Minutes appved. 1. The minutes of the proceedings of the last meeting were read and approved.

Minutes of Com. of Accts. appved. 2. Read and approved the minutes of the Committee of Accounts of this day, and signed an Order for payment of the following sum, viz.

Thos. Campbell, Esq. for Lectures in 1813. £105 —

Lectures decd. by Mr. Southey 3. Read a letter from Wm. Sotheby, Esq. stating that Mr. Southey declines lecturing at the Royal Institution.

Presents received.

4. The Honble. Robt. Clifford presented

A sketch of the Russian Campaign in 1812 —

The Managers of the London Institution presented

A Catalogue of their Library.

Ordered. That the thanks of the Managers be returned for these presents.

Seal of ~~the~~ Corporation affixed to Lease of 22 Albemarle St. for 57 years from 1813 renewable in 1823. 5. The Seal of the Corporation was affixed to a Lease granted by The Members of the Royal Institution to Jas. Spooner, Esq. of a Messuage or Tenement and Premises No. 22 Albemarle Street for the term of 57 years from 24th. June next at the yearly rent of £48 renewable at Midsummer 1823 for a fresh Term of 60 years & 3/4 and so at the end of every future 14 years on payment of a Fine to the Royal Institution of £240. — &

Countersigned in conformity with Chapter 12th. Art. 3d. of the Bye Laws of the Royal Institution, by

Tho. Murdoch — Chairman 1813.
Humphry Davy.
J. P. Auriol.
J. Guillemard.
Wm. Watson.
} a majority of the Managers present.
& Thomas Harrison — Secretary.

Recomdation from Visitors — 6. Read a reference from the Committee of Visitors entered in their proceedings of the 7th Inst:

Ordered. That Mr. Harris be desired to complete the Inventory of the Furniture with as little delay as possible. —
Resolved. That the consideration of painting the Staircase & Hall be deferred.
Ordered. That the Oil Cloth in the anti-room and at the entrance of the small Library be taken up & the boards — .. cleaned.
Ordered. That the green baize for the feet in the front of the seats in the lecture room be removed and the boards cleaned.
Ordered. That Mr. Royce be directed to examine the roof of the house and report the state of the same.

7. Ordered. That the wages of the Housekeeper be raised from £60 to £65 ℔ ann:
& — That the wages of the two Housemaids respectively be raised from £40 to £42 ℔ ann: to commence from the 30th Inst: —

Scientific Committees. 8. Josh. Fincher reported, that the following Committees met on the 15th 16th & 17th Inst: respectively and made their Elections viz.

Mathematics, Mechanics & Mechanical Inventions.
Honble. Robt. Clifford 1st Chairman
Revd. Wm. Pearson — 2d Chairman
Dr. Buchan — Secretary.

Chemistry Geology & Mineralogy.
Sir Gilbert Blane, Bart: M.D. 1st Chairman
Jn: George Children, Esqr. —— 2d Chairman
Wm. Hasledine Pepys, Esqr. — Secretary.

June 21.

General Science Literature & the Arts.
Rt. Honble. Sir Jno. Nicholl. 1st Chairman
John Disney, Esqr. —— 2d Chairman
John Hinckley Esqr. —— Secretary.

9. Josh. Fincher, reported that he had seen Mr. Woollett who stated that he could not supply the Royal Institution with the best Walls end Coals under 60s. ℔ Chaldron free of all other expenses and that the Coals at 56s 9d. were of very inferior quality.
Ordered. That Mr. Dickins supply the Institution with 30 Chaldrons of best walls-end Coals between this time and the 1st Octr. next giving 1 day's notice of the time and place when and where the Coals are begun to be loaded into the Sacks. —

Humphry Davy.

1813.

Managers Meeting 28 June 1813.

At a Meeting of the Managers of the Royal Institution of Great Britain held at the house of the Institution on the 28th of June 1813 —

Present
Sir Humphry Davy, V.P. in the Chair.
George Duckett, Esqr.
Wm. Watson, Esqr.
Scrope Bernard Morland, Esqr. Treasurer.
Alexr. Marcet, M.D.
Thos. Harrison, Esqr. Secretary.

Minutes appvd. 1. The minutes of the proceedings of the last meeting were read and approved. —

Minutes of Com. of Accts appvd. 2. Read and approved the minutes of the Committee of Accounts of this day. —

3. Mr. Royce reported that he had examined the roof of the house and found a few Slates disturbed by the wind which he had replaced so as to require no expense.

Lectures 4. Resolved. That Sir Humphry Davy be requested to write to Mr. Sotheby on the subject of Mr. Campbell's giving Lectures at the Royal Institution for the ensuing Season. —

J P Auriol

Managers Meeting 5th July, 1813

At a Meeting of the Managers of the Royal Institution of Great Britain held at the house of the Institution on the 5th of July 1813. —

Present
James Peter Auriol, Esqr. in the Chair.
William Watson, Esqr.
John Walker, Esqr.
Wm. Sotheby, Esqr.
Thos. Harrison, Esqr. Secretary.

Minutes appved. 1. The minutes of the proceedings of the last meeting were read and approved.

Minutes of Com. of Accts appved. 2. Read and approved the minutes of the Committee of Accounts of this day, & Signed an Order for payment of the following sums, vizt.

— Mr. Stuart 1/4 years Salary to 30th June.	£22. 10. —
Revd. Mr. Powell for wages to Professr. Farish's Man & Carr: of Apparatus back to Cambridge. —	11. 16. 8
Josh. Fincher on acct. of the Royal Institution	40. — . —
	£74. 6. 8

Mr. Campbell's Lectures. 3: Read a letter from Thos. Campbell, Esqr. addressed to Wm. Sotheby, Esqr. stating "that provided the Institution will allow Mr. Campbell to lecture at the time of the Season to be chosen by himself he shall be proud to enlist again in their service for 2 lectures on the terms mentioned by Mr. Sotheby.
Resolved.
That Mr. Campbell's offer be accepted on the terms mentioned by Mr. Sotheby namely 50 guineas. —

Present. Present received.
4 Col. de Serre presented
Essais Historiques et Critiques sur la Marine de France de 1661 a 1789, 8vo.
Ordered. —
That the thanks of the Managers be returned for this present. —

The

5. The following person having been duly recommended was admitted as an Annual Subscriber viz.t
Jos.h Ritchie, Esq.r of Lock Hospital Grosv.r Place.

Adjourned to Monday the 26th Inst:
J P Auriol

Managers Meet.g 26th July 1813.

At a Meeting of the Managers of the Royal Institution of Great Britain, held at the house of the Institution on the 26th of July 1813 —

Present
James Peter Auriol, Esq.r in the Chair.
Thomas Murdoch, Esq.r
John Walker, Esq.r
Tho.s Harrison, Esq.r Secretary.

Minutes app.ved 1. The minutes of the proceedings of the last meeting were read and approved.

Minutes of Com. of Acts app.ved 2. Read and approved the minutes of the Committee of Accounts of this day.—

Presents. Presents received.

3. The Commissioners of the public Records, presented
Inquisitionum Retornatarum Abbreviatio, 2 vol.
Rotuli Hundredorum, vol 1.
Catalogue of the Harleian Manuscripts, vol 4.
Reports from the Commissioners of the Public Records.
The Statutes of the Realm, vol 1.
Valor Ecclesiasticus temp. Hen. VIII vol 1.
Sir George Paul, Bart: presented
Tracts on the Prisons of Gloucestershire.
Ordered.
That the thanks of the Managers be returned for these presents.—

4. Read the minutes of the last General Monthly Meeting.

5. Resolved.
That the two Rooms lately occupied by Sir H. Davy be painted, papered and whitewashed for the use of Professor Brande on the terms mentioned in Mr. Royce's estimate amounting to £17.19.—

Charles Hatchett

Managers Meeting Aug. 2d 1813.

At a Meeting of the Managers of the Royal Institution of Great Britain, held at the house of the Institution, on the 2nd of August, 1813 —

Present
Charles Hatchett, Esq.r in the Chair.
William Watson, Esq.r
George Duckett, Esq.r
Alex.r Marcet, M.D.
James Peter Auriol, Esq.r
Thomas Murdoch, Esq.r
Thomas Harrison, Esq.r Secretary.

Minutes app.ved 1. The minutes of the proceedings of the last meeting were read and approved.

Minutes of Com. of Acts. app.ved 2. Read and approved the minutes of the Committee of Accounts of this day, & Signed an Order for payment of the following sums, viz.t
Mr Harris 1/4 years Salary to 30th March £45.—
Mr Brooks Collector 1/4 year's d.o to d.o — 13. 2. 6
Jos.h Fincher on account of the Royal Inst.n — 70.—
£128. 2. 6

Mr Feetham's present of a stove for Lib.y 3. Read the following letter from The Rev.d Edward Forster addressed to the Secretary of the Royal Institution.—

"About 2 years & 1/2 ago in consequence of various complaints of the coldness of the Library in Winter I proposed a plan for warming it which was not adopted.
I have since had various communications with Mr Feetham Ironmonger, and I have seen an improved Grate &c of his which he says will most effectually warm every part of it, and which in consideration of the favors he has received from the Royal Institution, he is willing to present to the Library and fit it completely up free of expense.—
I have seen the plan and think it will succeed and should the Managers approve of it I shall be happy to superintend the fitting it up.— It cannot in the least affect the Insurance.—

Resolved. That the thanks of the Managers be returned thro' Mr Forster to Mr Feetham for his offer and that the subject be referred to Sir Humphry Davy & Mr Brande.—

Aug. 2.

4. Ordered — That Mr Royce be directed to commence the repairs of the Rooms (to be occupied by Professor Brande) on Thursday next. And that this Order be communicated to Mr Royce by the Superintendent of the House who is to see to the execution thereof.—

5. The following person having been duly recommended was admitted as an Annual Subscriber, viz.t —
Timothy Macnamara, Esq.r Blenheim Hotel

6. Adjourned to Monday the 18th October.—

W Watson

Special Managers Meeting Oct 4th. 1813.

1813.

At a Special Meeting of the Managers of the Royal Institution of Great Britain, held at the house of the Institution on the 4th October 1813.—

Present
Sir Humphry Davy. V.P. in the Chair.
Charles Hatchett, Esqr.
George Duckett, Esqr.
J. P. Auriol, Esqr.

1. The following Notice having been sent to the Managers according to Chap: VI Art: 6 of the Bye Laws was read from the Chair.—

Prof. of Chemistry, (Mr. Brande) appted. Superintendent of the house—

"By appointment of a Vice President A Special Meeting of the Managers will be held on Monday the 4th. of October next at One o'Clock, precisely to consider on the propriety of appointing the Professor of Chemistry Superintendent of the House; and on matters connected with that appointment.—

2. Resolved. That Professor Brande be appointed Superintendent of the House & Establishment of the Royal Institution. And that he be invited to attend the Meetings of the Managers.—

Chemical School.

3. An idea having been started at the suggestion of the Professor of Chemistry of the possibility of transferring the Chemical School in Windmill Street to the Laboratory of the Royal Institution, It was Resolved That the idea be recommended for consideration at a future meeting of the Managers. At the same time the Managers present were unanimously of opinion that the measure if it can be accomplished will be highly beneficial to the interests of Science and of the Institution.
Resolved.
That Mr. Brande be requested to put his ideas on this subject on paper.—

Mr. Faraday's resignation, to attend Sir H. Davy.

4. Michael Faraday having expressed a wish of accompanying Sir Humphry Davy on his Scientific travels, Sir H. Davy

Oct. 4.

reported this to the Managers, stating that He would not engage him if the Professor of Chemistry considered his services as at all essential to the Institution or if the Managers had the slightest objection to the measure.
Mr. Brande having reported that he conceived arrangements could be made which would prevent Mr. Faraday's resignation being felt by the Body.
Ordered.—That Michael Faraday be permitted to resign his situation.—

[...]mation to Mr. Faraday

5. Mr. Brande reported that Michael Faraday had shewn considerable diligence and attention in cleaning and arranging the Mineral Collection, which not being immediately his duty in the Institution, he recommended his services to the Managers attention.
Ordered in consideration of these services that Michael Faraday be paid a Month's wages on the day of his departure.—

6 Ordered. That Mr. Brande be requested to engage a suitable person in his place at the same stipend.—

W. Watson

Managers' Meeting 18 Oct. 1813.

1813.

At a Meeting of the Managers of the Royal Institution of Great Britain, held at the house of the Institution on the 18th of October 1813.—

Present
William Watson, Esqr. in the Chair.
J. P. Auriol, Esqr.
Alexr. Marcet, M.D.

Minutes approved. 1. The minutes of the proceedings of the Meeting of the 2d. August, and also of a Special Meeting held on the 4th. Instant were read and approved.

Minutes of Com. of Accts. approved. 2. Read and approved the minutes of the Committee of Accounts of the 9th. 16th. 23d. & 30th. August, the 6th. 13th. 20th. & 27th. Septr. the 4th. & 11th. October and also the Committee of Accounts of this day, And Signed an Order for payment of the following sums,

Mr. Sturt for 1/4 year's Saly. to 30th. Septr.	£22.10 —
Josh. Fincher on acct. of the Royal Instn.	80. —
	£102.10 —

Presents. 3

Presents received.
George Wilson Meadley, Esqr. presented,
Memoirs of Algernon Sydney 8vo.
Le Chevalier De Sade, presented.
De la Tydologie, ou de la Science des Marées 2 vol. 8vo.
Sir Francis D'Ivernois, presented.
Exposé de l'Exposé de la situation de l'Empire Francais.
Ordered. That the thanks of the Managers be returned for these presents.

Commee. of Visitors. 4. Read a communication from the Committee of Visitors entered in their proceedings of the 11th. Instant.

Delay in Inventory of Furniture.

Mr. Harris being called in – stated that the reason for his delaying to make out the Inventory of the furniture was on account of the changes taking place in the different parts of the Institution and the necessity of removing the Furniture from Room to Room ~~to enable the Workmen~~ to enable the Workmen to carry into effect the

Oct. 18.

repairs as ordered by the Managers on the 2d. August last But that he would proceed now to make it out as soon as possible.—

5. Mr. Royce reported that he had examined the roof of the House and found several Slates broken, and some disturbed by the Wind and part of the Mortar removed from the edges and sides of the Slates.—the probable expence of the repairs may be £3.—
Ordered.
That Mr. Royce be directed to complete the repairs at an expense not exceeding £3.—

J. P. Auriol

1813.

Managers Meet^g. Novr. 1st. 1813.

At a Meeting of the Managers of the Royal Institution of Great Britain, held at the house of the Institution on the 1st. of November 1813. —

Present
J. P^r. Auriol, Esqr. in the Chair.
George Duckett, Esqr.
Thos. Harrison, Esqr. Secretary.

Minutes app^{d}. 1. The minutes of the proceedings of the last meeting were read and approved. —

Minutes of Comm^e Accts. app^{ved} 2. Read and approved the minutes of the Committee of Accounts of this day. —

Presents. 3.

Presents received.

Chas. Etter, Esqr. from S^t. Petersburgh, presented the following Specimens to the Museum of the Royal Institution. —

1. Avanturin felspar from Katherinenburg in Siberia
2. Dendritic Jasper, Siberia.
3. A Specimen of red Hornstone with which the new Cathedral at Petersburgh is paved.
4. A Specimen of sky blue felspar from the vicinity of the lake Baikal in Siberia.
5. Chromate of Lead – Siberia.
6. A Specimen of ferruginous fossil wood. —
7. 8. 9. 10. 11. 12. 13. 14. Geological Specimens chiefly Swedish & Siberian.

The Commissioners of the Public Records presented.
A Catalogue of the Harleian Manuscripts in the British Museum – 3 vol^s. folio. —

Ordered

That the thanks of the Managers be returned for these presents. —

Thos Murdoch

Managers Meet^g. 8th. Nov. 1813.

At a Meeting of the Managers of the Royal Institution of Great Britain, held at the house of the Institution on the 8th. of November 1813. —

Present
Thomas Murdoch, Esqr. in the Chair.
Leonard Horner, Esqr.
Scrope Bernard Morland, Esqr. Treasurer.
James Peter Auriol, Esqr.
Alex^r. Marcet, M.D.
Thos. Harrison, Esqr. Secretary.

Minutes approved 1. The minutes of the proceedings of the last meeting were read and approved.

Minutes of Com. of Acct^s. app^{ved} 2. Read and approved the minutes of the Committee of Accounts of this day.

3. Read a letter from Chas. Etter Esqr. requesting a Copy of the Bye Laws. — Ordered That a copy be sent him. —

4. Read the minutes of the last General Monthly Meeting.

Assist^t. in Lab^y &c. 5. M^r. Brande reported that he had engaged Stephen Slatter the Assistant Porter as Assistant in the Laboratory, upon the same terms as Michael Faraday, and had engaged as Assistant Porter Blacket Thomas Wallis. —

Ordered.

That Stephen Slatter be appointed Assistant in the Laboratory, on the same terms as Michael Faraday. — And that the Clerk furnish him with a Copy of the Order relating to his duties.

Ordered. That Blacket Thos. Wallis, be engaged as Assistant Porter on the same terms as Step^n. Slatter, the late Porter vizt. £1 p^r Week. —

Inventory of Furniture 6. M^r. Harris the Assistant Superintendent of the house, laid the Inventory of the Furniture made by him upon the Table. —

7. Resolved. That M^r. Brande be requested to get the

1813.

Mineralogical Room better heated, taking care that nothing is done to invalidate the Insurance.

Scrope Bernard Morland

Managers Meet^g. 15th. Novr. 1813.

At a Meeting of the Managers of the Royal Institution of Great Britain, held at the house of the Institution on the 15th. of November, 1813. —

Present
Scrope Bernard Morland, Esqr. Treas^r. in the Chair.
John Guillemard, Esqr.
William Watson, Esqr.
Chas. Hatchett, Esqr.
George Duckett, Esqr.
Alex^r. Marcet, M.D.
W^m. Blake, Esqr.
Thos. Harrison, Esqr. Secretary.

Minutes app^{ved} 1. The minutes of the proceedings of the last meeting were read and approved. —

Minutes of Com. of Acct^s. app^d. 2. Read and approved the minutes of the Committee of Accounts of this day. And Signed an Order for payment of the following sum —

Jos^h. Fincher on acc^t. of the Royal Inst^n. £60. —

Lectures 3. Resolved. That M^r. Guillemard be requested to make enquiries for a Lecturer on Natural Philosophy.

Notice to Prop^{rs} of "Clarendon Hotel" concerning Trees 4. Ordered. That M^r. Harris be directed to represent to the Owners of the Clarendon Hotel that the Trees are a nuisance to the Royal Institution and to request that they may be cut down or Lopped and to report hereon to the Managers at their next Meeting. —

Charles Hatchett

1813.

Managers' Meeting. 22nd Nov. 1813

At a Meeting of the Managers of the Royal Institution of Great Britain, held at the house of the Institution on the 22d of November, 1813 –

Present

Charles Hatchett, Esq.r in the Chair.
John Walker, Esq.r
George Duckett, Esq.r
W.m Blake, Esq.r
Ja.s Peter Auriol, Esq.r
Alex.r Marcet, M.D.
Tho.s Murdoch, Esq.r
Tho.s Harrison, Esq.r Secretary

Minutes app.d 1. The minutes of the proceedings of the last meeting were read and approved.

2. Read and approved the minutes of the Committee of Accounts of this day.

Reply from Mr. Jacquier "of Clarendon" engaging to clear Leads of R.I. of litter from Trees. 3. M.r Harris reported, that he waited on M.r Jacquier, the Owner of the Clarendon Hotel, with the Managers request, that he would permit the Trees in his garden to be cut down or lopped, as they are injurious to the buildings of the Royal Institution; M.r Jacquier stated that the Trees were of such great advantage to his House, that he would not lose them for £200; that they are specified in his lease, and he is answerable for their preservation to the Duke of Grafton. M.r Jacquier further stated, that he would undertake to have the leads of the Building kept clean by his own Porter whenever necessary.

4. Ordered. That the notice for announcing the Lectures for the ensuing Season be ready for printing by the next meeting. –

W.m Blake

Managers' Meet.g Nov. 29. 1813

At a Meeting of the Managers of the Royal Institution of Great Britain, held at the house of the Institution, on the 29th of November, 1813 –

Present

William Blake, Esq.r in the Chair.
Ja.s Peter Auriol, Esq.r
Alex.r Marcet, M.D.
George Duckett, Esq.r
Tho.s Murdoch, Esq.r
Scrope Bernard Morland, Esq.r Treasurer
Tho.s Harrison, Esq.r – Secretary.

Minutes app.ved 1. The minutes of the proceedings of the last meeting were read and approved. –

Minutes of Com. of Acc.ts app.ved 2. Read and approved the minutes of the Committee of Accounts of this day. –

Presents

Presents received

3. M.r Brande reported that M.r Feetham had presented a very handsome Grate for the Library which has been put up and answers the purpose very well. –
Edward Strachey, Esq.r presented
Bija Ganita, or the Algebra of the Hindus.
Ordered. That the thanks of the Managers be returned for these presents.

Lectures 4. Resolved. That the Lectures begin on Saturday the 22d of January next.

5. Resolved. That the notice for announcing the Lectures be deferred till this day week. –

Charles Hatchett.

1813

Managers' Meet.g Dec. 6. 1813.

At a meeting of the Managers of the Royal Institution of Great Britain, held at the house of the Institution on the 6th of December, 1813. –

Present

Cha.s Hatchett, Esq.r in the Chair.
J. P. Auriol, Esq.r
George Duckett, Esq.r
Jn.o Walker, Esq.r
W.m Blake, Esq.r
Tho.s Murdoch Esq.r
Alex.r Marcet M.D.

Minutes app.ved 1. The minutes of the proceedings of the last meeting were read and approved.

Minutes of Com. of Acc.ts app.ved 2. Read and approved the minutes of the Committee of Accounts of this day. –

Lectures 3. M.r Blake reported that Tho.s Simpson Evans, Esq. F.L.S. Master of the Royal Mathematical School at Christ's Hospital is willing to give 10 Lectures on Astronomy at the Royal Institution for the ensuing Season at such times as the Managers shall appoint.

4. Ordered. That the Card announcing the Lectures be sent round to the Members & Subscribers as soon as possible.

Leases. 5. Read a letter from M.r Seymour, Sol.r to the Royal Institution stating, that he had the pleasure of sending An Abstract of the Title of M.r Spooner to the Lease dated 6th July 1782 granted by W.m Mellish, Esq.r to M.r Swinton – also The original Lease of 6th July 1782 – also – An Attested Copy of an Assignment from the Executors of M.r Swinton to M.r Spooner, & the Counterpart Lease of the 31st May 1813 from the Members of the Royal Inst.n to M.r Spooner, and in exchange he wanted the Counterpart Lease of 6th July 1782 – Mellish to Swinton, in order that the same may be cancelled, he also had occasion for the Counterpart Lease of 2d Nov.r 1776 in order to enable him to surrender such Counterpart in exchange for the original Lease – He also stated that he should send a Lease from

Dec. 6

the Members of the Royal Institution to M.r Paynter of Premises demised by Mellish to Gray at the yearly rent of £4 requested that the Board of Managers might be summoned on the 13th Inst. to affix the seal of the Corporation thereto. –
Ordered. – That M.r Seymour be furnished with the Deeds mentioned in his letter. And that the meeting be summoned on the 13th Inst. to affix the seal of the Corporation to a Lease from the Royal Institution to M.r Paynter. –

J P Auriol

406

1813.

Managers' Meeting 13 Dec. 1813

At a meeting of the Managers of the Royal Institution of Great Britain, held at the house of the Institution on the 13th of December, 1813 —

Present
Jas. Peter Auriol, Esqr. in the Chair.
Thos. Murdoch, Esqr.
Jn.o Guillemard, Esqr.
Wm. Blake, Esqr.
Alexr. Marcet, M.D.
George Duckett, Esqr.
Thomas Harrison, Esqr. Secretary.

Minutes appd. 1. The minutes of the proceedings of the last meeting were read and approved.

Minutes of Com. of Acct. appd. 2. Read and approved the minutes of the Committee of Accounts of this day, and Signed an Order for an advance to the Clerk of £70. —

Lectures. Dr. Roget's. 3. Read the following letter from Dr. Roget, addressed to the Committee of Managers —

"As I am ignorant whether any arrangement has been made with regard to the days of the week on which the different Lectures are to be given this season I must beg leave to state those on which it would be most convenient for me to lecture, which are Wednesdays, or Saturdays, but of course, am, desirous of accomodating myself to the general convenience of others. — As I am preparing an entirely new course, I do not think I shall be ready to commence the Lectures till the middle of February, which I trust will meet with your approbation. —

4. Ordered. That letters be written to
Dr. Roget.
Mr. Evans.
Revd. Edward Forster.
Mr. Landseer.
on the subject of commencing their lectures. —

5 Read the minutes of the last General Monthly Meeting

407

Dec. 13. 6. The Secretary reported, that he had delivered to Mr. Seymour, Solicitor to the Royal Institution, the Counterpart Lease of 6th July, 1782, Mellish to Swinton, and also the Counterpart Lease of 2d Novr. 1776 as ordered to be delivered by him at the last Board — And that he had received the Deeds &c. mentioned in Mr. Seymour's letter and had locked them up in the Iron Chest of the Royal Institution. —

Lease for 57 years to Mr. S. Raynter. 7. The Seal of the Corporation was affixed to a Lease granted by The Members of the Royal Institution with the consent of The Extrix & Exors of Mr. Edwd. Gray deceased to Mr. Saml. Raynter of a piece of Ground and part of a Messuage and Premises No. 16 in Grafton St. for the term of 57 years from 24th June last, at the yearly rent of £4 renewable at Midsummer 1823 for a fresh term of 60 years ¾ and so at the end of every future 14 years on payment of a Fine to the Royal Institution of £20. —

8. Countersigned in conformity with Chap: 12 Art: 3 of the Bye Laws, by
J. P. Auriol — Chairman.
Tho. Murdoch }
Wm. Blake }
Alex. Marcet } The Managers present
Geo Duckett }
J. Guillemard }
& Thomas Harrison — Secretary. —

Subscribers. 9 The following persons having been duly recommended were admitted as Subscribers viz. —

To the Lectures.
Mrs. Polls — Harley Street. —

Annual Subscribers.
Mr. da Costa — 3. Lower John St. Golden Sqr.
Frederick Daniel, Esqr. Lincolns Inn fields.
Revd. Dr. Knox — 1 Adam Street, Adelphi.
John Medley, Esqr. 7 Arundel St. Strand.
John Wade, Esqr. upper Marylebone Street.
Revd. J. R. Roberts — 5 Portman Street

408

1813.

Eustace Barron, Esqr. 20 Holles Street
Dr. Anderson, 25 Oxendon St.
Jn.o Polls, Esqr. 50 Harley St.
Jn.o Thos. Claridge, Esqr. 8 Millman St.
Thos. Roberts, Esqr. 12 Paper Build. Temple
Mr. F. Ronalds — Apothecaries Hall.
Sir Francis Henry Drake, Bart.

W Watson

409

Dec 20.

Managers' Meetg. 20th Dec. 1813.

At a meeting of the Managers of the Royal Institution of Great Britain, held at the house of the Institution, on the 20th of December, 1813 —

Present
William Watson, Esqr. in the Chair.
Chas. Hatchett, Esqr.
Jas. Peter Auriol, Esqr.
George Duckett, Esqr.
Jn.o Guillemard, Esqr.
Alexr. Marcet, M.D.
Thos. Murdoch, Esqr.

Minutes appd. 1. The minutes of the proceedings of the last meeting were read and approved. —

Minutes of Com. of Acct. appd. 2. Read and approved the minutes of the Committee of Accounts of this day. —

Mr. Flaxman declines to Lecture 3. Read a letter from Mr. Flaxman declining for reasons therein stated to deliver the lectures promised by him and announced in the General Card of 6th Decr. 1813. —

Resolved.

Lectures. That Mr. Hatchett be requested to communicate the above letter to the President, Earl Spencer and that the further consideration of it be postponed. —

4. Mr. Brande reported, that Mr. Evans will be prevented from lecturing on Tuesdays by business and that he should prefer Monday.

Ordered. That Monday be assigned to him and that he be desired to deliver his first lecture on Monday the 24th of January.

Dr. Roget. 5. Read a letter from Dr. Roget stating that he shall be able to commence his course of lectures on Wednesday the 9th February.

6. Read letters from The Revd. Edward Forster, and Mr. Landseer, in which they appear desirous to postpone the commencement of their lectures till after the Easter recess.

Ordered. That they be written to with an intimation that it is extremely desirable that their lectures should commence before that period. –

Presents 7. **Presents received.**

Revd. Edward Balme, presented.
Dr. Richd. Hey's Dissertations on Gaming, Duelling, & Suicide, 8vo.
Mr. Mawe of the Strand presented.
A specimen of Hydrargillite to the Museum of the Institution.
Ordered. That the thanks of the Managers be returned for these presents. –

Subscribers. 8. The following persons having been duly recommended were admitted as Subscribers, viz.t Annual Subs.rs

William Vaughan, Esq.r 7 Hinde St. Manch.r Sq.e
The Hon.ble Mr. Plunkett, Somerset Street
Robt. Wm. Hay, Esq.r 39 St. James's place.
Mr. Geo: Tebbs – 172 New Bond St.
Augustus B. Granville, Esq.r 16 Charles St. Grosv.r Sq.e
Francis Ronalds, Esq.r of the upper Mall, Hammersmith
Revd. W. H. Ward – 2, Vere St. Cavendish Square.
Mr. Edward Lydiatt 2 North St. Fitzroy Square

Charles Hatchett

Managers Meet.g 27th Dec. 1813

At a meeting of the Managers of the Royal Institution of Great Britain, held at the house of the Institution, on the 27th of December, 1813. –

Present
Charles Hatchett, Esq.r in the Chair.
Wm. Watson, Esq.r
George Duckett, Esq.r
J. P. Auriol, Esq.r
Wm. Blake, Esq.r
Thos. Harrison, Esq.r Secretary

Minutes appr.ved 1. The minutes of the proceedings of the last meeting were read and approved. –

Minutes of Com. of Acc.ts approv.d 2. Read and approved the minutes of the Committee of Accounts of this day & Signed an Order for payment of the following sums, viz.t

Mr. Stuel for 1/4 year's Salary to 31st Inst:	£22.10.0
Jos: Fincher on account of the Royal Inst.n	30. –
	£52.10 –

Lectures 3. **Letters received.**

Read a letter from Mr. Evans wishing for various reasons to postpone the commencement of his lectures a few weeks beyond the period which the Managers had requested him to commence.
Resolved.
That the Secretary inform Mr. Evans that they concur, for the reasons mentioned in his letter, for the postponement of his lectures and that his first lecture may be delivered on Monday the 14th. February next.

Mr Landseer's 4. Read a letter from Mr. Landseer stating that he will comply with the wishes of the Board in lecturing before the Easter recess and that he should prefer Thursdays.
Resolved.
That the Secretary inform Mr. Landseer that the Managers will be happy to accomodate him with Thursday in each week as long as they can but that they may be obliged to alter the day before the close of the course and if it should occur due notice will be given him. –

5. The following persons having been duly recommended were admitted as Subscribers, viz.t

Subscribers Annual Subscribers.

Gen.l Browne. 26 York place Portman Sq.e
Major Wm. Morgenthall, 27. Hand Court
Henry Penny, Esq.r 72 upper Norton St.

Adjourned to Monday the 17th Jany.

W. Watson

Managers Meeting Jan 17 1814.

At a meeting of the Managers of the Royal Institution of Great Britain, held at the house of the Institution, on the 17th of January, 1814. –

Present
William Watson Esq.r in the Chair.
Wm. Blake, Esq.r
Scrope Bernard Morland, Esq.r Treasurer.
George Duckett, Esq.r
Alex.r Marcet, M.D.
Jn.o Guillemard Esq.r

Minutes approved. 1. The minutes of the proceedings of the last meeting were read and approved. –

Minutes of Com. of Acts approved 2. Read and approved the minutes of the Committee of Accounts of this day. –

Presents. 3. **Presents received.**

Revd. Stephen Weston, B.D. F.R.S &c. presented.
Persian Distichs, 8vo.
The Editors of the London Medical Surgical & Pharmaceutical Repository presented No. 1, of their Repository.
Ordered. – That the thanks of the Managers be returned for these presents. –

Repairs. 4. Ordered. That the Door of the Newsroom be listed and fitted up with a spring so as to close of itself. –

5. Ordered. That Mr. Royce make an estimate of the expense of making two inner Skylights to the Mineralogical room

6. Ordered. That notice be given to the Surveyor of the Sun Fire Office that new Stoves are about to be erected in the House of the Institution and that he be requested to inspect the same. –

Subscribers 7. The following persons having been duly recommended were admitted as Annual Subscribers, viz.t

Archibald Mearns, Esq.r 55 Curzon St.
Capt.n Smith, of the Madras Cavalry.

414

1814.

John Turner, Esqr 40 Wimpole S.^t
Revd Henry Penny, N.º 2 Cadogan Terrace
Henry Earle, Esqr 12 Hanover Square.
Revd J. T. G. Riddell, 7 Maddox S.^t
Arthur Crichton, Esqr 22 upper Gower S.^t
Duncan Stewart. N.º 3 Brick Court Temple
Anth.^y Vandam Searle, 16 Guilford S.^t –
Captn Broughton of Bengal.
John Fearn, Esqr 79 Newman S.^t
Dr Franklin, 24 Charlotte St Portland place
To the Lectures. – Miss Beeby – 48 Welbeck Street. –

Resolved. That the election of 7 Ladies recommended this day as Subscribers to the Lectures by two Annual Subscribers be postponed in order that a proper recommendation be obtained. –

Adjourned in consequence of the foregoing Resolution to Friday the 21.st Inst.^t at One o'Clock –

Geo Duckett

415

Managers Meeting 21st Jan. 1814.

At an adjourned meeting of the Managers of the Royal Institution of Great Britain, held at the house of the Institution on Friday the 21.st of January 1814 –

Present
Thos Murdoch, Esqr in the Chair.
Wm Blake, Esqr
George Duckett, Esqr
Scrope Bernard Morland, Esqr Treasurer
Jno Guillemard, Esqr

1. Read the resolution of Adjournment from Monday last to this day –

Subscribers.

2. The following persons having been duly recommended were admitted as Subscribers, vizt

To the Lectures.
Mrs A. Macnab
Miss A. Macnab } 8 Alfred place Bedford Square
Miss C. Macnab
Miss Ann Stewart
Mrs Wright
Miss Watson } 16 upper Cadogan Place
Miss Kent
Mrs Strutt, 28 upper Berkeley S.^t
Miss Drummond – Sloane S.^t
Mrs Richards, Park Street
Mrs Arnold, 13 Manchester Square.

Annual Subscribers.
Captn Wm Davidson of Maida-hill –
Jno Baker Richards, Esqr Park S.^t
Ja.^s Burton Esqr of Tanfield Court Temple.
Henry Metcalf, Esqr – Albany.
Captn Jos.^h Packwood, R. N.^l East St Queen Sq.
Peter Brown, Esqr Welbeck S.^t
Chas Agnew, Esqr 42 Russell Square
Mr Geo.^s Walter Hume, Long Acre.

G Duckett

416

1814.

Managers Meeting Jan. 24th, 1814

At a meeting of the Managers of the Royal Institution of Great Britain held at the house of the Institution on Monday the 24.th January, 1814.

Present
George Duckett, Esqr in the Chair.
J. P. Auriol, Esqr
Wm Watson, Esqr
Jno Guillemard, Esqr

Minutes app.^ved 1. The minutes of the proceedings of the 2 last meetings were read and approved.

Minutes of Com. of Acc.^ts approved. 2. Read and approved the minutes of the Committee of Accounts of this day, & Signed an Order for an advance to the Clerk of £60. –

Present. Present received.

3. John Grenfell, Esqr presented "Observations on the Copper Coinage."
Ordered. That the thanks of the Managers be returned for this present. –

4. Ordered. That Arrowsmith's new Physical map of Europe be procured and made to draw up and down on spring rollers.

Skylight. 5. Ordered. That the Skylights in the Mineralogical Room be completed immediately at an expence not exceeding Mr Royce's Estimate of £42 –

6. Resolved. That the question of Insurance be taken up at the next meeting. –

Subscribers 7. The following persons having been duly recommended were admitted as Subscribers, vizt

To the Lectures. – Miss Wells of Crawford St Portman Sq.
Miss Julian Powers – 45 Queen Ann S.^t –

Annual Subscribers.
Earl of Tyrconnel 53 George St Portman Sq.
Jno Wilton, Esqr Chandos S.^t –
M.^r De Monte 7 Grafton Street Fitzroy Sq.

J P Auriol

417

Managers Meet.^g 31st Jan. 1814.

At a meeting of the Managers of the Royal Institution of Great Britain, held at the house of the Institution on 31.st January 1814. –

Present
Ja.^s Peter Auriol, Esqr in the Chair.
Wm Watson, Esqr
Jno Guillemard, Esqr
Alex.^r Marcet, M. D.
Wm Blake, Esqr
Thos Harrison, Esqr Secretary.

Minutes app.^ved 1. The minutes of the proceedings of the last meeting were read and approved.

Minutes of Com. of Acc.^ts approved 2. Read and approved the minutes of the Committee of Accounts of this day.

Lectures by Rev. E. Forster 3. Read a letter from The Revd Edward Forster proposing to begin his lectures in February or the first week in March and read one lecture in each week until Passion week which would make five lectures before and five after Easter and that he should wish to read them on a Friday and have the same day for the whole course.

Ordered. That the Secretary inform him that the Managers can appoint Friday till after Easter but cannot be certain afterwards and request him to begin on the first Friday in March. –

4 Mr Hatchett communicated a letter from Mr Flaxman to the President of the Royal Institution, stating it to be quite impracticable for him to deliver any Lectures at the Institution this Season. –

Insurance of House, &c. 5 Resolved. That a policy to the following amount be forthwith insured and notice thereof given to the Treasurer and that he be requested to pay in future upon the new policy.

Buildings ———— £10,000

 1814

Household Goods, wearing apparel Plate & Linen	£ 300
Shelves in the Library & Glass cases containing Minerals & Models	500 —
Books	10,000
Chemical Apparatus, Mathematical, & Philosophical Instruments & Models	1200
Minerals	300 —
	£ 22,300. —

6\. Mr. John Hinckley having applied for the repayment of the sum of £25 advanced by him for himself and Brother and it appearing to the Managers that Mr. John Hinckley himself only advanced £12 . 10 — on the sum subscribed by him of £250 by way of Loan.
Ordered. That the Treasurer be requested to pay to Mr. John Hinckley £12 . 10. — taking his receipt and place the same to the Loan Account of the Royal Institution. —

Committee Room,
7\. Ordered. That the Secretary of the Committee of General Science be informed, that there being a Room appropriated for the meeting of the Committees of the Institution, no more meetings of that Committee can take place in the Library.

8\. Resolved. That for a Month from this day the reading room be lighted by wax Candles instead of Lamps for the purpose of ascertaining the comparative expense of the two modes of lighting that Room.

Subscribers
9\. The following persons having been duly recommended were admitted as Subscribers viz.

To the Lectures.

Mrs. Bale 6 Glasshouse St. Piccadilly.
Miss Anna Dilkes 6 upper Seymour St.
Mrs. Chevallier 121 Gt. Russell St.
Mrs. Doyly — 10 Fludyer St. Westmr.
Mrs. Benyon } of Russell Square
Miss Benyon }

Jan. 31.
Subscribers

Annual Subscribers.

Mr. John Litchfield } at Willis's Coffee-house
Mr. Chas. Moore }
Jno. Chevallier Esqr. M. D. 121 Gt. Russell St.

Spencer

 1814

Managers' Meetg 7th Feb. 1814.

At a meeting of the Managers of the Royal Institution of Great Britain held at the house of the Institution, on the 7th of February, 1814. —

Present

Earl Spencer, President in the Chair.
Chas. Hatchett, Esqr.
William Watson, Esqr.
Jno. Guillemard, Esqr.
Scrope Bernard Morland, Esqr. Treasurer.
George Duckett, Esqr.
J. P. Auriol, Esqr.
Wm. Blake, Esqr.
Thos. Murdoch, Esqr.
Thos. Harrison, Esqr. Secretary.

Minutes approved
1\. The minutes of the proceedings of the last meeting were read and approved —

Minutes of Com. of Accts approved.
2\. Read and approved the minutes of the Committee of Accounts of this day.

Audit of Treasurer's acct of 1813.
3\. Audited the Treasurers Account up to the 31st Decr. last, and found a balance in favor of the Institution up to that day of £66 . 6 . 1. —

4\. A letter from Mr. Jno. Hinckley, respecting the Library was read.
Ordered. That the Secretary inform him that his letter has been laid before the Board.

5\. Read a letter from Mr. Henry Hinckley, stating that his Brother Mr. John Hinckley had advanced in his name £12 . 10 . by way of Loan to the Royal Institution & requesting that sum to be paid to him.
Ordered.
That the Treasurer be requested to pay to Mr. John Hinckley £25 for his own and his Brother's Instalment of £5 ℘Cent on £250 each, advanced by way of Loan to the Royal Institution taking his receipt and place the same to the Loan Acct. of the Royal Institution

Feb. 7.

Mr. Forster's Lectures.
6\. The Secretary reported that Mr. Forster will commence his Lectures on Oratory, on Friday the 4th March, and give 5 Lectures before Easter, and 5 afterwards. —
Ordered.
That his Lectures be announced to begin on Friday the 4th March on the next Cards. —

Insurance.
7\. The Secretary having reported that he had communicated with the Sun Fire Office respecting the increased Insurance, and was informed that they could not give an answer for a week or two, and that he had not yet received their answer but was informed that if £20 were deposited, the Insurance would be in the mean time extended to the sum of £22,300 — it being understood that if hereafter the Managers do not approve of the new rate of Insurance the Office will return the £20.
Resolved.
That the Treasurer be requested to pay the sum of £20 to the Office taking their accountable receipt for that sum.

Subscribers
8\. The following persons having been duly recommended were admitted as Subscribers, viz.

To the Lectures.

Mrs. Aust, Whitehall
Mrs. Maberley Grosvenor Square
Miss Caroline Greville, 13 Bruton St.

Annual Subscribers.

Hon: Fredk. Kinnaird St. Albans St.
Jno. Ricketts, Esqr. — Princes St.
Fras. Hartwell, Esqr. 35 York place.

Charles Hatchett

422 | 1814

Managers Meeting 14th Feb. 1814

At a meeting of the Managers of the Royal Institution of Great Britain, held at the house of the Institution, on the 14th February, 1814.

Present
Earl Spencer, President. in the Chair.
Charles Hatchett, Esq.r
W.m Blake, Esq.r
J. P. Auriol, Esq.r
George Duckett, Esq.r
John Guillemard, Esq.r

Minutes app.ved 1. The minutes of the proceedings of the last meeting were read and approved.

Minutes of Com. of Acc.ts app.ved 2. Read and approved the minutes of the Committee of Accounts of this day, and signed an Order for payment of the following sums, viz.t

M.r Pond for Lectures in 1812, in balance of account	£35. –
Mess.rs Jones & Yarrell Newspapers in 1812	99. 17. 8
Feetham for Ironmongery in 1812	34. 8. 7
Parker for Glaziery in d.o	19. 2. 8
Allen for Chemicals in d.o	41. 9. 5
M.r Newman for Smeaton's Machine	5. 5. –
Dickins & C.o Coals in 1813	159. 3. 4
Reeve 3/4 years Land, Property & Assessed taxes up to Xmas 1813	205. –. 4
Butler 3/4 years Parochial rates to d.o	38. 15. –
Sun Fire Office remainder of Insurance on £22,300 up to Lady-day 1815	42. 4. 7
	£700. 6. 7

Insurance – 3. Ordered. That the Insurance of £22,300 be effected on the terms offered by the Sun Fire Office. And that the Treasurer be directed in future to pay on the new Policy £58. 5. 0 ℘ Annum, commencing from Lady-day 1815.

4 Read the minutes of the last General Monthly meeting

Feb. 14 | 423

5. M.r Sturt having applied to receive his Salary Monthly. –
Ordered. That his request be complied with, and that M.r Fincher be authorized to pay him accordingly.

London Institution 6. M.r Brande having stated that he had applied to the Managers of the London Institution for the loan of a Model of a Steam Engine belonging to that Society, and that they had very politely acceded to his request.
Resolved. That the thanks of the Managers of this Institution be returned forthwith to the Managers of the London Institution for their liberal compliance with the request of the Professor of Chemistry.

Subscribers 7. The following persons having been duly recommended were admitted as Subscribers viz.t –

Life subscriber – Life Subscriber.
Peter Tho.s Westcott, Esq.r of Walworth.

Annual – Annual Subscribers.
George B. Kelly, Esq.r 33 Piccadilly.
Gen.l Browne, 37 Sackville Street.
George Silvertop, Esq.r 29 Conduit S.t
M.r W.m Newberry S.t Pauls Church Yard.
M.r W.m Smith Dickinson 22 Doughty S.t
W.m Douglas, Esq.r M.P. Albany.

To the Lectures.
M.rs Ewart } 34 Clarges Street.
Miss Ewart }
M.rs E. Hopkinson, 25 Montagu Square
Miss Hamilton 19 Orchard Street.
Miss Lister Lincolns Inn fields.
Lady Leigh } 18 Montagu Square.
Miss Downes }
Miss E. Prinsep – 28 Argyll Street.

J. P. Auriol

424 | 1814

Managers Meeting 21st Feb. 1814

At a meeting of the Managers of the Royal Institution of Great Britain, held at the house of the Institution, on the 21st of February, 1814. –

Present
Earl Spencer, President. in the Chair.
J. P. Auriol, Esq.r
W.m Watson, Esq.r
George Duckett, Esq.r
John Guillemard, Esq.r
Alex.r Marcet, M.D.
W.m Blake, Esq.r
Tho.s Harrison, Esq.r Secretary.

Minutes app.ved 1. The minutes of the proceedings of the last meeting were read and approved.

Minutes of Com. of acc.ts appr.ved 2. Read and approved the minutes of the Committee of Accounts of this day, and signed an Order for an advance to the Clerk of £80. –

Lectures 3. Read a letter from M.r Elmes of Chichester requesting to read a Course of Lectures upon Architecture.
Resolved. That the consideration of his proposition be deferred till the next Board of Managers. –

Present. Present received.
4 Rob.t Richards, Esq.r M.P. presented his Speech on the renewal of the Charter of the East India Comp.y with Appendix.
Ordered. That the thanks of the Managers be returned for this present.

Hon.y ~~Member~~ Admission. 5. Ordered. That a letter of admission be written by the Secretary to Baron Dornberg, Envoy Extraordinary from Hesse Cassel according to Art: 16 Chap: VI of the Bye Laws. –

6. The following persons having been duly recommended were admitted as Subscribers, viz.t
Annual Subscribers.
William Curran, Esq.r 47 upper Norton S.t

Feb. 21. | 425

Frederick Webb, Esq.r of Albany
Rev.d Noel Digby – 62 Park Street.

To the Lectures.
M.rs Young – 48 Welbeck S.t
Miss Clitherow } 30 Wimpole S.t
Miss S. Clitherow }
Miss Annabel Pole Carew. New Cavendish Street.

Geo Duckett

Managers' Meeting, Feb. 28, 1814

At a meeting of the Managers of the Royal Institution of Great Britain, held at the house of the Institution, on the 28th of February, 1814.—

Present
Earl Spencer, President in the Chair.
Wm Blake, Esqr
Charles Hatchett, Esqr
John Guillemard, Esqr
Thomas Murdoch, Esqr
George Duckett, Esqr
Wm Watson, Esqr
Thos Harrison, Esqr Secretary.

Minutes appved 1. The minutes of the proceedings of the last meeting were read and approved.—

Minutes of Com. of Accts appved 2. Read and approved the minutes of the Committee of Accounts of this day.

3. Ordered. That the Secretary inform Mr Elmes that as the Lectures are all arranged for this year the Managers must decline his offer.

4. The following persons having been duly recommended were admitted as Subscribers, vizt

Life Subscriber — Life Subscriber.
Mr K. R. Hedges 31 Dean St Soho

Subscribers — Annual Subscribers.
Mr Briscoe — Twickenham.
Captn Thurston. R.N. 70 Gt Russell St

To the Lectures.
Miss Nesbitt. 27 upper Brook St
Mrs Jas Alexander. 37 upper Brook St
Mrs Fawcett. } 47 Portland place
Miss Fawcett. }
Mrs Simson. } 37 Portland place.
Miss Simson. }

Spencer.

Managers' Meeting 7th March, 1814.

At a meeting of the Managers of the Royal Institution of Great Britain, held at the house of the Institution, on the 7th of March, 1814.—

Present
Earl Spencer, President in the Chair.
Wm Watson, Esqr
Charles Hatchett, Esqr
Jno Guillemard, Esqr
Thos Murdoch, Esqr
Alexr Marcet, M.D.
Wm Blake, Esqr
Thos Harrison, Esqr Secretary.

Minutes approved. 1. The minutes of the proceedings of the last meeting were read and approved.

Minutes of Com. of Accts approved. 2. Read and approved the minutes of the Committee of Accounts of this day & signed an Order for payment of the following sum, vizt

Mr Harris Librarian 3/4 years Salary to 31st Decr last } £135. — —

Mr Landseer's Lectures. 3. Read a letter from Mr Landseer to Mr Fincher stating that he is ready to deliver two of his lectures before Easter — and that in consequence of being hard of hearing he would wish to be allowed to introduce a friend to the lectures without himself being personally present
Resolved. That his 2 Lectures be given on two Thursdays before Easter, and that in consideration of his affliction from deafness he be allowed to introduce a friend to the Lectures without being personally present.—

4. The following persons having been duly recommended were admitted as Subscribers, vizt

Subscribers — Annual Subscriber.
Whitshed Keene, Esqr M.P. Sackville St

To the Lectures.
Mrs Jeddere Fisher, 31 upper Grosvr St
Miss Adamson, 25 Hill St
Miss L Grant — 35 Dover Street

Spencer.

Managers' Meeting 14th March 1814

At a meeting of the Managers of the Royal Institution of Great Britain, held at the house of the Institution on the 14th of March, 1814.—

Present
Earl Spencer, President in the Chair.
Chas Hatchett, Esqr
Scrope Bernard Morland, Esqr Treasurer.
Wm Watson, Esqr
George Duckett, Esqr
Marquis of Lansdowne V.P.—
Alexr Marcet, M.D.
Thos Harrison, Esqr Secretary.

Minutes appved 1. The minutes of the proceedings of the last meeting were read and approved.—

Minutes of Com. of Accts appved 2. Read and approved the minutes of the Committee of Accounts of this day & signed an Order for payment of the following sums, vizt—

Mr Brooks 3/4 years Saly to 31st Decr last	£39. 7. 6
Gardner for Oil in 1813	57. 12. —
Thompson for Soap & Candles incd	38. 19. 6
Mr Hole for Dial &c	12. 16. 6
Mrs Novill Sundrs for the house	7. 8. 3
Henry, Charcoal in 1813	4. 1. —
	£160. 4. 9

Life Subscribers 3. In pursuance of nominations under the Act, the following persons were admitted to the privileges of Life Subscribers to the Royal Institution.
Archibald Home, Esqr of Sackville St by his father, Sir Everard Home, Bart.
& — Walter George Stirling Esqr of Hertford St by his Grandfather George Trenchard Goodenough, Esqr

4. Read the minutes of the last General Monthly Meeting.—

5. The following persons having been duly recommended

430

Subscribers — were admitted as Subscribers, viz.t — 1814.

Annual Subscribers.

Chas. Clowes, Esqr. of Dellaford Park Bucks
The Honble John Calthorpe Gros.r Sqe.
Saml. P. Wright, Esqr. Wood Green Tottenham.
S. M. Thriepland, Esqr. 32 St. Albans St.

To the Lectures.

Mrs. Rock 18 Albemarle St.
Miss Barker — } 39 Harley St.
Miss Elizth. Barker }
Miss Telfer, 77 upper Berkeley St.

Charles Hatchett.

431

Mar. 21.

Managers' Minutes 21st. March 1814

At a meeting of the Managers of the Royal Institution of Great Britain, held at the house of the Institution, on the 21st of March, 1814. —

Present

Chas. Hatchett, Esqr. in the Chair.
Jno. Walker, Esqr.
George Duckett, Esqr.
J. P. Auriol, Esqr.
Thos. Murdoch, Esqr.
Wm. Sotheby, Esqr.
Alexr. Marcet, M.D.
Earl of Selkirk, V.P.
Thomas Harrison, Esqr. Secretary.

Minutes appved. 1. The minutes of the proceedings of the last meeting were read and approved. —

Minutes of Com. of Accts appved. 2. Read and approved the minutes of the Committee of Accounts of this day, and Signed an Order for payment of the following sums, viz.t —

Mr. Newman Sundries for the Laboratory	£61.1.10
Josh. Fincher on acct. of the Royal Instn.	70.–.–
	£131.1.10

Lectures Mr. Landseer. 3. Read a letter from Mr. Landseer, to Mr. Fincher, requesting the privilege of being allowed, by the Committee of Managers, to introduce more than 4 persons to each of his own lectures. —
Resolved. That the Secretary inform him, that he cannot be allowed to introduce more than 4, the other Lecturers having no greater privilege.

Dr. Evans 4 Read a letter from Dr. Evans requesting permission of the Board to conclude the present Course of his Lectures with the 6th. Lecture. —
Resolved. That the Secretary inform him that the Managers are so pledged to the Public, that they must request him to deliver the 10 Lectures which he undertook to do —

432

1814.

Presents 5. Presents received.
Mr. D'Israeli, presented Quarrels of Authors 3 vol. 8vo.
Mattw. Raper, Esqr. presented.
Index ad Specimen Geographicum.
Ordered.
That the thanks of the Managers be returned for these presents.

Foreign Ambassadors 6 Dr. Marcet having informed the Board that the Count de la Gardie Ambassador from the King of Sweden to the Spanish Government was desirous of visiting the Royal Institution during his residence here.
Resolved.
That a letter of admission be written by the Secretary to him according to Art: 16 Chap VI of the Bye Laws.

Subscribers. 7. The following persons having been duly recommended were admitted as Subscribers, viz.t

Annual Subscribers.

Mr. Bertin, 63 Berners St.
Chas. Bosanquet, Esqr. 60 Lower Gros.r St.

To the Lectures

Miss Galton, 24. Gloster Place.

Charles Hatchett.

433

Managers Meeting March 28 1814

At a meeting of the Managers of the Royal Institution of Great Britain, held at the house of the Institution; on the 28th of March, 1814. —

Present

Chas. Hatchett, Esqr. in the Chair.
George Duckett, Esqr.
Jno. Guillemard, Esqr.
J. P. Auriol, Esqr.
Wm. Blake, Esqr.
Scrope Bernard Morland, Esqr. Treasurer.
Wm. Sotheby, Esqr.
Thos. Harrison, Esqr. Secretary.

Minutes appved. 1. The minutes of the proceedings of the last meeting were read and approved. —

Minutes of Com of Accts 2. Read and approved the minutes of the Committee of Accounts of this day.

Presents 3. Presents received.
Mr. J. J. Stockdale the Publisher presented.
The Origin of the Constitution of the Church & State in Great Britain.
& The Character of the Irish Roman Catholics.
Ordered.
That the thanks of the Managers be returned for these presents. —

Subscriber 4. The following person having been duly recommended was admitted as an Annual Subscriber, viz.t
Stephen Cave, Esqr. 3 Bolton row.

Spencer.

Managers Meeting, 4th April, 1814.

1814

At a meeting of the Managers of the Royal Institution of Great Britain, held at the house of the Institution on the 4th of April, 1814. –

Present
Earl Spencer, President in the Chair.
Charles Hatchett, Esqr.
Wm. Watson, Esqr.
George Duckett, Esqr.
J. P. Auriol, Esqr.
Thos. Murdoch, Esqr.
Jno. Walker, Esqr.
Marquis of Lansdowne, V. P.
Scrope Bernard Morland, Esqr. M.P. Treasurer
Alexr. Marcet, M.D.
Thos. Harrison, Esqr. Secretary.

Minutes appved. 1. The minutes of the proceedings of the last meeting were read and approved.

Minutes of Com. of Accts. 2. Read and approved the minutes of the Committee of Accounts of this day, and Signed an Order for payment of the following sum, vizt.

One years Ground Rent to the City of London to 25th March last } £135.1.8

Dr. Smith's Lectures. 3 Read a letter from Dr. Smith requesting to have 100 guineas for his Course of 12 Lectures on Botany, and to have 2 days together in each week for his Lectures.
Resolved.
That the Secretary do inform him that the Managers are willing to give him 100 guineas for 12 Lectures and to accomodate him as to his days of lecturing as far as they can consistently with their arrangements with the other Lecturers, and to request that he will give his first and second Lectures on Tuesday the 10th & Wednesday the 11th May.

Presents. Presents received.
4. Earl Spencer presented "Bibliatheca Spenceriana" 3 vol.

April 4 and James Park, Esqr. presented
"The Topography & Natural History of Hampstead" 8vo.
Ordered.
That the thanks of the Managers be returned for these presents. –

Subscriber. 5. The following person having been duly recommended was admitted as an Annual Subscriber, vizt.
Mr. Richard Rozea, 46. Paddington Street.

Adjourned to Monday the 18th Instant.

J. P. Auriol

Managers Meeting 18th Ap. 1814

1814

At a meeting of the Managers of the Royal Institution of Great Britain, held at the house of the Institution, on the 18th April, 1814. –

Present
J. P. Auriol, Esqr. in the Chair.
George Duckett, Esqr.
Thomas Harrison, Esqr. Secretary.

Minutes appd. 1. The minutes of the proceedings of the last meeting were read and approved. –

Minutes of Com. of Accts. appved 2 Read and approved the minutes of the Committee of Accounts of this day, and Signed an Order for payment of the following sums vizt. –

Mr. Searle for Bookbinding in 1813. £12.1.6
Josh. Fincher on account of the Instn. 70.–.–
£82.1.6

Lectures 3. Read a letter from Dr. Smith presenting his best thanks to the Managers for their acquiescence in his proposal and that he shall accomodate his time to theirs – His 1st & 2d Lectures will therefore be given on the 10th & 11th May. –

4. Read the minutes of the last General Monthly Meeting. –

Coal. 5 Ordered. That 35 Chaldrons of best Wallsend Coals be laid in at the Royal Institution at such time between this day and the 1st of October next as shall appear to Mr. Dickins to be most œconominal for the Institution. –

6. Read the Annual report of the Visitors on the state of the Institution when a few alterations were suggested and adopted by the Visitors.

7. The following persons having been duly recommended

Apl. 18 Subscribers were admitted as Subscribers, vizt.
Annual Subscribers.
John Calcraft, Esqr. M.P. 7 Clifford Street.
John Bedingfield Esqr. of the Navy Pay Office

Adjourned to Saturday next the 23d Instt. –

J. P. Auriol

Managers Meeting 23rd April 1814.

At an adjourned meeting of the Managers of the Royal Institution of Great Britain, held at the house of the Institution on Saturday, the 23rd of April 1814.—

Present
Thos. Murdoch, Esqr. in the Chair.
Jn. Walker, Esqr.
J. P. Auriol, Esqr.
Chas. Hatchett, Esqr.
George Duckett, Esqr.
Jn. Guillemard, Esqr.
S. Bernard Morland, Esqr. Treasurer.
Thos. Harrison, Esqr. Secretary.

House List. Ordered. That the House Lists be filled up as follows.—

President. Earl Spencer, K. G. F.R.S. &c. &c.

Treasurer.— S. Bernard Morland Esqr. M.P. L.L.D.

Secretary.— Thomas Harrison Esqr. F. R. S.—

Managers.
J. P. Auriol Esqr. F. R. S.
Sir Thos. Bernard Bart. L.L.D.
Earl of Chichester. F.R.S.
Daniel Giles, Esqr.
Charles Hatchett, Esqr. F.R.S. & F. A. S.
Sir Benjn. Hobhouse, Bart. F.R.S.
Henry Holland, Esqr.
Marquis of Lansdowne F.R.S.
Earl of Morton, K.T. F.R.S.
Daniel Moore Esqr. F.R.S. F.A.S. & F.L.S.
Rt. Honble. Sir Jn. Nicholl, L.L.D. F.R.S. &c.
Matthw. Raper Esqr. F.R.S. V. Pres A.S.
Thos. Snodgrass, Esqr.
William Watson, Esqr. F.R.S.
Earl of Winchilsea & Nottingham. K.G. &c. &c.

Apl. 23

Visitors.
George Aust, Esqr. F.R.S. & F.A.S.
Revd. Chas. Burney D.D. F.R.S. & F.A.S.
Rt. Hon: Reginald Pole Carew. M.P. F.R.S. &c.
Earl of Darnley F.R.S.
Wm. Dunlop, Esqr.
Rt. Hon: Lord Grenville
Sir J. C. Hippisley, Bart. M.P. L.L.D. F.R.S. &c.
Rt. Hon: Lord Holland.
Wm. Hasledine Pepys, Esqr. F.R.S.
Louis Hayes Petit, Esqr. F.R.S. &c.
Charles Pieschell, Esqr.
Claude Scott, Esqr.
Richd. Horseman Solly, Esqr. F.R.S. & F.A.S.
Lt. Genl. William Thornton.
Saml. Thornton, Esqr. F.A.S.

J P Auriol

Managers Minutes

At a meeting of the Managers of the Royal Institution of Great Britain held at the house of the Institution, on Monday the 25th of April, 1814.—

Present
J. P. Auriol, Esqr. in the Chair.—
Chas. Hatchett, Esqr.
George Duckett, Esqr.
Leonard Horner, Esqr.
Jn. Guillemard, Esqr.
Thos. Harrison, Esqr.—Secretary.—

Minutes appd. 1. The minutes of the proceedings of the last meeting were read and approved.—

Minutes of Comee. of Acts. 2. Read and approved the minutes of the Committee of Accounts of this day.—

Presents Presents received.—

3 The Society for the Encouragement of Arts presented Vol 31 of their Transactions.—
John Britton Esqr. presented "The Rights of Literature".
Ordered.
That the thanks of the Managers be returned for these presents.—

New Policy of Insurance 4 The renewed and extended Policy of Insurance having been received from the Sun Fire Office was—deposited in the Iron Chest of the Royal Institution.—

Spencer

Managers Meeting

At a meeting of the Managers of the Royal Institution of Great Britain, held at the house of the Institution, on Monday the 2d of May, 1814.—

Present
Earl Spencer P. in the Chair.
Jn. Guillemard, Esqr.
George Duckett, Esqr.
Earl of Winchilsea, V.P.
Wm. Watson, Esqr.
J. P. Auriol, Esqr.
Thos. Murdoch Esqr.
S. Bernard Morland, Esqr. Treasurer
Thos. Harrison, Esqr. Secretary.

Minutes approved 1. The minutes of the proceedings of the last meeting were read and approved.

Minutes of Comee. of Acts. approved 2 Read and approved the minutes of the Committee of Accounts of this day, and signed an Order for payment of the following sums viz.:

Mr. Royce for Repairs of the House in 1813	£88. 15. 11
Mr. Sheppard for Upholstery in 1813	60. 18. 2.
	£149. 14. 1

Moniteur ordered 3 Read a letter from Mr. Stanhope of the Foreign Post Office, addressed to Mr. Fincher, stating that the Clerks of his Office will now have it in their power to supply the Royal Institution with the French Papers.
Ordered.
That the Moniteur be taken.

Subscribers 4. The following persons having been duly recommended were admitted as Subscribers viz.:
To the Lectures.
Mrs. Henry Clive Lower Brook St.
Rt. Hon: Lady Elizth. Thynne Grosvenor Square

5 Ordered. That letters of admission be written by the

Secretary to Prince Gagarin and Count Orloff inviting their attendance upon the lectures of the Royal Institution. –

Spencer

Managers Meeting May 9th, 1814.

At a meeting of the Managers of the Royal Institution of Great Britain, held at the house of the Institution on Monday the 9th of May, 1814. –

Present
Earl Spencer Pt. in the Chair. –
Earl of Morton.
Wm. Watson, Esqr.
S. Bernard Morland, Esqr. Treasurer.
Chas. Hatchett, Esqr.
Thos. Snodgrass, Esqr.
James Peter Auriol, Esqr.
Sir Thos. Bernard, Bart.
Daniel Moore, Esqr.
Henry Holland, Esqr.
Thos. Harrison, Esqr. Secretary.

Minutes appved. 1. The minutes of the proceedings of the last meeting were read and approved.

Minutes of Com. of Accts. appved. 2. Read and approved the minutes of the Committee of Accounts of this day.

Vice Presidents 3. The President nominated as Vice Presidents for the ensuing year. –

The Marquis of Lansdowne.
Earl of Winchilsea.
Earl of Morton
Rt. Honble. Sir John Nicholl.
Sir Thos. Bernard, Bart.
J. P. Auriol, Esqr.

4. The following persons having been duly recommended were admitted as Subscribers, viz.:

Life Subscriber
Life Subscriber.
William Gaussen, Esqr. 12 Paper Buildgs. Temple
Annual Subscriber.
Saml. Folfrey, Esqr. Hans Place Sloane St.

Spencer

Managers Meeting 16th May 1814.

At a meeting of the Managers of the Royal Institution of Great Britain, held at the house of the Institution on the 16th of May, 1814.

Present
Earl Spencer, President in the Chair.
Earl of Morton, V.P.
Charles Hatchett, Esqr.
Wm. Watson, Esqr.
Daniel Moore, Esqr.
J. P. Auriol, Esqr. V.P.
Rt. Hon: Sir John Nicholl, V.P.
Thos. Snodgrass, Esqr.
Henry Holland, Esqr.
S. Bernard Morland, Esqr. Treasurer.
Thos. Harrison, Esqr. Secretary.

Minutes appd. 1. The minutes of the proceedings of the last meeting were read and approved.

Minutes of Com. of Accts. appved. 2. Read and approved the minutes of the Committee of Accounts of this day. –

Present

Present received.

3 The Lord Bishop of Durham presented.
The Political Life of Viscount Barrington, 4to.
Ordered.
That the thanks of the Managers be returned for this present.

4. Read the minutes of the last General Monthly Meeting

Mr. Lee's Lectures. 5. Mr. Moore reported that Mr. Lee the Clerk to the Royal Society will give a course of Lectures on Astronomy during the next Season.
Resolved. That Mr. Moore be requested to state to Mr. Lee that the Managers cheerfully accept of his offer and will allow him 5 guineas for each lecture – The course to consist of 12 Lectures or thereabouts. –

6 Resolved. That Mr. Hatchett be requested to apply to

Dr. Roget & the Revd. Edward Fowter to know whether they are disposed to give Lectures at the Royal Institution the next Season. –

7 Read a nomination from Sir Thos. Clarges Bart. of his Cousin Captn. Bagnold of Knightsbridge to be admitted to the privileges of a Life Subscriber to the Royal Institution
Resolved.
That the Secretary do inquire of Sir Thos. Clarges whether the Relationship is such as is required by the Act of Parliament. –

Spencer

446.

Managers Meeting 23rd May 1814.

1814.

At a meeting of the Managers of the Royal Institution of Great Britain, held at the house of the Institution, on the 23.d of May, 1814 —

Present

Earl Spencer, President in the Chair.
Ja.s Peter Auriol, Esq.r V. P.
Cha.s Hatchett, Esq.r
R.t Hon: Sir Jn.o Nicholl, V.P.
Dan.l Moore Esq.r
Henry Holland, Esq.r
Tho.s Snodgrass, Esq.r
Sir Tho.s Bernard, Bart. V.P.
Tho.s Harrison, Esq.r Secretary.

Minutes app.ved 1. The minutes of the proceedings of the last meeting were read and approved.

Minutes of Com. of Acct.s app.ved 2. Read and approved the minutes of the Committee of Accounts of this day.

Life Subscriber. 3. In pursuance of a nomination under the Act by Robert Holford, Esq.r his Neice Miss Mary Ann Bosanquet of Grosvenor S.t was admitted to the privileges of a Life Subscriber to the Royal Institution. —

Presents.

Presents received.

4 The Society of Antiquaries, presented Archaeologia vol 17. —
Society for bettering the Condition of the Poor, presented N.os 35 & 36 of their Reports. —
Sam.l Ware Esq.r M.R.I. presented Observations on Vaults. —
Ordered. That the thanks of the Managers be returned for these presents —

5 The Secretary reported that he had written to inquire of Sir Tho.s Clarges the Relationship between him and Capt.n Bagnold but had not yet received an answer.

447.

May 23. 6 M.r Hatchett reported that he had applied to D.r Roget to know whether he could give a Course of Lectures on Comparative Physiology at the Royal Institution the next Season but that it interfered so much with his Professional duties that he could not venture to engage giving a Course next Season. —

Lectures. 7. M.r Hatchett stated that M.r Wood would give a Course of Lectures on Perspective next Season gratuitously. Resolved. That M.r Wood's offer be accepted.

8. M.r Moore reported that M.r Lee would give 12 — Lectures at 5 guineas each Lecture.

Contract for Entertaining the Duchess of Oldenburg. 9. Her Imperial Highness the Grand Duchess of Oldenburgh having signified her desire of visiting the Royal Institution, Received the following estimate from M.r Gunter of Berkeley Square for a Cold Collation —

"As under, I offer for your consideration, the Bill of Fare, the expence of which, the hire of Silver waiters, Tea & Coffee Urns, Knives, Forks, Spoons, Desert d.o & Tea spoons & tongs, Cut goblets, wines, sweet wines, Decanters, Carofts, fine China Dishes, Plates Tea & Coffee China, Table Cloaths, napkins, Tea Coffee, Sugar, Sugar Candy, Cream, Rolls, Muffins, wine, Noyeau, Ices, persons to attend and every requisite, 20 guineas

Resolved. That it be accepted & that the expence be defrayed by the Managers

Hon. Admiss.n 10. The Count Beraldingen Envoy Extraordinary & Minister Plenipotentiary from the Court of Wirtemburg & Mons.r Wagner his Secretary of Legation having respectively applied for admission to the Royal Institution.
Ordered. That the Secretary do send them Letters of admission. —

Present. 11. M.r Harrison as a Member of the Council of the Royal Society having communicated that the Council had resolved to present the Royal Institution with such numbers of their Transactions as the Royal Institution are not in possession of
Ordered. That the thanks of the Managers be returned to the President & Council of the Royal Society for these presents. —

Spencer.

448

Managers Meeting 30th May 1814.

1814.

At a meeting of the Managers of the Royal Institution of Great Britain, held at the house of the Institution; on the 30.th May, 1814. —

Earl Spencer, President in the Chair
Cha.s Hatchett, Esq.r
W.m Watson, Esq.r
Earl of Morton, V.P.
Matthew Raper, Esq.r
S. Bernard Morland, Esq.r Treasurer.
R.t Hon: Sir John Nicholl V.P.
Tho.s Snodgrass, Esq.r
J. P. Auriol, Esq.r
Daniel Moore, Esq.r
Henry Holland, Esq.r
Sir Tho.s Bernard, Bart. V.P.
Tho.s Harrison, Esq.r Secretary.

Minutes app.ved. 1. The minutes of the proceedings of the last meeting were read and approved.

Minutes of Com. of Acct.s approved, 2. Read and approved the minutes of the Committee of Accounts of this day, & Signed an Order for an advance to the Clerk of £80. —

Hon.y admissions 3. M.r Manning having stated that E. Bourke Esq. Minister Plenipotentiary from the Court of Denmark is desirous of attending the Royal Institution.
Resolved
That the Secretary do invite M.r Bourke to attend. —

Presents. 4. Presents received
D.r Jos.h Adams presented "Treatise on Diseases."
M.r John Martin, the publisher, presented.
"The Inquirer or Literary Miscellany N.o 1."
Ordered.
That the thanks of the Managers be returned for these presents.

5 The following persons having been duly recommended

449

May 30. Subscribers were admitted as Subscribers, viz.t
To the Lectures.
Marchioness of Sligo
Hon: Miss Curzon
Hon: Miss A Cust
M.rs Wildman
Miss Wildman } 7 upper Cumberland St.
Miss Langston }

Spencer.

Minutes of Managers' Meetings
1799–1900

Volume VI

With an introduction by

Morris Berman
Assistant Professor of History, Rutgers University

INTRODUCTION

"No Newton," wrote Thomas Carlyle in *The Edinburgh Review* in 1829,

> by silent meditation, now discovers his system of the world from the falling of an apple; but some quite other than Newton stands in his Museum, his Scientific Institution, and behind whole batteries of retorts, digesters, and galvanic piles, imperatively "interrogates Nature,"—who, however, shows no haste to answer.

The age of the independent scientist had not yet passed when Carlyle wrote these lines, and in fact the scientific amateur remained on the English scene long after the industrial age had arrived; but it was nonetheless significant that Carlyle had entitled the essay in which these words appeared, "Signs of the Times". By 1829 the scientific leadership of the Royal Institution had been, for already more than a decade, in the hands of William Thomas Brande, whose interests in chemistry, unlike those of his predecessor, literally never strayed into purely intellectual areas. Davy saw his practical investigations as an obligation to present and future patrons, a way of selling science so that research, his "real" work, could proceed. Brande, as Davy's brother once remarked, had no such lofty aims. The "real" work of science was for him its ability to solve short-range, practical problems, and his businesslike approach led to important modifications in the social function of the R.I.

That Brande, an apothecary, could assume and maintain leadership of the Royal Institution, is a development that was presaged in the formal electoral changes of 1810, described in the introduction to the previous volume. The Institution's original fortunes, and Davy's brilliant career, had been closely tied up with the semi-official Board of Agriculture, whose members dominated the R.I.'s governorship for many years. That monopoly of control had led to a financial crisis for the R.I., requiring it to open its doors to wider participation, specifically to more middle-class and professional groups. In 1822 the already ineffective Board dissolved itself, and shortly after that control of the R.I. passed out of agricultural hands. Between 1821 and 1826 twenty-five per cent of the governors still belonged to the landed interest, but thirty per cent could now be classified as doctors, lawyers, or scientific administrators. It was this change in composition that made Brande's career, and the influence he exerted upon the Royal Institution, possible.

For this reason, it would be unfair to apply Carlyle's closing phrase too rigidly to Davy's successor: it was not the case that with Brande, Nature "showed no haste to answer". Rather, he was asking different questions. Brande's interest lay in gathering information relevant to medical practice, and his earliest researches were on subjects such as respiration, teeth, urine, blood, albumen, and calculi. Brande had done some informal lecturing on pharmaceutical chemistry as far back as 1808, and he was appointed Professor of Chemistry and Materia Medica at the Apothecaries' Society the same year that he began his professorship at the Royal Institution. Although he branched out into various industrial and technical areas, in many ways emulating the work of his Scottish colleague and friend, Andrew Ure, Brande's particular role, as Aubrey Tulley has written in his biographical sketch, "was that of a link-man between chemistry and medicine". In 1825 he published *A Manual of Pharmacy* and in 1839 his *Dictionary of Materia Medica and Practical Pharmacy*. In 1851 he was elected Master by the Society of Apothecaries. His interest in the ancillary value of chemistry to medicine proved to be life-long.

Brande's medical orientation was no accident. He had been born into a family of apothecaries, and it was only natural that he continue the tradition. But that this tradition should have had an impact upon scientific organisation *per se* was hardly to be expected, and it would have been unlikely without more general developments taking place *vis-à-vis* the social role of the traditional professions. Up to about 1750 the professions were marginal to the economy; physicians, for example, normally served only a small number of wealthy patients. The doctor was first and foremost a gentleman, with technical expertise playing a minor, and knowledge of the classics a major, role in his education. Within fifty years, however, the situation had

changed considerably. Urbanization, a rapid increase in population, and problems of overcrowding and sanitation created a middle-class and lower-middle-class demand for medical care, and thus a widening opportunity for surgeons, apothecaries, and druggists. In 1806 the apothecaries began to organise to take advantage of these opportunities as well as to prevent encroachment by quacks, and in 1815 "the Act for better regulating the Practice of Apothecaries throughout England and Wales" (55 Geo. III, c. 194) established a system of qualification and registration along modern lines. To be licensed by the Society the student had to attend lectures on anatomy, medicine, physiology, chemistry, and materia medica, and thus the demand for medical—i.e. scientific—education soon began to increase.

Brande's involvement in these issues predated this legislation. He first proposed using the Institution's laboratory as a classroom for medical students in October of 1813 (Managers' Minutes, V, 394), and with the passage of the Apothecaries' Act the Managers agreed (p. 77)

> that it would be highly advantageous as well as hono[u]rable to the Institution, to give authority to M^r^. Professor Brande to transfer his Course of Lectures on Chemistry from the room in Windmill Street to the laboratory of the Royal Institution . . . .

What had been an agricultural laboratory during the early years was now to become, at least in part, a place of medical instruction. Three times a week Brande, and later Faraday (his re-appointment is noted on p. 58) would lecture on chemistry to the Windmill Street students and also to the students of St. George's Hospital (see the Managers' Minutes, IX, 416) and give examinations in the course as well. In both teaching and publication, Brande and Faraday worked as a team. Brande's *Manual of Chemistry* (1819), which acknowledged Faraday's assistance, was written to give the R.I. medical student a general working knowledge of the subject, while Faraday's *Chemical Manipulation* (1827), surprisingly the only book he ever wrote, was designed to meet the needs of students seeking to acquire sound laboratory technique. Chemistry was particularly emphasized by the new generation of surgeons and apothecaries because of its supposed importance for drug preparation, and Thomas Wakley editorialized in *The Lancet* that the subject was "now justly regarded as the ground-work of all medical knowledge". In October 1827 *The Lancet* began an extensive, possibly verbatim, coverage of the chemistry lectures being given at the R.I. In a later issue Wakley listed the Institution under a column entitled "The Medical and Surgical Schools of London", and described the course as including "a full examination of pharmaceutical chemistry; the processes of the pharmacopoeiae will be similarly described and compared with those adopted by manufacturers" (*The Lancet*, 26 September 1829, pp. 11–12). In this way the Royal Institution was becoming publicly identified as serving the medical community, but the reverse was also true. Brande's lectures provided a regular source of income for the Institution (p. 81), and the Managers recognised that they "have been productive of great advantage to the Royal Instit^n^. and have greatly raised its character as a school of Chemistry . . ." (Minutes, VII, 54). If the Institution contributed to new trends in medical professionalization, it is also the case that much of the prestige and organisational consolidation of the R.I. can be traced to its orientation toward middle-class concerns. Science, it might be suggested, did not get professionalized in terms of itself, but rather in terms of its ability to accelerate the professionalization of already established occupations.

A second professional development initiated by Brande is also recorded in this volume, namely his decision to edit and publish a journal under Institution auspices (pp. 78–79). *The Journal of Science and the Arts* ran (under various titles) from 1816 to 1831, and reflected, to some extent, Brande's particular bias toward medicine and industrial chemistry. It also served as an advertisement for the Royal Institution, carrying the Institution's seal and giving prominent attention to lectures and laboratory work in progress. Its importance to medicine did not go unnoticed, and *The Lancet* commented approvingly that the "medical man *ought* to be a scientific man". But Brande's scope was in fact more catholic than this. He saw the *Journal* as an essentially new enterprise, unlike the publications of, say, David Brewster or William Nicholson. The *Journal* was intended as a co-ordinator of information relevant to the concerns of a new society, reviewing the latest publications and keeping the technical reader up-to-date on the most recent developments. It was to be a tool and not, like the *Philosophical Transactions of the Royal Society*, an archive. Brande was a new type of scientist, as Thomas Carlyle had recognised; a sophisticated mixture of laboratory analyst and civil servant. His career was characterized by medical research, lectures, textbooks, handbooks, and consultant work, and

the *Journal* was another aspect of his desire to organise factual information to meet modern professional requirements.

Yet the Royal Institution continued to cater to polite society, as would be expected. The chemistry classes were, after all, held in the laboratory rather than the lecture theatre, and the *Journal* retained, if only out of financial necessity, a lively interest in meteorology, natural history, and other traditional concerns of lay scientists. But more and more the Institution began to take on the general appearance of the larger society: a staid and elegant aristocratic shell with a dynamic entrepreneurial-professional kernel. In English science, amateurism and expertise, research and professionalization, were by definition mutually exclusive, and the R.I. is at least one model of how these distinctions were maintained. If polite society continued to enjoy popular lectures, it was the more utilitarian needs of professional society to which Brande felt his allegiance. By the 1820's Brande was considered the leading chemist of the metropolis and the R.I. London's leading laboratory. All kinds of routine analyses were turned over to Brande and Faraday by government, industry, and private individuals, some of whom apparently were disrupting the work of the laboratory by using its facilities for their own experiments (p. 349). In 1816 Brande testified before a Parliamentary commission on leather manufacture, and in 1823 undertook an investigation on metal dyes for coinage for the Mint. Faraday's best known work in this vein was his research on steel for one of the Institution's governors, James Stodart, a surgical instrument maker. Given this activity it is not surprising that Davy resented Brande's influence, despite his own earlier promotion of practical chemistry. Thus we read, at a meeting at which Davy was present, the Managers' resolution that "due attention [be given] to preserving the Laboratory as a place of Experimental research on Chemistry" (p. 81). But it was not until the 1830's that Faraday had time to return to such research, and despite his spectacular results the Managers found Brande's businesslike thinking more in line with their own.

There are other issues in this volume that deserve our attention, and which probably merit further investigation. Even with the changing composition and scientific direction of the Institution, it nevertheless remained in debt throughout the period of this volume. As early as page 3 we read of a voluntary subscription to discharge debts and provide for the Institution's future support. In 1816 the Managers temporarily suspended the purchase of books (pp. 124–25), revealing their concern with "the present depressed state of the Finances of the Institution . . ." (p. 127). In 1818 John Fuller loaned the R.I. £1000, but by 1823 the Managers were contemplating selling the stock purchased by the loan to meet certain expenses (pp. 183 and 394). The Managers also found it necessary to shorten the hours of the Reading Room and discharge the librarian (pp. 398 and 400). Further discussion of finances and this "system of retrenchment", as they called it, occurs on pages 402–4, and on page 408 we learn that, as in past years, solvency was maintained by loans from the governing group. A separate subscription for the laboratory was raised by soliciting Members to contribute one guinea per annum (pp. 415 and 425).

If the Institution was financially shaky, its scientific reputation was quite secure. Its lecturers were occasionally asked to repeat their course elsewhere, as in the case of John Millington, who was sought after by the Surrey Institution (p. 74). A number of gratis lectures are also recorded, for example by John Soane on architecture and John Frost, Professor of Botany to the Medico-Botanical Society (pp. 157, 160, 262, 295–96, and 378). Lecturing at the R.I. had apparently become a way of establishing or enhancing one's reputation, since the Institution customarily recruited its lecturers from among England's leading scientific and literary figures. Charles Babbage, for example, was engaged to lecture on astronomy (p. 30), and page 366 records his presentation of "A letter to Sir Humphry Davy, Bart. on the application of Machinery to the purpose of calculating and printing Mathematical tables".

One of the least attractive episodes in the Institution's history, the probably fraudulent charges brought against Frederick Accum for mutilating library books, which followed soon after his muckraking treatise on food adulteration that incensed a number of influential citizens, is also recorded in this volume (pp. 303–5, 320–21, and 352). Accum, at one time Davy's chemical assistant, was principally known for his interest in commercial chemistry and especially gas illumination, and presented his "Practical Treatise on Gas-Light" to the R.I. in 1815 (p. 58). Brande and Faraday had also done research on coal gas, and modern illumination was a favourite topic in the *Journal*. Brande had directed the installation of gas lighting at Apothecaries' Hall, and a gas apparatus at the R.I. laboratory and lecture theatre, in 1816 (pp. 79–80). Gas lighting for the Institution was first considered in 1818 (p. 225). Brande's

involvement in these matters may have led to his subsequent discovery of naphthalene in coal tar, which was first analysed by Faraday.

During these years, Sir Joseph Banks' long presidency of the Royal Society, and his domination of the English scientific scene, came to a close. He bequeathed a substantial section of his library to the R.I. in 1819 (p. 234), and died the following year. Another founder of the R.I., Sir Thomas Bernard, also died during this time, and his posthumous autobiography was presented to the Institution by the Reverend James Baker (p. 258).

Rutgers University
April, 1973

Minutes

of the

Meetings of Managers.

Vol. VI.

(June, 1814 to Dec. 1824.)

1

Managers' Meeting. 6th June 1814.

At a Meeting of the Committee of Managers of the Royal Institution of Great Britain, held at the house of the Institution on Monday the 6th June 1814. —

Present

Earl Spencer President in the Chair.
James Peter Auriol, Esq.r V.P.
Cha.s Hatchett, Esq.r
Matt.w Raper, Esq.r
Wm Watson, Esq.r
Sir Tho.s Bernard, Bart. V.P.
Tho.s Snodgrass, Esq.r
Daniel Moore, Esq.r
Daniel Giles, Esq.r
Earl of Morton, V.P.
Sir Jn.o Nicholl, V.P.

Minutes approved. 1. The minutes of the proceedings of the last meeting were read and approved.

Minutes of Com. of Acct.s approved. 2. Read and approved the minutes of the Comm.t of Accounts of this day.

Mr Craig's offer of Lectures on Drawing gratuitously. 3 Read a letter from W. M. Craig, Esq.r offering to read gratuitously in the Lecture Room A Course of Eight Lectures on the principles and practice of Drawing and Painting as branches of elegant Education. —

Resolved.

That Mr Craig's offer be accepted with the thanks of the Managers.

Life Subscriber 4 In pursuance of a nomination under the Act by Thomas Warre, Esq.r his Nephew
John Ashley Warre, Esq.r M.P.
was admitted to the privileges of a Life Subscriber to the Royal Institution. —

J P Auriol

2

June 13. 1814.

At a Meeting of the Managers of the Royal Institution of Great Britain held at the house of the Institution, on the 13th of June 1814. –

Present

James Peter Auriol, Esqr. V.P. in the Chair.
Charles Hatchett, Esqr.
Matthew Raper, Esqr.
Wm. Watson, Esqr.
Daniel Moore, Esqr.
Henry Holland, Esqr.
Sir Thos. Bernard, Bart. V.P.
S. Bernard Morland, Esqr. Treasurer.

1. The minutes of the proceedings of the last meeting were read and approved.

Payments. 2. Read and approved the minutes of the Committee of Accounts of this day, and Signed an Order for payment of the following sums, vizt.

Upton & Co. for Oil in 1813 ——	£ 48. 7. 6
Allen — Chemicals " ——	47. 14. 3
Knight & Co. Chemicals in 1813 ——	13. 3. 4
Bulmer – Printing . . " " ——	110. 15. 6
Dalton & Keys – Cards " " ——	69. 5. 6
Jones & Yarrell Newspapers . . ——	100. 18. 5
Wrangham — Do. . . ——	38. 13. 2
Wrangham . — Stationary . . ——	35. 17. 10
Feetham — Ironmongery ——	50. 7. 9
Mr. Howis – Sundt. for the house ——	21. 1. 7
Mr. Hammond – Linendrapery in 1813 ——	12. 13. 5
Mr. Newman Sundt. for the Lectures ——	45. 16. 2
Mr. Parker Glaziery in 1813 ——	10. 18. 9
De Boffe – Foreign Books in 1813 ——	21. 6. 6
	£ 626. 19. 8

3. Presents received.

The Commissioners of the Public Records presented

3

June 13. Rotuli Scotiæ vol 1. folio –
Valor Ecclesiasticus temp. Hen. VIII vol 2. folio.
Ordered. That the thanks of the Managers be returned for these presents. —

Scientific Committees. 4. Ordered. That the following Committees be requested to meet for the election of two Chairmen and a Secretary for each Committee viz!

Chemistry, Geology &c. on Thursday the 16th. at 1 o'Clock
General Science Literature &c. Friday 17th. do. —
Mathematics, Mechanics &c. — Tuesday 21st. do. —

List of Benefactions 5 Ordered. That a List of the "Benefactions to the Fund for discharging the Debts of the Royal Institution and for providing for its future support" be printed and sent round on the two next Saturdays to the Members and Subscribers. –

6. Read the minutes of the last General Monthly Meeting. –

Spencer.

4

June 20. 1814.

At a Meeting of the Managers of the Royal Institution of Great Britain, held at the house of the Institution; on the 20th of June, 1814. –

Present

Earl Spencer, President in the Chair.
Chas. Hatchett, Esqr.
Thos. Snodgrass, Esqr.
Sir Thos. Bernard, Bart. V.P.

1. The minutes of the proceedings of the last meeting were read and approved.

2. Read and approved the minutes of the Committee of Accounts of this day, and Signed an Order for payment of the following sums, vizt.

Payments

Dr. Smith for a Course of Lectures in 1814	£ 105. —
Coachhire & expt. Collecting Specimens ——	6. 10. 6
Mr. Reeve for ¼ years Assessed taxes to Ladyday	35. 17. 9
Mr. Butler for do. – Parochial " — " ——	12. 18. 4
	£ 160. 6. 7

Present received.

3. Prince Hoare, Esqr. M.R.I. presented
Epochs on the Arts including Hints on the use & progress of Painting & Sculpture in Great Britain. 8vo.
Ordered.
That the thanks of the Managers be returned for this present.

4. Read a Recommendation from the Visitors entered in their proceedings of the 13th. Inst. —

Lectures. Rev. E. Forster 5. Mr. Hatchett reported that the Revd. Edward Forster will give a Course of Lectures in the next Season upon Oratory, or upon the Poetry of Milton, at the option of the Managers. –
Resolved. That Mr. Hatchett be requested to inform

5

June 20. Mr. Forster that the Managers would prefer a Course of Lectures on the Poetry of Milton. –

JP Auriol

June 27. | 1814.

At a Meeting of the Managers of the Royal Institution of Great Britain, held at the house of the Institution on the 27th June, 1814.

Earl Spencer, President, in the Chair.
J. P. Auriol, Esqr. V.P.
Chas. Hatchett, Esqr.
Sir Thos. Bernard, Bart. V.P.
Wm. Watson, Esqr.
Daniel Moore, Esqr.
S. B. Morland, Esqr. Treasurer.

1. The minutes of the proceedings of the last Meeting were read and approved.

2. Read and approved the minutes of the Committee of Accounts of this day, and signed an Order for payment of the following sums, viz.

Payments

Wm. Thos. Brande, Esqr. 1 years Salary as Professor of Chemy	£200 . – . –
Dr. Roget – for a course of Lect. in 1814	105 . – . –
Revd. Edwd. Forster – do.	105 . – . –
Thos. Campbell, Esqr. do.	52 . 10 –
Dr. Evans do.	52 . 10 –
J. Landseer, Esqr. do.	52 . 10 –
Josh. Fincher on acct. of the Royal Instn.	80 . – –
	£647 . 10 –

3. In pursuance of a nomination under the Act by Sir Mark Wood Bart, his Son
Mark Wood, Esqr. of Pall Mall
was admitted to the privileges of a Life Subscriber to the Royal Institution.

Easter Offerings

4. The Revd. R. Hodgson Rector of this Parish having applied for the Easter Offerings.
Resolved. That £2.2.0 be paid to his Collector and that the same be continued annually. –

Spencer

July 4.

At a Meeting of the Managers of the Royal Institution of Great Britain, held at the house of the Institution on the 4th July, 1814. –

Present
Earl Spencer, President in the Chair.
J. P. Auriol, Esqr. V.P.
Daniel Moore, Esqr.

1. The minutes of the proceedings of the last meeting were read and approved. –

2. Read and approved the minutes of the Committee of Accounts of this day, and Signed an Order for payment of the following sums, viz.

Payments

Mr. Harris for ½ years Salary to 30th. ult.	£90 . . –
Mr. Brooks for do. to do.	26 . 5 . –
Mr. Reeve for Fares to 25th. March	32 . 8 . 4
Josh. Fincher on acct. of the Royal Instn.	80 . . –
	228 . 13 . 4

Lectures.

3. Read a letter from The Revd. Edward Forster, to Chas. Hatchett, Esqr.; stating that he shall be happy to comply with the wishes of the Managers in reading a Course of Lectures on the Poetry of Milton which He will endeavour to comprize within 6 Lectures and that He shall consider the remuneration of 60 guineas as ample.
Agreed to. –

Present received.

4 Revd. Stepn. Weston presented
The Chinese Tale of Fan – Hy – Chew.
Ordered. That the thanks of the Managers be returned for this present.

5 Adjourned to Monday the 1st of August.

JP Auriol

Aug. 1. | 1814.

At a Meeting of the Managers of the Royal Institution of Great Britain; held at the house of the Institution, on the 1st of August, 1814. –

Present
J. P. Auriol, Esqr. V.P. in the Chair.
Thomas Snodgrass, Esqr.
Henry Holland, Esqr.

1. The minutes of the proceedings of the last meeting were read and approved.

2. Read and approved the minutes of the Committee of Accounts of this day, and Signed an Order for payment of the following sums, viz.

Payments.

Mr. Newman, Sunds. for the Lectures on Chemy.	£25 . 3 . 2
Ditto do. do. on Astrony.	7 . 7 . –
Josh. Fincher on acct. of the Royal Instn.	90 . . –
	£122 . 10 . 2

Lectures.

3. Read a letter from Thos. Campbell, Esqr. dated 30th. July 1814 addressed to the Secretary of the Royal Institution, on the subject of his Lectures on Poetry for the ensuing Season. –

Resolved. The Managers present feel themselves very desirous of continuing to benefit the Institution by the ability and talents of Mr. Campbell and hope that an arrangement may be made with him for delivering a Course of Lectures in the ensuing Season. Ordered therefore that the Clerk do request that Mr. Campbell will propose a plan for their consideration at the next Meeting on the 7th of Novr. next.
Resolved. That the consideration of the other parts of Mr. Campbell's letter be deferred till that Meeting.

Estimate of Repairs.

4. Read the following Estimate made by Mr. Royce of the expence of painting and repairing the House of the Institution. –

Repairing Ceiling of Lecture Room washing & whitewashing do.

Aug. 1.

the passages round the covered way of do. the Staircase leading to Gallery water closet and erecting Scaffolding to do.	£60 . . – . –
Repairing the Ceiling of Lanthern, which lets down to darken the Lecture room	10 . – . –
Painting Lecture room Gallery & Passages round do. the Stairs leading to Gallery & Apparatus room adjoining	78 . . –
Entrance Hall best Stair-case & Passages leading to Rooms, Water Closet on Ground floor, the Ceiling distempered of front Hall	68 . – . –
Reading Room adjoining Hall	11 . 12 . –
Library on Ground floor	16 . 14 . –
Newspaper Room & Clerks' Office	16 . 19 . –
Library of Reference. Painting all those parts which are now white & whitewashg. Ceiling & Cornices	13 . 10 . –
Whitg. Ceilings & Cornice of Mineral rooms & painting Walls & Doors of Rooms	5 . – . –
Managers Room – painting walls windows & Glass cases & whitewashg. Ceiling & Cornices	10 . 2 . –
Passage leading to do. being part of Model room	6 . –
The Back Staircase & passages leading to Mr. Brande's Apartments – painting all those parts formerly painted – Colouring walls of do. – white-washing Ceilings & Stairs; & painting window frames &c.	36 . 8 . –
Painting the walls of back Staircase in Oil colour 5 ft. high	10 . – –
Painting & whitewashing Housekeeper's room Kitchen & Passages leading to do.	15 . 10 –
Carrd. forwd.	£357 . 15 . –

10 1814.

Brot. forward	357. 15 –
Painting outside of House front & back – all the Iron & wood work Skylights &c. rails	47 –
Repairing slating of Roof & mending the Gutters Repairing Chimnies, Carpenters work to sundries & Lacquering door furniture of Locks and window Latches &c.	20. –
Fitting up Apparatus in water closet	16 –
Altering the Library on Ground floor by convertg it into a Newspaper Room, removing the Book-cases to the Room proposed to be converted into the Lower Library	45. –
The above cases lengthened to reach the ceiling of the aforesaid Room	30. –
Fitting up Shelves for the bound & unbound Newspapers in the new Newspaper Room	25. –
	£ 540. 15 –

Repairs ordered. **Resolved**. – That Mr. Royce be directed to carry the same into execution at an expence not exceeding £540.15 –

Holyday to Librarian. 5. Mr. Harris requested of the Managers leave for a months absence after the present week. –
Resolved
That Mr. Harris's request be complied with.

Adjourned to Monday the 7th. of Novr. –

J. Nicholl

11

Nov. 7.

At a Meeting of the Managers of the Royal Institution of Great Britain, held at the house of the Institution on the 7th. of November, 1814. –

Present
Rt. Honble. Sir John Nicholl, V.P. in the Chair.
William Watson, Esqr.
Daniel Moore, Esqr.
Sir Benjn. Hobhouse, Bart.
Thomas Snodgrass, Esqr.
Thomas Harrison, Esqr. Secretary.

1. The minutes of the proceedings of the last meeting were read and approved.

2. Read and approved the minutes of the Committee of Accounts of this day, & Signed an Order for an advance to the Clerk of £s.70. –

Lectures. 3. Read a letter from Mr. Rippingham, addressed to the Secretary, proposing to deliver a Course of Lectures at the Royal Institution on the Art of Extempore Eloquence.
Resolved.
That it be referred to the Secretary to make the necessary enquiries. –

Presents received.

4. The Royal Society presented
Philosophical Transactions, for 1809 part 2. to 1814 part 1, inclusive.
The Revd. Johnson Grant, presented
A Summary of the History of the English Church, vol. 2d. –
The African Institution, presented.
Lettre au Prince de Talleyrand, par W. Wilberforce.
De l'Interêt de la France a l'egard de la traite des Negres par Simonde de Sismondi,
The Prayer Book & Homilly Society presented

12 1814.

Two Sermons preached at their Anniversary.
Ordered. That the thanks of the Managers be returned for these presents. –

Irish Newspaper. 5. General Thornton having represented that it would be expedient to change the Dublin Journal which is published 3 times a week for the Irish Farmers Journal a weekly Paper. –
Resolved. That the Dublin Journal be discontinued and the other substituted in its stead. –

6. The Room No. 15 in the Institution being vacant & the Secretary having requested to be allowed to have it for a Bed room. –
Resolved. That his request be complied with.

Thomas Snodgrass.

13

Nov. 14.

At a Meeting of the Managers of the Royal Institution of Great Britain, held at the house of the Institution on Monday the 14th. of November, 1814. –

Present
Thomas Snodgrass, Esqr. in the Chair.
Henry Holland, Esqr.
Daniel Moore, Esqr.
William Watson, Esqr.
Thomas Harrison, Esqr. Secretary.

1. The minutes of the proceedings of the last meeting were read and approved.

2. Read and approved the minutes of the Committee of Accounts of this day, & Signed an Order for payment of the following sum, vizt.

Payment for Coals. Messrs. Dickins & Co. for Coals in 1814. £110.5. –

3. **Present received**
The Author presented – A slight sketch of Paris in its improved state, since 1802. –

4 Read a recommendation from the Visitors entered in their proceedings of the 11th. Instant. –
Resolved. That the consideration of it be deferred to the next Board.

Lectures. 5 **Ordered**. That the Secretary be requested to write to Sir James Edward Smith, to ascertain whether he will give a Course of Lectures on Botany in the ensuing Season.

6 Read a letter from Mr. Wood stating that he should be ready to commence his lectures any time after April the 1st. –
Resolved. That Mr. Wood's wishes be complied with, and that the Secretary communicate it accordingly. –

7. The following persons having been duly recommended were

admitted as Annual Subscribers to the Royal Institution for the year 1815.—

Ann^{l}. Subscribers.

Vere Fane, Esqr. 2 Alsops Place upper Baker S^{t}.
Isaac Golledge, Esqr. 41 Jermyn Street.
Thos. George Apreece, Esqr. 15 Dover Street.
M^{r}. H de Saran, Sloane Street.
D^{r}. John Davy, 7. Old Lisle Street
D^{r}. Yeats – 3 King Street, S^{t}. James's Square.
W^{m}. Lloyd, Esqr. 15 Southampton S^{t}.—

Spencer

Nov. 21.

At a Meeting of the Managers of the Royal Institution of Great Britain, held at the house of the Institution, on Monday the 21st. of November, 1814.—

Present

Earl Spencer, President, in the Chair.
Charles Hatchett, Esqr.
Sir Benjamin Hobhouse, Bart.
William Watson, Esqr.
Sir Thos. Bernard, Bart. V. P.
Thomas Snodgrass, Esqr.
Daniel Moore, Esqr.

1. The minutes of the proceedings of the last meeting were read and approved.—

2. Read and approved the minutes of the Committee of Accounts of this day.

Lectures.

3. Read a letter from Thos. Campbell Esqr, addressed to the Sec.^{y} of the R. I, stating that He has sketched the heads of 5 Lectures which he shall be able to deliver at the R. I.—2 before, and 3 after Easter,—the terms of which are 100 guineas.—

Resolved.

That M^{r}. Campbell's offer of giving 5 lectures on Poetry at the R. I. two before and three after Easter be acceded to, upon the terms proposed by him as a remuneration—namely 100 guineas.—

4. Presents received.

James Payting Gyles, Esqr. M. R. I. presented
Elements of Hebrew Grammar, 8vo.
The Society for the distribution of Tracts in defence of the Church of England, presented
The Churchman armed against the errors of the time, 3 vol^{s} 8vo.

Ordered. That the thanks of the Managers be returned for these presents.—

Estimate for Book Case

5. Read the recommendation from the Visitors—referred for consideration from the last Board.—

Resolved.

That M^{r}. Royce be directed to prepare an estimate of the expence of making a Book-case to be placed at the further-end of the present Newsroom.—

Proposal respecting Syllabus of Lectures, declined.

6. Read the following regulation proposed by The Hon: Robert Clifford.—

In consequence of several applications made to M^{r}. Robt. Clifford he begs leave to submit to the Managers whether it would not be convenient to publish a Syllabus of each course of lectures previous to their beginning; such syllabus not to exceed one sheet of paper in 8vo. or 16 pages of print, in which the lecturers should detail their general view of the subject, and also point out any parts of the lectures to which they more particularly wished to draw the attention of their hearers—this would not interfere with the usual Cards.—

Resolved.—

That the Managers having considered the above proposal are of opinion that it is not advisable to adopt it.—

Cushions for Lecture-room.

7. Read an Estimate from Messrs. Pringle & Sons Upholsterers offering to make new Cushions, to the seats in the Lecture room, of Scarlet Morine (as ℘ pattern) upon the same construction as the present ones with a plain flounce 7 inches deep the whole stuffed with the best flocks (as ℘ sample) at £82. 15—the old flocks to be taken in exchange, and the bottoms of the present Cushions to be used.—

Resolved,

That Messrs. Pringle & Sons complete the same at the above sum, under the direction of M^{r}. Brande with the assistance of M^{r}. Fincher.

Matting for Lecture room.

8 Read an estimate from M^{r}. Royce stating that the expence of laying down fine matting round the passage of Lecture room and the centre part of Floor in ditto will cost £13.

Ordered. That M^{r}. Royce be directed to complete the same at an expence not exceeding the above sum under the

Nov. 21.

direction of M^{r}. Brande.—

9. The following persons having been duly recommended were admitted as Annual Subscribers to the Royal Institution for the year 1815.—

W. H. Fisher, Esqr. of Conduit Street.
Sir Egerton Brydges, M.P. 11 Hinde S^{t}.—

J. Nicholl

18 1814.

Nov. 28.

At a Meeting of the Managers of the Royal Institution of Great Britain, held at the house of the Institution, on the 28th. of November, 1814.—

Present

Rt. Hon: Sir John Nicholl, V.P. in the Chair.
Thomas Snodgrass, Esqr.
Daniel Moore, Esqr.
Sir Benjamin Hobhouse, Bart.
Henry Holland, Esqr.
William Watson, Esqr.
Thomas Harrison, Esqr. Secretary.—

1. The minutes of the proceedings of the last meeting were read and approved.

2. Read and approved the minutes of the Committee of Accounts of this day.—

Lectures. 3. Read a letter from Sir J. E. Smith, addressed to the Sec: of the R. I, assuring the Managers that He shall be happy to be employed in their service in the ensuing Spring on the same terms as the last.—

Book-case 4. Read an estimate from Mr. Royce stating that the expence of fitting up a new Book-case at the South end of the present News-room 8 feet 6 inches high and of the same form as those in the Library of Reference will amount to £38.—
Resolved. That it is not expedient at present to carry the above into effect.—

Iron-gate 5. Ordered. That an Iron Gate be fixed in the Area of the house of the Institution, under the direction of Mr. Brande.—

6. The following persons having been duly recommended were admitted as Annual Subscribers to the Royal

Nov. 28. Institution for the year 1815.— 19

Annl. Subs.rs.
George Allan, Esqr. M.P. 24 Pall Mall.
Alexander Scott, Esqr. Portland Hotel, Portland St.

J. Nicholl

20 1814.

Dec. 5.

At a Meeting of the Managers of the Royal Institution of Great Britain held at the house of the Institution on the 5th. of December, 1814.—

Present

Rt. Hon: Sir John Nicholl V.P. in the Chair
Sir Benjamin Hobhouse, Bart:
Henry Holland, Esqr.
Thomas Snodgrass, Esqr.
Daniel Moore, Esqr.
Thomas Harrison, Esqr. Secretary.—

1. The minutes of the proceedings of the last meeting were read and approved.

Payments. 2. Read and approved the minutes of the Committee of Accounts of this day, and signed an Order for payment of the following sums.—viz.

John Bird for Bricklayers work in 1814.	£43. 13. —
Josh. Fincher on account of the Royal Instn.	70. —. —
	£113. 13. —

Presents received.

3. Mr. Lindley Murray, presented
The power of Religion on the Mind, 8vo.
The Secy. of the African Institution, presented
Eighth Report of the Directors of the African Institution.—
Ordered.—
That the thanks of the Managers be returned for these presents.—

Lectures. 4. The Secretary reported that Mr. Lee will be ready to commence his Lectures in the beginning of February next.

J. Nicholl

21

Dec. 12.

At a Meeting of the Managers of the Royal Institution of Great Britain; held at the house of the Institution, on the 12th. of December, 1814

Present

Rt. Hon: Sir John Nicholl, V.P. in the Chair.
William Watson, Esqr.
Thomas Snodgrass, Esqr.
Daniel Moore, Esqr.

1. The minutes of the proceedings of the last meeting were read and approved.

2. Read and approved the minutes of the Committee of Accounts of this day.—

Bibliothèque Britannique to be completed 3. Read a letter from Dr. Marcet M.R.I. addressed to the Secretary, suggesting to the Managers the expediency of completing the Bibliothèque Britannique, which will be at an expence of not more than 5 or 6 guineas as the Institution are in possession of 75 vols. and the whole series 126 vols.
Resolved.
That the Board of Managers think it advisable to complete the Bibliothèque Britannique, if it meets with the approbation of the Patrons of the Library — And that this resolution, be communicated, by the Clerk, to the Patrons at their next Meeting.—

Presents received.

4 Sir Egerton Brydges, M.P. presented
Poems, 4to. — printed at Lee Priory
Dr. Buchan, presented
Kunopædia: An Essay on breaking or training the English Spaniel or Pointer 8vo.
Ordered.
That the thanks of the Managers be returned for these presents —
The 1st. Report of a Society for preventing accidents in Coal Mines — By Mr. Wilkinson of Pumps Ct. Temple

5. Read the Minutes of the last General Monthly Meeting. —

6. Resolved. That the Lectures be announced for the ensuing Season under the direction of the Secretary. —

I. Nicholl

Dec. 19

At a Meeting of the Managers of the Royal Institution of Great Britain held at the house of the Institution on the 19th of Decemr. 1814.

Present

Rt. Hon: Sir Jno. Nicholl V.P. in the Chair.
Charles Hatchett, Esqr.
William Watson, Esqr.
Thomas Snodgrass, Esqr.
Henry Holland, Esqr.
Daniel Moore, Esqr.
Matthew Raper, Esqr.
Thomas Harrison, Esqr. Secretary.

1. The minutes of the proceedings of the last meeting were read and approved.

2. Read and approved the minutes of the Committee of Accounts of this day, and Signed an Order for payment of the following sums, vizt.

Payments.

Mr. Butler 3/4 yrs Parochial rates to Xmas 1814	£ 40. 12. 6
Mr. Reeve 3/4 yrs Assessed taxes to do.	204. 19. 6
Wm. Harris Librarian 1/2 yrs Salary to 31st. Decr. 1814.	90. –. –
David Brooks Collr. 1/2 yrs Saly. to do.	26. 5. –
Josh. Fincher on acct. of the R.I.	60. –. –
	421. 17. –

Ordered.
That the Taxes and Salaries be paid as they become due.

Letter from Mr. Lee declining to Lecture.

3 Read the following letter from Mr. Lee, addressed to the Secretary of the R.I. dated Royal Society 16th. December 1814.

Sir,

It is truly mortifying to me, but I am under the necessity of relinquishing my engagement with the

Royal Institution, for Sir Joseph Banks has, in the most decided manner, expressed his opinion, that my delivering Lectures there would be totally incompatible with my situation in this Society.

I do assure you, I had not the slightest suspicion of any opposition till late last night. I called at the African Institution this morning to see you on the subject, and mean to call again to-morrow, that I may have an opportunity of more fully explaining the business to you, and which can be better done in conversation than in writing.

I have the honor to be &c.
(signed) Stephen Lee

Resolved. That the Chairman and Secretary do wait upon Sir Joseph Banks, and endeavour to prevail upon him under the particular circumstances of the case still to allow Mr. Lee to give a Course of Lectures on Astronomy at the Royal Institution during the ensuing Season.

Mr. Forster's Lectures.

4 Read a letter from The Reverend Edward Forster, addressed to the Secretary of the R.I. dated Paris 2d. Decr. 1814. stating, that He is fearful it will not lie in his power to read the short course of lectures which he proposed doing this winter, as his stay upon the Continent is prolonged much more than he expected. Should he however return in time, he shall be happy to do so towards the end of the Season.

Resolved.
That the Secretary do write to Mr. Forster and state to him the necessity of saying when he will be able to commence his Course.

5. The following persons having been duly recommended were admitted, as Annual Subscribers to the Royal Institn. for the year 1815. —
Dr. Bain 10 Curzon St.
Henry Peters 2 Park Street.

6 Adjourned to Monday the 16th. January 1815

I. Nicholl

Jan. 16.

At a Meeting of the Managers of the Royal Institution of Great Britain held at the house of the Institution, on the 16th January, 1815. —

Present

Rt. Honble. Sir Jno. Nicholl, V.P. in the Chair.
Sir Benjamin Hobhouse, Bart.
Charles Hatchett, Esqr.
Wm. Watson, Esqr.
Daniel Moore, Esqr.
Matthew Raper, Esqr.
Thomas Snodgrass, Esqr.
Henry Holland, Esqr.
Sir Thos. Bernard, Bart. V.P.
Thomas Harrison, Esqr. Secretary.

1. The minutes of the proceedings of the last meeting were read and approved.

2. Read and approved the minutes of the Committee of Accounts of this day.

Mr. Lee's Lectures.

3 The Chairman & Secretary reported, that they had had, several interviews with Sir Joseph Banks and found Mr. Lee so circumstanced with the Royal Society, as to prevent him giving his Course of Lectures on Astronomy at the Royal Institution during the ensuing Season.

Presents received.

4. The Royal Society, presented
Philosophical Transactions 1814. — part 2.
Sir S. Egerton Brydges, Bart: M.P. presented
Stanzas, by the Author of Dunlme Castle, printed at Lee Priory.
Sir Humphry Davy, presented
Elements of Agricultural Chemistry, new Edition in 8vo.

Ordered.
That the thanks of the Managers be returned for these presents. —

1815.

5\. Mr. Hatchett proposed to the Board, that Baron Pfeffel, the Minister from the Court of Bavaria, should have admission to the Institution in conformity with Art: 16 Chap: VI of the Bye Laws. — Hon. Admission.

Ordered.

That the Secretary do transmit to him the usual letter of admission to the Lectures, Library and Reading Rooms of the Royal Institution.

6\. Sir Thomas Bernard stated to the Board that Mr. Brooks in consequence of his increased duties at the Alfred would not be able to attend the Lecture room-door in future but that he should continue his office of Collector to the Royal Institution. — Resp. Collector.

7\. The following persons having been duly recommended were admitted as Subscribers to the Royal Institution for the year 1815.—

Annual Subscribers.

Mr. Manuel Palacio — 52 Welbeck Street
Charles le Tavernier, Esqr. 33 Mount Street
Rt. Revd. Lord Bishop of Cloyne. Montagu Square.
Jno. Wm. Searle, Esqr. 30 Welbeck Street.—
John Vivian, Esqr. — 26 Portland Place
Richd. Arnold, Esqr. 25 New Milman St.
Hugh Leycester, Esqr. New St. Spring Gardens.
G. M. Burrows, Esqr. Brompton-row.
Revd. Thos. Fuller — 47. York Street.
John Joseph Dillon, Esqr. Albany.—

To the Lectures.

Mrs. Pemberton }
Miss. C. Pemberton } George St. Hanover Square.

J. Nicholl

Jan. 23.

At a Meeting of the Managers of the Royal Institution of Great Britain held at the house of the Institution, on the 23d. January, 1815.—

Present

Rt. Hon: Sir John Nicholl, in the Chair.
Sir Benjamin Hobhouse Bart:
Daniel Moore, Esqr.
Thomas Snodgrass, Esqr.
Henry Holland, Esqr.
S. Bernard Morland, Esqr. Treasurer.
William Watson, Esqr.

1\. The minutes of the proceedings of the last meeting were read and approved.

2\. Read and approved the minutes of the Committee of Accounts of this day, and signed an Order for an Advance to the Clerk of £60.—

3\. In pursuance of a nomination under the Act by John Hinckley Esqr. deceased, his Sister — Life Subscr.

Esther Hinckley, of Guilford Street.

was admitted to the privileges of a Life Subscriber to the Royal Institution.—

Present received.

4\. The African Institution, presented 8 nos. of their Reports.

Ordered. That the thanks of the Managers be returned for this present.—

Honry. Admissions 5. Hon: Robt. Clifford proposed to the Board of Managers, that Mr. Ed: de Bourke, Minister from His Danish Majesty, be admitted to the Royal Institution under Art: 16 Chap: VI of the Bye Laws.—

& Mr. Brooks the Collector stated, that Count St. Martin D'Aglie, the Sardinian Ambassador, who annually subscribed to the Institution till he was appointed Ambassador, wished to have admission to the Royal Institution in conformity

with Art: 16. Chap: VI of the Bye Laws — 1815.

Ordered.

That the Secretary do transmit to them the usual letters accordingly.—

Resolved.

6\. That Carpets be provided for the Newspaper Room extending about a Chairs breadth or upwards beyond the edge of each Table, and that the sides of the Rooms and the intervals between the Carpets be covered with Floor Cloth. — Carpets.

That a Turkey Carpet be provided for the Reading room of a size nearly to cover the whole Floor and that the fitting up of the Reading Room be forthwith compleated and announced to the Members and Subscribers and that the carrying of this Resolution into effect be under the direction of Mr. Brande with the assistance of Mr. Fincher.—

7\. Resolved. That Thomas Slatter be appointed Door-keeper to the Lecture Room at a Salary of £10.10.— pr. Ann: to commence from the 1st. January last.— — Doorkeeper.

8\. The following persons having been duly recommended were admitted as Annual Subscribers to the Royal Institution for the year 1815.

Annual Subscribers.

Genl. Bude 8 Henrietta St. Cavendish Square
Edward Jas. Jarvis Esqr. Inner Temple.
Dr. Chambers, Upper Bedford Place.

To the Lectures.

Miss Trotter, 7 Soho Square
Mrs. Boyd, Princes St. Cavendish Square
Mrs. Tolfrey — 24 Hans Place
Miss Kranter — 10 Arlington Street.
Miss K. Nicholl — Bruton Street.

J. Nicholl

Jan. 30.

At a Meeting of the Managers of the Royal Institution of Great Britain held at the house of the Institution, on the 30th. January, 1815

Present

Rt. Hon: Sir John Nicholl, in the Chair.
Wm. Watson, Esqr.
Matthw. Raper, Esqr.
Thomas Snodgrass, Esqr.
J. P. Auriol, Esqr. V.P.
Daniel Moore, Esqr.
Thomas Harrison, Esqr. Secretary.

1\. The minutes of the proceedings of the last meeting were read and approved.—

2\. Read and approved the minutes of the Committee of Accounts of this day.

Lectures. Mr. J Millington. 3. Read a letter from Mr. John Millington, addressed to the Managers of the Royal Institution, stating that as he understood there was a vacancy for a Course of Lectures, He begged leave to offer himself to their notice; and that as he had been in the habit of delivering lectures on general Philosophical subjects for some years he should be happy to undertake a Course of Lectures at the Royal Institution — He also begged to state that he had it in contemplation to get up a Course of Lectures on the rise progress and present state of the Steam Engine and its applications; And also on the various manufacturing processes of this Country which his profession had made him acquainted with, and their dependance on the several Sciences, And to illustrate the whole by working models, but hitherto his Professional engagements, and the great expence attendant upon Model-making had prevented his carrying this project into effect — He was however persuaded it would carry considerable interest with the Public, and if such a Course would be desirable at the Royal Institution, and he could have its patronage in carrying it into

30 — 1815.

effect he would lose no time in completing it.

Mr. Millington being in attendance, was called in and stated to the Meeting that he would be happy to deliver a Course of Lectures on Natural Philosophy. – the Course to consist of about 12 Lectures at 3 guineas for each lecture.

<u>Resolved</u>. That Mr. Millington be engaged to give a Course of Lectures on Natural Philosophy on his own terms – namely 3 guineas p. lecture. –

Lectures. Mr. Babbage

4 Mr. Moore having stated that Mr. Babbage would be willing to deliver a Course of Lectures on Astronomy He was introduced by Mr. Moore to the Board and stated that he would be ready to commence a Course of Lectures on Astronomy, on Thursday the 16th. of February next – the terms of which are 12 Lectures for 50 guineas and after that rate in case the Course did not consist of that number. –

<u>Resolved</u>. That Mr. Babbage be engaged to give a Course of Lectures on Astromy on the above terms.

Model of Steam-Engine – – C. Broderip, Esq

5. Mr. Brande reported that Charles Broderip Esq.r M.R.I. had offered to direct the construction of a Model of the Steam Engine of the value of £120 – and to present the same to the Institution provided the Managers would defray a portion of the expence not exceeding 50 guineas.

<u>Resolved</u>. That Mr. Broderip's proposal be accepted and that the thanks of the Managers be returned for his liberal offer.

6. <u>Presents received</u>.

Sir Humphry Davy presented Specimens of Minerals from Mount Vesuvius. –

<u>Ordered</u>. That the thanks of the Managers be returned for this present.

——— 2 Copies of a Report of the General Committee of the London Infirmary for curing diseases of the Eye.

7. Jos.h Fincher had the honor to report to the Managers

31

Jan 30. completion of Bibliotheque Britannique

that he laid the Resolution in their Minutes of the 12th. December last, before the Patrons of the Library at their Meeting on Wednesday the 25th. Instant, upon which they Resolved.

"That the Patrons highly approve ~~of the compleation~~ "of the completion of the Bibliotheque Britannique, and "return their thanks to the Managers for their communication "and intended assistance for that purpose." –

<u>Resolved</u>.

That Mr. Harris enquire and report to the next Board the expence of completing the series of the Bibliotheque Britannique. –

8. The following persons having been duly recommended were admitted as Subscribers to the Royal Institution for the year 1815

<u>Annual Subscribers</u>.

Major Gen.l Madden 11 Cleveland row
Philip Cipriani, Esq.r 18 Wimpole St.
Hon: Wm. Grey – 13 Hertford St.
Dr. O Neil – 41 Duke St. Manchester Sq.e
David Ricardo, Esq.r 56 upper Brook St.
Barrington Pope Blackford Esq.r Admiralty.
Rev.d Rob.t Jones 13 Arundel St. Strand
Rev.d Edward Smedley – Sanctuary Westm.r

<u>To the Lectures</u>.

Miss Cotton } York St. Portman Sq.e
Miss Maryanne Bonomi }
Mrs. Cooke & 3 Miss Cookes – 21 Hertford St.
Miss Caroline Maxwell – 62 Harley St.

J P Auriol

32 — 1815.

Feb. 6.

At a Meeting of the Managers of the Royal Institution, of Great Britain held at the house of the Institution, on the 6th. of February, 1815. –

Present

J. P. Auriol, Esq.r V.P. <u>in the Chair</u>.
Thomas Snodgrass, Esq.r
Matthew Raper, Esq.r
Daniel Moore, Esq.r

1. The minutes of the proceedings of the last meeting were read and approved.

2. Read and approved the minutes of the Comm.e of Accounts of this day. –

Bibliotheque Britannique

3. Mr. Harris the Librarian reported, that Mr. De Boffe not having received any numbers of the Bibliotheque Britannique since the communication with the Continent has been open, cannot say what the price will be at present, but supposes it will be about 3 guineas & a half per year, which will amount to about £24 to compleat the 6 years & 1/2 wanting to compleat the set at the Royal Institution.

<u>Resolved</u>.

That Mr. Harris be desired to write to Dr. Marcet to enquire of him whether he could put the Institution in the way of procuring the Bibliotheque Britannique and what would be the actual expense.

Mr. Harris further reported, that an Alphabet of Capital letters is wanting to be placed over the cases in the Lower Library, those at present not being regular and too small – A key also is wanting for the Door.

<u>Resolved</u>.

That the above application be complied with under the direction of Mr. Brande and that he be requested in future to order such triffling necessaries as appear to him to be immediately requisite, reporting the same to the next Board.

33

Feb. 6.

4. The following persons having been duly recommended were admitted as Subscribers to the Royal Institution for the year 1815. –

<u>Annual Subscribers</u>.

R.t Hon: L.d Kilworth 10 Stratford Place.
George Daniell, Esq.r Lincolns Inn fields.
Wm. Noble Rule, Esq.r Montagu Place.

<u>To the Lectures</u>.

Mrs. Brett, 7 Hanover Square
Miss Warrell – Albemarle St.
Miss C. Warrell d.o –
Mrs. Bell } Dean St. Soho
Miss Bell }

J. Nichols

Feb. 13. 1815.

At a Meeting of the Managers of the Royal Institution of Great Britain, held at the house of the Institution, on the 13th. of February, 1815. –

Present

Rt. Hon: Sir John Nicholl, V.P. in the Chair.
Charles Hatchett, Esqr.
Sir Benjn. Hobhouse, Bart. M.P.
Matthew Raper, Esqr.
Henry Holland, Esqr.
Thomas Snodgrass, Esqr.
S. Bernard Morland, Esqr. Treasurer.
J. P. Auriol, Esqr. V.P.
Daniel Moore, Esqr.

1. The minutes of the proceedings of the last meeting were read and approved. –

2. Read and approved the minutes of the Committee of Accounts of this day, and signed an order for payment of the following sums, viz:

Payments;

Mr. Royce for Painting & repairing the House of the Institution in 1814	£626 - 3 - 6
Mr. Pringle for Upholstery – do.	142 - 14 - 4
	768 - 17 - 10

Presents received.

3. The President & Committee of the Liverpool Library presented 4 Copies of their Catalogue.
and Joseph Adams M.D. presented "A Treatise on Hereditary Diseases," and 2 Cops. of an Illustration of Dr. Hunter's Doctrine concerning the Life of the Blood.
Ordered. That the thanks of the Managers be returned for those presents.

4. Read the minutes of the last General Monthly Meeting. –

Feb. 13.

5. The following persons having been duly recommended were admitted as Subscribers to the Royal Institution for the year 1815.

Annual Subscribers.

Hon: Jas. Somers Cocks 22 George St.
George Townley Esqr. 59 Welbeck St.
Count Ludolf – Bury St.
Philip Vaughan Esqr. – 27 York Place
Chas. Shaw Lefevre junr. Esqr. New St. Spring-gardens
Anthony Bertolacci Esqr. 1 High row Knightsbridge.

To the Lectures.

Mrs. Bertolacci – High row Knightsbridge.
Lady Frances Vane Tempest }
Mrs. Cade — } No. 1. Cadogan Place
Miss Ballatine – – }
Mrs. Vaughan } 27 York Place Baker St.
Miss Vaughan }

J. Nicholl

Feb. 20. 1815.

At a Meeting of the Managers of the Royal Institution of Great Britain, held at the house of the Institution, on the 20th of February, 1815. –

Present

Rt. Hon: Sir John Nicholl M.P. V.P. in the Chair
Sir Benjn. Hobhouse, Bart. M.P.
Daniel Moore, Esqr.
Thos. Snodgrass Esqr.
S. Bernard Morland, Esqr. M.P. Treasurer.
Thomas Harrison, Esqr. Secretary.

1. The minutes of the proceedings of the last meeting were read and approved.

2. Read and approved the minutes of the Committee of Accounts of this day, and signed an Order for an advance to the Clerk of £60. –

Hon. Admission.

3. Read a letter from William Watson Esqr. Manager submitting to the Board that M. De Freudenreich, Envoy Extraordinary from the Republic of Berne, and M. De Haller the Counsel of Legation be admitted to the Royal Institution under Art: 16 Ch: VI of the Bye Laws. –
Resolved.
That the Secretary do transmit to them the usual letters accordingly. –

4. Audited the Treasurer's Account to the 31st. Decr. last. –

5. The following persons having been duly recommended were admitted as Subcribers to the Royal Institution for the year 1815. –

Annual Subscribers.

Robert Wilmot Esqr. – 23 Montagu Square
George Henry Jones, Esqr. 4 Eaton St. Pimlico
Duncan Campbell Esqr. 22 Alfred Place

Feb. 20.

Captn. Thos. Forster. R. N. No. 1. Bentinck St.
H. B. Sawbridge, Esqr. Green St. Grosvr. Sq.
Marm: Constable Maxwell, Esqr. 1 Arklow Place
Wm. Baillie Esqr. 72 Lower Grosvenor St.

To the Lectures.

Rt. Hon: Lady Hannah Ellice 1 New St. Spring gardens
Mr. John Buller 94 Gloster Place
Miss Baillie 72 Lower Gvr. St.
Dowager Lady Brownlow 30 Hill St. –

Morland

38

Feb. 27. 1815.

At a Meeting of the Managers of the Royal Institution of Great Britain, held at the house of the Institution, on the 27th of February, 1815. —

Present

Earl of Morton, V.P. in the Chair.
J. P. Auriol, Esqr. V.P.
Matthew Raper, Esqr.
Thomas Snodgrass, Esqr.
Henry Holland, Esqr.
S. B. Morland, Esqr. Treasurer.
Daniel Moore, Esqr.
Wm. Watson, Esqr. —
Thomas Harrison, Esqr. Secretary.

1. The minutes of the proceedings of the last meeting were read and approved.

2. Read and approved the minutes of the Committee of this day. —

Coals. 3 Read a letter from Messrs. Dickins & Co. addressed to the Clerk, stating that from the present low price of Coals they are enabled to supply the Royal Institution with the best Wallsend Coals at 58s. per Chaldron free of all expences. —
Ordered. That Messrs. Dickins & Co. supply the Institution with 20 Chaldrons at present, on the above terms. —

Presents received.

4. Memoirs of William Stevens Esqr. —
Mr. Mallison presented His 2d. Report of Assistance to Shipwrecked Mariners. —
Ordered. That the thanks of the Managers be returned to him for this present.

5 The Secretary reported that the price of the Bibliotheque Britannique, which as before reported, will be £3. 13. 6

Feb. 27. 39

per Ann: and as 6 years and a half are wanting the amount will be £24. —

Cushion for Hall. Mr. Fincher reported that the refitting + a Cushion for the Hall seat made of black leather stuffed with the materials of the present old Cushion, would be at an expence of 2 guineas.
Resolved. That it be carried into effect at a sum not exceeding 2 guineas.

The following persons having been duly recommended were admitted as Subscribers to the Royal Institution for the year 1815.

Life Subscriber.
James Agnew, Esqr. of Russell Square

Annual Subscribers.
Samuel Smith Esqr. 6 Park St. Westmr.
William à Beckett, Esqr. 20 Broad St. Golden Sq.

To the Lectures.
Lady Hamlyn Williams Upper Grosvr. St.
Mrs. Wilmot. 23 Montagu Sq.
Miss Emma Lemon. Whitehall
Hon: Miss Cocks, Gt. George St. Westmr.
Miss Horton, 22 Montagu Square.

Spencer

40

Mar. 6. 1815.

At a Meeting of the Managers of the Royal Institution of Great Britain held at the house of the Institution, on the 6th. March, 1815. —

Present

Earl Spencer, President, in the Chair.
Sir Benjn. Hobhouse, Bart. M.P.
Daniel Moore, Esqr.
J. P. Auriol, Esqr. V.P.
Thomas Snodgrass, Esqr.
Henry Holland, Esqr.
S. Bernard Morland, Esqr. M.P. Treasurer.
Thomas Harrison, Esqr. Secretary.

1. The minutes of the proceedings of the last meeting were read and approved. —

2. Read and approved the minutes of the Committee of Accounts of this day, and signed an Order for payment of the following sums vizt.

Wrangham for Newspapers in 1814	£31. 13. 3
Ditto — for Stationary — do. —	35. 17. 11
Thompson for Soap & Candles — do. —	48. 15. 8
Upton for Oil in 1814 —	74. 1. —
Howis Sundries for the house in 1814 —	25. 17. —
Wyatt — do. — do. —	12. 16. 6
Searle for Bookbinding — do. —	14. 15. 3
Hammond for Linen-drapery do. —	15. 1. 6
Mr. Hutchins for Sewers rate —	12. 10. —
	£271. 8. 1

Lectures — Mr. Forster. 3. Read a letter from The Revd. Edward Forster, addressed to the Secretary, stating that it will be totally out of his power to deliver his lectures this Season as he has been confined for the 3 last months to his Room by a violent attack of Rhematism which will render his stay on the Continent absolutely necessary for some time longer

Mar. 6. 41

4. In pursuance of a nomination under the Act by Richard Chicheley Plowden, Esq.; his Son
Charles Hood Chicheley Plowden, Esq. of Devonshire Place was admitted to the privileges of a Life Subscriber to the Royal Institution.

5. The following persons having been duly recommended were admitted as Subscribers to the Royal Institution for the year 1815. —

Annual Subscribers.

W. Rookes Crompton, Esq. of Esholt Hall Yorkshire.
Robert Woulfe, Esq. 25 Downing St.
Genl. H. M. Gordon, 4 Dean St. Park lane
Captn. Maxwell, R.N. 62 Harley St. —
Wm. Shrine Esq. 12 Baker St.

To the Lectures.
Miss Frances Lemon, Whitehall
Mrs. Rose 41 Beaumont St.
Miss Protheroe, Harley St. —

Spencer.

42 1815.

Mar. 13.

At a Meeting of the Managers of the Royal Institution of Great Britain held at the house of the Institution on the 13th of March, 1815; –

Present

Earl Spencer. President. in the Chair.
Rt. Hon: Sir John Nicholl. M.P. V.P.
Henry Holland, Esqr.
Daniel Moore, Esqr.
William Watson, Esqr.
Thomas Snodgrass, Esqr.
Sir Benjn. Hobhouse, Bart. M.P.
Matthew Raper, Esqr.
Daniel Giles, Esqr. –
Thomas Harrison, Esqr. Secretary

1. The minutes of the proceedings of the last meeting were read and approved. –

2. Read and approved the minutes of the Committee of Accounts of this day, and signed an Order for an advance to the Clerk of £60. –

3. Read the minutes of the last General Monthly Meeting. –

4. The following persons having been duly recommended were admitted as Subscribers to the Royal Institution for the year 1815. –

Annual Subscribers.
Joseph Langstaff, Esqr. 8 Collen St. Hanover Sqr.
Captn. F. V. Vernon, R.N. 40 Grosvenor Sqr.
W. S. Fuller, Esqr. R.N. – 7. Haymarket

To the Lectures.
Hon: Miss Bassett – 18 upper Grosvenor St.

Spencer.

43

Mar. 20.

At a Meeting of the Managers of the Royal Institution of Great Britain, held at the house of the Institution, on the 20th of March, 1815.

Present

Earl Spencer, President. in the Chair.
Sir Thos. Bernard, Bart. V.P.
Daniel Moore, Esqr.
J.P. Auriol Esqr. V.P.
Thomas Harrison, Esq. Secretary.

1. The minutes of the proceedings of the last meeting were read and approved.

2. Read and approved the minutes of the Committee of Accounts of this day

Present received.

3. Dr. Curry – M.R.I. presented
Observations on Apparent Death.
Ordered. That the thanks of the Managers be returned for this present. –

Insurance. 4. The Secretary reported that the Insurance of the House amounting to £58. 5. 0 would become due at Lady day.
Resolved.
That the Treasurer be requested to pay that sum when it becomes due, and signed an Order accordingly

5. The following persons having been duly recommended were admitted as Subscribers to the Royal Institution for the year 1815. –

Annual Subscriber.
George William Tapps, Esqr. Cumberland Place

To the Lectures.
Miss E. Agnew, Russell Square.
Miss Hamlyn 32 upper Grosvenor St.

6 In pursuance of a nomination under the Act by

44 1815.

Life Subscriber. Isaac Walker Esqr. – his Son John Walker Esq. of Bedford Square was admitted to the privileges of a Life Subscriber to the Royal Institution.

7. Adjourned to Monday the 3d. of April.

Spencer.

45

April 3.

At a Meeting of the Managers of the Royal Institution of Great Britain, held at the house of the Institution, on the 3d of April, 1815.

Present

Earl Spencer, President in the Chair.
Marquis of Lansdowne, V.P.
J. P. Auriol, Esqr. V.P.
Henry Holland, Esqr.
Daniel Moore, Esqr.
Thos. Snodgrass, Esqr.
Sir Benjn. Hobhouse, Bart. M.P.
Wm. Watson, Esqr.
Thomas Harrison, Esqr. Secretary.

1. The minutes of the proceedings of the last meeting were read and approved.

2. Read and approved the minutes of the Commee. of Accounts of this day, and signed an Order for an advance to the Clerk of £60. –

Lectures. 3. Read a letter from John Roche, M.D. addressed to the Secretary of the R.I. dated Belfast 20th. March 1815, offering to read a Course of Lectures on Political Economy at the Royal Institution during the present Season.
Resolved.
That the Secretary inform Dr. Roche that the arrangements for lectures for the present Season are full.

Presents received.

4. Mr. Payne presented 4 nos. of the Commercial Magazine. –
Ordered. That the thanks of the Managers be returned for this present.

5. Read a recommendation from the Visitors entered in their proceedings of the 25th. of March last
Resolved That Mr. Royce inspect the Wall of the Laboratory

Repairs &c. next the Garden of the Clarendon Hotel, under the direction of Mr. Brande, and lay before the next Board an estimate of the expence of rendering the Wall and Laboratory dry. —

Resolved. That the Floor Cloth in the Passage leading to the Reading Room be renewed under the direction of Mr. Brande.

Resolved. That the Visitors be invited to a conference with the Managers on the subject of the Anniversary Dinner on Monday next at 2 o'Clock p.m.

6. The following persons having been duly recommended were admitted as Subscribers to the Royal Institution.

Life Subscriber

Samuel Moulton Barrett, Esqr.

Annual Subscriber.

Benjn. Hawes jun. Esqr. Old Barge house Christ-Church

Chichester

April 10.

At a Meeting of the Managers of the Royal Institution of Great Britain, held at the house of the Institution, on the 10th of April, 1815. —

Present

Earl of Chichester, in the Chair.
Charles Hatchett, Esqr.
Thomas Snodgrass, Esqr.
S. Bernard Morland, Esqr. Treasurer
J. P. Auriol, Esqr. V.P.
Thomas Harrison, Esqr. Secretary.

1. The minutes of the proceedings of the last meeting were read and approved.

2. Read and approved the minutes of the Committee of Accounts of this day.

3. Presents received.

Presented by ———

A Letter to W. Wilberforce Esqr. M.P. containing remarks on the reports of the Sierra Leone Company & African Institution 3rd Edit.

John Weyland jun. Esqr. F.R.S. presented

The principle of the English Poor Laws, and on the connexion observed in Scotland between the price of Grain & the wages of Labor 8vo.

Ordered. That the thanks of the Managers be returned to Mr. Weyland for this present.

4. Read the minutes of the last General Monthly Meeting. —

Repairs 5. Mr. Royce reported that the expence of rendering the Wall of the laboratory dry, would amount to between £30 & £35.

Ordered. That it be done under the inspection of Mr. Brande at an expence not exceeding £35. —

Annl. Dinner 6 The Visitors having conferred with the Managers on the subject of the Anniversary Dinner — It was Resolved.

That the further consideration of it be deferred till the next General Monthly Meeting on the 8th of May.

Minerals. 7. Ordered. That the duplicates of the Minerals be disposed of under Mr. Brandes direction. —

8. The following persons having been duly recommended were admitted as Subscribers to the Royal Institution for the year 1815.

Annual Subscribers.

John Frere, Esqr. 19 Sackville St.
Benjn. Harper, Esqr. Albany
Rowland Stephenson, Esqr. Albany.
Charles Hutchinson Esqr.

To the Lectures.

Lady & Miss Buller 9 St. James's Place

JP Auriol

April 17.

At a Meeting of the Managers of the Royal Institution of Great Britain held at the house of the Institution, on the 17th of April 1815. —

Present

J. P. Auriol, Esqr. V.P. in the Chair.
William Watson, Esqr.
Matthew Raper, Esqr.
Daniel Moore, Esqr.
Sir Benjn. Hobhouse, Bart. M.P.
Thomas Snodgrass, Esqr.
Rt. Hon: Sir John Nicholl V.P.
S. Bernard Morland, Esqr. M.P. Treasurer.
Daniel Giles, Esqr.
Thos. Harrison, Esqr. Secretary.

1 The minutes of the proceedings of the last meeting were read and approved. —

2. Read and approved the minutes of the Committee of Accounts of this day.

Lectures. 3. Read a letter from Mr. Stackhouse offering to read a Course of Lectures on the Architectural & other remains of Britain during the present Season.

Resolved.

That the Secretary inform him that the arrangements for lectures for the present year are full.

4. Ordered. That the Apothecary's bill on Mrs. Greenwood the Housekeeper amounting to £3.14. — be paid. —

5. The following person having been duly recommended was admitted an Annual Subscriber to the Royal Institution for the year 1815.

Annl. Subsr. Richard Bean, Esqr. of Albemarle Street.

J. Nicholl

50

April 24.

1815.

At a Meeting of the Managers of the Royal Institution of Great Britain held at the house of the Institution, on the 24th of April, 1815.

Present
Rt. Hon: Sir John Nicholl, V.P. in the Chair
William Watson, Esqr.
S. Bernard Morland, Esqr. M.P. Treasurer
Thomas Snodgrass, Esqr.
J. P. Auriol, Esqr. V.P.
Daniel Moore, Esqr.
Thomas Harrison, Esqr. Secretary.

1. The minutes of the proceedings of the last meeting were read and approved.

2. Read and approved the minutes of the Committee of Accounts of this day, & Signed an Order for an advance to the Clerk of £60. –

On Property-tax. – Cooperation with London Instn. declined.

3. Read a letter from: Robert Stevens Esqr. Secy. to the London Institution, addressed to the Secretary, dated 19th April, stating that he believes it is the intention of the Board of Management of the London Institution in case of the renewal of the Property tax to make application to the Chancellor of the Exchequer for a Clause of exemption in favor of the Institution on the ground of its being a Chartered Literary Establishmt. and requesting to know if any such intention exists in the Managers of the Royal Institution, with whom if such intention existed, he has no doubt the Managers of the London Institution would gladly co-operate
Resolved.
That the Secretary inform Mr. Stevens that the Managers decline the offer of co-operating with the London Institution.

Life-Subscriber.

4. In pursuance of a nomination under the Act by Sir John Cox Hippisley Bart – his Son John Stuart Hippisley Esqr. of Grosvenor Street was

April 24.

51

admitted to the privileges of a Life Subscriber to the Royal Institution. –

5. The following persons having been duly recommended were admitted as Subscribers to the Royal Institution for the year 1815. –

Annual Subscriber.
Charles Hornyold, Esqr. Cecil Street Strand.

To the Lectures.
Countess of Dartmouth } 12 Berkeley Square
Rt. Hon: Lady Caroline Legge }
Mrs. Paltmer } 26 Gloster Place.
Miss Dawson }
Mrs. George Weld – 30 Welbeck Street.

6. Ordered. That the House Lists be filled up as follows.

President. Earl Spencer K.G. F.R.S. &c. &c. &c.

Treasurer. Scrope Bernard Morland Esqr. M.P. &c. –

Secretary. Thomas Harrison, Esqr. F.R.S.

Managers.
J. P. Auriol, Esqr. F.R.S.
Sir Humphry Davy, L.L.D. F.R.S. &c. &c.
Duke of Devonshire.
George Duckett, Esqr. F.R.S. & F.A.S
Lord Dundas, F.R.S. & F.A.S
Davies Giddy, Esqr. M.P. F.R.S & F.L.S.
Earl of Hardwicke, K.G. F.R.S & F.A.S.
Benjamin Hall, Esqr. M.P. F.R.S.
Sir Benjamin Hobhouse, Bart. M.P. F.R.S.
Henry Holland, Esqr.
Alexr. Marcet, M.D. F.R.S.
Daniel Moore, Esqr. F.R.S. F.A.S. & F.L.S
Rt. Hon. Sir John Nicholl, M.P. &c. &c.
Thomas Snodgrass, Esqr.
William Watson, Esqr. F.R.S.

52

Visitors. –

1815.

Timothy Brent, Esqr. –
Revd. Chas. Burney, D.D. F.R.S. & F.A.S.
William Manning, Esqr. M.P.
Wm. Hasledine Pepys, Esqr. F.R.S.
Louis Hayes Petit, Esqr. F.R.S. F.A.S. & F.L.S
Chas. Pieschell, Esqr.
Claude Scott, Esqr.
Sir John Saunders Sebright, Bart. M.P.
Richd. Horsman Solly, Esqr. F.R.S & F.A.S
Duke of Somerset F.R.S.
Marquis of Stafford. K.G.
Lord Teignmouth
Lt. General Wm. Thornton M.P.
John Walker, Esqr. F.R.S F.A.S. & F.L.S.
Rt. Hon: Wm. Wickham.

J. Nicholl

53

May 1.

At a Meeting of the Managers of the Royal Institution of Great Britain, held at the house of the Institution, on the 1st. of May, 1815. –

Present
Rt. Hon: Sir John Nicholl. in the Chair.
Daniel Moore, Esqr.
S. Bernard Morland, Esqr. M.P. Treasurer.
Wm. Watson, Esqr.
Thos. Snodgrass, Esqr.
J. P. Auriol, Esqr.
Sir Benjamin Hobhouse, Bart. M.P.
Thomas Harrison, Esqr. Secretary.

1. The minutes of the proceedings of the last meeting were read and approved.

2. Read and approved the minutes of the Committee of Accounts of this day, and Signed an Order for payment of the following sums, vizt.

Payments.

1 years Ground rent to the City of London to 25th. March last. –	£135. 1. 8
¼ yrs. Land Property & Assessed taxes to 25th. March last.	68. 6. 9
¼ yrs. Poor rates to do.	13. 10. 10
¼ yrs. Salary to Mr. Harris to 30th March	45. –. –
	£ 261. 19. 3

Presents received.

3. Chas. Kelsall, Esqr. presented
Phantasm of an University, 4to. –
The Society of Arts, Manufactures &c. presented
Transactions of the Society, vol 32.
Miss Porden presented, A Poem called the Veils, or the Triumph of Constancy. 8vo. elegantly bound in red morocco.
Ordered. That the thanks of the Managers be returned.

for these presents. –
7 The pleasures of Anarchy, By ——

4. Read the following Resolution from the Patrons of the Library. — *Library Hours.*
Resolved. That it be recommended to the Managers to order the Library of Reference to remain open from the 1st of March to the 1st of September till 6 o'Clock in the Afternoon instead of closing as it now does at 4 –
Ordered accordingly. –

5. The following persons having been duly recommended were admitted as Subscribers to the Royal Institution for the year 1815. –

Annual Subscribers.
Sir John English, 2. Stafford Street
Anthony Hill, Esqr. –
Robt. Burrow, Esqr. 3. Garden Court Temple

To the Lectures.
Rt. Hon: Lady Charlotte Legge, of Berkeley Sqr.
Miss Louisa Ricketts, 28 Gloster Place

J.P. Auriol

May 8.

At a meeting of the Managers of the Royal Institution of Great Britain, held at the House of the Institution, on the 8th of May, 1815. —

Present
J. P. Auriol, Esqr. in the Chair.
George Duckett, Esqr.
Daniel Moore, Esqr.
Thomas Snodgrass, Esqr.
Thomas Harrison Esqr. Secretary.

1. The minutes of the proceedings of the last meeting were read and approved. –

2. Read and approved the minutes of the Committee of Accounts of this day. –

3. Read the following letter from the President of the Royal Institution, addressed to the Secretary, dated St. James's Place 5th. May 1815. — *Letter from Earl Spencer nominating Vice Presidents*

Sir,

My health not allowing me to attend at the next General Monthly Meeting of the Royal Institution, I am under the necessity of troubling you to inform the Managers that in pursuance of the authority vested in me as President I nominate the following Managers to be Vice Presidents for the year ensuing. –
J. P. Auriol, Esqr.
Sir Humphry Davy.
Lord Dundas.
The Earl of Hardwicke.
The Rt. Hon: Sir John Nicholl.
And I desire that this nomination may be regularly notified from the Chair at the General Monthly Meeting.
I remain Sir, &c. &c.
(signed) Spencer.
Thomas Harrison, Esqr. –

4. Read the minutes of the General Annual Meeting of Members

5. The following persons having been duly recommended were admitted as Subscribers to the Royal Institution for the year 1815. –

Annual Subscribers.
Chas. Kelsall, Esqr. 24 Sackville St.
Henry Peters, junr. Esqr. 11. Saville row
The Hon: Basil Cochrane, 12 Portman Sqr.

To the Lectures.
Rt. Hon: Lady Catherine Fielding – Upper Wimpole St.

Dundas

~~May 8~~ May 15.

At a Meeting of the Managers of the Royal Institution of Great Britain, held at the house of the Institution, on the 15th of May, 1815. —

Present
Rt. Hon: Lord Dundas, V.P. in the Chair.
Sir Humphry Davy, V.P.
William Watson, Esqr.
Thomas Snodgrass, Esqr.
Daniel Moore, Esqr.
S. Bernard Morland, Esqr. M.P. Treasurer.
Rt. Hon: Sir John Nicholl, M.P. – V.P.
George Duckett, Esqr.
Henry Holland, Esqr.
Alexr. Marcet, M.D.
J. P. Auriol, Esqr. V.P.
Thomas Harrison, Esqr. Secretary.

1. The minutes of the proceedings of the last meeting were read and approved.

2. – Read and approved the minutes of the Committee of Accounts of this day, and signed an Order for payment of the following sums, vizt.

Payments.

Mr. E Odell for Plumbing & Glazing &c. in 1814	£29 . 9 . 5
B. Shepard for Upholstery in 1814	6 . – . 6
	35 . 9 . 11

3. In pursuance of nominations under the Act, the following persons were admitted to the privileges of Life Subscribers to the Royal Institution. —
Life Subscribers
Samuel Turner, Esqr. of Nottingham Place, by the Will of his deceased Father Samuel Turner, Esqr. and – The Honorable Lawrence Dundas of Hertford St. May Fair by His Father,
The Rt. Hon: Lord Dundas. –

4. Hon: Robt. Clifford proposed to the Board of Managers

Hon Admissions — that Professor Esmark of Norway be admitted to the Royal Institution under Art. 16 Chap VI of the Bye laws.

Ordered.

That the Secretary do transmit to Professor Esmark the usual letter accordingly.

Present received.

5 Mr. Frederick Accum, presented

A Practical Treatise on Gas-Light, 8vo.

Ordered.

That the thanks of the Managers be returned for this present. —

6. Read the minutes of the last General Monthly Meeting. —

Re-engagement of M. Faraday. — 7. Mr. Brande having stated to the Board that he wanted assistance in the Laboratory, And that Michael Faraday who was formerly Assistant in the Laboratory, would be willing to resume his situation.

Resolved.

That Blacket Thos. Wallis the Assistant Porter be discharged on Saturday next, with a Month's wages in advance; And that Stephen Slatter be restored to his former situation as Assistant Porter at £1 pr week.

Resolved further.

His salary and duties. — That Michael Faraday be engaged as Assistant in the Laboratory and Mineralogical Collection, and Superintendant of the Apparatus at a Salary of 30s. pr week. And that He be accomodated with Apartments in the house under the superintendance of Mr. Brande.

8. The following persons having been duly recommended were admitted as Subscribers to the Royal Institution for the year 1815.

Annual Subscribers.

James Henry Slater Esqr. 77 Lower Grosvr. St.
Lt. Genl. Hart. M.P. 33 St. Albans Street.
Monseigneur L'Evêque de Sisteron, 18 Queen St. Brompton

Spencer

May 22.

At a Meeting of the Managers of the Royal Institution of Great Britain, held at the house of the Institution, on the 22d of May 1815. —

Present

Earl Spencer, President, in the Chair.
Earl of Hardwicke. V.P.
George Duckett, Esq.
Thomas Snodgrass, Esq.
Henry Holland, Esqr.
Daniel Moore, Esqr.
J. P. Auriol, Esqr. V.P.
Davies Giddy, Esq. M.P.
Alexr. Marcet, M.D.
Sir Humphry Davy. V.P.
Thomas Harrison, Esqr. Secretary.

1. The minutes of the proceedings of the last meeting were read and approved. —

2. Read and approved the minutes of the Committee of Accounts of this day, and signed an Order for an advance to the Clerk of £60. —

Present received.

3. James Moore, Esqr. presented

The History of the Small Pox, 8vo. —

Ordered.

That the thanks of the Managers be returned for this present. —

Loan of Steam Engine. — 4. Ordered. That the thanks of the Managers be returned to Mr. Allen for the loan of his Steam Engine — And that the thanks of the Managers be returned to the London Institution for the loan of models of Steam Engines to the Royal Institution.

Annl. Subscr. — 5 The following person having been duly recommended was admitted as an Annual Subscriber to the Royal Institution.

Rt. Hon: Lord Graves, Uxbridge-house.

Spencer

May 29.

At a Meeting of the Managers of the Royal Institution of Great Britain, held at the house of the Institution, on the 29th of May, 1815. —

Present

Earl Spencer, President. in the Chair.
William Watson, Esq.
George Duckett, Esqr.
Thomas Snodgrass, Esq.
Henry Holland, Esq.
J. P. Auriol, Esqr. V.P.
Benjamin Hall, Esq. M.P.
Davies Giddy, Esq. M.P.
Sir Humphry Davy. V.P.
Rt. Hon. Sir Jn. Nicholl, M.P. V.P. —
Thomas Harrison, Esq. Secretary

1. The minutes of the proceedings of the last meeting were read and approved.

2. Read and approved the minutes of the Committee of Accounts of this day.

Lectures. — 3. Read a letter from Mr. Landseer to the Secretary stating that he should be disposed and prepared to read more Lectures this Season if the Managers think fit. —

Resolved. That the Secretary do inform him that the Managers will be perfectly satisfied if he will read 2 more Lectures this Season, in consideration of his having delivered an extra Lecture last year.

Read a letter from Mr. Campbell to the Secretary stating that as the Season is so far advanced he would not wish to resume his lectures.

4. Presents received.

The African Institution presented.

"Special Report of the Directors of the African Institution in answer to Dr. Thorpe's letter." —

May 29.

8 The Revd. Andrew Irvine, A.M. presented

Reflections on the Education of the Poor.

Ordered.

That the thanks of the Managers be returned for these presents. —

Spencer

62 1815.

June 5.

At a Meeting of the Managers of the Royal Institution of Great Britain, held at the house of the Institution, on the 5th. of June, 1815. –

Present

Earl Spencer, President. in the Chair.
Lord Dundas V.P.
Thomas Snodgrass, Esqr.
Henry Holland, Esqr.
J. P. Auriol, Esqr. V.P.
Sir Benjamin Hobhouse, Bart: M.P.
Daniel Moore, Esqr.
S. Bernard Morland, Esqr. M.P. Treasurer.
Sir Humphry Davy. V.P.
William Watson, Esqr.
Thomas Harrison, Esqr. Secretary.

1. The minutes of the proceedings of the last meeting were read and confirmed. –

2. Read and approved the minutes of the Committee of Accounts of this day, & Signed an Order for payment of the following sum, vizt.

Payment

Mr. Babbage for a Course of Lectures on Astronomy in 1815 — £32.10.0

Presents received.

3 The Royal Society presented, Philosophical Transactions 1815, part 1.
Mr. Lelyveld, presented, A Collection of State Papers, Treaties &c. of the United Provinces, as p List. –
Ordered.
That the thanks of the Managers be returned for these presents.

4. The Managers having occasion for the Room No. 26 which is now occupied by Miss Harris in pursuance of an Order of the Managers of the 6th of July, 1812 on

June 5 63

condition that no inconvenience be found to arise therefrom.
Resolved.
That Miss Harris be accomodated with the Room No. 21 instead of No. 26, on the same conditions.
Resolved further.
That Michael Faraday be accomodated with the Rooms No. 25, & 26 which he formerly occupied.

5. The following person having been duly recommended was admitted as an Annual Subscriber to the Royal Institution for the year 1815. –

Annl. Subsr. Captn. Elton R.N. – 37. Suffolk Street

Humphry Davy V.P. —

64 1815.

June 12

At a Meeting of the Managers of the Royal Institution of Great Britain, held at the house of the Institution, on the 12th. of June, 1815. –

Present

Sir Humphry Davy, V.P. in the Chair.
Sir Benjn. Hobhouse, Bart: M.P.
George Duckett, Esqr.
Thos. Snodgrass, Esqr.
Henry Holland, Esqr.
Thomas Harrison, Esqr. Secretary.
J. P. Auriol, Esqr. V.P.
S. Bernard Morland, Esqr. M.P. Treasurer.
Daniel Moore, Esqr.

1. The minutes of the proceedings of the last meeting were read and approved.

2. Read and approved the minutes of the Committee of Accounts of this day, and Signed an Order for payment of the following sums, vizt.

Payments.

Wm. Thos. Brande, Esqr. 1 years Salary as Professor of Chemy. in 1815	£200. – –
Mr. Millington for a Course of Lectures in 1815	47. 5. –
Mr. Landseer for do. – in do.	52. 10. –
Sir James Edward Smith – for do.	105. – –
Josh. Fincher on acct. of the R.I.	60. – –
	£464. 15. –

Present received.

3. Revd. W. Dealtry B.D. presented.
2 Copies of A Review of Mr. Norris' Attack upon the British & Foreign Bible Society.
Ordered.
That the thanks of the Managers be returned for this present. –

June 12. 65

Application from Mr. Harris declined.

4. Mr. Harris having addressed the Board on the subject of the Room No. 21 granted to his Niece by the last Board, and proposing with their consent at his own expence to divide the Room by a partition, and to have the large high window closed up, with such whitewashing and painting as may be necessary for the accomodation of his Niece. –
Resolved.
That the Board, after mature deliberation, and with every disposition to accommodate Mr. Harris, finding that the Room No. 21 may be wanted at no distant period for the service of the Institution, cannot consent to any alteration in their former Minute.

Lectures Mr. Millington.

5. Mr. Millington being in attendance, was called in, and stated to the Board, that he should be happy to give, in the next Season, a Course of Lectures on Natural Philosophy & Astronomy, to consist of 18 Lectures at the sum of 100 guineas.
Resolved accordingly.

6. Read the minutes of the last General Monthly Meeting.

7. Ordered. That the Scientific Committees be summoned to meet for the purpose of electing 2 Chairmen & a Secretary for each Committee.

8. Ordered. That all the Scientific Journals be placed in the Reading Room instead of the Newspaper Room.

Holiday to Housekeeper.

9. Mrs. Greenwood, the Housekeeper, having applied to the Board for permission to go into the Country, in consequence of her illness.
Resolved. That leave be granted her, under the direction of Mr. Brande.

Adjourned to Monday the 3d. of July.

J. P. Auriol

66

July 3.

1815.

At a Meeting of the Managers of the Royal Institution of Great Britain held at the house of the Institution on the 3rd of July, 1815.—

Present
J. P. Auriol, Esqr. V.P. in the Chair.
Thomas Snodgrass, Esqr.
William Watson, Esqr.
Daniel Moore, Esqr.
S. Bernard Morland, Esqr. M.P. Treasurer.
George Duckett Esqr.
Davies Giddy, Esqr. M.P.

1. The minutes of the proceedings of the last meeting were read and approved.

2. Read and approved the minutes of the Committee of Accounts of this day, and signed an Order for an advance to the Clerk of £70.—

3. Read a letter from Mr. Farish, addressed to Thomas Harrison Esqr. Sec: R. I., declining for the reasons therein assigned to lend his Apparatus ~~to lend his Apparatus~~ to the Royal Institution for the present.—

Resolved. That the letter be returned to Mr. Harrison with the thanks of the Managers for the communication.—

Presents received.

4. Sir James Edward Smith, P. L. S. presented
Traité du Citrus, par Georges Gallesio, 8vo.
Revd. Stephen Weston, B.D. F.R.S. presented
Episodes from the Shah Nameh, 8vo.
The Society for bettering the condition of the Poor, presented the 37th. & 38th. Reports of the Society, 2 cops. of each.
Ordered.
That the thanks of the Managers be returned for these presents.—

July 3. 67

Scientific Committees — 5. Joseph Fincher reported that the Committee of Mathematics Mechanics & Mechanical Inventions, met on Monday the 19th. of June, and made their election as follows.
Hon. Robert Clifford — 1st Chairman.
Sir Gilbert Blane Bt. M.D. 2d.– Chairman.
Dr. Buchan ——— Secretary.

6. Read a communication from the Visitors entered in their proceedings of the 12th. of June.
Mr. Brande reported that the apertures have been made and that the Partition answers the design completely.

Mr Harris applies to reside out of the house. — 7. Read the following request from Mr. Harris the Librarian
W. Harris presumes the Managers of the Royal Institution cannot have any objection to his residing out of the house as he shall always attend the Library hours as usual.
Resolved. That the Managers have no objection to Mr. Harris residing out of the house provided it be attended with no additional expense to the Institution

Chest of Drawers for Mr Faraday. — 8. Ordered. That a Chest of Drawers be provided for Mr. Faraday at an expense not exceeding 5 guineas.

Adjourned to Monday the 7th. of August.— Thomas Snodgrass.

68

Aug. 7.

1815.

At a Meeting of the Managers of the Royal Institution of Great Britain, held at the house of the Institution on Monday the 7th. of August, 1815.—

Present
Thomas Snodgrass, Esqr. in the Chair.
Daniel Moore, Esqr.
J. P. Auriol, Esqr. V.P.

1. The minutes of the proceedings of the last meeting were read and approved.—

2. Read and approved the minutes of the Committee of Accounts of this day, and Signed an Order for an advance to the Clerk of £140.—

Presents received.

3. Sir Humphry Davy, presented
Peintures de Vases Grecs, par, J. V Millingen, folio.
Samuel Boddington, Esqr.– presented
A plan of Munich,
Dr. George Henning, of Bristol, presented
A Critical Inquiry into the pathology of Scrofula, 8vo.
The publishers of the British Review presented
The British Review, No. 11.
M. Haüy, Paris – presented
Memoire sur une Loi de la Cristallisation, appelée Loi de Symétrie,
The American Academy of Arts & Sciences presented the 3d. Volume of their Memoirs – part 1. & 2.
Ordered.
That the thanks of the Managers be returned for these presents.

Dusting Books during Recess. — 4. Mr. Sturt reported that the Books will require dusting during the recess, and that he would wish to have the assistance of the Porters for that purpose.
Ordered accordingly.

Aug. 7. 69

5. The following person having been duly recommended was admitted an Annual Subscriber to the Royal Institution for the year 1815.

Annl. Subsr. — Dr. Robert Richardson – 27 Rathbone Place.

Coals. — 6. Ordered. That Messrs. Dickins & Co. supply the Royal Institution with 20 Chaldrons of the best Wallsend Coals.

Holyday to Asst. Sec. — 7. Joseph Fincher requested leave of absence for 2 or 3 weeks during the recess, which was granted.

8. Adjourned to the first Monday in November.—

J P Auriol

Nov. 6.

At a Meeting of the Managers of the Royal Institution of Great Britain held at the house of the Institution on Monday the 6th of November, 1815.

Present

J. P. Auriol, Esqr. V.P. in the Chair.
Sir H. Davy V.P.
Thomas Snodgrass, Esqr.
Henry Holland, Esqr.
Daniel Moore, Esqr.
George Duckett, Esqr.
Sir Benjn. Hobhouse, Bart. M.P.
Doctor Marcet.
Thomas Harrison, Esqr. Secretary.

1. The minutes of the proceedings of the last meeting were read and approved.

2. Read and approved the minutes of the Committee of Accounts of this day, and signed an Order for an advance to the Clerk of £40.

3. *Lectures on Political Economy.* Read letters from Dr. Roche addressed to the Secretary offering to give a Course of Lectures, at the Royal Institution, on Political Economy.
Resolved. That as the subject does not come within the scope of the Institution Dr. Roche's offer be declined.

Invention of Universal Standard of Measure by J. Street. Read letters from J. Street addressed to the Secy. conceiving he had discovered a simple mode of finding a measure which possesses the accuracy sought for in an universal standard, with a drawing illustrating the principle he adopts, and requesting to lay the principle of his invention before a Committee of the Royal Institution.
Ordered
That Mr. Street receive the thanks of the Managers and be informed that his communication has been laid before them.

Presents received.

4. The Rt. Hon: Earl Spencer President – presented.
Bibliotheca Spenceriana, vol. 4.
Le Chevalier De Sade, presented
Dialogues Politiques sur les principales opérations du Gouvernment François, depuis la Restauration, et sur leurs consequences necessaires 8vo.
M. J. R. Duval, Membre des College et Academie Royale de Chirurgie de Paris. presented
Collection d'Opuscules sur l'Art du Dentiste. 8vo.
Francis Balfour, M.D. presented
A Collection of Treatises on the effects of Sol-Lunar Influence in Fevers 8vo.
George Man Burrows, Esqr. Member of the Royal College of Surgeons, presented
An Account of two cases of Death from eating Mussels: with observations on Fish Poison 8vo.
Samuel Tolfrey, Esqr. presented
A narrative of events which have recently occurred in the Island of Ceylon, 8vo.
The Commissioners of the Public Records presented
The Acts of the Parliament of Scotland, vols. 2, & 3, folio
Registrum Magni Sigilli Regum Scotorum, folio.
Reports respecting the Public Records of Ireland, folio.
John Fearn, Esqr. presented.
A demonstration of the principles of Primary vision with the consequent state of Philosophy in Great Britain 4to. – 2 Cops.

Ordered. That the thanks of the Managers be returned for these presents.

5. *Hon. Admission.* Mr. Ramadini Charge d'Affaires de la S. P. Ottomane having applied to be admitted to the Royal Institution under Art. 16 Chap. VI of the Bye Laws.
Ordered. That the Secretary do transmit him the usual letter accordingly.

Humphry Davy

Nov. 13.

At a Meeting of the Managers of the Royal Institution of Great Britain, held at the house of the Institution, on Monday the 13th November, 1815.

Present

Sir Humphry Davy, V.P. in the Chair.
J. P. Auriol, Esqr. V.P.
William Watson, Esqr.
Thomas Snodgrass, Esqr.
George Duckett, Esqr.
Rt. Hon: Sir John Nicholl. M.P. V.P.
Alexander Marcet. M.D.
Sir Benjamin Hobhouse, Bart. M.P.

1. The minutes of the proceedings of the last meeting were read and approved.

2. Read and approved the minutes of the Committee of Accounts of this day.

3. Read the Minutes of the last General Monthly Meeting.

The following persons having been duly recommended were admitted as Annual Subscribers to the Royal Institution for the year 1816.

Payments.
Captn. Thos. Wilson, R.N. No. 10 Knightsbridge
Benjamin Hanson, Esqr. – 2 Bruton St.
Captn. Dilkes, R.N. No. 6 Upper Seymour St.
Revd. James Vaughan, 27 York Place.
John Pee Esqr. No. 1. Fig Tree Court Temple
George Finch, Esqr. 32 South St. Park lane.

To the Lectures.
Rt. Hon: Lady Ellenborough
Hon: Miss Law
Miss Ann Law.
Miss Elizth. Law.
} St. James Square

J P Auriol

Nov. 20.

At a Meeting of the Managers of the Royal Institution of Great Britain, held at the house of the Institution, on Monday the 20th of November 1815.

Present

J. P. Auriol, Esqr. V.P. in the Chair.
Wm. Watson, Esqr.
Thomas Snodgrass, Esqr.
George Duckett, Esqr.
Daniel Moore, Esqr.
Thos. Harrison, Esqr. Secy.

1. The minutes of the proceedings of the last meeting were read and approved.

2. Read and approved the minutes of the Committee of Accounts of this day.

Present received.

3. The President & Council of the Society of Antiquaries, presented
The first part of the 18th volume of Archaeologia.
Ordered. That the thanks of the Managers be returned for this present.

J. Nicholl

74

Nov. 27. 1815.

At a Meeting of the Managers of the Royal Institution of Great Britain, held at the house of the Institution, on the 27th of November, 1815.—

Rt. Honble Sir John Nicholl, M.P. V.P. in the Chair
Sir Benjamin Hobhouse, Bart. M.P.
S. Bernard Morland, Esqr. M.P. Treasurer.
William Watson, Esqr.
Thomas Snodgrass, Esqr.
Daniel Moore, Esqr.
Thomas Harrison, Esqr. Secretary.
J. P. Auriol, Esqr. V.P.

1\. The minutes of the proceedings of the last meeting were read and approved.

2\. Read and approved the minutes of the Committee of Accounts of this day, and signed an Order for an advance to the Clerk of £40.—

Mr. Millington permitted to re-deliver Lectures at Surry Institution.

3 Read a letter from Mr. Millington, enclosing an application made to him, from the Surry Institution, to deliver the same Course of Lectures there, after Christmas in the Evenings from 7 to 8 o'Clock, which he gave at the Royal Institution last Season, and requesting the sentiments of the Managers on that subject.

Resolved.

That the Secretary do inform Mr. Millington that the Managers have no objection to his delivering to so respectable a body as the Surry Institution after Christmas next in the Evenings from 7 to 8 o'Clock the same Course of Lectures which he gave last Season at the Royal Institution.

Lectures Mr. Craig

4\. Read a letter from Mr. W. M. Craig offering to read gratuitously a Course of Lectures on the principles and practices of Drawing and Painting which he should be ready to commence the first week in April.

75

Nov. 27.

Resolved. That Mr. Craig's liberal offer be accepted.

Presents received.

5 The President & Council of the Royal Society, presented Philosophical Transactions, for 1815 part 2.
Sir John Sinclair, Bart. presented
Account of an improved mode for raising Crops of Grain by means of a Drill-Barrow.
The Association for a general supply of Fish – their second Report.
Ordered. That the thanks of the Managers be returned to the President & Council of the Royal Society for the above present.

Transfer of Mr. Brande's Lectures at Windmill St.

6\. Mr. Brande having proposed to transfer the Lectures on Chemistry delivered by him in Windmill Street to the Laboratory of the Royal Institution.
Resolved. That the Visitors be invited – for Monday next at One o'Clock to confer with the Managers on the above proposition.

7\. The following person having been duly recommended was admitted as an Annual Subscriber to the Royal Institution for the year 1816.—

Annl. Subscr.

Revd. John Powell of Trinity College, Cambridge

J. Nicholl

76

Dec. 4. 1815.

At a Meeting of the Managers of the Royal Institution of Great Britain, held at the house of the Institution on the 4th of December, 1815.—

Present
Rt. Hon: Sir John Nicholl, M.P. V.P. in the Chair
Sir Humphry Davy, V.P.
George Duckett, Esqr.
James Peter Auriol, Esqr. V.P.
Thomas Snodgrass, Esqr.
Scrope Bernard Morland, Esqr. M.P. Treasurer.
Daniel Moore, Esqr.
Thomas Harrison, Esqr. Secretary.

1\. The minutes of the proceedings of the last meeting were read and approved.

2\. Read and approved the minutes of the Committee of Accounts of this day.

Presents received.

3\. Dr. Young, presented
A practical and historical Treatise on Consumptive Diseases. 8vo.
The College of Physicians presented
Medical Transactions, vol 5. 8vo.
Sir John Sinclair, Bart. presented
Account of some experiments to promote the improvement of Fruit Trees by peeling the bark.
Sir William Adams presented
A Collection of Documents in justification of the conduct of Governor Ainslie – 2 copies.
Ordered.
That the thanks of the Managers be returned to Dr. Young, and to the Registrar of the College of Physicians for the above presents.

4\. In pursuance of the invitation of the Managers at

75a

Dec. 4.

Removal of Mr. Brande's Lectures at Windmill st.

the last Board, the following Visitors attended.
The Revd. Dr. Burney.
Richd. Horsman Solly Esqr.
and the subject having been discussed between them and the Managers.
Resolved. That Mr. Brande's proposition of delivering Lectures in the Laboratory be recommended to the consideration of the Committee appointed to revise the Bye laws.—

5\. Dr. Crotch having offered thro' Mr. Brande to give a Course of Lectures on Music during the ensuing Season
Resolved. That his offer be declined.

6\. The following person having been duly recommended was admitted as an Annual Subscriber to the Royal Institution for the year 1816.

Annl. Subscr.

Mr. Frederic Brande, No. 10 Arlington Street.

J. Nicholl

76+ 1815.

Dec. 11.

At a Meeting of the Managers of the Royal Institution of Great Britain, held at the house of the Institution, on the 11th December, 1815.—

Present

Rt. Hon: Sir John Nicholl, M.P. V.P. in the Chair.
Sir Benjn. Hobhouse, Bart. M.P.
Thomas Snodgrass, Esqr.
William Watson, Esqr.
Sir Humphry Davy, V.P.
George Duckett, Esqr.
S. Bernard Morland, Esqr. M.P. Treasurer.
Alexr. Marcet, M.D.

1. The minutes of the proceedings of the last meeting were read and approved.

2. Read and approved the minutes of the Committee of Accounts of this day, and signed an Order for an advance to the Clerk of £56.—

3. Read the minutes of the last General Monthly Meeting.—

Adjourned to Monday the 8th of January 1816.—

J. Nicholl

1816. 77

Jan. 8.

At a Meeting of the Managers of the Royal Institution of Great Britain, held at the house of the Institution, on the 8th of January, 1816.

Present

Rt. Hon: Sir John Nicholl, M.P. V.P. in the Chair.
Sir Humphry Davy, V.P.
Thomas Snodgrass, Esqr.
J. P. Auriol, Esqr. V.P.
Daniel Moore, Esqr.
Alexr. Marcet, M.D.
Thomas Harrison, Esqr. Secretary.

1. The minutes of the proceedings of the last meeting were read and approved.

2. Read and approved the minutes of the Committee of Accounts of this day, and signed an Order for an advance to the Clerk of £60.

Present received.

3. Robert Crowe Esqr. – presented.
"A Chemical Chart or Table".
Ordered. That the thanks of the Managers be returned for this present.—

Mr Brande's Lectures at Windmill Street transferred to R.I.

4. Read the following resolution of the Committee appointed to revise the Bye laws of the 20th Decr. last pursuant to the Resolution of the Managers of the 4th December, recommending to that Committee Mr. Brande's proposition to transfer the lectures on Chemistry delivered by him in Windmill Street to the Laboratory of the Royal Institution

"That it is the opinion of this Committee that it would be highly advantageous as well as honorable to the Institution, to give authority to Mr. Professor Brande to transfer his Course of Lectures on Chemistry from the room in Windmill Street to the Laboratory of the Royal Institution, to give which authority the Managers are

78 1816.

fully competent by Chap XIII Art 8, & Chap XXI Art. 1 of the Bye laws. It being understood that such Course of Lectures is to be attended gratuitously by the Members, and ~~that Annual Subscribers and~~ that Annual Subscribers and others, may have admission to these lectures on terms to be settled between the Managers and the Professor."

Resolved.

That the Managers do concur in this report.

That persons who are not Subscribers to the Institution attending those lectures, "shall pay their Subscriptions to the Professor", And that Life Subscribers and Annual Subscribers, shall be admitted thereto upon severally paying 2 guineas for each Course to the Institution.

not confirmed J N

Resolved further. That copies of the following Minutes and Resolutions relating to this subject be sent round to the Managers not present at this Board with a request that they would meet the Committee on Monday next the 15th Instant at One o'Clock precisely when the minutes of this day will stand for Confirmation. [Minutes of Managers of 4th October 1813, 27th November & 4th December 1815, and of this day.]

Publication of Quarterly Journal.

5. Read the resolution, of the Committee appointed to revise the Bye Laws, of the 6th Inst. recommending Mr. Brande's prospectus of a Journal of the Royal Institution to be laid before the Managers.

Mr. Brande laid on the table a prospectus of a Journal of the Royal Institution to be published 4 times in each year.

The Bye Law respecting publications having been read.

Resolved. That in execution of the said Bye Law a Journal to the effect above mentioned be regularly published to commence as soon as convenient.

That the said Prospectus be forthwith printed and a Copy sent to each Manager, and also to each Member of the Committee appointed to revise

Jan. 8. 79

the Bye Laws in the course of the present week, and that the further consideration thereof be adjourned to Monday next.

That Mr. Brande be requested to undertake the office of Editor, and to employ such Assistants as he may find requisite.

6. Read the following resolution of the Committee appointed to revise the Bye Laws of the 6th Inst.t "That Mr. Brande's proposal for raising a Mineralogical Fund be laid before the Managers"

Proposal for raising a Mineralogical Fund by subscription.

The following proposal for raising a Mineralogical fund was read to the meeting by Mr. Brande.

The Geological Collection of the Royal Institution having been arranged and a Catalogue descriptive of its objects and contents being now in the press and nearly ready for publication, it appears highly desirable that the Mineralogical Collection should be brought into the same state of order and arrangement And that it should also be illustrated by a descriptive Catalogue — Notwithstanding several Members, Subscribers, and other Gentlemen liberally contributed towards the Mineralogical Collection, it is still in a very imperfect and incomplete state, and its objects for Mineralogical lectures, and for general reference cannot at present be fulfilled, it is therefore hoped that the Members and Subscribers and others will contribute to its completion, either by assisting a Fund for the purchase of Minerals or by the donation of specimens.

Resolved.

That a Book be opened in the Clerks' Office for receiving subscriptions and donations to the Mineralogical Collection.

on Erection of Gas Apparatus

7. Mr. Brande requested permission of the Managers to erect a small Gas Apparatus at private expence in the Laboratory of the Institution, and to carry tubes into the lecture room at the expence of the Institution

Resolved. That Mr. Brandes application be sanctioned with the inspection of the Surveyor of the Sun Fire Office.

8 The following persons having been duly recommended were admitted as Subscribers to the Royal Institution.

Life Subscriber.

Mr. Thomas William Chevalier – South audley St.

Annual Subscribers.

Major James Jones, 29 Thayer Street
James Goding, Esqr. Knightsbridge.
Dr. Coull, 57 Arundel St. Panton Square
Rt. Hon: Chas. Yorke, 28 Bruton Street
James Prodgers, Esqr. 11 Everett St. Brunswick Sqr.
Revd. J. B. F. Paquiet, 57 Bedford Place.
Dr. David Plenderleath, Upper Berkeley St.
B. Montcagndo, Esqr. 21 Nassau St.
Thomas Beckett, Esqr. 6 Southampton St. Covent Garden
Joshua Milne, Esqr. 3 Verulam Buildings.
Alexr. Mackenzie, 42 Portland Place.
Captn. J. W. Manby 5 Lyons Inn.
Ralph Green, Esqr. 19, Suffolk Street.

To the Lectures.

Mrs. Blake, Portland Place

J. Nicholl

Jan. 15.

At a Meeting of the Managers of the Royal Institution of Great Britain, held at the house of the Institution, on the 15th of January, 1816. –

Present

Rt. Hon: Sir John Nicholl, M.P. V.P. in the Chair.
Sir Humphry Davy, V.P.
Alexander Marcet, M.D.
Thomas Snodgrass, Esqr.
James Peter Auriol, Esqr. V.P.
Daniel Moore, Esqr.
S. Bernard Morland, Esqr. M.P. Treasurer.
Thomas Harrison, Esqr. Secretary.

1. The minutes of the proceedings of the last meeting were read and confirmed, except the words "that persons attending those lectures, to be delivered in the Laboratory, shall pay their Subscriptions to the Professor". –

Resolved.

Mode of payment to Mr. Brande for his Addl. Lectures — That instead thereof all such subscriptions be received on account of the Institution; and that two thirds of the whole amount be paid to Mr. Brande, and the remaining one third be retained by the Institution.

Regulations. — That all the expences of the said Lectures be borne by the Institution. –

That the said Lectures or Demonstrations on practical Chemistry, be given either in the Laboratory or the Lecture Room so that the same be conducted with due attention to preserving the Laboratory as a place of Experimental research on Chemistry.

That the Subscribers to the said Lectures be not as such admissible to the Laboratory, Lecture Room or other parts of the Institution, except for the purpose of attending the said Lectures.

That notice of the said Lectures and the terms of attending them be forthwith sent round to the Members and Subscribers. –

2. Read and approved the minutes of the Committee of Accounts of this day, and signed an Order for payment of the following sums, viz:

Payments.

Mr. Harris Librarian 3/4 yrs Salary to 31st. December last	£135. –
Mr. Brooks, Collector 1 years Salary to 31st. Decr. last –	31. 10. –
Thos. Slatter Keeper of Lecture-room Door, 1 years Salary to do.	10. 10. –
	£177. –

Presents received.

3. Mr. Joshua Milne, Annl. Subr. to R. I. presented
A Treatise on the valuation of Lives and Survivorships, 2 vol 8vo. –
Revd. Dr. Weston, presented
Ode to Her Imperial Majesty Catherine the Great 8vo. –
presented by ———
A Letter on the Game Laws, 8vo.
An Inquiry into the right and duty of compelling Spain to relinquish her Slave Trade, 8vo. –

Ordered.

That the thanks of the Managers be returned to The Revd. Dr. Weston, & to Mr. Milne, for the above presents. –

4. The following persons having been duly recommended were admitted as Subscribers to the Royal Institution.

Annl. Subrs.: Henry Broadwood, Esqr. – Piccadilly.
Saml. Henry Russell, Esqr. 27 Old Bond St.
Col. Mac Gregor, 27 Golden Square

To the Lectures.

Mr. George Nicholson, 63 Guilford St.

J. Nicholl

Jan. 22.

At a Meeting of the Managers of the Royal Institution of Great Britain, held at the house of the Institution, on the 22d. January, 1816. –

Present

Rt. Hon: Sir John Nicholl, M.P. V.P. in the Chair.
Sir Benjamin Hobhouse, Bart. M.P.
Sir Humphry Davy, V.P.
William Watson, Esqr.
Thomas Snodgrass, Esqr.
George Duckett, Esqr.
S. Bernard Morland, Esqr. M.P. Treasurer.
Alexander Marcet, M.D.
Thomas Harrison, Esqr. Secretary.

1. The minutes of the proceedings of the last meeting were read and confirmed.

2. Read and approved the minutes of the Committee of Accounts of this day.

Presents received.

3. Dr. Yeats. Annl. Subr. to the R. I. presented, A Treatise On the early symptoms which lead to water in the Brain 8vo. –
Sir Thomas Bernard, Bart. presented
"The Barrington School" – 3d. Edition. –
Sir H. Davy, presented,
"On the Fire-Damp of Coal Mines".
The Proprietors of the Asiatic Journal presented
The Asiatic Journal No. 1. –

Ordered.

That the thanks of the Managers be returned for these presents. –

Repairs. — 4. Read an estimate, from Mr. Royce for repairing, whitewashing & covering with green baise the Seats in the Laboratory, amounting to £11.18. –

Ordered.

That Mr. Royce be directed to complete the same immediately as p estimate. –

5. The following persons having been duly recommended were admitted as Subscribers to the Royal Institution.

Life Subscriber.

Mr. John Sweatman, 20 South Audley Street.

Annual Subscribers.

Earl of Shannon, 7 Hamilton Place.
Revd. Edwin Prodgers - Stockwell Surrey.

J. Nicholl

Jan 29.

At a Meeting of the Managers of the Royal Institution of Great Britain, held at the house of the Institution, on the 29th of January, 1816.—

Present

Rt. Hon: Sir John Nicholl, M.P. V.P. in the Chair.
Thomas Snodgrass, Esqr.
George Duckett, Esqr.
Alex: Marcet, M.D.
Thomas Harrison, Esqr. Secretary.

1. The minutes of the proceedings of the last meeting were read and approved.

2. Read and approved the minutes of the Committee of Accounts of this day, and signed an Order for payment of the following sums vizt.

Payments.

Wm. Bulmer for Printing in 1814	£180. 19. —
Josh. Fincher on acct. of the Royal Institution	50. —. —
	£230. 19. —

3. The following persons having been duly recommended were admitted as Subscribers to the Royal Institution.

Annual Subscribers

Dr. George Edwards, 40 Suffolk St. Charing Cross
Wm. Mervin Dillon, Esqr. 26 Bryanstone St.
The Hon: Captn. G. G. Waldegrave, R.N. Portland place
C. P. Meyer, Esqr. Forty Hall, near Enfield.
James P. Lind, Esqr. 13 Duke St. St. James'
Charles Harrison, Esqr. Treasury Chambers.

To the Lectures.

Mrs. Meyer } Forty Hall Enfield
Mrs. C. P. Meyer }
Mr. Bainbridge 45 Bread St. Cheapside.

Spencer

Feb 5.

At a Meeting of the Managers of the Royal Institution of Great Britain, held at the house of the Institution, on the 5th of February, 1816.—

Present

Earl Spencer, President in the Chair.
Sir Humphry Davy, V.P.
Rt. Hon: Sir John Nicholl, M.P. V.P.
James Peter Auriol Esqr. V.P.
William Watson, Esqr.
George Duckett, Esqr. —
Thomas Snodgrass, Esqr.
S. Bernard Morland, Esqr. M.P. Treasurer.
Daniel Moore, Esqr.
Alex: Marcet, M.D.
Thomas Harrison, Esqr. Secretary.

1. The minutes of the proceedings of the last meeting were read and approved.

2. Read and approved the minutes of the Committee of Accounts of this day.

Presents received.

3. The Publishers of the British Review presented The British Review No. 13.
The Publishers of the European Magazine, presented The European Magazine for January.
Ordered. That the thanks of the Managers be returned for these presents.

Lectures.

4. The Secretary reported that Dr. Buchan would be willing to give 3 Lectures on the Functions of Animal Life, during the present Season.
Resolved, That the said offer be declined.

5. Read the report from the Committee of Bye Laws proposed to be made to the General Meeting.—

Hon. Admission.

6 William Watson, Esqr. Manager proposed to the Board M. de Haller Envoye Extraordinaire from the Canton of Berne to be admitted to the Royal Institution, under Art. 16 Chap VI of the Bye Laws.
Ordered, That the Secretary do transmit M. de Haller the usual letter accordingly.—

7. The following persons having been duly recommended were admitted as Subscribers to the Royal Institution.

Annual Subscribers.

Osborne Markham, Esqr. 20 Upper Brook Street
Matthew Harrison junr. Esqr. Treasury Chambers.
Dr. Wm. Somerville 5 Berkeley Street.
F. H. A Clauswitz Esqr. 7 Hinde St. Manchester Sqr.
Major Snow, No. 2 Cork Street.
W. S. Northcote, Esqr. 23 Portland place
Revd. Thos. Raddish, Winchelsea, Sussex.

To the Lectures.

Miss Harriet Pemberton, George St. Hanover Square
Mr. & Miss Davis. 7 Portland Place
Mrs. Kerrison, 12 New Burlington St.
Mrs. Northcote 23 Portland place
Mr. Jn. Lambe, 25 George St. Hanover Square
Miss Copley — do.
Mr. Jeffreys — Clarges Street.
Mrs. J. Folly & 2 Miss Follys — Leyton Essex

Spencer

88

Feb. 12.

1816.

At a Meeting of the Managers of the Royal Institution of Great Britain held at the house of the Institution, on the 12th of February, 1816.

Present
Earl Spencer, President in the Chair.
Sir H Davy, V. P. –
Thomas Snodgrass, Esqr.
George Duckett, Esqr.
Alexr. Marcet, M.D.
S. Bernard Morland, Esqr. M.P. Treasurer.
Daniel Moore, Esqr.
Thomas Harrison, Esqr. Secretary.

1. The minutes of the proceedings of the last meeting were read and approved. –

2. Read and approved the minutes of the Committee of Accounts of this day, and signed an Order for payment of the following sums, vizt. –

Payments,

Reeve for 3/4 yrs Kings & Assessed taxes to Xmas 1815	£205 . 1 . 9
Butler 3/4 yrs Poor rates to do.	39 . 7 . 6
Jones & Yarrell for Newspapers in 1814,	109 . 9 . 6
Newman for Smith's bill for paint Geological pictures – Canvas &c.	27 . 5 . 6
Joseph Fincher on account of the Royal Institution	60 . – . –
	£441 . 4 . 3

3. Audited the Treasurer's Account for the year 1815. –

4. Read the minutes of the last General Monthly Meeting. –

5. In consideration of the additional labour caused to Michael Faraday by the Lectures in the Laboratory.

Feb. 12
Salary of M. Faraday raised.

89

Resolved.

That his Salary be £100 pr Anm. to commence from the 1st January last. –

6. The following persons were admitted as Subsrs. to the R. Institn.

Annual Subscribers.
Sir John Anstruther, Bart 10 Hanover St.
John Osborn, Esqr. M.P. – Admiralty.
Mr. J. Binns, 147 Tottenham Court Road.
Francis Downing, Esqr. 10 Guilford Street.
Mr. Francis Bramah, Grosvenor St. Pimlico. –
The Hon: Captn. Spencer, R. N. St. James's Place.
Wm. Coesfelt, Esqr. Upper Brook St. –

To the Lectures.
Mrs & Miss Atlee – Wandsworth.
Mrs. Osborn, of the Admiralty.
Miss Chanson, 27 York Place.
Mrs Hanson, 3 Burton St.
Miss Holford, Hampstead.
Mrs Pilgrim Kensington.

Spencer.

90

Feb. 19.

1816.

At a Meeting of the Managers of the Royal Institution of Great Britain, held at the house of the Institution, on the 19th of February, 1816. –

Present
Earl Spencer, President in the Chair.
Rt. Hon: Sir John Nicholl, M.P. V.P.
Sir Humphry Davy V. P.
J. P. Auriol, Esqr. V. P.
Daniel Moore, Esqr.
Thomas Snodgrass, Esqr.
George Duckett, Esqr.
Alexr. Marcet, M.D.
S. Bernard Morland, Esqr. M.P. Treasurer.
Thomas Harrison, Esqr. Secretary.

1. The minutes of the proceedings of the last meeting were read and approved.

2. Read and approved the minutes of the Committee of Accounts of this day, and signed an Order for payment of the following sums, vizt. –

Allen for Chemicals in 1814 –	£20 . 3 . 0
De Boffe for Foreign Books do. –	27 . 0 . 9
Dalton & Keys, for Cards – do. –	57 . 15 . –
	104 . 18 . 9

Letters received.

Withdrawal.

3. Read a letter from Francis Downing Esqr. Dy. Inspector of Hospitals – elected an Anl. Subr. on the 12th Inst. requesting permission of the Managers to withdraw his name as he has received an order to hold himself in readiness to embark for Foreign service. –

Resolved.

That under the particular circumstances of this case, and Mr. Downing having derived no benefit from his election, he be excused from paying his subscription for the present year. –

91

Feb. 19,
Borrowing Models from Socy. of Arts.

Read a letter from Mr. Millington, suggesting to the Managers that an application be made to the Society of Arts to lend some of their models to be used in the Lectures to be delivered by him on practical Mechanics with their applications to the Arts and to Manufactures, and likewise suggesting that the application be made with the following observations. That in the Lectures the models will be laid before a large assemblage of persons who are most likely to profit by them, and who perhaps otherwise might not be aware of their existence. That but a very small portion of them will be required at any one time, and that only once a week from Easter to the middle of June. – That the models lent shall not be retained or kept out of the Society's repository longer than from 12 till 6 o'Clock on the days on which they may be required. That they shall be moved with the greatest care and without any expence or trouble to the Society and that the Royal Institution will be answerable for their safe and punctual return, and for all damage, if any, which may happen to them by means of their removal.

In order to enable the Managers of the Royal Institution to enter into the above pledge, Mr. Millington undertakes that he will personally attend to the removal of such Models as may be so lent, both in going and coming, and that he will use every care and diligence in his power to prevent them being injured or mutilated, upon the Managers allowing him such expences of Coachhire Porterage, or other conveyance as may be necessary. –

Resolved.

That the Secretary be directed forthwith respectfully to apply to the Society of Arts, Manufactures and Commerce, thro' their Secretary Dr. Taylor, for the loan of such Models as Mr. Millington may point out upon the above mentioned conditions. –

Lectures

Read a letter from Mr. Stackhouse dated 17th Instant, offering to give a short course of Lectures gratuitously, on the earliest periods of British History illustrated by drawings from the Aboriginal Roman, Saxon and

92

Norman remains still extant.— 1816.

Resolved. That the said offer be declined.

Hon.y Admission. 4. On the application of Henry Warburton, Esq.r— Mons.r Coquebert Montbret, a Foreigner of distinguished Scientific and Literary acquirements was admitted to the Royal Institution, under Art. 16 Chap VI of the Bye laws.—

Presents received.

5. W. Reid Clanny, M.D. presented
A Treatise on the Mineral waters of Gilsland, 8vo.
Mr. Beaumont, presented
Letters on public-house Licencing.—

Ordered. That the thanks of the Managers be returned for these presents.—

Lectures. 6. The Clerk having informed the Managers that he had received a letter from Sir James Smith offering to deliver 2 Lectures on Botany gratis, during the ensuing Season, or by any future service to shew his respect for the Royal Institution, and its Managers.

Resolved.

That the Secretary do return the thanks of the Managers to Sir James Edward Smith for his communication and inform him that they are happy to accept his offer of giving such lectures.

Payment of Apothecary's Bill for Mrs Greenwood. 7. Ordered. That the Apothecary's bill on Mrs Greenwood the Housekeeper, amounting to £6.17.6 be paid.—

8. The following persons having been duly recommended were admitted as Subscribers to the Royal Institution.

Annual Subscribers.
Rev.d Edward Owen, 8 Mount St.
Rich.d Jas. Lawrence Esq. 49 Closter Place
Capt.n Henry E. Reginald Baker R.N. 28 Pulteney St.

To the Lectures
Miss Whitbread. Dover Street

Feb. 19. 93

To the Lectures.
Mrs Walker & Miss Walford. 39 Portland place
Mr Chas. Morris
Miss Morris & Miss E. Morris } Dover Street
Miss Doville — Balham Clapham
Miss Clarke — 46 Park Street.

Spencer.

94

1816. Feb. 26. 1816.

At a meeting of the Managers of the Royal Institution of Great Britain, held at the house of the Institution, on the 26th. of February, 1816. —

Present
Earl Spencer, President in the Chair.
Sir Humphry Davy, V.P.
Sir Benjamin Hobhouse, Bart. M.P.
Thomas Snodgrass, Esq.r
James Peter Auriol, Esq.r V.P.
George Duckett, Esq.
Thomas Harrison, Esq.r Secretary.

1. The minutes of the proceedings of the last meeting were read and approved.

2. Read and approved the minutes of the Committee of Accounts of this day, and signed an Order for an advance to the Clerk of £80.—

Letter resp.g Models from Soc. of Arts. 3. Read the following letter from Dr. Taylor, Sec.y to the Society of Arts, in answer to the application made to him by the Secretary, pursuant to the resolution of the last Board.—

Society of Arts &c. Adelphi
Feb.y 22d 1816.

Your letter respecting the loan of the Models of this Society to the Royal Institution, has been laid before the Society, and will be taken into consideration on Wednesday next. As soon as their determination is known, I will do myself the honor to acquaint you with the result.

I remain respectfully Sir.
Your obed.t Servant
Charles Taylor M.D. Sec.y

Thomas Harrison, Esq.
Sec. R. I. &c. &c.

Feb. 26. 95

4. In pursuance of a nomination under the Act by Sir Charles William Rouse Boughton, Bart. his Son

Life Subs.r William Edward Rouse Boughton Esq. F.R.S. of Devonshire Place — was admitted to the privileges of a Life Subscriber to the Royal Institution.—

Presents received.

5. Thomas Moore, Esq. Ann. Subs. to the R. I. presented
Trial in King's Bench between Moore & Adam.
Earl Spencer — presented
The Chevalier de Mac Carthy's Catalogue of his Library
W. T. Brande, Esq.r presented
A descriptive Catalogue of the British Specimens deposited in the Geological Collection of the Royal Institution —

Ordered. That the thanks of the Managers be returned for these presents.

6. The following persons having been duly recommended were admitted as Subscribers to the Royal Institution

Annual Subscriber.
Edward Stanley, Esq.r 13, Hanover Street.

To the Lectures.
Miss Coesfelt 16. Upper Brook St.
Mr & Miss Searancke — 26. Montagu Place.

J. Nicholl

96

Mar. 4. — 1816.

At a Meeting of the Managers of the Royal Institution of Great Britain, held at the house of the Institution, on the 4th of March, 1816. —

Earl Spencer, President —
R^t Hon: Sir Jno Nicholl, M.P. V.P in the Chair
Sir Humphry Davy, V.P.
J. P. Auriol, Esqr V.P.
William Watson, Esqr
Thomas Snodgrass, Esqr
George Duckett, Esqr
Thomas Harrison, Esqr Secretary.

1. The minutes of the proceedings of the last meeting were read and approved.

2. Read and approved the minutes of the Committee of Accounts of this day, and signed an Order for payment of the following sum, vizt

Payment. Messrs Dickins & Co for Coals in 1815. — } £118. —

3. On the application of Sir Humphry Davy, Admiral Chichagou, was admitted to the Royal Institution, under Art: 16 Chap VI of the Bye laws.

Hon. Admission.

Present received.

T. S. Evans, L.L.D. presented, "Historical Memoranda, respecting Experiments intended to ascertain the Calorific powers of the different Prismatic Rays."

Ordered. That the thanks of the Managers be returned for this present. —

4. The following persons having been duly recommended were admitted as Subscribers to the Royal Institution.

Life Subscriber.

Daniel Tighe, Esqr 18 St James's Place

97

Mar. 4.

Annual Subscribers.

Admiral Sir Pulteney Malcolm, K.C.B. N° 2. Upper Harley Street.
Ashhurst Majendie, Esqr 16 Park lane.

To the Lectures.

Mrs Dixon } 130 Park Street.
Miss Sisson }
Mrs Hoare, — 24 Chapel S^t

Spencer

98

Mar. 11. — 1816.

At a Meeting of the Managers of the Royal Institution of Great Britain, held at the house of the Institution on the 11th of March, 1816. —

Present

Earl Spencer, K.G. President. in the Chair
R^t Honble Sir John Nicholl, M.P. V.P.
J. P. Auriol, Esqr V.P.
Thomas Snodgrass Esqr
William Watson, Esqr
S. Bernard Morland, Esqr M.P. Treasurer
Daniel Moore, Esqr
Alex^r Marcet, M.D.
Thomas Harrison, Esqr Secretary.

1. The minutes of the proceedings of the last meeting were read and approved.

2. Read and approved the minutes of the Committee of Accounts of this day, and signed an Order for an advance to the Clerk of £150. —

3. Read a letter from Mr Stackhouse, stating that he is desirous of giving one, two, or three Lectures "On the Architectural and other remains of Aboriginal, Roman, Saxon, and Norman Britain illustrated by numerous large Drawings" in the Royal Institution. He wishes it to be clearly understood that he proposes to deliver the said Lectures gratuitously, and that he shall consider the Committee as perfectly free from any engagement as to the future in consequence of the present Lectures.

Lectures.

Resolved.

That the Managers consent to his reading one lecture on this subject during the present Season upon the conditions specified in his letter. —

Presents received.

99

Mar. 11.

4. The London Institution, by their Secretary, presented, The Inaugural Oration, by Chas Butler Esqr on laying the first Stone of the London Institution.

Robert Morgan Tighe, Esqr M.R.I. presented, Sermons from Massillon & Bourdalone, by Wm Parnell.

Ordered.

That the thanks of the Managers be returned for these presents.

5. Read the minutes of the last General Monthly Meeting. —

6. The following persons having been duly recommended were admitted as Subscribers to the Royal Institution

Annual Subscribers.

William Oxcatt, Esqr 380. Strand.
Chas Kerr, Esqr Marshalls near Romford.
Captn Simeon, R.N. 50 Queen Ann S^t

To the Lectures.

Lady Codrington, Charles St Berkeley Square.
M^{rs} & Miss Monke, 30 Hertford S^t
Miss Roberts, 91, Gloster Place
Miss Wilson, Marshalls near Romford.
Viscountess Althorp, St James's Place.

Spencer

Mar. 18. 1816.

At a Meeting of the Managers of the Royal Institution of Great Britain, held at the house of the Institution, on the 18th of March, 1816.—

Present
Earl Spencer, K.G. President in the Chair.
Rt. Hon Sir John Nicholl, M.P. V.P.
J. P. Auriol, Esqr. V.P.
Lord Dundas, V.P.
George Duckett, Esqr.
Thomas Snodgrass, Esqr.
Alexr. Marcet, M.D.
William Watson, Esqr.
Daniel Moore, Esqr.
Thomas Harrison, Esqr. Secretary.

1. The minutes of the proceedings of the last meeting were read and approved.

2. Read and approved the minutes of the Committee of Accounts of this day, and signed an Order for an advance to the Clerk of £120.— And that he pay the following bills thereout.—

Payments.

Thompson for Soap & Candles in 1815	£26 . 1 . 6
Allnutt for Matting for Lecture room	13 . 11 . –
Fisher for Chemicals in 1815	23 . 2 . 4
	62 . 14 . 10

Application from E. B. Impey, Esq.

3. Read a letter from Elijah Barwell Impey, Esqr. addressed to the Secretary, stating that it has lately come to his knowledge upon enquiry into the privileges to which he is entitled, by his late Father's bequest in the Royal Institution; that the resolution of the Managers which reduced his interest in the establishment from that of a Member to a Life Subscriber is founded upon a statement which is not correct, as it had been reported to the Managers that the day of his Father's

Mar. 18.

decease was the 1st of October 1810, some months subsequent to the date of the Act of Parliament — and enclosing a Certificate from the Register of Burials shewing that his Father died in October 1809 anterior to the passing of the Act.

Resolved. That the above letter and Certificate, together with a copy of the Minute of the Managers of the 4th October 1811, be sent to Mr. Seymour, Solicitor to the Royal Institution.—

Presents received.

4. George Hibbert, Esqr. presented,
"Further proceedings of the Honorable House of Assembly of Jamaica, relative to a Bill for preventing the unlawful Importation of Slaves in the British Colonies.
Gibbes Walker Jordan, Esqr. Agent for Barbadoes, presented
Copies of a Letter, and of an Answer, respecting the state of the Silver and Copper Coins in Barbadoes.
Presented by the Author
Histoire de Napoleon Buonaparte
par M. De Chateauneuf — premiere partie.
Ordered.
That the thanks of the Managers be returned for these presents.

Coals.

5. Ordered. That 40 Chaldrons of Adairs Main Coals be delivered by Messrs. Dickins & Co. into the Cellar of the Royal Institution, at 50s. ℔ Chaldron, Metage, Shooting, and all expences included.—

6. The following persons having been duly recommended, were admitted as Subscribers to the Royal Institution.
Annual Subscriber
David Dewar Esqr. 79 New Bond St.
To the Lectures.
Mrs. Rogers St. Swithins lane
Honble. Mrs. Waldegrave, Bolton Street.
Miss Vaughan 27 York Place.

Benjn. Hobhouse

Mar. 25. 1816.

At a Meeting of the Managers of the Royal Institution of Great Britain, held at the house of the Institution, on the 25th of March, 1816.—

Present
Sir Benjamin Hobhouse, Bt. M.P. in the Chair
S. Bernard Morland, Esqr. M.P. Treasurer
Thomas Snodgrass, Esqr.
Daniel Moore, Esqr.
George Duckett, Esqr.
Thomas Harrison, Esqr. Secretary.

1. The minutes of the proceedings of the last meeting were read and approved.

2. Read and approved the minutes of the Committee of Accounts of this day, and signed an Order for an advance to the Clerk of £50.— And that he pay the following sums thereout.

Payments.

Accum, Sundrs. for the lectures	£22 . 0 . 4
Howis Sundrs. for the house	26 . 1 . 4
	£48 . 1 . 8

Letter from Mr. Seymour respg. Mr. E. B. Impey's Claim.

3. Read the following letter from Wm. Seymour, Esqr. Solicitor to the Royal Institution.

My Lords & Gentlemen.

When Mr. E. B. Impey's Claim was referred to me in 1811, it was stated by Mr. Boibonds who represented him, that Sir Elijah died on the 1st October 1810, which was the foundation of my opinion given in my letter to the Managers under date of the 4th Oct. 1811. — It now appears by the Certificate of the Minister of the Parish of Hammersmith, that Sir Elijah died on the 10th Oct. 1809, and inasmuch as Sir Elijah devised his share to his Son the said Elijah Barwell Impey (who is now living) such Son was a Member

Mar. 25.

of the Royal Institution on the 10th April 1810 — If the Devisee did not within three Calendar Months after the 10th April 1810, signify according to the 4th Sect. of the 51st Geo: 3d. Ch. 51, that he would not agree to relinquish his Hereditary and Saleable share, then of course he is not now entitled to compensation, but he will be deemed a Member for life, with the privileges conferred by Section 3 of the same Act.

I have the honor to be
&c. &c.
(signed) Wm. Seymour

To the Managers of the
Royal Institution.

Mr. Impey to be admitted as Member.

Resolved.
That Elijah Barwell Impey, Esqr. be considered as Member for life of the Royal Institution with the privileges conferred by Sect. 3. (pa: 21) of the Act of Parliament.—

Insurance — Erection of Steam boiler.

4. Read a letter from Edward Griffin, Esqr. Secy. of the Sun Fire Office, addressed to the Secretary, stating that in consequence of the erection of a Steam Boiler on the premises in Albemarle Street it will be necessary to insert the following Warrantee in their Policy No. 893167 otherwise the Insurance will not stand good.

Warrantee.
This Office to be free from all loss or damage occasioned by explosion, and from all consequences thereof.

Resolved.
That the Clerk make enquiries at several of the principal Fire Offices on this subject, & report to the next Board.

Presents received.

5. James Stephen, Esqr. presented
"A Defence of the Bill for the registration of Slaves.
"An exposure of misstatements and misrepresentations contained in Mr. Marryats Pamphlet." — By ——
Ordered.
That the thanks of the Managers be returned to James Stephen, Esqr. for this present.

6 The following persons having been duly recommended were admitted as Subscribers to the Royal Institution.

Annual Subscribers.

Hon.ble Matt.w Fortescue Devonshire place
The Marquis of Salisbury Arlington St.

To the Lectures.

Mrs Leslie Foster 24 Gt George Street
Hon.ble Mrs Fortescue } Devonshire place
Miss Acland
Lady Acland, 85 Pall Mall
Mrs Robinson, 47 Weymouth St.
Miss Hinchley, 47 Weymouth St.

Spencer

April 1.

At a Meeting of the Managers of the Royal Institution of Great Britain, held at the house of the Institution, on the 1st of April, 1816.—

Present
Earl Spencer, K.G. President in the Chair.
Lord Dundas, V.P.
Sir Humphry Davy V.P.
J. P. Auriol, Esq. V.P.
Thomas Snodgrass, Esq.
Sir Benj.n Hobhouse, Bart. M.P.
George Duckett, Esq.
William Watson, Esq.
S. Bernard Morland, Esq. M.P. Treasurer.
Daniel Moore, Esq.

1. The minutes of the proceedings of the last meeting were read and approved.

2. Read and approved the minutes of the Committee of Accounts of this day, and signed an Order for payment of the following sums, viz.t

Payments.

1 years Insurance in the Sun Fire Office to 25th March 1817	£64. 1.—
Joseph Fincher on account of the Royal Institution	50.—
	114. 1

On Erection of Steam Boiler

3. The Clerk reported that he had made enquiries at several of the principal Fire Offices, and was informed that, though on an inspection of the Institution, by their Surveyors, they might consent to an Insurance being effected without inserting a warrantee, yet if any loss or damage was occasioned to the property insured, by explosion or from the consequence thereof, they would contest the Insurance.

Resolved. That the Insurance be continued in the Sun Fire Office the Board consenting to the warrantee being inserted in their Policy.—

Life Subs.rs

4. In pursuance of a nomination under the Act by Benjamin Harrison, Esq. of Guys Hospital, his Nephew Matthew Harrison, Esq. of Queen Square was admitted to the privileges of a Life Subscriber to the Royal Institution.—

5 Present received.
Miss Benett presented
Silicious Madrepore from Wiltshire.
Ordered.
That the thanks of the Managers be returned for this present.

Journal – No 1.

The Journal of the Royal Institution, No 1, was laid on the table.

6. The following persons having been duly recommended were admitted as Subscribers to the Royal Institution.

Annual Subscribers.
Capt.n Adam Brown, R.N. Berkeley Sq.
Dr Compton No 1. Burlington Gardens.
Admiral Sir H. Hotham 23 Harley St.

To the Lectures
Miss Arabella Hoare } 16, Devonshire place
Miss Frances Hoare

J P Auriol

April 8.

At a Meeting of the Managers of the Royal Institution of Great Britain held at the house of the Institution, on the 8th of April, 1816.—

Present
James Peter Auriol, Esq. V.P. in the Chair.
Sir Humphry Davy, V.P.
Thomas Snodgrass, Esq.
George Duckett, Esq.
Daniel Moore, Esq.
Alex.r Marcet, M.D.
Thomas Harrison, Esq.r Secretary.

1. The minutes of the proceedings of the last meeting were read and approved.

2. Read and approved the minutes of the Committee of Accounts of this day, and signed an Order for payment of the following sums, viz.t

Payments.

Mess.rs Upton & Co for Oil to 27th February last	£84.—
Jn.o Bird for Bricklayers work in 1815	25.—
Parker for Plumbing & Glaziery do	16. 4. 3
Odell for do — do	17. 7. 1
	142. 11. 4

Letter from Mr Stackhouse resp.g Lectures

3. Read the following letter from Mr Stackhouse, addressed to the Managers of the Royal Institution.

Gent.n

"At the conclusion of my lecture on Tuesday the 26th Ult.o I was asked by several of the Subscribers when I should give my next lecture: Upon my informing them that I believed the previous arrangements of the Royal Institution were so made as to preclude my giving any more lectures at this time, they expressed very great regret, and requested that an application should be made to you, that I might give two more lectures this Season. Should you accede to this their

request, I shall on my part most willingly deliver two more lectures gratuitously on the following days, — Monday April 22^{d} — and
Monday April 29th.
In these lectures the entire subject will be epitomized.
I remain, Gentlemen
Yours respectfully
(signed) T. Stackhouse

Resolved. That this offer be accepted.

4. In pursuance of nominations under the Act, the following persons were admitted to the privileges of Life Subscribers to the Royal Institution.

Life Subscribers. Samuel S. Boddington, Esqr. of Upper Brook S^{t}. nominated by his Father, Saml. Boddington, Esqr. and — Mrs Dickinson of N° 11 Stafford row Buckingham Gate by her Husband, Charles Dickinson, Esqr.

Presents received.

5. The Society of Antiquaries presented, by Nicholas Carlisle, Esqr. their Secretary.
A Catalogue of the printed Books & Manuscripts in the Library of the Society of Antiquaries.
M^{r}. Richard Phillips, by Professor Brande, presented
Remarks on the Editio altera of the Pharmacopæia Londinensis.
The Rev^{d}. Stephen Weston, F.R.S. & F.A.S. presented
A Chinese Poem inscribed on Porcelain, with a Translation & Notes.
Patrick Colquhoun, Esqr. presented,
Epitome of a scheme of Finance.
Ordered.
That the thanks of the Managers be returned for these presents.

6. Read the minutes of the last General Monthly Meeting.

7. The following person having been duly recommended was admitted a Subscriber to the Royal Institution.
Annual Subscriber.
Samuel Marsham, Esqr. 49 John S^{t}. Fitzroy Sq^{re}.

8. Adjourned to Monday the 22^{d} April.

Dundas

At a Meeting of the Managers of the Royal Institution of Great Britain held at the house of the Institution, on the 22^{d} of April, 1816. —

R^{t}. Hon: Lord Dundas V.P. in the Chair.
J. P. Auriol, Esqr. V.P.
Henry Holland, Esqr.
Thomas Snodgrass, Esqr.
Sir Benjamin Hobhouse, Bart. M.P.
George Duckett, Esqr.
Thomas Harrison, Esqr. Secretary.
Daniel Moore, Esqr.

1. The minutes of the proceedings of the last meeting were read and approved.

2. Read and approved the minutes of the Committee of Accounts of this day, and signed an Order for an advance to the Clerk of £60. —

Hon^{y} Admission. 3. On the application of John Searle, Esqr an Annual Subs^{r}. — Chev^{r}. Reys, Secretaire General & Conseiller de Legation de S. A. R. le Prince Regent de Portugal, was admitted to the Royal Institut^{n}. under Art: 15. Ch: VI of the Bye Laws.

Presents received.

4 Henry Goodwyn, Esqr. M.R.I. presented
The first Centenary of Tables of all the complete Decimal Quotients, from 1 to 1024,
The Society for the Encouragement of Arts &c. presented
Their Transactions, vol. 33.
I. D'Israeli, Esqr. presented
An Inquiry into the Literary & Political Character of James I.
M^{r}. J. Harding, presented
Robert Bakewell's Introduction to Geology 2^{d} Edition
The Author of Three Familiar Lectures on Craniological Physiognomy.
Ordered. That the thanks of the Managers be returned

for these presents.

5. Read a recommendation from the Visitors entered in their proceedings of the 8th Instant.
Repairs. Resolved. That M^{r}. Royce prepare an estimate of the expence of white-washing the three Rooms belonging to the Housekeeper and Servants, and that it be laid before the next Board. —

6. The following persons having been duly recommended were admitted as Subscribers to the Royal Institution.
Annual Subscribers.
Thomas Thompson, Esqr. Chapel S^{t}. Gros^{r}. Place
W. Harkness, Esqr. 33 Conduit S^{t}. —
Arthur Chichester Esqr. Duke S^{t}. Man^{r}. Square
D. Taylor, Esqr. Tavistock Hotel
John Cripps, Esqr. 13 John S^{t}. Adelphi.

Dundas

112

April 29 — 1816.

At a Meeting of the Managers of the Royal Institution of Great Britain, held at the house of the Institutn. on the 29th of April, 1816. —

Rt. Hon: Lord Dundas, V. P. in the Chair
Sir Humphry Davy V. P.
J. P. Auriol, Esqr. V. P.
George Duckett, Esqr.
Daniel Moore, Esqr.
Thomas Snodgrass, Esqr.
S. Bernard Morland, Esqr. M. P. Treasurer.
Alexander Marcet, M. D.
Thomas Harrison, Esqr. Secretary.

1\. The minutes of the proceedings of the last meeting were read and approved.

2\. Read and approved the minutes of the Committee of Accounts of this day, and signed an Order for payment of the following sums, vizt.

Payments,

Newman for Chemicals in 1815	£67. 8. –
Pringle & Co. for Upholstery, &c.	28. 8. 2.
	95. 16. 2

Life Subscriber,

3\. In pursuance of a nomination under the Act by Lord Dundas, his Grandson
Thomas Dundas Esqr. of Arlington Street
was admitted to the privileges of a Life Subscriber to the Royal Institution.

Hon. Admissions.

4\. On the application of Sir Humphry Davy,
Mr. Camille Jordan – of Lyons –
& M. Benjamin de Constant
Foreigners distinguished for Literary acquirements were admitted to the Royal Institution under Art: 16. Chap VI of the Bye Laws. –

113

April 29.

Repairs.

5\. Read an estimate from Mr. Royce of the expence of whitewashing the Rooms, and colouring the Staircase and Passage leading to them, amounting to £9. 1. 0 –
Ordered. That it be done forthwith at an expense not exceeding £9. 1. 0 –

Coals.

6\. The Clerk reported that the Adairs Main Coals are of a very inferior quality to those formerly used in the Institution and as 5 Chaldron only, had been delivered, He had desired that the remaining 15 Chaldr. be not sent till further orders.
Resolved. That for the present 5 Chaldron of the best Wallsend Coals be laid in at 55s. ₩ Chaldn. free of all expences. –

6\. The following persons having been duly recommended were admitted as Subscribers to the Royal Institution.

Annual Subscribers.
Rt. Honble Lord Clifford, 3 Portman Square
Dr. Sigmond – 12 Dover Street
Wm. Vernon, Esqr. 40 Grosvenor Square
John Halkett, Esqr. 21. New Norfolk St.
Viscount Gage – Royal Hotel.
Henry Seymour, Esqr. 22 Cadogan Place.
M. L. Este, Esqr. of the 1st Life Guards Hyde Park.
Major Walwyn No. 16 Cork Street.

To the Lectures.
Miss Grey 50 Bedford Square
Miss Moula, Miss Cooper } 15 Upper Gower Street.
Mrs. Ashton, Miss Ashton, Miss F. Ashton } 21 Grafton Street

7\. Ordered. That the House Lists be filled up as follows.

President Earl Spencer K. G. &c. &c.

Treasurer. Scrope Bernard Morland, Esqr. M. P. &c.

Secretary. Thomas Harrison, Esqr. F. R. S.

114

Managers. — 1816.
J. P. Auriol Esqr.
John Frederick Daniell, Esqr.
Sir Humphry Davy
George Duckett, Esqr.
Charles Hatchett, Esqr.
Sir John Cox Hippisley, Bart. M. P.
Marquis of Lansdowne
Daniel Moore, Esqr.
Earl of Morton, K. T.
Rt. Hon: Sir John Nicholl
John Sanders, Esqr.
Thomas Snodgrass, Esqr.
Samuel Solly, Esqr.
William Watson, Esqr.
Earl of Winchilsea

Visitors.
George Aust, Esqr.
Revd. Charles Burney D. D.
Earl of Darnley.
William Hasledine Pepys, Esqr.
Sir Christr. Hawkins, Bart.
Lord Holland, F. R. S.
Gibbes Walker Jordan, Esqr.
William Morgan, Esqr.
Charles Pieschell, Esqr.
Peter Mark Roget, M. D. F. R. S.
Richd. Horsman Solly, Esqr.
Duke of Somerset, F. R. S.
Marquis of Stafford, K. G.
Sir Jno. Thomas Stanley, Bart.
Lt. Genl. Wm. Thornton, M. P.

Spencer.

115

May 6.

At a Meeting of the Managers of the Royal Institution of Great Britain, held at the house of the Institution, on the 6th of May, 1816. —

Earl Spencer, K. G. President, in the Chair.
J. P. Auriol, Esqr.
William Watson, Esqr.
Daniel Moore, Esqr.
Samuel Solly, Esqr.
Rt. Honble Sir John Nicholl, M. P.
Thomas Snodgrass, Esqr.
Sir Humphry Davy
John Sanders, Esqr.
S. Bernard Morland, Esqr. M. P. Treasurer.
George Duckett, Esqr.
J. F. Daniell, Esqr.
Thomas Harrison, Esqr. Secretary.

1\. The minutes of the proceedings of the last meeting were read and approved.

2\. Read and approved the minutes of the Committee of Accounts of this day.

Present received.

3\. Gibbes Walker Jordan, Esqr. presented
Examination of the principles of the Slave Registry Bill.
Ordered. That the thanks of the Managers be returned for this present.

Committee on Expenditure.

4\. Resolved. That Mr. Auriol, Sir H. Davy, and Mr. Solly, be a Sub-Committee to examine into the expenditure of the Institution and report their opinion thereon to the Board – the same to be an open Committee, and meet immediately after the breaking up of the Board on Monday next.

5\. The President nominated as Vice Presidents for

116 1815.

the ensuing year,
The Marquis of Lansdowne
Earl of Morton.
Earl of Winchilsea
Rt. Honble Sir John Nicholl
Sir Humphry Davy
James Peter Auriol, Esqr.

6. The following persons having been duly recommended were admitted as Subscribers to the Royal Institutn.

Annual Subscriber
John William Bouden, Esqr. of Fulham.

To the Lectures.
Mrs. Otway } 5 Hill Street.
Miss Otway }
Miss Randall, 35 Upper Berkeley St.
Miss Alleyne, 10 Upper Harley Street,

Spencer.

117

May 13.

At a Meeting of the Managers of the Royal Institution of Great Britain, held at the house of the Institution, on the 13th of May 1816. —

Earl Spencer, K.G. President in the Chair.
Sir Humphry Davy V.P.
J. P. Auriol, Esqr. V.P.
Rt. Hon: Sir John Nicholl, M.P. V.P.
Charles Hatchett, Esqr.
Samuel Solly, Esqr.
Thomas Snodgrass, Esqr.
Daniel Moore Esqr.
J. F. Daniell, Esqr.
John Sanders, Esqr.
George Duckett, Esqr.
Thomas Harrison, Esqr. Secretary.

1. The minutes of the proceedings of the last meeting were read and approved.

2. Read and approved the minutes of the Committee of Accounts of this day.

Present received.

3. John Weyland junr. Esqr. F.R.S. presented
The principles of population and production, as they are affected by the progress of Society, with a view to moral and political Consequences, 8vo.
Ordered. That the thanks of the Managers be returned for this present.

4. Read the minutes of the last General Monthly Meeting.

5. The following person having been duly recommended was admitted as an Annual Subscriber to the Royal Institution. —
Annl. Subsr. Robert Coombe, Esqr. 27 Gt. Pulteney Street.

Spencer

6. The Sub-Committee appointed by the last Board having

118 1816.

~~examined into the expenditure of the Institution recommended the following diminution in the Expenditure.~~
~~That the sending round the weekly notices to the Managers be discontinued. —~~
~~That the printed Cards of the lectures be discontinued and that in lieu thereof the lectures be advertized once every fortnight in the ~~Times~~ Evening Post and Morning Chronicle.~~
~~That every Tradesman who furnishes any thing to the Institution insert the same together with the price thereof in a Book to be left at the Institn. when the articles are delivered.~~
~~which was resolved accordingly.~~

119

May 20.

At a Meeting of the Managers of the Royal Institution of Great Britain, held at the house of the Institution, on the 20th of May, 1816.

Earl Spencer, K.G. President in the Chair.
Sir Humphry Davy V.P.
J. P. Auriol, Esqr. V.P.
Charles Hatchett, Esqr.
George Duckett, Esqr.
Thomas Snodgrass, Esqr.
John Sanders, Esqr.
Daniel Moore, Esqr.
John Frederick Daniell, Esqr.
Thos. Harrison, Esqr. Secretary.

1. The minutes of the proceedings of the last meeting were read and approved.

2. Read and approved the minutes of the Committee of Accounts of this day, and signed an order for payment of the following sums, vizt.

Payments.

1 years Ground rent to the City of London to Lady-day last	£135. 1. 8.
Joseph Fincher on account of the Royal Institution	50. —. —
	185. 1. 8

Presents received.

3. Sir Egerton Brydges, Bart. M.P. presented
Bertram — a poetical Tale, 12mo.
John Symmons, Esqr. F.R.S. M.R.I. presented
"Remedies proposed for the relief of our present Embarrassments." —
Ordered. That the thanks of the Managers be returned for these presents.

4. The following persons having been duly recommended

were admitted as Subscribers to the Royal Institution

Annual Subscribers.

Edward Price, Esqr. No. 119 Bond Street.
George Pocock, Esqr. 39 Charles St. Berkeley Sqr.

5. Recommendations of Comee. on Expenditure. The Secretary from the Sub-Committee appointed to examine into the Expenditure of the Institution, brought up the following report.

The Sub-Committee having examined into the expenditure of the Institution recommend the following diminution in the Expenditure.

Notices of Meetings & Lectures discontinued. That the sending round the weekly notices to the Managers be discontinued.

That the printed Cards of the Lectures be discontinued, and that in lieu thereof the Lectures be advertized once every fortnight in the Morning Post and Morning Chronicle.

That every Tradesman who furnishes any thing to the Institution insert the same together with the price thereof in a Book to be left at the Institution when the Articles are delivered.

Resolved.

That the same be adopted by this Board.

That the following notice be forthwith given to the Members and Subscribers.

Advertisements in Papers ordered. The Members and Subscribers are informed that for the future the printed Cards announcing the Lectures will be discontinued, and in stead thereof due notice of the Lectures will be given in the Morning Chronicle and Morning Post Newspapers.

Humphry Davy

May 27.

At a Meeting of the Managers of the Royal Institution of Great Britain, held at the house of the Institution, on the 27th. of May, 1816.

Sir Humphry Davy, V.P. in the Chair.
Samuel Solly, Esqr.
John Sanders, Esqr.
Thomas Snodgrass, Esqr.
Sir John Cox Hippisley, Bart. M.P.
Charles Hatchett, Esqr.
Thomas Harrison, Esqr. Secretary.

1. The minutes of the proceedings of the last meeting were read and approved.

2. Read and approved the minutes of the Committee of Accounts of this day.

3. The following person having been duly recommended was admitted as an Annual Subscriber to the Royal Institution.

Annl. Subsr. G. H. Nochden L.L.D. H. 2. Albany.

Spencer

June 3.

At a Meeting of the Managers of the Royal Institution of Great Britain held at the house of the Institution, on the 3d. of June, 1816.

Earl Spencer, President, in the Chair.
Sir Humphry Davy, V.P.
James Peter Auriol, Esqr. V.P.
Samuel Solly, Esqr.
Charles Hatchett, Esqr.
John Frederick Daniell, Esqr.
Thomas Snodgrass, Esqr.
George Duckett, Esqr.
Daniel Moore, Esqr.
Sir J. C. Hippisley, Bart. M.P.

1. The minutes of the proceedings of the last meeting were read and approved.

2. Read and approved the minutes of the Committee of Accounts of this day, and signed an Order for an advance to the Clerk of £50.—

3. Letter from Mr. Millington respt. Lectures. Read the following letter from Mr. Millington, addressed to the Managers of the Royal Institution.

My Lords & Gentlemen,

Having been informed by Mr. Solly, that it is your wish that I should continue my Lectures upon Practical Mechanics in the ensuing Season for 1817,—I beg to state that I shall have great pleasure in doing so, on the same terms as at present, provided I have early intimation of your wishes, so as to plan my arrangements accordingly.

The Course which I am now upon, cannot be completed in less than 24 Lectures, and perhaps may extend to 26, and as nine of these will be delivered in the present Season, from 15 to 17 Lectures will be left for the ensuing one.—

The satisfaction which has been expressed of my late

June 3.

attempts both by the Managers and attendants on the Royal Institution, cannot be otherwise than highly gratifying, and demands my utmost exertion to make my future efforts worthy of that approbation, which I assure you, will always be my proudest wish to obtain. And from the many flattering promises of support which have been made to me, not only for the loan of models, but supply of materials, I confidently hope that the part of my Course which is reserved for the next year, will not be found altogether devoid of interest, information, novelty, and amusement.

I am, My Lords and Gentlemen,
your obliged and very humble Servant
(signed) John Millington.

Upper Mall Hammersmith
1st. June 1816.

Resolved. That Mr. Millington's offer be accepted.

4. Sir J. E. Smith declines lecturing. Read a letter from Sir James Edward Smith, stating that it is with great regret he is obliged to make his apologies to the Managers of the Royal Institution, his health not allowing him to perform his promise of lecturing here this Season, but if the Managers will allow him, he will consider his promise holding good for the Spring of 1817.

Resolved. That the Board regret the cause that prevents Sir James Edwd. Smith delivering his lectures on Botany this Season, and accept with their thanks his offer of delivering them in 1817.—

Present received.

5. Charles Scudamore, M.D. presented

An Analysis of the Mineral Water of Tunbridge Wells.

Ordered. That the thanks of the Managers be returned for this present.—

6. The following person having been duly recommended was admitted as a Life Subscriber to the Royal Institution.

Life Subsr. G. H. Longridge, Esqr. 35 Leicester Square.

Spencer

124

June 10.

1816.

At a Meeting of the Managers of the Royal Institution of Great Britain held at the house of the Institution, on the 10th of June, 1816. –

Present
Earl Spencer, K. G. President, in the Chair
Sir Humphry Davy V. P.
Samuel Solly, Esqr.
William Watson, Esqr.
Thomas Snodgrass, Esqr.
George Duckett, Esqr.
Sir J. C. Hippisley, Bart: M.P.

1. The minutes of the proceedings of the last meeting were read and approved.

2. Read and approved the minutes of the Committee of Accounts of this day.

Present received.

3 Samuel Solly, Esqr. presented
"Edda, Rythmica seu Antiquior, vulgo Sæmundina dicta, 4to. –
Ordered. That the thanks of the Managers be returned for this present.

Hon Admission. 4. On the application of Sir Humphry Davy,
Count Jules de Polignac of Paris
was admitted to the Royal Institution under Art: 16 Chap: VI of the Bye Laws.

5. Read the minutes of the last General Monthly Meeting.

Suspension of Purchase of Books. 6. Resolved. That the Patrons be informed that under the present circumstances of the Royal Institution, the Managers find it impracticable to supply the usual Funds to the Library, they therefore request that the

June 10.

125

Patrons will suspend for the present their purchases of Books, and employ their Funds as far as they can be applied to the payment of the Salary of the Keeper of the Library.

Annl. Subsrs. 7. The following persons having been duly recommended were admitted as Annual Subscribers to the Royal Institution for the year 1816.
Robert Chappell, Esqr. } 17 Caroline St. Bedford Square
Henry Baldwin, Esqr. }

Adjourned to Monday the 1st. of July.

Humphry Davy

126

July 1.

1816.

At a Meeting of the Managers of the Royal Institution of Great Britain held at the house of the Institution, on the 1st of July, 1816. –

Present
Sir Humphry Davy, V. P. in the Chair.
Rt. Honble Sir John Nicholl, V.P.
J. P. Auriol, Esqr. V. P.
Earl of Winchilsea, V.P.
Thomas Snodgrass, Esqr.
Charles Hatchett, Esqr.
William Watson, Esqr.
Earl Spencer K. G. President

1. The minutes of the proceedings of the last meeting were read and approved.

2. Read and approved the minutes of the Committee of Accounts of this day, and signed an Order for payment of the following sums, vizt.

Payments.

William Thos. Brande, Esqr. 1 years Salary, as Professor of Chemistry, to May 1816 –	£ 200 . – . –
John Millington, Esqr. for Lectures in 1816.	105 . –
Sundry expences for do.	6 . 3 . –
Joseph Fincher on acct. of the Royal Institution	210 . –
Mr. Wm. Harris ¼ yrs Salary to 30th. March 1816	45 . – . –
Mr. Reeve ¼ yrs Kings & Assessed taxes to Lady day 1816	68 . 7 . 5
Mr. Butler ¼ yrs Parochial do.	13 . 2 . 6
	£ 647 . 12 . 11

Present received.

3 The Committee of the Society for diffusing knowledge

July 1.

127

respecting the punishment of Death, presented
Montagu's Selections of Opinions of different Authors upon the punishment of Death, 3 vol. 8vo.
Ordered. That the thanks of the Managers be returned for this present.

The Journal of the Royal Institution No. 2, was laid on the table.

Commn. from Patrons of Library. 4. Read the following Resolution from the Patrons of the Library.
The Patrons of the Library having considered the communication made to them by the Board of Managers are inclined to hope that it will be in their power during the present depressed state of the Finances of the Institution to contribute One hundred Pounds annually from this period towards the Salary of the Keeper of the Library, but conceive the surplus of their Receipts beyond that sum will be absolutely necessary for unavoidable Expences attending their Establishment.

5. Ordered. That 30 Chaldrons of the best Wallsend Coals be laid in at 55s. ℔ Chaldron, Metage, Shooting, and all Expences included.

New Edn. of Bye Laws. 6. Ordered. That a new edition of the Bye Laws be printed under the direction of Mr. Brande.

Adjourned to the 4th. of November. –

S Solly

1816

Nov. 4.

At a Meeting of the Managers of the Royal Institution of Great Britain held at the house of the Institution, on the 4th of November, 1816. —

Present
Samuel Solly, Esq.r in the Chair.
Thomas Snodgrass, Esq.r
Daniel Moore, Esq.r
John Frederick Daniell, Esq.r
Thomas Harrison, Esq.r Secretary.

1. The minutes of the proceedings of the last meeting were read and approved.

2. Read and approved the minutes of the Committee of Accounts of this day, and signed an Order for an advance to the Clerk of £50.

3. In pursuance of nominations under the Act, the following persons were admitted to the privileges of Life Subscribers to the Royal Institution.

Life Subs.ers nominated Roderick Macleod, M.D. of 70 Berwick St. Soho by his Cousin Sir John Macpherson, Bart.
& Captain Richard Peacocke of the Royal Navy. by his Brother Major Gen.l Sir Warren Peacocke

4. Read a letter from Mr. T. Stackhouse proposing to deliver in the Institution the ensuing Season, A Course of 5 Lectures on the Architectural and other remains of Aboriginal Roman, Saxon, & Norman Britain, accompanied with numerous large Drawings illustrative of the progress of Architecture in this Country.

Lectures. Resolved. That the consideration of this proposal be deferred for the present.

Presents received.

5 John Stackhouse, Esq.r F.L.S. presented,
Theophrasti Eresii de Historia Plantarum Libri X Graece, 2 vol.s 8vo —

Nov. 4.

Sir Thomas Bernard, Bart: presented,
Spurinna: or, the Comforts of Old Age, 8vo
Thomas N. Parker, Esq.r M.A. presented
An Essay on the construction & hanging of a Turnpike Gate, Turn wicket, &c. 8vo
The Publishers of Conversations on Political Economy, presen.d
"Conversations on Political Economy; in which the elements of that Science are familiarly explained," 8vo
Another Copy of the same Book, from the Editors.
Charles Scudamore, M.D. &c. &c. presented
An analysis of the Mineral water of Tunbridge Wells, 8vo
Mr. Robert Owen, of New Lanark, presented
An address delivered to the Inhabitants of New Lanark
& A new view of Society, or Essays on the formation of the human Character 8vo
Earl Spencer K.G. President of the Royal Institution presented,
M. Lutheri Opera omnia, 7 vol.s folio.
P. Melancthonis Opera omnia, 4 vol.s folio.
J. Huss et Hieron Pragensis Opera, 2 vol.s folio.
The Publishers of the Colonial Journal, presented
The Colonial Journal, No. 2.
The Editor of the Magnetiser's Magazine,
The Magnetisers Magazine No.s 1. & 2.
No. 3 of the Journal of the Royal Institution has been laid on the table of the Library.
Ordered.
That the thanks of the Managers be returned for the above presents. —

6. The following persons having been duly recommended were admitted as Annual Subscribers to the Royal Institution for the year 1817. —

Ann.l Subs.ers
Rev.d A. Montgomery Campbell 36 York St. Portman Sq.
George Oswald Sym, Esq.r — 63 Baker Street
Joseph Green, Esq.r of South Lambeth
Samuel Reed, Esq.r — 4 Berkeley Street
Colonel Catinelli, 20 Portman Street

1816.

James Gregory, Esq.r 8 Adam St. Adelphi.
Sir Christ.r Cogge Upper Grosvenor Street.
James Morier, Esq.r 3 Wimpole Street
Henry Jenkinson Esq.r — Middlesex Hospital

7 Adjourned to Monday the 2d of December.

Dan Moore

Dec. 2.

At a meeting of the Managers of the Royal Institution of Great Britain, held at the house of the Institution, on the 2d of December 1816. —

Present
Daniel Moore, Esq.r in the Chair.
Samuel Solly, Esq.r
George Duckett, Esq.r
John Sanders, Esq.r
Thomas Snodgrass, Esq.r
J. P. Auriol, Esq.r V.P.
S. Bernard Morland, Esq.r M.P. Treasurer.
Thomas Harrison, Esq.r Secretary.

1. The minutes of the proceedings of the last meeting were read and approved.

2. Read and approved the minutes of the Committee of Accounts of this day, and signed an Order for an advance to the Clerk of £60.

Letters received.

3. Read a letter from Mr. Stackhouse proposing to deliver the 5 Lectures mentioned in his former letter, for 30 guineas.

Lectures. Resolved.
That the Secretary do request him to give 6 Lectures for 30 guineas, or 5 Lectures for 25 guineas.

Read a letter from W. M. Craig, Esq.r offering, gratuitously, to read a Course of Lectures on Drawing & Painting in the ensuing Season at the Royal Institution.
Resolved, That Mr. Craigs offer be accepted.

Offer of Books as Present by J. Guillemard, Esq.

Read the following letter from John Guillemard, Esq.r to the Sec.y of the Royal Institution dated 30th June 1816. —
"I have a small collection of Books relating to the French and American Revolutions which I considered as a nucleus round which I might form a compleat and full one: The state of my health requires me to seek a warm Climate and in the mean time my Books lie in a Closet,

what can I do better than present them, should they be acceptable to the Royal Institution, with one condition, that in case I wish myself to borrow them out of the House I might be allowed that privilege."

Resolved. That Mr. Guillemard's offer be accepted, and that the thanks of the Board be returned to him for the same.

Presents received.

4. George Henry Nochden, L.L.D. &c. &c. presented,
A Grammar of the German Language.
Samuel Solly, Esqr. presented
Augusteum Dresden's Antike Denkmäler Enthaltend folio
The Geological Society, presented,
Transactions of the Geological Society vol. 3.

Ordered.
That the thanks of the Managers be returned for these presents.

5 Resolved. That the Journals of the Royal Institution be presented to the Geological Society, as they are published.

Journal of R.I. to be presented to Societies presenting.

That where public Bodies are in the habit of presenting their periodical works to the Royal Institution, the Journals of the Institution should be presented to them in return, at the expence of this Institution."

6. The following persons having been duly recommended were admitted as Annual Subscribers to the Royal Institution for the year 1817. —

Annl. Subscrs.

Mr. George Wilkinson Tottenham Green.
Honble Captn. Dawson R.N. 17 Bruton Street.
John Claudius London, Esqr. Hermitage house Bayswater.

7. Adjourned to Monday the 16th Instant.

Humphry Davy

Dec. 16.

At a Meeting of the Managers of the Royal Institution of Great Britain, held at the house of the Institution, on the 16th of December, 1816. —

Present
Sir Humphry Davy, V.P. in the Chair.
Charles Hatchett, Esqr.
Daniel Moore, Esqr.
S. Bernard Morland, Esqr. M.P. Treasurer.
John Frederick Daniell, Esqr.
Thomas Harrison, Esqr. Secretary.

1. The minutes of the proceedings of the last meeting were read and approved.

2. Read and approved the minutes of the Committee of Accounts of this day, and signed an Order for an advance to the Clerk of £60 for the Current expences of the present year, and also a further advance of £50 in Jany. next.

3. Read a letter from Mr. Stackhouse, addressed to the Secretary, in answer to the Resolution of the last Board, stating, that he is willing to give 6 Lectures for 30 guineas, and should the subject lead him to a 7th Lecture, He will consider it as included in the above stipulation of 30 guineas. —
Agreed to.

Lectures Mr. Stackhouse

4. In pursuance of a nomination under the Act, the following person was admitted to the privileges of a Life Subscriber to the Royal Institution.

Life Subscriber

Mr. Robert Taylor of No. 61 Harley Street nominated by his Father Robert Taylor, Esqr.

Presents received.

5. The Corporation of London, presented
An Account of the Entertainment given in the Guildhall to His Royal Highness the Prince Regent, and his Illustrious Guests, their Imperial & Royal Majesties the Emperor of all the Russias & the King of Prussia, & also the Entertainment to His Grace the Duke of Wellington.

The Royal Society, presented
Philosophical Transactions for 1816, 2 parts.
W. A. Cadell, Esqr. F.R.S. Lond: & Edin: presented
"On the lines that divide each semidiurnal Arc into six equal parts, 4to. —
Mr. Ackermann – the Publisher, presented
Repository of Arts for 1816.
Mr. Asperne – the Publisher, presented
European Magazine for 1816.
The Geological Society, presented.
The 2d. volume of their Transactions.

Ordered.
That the thanks of the Managers be returned for these presents.

6. Read the minutes of the last General Monthly Meeting.

7. The following persons having been duly recommended were admitted as Annual Subscribers to the Royal Institution for the year 1817.

Annl. Subscrs.

John Mansfield, Esqr. 61 Wimpole Street
David Majoribanks, Esqr. 4 York St. St. James's Sqe.
Alexander Grant, Esqr. 17 Piccadilly.

Adjourned to Monday the 27th of January, 1817. —

Humphry Davy

1817, Jan. 27.

At a Meeting of the Managers of the Royal Institution of Great Britain, held at the house of the Institution, on the 27th January 1817. —

Present
Sir Humphry Davy V.P. in the Chair.
George Duckett, Esqr.
Samuel Solly, Esqr.
James Peter Auriol Esqr. V.P.
S. Bernard Morland, Esqr. M.P. Treasurer.
Thomas Snodgrass, Esqr.
John Sanders, Esqr.
Daniel Moore, Esqr.
Charles Hatchett, Esqr.
Thomas Harrison, Esqr. Secretary.

1. The minutes of the proceedings of the last meeting were read and approved.

2. Read and approved the minutes of the Committee of Accounts of this day, and signed an Order for an advance to the Clerk of £70 —

3. Read the following letter from Peter Evan Turnbull, Esqr. dated 14th January 1817, addressed to the Secretary.

Letter: P. E. Turnbull's application to be a Life Subscr.

Sir,
I beg to request that you will be pleased to lay before the Committee of Managers, the following circumstances, which I am willing to hope that they will have the goodness to take into their favorable consideration.

"My Father, John Turnbull, Esqr. of Guilford St. one of the early Proprietors of the Royal Institution, expired at Geneva in August last. — He had of late years resided much on the Continent, or at a distance from London, and this and other circumstances had prevented his giving timely attention to the provisions of the Act of Parliament by which the Proprietors were converted into Members with the allowance of a certain pecuniary compensation for the Cession of their hereditary Rights, and the privileges of appointing one of their Relatives

by Blood, to be a Life Subscriber – By this omission he was deprived of the opportunity of applying for, or receiving the pecuniary compensation so allowed – and as the time limited by the Act is now so long gone by, I presume not to solicit any indulgence in this respect. With regard to the Life Subscribership however, I beg leave to submit to the Managers that I have a letter written by my Father a very few days before his decease, wherein he expressly says that he will give to me, (his eldest Son) the nomination in question, whensoever I please – and had he not been called from this state of existence in a very sudden and unexpected manner, this, his intention, would doubtless have been formally carried into effect.

I venture to flatter myself, Sir, that under these peculiar circumstances, the Managers will consider this intimation of my Father's intention to be a sufficiently valid nomination – and that they will have the goodness to allow my name to be admitted accordingly into the List of Life Subscribers.'

Resolved. That the Secretary be directed to write to Mr. Turnbull requesting his attendance here on Monday next at One o'Clock, with his Father's letter expressing his intention of nominating him as a Life Subscriber, and with some person able to identify his Fathers' hand writing.

Model of Orrery.

4. Read a letter from Mr. Thomas Yeates, dated 23d. Jany 1817. addressed to the Managers, stating that he had prepared a set of Drawings for an Orrery on improved principles and if a Model were esteemed worthy of a place among the Philosophical Apparatus of the Royal Institution, he should hope to see it realized for the benefit and advancement of the Science.

Resolved.

That it be referred to Daniel Moore Esqr. and that he be requested to call in such assistance as he may think proper.

Jan. 27.

Presents received.

5. Anthony Bertolacci, Esqr. presented
A view of the Agricultural, Commercial, and Financial interests of Ceylon 8vo.
Dr. Elliotson, presented
Institutions of Physiology by J. F. Blumenbach, 8vo.
Edward Brooke, Esqr. presented
Le Vray et Methodique Cours de la Physique resolutive, vulgairement dite Chymie, 4to. 1657.
The Society for bettering the condition of the Poor, presented
The Society's 39th. Report – 2 Copies.
The Author of Chronique de Paris, presented
Chronique de Paris, Nos. 1 & 2.
The Geological Society, presented
Vol 4 part 1st of their Transactions.
Charles Hatchett Esqr. F.R.S. M.R.I. presented
A Letter to Sir Joseph Banks Bart. on purifying Corn tainted with Must.
George Allan, Esqr. M.P. presented
Collectanea ad Statum Civilem et Ecclesiasticum Comitatus Dunelmensis, 4to.

Ordered.

That the thanks of the Managers be returned for these presents.

6. Read a recommendation from the Visitors entered in their proceedings of the 13th. Instant.

Resolved.

That the consideration thereof be referred to Mr. Solly, and that he be requested to report thereon to the Managers as soon as convenient.

Hon. Admission.

7. General de Neuffer Envoy Extraordinary and Minister Plenipotentiary of His Majesty the King of Wurttemberg having applied to be admitted to the Royal Institution under Art 16. Chap: VI of the Bye laws.

Ordered.

That the Secretary do transmit him the usual Letter accordingly.

8. It having been communicated from Sir Wm. Congreve,

Request of ~~to~~ Grand-duke Nicholas to visit R. I.

through Sir Humphry Davy to this Board, that the Grand Duke Nicholas of Russia is desirous of seeing this Institution.

Resolved. That the Secretary be directed to inform him that the Managers will feel much gratified by having the honor of receiving him and that they should esteem it a further honour to propose His Imperial Highness as an honorary Member at the General Monthly which will be held on Monday next the 3d. of February.

9. The following persons having been duly recommended were admitted as Annual Subscribers to the Royal Institution for the year 1817.

Annual Subscribers.

Revd. Gilbert Parke 9 Alfred place.
John Lane, Esqr. 18 Lower Thornhaugh Street
Giles Prodgers 11 Everett Street
Dr. Thomas Donahoo 337 Strand.
Thomas Potter Macqueen Esqr. M.P. 28 Spring gardens
Major Dance 29 Upper Gower Street
Lt. General Mitchell
George Lawrence, Esqr. 8 Montagu Place.

To the Lectures.

Miss Children & Miss Bullen 28. Mercer St.
Mrs. Gregson 2 Charles St. Grosvenor Square
Mrs. Reed & Miss Hendrie 4 Berkeley Street
Mrs. Dance 29 Upper Gower St.
Mrs. Jenyns & two Misses Jenyns 6 Connaught place
Miss Heberden 2 Upper Grosvenor Street.
Miss Burroughs, 42 Berkeley Square
Mrs. Poynder 2 Montagu Place
Countess Powis & Lady Charlotte Clive Berkeley Square
Mrs. Alcock, Miss Alcock } 27 Queen Ann Street
Miss Maria Alcock & Miss Jane Alcock }

J. Nicholl

Feb. 3.

At a Meeting of the Managers of the Royal Institution of Great Britain, held at the house of the Institution, on the 3d. of February 1817,

Present

Rt. Honble Sir John Nicholl, M.P. V.P. in the Chair
Earl Spencer, K.G. President.
J. P. Auriol, Esqr. V.P.
Sir H. Davy, V.P.
William Watson, Esqr.
Samuel Solly, Esqr.
J. F. Daniell, Esqr.
John Sanders, Esqr.
Daniel Moore, Esqr.
George Duckett, Esqr.
Thomas Harrison, Esqr. Secretary.
S. Bernard Morland, Esqr. M.P. Treasurer.

1. The minutes of the proceedings of the last meeting were read and approved.

2. Read and approved the minutes of the Committee of Accounts of this day, and signed an Order for payment of the following sums, viz:

Payments.

3/4 yrs Assessed taxes to Xmas last	£118 . 13 . 10
3/4 yrs Poor rates to do.	40 . —
Mr. Harris 3/4 yrs Salary to 31st. Dec last	135 . —
Mr. Brooks 1 year to do.	31 . 10 . —
Thomas Slatter Lecture room Door keeper do.	10 . 10 . —
	335 . 13 . 10

Present received.

3. Messrs. Longman & Co. the Publishers presented "The Correspondent No. 1."

Ordered. That the thanks of the Managers be returned for this present.

4. In pursuance of a nomination under the Act, the following person was admitted to the privileges of a Life Subscriber to the Royal Institution – Thomas Pitt Esqr. of Wimpole Street nominated by his Father Thomas Pitt Esqr. –

Life Subscr.

1817.

Case of Mr. P. E. Turnbull.

5\. Agreeably to the Resolution of the last Board, Mr. Peter Evan Turnbull attended with his Brother Mr. Thomas Turnbull who proved the hand writing of his Father to the letter of the 4th of August last, which he himself directed, and he stated that his Father died on the 8th having in the interim declared to him that he meant to nominate his Son, Mr. P. E. Turnbull as a Life Subscriber to the Royal Institution. The letter contained the following passage.

"I will with pleasure appoint you to the Royal Institution, but I think I should not lose the £45."

and Mr. P. E. Turnbull stated that he had proved his Fathers Will and that it did not contain a nomination to any Relative by Blood of the privileges of a Life Subscriber to this Institution, and that there was no Codicil thereto.

Resolved. That the Case of Mr. Peter Evan Turnbull be specially reported to the General Monthly Meeting of this day.

6\. The following persons having been duly recommended were admitted as Subscribers to the Royal Institution for the year 1817.

Annual Subscribers.

Mr. Murray 19 Upper Harley Street
Mr. Stirling. 2 Clarges Street.
Capt. Milligan, Lower Grosvenor Street
Henry James Barchard, Esqr. of Wandsworth
Dr. Bollman, 16 Buckingham St. Strand.
Mr. Walker 29 East St. Lambs Conduit St.
Mr. John Bramah - Pimlico.
Mr. Robert Richmond, 15 John St. Bedford row

To the Lectures.

Miss Blake - Portland Place
Mrs. Twiss 18 Cadogan Place
Mrs. Lawes } 27 Montague Street
Miss Burrell }
Miss Lucy Nicholls - Kensington.
Mrs. Robert Milligan. Lower Grosvenor Street

Feb 3.

Miss A. I. Jordan 21 Abingdon Street
Miss Caroline Bollman } 16 Buckingham St. Strand.
Miss Elizth. Bollman }
Mrs. John Reid } 24 Upper Harley Street
Miss Reid }
Countess Munster 21 Piccadilly.
Countess of Glengall Lower Grosvenor Street.
Miss Dovall
Mrs. Burrowes } of Hill Street
Miss Burrowes }
Miss Holford, York Place Portman Square.

Spencer.

1817.

Feb 10.

At a Meeting of the Managers of the Royal Institution of Great Britain, held at the house of the Institution, on the 10th of February, 1817. —

Present

Earl Spencer, K. G. President, in the Chair
Sir Humphry Davy V. P.
William Watson, Esqr.
George Duckett, Esqr.
Charles Hatchett, Esqr.
John Sanders, Esqr.
John Frederick Daniell, Esqr.
James Peter Auriol, Esqr. V. P.
Thomas Harrison, Esqr. Secretary

1\. The minutes of the proceedings of the last meeting were read and approved.

2\. Read and approved the minutes of the Committee of Accounts of this day, and signed an Order for payment of the following sums, viz.

Payments.

Bulmer for Printing in 1815.	£126 . 11 . 6
Feetham Ironmongery in do.	124 . 14 . 11
Joseph Fincher on account of the Royal Institution	50 . - . -
	301 . 6 . 5

on Present of Books. by Mr. Guillemard.

3\. Read a letter from John Guillemard Esqr. addressed to the Secretary, dated Paris 3rd Feby. 1817, stating that he had received with great satisfaction, the acquiescence of the Board of Managers of the Royal Institution, in his proposal of offering to its Library his Collection of Books on the French and American Revolutions, but that on leaving London in August last as the Board was not sitting and its answer uncertain, he had packed up in Chests a considerable part of those Books along with others of a different nature and it will not be possible

Feb 10.

to separate them without great inconvenience till his return to England, He should give Orders however for all those which are disposable to be sent to the Royal Instn. as soon as possible.

4\. Present received.

William Luxmoore Esqr. M. R. I. presented,
Essays on Hypochondriacal and other Nervous affections by John Reid, M. D. 8vo.

Ordered. That the thanks of the Managers be returned for this present.

5\. Read the minutes of the last General Monthly Meeting.

6\. The following persons having been duly recommended were admitted as Subscribers to the Royal Institution for the year 1817. —

Annual Subscribers.

Mr. Aylmer 53 Harley Street
Mr. Robert Ware of Clapham.
Henry J. Williams, Esqr. Lincolns Inn fields.
Mr. Thos. Josling, 53 Bread St. Buildings.
Colonel St. George 6 Upper Gloucester Street.
Rt. Honble. Wm. Sturges Bourne, M. P. Spring Gardens
Colonel Turner. 15 Edward St. Portman Sq.

Subscribers to the Lectures.

Miss Wollaston & Miss K. Wollaston 26 Sackville St.
Miss Frances Griffin & Miss Jane Griffin - 21 Bedford place
Mrs. Ware, Bridge Street.
Miss Mary Milligan, Wimpole Street.
Miss Alexander, Lincolns Inn Fields.
Miss Cotton.
Miss Frances Lemon - Whitehall.
Honble. Mrs. Vernon & Miss Vernon Grosvenor Square
Miss Anna Wilbraham 57 Upper Seymour St.
Mrs. Marx, Bedford Place. —

Spencer.

144

Feb. 17.

1817.

At a Meeting of the Managers of the Royal Institution of Great Britain held at the house of the Institution, on the 17th of February, 1817.

Present

Earl Spencer, K.G. President, in the Chair
Sir Humphry Davy V.P.
J. P. Auriol, Esqr. V.P.
George Duckett, Esqr.
Daniel Moore, Esqr.
John Sanders, Esqr.
J. F. Daniell, Esqr.
Thomas Harrison, Esqr. Secretary.

1. The minutes of the proceedings of the last meeting were read and approved.

2. Read and approved the minutes of the Committee of Accounts of this day, and signed an Order for payment of the following sums, vizt.

Payments,

Jones & Yarrell for Newspapers in 1815	£109 . 8 . 6
Mr. Royce for Repairs in do.	77 . 8 . –
	186 . 16 . 6

3. Audited the Treasurer's Account for the year 1816. –

4. Mr. Moore reported that he and Mr. Millington had examined the Orrery made by Mr. Yeates, and were of opinion that it would not be expedient for the Institution to purchase a Model of it

Model of Orrery declined

Resolved.

That the Secretary do inform Mr. Yeates that the Board are under the necessity of declining his offer.

Ordered. That after the 25th of March next, Mr. Hole's services in winding up the Clocks be dispensed with.

145

Feb 17.

The following persons having been duly recommended were admitted as Subscribers to the Royal Institution for the Year 1817. –

Annual Subscribers.

Rear Admiral Sir Josias Rowley – Albany
Alexander Spearman, Esqr. 38 Davies St. Berkeley Sq.
Francis S. Bayley, Esqr. 3 New Square Lincolns Inn

To the Lectures.

Mrs. Blackburn 44 Charles St. Berkeley Square
Miss Protheroe – Harley Street.

Spencer

146

Feb. 24.

1817.

At a Meeting of the Managers of the Royal Institution of Great Britain, held at the house of the Institution, on the 24th of February, 1817.

Present

Earl Spencer, K.G. President in the Chair
Sir Humphry Davy, V.P.
J. P. Auriol, Esqr. V.P.
Charles Hatchett, Esqr.
Samuel Solly, Esqr.
John Sanders Esqr.
George Duckett, Esqr.
Thomas Harrison, Esqr. Secretary.

1. The minutes of the proceedings of the last meeting were read and approved.

2. Read and approved the minutes of the Committee of Accounts of this day, and signed an Order for an Advance to the Clerk of £50.

3. Mr. I. Sturt, the Assistant Librarian, stated to the Board that about 100 Volumes of boarded Books in the Reading-Room are in a very bad condition and require to be re-boarded or fresh covered with coloured Paper which he will undertake to get done for one guinea.

Repairing Books.

Resolved.

That the above proposal be accepted.

4. The following persons having been duly recommended were admitted as Subscribers to the Royal Institution for the year 1817.

Annual Subscribers.

Chas. Greenwood, Esqr. 14 Great Queen St. Westmr.
Octavius Vernon Esqr. Grosvenor Square.
Henry Gally Knight, Esqr. 16 Henrietta Street

To the Lectures.

Mrs. Weekes – 7 Bryanston Square
Lady Francis St. James's Square.

Spencer.

147

Mar. 3.

At a Meeting of the Managers of the Royal Institution of Great Britain, held at the house of the Institution, on the 3d. of March, 1817. –

Present

Earl Spencer, K.G. President, in the Chair.
Sir Humphry Davy V.P.
J. P. Auriol Esqr. V.P.
Charles Hatchett, Esqr.
Samuel Solly, Esqr.
J. F. Daniell, Esqr.
Rt. Honble Sir John Nicholl, M.P. V.P.
Daniel Moore, Esqr.
John Sanders, Esqr.
Thomas Snodgrass, Esqr.

1. The minutes of the proceedings of the last meeting were read and approved.

2. Read and approved the minutes of the Committee of Accounts of this day, and signed an Order for payment of the following sum, vizt.

Messrs. Keys for Cards in 1815 – £83 . 17 . 9. –

Presents received.

3. I. D'Israeli, Esqr. presented,
Curiosities of Literature, vol. 3d.
Lt. Genl. Thornton, M.P. – presented
Essays, Religious & Moral, by Isaac Hawkins Browne Esqr.

Ordered.

That the thanks of the Managers be returned for these presents.

4. Ordered. That 12 black Chairs be provided for the Newspaper Room.

Chairs.

5. The following persons having been duly recommended were admitted as Subscribers to the Royal Institution.

Life Subscriber.

Charles Robinson, Esqr. 36 Highbury Place Islington.

Annual Subscribers. 1817.

Honble P. B. Blaquiere, 46 Bryanston Square

Mr Wigsel – 7 Conduit Street

To the Lectures.

Lady Strange

Mrs Wigsel } 7 Conduit Street

Mrs Evans }

Spencer.

Mar. 10.

At a Meeting of the Managers of the Royal Institution of Great Britain held at the house of the Institution, on the 10th of March, 1817. –

Present

Earl Spencer, K. G. President, in the Chair

J. P. Auriol, Esqr. V. P.

Sir Humphry Davy, V. P.

Samuel Solly, Esqr.

William Watson, Esqr.

George Duckett, Esqr.

Thomas Snodgrass, Esqr.

Charles Hatchett, Esqr.

J. F. Daniell, Esqr.

Daniel Moore, Esqr.

Sir J. C. Hippisley, Bart. M.P.

Thomas Harrison, Esqr. Secretary.

1. The minutes of the proceedings of the last meeting were read and approved.

2. Read and approved the minutes of the Committee of Accounts of this day, and signed an Order for payment of the following sums viz.

Payments.

Wrangham for Stationary	in 1815.	£37 . 4 . 2
Ditto – Newspapers	in do.	30 . 9 . 4
Allen – Chemicals	in do.	9 . 2 . 7
Newman Chemicals	in 1816	168 . 5 . 10
Searle for Bookbinding	in 1816	12 . 6 . 6

Presents received.

3. Presented by the Author,

Two Sketches of France, Belgium, & Spa, in two Tours during 1771 & 1816. 8vo. –

Mr Mitchell presented

"Genealogy of the Kings of England."

Ordered. That the thanks of the Managers be returned for this present.

1817.

4. Read the minutes of the last General Monthly Meeting.

Illness of Mr Fincher.

5. Resolved. That during the continuance of Mr Fincher's present Indisposition, the Institution will defray the expences of his Coachhire on their business! –

6. The following persons having been duly recommended were admitted as Annual Subscribers to the Royal Institution for the year 1817. –

Captn. Henry Walpole 36 Half Moon Street

Edward Augustus Kendall, Charlotte St. Fitzroy Sqe.

Mr Jonte –

Spencer.

Mar. 17.

At a Meeting of the Managers of the Royal Institution of Great Britain, held at the house of the Institution, on the 17th of March, 1817. –

Present

Earl Spencer, K. G. President, in the Chair.

Charles Hatchett, Esqr.

George Duckett, Esqr.

J. P. Auriol, Esqr. V. P.

Daniel Moore, Esqr.

Thos. Snodgrass, Esqr.

John Sanders, Esqr.

William Watson, Esqr.

Earl of Morton, V. P.

J. F. Daniell, Esqr.

Thomas Harrison, Esqr. Secretary.

1. The minutes of the proceedings of the last meeting were read and approved.

2. Read and approved the minutes of the Committee of Accounts of this day, and signed an Order for payment of the following sums, viz.

Josh. Fincher on account of the Royal Institution		£50 . – . –
Thompson, Soap & Candles	in 1816	27 . 12 . 11
Wyatt, Sundries for the house	do.	7 . 19 . 10
Odell, Plumbing & Glazing	do.	9 . 8 . 4
Howis Sundries for the house	do.	26 . 8 . 10
Pringle Upholstery	do.	19 . 14 . 2
Fisher, Chemicals	do.	34 . 13 . 3
		£175 . 17 . 4

Election of Grand-duke Nicholas of Russia as Hon. Mem.

3. Earl Spencer, President, reported that he had written a letter to the Grand Duke Nicholas of Russia, announcing to His Imperial Highness, his election as an honorary Member of this Institution, and that he had received an answer from Baron Nicolay of which the following is a Copy.

152

St. Albans House 1817.
March 12th 1817. –

Letter from Baron E Nicolay.

My Lord,

By order of His Imperial Highness the Grand Duke of Russia, I have to request your Lordship to express to the Royal Institution, the satisfaction which His Imperial Highness has felt from the honour of becoming Honorary Member of that Society. The general utility of their pursuits has inspired the Grand Duke with a sincere esteem for the supporters of it. — I am further commanded to convey to your Lordship the expression of the particular pleasure the Grand Duke has felt in receiving this mark of attention from the Royal Institution through your Lordship, as it affords His Imperial Highness an opportunity of testifying to you his sentiments of esteem.

I have the honor to be,
with high consideration,
My Lord,
Your obedient humble Servant
(signed) Baron E Nicolay

The Earl of Spencer
&c. &c. &c.

4 **Presents received.**

The Author, presented,
Sketch of a plan for a Reformation in the system of Provincial Banking.
Wm Thos Brande, Esqr Professor of Chemistry R.I.
Outlines of Geology, 8vo
Ordered. That the thanks of the Managers be returned for this present.

5. The following persons having been duly recommended were admitted as Subscribers to the Royal Institution.
Annual Subscribers.
W. Amsinck, Esqr 4 Vigo Lane Sackville Street
John Lewis, Esqr – 11 Powis Place
To the Lectures
Miss Matilda Green, 29 Upper Gower Street.

Spencer

153

Mar. 24.

At a Meeting of the Managers of the Royal Institution of Great Britain, held at the house of the Institution, on the 24th of March, 1817. –

Present
Earl Spencer, K.G. President, in the Chair.
Sir Humphry Davy, V.P.
J. P. Auriol, Esqr V.P.
Samuel Solly, Esqr
Thomas Snodgrass, Esqr
John Sanders, Esqr
J. F. Daniell, Esqr
Sir John Cox Hippisley Bart. M.P.
Thomas Harrison, Esqr Secretary.

1. The minutes of the proceedings of the last meeting were read and approved.

2. Read and approved the minutes of the Committee of Accounts of this day, and signed an Order for paymt of the following sums, viz! –

Payments

1 years Insurance in the Sun Fire Office to 25th March 1818 ——	63 . 16 . 6
Mr Stackhouse for a Course of Lectures delivered in 1817 ——	31 . 10 .
	95 . 6 . 6

Present received.

3. Samuel Solly, Esqr presented,
2 numbers of the German Encyclopædia.
Ordered. That the thanks of the Managers be returned for this present. –

Liby Hours

4. Ordered. That the following notice be painted on a Board and placed at the Door of the Library of Reference.
"Open from the 1st of March to the 1st of September from 10 till 6, and during the rest of the year till 4 in the Afternoon."

5. The following persons having been duly recommended

154

1817.

were admitted as Annual Subscribers to the Royal Institution for the year 1817.
Henry Cowper, Esqr Old Palace yard.
Lieutenant Thomas Stopford, 63 Welbeck St.

To the Lectures.
Lady Sitwell 66 Upper Mount Street.

Adjourned to Monday the 7th of April next. –

Spencer

155

April 7.

At a Meeting of the Managers of the Royal Institution of Great Britain held at the house of the Institution, on the 7th of April, 1817. —

Present
Earl Spencer, K.G. President in the Chair.
James Peter Auriol, Esqr V.P.
Thomas Snodgrass, Esqr
Samuel Solly, Esqr
J. F. Daniell, Esqr
John Sanders, Esqr
S. Bernard Morland, Esqr M.P. Treasurer.
Daniel Moore, Esqr
Sir John Cox Hippisley, Bart. M.P.
Thomas Harrison, Esqr Secretary.

1. The minutes of the proceedings of the last meeting were read and approved.

2. Read and approved the minutes of the Committee of Accounts of this day. –

3. The following persons having been duly recommended were admitted as Subscribers to the Royal Institution for the year 1817. –
Annual Subscribers.
Lieut. Col. Ansley, of Otto House North End.
George Dickson, Esqr 15 Chester Place Kennington
George S. Ford, Esqr No 1 Mecklenburgh Square
Captn J. G. Peters, No 18 Charles St. Berkeley Sq.

To the Lectures.
Mrs Boswell, 13 Grafton Street
Miss Robarts, Grosvenor Street.

Spencer

156

April 14.

1817.

At a Meeting of the Managers of the Royal Institution of Great Britain, held at the house of the Institution, on the 14th of April, 1817. –

Present
Earl Spencer K.G. President in the Chair.
Charles Hatchett, Esq.r
Samuel Solly, Esq.r
J. P. Auriol, Esq.r V.P.
Thomas Snodgrass, Esq.r
Daniel Moore, Esq.r
John Sanders, Esq.r
Thomas Harrison, Esq.r Secretary. –

1. The minutes of the proceedings of the last meeting were read and approved.

2. Read and approved the minutes of the Committee of Accounts of this day, and signed an Order for an Advance to the Clerk of £50. –

Presents received.

3. Narrative of occurrences in the Indian Countries of North America, since the connexion of the Earl of Selkirk with the Hudson's Bay Company. – 2 Copies. –
G. M. Burrows, M.D. &c. &c. presented,
Cursory Remarks on a Bill now in the House of Peers for regulating of Mad-houses – 2 Copies.
John Mawe, Esq.r presented the following Specimens,
Octoedral fluor —— Devonshire.
Derby Marble —— Derbyshire.
Tourmaline with Phosphate of Lime – Devonshire.
Ordered.
That the thanks of the Managers be returned for these presents. –

4. Read the minutes of the last General Meeting.

157

April 14.

Lectures, Mr Soane

Mr. Sanders having communicated to the Board that it was probable Mr. Soane would be willing to deliver a few lectures on Architecture gratuitously, if the Managers should think proper to make any application to him upon the subject.
Resolved.
That Mr. Sanders, Mr. Brande and Mr. Moore be requested to wait on Mr. Soane to express the great satisfaction with which a proposal of this nature would be received by the Managers.

The following persons having been duly recommended were admitted as Subscribers to the Royal Institution for the year 1817. –
Annual Subscriber.
Capt.n John Bowen R.N. No. 1. Dover Street
To the Lectures.
Miss Emma Milford }
Miss Caroline Milford } Clifford Street

Spencer.

158

April 21.

1817.

At a meeting of the Managers of the Royal Institution of Great Britain, held at the house of the Institution, on the 21st of April, 1817. –

Present
Earl Spencer, K.G. President, in the Chair.
Rt. Hon: Sir John Nicholl, M.P. V.P. –
Thomas Snodgrass, Esq.r
George Duckett, Esq.r
Samuel Solly, Esq.r
John Sanders, Esq.r
J. F. Daniell, Esq.r
Daniel Moore, Esq.r

1. The minutes of the proceedings of the last meeting were read and approved.

2. Read and approved the minutes of the Committee of Accounts of this day.

Present received.

3. Ant.y Bertolacci, Esq.r M.R.I. presented,
"Inquiry into several questions of Political Economy."
Ordered.
That the thanks of the Managers be returned for this present.

Coals. 4. Ordered. That 40 Chaldrons of the best Walls-End Coals be laid in at the Royal Institution, at 54s/6d. ℔ Chaldron, Metage, Shooting, and all expences included.

5. The following persons having been duly recommended were admitted as Subscribers to the Royal Institution for the year 1817.
Annual Subscribers
Martin Tupper, Esq.r 13 New Burlington Street
George C. Harvey, Esq.r

To the Lectures
Miss H. Sophia Isabella Dillon. 15 Quebec Street

159

April 21.

Ordered. That the House Lists be filled up as follows.

President. Earl Spencer, K.G. F.R.S. and F.A.S.

Treasurer. Scrope Bernard Morland, Esq.r M.P. L.L.D.

Secretary. Thomas Harrison Esq.r F.R.S.

Managers.
John Frederick Daniell, Esq.r F.R.S.
Sir Humphry Davy L.L.D. F.R.S. &c. &c.
George Duckett, Esq.r F.R.S. & F.A.S.
Lord Dundas F.R.S. & F.A.S.
Earl of Hardwicke, K.G. L.L.D. F.R.S. &c.
Charles Hatchett, Esq.r F.R.S.
Sir Benjamin Hobhouse Bart. F.R.S. and F.A.S.
Charles Holford, Esq.r
Daniel Moore, Esq.r F.R.S. F.A.S. and F.L.S.
Louis Hayes Petit, Esq.r A.M. F.R.S. &c.
Claude Scott, Esq.r
Thomas Snodgrass, Esq.r
Samuel Solly, Esq.r
James Stodart, Esq.r
Giles Templeman, Esq.r

Visitors.
Sir Edmund Antrobus, Bart.
George Aust, Esq.r F.R.S. and F.A.S.
Timothy Brent, Esq.r
Sir William Burroughs, Bart.
John Chamier, Esq.r
Earl of Chichester, F.R.S.
Henry Tho.s Colebrooke, Esq.r
Lord Erskine F.R.S.
Gibbes Walker Jordan, Esq.r A.M. F.R.S.
William Morgan, Esq.r F.R.S.
Wm. Hasledine Pepys, Esq.r F.R.S.
Rich.d Horsman Solly, Esq.r F.R.S. and F.A.S.
Duke of Somerset, F.R.S.
Lord Teignmouth.
Lt. Gen.l Wm. Thornton, M.P.

J P Auriol

160

April 28. 1817.

At a Meeting of the Managers of the Royal Institution of Great Britain, held at the house of the Institution, on the 28th of April, 1817. —

Present

James Peter Auriol Esqr. V.P. in the Chair.
Charles Hatchett, Esqr.
J. F. Daniell, Esqr.
Samuel Solly Esqr.
S. Bernard Morland, Esqr. M.P. Treasurer.
Daniel Moore, Esqr.
Thos. Harrison, Esqr. Secretary.

1. The minutes of the proceedings of the last meeting were read and approved.

2. Read and approved the minutes of the Committee of Accounts of this day, and signed an Order for payment of the following sum, vizt.

Payment — 1 years Ground to the City of London from Lady day 1816 to Lady day 1817 — £150 . —

3. Read the following letter addressed by John Soane Esqr. to Mr. Brande, Mr. Moore and Mr. Fanders in consequence of the Resolution of the Managers of the 14th Instant.

Letter from Mr. Soane.

Gentlemen — Lincolns Inn Fields 28th April 1817.

I am much obliged by the flattering manner in which you were pleased to communicate to me the wishes of the Committee of Managers of the Royal Institution respecting my Lectures on Architecture; — I will thank you to assure the Committee that on all occasions I shall have great pleasure in exerting my best endeavours for the advancement of Architectural knowledge, and altho' I am fearful that any thing I can say on the subject will fall very short of the expectations you have so kindly expressed I am ready to make the attempt in one or two Lectures

I am, &c &c. (signed) John Soane.

Resolved. That the Secretary be directed to return the thanks of the Board to Mr. Soane for the communication & to request that he will arrange the days for his Lectures with Mr. Brande.

161

April 28.

Presents received.

4. M. M. A. Jullien of Paris, — by Hugh Hammersley Esqr. M.P. presented,

Essai Général d'Education Physique, Morale et Intellectuelle par M. A. Julien 4to. —

Discours servant d'Introduction à la seconde édition de l'Essai sur l'emploi du Tems, 8vo.

Esquisse d'un Ouvrage sur l'Education comparée, et Series de Questions sur l'Education, 8vo.

Sir Egerton Brydges, Bart. M.P. presented,

Arguments in favour of the practibility of relieving the able bodied Poor, by finding employment for them.

Isaac Hawkins Browne Esqr. presented

A Treatise on the Records of the Creation, and on the moral attributes of the Creator, 2 vols. 8vo.

The Horticultural Society, presented,

A set of the Transactions of their Society.

Ordered.

That the thanks of the Managers be returned for these presents.

5 In pursuance of a nomination under the Act, the following person was admitted to the privileges of a Life Subscriber to the Royal Institution

Life Subscr. — John Hopkins Bradney, Esqr. of Mount Street Grosvenor Sq. nominated by his Father Joseph Bradney, Esqr.

6. The following persons having been duly recommended were admitted as Subscribers to the Royal Institution for the year 1817.

Annual Subscriber.

Lord Prudhoe — Northumberland house

To the Lectures

Miss Sebright 31 Argyll Street

Spencer.

162

May 5. 1817.

At a Meeting of the Managers of the Royal Institution of Great Britain held at the house of the Institution on the 5th of May, 1817. —

Present

Earl Spencer, K.G. President, in the Chair.
Lord Dundas.
J. F. Daniell, Esqr.
Samuel Solly, Esqr.
Thomas Snodgrass, Esqr.
Claude Scott, Esqr.
Sir Humphry Davy.
Daniel Moore, Esqr.
James Stodart, Esqr.
Sir Benjamin Hobhouse, Bart. M.P.
George Duckett, Esqr.
S. Bernard Morland, Esqr. M.P. Treasurer.
Thomas Harrison, Esqr. Secretary.

1. The minutes of the proceedings of the last meeting were read and approved.

2. Read and approved the minutes of the Committee of Accounts of this day, and signed an Order for an advance to the Clerk of £50. —

3. Read a letter from Viscount Clive, addressed to the Secy., requesting that permission be granted to Monsieur de Jouffroy Chargé d'Affaires of His Majesty the King of Prussia to avail himself of the Library of the Royal Institution in the manner usually granted to Foreigners of distinction while resident in this Country.

Hon. Admission

Ordered. That the Secretary do transmit to M. de Jouffroy the usual letter accordingly.

4 Read a letter from R. Fidler addressed to the Managers stating his distress and requesting the Board to furnish him with £15 out of the Subscription which he understood was going on in his behalf. —

Application from ~~Subscription~~ R. Fidler.

163

May 5.

Resolved.

That under all the circumstances of his case the Board cannot entertain his application. —

5. In pursuance of a nomination under the Act, the following person was admitted to the privileges of a Life Subscriber to the Royal Institution.

Life Subscr. — Sir Thomas Strange of Great George St. Westmr. nominated by his Brother James Strange Esqr. —

Present received.

6. By the Publishers,

The Colonial Journal, No. 3.

Ordered.

That the thanks of the Managers be returned for this present. —

7. Read the minutes of the General Annual Meeting of the 1st Instant.

8. The President nominated as Vice Presidents for the ensuing year.

Vice Presidents

Sir Humphry Davy.
Lord Dundas.
Lord Erskine K.T.
Earl of Hardwicke K.G.
Charles Hatchett, Esqr.
Sir Benjn. Hobhouse Bart. M.P.

9. Ordered. That Mr. Royce do put up in the Reading room a few Shelves under the direction of Mr. Harris.

Dundas

May 12.

1817.

At a Meeting of the Managers of the Royal Institution of Great Britain held at the house of the Institution, on the 12th of May, 1817. —

Present
Lord Dundas, V.P. in the Chair.
Charles Hatchett, Esqr. V.P.
Sir Benjamin Hobhouse, Bart. M.P. V.P.
George Duckett, Esqr.
Samuel Solly, Esqr.
J. F. Daniell, Esqr.
Thomas Snodgrass, Esqr.
Daniel Moore, Esqr.
Sir Humphry Davy, V.P.
Earl Spencer, K.G.
Lord Erskine, K.T.
Thomas Harrison, Esqr. Secretary.

1. The minutes of the proceedings of the last meeting were read and approved.

2. Read and approved the minutes of the Committee of Accounts of this day. —

Presents received.

3. His Majesty's Commissioners of the Public Records, presented,
Statutes of the Realm, vol. 2 & 3 folio.
Rymeri Foedera, vol. 1 - 2 parts — new edit. folio.
Domesday Book, vol. 4 - Supplement folio.
Domesday Book, General Introduction to the Indexes.
Inquisitionum ad Capellam Domini Regis Retornatarum, vol. 3 and Supplement.
Henry Thomas Colebrooke, Esqr. M.R.I. presented,
Algebra, with Arithmetic & Mensuration, from the Sanscrit of Brahmegupta & Bhascara, 4to.
Mr. Thomas Preston,
A new system of Commercial Arithmetic; for Monies, Weights and Measures, 12mo.

May 12.

Ordered. That the thanks of the Managers be returned for these presents.

4. Read the minutes of the last General Monthly Meeting.

Committee on Form.

5. Resolved. That the following Select Committee be appointed to consider of a proper form of a Letter to be sent to Foreign Members announcing their Election viz.
Sir H. Davy.
Charles Hatchett, Esqr.
Thomas Harrison, Esqr.
And that Mr. Brande be requested to attend the said Committee. —

Spencer.

May 19.

1817.

At a Meeting of the Managers of the Royal Institution of Great Britain held at the house of the Institution on the 19th of May, 1817. —

Present
Earl Spencer, K.G. President in the Chair.
Sir Humphry Davy V.P.
J. F. Daniell, Esqr.
Samuel Solly, Esqr.
Lord Dundas V.P.
Thomas Snodgrass Esqr.
James Stodart, Esqr.
George Duckett, Esqr.
Sir Benjamin Hobhouse, Bart. M.P. V.P.
Giles Templeman, Esqr.
Daniel Moore, Esqr.
Thomas Harrison, Esqr. Secretary.

1. The minutes of the proceedings of the last meeting were read and approved.

2. Read and approved the minutes of the Committee of Accounts of this day. —

Spencer.

May 26.

At a Meeting of the Managers of the Royal Institution of Great Britain, held at the house of the Institution, on the 26th of May, 1817. —

Present
Earl Spencer, K.G. President in the Chair.
Sir Benjamin Hobhouse, Bart. M.P. V.P.
Samuel Solly, Esqr.
Charles Hatchett, Esqr. V.P.
Giles Templeman, Esqr.
James Stodart, Esqr.
J. F. Daniell, Esqr.
Thomas Snodgrass, Esqr.
Daniel Moore, Esqr.
Sir Humphry Davy V.P.
Thomas Harrison, Secretary.

1. The minutes of the proceedings of the last meeting were read and approved.

2. Read and approved the minutes of the Committee of Accounts of this day, and signed an Order for payment of the following sums, viz.

Payment.

Wm. Thos. Brande Esqr. 1 years Salary as Prof. of Chemistry to May 1817	£200 —
Jas. Fincher on acct. of the Royal Instn.	50 —
	£250 —

3. Presents received.

— Telford, Esqr. presented,
A Design for a Bridge over the River Mersey.
Ordered.
That the thanks of the Managers be returned for this present.

Diploma for Hon. Members.

4. The Select Committee appointed to consider of the proper form of a letter to be transmitted to such distinguished Foreigners

as may be elected Honorary Members of the Institution by virtue of Chap 3 Art: 1 of the Bye Laws, recommended to the Board the adoption of the following form.

"Præses, Directores, et Socii Regalis Institutionis Magnæ Britanniæ omnibus et singulis ad quos præsentes pervenerint, Salutem.

Quoniam

singularem suum in dictæ Institutionis conatus & studia affectum uberrime fuerit testatus, suisque meritis egregiis Rem literariam, et solidam imprimis Philosophiam augere and ornare pro virili satagat, dicta societas inclytum

die instantis anni solenni Consessu in societatem suam cooptavit, inque hujus rei testimonium sigillum suum præsentibus affigi curavit.

Datum Londini Anno Christi MDCCC et

Regni Georgii tertii Britanniarum Regis dictæ Institutionis Patroni illustrissimi.

Resolved.

That the Secretary do transmit a Copy hereof to Dr Burney with a request that he will have the goodness to peruse and correct the same.

Spencer.

June 2.

At a Meeting of the Managers of the Royal Institution of Great Britain, held at the house of the Institution on the 2d of June, 1817.

Present

Earl Spencer K.G. President, in the Chair.
Sir Humphry Davy, V.P.
J. F. Daniell, Esqr.
Thomas Snodgrass, Esqr.
Daniel Moore, Esqr.
Sir Benjn. Hobhouse Bart. M.P. V.P.
James Stodart, Esqr.
Claude Scott, Esqr.
S. Bernard Morland, Esqr. M.P. Treasurer
Thomas Harrison, Esqr. Secretary.

1. The minutes of the proceedings of the last meeting were read and approved.

2. Read and approved the minutes of the Committee of Accounts of this day.

Lectures.

3. Read a letter from W. M. Craig, Esqr. addressed to the Managers offering to read gratuitously the next Season at the Royal Institution a Course of Lectures on the connexion between Poetry and Painting.

Resolved.

That Mr. Craigs offer be accepted.

4 Read a letter from Mr. Hindmarsh offering to read Rhetorical lectures at the Royal Institution.

Resolved. That his offer be declined.

Life Subscr.

5 In pursuance of a nomination by a Testamentary Paper of The Honble Robert Clifford deceased, his Nephew

The Honble Robert Clifford

was admitted to the privileges of a Life Subscriber to the Royal Institution.

Present received.

6. M. Dupin presented "Memoire Descriptif."

Ordered. That the thanks of the Managers be returned for this prese[nt]

7 Resolved.

That Mr. Faraday be allowed leave of absence for such time as Mr. Brande may think expedient.

Request of Mrs Greenwood.

8. Sir H. Davy having stated to the Board that Mrs Greenwood continued to be in a bad state of Health requested permission for her Son to sleep in the House for a short period.

Resolved. That under the special circumstances of her case this permission be granted subject to the superintendance of Mr. Brande.

Adjourned to Monday the 16th. Inst.

Spencer.

June 16.

At a Meeting of the Managers of the Royal Institution of Great Britain, held at the house of the Institution, on the 16th. of June, 1817.

Present

Earl Spencer, K.G. President, in the Chair.
Charles Hatchett, Esqr. V.P.
Sir Benjamin Hobhouse, Bart. M.P. V.P.
Thomas Snodgrass, Esqr.
Claude Scott, Esqr.
Samuel Solly, Esqr.
George Duckett, Esqr.
Daniel Moore, Esqr.
S. Bernard Morland, Esqr. M.P. Treasurer
Charles Holford, Esqr.
James Stodart, Esqr.

1. The minutes of the proceedings of the last meeting were read and approved.

2. Read and approved the minutes of the Committee of Accounts of this day, and signed an Order for payment of the following sums viz:

Payments.

John Millington Esqr. for a Course of Lectures in 1817	£105. —
Sundry expences for do.	5. 5 —
Joseph Fincher on acct. of the Royal Institution	50. — —
	£ 160. 5. —

Presents received.

3. Earl Spencer President of the Royal Institution — presented, An old Edition printed in the 15th. Century of the Book entitled Aggregator Practicus with figures of Plants.

John Landseer Esqr. F.S.A. presented,

Observations on the Engraved Gems, brought from Babylon to England by Abraham Lockett, Esqr.

Le Chevalier de Sade, presented,

Letter on the revision of the French Code,

A free Critique on Dr. Chalmers's Discourses on Astronomy).

172

Ordered. 1817.

That the thanks of the Managers be returned for these presents. –

4. Read the minutes of the Committee of Visitors entered in their proceedings of the 9th Instant.

5. Resolved. That the further consideration of the Draft of the Diploma be adjourned till the next meeting. –

Adjourned to Monday the 7th of July.

Charles Hatchett.

173

July 7.

At a Meeting of the Managers of the Royal Institution of Great Britain, held at the house of the Institution, on the 7th of July, 1817. –

Present

Charles Hatchett, Esqr. V.P. in the Chair.
Samuel Solly, Esqr.
Charles Holford, Esqr.
James Stodart, Esqr.
Thomas Snodgrass Esqr.
Daniel Moore, Esqr.
George Duckett, Esqr.

1. The minutes of the proceedings of the last meeting were read and approved.

2. Read and approved the minutes of the Committee of Accounts of this day, and signed an Order for payment of the following sums, viz.t

Payments.

Reeve for ¼ yrs King's & Assessed taxes to 25th March last	39 . 13 . 1
Butler for ¼ yrs Parochial rates to do.	15 . —
W. Harris Librarian ¼ yrs Salary to 31st. March last	45 . —
Joseph Fincher on account of the Royal Institution	200 . —
	£ 299 . 13 . 1

Lectures

3. Read a letter from John Landseer Esqr., dated 5th. July 1817. addressed to Earl Spencer & transmitted by His Lordship, expressing his willingness to read the number of Lectures that might be required of him at the Royal Institution in the ensuing Season. Resolved.

That the arrangements for the next Season having been completed the Managers are under the necessity of declining Mr. Landseer's offer for that Season.

Presents received.

4. The Society for the Encouragement of Arts, Manufactures

174

1817.

and Commerce presented,

Transactions of the Society, volume 34

Mr. Aikin's Address at the distribution of the Rewards adjudged by the Society. –

Mr. J. C. London F.L.S. presented

Remarks on the Construction of Hothouses 4to.

Ordered. That the thanks of the Managers be returned for these presents.

5. Read the minutes of the last General Meeting.

6. Resolved. That the consideration of the Diploma be further adjourned till the next meeting.

7. The following person having been duly recommended was admitted as an Annual Subscriber to the Royal Institution for the year 1817.

Ann.l Subs.r

Henry Pearse, Esqr. 17. Spring Gardens.

Leave of Absence granted to Ass.t Librarian

8. Mr. Sturt the Assistant Librarian, having requested leave of absence from the Institution for three weeks during the recess at two different periods – one week in the month of August, and a fortnight in September. Resolved,

That his request be complied with, provided that due attention be paid to the Library during his absence.

Adjourned to Monday the 3rd. of Nov.r

Humphry Davy

175

Nov. 3.

At a Meeting of the Managers of the Royal Institution of Great Britain held at the house of the Institution, on the 3rd of November, 1817. –

Present

Sir Humphry Davy V.P. in the Chair.
Charles Hatchett, Esqr. V.P.
Giles Templeman, Esqr.
Daniel Moore, Esqr.
Thomas Snodgrass, Esqr.
J. F. Daniell, Esqr.
Charles Holford, Esqr.
S. Bernard Morland, Esqr. M.P. Treasurer.
Thomas Harrison, Esqr. Secretary

1. The minutes of the proceedings of the last meeting were read and approved.

2. Read and approved the minutes of the Committee of Accounts of this day, and signed an Order for payment of the following sums, viz.t

Payments.

Joseph Fincher on acc.t of the Royal Inst.n	£60. —
Mess.rs Upton & Co. for Oil in 1816	56. —
	£116. —

Withdrawal.

3. Read a letter from Leonard Horner, Esqr., addressed to Thos. Harrison Esqr. Sec.y R.I., dated Edinburgh 15th. Oct.r 1817 stating that as he is not likely to visit London, except during the Summer Months, he begs leave to offer to the Managers his resignation as a Member of the Royal Institution. –

Resolved. That Mr. Horner's resignation be accepted and that his name be omitted in the succeeding Lists of Members. – The Bond for his annual payments was cancelled by the Secretary. –

Presents received.

4. The Royal Society, presented, Philosophical Transactions 1817 part 1.

A parallel drawn between the two intended Chinese Dictionaries, 4to. – By the Author.

The Editor of the Colonial Journal, presented,
The Colonial Journal No. 5 and 6.

Mr. Richard Rawlins, Surgeon, Oxford, presented,
A Dissertation on the structure of the Obstetric Forceps; with several new forms, 8vo. –

Mr. John Marshall, Surgeon, London presented
Remarks on Arsenic, considered as a Poison & a Medicine, 8vo.

Mr. Frederick Accum, presented,
Chemical Amusement:– Instructive Experiments in Chemy. 12mo. –

Charles Carpenter Bompass, Esqr. presented
An Essay on the nature of Heat, Light and Electricity, 8vo. –

The Geological Society, presented,
Transactions of the Geological Society vol. 4 part 2.

Dr. Marshall Hall, presented,
The principles of Diagnosis 8vo.

Dr. Alexander Marcet F.R.S. M.R.I. &c. presented,
An Essay on the Chemical History and Medical Treatment of Calculous Disorders, 8vo. –

Mr. John Sadler, presented,
Narrative of the Aerial Voyage of Mr. Sadler across the Irish Channel Oct: 1st. 1812 –
Narrative of the Aerial Voyage of Mr. Windham Sadler across the Irish Channel July 22d. 1817 –

The Revd. Dr. Abauzit, presented,
The Geneva Catechism prepared by the Pastors of Geneva, 12mo.

His Majesty's Commissioners of the Public Records, presented,
Valor Ecclesiasticus temp: Hen: VIII vol. 3 folio.

Sir William Adams, presented,
A Letter on the measures of the Medical Officers of the London Eye Infirmary.

Dr. B. De Sanctis presented,
Lusus Naturæ Londini observatus descriptus Tabula & Notis insuper illustratus.

William Tooke, Esqr. presented,
Report of the Committee of the House of Commons on the employment of Boys in sweeping of Chimnies.

John Fearne Esqr. presented – "A letter to Professor Stewart

on the objects of general terms and on the Axiomatical laws of Vision."

Ordered. That the thanks of the Managers be returned for these presents. –

5. Read the minutes of the Committee of Visitors entered in their proceedings of the 13th. of October last.

Carpeting for Library.

Resolved.
That Scrope Bernard Morland, Esqr. M.P. Treasurer, Thomas Harrison, Esqr. Secretary, and William Thos. Brande, Esqr. Superintendant of the House, be authorized to provide Carpeting for the Library of Reference.

Diploma

6 Resolved. That the consideration of the Diploma be deferred till Lord Spencer shall attend.

Porter.

7. Resolved. That William Bickley be engaged as Assistant Porter in the room of Stephen Statter.

8. The following persons having been duly recommended were admitted as Annual Subscribers to the Royal Institution.

Annl. Subscrs.

For the year 1817. –
Thomas Sharp Smith, Esqr. of Albany.
For the year 1818. –
Thomas Calvert, Esqr. 16 Grafton Street.
Thos. Taunton Wright, Esqr. 1. Kings Bench Walk Temple
Dr. Ord – Berkeley Square.
James Evans, Esqr. 10 St. Georges row Oxford Road.
Mr. George Green, 11 High row Knightsbridge
Robert Hendrie, Esqr. 12 Titchborne Street Haymarket
William Greenwood, Esqr. 37 Carey St. Lincolns Inn.

To the Lectures.
Mrs. Hart of Mecklenberg Square.
Mrs. C. Fulk of 3 Dorset Square.
Mrs. Watts Russel 33 Portland Place.

Adjourned to Monday the 1st. of December.

Benjamin Hobhouse V.P.

Dec. 1

At a Meeting of the Managers of the Royal Institution of Great Britain, held at the house of the Institution, on the 1st. of December, 1817. –

Sir Benjamin Hobhouse Bart. M.P. V.P. in the Chair.
Earl Spencer. President.
J. F. Daniell, Esqr.
Thomas Snodgrass, Esqr.
James Stodart, Esqr.
Sir Humphry Davy V.P.
George Duckett, Esqr.
S. Bernard Morland, Esqr. M.P. Treasurer.
Thomas Harrison, Esqr. Secretary.

1. The minutes of the proceedings of the last meeting were read and approved.

Payments.

2. Read and approved the minutes of the Committee of Accounts of this day, and signed an Order for an advance to the Clerk of £100 for the current expences of the present year, and also a further advance of £50 on the 1st. of January next.

Hon. Admission.

3. On the application of Alex: Marcet M.D. Baron de Humbolt Envoy Extraordinary and Minister Plenipotentiary of His Majesty the King of Prussia, was admitted to the Royal Institution under Art: 16 Chap VI of the Bye Laws.

Carpeting of Library.

4. The Secretary, from the Committee appointed to provide Carpeting for the Library of Reference, reported that they had given orders that the whole of the Floor should be covered with Hempen Carpeting, which they have reason to believe likely to be durable and may be laid down at an expence to the Institution of little more than £20.

Resolved.
That this report be confirmed.

5. Presents received.

Thomas Forster, Esqr. presented,
Observations on the Phenomena of Insanity.

Samuel Ware, Esqr. M.R.I. presented,
Observations on the Origin of Decorative Architecture.

The Society of Antiquaries presented,
Archaeologia, vol. 18 part 2.

The Royal Society presented,
Philosophical Transactions 1817 part 2.

Horticultural Society, presented
vol. 2 part 6 of their Transactions.

Ordered,
That the thanks of the Managers be returned for these presents.

6. Read the minutes of the last General Monthly Meeting. –

Diploma for Hon. Mems.

7. Earl Spencer being now present, the consideration of the Diploma for Honorary Members was resumed, and the following form adopted by the Board viz! –

Præses, Moderatores, et Socii Regalis Institutionis Magnæ Britanniæ
Omnibus et singulis salutem:
Quum
affectum, erga Regalis Institutionis conatus et studia, Philosophiam pro virili parte, in urbes et domos introducendo, singularem testatus est, Illum, solenni consessu in Societatem nostram cooptavimus; et, in hujus cooptationis testimonium, his Litteris præsentibus, Nos, Institutionis antea dictæ Præses, Moderatores et Socii, sigillum commune nostrum apponi curavimus: Datum Londini, die
Anno Christi
MDCCC ; et Regni Britanniarum
Regis, Institutionis Regal: illustrissimi Patroni

Resolved.
That the Secretary do cause the same to be neatly engraved on a Copper Plate. –

1817.

8. The following persons having been duly recommended were admitted as Annual Subscribers to the Royal Institution for the year 1818. –

Ann. Subscrs.

Captn. Steell 13 Maddox St. Hanover Square
Robt. George Clarke, Esq. 5 Craven St. Strand.
Samuel Young, Esq. 39 Lower Brook Street
John Allen, Esq. 8 Montagu Street

Adjourned to Monday the 26th. of January 1818. –

Charles Hatchett V.P.

~~Dec.~~ 1818 Jan. 26

At a Meeting of the Managers of the Royal Institution of Great Britain, held at the house of the Institution, on the 26th. of January, 1818. –

Present
Charles Hatchett, Esqr. V.P. in the Chair.
Sir Humphry Davy. V.P.
Daniel Moore, Esqr.
Giles Templeman, Esqr.
James Stodart, Esqr.
George Duckett, Esqr.
Thomas Snodgrass, Esqr.
J. F. Daniell, Esqr.
Samuel Solly, Esqr.
S. Bernard Morland, Esqr. M.P. Treasurer.
Thomas Harrison, Esqr. Secretary.

1. The minutes of the proceedings of the last meeting were read and approved.

2. Payment. Read and approved the minutes of the Committee of Accounts of this day, and signed an Order for an advance to the Clerk of £60.

3. Proposal of Communication between Academy at Dijon & R.I. Read a letter from C. Moreau, Esqr. Correspondent of the Academy at Dijon stating, that in virtue of the power invested in him, and the instructions he had received, he hastened to make known to the Members of the Royal Institution the wish of the Academy, who desired the honour of associating itself with the Royal Institution, so that in cases of importance to open a correspondence on every species of Sciences, Arts, and useful discoveries, and that there shall exist a perfect reciprocity on both sides.

Resolved.

That the Secretary be directed to acknowledge the receipt of the above, and to state that the Managers receive with pleasure the proposition made on the part of the Academy of Dijon, and shall be happy to be

1818.

favoured with any communications from the Fellows of that learned body, and to transmit them an account of any discoveries that may be made in the Royal Institution connected with the progress of Science or the useful Arts.

4 Presents received.

Dr. A. P. Wilson Philip, presented
An experimental Inquiry into the laws of the vital Functions,
Sir Thomas Bernard, Bart. M.R.I. presented
Case of the Salt Duties, with proofs & illustrations
Sir G. O. Paul, Bart.
Observations on the alarming progress of the Gaol or Typhus Fever.
W. Wadd Esq. Surgeon Extraordy. to the Prince Regent
Cases of diseased Prepuce & Scrotum.
Cases of diseased Bladder & Testicle.
Companion for Caraboo: A Narrative of the conduct and adventures of Henry Frederic Moon.

Ordered.

That the thanks of the Managers be returned for these presents.

5 Read the minutes of the last General Monthly Meeting.

6. Read the minutes of the Committee of Visitors entered in their proceedings of the 12th. Instant

Present of Patent Door-spring.

The Clerk reported that he had received the following letter from Messrs. E. & J. Dowson, the makers of a new invented Single Door Spring, in consequence of an application made to them by Richard Horsman Solly, Esq. Visitor –

Sir,

Mr. Solly having called & stated that a new Spring was wanted for the News-room of the Royal Institution; we beg to inform you that we will, with pleasure, furnish one of our new improved single Door Springs free of expence, if the Members of the Institution will do us the favour to accept it.

6 Welbeck Street 22d. Jany. 1818. (signed) E. & J. Dowson

Jan. 26.

Resolved. That Messrs. Dowson's liberal offer be accepted with the thanks of the Managers.

7 Diploma. The Secretary reported, that in order to save the Institution the expence of a Copper Plate Engraving for the Diploma which was found to be at an amount of upwards of £5 Joseph Fincher had been so good as to undertake to make a sufficient number of copies in a neat hand upon Parchment

8 Carpeting of Library. The Secretary, from the Committee appointed to provide Carpeting for the Library of Reference, reported that the Carpet had been laid down at an expence of £24.3.2.

9 Loan of £1000 by Mr Fuller. John Fuller Esq. M.R.I. having communicated thro' Daniel Moore, Esq. to the Managers of the Royal Instn. that feeling the state of its finances, and desirous to promote its objects he desires to place in their hands the sum of £1000 – the interest to be applied for the purposes of the Institution and the principal so secured to him that it may never become liable to the Debts of the Institution.

Resolved.

That the Managers receive with great pleasure the very liberal offer of Mr. Fuller; – that their thanks be returned to him by the President, and that Mr. Fuller be invited to confer with the Managers at the next meeting of the Board on the manner in which it may be most agreeable to him to have the donation secured.

10 The following persons having been duly recommended were admitted as Subscribers to the Royal Institution.

Life Subscriber.

Lt. Genl. Mitchell of Fisherton near Salisbury.

Annual Subscribers.

Edward Dodwell, Esq. Wimpole Street
William Ogilvy, Esq. 6 Albany Chambers
George Pocock, Esq. 39 Charles St. Berkeley Sq.
Colonel Wraw 11 Park Place St. James's
Lt. Col. Colston 57 Upper Berkeley Street
Lt. W. G. Ferry of the 1st. Life Guards – Hyde Park Barracks

184

Annual Subscribers contd. 1818.

Revd. Calvert Moore, 99 Park St.
Captn. Jones, 5 upper Fitzroy Street
George Paterson, Esqr. 14 Princes St. Hanover Sqr.
Sir George Prescott, Bart.
David Caddell, Esqr. Salisbury Square

To the Lectures.

Mrs. Shadwell 9 upper Gower Street
Mrs. Mills 20 Finsbury Place.

Benjamin Hobhouse V.P.

185

Feb 2

At a Meeting of the Managers of the Royal Institution of Great Britain, held at the house of the Institution, on the 2d. of February, 1818. —

Present

Sir Benjamin Hobhouse, Bart. M.P. V.P. in the Chair.
James Stodart, Esqr.
J. F. Daniell, Esqr.
Charles Hatchett, Esqr. V.P.
Sir Humphry Davy, V.P.
Samuel Solly, Esqr.
Daniel Moore, Esqr.
Thomas Harrison, Esqr. Secretary.

1. The minutes of the proceedings of the last meeting were read and approved.

2. Read and approved the minutes of the Committee of Accounts of this day, and signed an Order for payment of the following sums, vizt.

Payments.

Dickins & Co. for Coals in 1816	£108 . 15 . —
Jos. Fincher for the payment of small bills in 1816	100 . — . —
	£208 . 15 . —

3. Audited the Treasurers' Account for the year 1817. —

Presents received.

4. E. A. Kendall, Esqr. F.A.S. presented
An Argument for construing largely the right of an Appellee of Murder to insist on Trial by Battle, and also for abolishing Appeals.
E. Thos. Svedenstierna, Esqr. of Stockholm, presented,
Specimens of Minerals
Ordered. That the thanks of the Managers be returned for these presents.

5 John Fuller, Esqr. attended, and was thanked by the Chairman

186

1818.

Loan by Mr. Fuller.

for his very liberal offer; which he now extended so far, as to allow the Institution, upon certain conditions, to avail themselves of the principal sum of £1000, in discharge of their Debt, and it was agreed that Mr. Fuller should instruct Mr. Moore to lay those conditions before the next meeting of the Board.

Communication from Patrons. Deficiencies.

6. Read the following Resolution of the Patrons of the Library — vide their minutes of the 28th. Jany. 1818. —
Resolved.
That a statement of the Funds of the Library for the last year, be laid before the Managers at their next meeting, requesting that they would be pleased to allow the deficiency to be supplied out of the general Funds of the Institution.
Resolved. That the Patrons of the Library be informed that the general Funds of the Institution, are not at present in a state to enable the Managers to assist the Patrons to discharge the deficiency of their account.

7. The following persons having been duly recommended were admitted as Subscribers to the Royal Institution for the year 1818. —

Annual Subscribers.

Revd. John Abbiss, 23 Grosvenor Place.
James Warington, Esqr. 57 Upper Berkeley Street
Lord Frederick Montagu, 14 Berkeley Square.
John Chatfield Tyler, Esqr. 39 Davies Street.

To the Lectures.

Miss Parkes — Goswell Street.
Miss Franks, Charles St. Berkeley Square.
Lady Rumbold.
Mrs. Thos. Farrer, Bedford Place.
Miss Farrer. Lincolns Inn.
Mrs. & Miss MacTaggart, of Leighton.

Spencer

187

Feb. 9.

At a Meeting of the Managers of the Royal Institution of Great Britain, held at the house of the Institution, on the 9th. of February, 1818.

Present

Earl Spencer, K.G. President in the Chair.
Sir Benjn. Hobhouse, Bart. M.P. V.P.
Sir H. Davy, V.P.
James Stodart, Esqr.
Giles Templeman, Esqr.
George Duckett, Esqr.
J. F. Daniell, Esqr.
Daniel Moore, Esqr.

1. The minutes of the proceedings of the last meeting were read and approved.

2. Read and approved the minutes of the Committee of Accounts of this day, and signed an Order for payment of the following sums vizt.

Payments.

Mr. Harris 3/4 years Salary to 31st. Decr. 1817	£135 . — . —
Mr. Brooks 1 years do. to do.	31 . 10 . —
Thos. Hatter Lecture room door keeper do.	10 . 10 . —
Bird for Bricklayers work in 1816	44 . 11 . 6
Stone & Co. for Gas Apparatus in do.	36 . 3 . 11
Royce — for repairs of the House in do.	33 . 11 . 3
Joseph Fincher on acct. of the Royal Instn.	60 . — . —
	351 . 6 . 8

3. Read the minutes of the last General Monthly Meeting.

Mr. Fuller's Loan.

4. Mr. Moore reported that he had not yet received Mr. Fuller's instructions on the subject referred to him by the last Board; — the consideration of it was therefore adjourned to the next Meeting.

5. The following persons having been duly recommended

188

1818.

were admitted as Subscribers to the Royal Institution for the year 1818.

Annual Subscribers.

Patrick Paterson, Esqr. 81 Jermyn Street.
George Pring, Esqr. of Hammersmith.
Revd. Dr. Randolph, Rectory Covent Garden
Henry Harrison Briscoe, Esqr. 12 Rathbone place
Lt. Col. Johnson, 18 upper Seymour Street.
Dr. Heberden, 2 upper Brook Street
C. Dering Esqr. 9 Clarges St.
James Henry Earle, Esqr. 11 Hinde St. Man. Sqe.
Captn. Thos. Garth, R.N. Kings Mews.
Col. Newbery, 19 Nottingham Place
Hon: E. Lascelles, Harewood Place
Joseph Alcock, Esqr. 27 Queen Ann St.
John Shaw Lefevre, Esqr. New St. Spring Gardens
Lucius O'Brien, Esqr. 1 Fig Tree Court Temple.

To the Lectures.

Miss Frances Hamilton } 18 Orchard St.
Miss Emily Fitzhugh }
Viscountess Lascelles } Harewood Place
Miss Frances Lascelles }
Miss Lascelles }
Miss Colebrooke — Branch Hill Lodge Hampstead.
Miss Jane Barrow } 14 Orchard Street
Miss Margaret Barrow }
Mrs. George Tritton } West Hill Wandsworth.
Miss Tritton }
Miss E. Tritton }

Spencer.

189

Feb. 16.

At a Meeting of the Managers of the Royal Institution of Great Britain, held at the house of the Institution, on the 16th of February, 1818.—

Present
Earl Spencer K.G. President, in the Chair.
Sir Benjn. Hobhouse, Bart: M.P. V.P.
Sir Humphry Davy, V.P.
Charles Hatchett, Esqr. V.P.
Thomas Snodgrass, Esqr.
George Duckett, Esqr.
Giles Templeman, Esqr.
James Stodart, Esqr.
S. Bernard Morland, Esqr. M.P. Treasurer.
Daniel Moore, Esqr.
Thomas Harrison, Esqr. Secretary.

1. The minutes of the proceedings of the last meeting were read and approved.

2. Read and approved the minutes of the Committee of Accounts of this day, and signed an Order for an Advance to the Clerk of £200.

Resign. of Porter. 3. Mr. Brande stated to the Board, that William Bickley the Assistant Porter wished to resign.
Resolved. That his resignation be accepted, and that a new Porter be looked out for to supply his situation.

4. The following persons having been duly recommended were admitted as Subscribers to the Royal Institution for the year 1818.—

Annual Subscribers.
Lewis George Dive Esqr. Tavistock St. Bedford Sq.
Captn. Chas. Grant R.N. 23 upper Harley St.
James South, Esqr. High Street Borough.
Charlton B. Wollaston Esqr. Paper Build. Temple
John Ware, Esqr. Bridge St. Blackfriars.
Thomas Polhill, Esqr. 52 Harley Street.

190

1818

Henry Sich, Esqr. Chiswick.
Paul Fras. Benfield, Esqr. 13 Somerset St.
Henry Gahagan Esqr. 13 upper Grosvenor St.
Sir Chas. Hulse, Bart. Audley Square.

To the Lectures.

Miss Mary Anne Flaxman, Park St. Southwark
Miss Sarah Joy } Blackman St.
Mrs. James South }
Miss Swinburne } Grosvenor Place
Miss E. Swinburne }
Miss Harriet Grant 23 upper Harley St.
Miss Foyster } 45 Charlotte Street Fitzroy Square
Miss Maria Foyster }
Miss Harriet Foyster }
Miss Cecilia Markham 17 Curzon Street
The Honble Mrs. Abercromby 6 New St. Spring Gardens
Miss Dawes } 36 Foley Place.
Miss Noble }

Spencer.

191

Feb. 23.

At a Meeting of the Managers of the Royal Institution of Great Britain, held at the house of the Institution on the 23d. of February, 1818.—

Present
Earl Spencer, K.G. President, in the Chair.
Sir Benjamin Hobhouse, Bart. M.P. V.P.
Sir Humphry Davy, V.P.
George Duckett, Esqr.
James Stodart, Esqr.
J. F. Daniell, Esqr.
S. Bernard Morland, Esqr. M.P. Treasurer.
Thomas Harrison, Esqr. Secretary.

1. The minutes of the proceedings of the last meeting were read and approved.—

2. Read and approved the minutes of the Committee of Accounts of this day, and signed an Order for an advance to the Clerk of £120.

Lectures. 3. Read a letter from Mr. B. H. Smart offering to deliver gratuitously at this Institution, 4 lectures on Elocution.
Resolved.
That Mr. Smart's offer be declined, the arrangements for the present Season having been completed.

Present received.

4. Henry Goodwyn, Esqr. M.R.I. presented,
The first Centenary of a series of Tables of complete Decimal Quotients —
Ordered. That the thanks of the Managers be returned for this present

5. Ordered. That Mr. Bramah's Press be purchased at the sum of 35 guineas.

Porter. 6 Resolved. That William Dowling be appointed Assistant Porter at the usual Salary of £1 per week.—

1818.

The Clerk reported that Mr. Hole the Watchmaker will undertake to wind clean and repair all the Clocks in the Institution at 2 guineas a year.

Resolved. That he be engaged at that sum.

The following persons having been duly recommended were admitted as Subscribers to the Royal Institution.

Life Subscriber.

George Townley, Esqr. – 61 South Molton St.

Annual Subscribers.

George Hallam, Esqr. White Barns Buntingford.
John Want, Esqr. 11 North Crescent Bedford Sqr.
Daniel Cloete, Esqr. Jeffries Square
Frederick Glover, Esqr. 5 Campden Place Peckham
William Green, Esqr. Milbank Street
Edward Abell, Esqr. 9 Fenchurch Buildings.

To the Lectures.

The Honble Miss Courtney } St. James's Palace
Miss Anne Courtney }
Mrs. Maxn. Kymer, Finsbury Square.

Spencer

Mar. 2.

At a Meeting of the Managers of the Royal Institution of Great Britain, held at the house of the Institution on the 2d. of March, 1818. –

Present.

Earl Spencer K.G. President, in the Chair.
Charles Hatchett, Esqr. V.P.
Sir Humphry Davy, V.P.
Sir Benjamin Hobhouse Bart. V.P.
Claude Scott, Esqr.
Thomas Snodgrass, Esqr.
James Stodart, Esqr.
Samuel Jolly, Esqr.
George Duckett, Esqr.
S. Bernard Morland, Esqr. M.P. Treasurer.
Daniel Moore, Esqr.

1. The minutes of the proceedings of the last meeting were read and approved.

2. Read and approved the minutes of the Committee of Accounts of this day, and signed an Order for payment of the following sums – vizt.

Payments.

Mr. Newman for Chemicals in 1817	£120 . 3 . 11
Joseph Fincher on account of the Royal Institution	50 . –
	£170 . 3 . 11

Present received.

3. Honble Francis Henry Egerton, presented Addenda & Corrigenda to his Edition of the Hippolytus Stephanephorus of Euripides.

Ordered. That the thanks of the Managers be returned for this present.

4. On the application of Sir H. Davy,

Hon. Admission.

His Excellency Count Ludolf the Neapolitan Ambassador was admitted to the Royal Institution under Art. 16 – Chap. VI of the Bye Laws. –

1818

5. Daniel Moore Esqr. laid before the Board a Draft of a Deed on the part of Mr. Fuller for carrying into effect the terms on which he intended to present to this Institution the sum of £1000 for its benefit which having been read, was ordered to be laid before the General Monthly Meeting of this day.

Mr. Fuller's loan.

6. The following persons having been duly recommended were admitted as Subscribers to the Royal Institution for the year 1818. –

Annual Subscriber.

John L. Anderdon, Esqr. 14 New St. Spring Gardns.

To the Lectures.

Miss Grant 40 Albemarle Street
Miss Taylor 61 Harley Street.

Spencer.

Mar. 9.

At a Meeting of the Managers of the Royal Institution of Great Britain, held at the house of the Institution, on the 9th. of March, 1818. –

Present

Earl Spencer, K.G. President, in the Chair.
Daniel Moore, Esqr.
Thomas Snodgrass, Esqr.
John Frederick Daniell, Esqr.
Samuel Jolly, Esqr.
Giles Templeman, Esqr.
S. Bernard Morland, Esqr. M.P. Treasurer
Thomas Harrison, Esqr. Secretary.

1. The minutes of the proceedings of the last meeting were read and approved.

2. Read and approved the minutes of the Committee of Accounts of this day, and signed an Order for payment of the following sums, vizt.

Payments.

Reeve 3/4 yrs. Assessed taxes to Xmas 1817.	£118 . 13 . 4
Chappell 3/4 yrs. Parochial rates to do. –	42 . 10 . –
	161 . 3 . 4

3. Read the minutes of the last General Monthly Meeting.

4. The following persons having been duly recommended were admitted as Subscribers to the Royal Institution.

Life Subscriber.

Vice Admiral Sir Thos. Williams 8 Hill St.

Annual Subscribers.

Bingley Garland, Esqr. 2 Verulam Buildgs.
Wm. George Carter, Esqr. Temple Chambers.
Gilbert Munro, Esqr. New Bank Buildings
Frederick Serjeant, Esqr. Treasury Chambers
Henry Thos. Liddell, Esqr. 51 Portland Place
Edwd. Polhill, Esqr. 36 York Place Portman Sqr.

196

Annual Subscribers continued – 1818.

Thos. Newman, Esqr. Grillions Hotel
Joseph Fisher, Esqr. 78 Jermyn Street

To the Lectures.

Mrs. Taylor, 61 Harley Street
Miss Hayles –
Miss Wollaston – } 5 Argyll Street
Miss A. Wollaston }
Miss Kymer – Streatham
Lady Williams 8 Hill Street

Spencer.

197

Mar. 16.

At a Meeting of the Managers of the Royal Institution of Great Britain, held at the house of the Institution, on the 16th. of March, 1818. –

Present

Earl Spencer, K. G. President, in the Chair.
J. F. Daniell, Esqr.
Charles Hatchett, Esqr. V.P.
Samuel Solly, Esqr.
Sir Humphry Davy, V.P.
Claude Scott, Esqr.
Thomas Snodgrass, Esqr.
Daniel Moore, Esqr.
S. Bernard Morland, Esqr. M.P. Treasurer.
James Stodart, Esqr.
Thomas Harrison, Esqr. Secretary.

1. The minutes of the proceedings of the last meeting were read and approved.

2. Read and approved the minutes of the Committee of Accounts of this day, and signed an Order for an Advance to the Clerk of £130.

Presents received

3. By the Author –
La Scava; or, some account of an excavation of a Roman Town; to which is added a Journey to the Simplon, 8vo.
Sketches of "Curvilinear Hothouses – By Mr. J. C. Loudon
Ordered.
That the thanks of the Managers be returned to Mr. Loudon for the above present.

4. The following persons having been duly recommended were admitted as Annual Subscribers to the Royal Institution for the year 1818. –

Ann Subrs.
Daniel Mussanden, Esqr. Fladongs Hotel
John Sayer, Esqr. Wick House near Worcester.

198

Annual Subscribers, contd. 1818.

Alexander Campbell, Esqr. Walton on Thames
Revd. Dr. Tournay, Warden of Wadham College Oxford
Col. Hughes M.P. 12 Gloster Place
Captn. Harvey, King Street St. James's

Adjourned to Monday the 6th of April.

Spencer

199

April 6.

At a Meeting of the Managers of the Royal Institution of Great Britain held at the house of the Institution, on the 6th. of April 1818. –

Present

Earl Spencer, K. G. President, in the Chair.
Sir Benjn. Hobhouse Bart. V.P.
Charles Hatchett, Esqr. V.P.
J. F. Daniell, Esqr.
Giles Templeman, Esqr.
Thomas Snodgrass, Esqr.
Claude Scott, Esqr.
S. Bernard Morland, Esqr. M.P. Treasurer.
George Duckett, Esqr.
Samuel Solly, Esqr.
Daniel Moore, Esqr.
Charles Holford, Esqr.
Thomas Harrison, Esqr. Secretary.

1. The minutes of the proceedings of the last meeting were read and approved.

2. Read and approved the Committee of Accounts of this day and signed an Order for payment of the following sums vizt.

Payments

1 years Insurance in the Sun Fire Office to 25th. March 1819 –	£63 . 16 . 6
J. Fincher on account of the Royal Institution	50 . – . –
	113 . 16 . 6

Letter from Sir J. E. Smith resp. Lectures.

3. Read the following letter from Sir E. Smith dated Cambridge April 5th. 1818 addressed to the Managers.

Gentlemen,

Being disappointed of giving a Course of Botanical Lectures in this University (as I had been requested to do by the Professor of Botany and the Vice Chancellor,) this term, I am at liberty to resume my old practice of

lecturing under your auspices. I therefore take the earliest opportunity of saying that I am ready immediately to commence a course of any convenient length that you may chose, and I wish the pecuniary remuneration to be such as may suit the finances of the Institution. – If they do not admit of payment for my lectures, I shall be ready to concur with those eminent Men who have formerly like myself, been engaged by the Managers, and have profited thereby, in giving gratuitously such a set of Courses of Lectures as may do credit to Science, and essentially benefit the Royal Institution in a pecuniary way. If this is not concurred in, and my first proposal cannot conveniently be complied with, I trust you will think me at liberty to lecture in any other place or manner that I may, for the present season, judge best, without any dissolution of those ties, which may be renewed on some future occasion, and which have always been found agreeable and honourable by Gent^n

your very faithful & obedient Servt.

(signed) J. E. Smith.

Resolved. That the Secretary do inform Sir J. E. Smith that the Managers accept with many thanks his proposal of delivering gratuitous Lectures.

Presents received.

4. The Society for the Encouragement of Arts &c. presented

The 35th. Volume of the Society's Transactions.

J. H. Curtis Esqr. presented

An introductory Lecture on the Diseases of the Ear.

Ordered.

That the thanks of the Managers be returned for these presents.

5. Read the minutes of the Committee of Visitors entered in their proceedings of the 25th. of March.

6. The following persons, having been duly recommended were admitted as Subscribers to the Institution for the year 1818. –

April 6.

Annual Subscribers.

Rev^d. M^r. Watkins N°. 2 Great Cumberland place
Thos. K. Hall, Esqr. Hollybush Staffordshire
Henry Jeffreyes, Esqr. 33 Clarges S^t.
J. P. Cox, Esqr. 17 Howard Street
Vice Admiral Sir George Martin 8 Hertford S^t.
Constable Maxwell, Esqr. 33 Hill S^t.

To the Lectures.

Mrs Scott Waring } Half Moon S^t.
Miss Hamilton }
Mrs Slater York Place
Miss Stanley Privy Gardens

Benjn. Hobhouse V.P.

April 13

At a Meeting of the Managers of the Royal Institution of Great Britain held at the house of the Institution, on the 13th. of April, 1818. –

Present

Sir Benjamin Hobhouse, Bart. M.P. V.P. in the Chair
Sir Humphry Davy, V.P.
Thomas Snodgrass Esqr.
Giles Templeman, Esqr.
James Stodart, Esqr.
Daniel Moore, Esqr.
Samuel Solly, Esqr.
George Duckett, Esqr.

1. The minutes of the proceedings of the last meeting were read and approved.

2. Read and approved the minutes of the Committee of Accounts of this day, and signed an Order for payment of the following sums vizt.

Payments

W. Bulmer for Printing in 1816	£148 . 7 . 0
W. Feetham for Ironmongery in d:–	103 . 19 . 2
Timothy Bramah, Esqr. for his Press	36 . 15 . –
	289 . 1 . 2

3. Read the minutes of the last General Monthly Meeting.

4. On the application of Sir Humphry Davy

Hon Admission.

His Excellency Richard Rush, Esqr.
Ambassador from the united States of America.
was admitted to the Royal Institution under Art. 15. Ch. VI of the Bye Laws.

5. The following persons having been duly recommended were admitted as Subscribers to the Royal Institution for the year 1818. –

April 13.

Annual Subscriber

Newman Smith, Esqr. Mount Street

To the Lectures.

Miss Clinton, Foley Place
Miss Churchill of Baker Street.

Spencer.

204

April 20. 1818.

At a Meeting of the Managers of the Royal Institution of Great Britain, held at the house of the Institution on the 20th of April, 1818. –

Present

Earl Spencer K. G. President. in the Chair.
Charles Hatchett, Esqr. V. P.
Sir Humphry Davy V. P.
Thomas Snodgrass, Esqr.
J. F. Daniell, Esqr.
James Stodart, Esqr.
George Duckett, Esqr.
Giles Templeman, Esqr.
Samuel Solly, Esqr.
S. Bernard Morland, Esqr. M.P. Treasurer.
Daniel Moore, Esqr.
Thomas Harrison, Esqr. Secretary.

1. The minutes of the proceedings of the last meeting were read and approved.

2. Read and approved the minutes of the Committee of Accounts of this day, and signed an Order for payment of the following sum, vizt.

Payments.

The Assignees of Dudley Adams for Mechanical Apparatus from 1802 to 1805 } £17 . 17 . 0

Presents received.

3. Lt. Col. W. M. Leake F.R.S &c. &c. presented
A Map of Egypt on two Sheets
Mr. John Hennen, Deputy Inspector of Military Hospitals "Observations on the practice of Military Surgery." presented
The Directors of the East India Company presented
Morrison's Chinese Dictionary part 1 & 2.
Dialogues &c. in Chinese and English,

Ordered.
That the thanks of the Managers be returned for these presents.

205

April 20

4. Read the minutes of the Committee of Visitors entered in their proceedings of the 13th of April.

List of Subscrs.

5. Ordered. That a List of Life Subscribers, Annual Subscribers and Subscribers to the Lectures be made out by the Clerk, and laid before the Managers.

6. The following persons having been duly recommended were admitted as Subscribers to the Royal Institution for the year 1818.

Annual Subscribers.
Lt. Philip Yorke 3d Regt. of Guards, 12 Portman St.
The Honble Charles John Shore 4 Portman Sqr.
Robert Holden, Esqr. 17 Montagu Square

To the Lectures.
Mrs. Bethell
Miss Bethell } 39 Grosvenor Place.

7. Ordered. That the House Lists be filled up as follows.

President Earl Spencer K. G. F.R.S and F.S.A.

Treasurer Scrope Bernard Morland, Esqr. M.P. L.L.D.

Secretary Thomas Harrison, Esqr. A. M. F.R.S.

Managers.
James Peter Auriol, Esqr. F.R.S
Sir Thos. Bernard Bart. L.L.D.
Earl of Bessborough.
John Frederick Daniell, Esqr.
Sir Humphry Davy L.L.D. F.R.S &c. &c.
Charles Hatchett, Esqr. F.R.S & F.S.A
Major Genl. Kyd
Marquis of Lansdowne L.L.D. F.R.S.
Daniel Moore, Esqr. F.R.S. F.S.A & F.L.S
Rt. Honble Sir John Nicholl M.P. L.L.D. F.R.S &c.
Thomas Snodgrass, Esqr.
Samuel Solly, Esqr. F.R.S.
Sir John Thos. Stanley Bart. F.R.S.

206

1818.

James Stodart, Esqr.
Robert Ward, Esqr.

Visitors
George Aust, Esqr. – F.R.S and F.S.A
Timothy Brent, Esqr.
Rear Admiral Sr. Edwd. Codrington, K.C.B
Henry Thos. Colebrooke, Esqr.
Earl of Darnley, F.R.S.
John Fuller, Esqr.
Davies Gilbert, Esqr. M.P. F.R.S.
Gibbes Walker Jordan Esqr. A.M. F.R.S
Charles Rieschell, Esqr.
Sir John S. Sebright, Bart. M.P.
John Soane, Esqr. F.S.A
Richd. Horsman Solly, Esqr. F.R.S & F.S.A.
Duke of Somerset F.R.S.
Marquis of Stafford K.G.
Lt. Genl. Wm. Thornton, M.P.

Humphry Davy

207

April 27.

At a Meeting of the Managers of the Royal Institution of Great Britain held at the house of the Institution, on the 27th of April, 1818.

Present

Sir Humphry Davy, V.P. in the Chair.
Earl Spencer, K. G. President
Charles Hatchett, Esqr. V. P.
Samuel Solly, Esqr.
Thomas Snodgrass, Esqr.
James Stodart, Esqr.
Daniel Moore, Esqr.
Giles Templeman, Esqr.
George Duckett, Esqr.
S. Bernard Morland, Esqr. M.P. Treasurer.
Thomas Harrison, Esqr. Secretary.

1. Read and approved the minutes of the last meeting.

2. Read and approved the minutes of the Committee of Accounts of this day, and signed an Order for an Advance to the Clerk of £50. –

Lectures Sir J. E. Smith.

3. Read the following letter from Sir J. E. Smith, addressed to the Secy., dated Norwich April 18th 1818. –

Sir,

Instead of proceeding to London, as I expected, I have been obliged by business to return home from Cambridge & shall not be in town for near so long a period as I intended – I have it not in my power therefore to give any Course of lectures there this year; but I shall be anxious, if I can, to present myself at the Royal Institution for a couple of lectures as last year, about which I shall have the honour of consulting the Managers when I come to town in May.

I remain with much respect &c. &c.
(signed) J. E. Smith.

208

1818.

Request of Mr Sturt resp[t] Dining.

4. Read a letter from Mr. Sturt, Assistant Librarian, stating that in consequence of his having from the 1st of March to the 1st of September, to attend every day in the Library on which the Institution is open from 10 o'clock in the Morning till 6 in the Evening without being allowed any fixed time to dine, He should esteem it a favor, if it met with the approbation of the Managers, to be allowed to dine in the middle of the day from 1 to 3, in consideration of the two extra hours which he is required to attend from the 1st. of March to the 1st. of Sept:. –

Resolved.

That his request be complied with.

5. In pursuance of a nomination under the Act, the following person was admitted to the privileges of a Life Subscriber to the Royal Institution.

Life Subs[cr].

William Edmund Hartopp, Esq[r] of Lincolns Inn nominated by his Father Sir Edmund Cradock Hartopp Bart.

6 Presents received.

I. D'Israeli Esq[r]. presented
The Literary Character illustrated.
John Ayrton Paris, M.D. F.L.S. presented
A Memoir of the Life & Scientific labours of the late Rev[d]. William Gregor, A.M.
John Vetch, M.D. presented
Observations relative to the treatment by Sir W. Adams of the Ophthalmic cases of the Army.

Ordered.

That the thanks of the Managers be returned for these presents.

7. Read the minutes of the last General Meeting.

Spencer

209

May 4.

At a Meeting of the Managers of the Royal Institution of Great Britain held at the house of the Institution, on the 4th of May, 1818. –

Present

Earl Spencer, K.G. President in the Chair.
Charles Hatchett, Esq[r].
J. P. Auriol, Esq[r].
Thomas Snodgrass, Esq[r].
Sir Humphry Davy.
Robert Ward, Esq[r].
J. F. Daniell, Esq[r].
Daniel Moore, Esq[r].
James Stodart, Esq[r].
S. Bernard Morland, Esq[r]. M.P. Treasurer.
Samuel Solly, Esq[r].
Major Gen[l]. Kyd
Thomas Harrison, Esq[r]. Secretary.

1. The minutes of the proceedings of the last meeting were read and approved.

2. Read and approved the minutes of the Committee of Accounts of this day, and signed an Order for payment of the following sums vizt.

Payments.

1 years Ground rent to the City of London to 25th. March 1818	£150. 1. 4
Messrs. Dicksons for Coals in 1817	109. –. –
	259. 1. 4

Presents received.

3. A. Barto-Vallé, Esqr. presented
Ittiolitologia Veronese, large folio.
T. H. Colebrooke, Esqr. presented
"On Import of Colonial Corn, 8vo."
Sir Frederick Baker, Bart. presented
"Medical Tracts by Sir George Baker, Bart."

Ordered.

That the thanks of the Managers be returned for these presents.

210

1818.

4 In pursuance of a nomination under the Act by Mrs. Mary Salter Dehany, under the share of her late Father Philip Dehany to which she became entitled + in November 1809.

Life Subs[cr].

her Cousin William Knight Dehany, of 1 Elm Court Temple Esq[r] was admitted to the privileges of a Life Subscriber to the Royal Institution.

5 Earl Spencer nominated as Vice Presidents for the ensuing year,
James Peter Auriol, Esq[r].
Sir Tho[s]. Bernard, Bart.
Earl of Bessborough.
Sir Humphry Davy.
Charles Hatchett, Esq[r].
Marquis of Lansdowne.
R[t]. Hon: Sir John Nicholl.
Sir John Tho[s]. Stanley, Bart.

6 Read the minutes of the General Annual Meeting.

Coals.

7. Ordered. That 40 Chaldrons of the best Walls-End Coals be laid in at the Royal Institution, at 54/6[d]. ℔ Chaldron, Metage, Shooting and all expences included.

Adjourned to Monday the 18th. of May

Spencer.

211

May 18.

At a Meeting of the Managers of the Royal Institution of Great Britain held at the house of the Institution, on the 18th of May, 1818. –

Present

Earl Spencer, K.G. President, in the Chair.
J. P. Auriol, Esq[r]. V.P.
Charles Hatchett, Esq[r]. V.P.
Thomas Snodgrass, Esq[r].
James Stodart, Esq[r].
Major Gen[l]. Kyd.
Daniel Moore, Esq[r].
Sir Humphry Davy, V.P.
Earl of Bessborough, V.P.
S. Bernard Morland, Esq[r]. M.P. Treasurer.
Thomas Harrison, Esq[r]. Secretary.

1. The minutes of the proceedings of the last meeting were read and approved.

2. Read and approved the minutes of the Committee of Accounts of this day, and signed an Order for payment of the following sums viz[t].

Payments

Joseph Fincher on acc[t]. of the Royal Institution. –	£60. –
Mainwaring for Apparatus in 1817	27. 2. 2
Fisher for Chemicals in 1817	14. 9. 7
	£101. 11. 9

Presents received.

3. Mr. Arnold Merrick, presented
L. J. Thenards' Treatise on the general principles of Chemical analyses 8vo.
Colonel De Bosset, by Col. Ansley M.R.I. presented
Essai sur les Médailles antiques des Iles de Cephalonie et d'Ithaque, 4to.
Sir James Edward Smith, M.D. presented
Considerations respecting Cambridge, more particularly relating to its Botanical Professorship.

1818.

Patrick Colquhoun, Esq: presented,
Considerations on the means of affording profitable employment to the redundant Population of Great Britain and Ireland &c.

Ordered.

That the thanks of the Managers be returned for these presents.

4. The following Gentlemen having been proposed to the Board as desirous of Visiting this Institution during their temporary residence in London, – viz.

Hon Admissions.

His Serene Highness the Prince Leppe
His Excell.y The Duke of San Carlos
Ambassador from the Court of Spain
His Excellency Count Shomborn, from Francfort
Adams Smith, Esq: Secy of Legation from the united States of America.
& Col. De Bossett,
a gentlemen of literary & scientific character.

Ordered.

That they be admitted to the Royal Institution under Art: 15 Chap VI of the Bye laws, and that the Secretary do transmit them the usual letter.

5 In pursuance of a nomination under the Act by Patrick Craufurd Bruce, Esq. his Son

Life Subs.er

Michael Bruce, Esq: of upper Grosvenor Street was admitted to the privileges of a Life Subscriber to the Royal Institution.

6. Read the minutes of the last General Monthly Meeting.

7. The following lady having been duly recommended was admitted as a Subscriber to the Lectures for the year 1818.

Subs.er

Mrs Langford Brooke of No: 22 Stratford Place.

8 Notice having been sent to the Managers agreeably to Art: 3 of Chap: XII of the Bye laws,
"At this Meeting the Seal of the Corporation

May 18.

Bond to Mr Fuller

was affixed to a Deed whereby the Members of the Royal Institution covenanted to pay to John Fuller his Executors Admors or Assigns the sum of £1000 in an event specified therein

It was also executed by Scrope Bernard Morland Esq: Treasurer and Thomas Harrison Esq: Secretary, and it was Countersigned by

Spencer – President

Bessborough
Charles Hatchett
A. Kyd
J. P. Auriol
Thomas Snodgrass
Dan: Moore
James Stodart

} A majority of the Managers present

Thomas Harrison Secretary.

Spencer.

1818.

May 25.

At a Meeting of the Managers of the Royal Institution of Great Britain held at the house of the Institution on the 25th of May, 1818.

Present
Earl Spencer, K.G. President, in the Chair.
Charles Hatchett, Esq.r V.P.
J. P. Auriol, Esq.r V.P.
Rt Hon: Sir John Nichell V.P.
Thomas Snodgrass, Esq:
Daniel Moore, Esq.r
Samuel Solly, Esq:

1. The minutes of the proceedings of the last meeting were read and approved.

2. Read and approved the minutes of the Committee of Accounts of this day

3. Read and approved the minutes of the last General Meeting.

Spencer.

June 1.

At a Meeting of the Managers of the Royal Institution of Great Britain, held at the house of the Institution, on the 1st of June, 1818.

Present
Earl Spencer, K.G. President, in the Chair
Charles Hatchett, Esq.r V.P.
Earl of Bessborough, V.P.
J. P. Auriol, Esq.r V.P.
Rt Hon: Sir John Nichell, M.P. V.P.
Thomas Snodgrass, Esq.r
James Stodart, Esq.
Daniel Moore, Esq.
S. Bernard Morland, Esq.r M.P. Treasurer.
Samuel Solly, Esq:
Thomas Harrison, Esq: Secretary.

1. The minutes of the proceedings of the last meeting were read and approved.

2. Read and approved the minutes of the Committee of Accounts of this day, and signed an Order for payment of the following sums, viz.t

Payments.

W. T. Brande Esq.r 1 year's Salary as Professor of Chem.y to May 1818	£200 —
J. Fincher on acc.t of the Royal Institution	80 —
	£280 —

on Loan by Mr Fuller.

3. Daniel Moore, Esq.r having stated to the Board that Mr Ainge the Solicitor to Mr Fuller had applied to him for the Declaration of Trust executed at the Board of Managers on the 18th of May last.

Resolved.

That the same be delivered to Mr Moore and that he take a receipt from Mr Ainge for the same, and the said Deed was delivered to Mr Moore accordingly.

Adjourned to Monday the 15th of June.

Charles Hatchett V.P.

216

June 15.

At a Meeting of the Managers of the Royal Institution of Great Britain held at the house of the Institution on the 15th of June, 1818. — 1818.

Present
Charles Hatchett, Esqr. V.P. in the Chair.
Major Genl. Kyd.
Samuel Solly, Esqr.
Daniel Moore, Esqr.
Thomas Snodgrass, Esqr.
J. P. Auriol, Esqr. V.P.
The Marquis of Lansdowne V.P.
Robert Ward, Esqr.
James Stodart, Esqr.
Thomas Harrison, Esqr. Secretary.

1. The minutes of the proceedings of the last meeting were read and approved.

2. Read and approved the minutes of the Committee of Accounts of this day, and signed an Order for payment of the following sums vizt.

Payments.

John Millington Esqr. for Lectures in 1818.	£105. —
Ditto expences for do.	5. 5. —
Joseph Fincher on account of the Royal Institution	50. —
	£160. 5. —

Presents received.

3 Mr. W. Holmes presented,
The Hours, a Poem by Henry Hudson Esqr.
Panthea; a Tragedy by W. Bennett, Esqr.
Dr. Geo: Man Burrows, presented
Strictures on the uses and defects of Parish Registers, and Bills of Mortality.
Report of the Committee of the London Infirmary for curing Diseases of the Eye,

217

June 15.

Sir W. Adams presented,
Reply to the Report of the London Infirmary.
Ordered.
That the thanks of the Managers be returned for these presents.

4 Read the minutes of the last General Monthly Meeting.

5. Read the minutes of the Committee of Visitors entered in their proceedings of the 8th of June.

6. The following persons having been duly recommended were admitted as Annual Subscribers to the Royal Institution for the year 1818.

Annl. Subscrs.
Archd. N. Armstrong Esqr. 8 Queen Sq. Westr.
George Ormerod, Esqr. 31 Welbeck St.

7. Adjourned to Monday the 6th of July at half past One o'clock.

Bessborough V.P.

218

July 6.

At a Meeting of the Managers of the Royal Institution of Great Britain held at the house of the Institution on the 6th of July, 1818. — 1818.

Present
Earl of Bessborough, V.P. in the Chair.
Charles Hatchett, Esqr. V.P.
James Peter Auriol, Esqr. V.P.
Thomas Snodgrass, Esqr.
Major Genl. Kyd.
James Stodart, Esqr.
Samuel Solly, Esqr.

1. The minutes of the proceedings of the last meeting were read and approved.

2. Read and approved the minutes of the Committee of Accounts of this day, and signed an Order for payment of the following sums, vizt.

Payments.

Reeve for ¼ year's Assessed taxes to Lady day last	£39. 12. 5
Chappell for do. Parochial rates to ditto	14. 3. 4
W. Harris Librarian ¼ yr's Salary to 31st. March last	45. —
Joseph Fincher on account of the Royal Institution	200. —
	£298. 15. 9

Presents received.

3. John Ayrton Paris, M.D. M.R.I. presented
Transactions of the Royal Geological Society of Cornwall, vol 1.
Sir Humphry Davy, presented
"On the safety Lamp for Coal Mines."
The Academy of Sciences of Philadelphia, through Mr. Wm. Allen, presented
Journal of the Academy, 6 numbers.
Ordered. That the thanks of the Managers be returned for these presents.

219

July 6.

Employment of W. Greenwood under Mr. Faraday.
4. Resolved. That William Greenwood be employed under the direction of Mr. Faraday to put the Laboratory in order, and that a remuneration of £5 be paid him for his services.

Holydays to Librns.
5. Mr. Harris and Mr. Stunt applied to the Board for leave of absence for one month each, during the recess.
Resolved. That their requests be granted, provided they be not both absent from the Library at the same time.

Map of London.
6. Ordered. That the large Map of London be put up in the Newspaper Room by Mr. Royce at an expence not exceeding £3 —

7. The following person having been duly recommended was admitted as a Life Subscriber to the Royal Institution.

Life Subscr.
Mr. John Prosser junr. of No. 9 Charing Cross

Adjourned to Monday the 2d of November next.

J P Auriol

220

Nov. 2 — 1818.

At a Meeting of the Managers of the Royal Institution of Great Brittain held at the house of the Institution on the 2d. of November, 1818.—

Present

James Peter Auriol, Esqr. V.P. in the Chair.
Thomas Snodgrass, Esqr.
Sir S. Bernard Morland, Bart. M.P. Treasurer
Major Genl. Kyd.
J. F. Daniell, Esqr.
James Stodart, Esqr.
Daniel Moore, Esqr.

1. The minutes of the proceedings of the last meeting were read and approved.

2. Read and approved the minutes of the Committee of Accounts of this day, and signed an Order for payment of the following sums vizt.

Payments

Messrs. Hayward & Furney for Hempen Carpeting for the Liby.	£24 . 3 . 2
Joseph Fincher on account of the Royal Institution —	60 . – . –
	84 . 3 . 2

Presents received.

3. His Majesty's Commissioners of the Public Records presented.

Foedera, vol. 2 part 1.
Placita de Quo Warranto temporibus Edw: I. II. III
Rotuli Hundredorum temp: Hen: III. & Edw: I vol. 2.

The Horticultural Society — vol. 2 part 7 — & vol. 3 part 1. of their Transactions.

Mr. Nodes Dickinson of the Royal College of Surgeons presented — "Remarks on Burns & Scalds." 8vo.

John Thomson, Esqr. — presented —
Observations Introductory to a work on English Etymology

Sir Richard Phillips, presented
Essays on the proximate Mechanical Causes of the general Phenomena of the Universe.

221

Nov. 2.

Lt. Genl. Wm. Thornton M.R.I. presented
His Speech in the House of Commons on the 7th. of May 1818 on his Motion to repeal the Declarations against the belief of Transubstantiation &c. 8vo.

Ordered. That the thanks of the Managers be returned for these presents.

Anonymous.
Studies of the Historic Muse, 4to.
Letters of Yorick in favour of the Established Church 3 parts, 8vo. —

Hon. Admission

4. On the application of The Rt. Honble Sir John Sinclair Bt. Baron de Stierneld, Envoy Extraordinary & Minister Plenipotentiary to His Majesty the King of Sweden, was admitted to the Royal Institution under Art 15 Chap: VI of the Bye laws. —

5. Read the minutes of the last General Monthly Meeting.

6. Read the minutes of the Committee of Visitors of the 12th. October last.

Oil.

7. Resolved. That Messrs. Gardner of the Strand, — who formerly supplied this Institution with Oil, be allowed to serve the Royal Institution with that Article instead of Messrs. Upton & Co.

Asst. Porter.

8. Ordered. That Richard Hall be appointed Assistant Porter at £1 pr. week.

Annl. Subsrs.

9. The following persons having been duly recommended were admitted as Annual Subscribers to the Royal Institution for the year 1819.

William George Bicknell, Esqr. 11 Queen St. May Fair
Andrew Trevor, Esqr. 50 upper Charlotte St. Fitzroy Sqr.
Charles Assey, Esqr. 57 Berners Street
Hugh Barclay Esqr. 46 Leicester Square.
Wm. Miller, Esqr. 2 Duchess St. Portland Place
Captn. Harris, Royal Engineers

222

1818.

Alexander Read, Esqr. 2 Dorset Square
George Arbuthnot, Esqr. 14 Wimpole Street
C. A. Mackenzie, Esqr. 55 Pall Mall.

Subscribers to the Lectures.

Mrs. & Miss Dashwood 41 Grosvenor Place.

Adjourned to Monday the 30th. of November.

A. Kyd

228

Nov. 30.

At a Meeting of the Managers of the Royal Institution of Great Britain, held at the house of the Institution on the 30th. of November, 1818.—

Present

Major General Kyd, in the Chair.
Thomas Snodgrass, Esqr.
James Stodart, Esqr.
Rt. Hon: Sir John Nicholl Bart. M.P. V.P.
Sir S. Bernard Morland, Bart. M.P. Treasurer
Thomas Harrison, Esqr. Secretary.

1. The minutes of the proceedings of the last meeting were read and approved.

2. Read and approved the minutes of the Committee of Accounts of this day, and signed an Order for an Advance to the Clerk of £50. —

Lectures

3. Read a letter from Mr. M. Ferstandiz offering to read a Course of Lectures on the Hebrew Language in this Institution.

Resolved. That the same be declined.

Presents received.

4. The Revd. Dr. Abauzit presented;
La Sainte Bible par les Pasteurs de l'Eglise de Geneve, 3 tom. en 2. 8vo.

Gibbes Walker Jordan, Esqr. M.R.I. presented
Report of a Select Committee of the House of Assembly in Barbados on the late Insurrection. 2 copies.

Ordered.
That the thanks of the Managers be returned for these presents.

5. Read the minutes of the last General Monthly Meeting.

6 Ordered. That the following periodical publications be taken in.

1818.

Periodicals ordered. The Antijacobin Review. — published Monthly
The new Monthly Magazine — d.o —
The Literary Panorama — d.o —
– and – The British Review. — Quarterly –

(Death of Sir T. Bernard.) 7. **Ordered**. That the House List for a Manager in the room of the late Sir Thomas Bernard Bart. be filled up as follows.

George Duckett, Esq.r F.R.S & F.S.A.

8. The following persons having been duly recommended were admitted as Annual Subscribers to the Royal Institution for the year 1819.

Rev.d E. A Broome – Brompton –
Rev.d Dr. Jennings 11 Glocester Place
Rev.d John Palmer Boteler, 7 Charles St.
Francis Brown, Esq.r 15 Old Cavendish St.
Rich.d Crozier Sherwood Esq.r 4 Bolton St.
and as a Life Subscriber to the Royal Institution.
W. T. Rose, Esq.r – Dean St. May Fair.

JP Auriol

Dec. 7.

At a Meeting of the Managers of the Royal Institution of Great Britain held at the house of the Institution on the 7th of December 1818. –

Present

James Peter Auriol, Esq.r V.P. in the Chair.
Daniel Moore, Esq.r
Major Gen.l Kyd.
Rob.t Ward, Esq.r
Thomas Snodgrass, Esq.r
James Stodart, Esq.r
Sir S. Bernard Morland, Bart. M.P. Treasurer
Thomas Harrison, Esq.r Secretary.
Samuel Solly, Esq.r
R.t Hon.ble Sir John Nicholl, M.P. V.P.

1. The minutes of the proceedings of the last meeting were read and approved.

2. Read and approved the minutes of the Committee of Accounts of this day, and signed an Order for an advance to the Clerk of £50 and also a further Advance of £50 on the 1st of January next

3. Mr. Brande, Superintendant of the Laboratory, having stated to the Board that assistance would be wanted in the Laboratory during the ensuing Season of Lectures, (Engagement of W. Greenwood.)
Resolved. That William Greenwood be employed at a Salary of 15s ℘ week to commence from the present time.

(Sub Comm.ee resp.g lighting R.I. with Gas.) 4. **Resolved**. That the following Sub-Committee be appointed to take into consideration the expediency of lighting the Institution with Gas instead of Oil
– William Tho.s Brande, Esq.r
– Samuel Solly, Esq.r
Rob.t Ward, Esq.r
James Stodart, Esq.r
Daniel Moore, Esq.r

Adjourned to 25th Jan.y 1819 Charles Hatchett V.P.

Jan. 25.

1819.

At a Meeting of the Managers of the Royal Institution of Great Britain held at the house of the Institution on the 25th of January, 1819,

Present

Charles Hatchett, Esq.r V.P. in the Chair.
J. F. Daniell, Esq.r
Daniel Moore, Esq.r
R.t Hon.ble Sir John Nicholl, M.P. V.P.
Earl of Bessborough V.P.
George Duckett, Esq.r
Sir S. Bernard Morland, Bart. M.P. Treasurer
James Stodart, Esq.r
Samuel Solly, Esq.r
Thomas Snodgrass, Esq.r

1. The minutes of the proceedings of the last meeting were read and approved.

2. Read and approved the minutes of the Committee of Accounts of this day, and signed an Order for payment of the following Sums, viz.t

Payments.

Mess.rs Upton & C.o for Oil in 1817	£ 80. 4. –
Ditto . . . in 1818	74. 16. –
Mr. Brooks Collector for 1 years Salary to 31st Dec.r last	31. 10. –
Joseph Fincher on account of the Royal Institution	50. –
	£ 236. 10. –

Lectures. 3. Read a letter from Dr. Bathie, addressed to Professor Brande, on the subject of delivering a Course of Lectures at this Institution on the state of Morals, Religion, and general Improvement.
Resolved.
That the thanks of the Managers be given to Dr. Bathie for his obliging proposal, but that as the Lectures for this Season have been arranged they beg leave to decline his offer.

Jan 25.

(Print.g Extracts from American MSS. ~~gran~~ permitted.) 4. Read a letter from E. A. Kendall, Esq.r Ann.l Subs.r to the Royal Institution, requesting permission to extract for printing in the Colonial Journal of which he is the Proprietor the whole or any part of the M.S. official Papers relating to the American War, contained in numerous volumes in the Library of the Royal Institution.
Resolved. That Mr. Kendall's request be granted.

(Life Subs.n) 5. In pursuance of a nomination under the Act by Robert Barclay, Esq.r –
his Grandson, Arthur Kett Barclay, Esq.r of Clapham was admitted to the privileges of a Life Subscriber to the Royal Institution.

6. Read the minutes of the last General Monthly Meeting.

(Stove in Lab.y) 7. Read the minutes of the Committee of Visitors entered in their proceedings of the 13th Instant
Resolved.
That the Stove in the Laboratory be removed under the direction of Mr. Brande, and that the other recommendations of the Visitors be taken into consideration at a future Board.

Presents received:

8. Dr. Marshall Hall – of Nottingham, presented,
On the Mimoses; a descriptive & practical Essay
William Cowper, Esq.r – presented
An English Hindoo, & Mahomedan Almanack for 1784
Mr. Robert Owen – presented
Tracts relative to the new view of Society.
The Rev.d Stephen Weston presented
Enchiridion Romæ: or Remarks on the Buildings &c. of Rome
Sir Gilbert Blane Bart. presented Elements of Medical Logick,
Le Chevalier de Sade, presented Le Correspondant, 5 tom.s 8.vo
L'Academie des Sciences de Dijon, presented
– Séance Publique tenue le Samedie 4 Juillet, 1818.
Ordered. – That the thanks of the Managers be returned for these presents.

228

1819. Jan 25

9 The following persons having been duly recommended were admitted as Subscribers to the Royal Institution for the year 1819.

To the Lectures

Mrs. H. J. Barchard – Wandsworth
Mrs. Chevalier
Miss Chevalier } 20 South Audley St.
Miss Frances Chevalier
Miss Mary Chevalier
Mrs. Manning } 14 New St. Spring Gardens
Miss Caroline Manning
Mrs. Gall
Miss Ross } 103 Gloucester Place
Miss Marian Ross
Mrs. Wild } 2 Gloucester Place
Miss Wilks
Mrs. Merriman } 133 New Bond St.
Miss Birchall
Mrs. Lee – Hampstead
Miss Innes } Gloucester Place
Miss Morse

Annual Subscribers.

Henry Hawley, Esqr. 41 Lower Cadogan Place
Danl. Mesman, Esqr. Trevor Place Knightsbridge
Lewis Wyatt, Esqr. Albany.
Wm. Mills, Esqr. 60 Upper Berkeley St.
Thos. Bent, Esqr. 21 Sackville St.
George Rose, Esqr. 25 Nottingham Place
Duncan Munro, Esqr. – do. –
John Bent, Esqr. – 21 Sackville St.
Wm. Grant, Esqr. 24 Manchester Square.
Chas. Jno. Manning, Esqr. New St. Spring Gardens
Major Gall, 103 Gloucester Place.
Richd. Powlett Wrighte Benyon, Esqr. —
Lt. Col. F. Browne, 63 Frith St.
— Deaken, Esqr. Fenchurch St.
Francis Walpole, Esqr. Half Moon St.
Captn. Wilson, R.N. – 10 Knightsbridge
Dr. Hugh Lee, 62 South Audley St.

229

Annual Subscribers, continued

Captn. Wm. Stopford, 63 Welbeck St.
M. Fitzgerald, Esqr. 3 Percy St. Bedford Sqr.

Bessborough V.P.

230

Feb. 1. 1819.

At a Meeting of the Managers of the Royal Institution of Great Britain held at the house of the Institution on the 1st. of February, 1819.

Present

Earl of Bessborough, V.P. in the Chair.
James Peter Auriol, Esqr. V.P.
Charles Hatchett, Esqr. V.P.
J. F. Daniell, Esqr.
Sir S. Bernard Morland, Bart. M.P. Treas.
Thomas Snodgrass, Esqr.
Robert Ward, Esqr.
George Duckett, Esqr.
Daniel Moore, Esqr.
Thomas Harrison, Esqr. Secretary.

1. The minutes of the proceedings of the last meeting were read and approved.

2. Read and approved the minutes of the Committee of Accounts of this day, and signed an Order for payment of the following sums, vizt.

Payments.

Jones & Yarrell for Newspapers in 1817.	£ 98 . 17 . 4
Feetham — for Ironmongery – do. —	55 . 1 . 5
Bulmer — for Printing — do. —	47 . 2 . 6
Reeve for 3/4 yrs Assessed taxes to Xmas last	118 . 13 . 9
Chappell for do. – Parochial rates to do.	50 . – –
William Harris 3/4 yrs Salary to 31 Decr.	135 . – –
Thos. Slatter Lecture room door keeper 1 years Salary to 31st. Decr. —	10 . 10 . –
	515 . 5 –

3. Audited the Treasurer's Account for the year 1818.

4. The following persons having been duly recommended were admitted as Subscribers to the Royal Institution for the year 1819.

231

Feb. 1.

Annual Subscribers.

Charles Collins Blane, Esqr. 4 Cleveland row
Francis Lunn, Esqr. – Clapham Common
George Brown, Esqr. 19 Baker St.
Captn. Clive, 37 Sackville Street.
George Pocock, Esqr. 39 Charles St. Berkeley Square
Henry Wilkinson, Esqr. Clapham Common
Wm. Bentinck Lethem Hawkins, Esqr. Gt. Marlboro' St.
James Lawson, Esqr. Lawn House Hanwell Middx.
John William Birch, Esqr. 5 Kings Bench Walk Temple
William Carlyon, Esqr. 47 Lamb's Conduit St.
William Heberden, Esqr. Upper Brook St.
Dr. Bostock, Coram St. Brunswick Square
H. J. Brooke, Esqr. Keppell St. Russell Square

To the Lectures.

Miss C. Blake – Portland Place
Miss de Ponthieu – 8 Stratton St.
Mrs. Briggs 30 Edgeware Road
Miss Bentley 21 Tavistock Square
Mrs. Arnold } 10 Chandos St. Portland place
Miss C. Arnold

Spencer

Feb. 8.

At a Meeting of the Managers of the Royal Institution of Great Britain, held at the house of the Institution, on the 8th of February, 1819.

1819.

Present
Earl Spencer, President, in the Chair
Charles Hatchett, Esqr. V.P.
George Duckett, Esqr.
J. F. Daniell, Esqr.
J. P. Auriol, Esqr. V.P.
James Stodart, Esqr.
Thomas Snodgrass, Esqr.
Daniel Moore, Esqr.
Thomas Harrison, Esqr. Secretary.

1. The minutes of the proceedings of the last meeting were read and approved.

2. Read and approved the minutes of the Committee of Accounts of this day, & signed an Order for payment of the following sums, vizt.

Payments

Newman for Chemicals for the year 1818	£168 . 18 . 10
John Deckins for Coals - do.	109 . - . -
Joseph Fincher on account of the Royal Institution	50 . - . -
	327 . 18 . 10

Life Subsr.

3. In pursuance of a nomination under the Act by William Wynch, Esqr. his Son The Revd. Peter B Wynch of Grosvenor Place, was admitted to the privileges of a Life Subscriber to the Royal Institution.

Hon. Admission

4. On the application of Charles Hatchett, Esqr. The Prince Wolkonsky was admitted to the Royal Institution under Art. 15 Chap VI of the Bye laws

Feb. 8.

5. Read the minutes of the last General Monthly Meeting.

6. The following persons having been duly recommended were admitted as Subscribers to the Royal Institution for the year 1819.

Annual Subscribers.
Matthew Munro, Esqr. 1 Lower Seymour Street
Chas. Jno. Clavering, Esqr. 22 Queen St. May Fair
John Round, Esqr. 25 Albemarle Street
Revd. Francis Thackeray 4 New Norfolk Street
Edward Deas Thomson, Esqr. Somerset Place

To the Lectures.
Miss Munro — No. 1 Lower Seymour St.
Miss Perry — 21 Tavistock Square
Miss Caroline Bathurst, 45 Wimpole St.
Miss Lewis 11 Powis Place
Mrs. Stirling } Lawn House Hanwell Middx
Miss Pope }
Miss Grace Hawkins Gt. Marlborough Street
Miss Bond } 13 Orchard Street
Miss Travers }

Spencer

Feb. 15.

At a Meeting of the Managers of the Royal Institution of Great Britain held at the house of the Institution on the 15th of February, 1819.

1819.

Present
Earl Spencer, K. G. in the Chair.
Thomas Snodgrass, Esqr.
John Frederick Daniell, Esqr.
Rt. Honble Sir John Nicholl, M.P. V.P.
Samuel Solly, Esqr.
James Peter Auriol, Esqr. V.P.
Charles Hatchett, Esqr. V.P.
George Duckett, Esqr.
Earl of Bessborough V.P.
James Stodart, Esqr.
Robert Ward, Esqr.
Sir S Bernard Morland, Bart M.P. Treasurer
Daniel Moore, Esqr.
Thomas Harrison, Esqr. Secretary.

1. The minutes of the proceedings of the last meeting were read and confirmed.

2. Read and approved the minutes of the Committee of Accounts of this day, and signed an Order for an Advance to the Clerk of £140. —

Presents received.

Present by Sir J. Banks.

3. Rt. Honble Sir Joseph Banks & Lady Banks presented 202 Volumes of valuable Books to the Library

Resolved.
That the special thanks of the Managers be returned to them by the Secretary for this their munificent present, and that Daniel Moore, Esqr. be requested personally to deliver this mark of the Managers' respect and gratitude.

Samuel Solly, Esqr. presented
Becker, Augusteum: ou Description des Monumens antiques qui se trouvent à Dresde, tome 3, folio.

Feb. 15.

The Honble Francis Henry Egerton, presented A Letter to the Parisians upon the Duke of Bridgewater's Navigation.

Ordered. That the thanks of the Managers be returned for these presents.

The following persons having been duly recommended were admitted as Subscribers to the Royal Institution for the year 1819.

Annual Subscribers.
Major Bristow No. 2 Windham Place Montagu Sq.
Samuel Good, Esqr.
Jeffries Allen, Esqr. Temple
James Henry Deacon, Esqr. 19 James Street
J. Irvine, Esqr. 7 Vere Street Oxford St.
John Nelson, Esqr. Somerset Place
Robert Selby junr. Esqr. Bryanston Street
Mr. Sharpe 44 Weymouth St.
Sir Chas. Mordaunt, Bart M.P. 9 Manchester Sq.

To the Lectures
Miss Walker } Wandsworth
Miss A. M Walker }
Lady Mordaunt } Manchester Square
The Honble Mrs. Eliot }
Mrs. Galiagan 13 Upper Grosvenor St.

Spencer

236

Feb. 22.

1819.

At a Meeting of the **Managers** of the Royal Institution of Great Britain held at the house of the **Institution** on the 22.d of February, 1819.

Present

Earl Spencer, K.G. President in the Chair
Charles Hatchett, Esq.r V.P.
Daniel Moore, Esq.r
J. P. Auriol, Esq.r V.P.
Thomas Snodgrass, Esq.r
Sir S. Bernard Morland, Bart. M.P. Treasurer
Thomas Harrison, Esq.r Secretary.

1. The minutes of the last meeting were read and approved.

2. Read and approved the minutes of the Committee of Accounts of this day; and signed an Order for an Advance to the Clerk of £130.

Presents received.

3. The Horticultural Society, presented Vol. 3 part 2 of their Transactions.
Ordered.
That the thanks of the Managers be returned for this present.

4. The following persons having been duly recommended were admitted as Subscribers to the Royal Institution for the year 1819.
Annual Subscribers.
John Jackson Esq.r 27 Chapel Street
J. R. Bowen, Esq.r – 16 Lower Seymour St.
Nath.l Barton, Esq.r 54 Baker Street
To the Lectures.
Baroness de Rolle 48 Upper Grosvenor St.
Mrs Jennings, 11 Gloucester Place

Bessborough V. P.

237

Mar. 1.

At a Meeting of the **Managers** of the Royal Institution of Great Britain held at the house of the Institution on the 1st of March, 1819.

Present

Earl of Bessborough V.P. in the Chair.
Cha.s Hatchett, Esq.r V.P.
J. F. Daniell, Esq.r
Thomas Snodgrass, Esq.r
Daniel Moore, Esq.r
James Stodart, Esq.r
Sir S. Bernard Morland, Bart. M.P. Treasurer
George Duckett, Esq.r

1. The minutes of the proceedings of the last meeting were read and approved.

2. Read and approved the minutes of the Committee of Accounts of this day; and signed an Order for an Advance to the Clerk of £50.

Present received.

3. Mr. E. Dowson, presented,
The Youth's explanatory Theological Dictionary.
Ordered.
That the thanks of the Managers be returned for this present.

Life Subs.cr

4. In pursuance of a nomination under the Act by the Earl of Breadalbane,
his Son Viscount Glenorchy of Park Lane,
was admitted to the privileges of a Life Subscriber to the Royal Institution.

Ann.l Subs.cr

5. The following persons having been duly recommended were admitted as Annual Subscribers to the R.I.
Joseph Marryat, Esq.r of the Albany.
Jn.o Fullarton Elphinstone, Esq.r 2 upper Harley St.
James Gellaway, Esq.r 34 Gower St.

Bessborough V. P.

238

Mar. 8.

1819.

At a Meeting of the **Managers** of the Royal Institution of Great Britain, held at the house of the Institution on the 8th of March, 1819.

Present

Earl of Bessborough, V.P. in the Chair.
James Peter Auriol, Esq.r V.P.
Cha.s Hatchett, Esq.r V.P.
Tho.s Snodgrass, Esq.r
J. F. Daniell, Esq.r
James Stodart, Esq.r
Tho.s Harrison, Esq.r Secretary.

1. The minutes of the proceedings of the last meeting were read and approved.

2. Read and approved the minutes of the Committee of Accounts of this day, and signed an Order for an Advance to the Clerk of £120.

3. Read the minutes of the last General Monthly Meeting.

Hon. Admissions to ~~the~~ Person conn.d with North.n Expedition.

4 In consequence of the Resolution of the General Meeting.
Resolved. That tickets of admission to the Lectures of this Institution be sent to the following Gentlemen about to sail on the Northern Expedition.
Lieut.t Parry
Lieut.t Liddon
Lieut.t Beechey
Lieut.t Hoppner
Capt.n Sabine R.A. (Astronomer &c.)
Mr. Edwards Surgeon
Mr. Fisher } Assist.t Surgeons
Mr. Beverley }
Mr. Hooper – Purser.
Mr. Ross – Midshipman.

239

Mar. 8.

Ann.l Subs.crs

5. The following persons having been duly recommended were admitted as Annual Subscribers to the Royal Institution for the year 1819.
Henry Gordon, Esq.r 15 Gower St.
Jos.h Barretto, jun.r Esq.r 15 Albemarle St.
Cha.s Bevan, Esq.r 10 Wimpole St.
Cha.s Grant, Esq.r 7 Henrietta St. Cavendish Sq.
Rev.d John Wynne, 45 Paternoster row
Mr. J. H. Maccan, 125 Jermyn Street

Charles Hatchett V P

Mar. 15.

1819.

At a Meeting of the **Managers** of the Royal Institution of Great Britain held at the house of the Institution on the 15th of March, 1819.

Present

Charles Hatchett, Esqr. V.P. in the Chair.
George Duckett, Esqr.
Daniel Moore, Esqr.
Samuel Solly, Esqr.
Sir S. Bernard Morland, Bart. M.P. Treasurer
Thomas Snodgrass, Esqr.
John Fredk. Daniell, Esqr.
James Stodart, Esqr.
James P. Auriol, Esqr. V.P.
Thomas Harrison, Esqr. Secretary.

1. The minutes of the proceedings of the last meeting were read and approved.

2. Read and approved the minutes of the Committee of Accounts of this day, and signed an Order for payment of the following sums, viz.

Payments.

Joseph Fincher, on account of the Royal Institution	£150. –
Library of the Royal Instn. Subsn. of Henry Hallam, Esq Member	5 - 5 –
	155 . 5 ..

Presents received.

3. John Fuller, Esqr. M.R.I. presented, Views in Sussex, Drawn by Turner & Engraved by Cooke No. 1. Imperial Paper.
The Rt. Honble John Trevor, presented Dr. Morton's Table of Alphabets

Ordered

That the thanks of the Managers be returned for these presents.

4. The following person having been duly recommended was admitted as an Annual

Mar. 15

Subscriber to the Royal Institution for the year 1819.

Annl. Subsr.

Stephen Tempest, junr. Esqr. 20 Duke St. Piccady

Spencer

Mar. 22.

1819.

At a Meeting of the **Managers** of the Royal Institution of Great Britain held at the house of the Institution, on the 22d of March, 1819.

Present

Earl Spencer, K.G. President, in the Chair
James Peter Auriol, Esqr. V.P.
Samuel Solly, Esqr.
J. F. Daniell, Esqr.
Chas. Hatchett, Esqr. V.P.
Thomas Snodgrass, Esqr.
James Stodart, Esqr.
Sir S. Bernard Morland, Bart M.P. Treasurer
Daniel Moore, Esqr.
Thomas Harrison, Esqr. Secretary.

1. The minutes of the proceedings of the last meeting were read and approved.

2. Read and approved the minutes of the Committee of Accounts of this day and signed an Order for an Advance to the Clerk of £120.

Presents received.

3. Dr. H. Salemi presented from the Abate Francesco Ferrara, Descrizione dell' Etna con la Storia delle Eruzioni e il Catalogo del Prodotti, 8vo.

Ordered.

That the thanks of the Managers be returned for this present

Window

4. Mr. Brande having stated a request from Mr. Robinson of No. 20 Albemarle Street respecting a Window belonging to the Royal Institution.

Ordered.

That Sir S. Bernard Morland, Bart Mr. Brande and Mr. Harrison be a Committee to inspect the Lease of the Institution, and the proposed alteration and report thereon to the Board.

Mar. 22.

5. The following persons having been duly recommended were admitted as Subscribers to the Royal Institution for the year, 1819.

Annual Subscribers.

Captn. Blake 30 East St. Red Lion Square
Henry Salemi M.D. 1 Sherrard Street, Golden Sqe.

To the Lectures.

Miss Raikes – 3 Portland Place

Spencer

1819.

Mar. 29.

At a Meeting of the Managers of the Royal Institution of Great Britain held at the house of the Institution on the 29th of March, 1819.

Present

Earl Spencer, K.G. President, in the Chair
Charles Hatchett, Esqr. V.P.
James Peter Auriol, Esqr. V.P.
Thomas Snodgrass, Esqr.
J. F. Daniell, Esqr.
Rt. Honble. Sir John Nicholl, M.P. V.P.
Sir S. Bernard Morland, Bart: M.P. Treasr.
James Stodart, Esqr.
Samuel Solly, Esqr.
Daniel Moore, Esqr.
George Duckett, Esqr.

1. The minutes of the proceedings of the last meeting were read and approved.

2. Read and approved the minutes of the Committee of Accounts of this day.

3. The following persons having been duly recommended were admitted as Annual Subscribers to the Royal Institution for the year, 1819

Annl. Subsrs.
Frederick Croft, Esqr. 15 Lower Brook St.
Thomas Goding jun: Esqr. Knightsbridge.

Spencer

April 5.

At a Meeting of the Managers of the Royal Institution of Great Britain, held at the house of the Institution on the 5th of April, 1819.

Present

Earl Spencer, K.G. President in the Chair
Charles Hatchett, Esqr. V.P.
Jas. Peter Auriol, Esqr. V.P.
J. F. Daniell, Esqr.
Robert Ward, Esqr.
Samuel Solly, Esqr.
Thos. Snodgrass, Esqr.
Sir S. Bernard Morland, Bart: M.P. Treasurer
James Stodart, Esqr.
Daniel Moore, Esqr.
Thos. Harrison, Esqr. Secretary.

1. The minutes of the proceedings of the last meeting were read and approved.

2. Read and approved the minutes of the Committee of Accounts of this day, and signed an Order for an Advance to the Clerk of £66.

Machine lent.

3. Read a request from R. Woodhouse, Esqr. one of the Commissioners of the Board of Longitude, through Mr. Gibbes Walker Jordan, to be allowed to borrow for a short time Smeaton's Machine for measuring the momentum of bodies in motion, Mr. Woodhouse insuring its safe return to the Institution.
Resolved – That the request of Mr. Woodhouse be granted on the above conditions.

Window.

4. The Committee appointed by the Board of the 22d. March viewed the Premises where Mr. Robinson desired leave to stop up a window belonging to the Royal Institution, they are of opinion that it would not be safe to permit the same to be done without special leave from the Corporation of London under whom the

1819.

House of the Institution is held.
Resolved.
That Daniel Moore be requested to apply to the City Solicitor on this subject.

5. The following persons having been duly recommended were admitted as Subscribers to the Royal Institution for the year 1819

Annual Subscribers.

Edward Jerningham, Esqr. 69 Gloucester Place.
Augustus Colfsold, Esqr. 68 Aldermanbury.

To the Lectures.

Mifs Clarke – 69 Gloucester Place

6. Adjourned to Monday the 19th. April.

JP Auriol

April 19.

At a Meeting of the Managers of the Royal Institution of Great Britain held at the house of the Institution on the 19th of April, 1819.

Present

James Peter Auriol, Esqr. V.P. in the Chair.
James Frederick Daniell, Esqr.
James Stodart, Esqr.
Thomas Snodgrass, Esqr.
Charles Hatchett, Esqr. V.P.
Robert Ward, Esqr.
Sir John Thos. Stanley, Bart.
Samuel Solly, Esqr.
Thomas Harrison, Esqr. Secretary.

1. The minutes of the proceedings of the last meeting were read and approved.

2. Read and approved the minutes of the Committee of Accounts of this day, and signed an Order for payment of the following sum, vizt.

Payment.
1 years Ground rent to the City of London to 25th. March last } £150.

Window.

3. Read a letter from Daniel Moore, Esqr. stating, that in consequence of the resolution of the last Board, he had consulted with Mr. Montagu, the City Surveyor, respecting the consent of that Corporation to close up a window belonging to the Royal Institution, and had been informed by him, that the Corporation will not interfere with any arrangements, that may be made by their Tenants for their own accomodation.
Resolved.
That Mr. Robinson be requested to attend the Committee in order to settling this busineſs.

Presents received.

4. The President & Council of the Royal Society, presented

1819.

Astronomical Observations made at Greenwich in the years 1803, to 1816 inclusive.
Philosophical Transactions, 1818 part 2.d
The American Academy, presented –
Memoirs of the American Academy vol 4 part 1.
Sir James Edward Smith, Pres. L. S. &c presented
A Defence of the Church and Universities of England.
John Vaillant, Esq.r M. A. presented
"Scripture compared with itself, in proof of the Doctrine of the Trinity."
Ordered. That the thanks of the Managers be returned for these presents.

5. Read the minutes of the last General Monthly Meeting.

6. The following persons having been duly recommended were admitted as Annual Subscribers to the Royal Institution for the year 1819.

Ann.l Subs.rs
Edward Segart, Esq.r 17 Charles St. S.t James's Sq.
Kenelm H.y Digby, Esq.r Trinity College, Cambridge.

7. Ordered. That the House Lists be filled up as follows.

President – Earl Spencer, K. G. F. R. S. & F. A. S.

Treasurer – Sir S. Bernard Morland, Bart M.P.

Secretary – Thomas Harrison, Esq.r F. R. S.

Managers.
L.t Col. Benj.n Ansley, F. S. A
James Peter Auriol, Esq.r F. R. S
John Chamier, Esq.r
Cha.s Madryll Cheere, Esq.r
Edmund Rob.t Daniell, Esq.r
George Duckett Esq.r F. R. S & F. S. A.
Earl of Hardwicke K. G. F. R. S & F. S. A
Charles Hatchett, Esq.r – F. R. S & F. S. A

April 19.

Daniel Moore, Esq.r F. R. S. F. S. A. & F. L. S
W.m Hasledine Pepys, Esq.r F. R. S
Louis Hayes Petit Esq.r F. R. S. F. S. A. & F. L. S.
Claude Scott, Esq.r
Thomas Snodgrass, Esq.r
Samuel Solly, Esq.r F. R. S
James Stodart, Esq.r

Visitors
Earl of Bridgewater, F. R. S. and F. S. A.
Earl of Brownlow F. R. S & F. S. A
Viscount Bulkeley
Henry Tho.s Colebrooke, Esq.r F. R. S
Earl of Darnley, F. R. S
John Fuller, Esq.r
Daniel Giles, Esq.r
Cha.s Holford, Esq.r
Sir John Cox Hippisley F. R. S. &c &c
Sir Benj.n Hobhouse, Bart. F. R. S. &c
Gibbes Walker Jordan, Esq.r A. M. F. R. S
John Soane, Esq.r F. S. A
Rich.d Horsman Solly, Esq.r F. R. S & F. S. A.
Sir George Staunton, Bart M. P. F. R. S & F. L. S
L.t Gen.l W.m Thornton.

8. Adjourned to the 3.d of May.

Spencer

May 3.

1819.

At a Meeting of the Managers of the Royal Institution of Great Britain, held at the house of the Institution on the 3.rd of May, 1819,

Present
Earl Spencer, K. G. President, in the Chair.
Edmund Rob.t Daniell, Esq.r
Louis Hayes Petit, Esq.r
Claude Scott, Esq.r
Charles Hatchett, Esq.r
Daniel Moore, Esq.r
George Duckett, Esq.r
Thomas Snodgrass, Esq.r
W.m Hasledine Pepys, Esq.r
James Stodart, Esq.r
James Peter Auriol, Esq.r
John Chamier, Esq.r
Cha.s Madryll Cheere, Esq.r
Samuel Solly, Esq.r
Sir S. Bernard Morland, Bart M.P. Treasurer
Thomas Harrison, Esq.r Secretary.

1. The minutes of the proceedings of the last meeting were read and approved.

2. Read and approved the minutes of the Committee of Accounts of this day; and signed an Order for an advance to the Clerk of £50.

Presents received.

Bramah's Press Presented
3. Mess.rs Timothy & Francis Bramah, presented through M.r Millington
The original Hydraulic Press invented & made by their late Father, M.r Joseph Bramah, being the first Machine that was constructed on a large and powerful scale.
Sir Christ.r Hawkins, Bart. presented, through Earl Spencer, Specimens of Gold which were found in streaming for Tin in the Parish of Ladock, County of Cornwall

May 3.

Ordered. That the special thanks of the Managers be returned for these valuable presents.

John Ring, Esq.r presented,
The Commemoration of Handel & other Poems;
M.r Brande, presented
A Manual of Chemistry – 8.vo
Capt.n Edward Sabine, presented,
Remarks on Capt.n Ross's Account of the late Voyage to Baffins' Bay.
Cha.s Wye Williams Esq presented
Exposure of the fallacies contained in the letter to R.t Honble R.t Peel, M.P.
Mons.r Hatchette, presented thro' M.r Millington,
Elements de Geometrie a trois dimensions, 8.vo
Second Supplement de la Geometrie descriptive, 4.to
Ordered,
That the thanks of the Managers be returned for these presents.

4. Read the minutes of the General Annual Meeting

Window.
5. The Committee appointed by the Board of the 22.d March last, reported that there appeared to them to be no objection to the substitution of a Skylight instead of a window as proposed by The Rev.d M.r Robinson.
Resolved.
That leave be granted to The Rev.d John Robinson to substitute a Skylight for the window which he proposed to block up, upon his undertaking to be at the whole expence of the alteration, and not to erect any Building which may overhang the Premises belonging to this Institution or darken the Skylight.

Hon. Admission
6. On the application of Earl Spencer,
Le Prof.eur Aldini
was admitted to the Royal Institution under Art: 15 Chap: VI of the Bye laws.

7. The President nominated as Vice Presidents for the

ensuing year,
James Peter Auriol, Esq.
George Duckett, Esq.
Earl of Hardwicke K.G.
Chas. Hatchett, Esq.
Claude Scott, Esq.

8. The following persons having been duly recommended were admitted as Annual Subscribers for the year 1819, (Annl. Subsers.)
Dr. Stephen Pellet, 49 Weymouth St.
Henry Cleaver, Esq. Fulham.

Spencer

May 10.

At a Meeting of the Managers of the Royal Institution of Great Britain held at the house of the Institution, on the 10th of May, 1819,

Present
Earl Spencer, K.G. President, in the Chair.
Samuel Solly, Esq.
Chas. Madryll Cheere, Esq.
Charles Hatchett, Esq. V.P.
John Chamier, Esq.
Sir S. Bernard Morland, Bt. M.P. Treasurer.
Thomas Snodgrass, Esq.
Daniel Moore, Esq.
James Stodart, Esq.
Lt. Col. Ansley.
James Peter Auriol, Esq. V.P.
George Duckett, Esq. V.P.
Thomas Harrison, Esq. – Secretary.

1. The minutes of the proceedings of the last meeting were read and approved.

2. Read and approved the minutes of the Committee of Accounts of this day.

Present received.

3. Mr. Chas. Bell, Surgeon – presented
An Essay on the circulation of the Blood.
Ordered. That the thanks of the Managers be returned for this present.

4. Read the minutes of the last General Monthly Meeting.

5. (Liby. Shelves) Ordered. That Mr. Royce be directed to fit up in the Library of Reference the Book-Shelves as mentioned in his estimate at an expence not exceeding £7.–

6. Resolved. That Mr. Moore & Mr. Solly be requested

forthwith to apply to Mr. Porden to give a Course of Lectures on Architecture at the Royal Institution during the present Season.

Spencer

May 17.

At a Meeting of the Managers of the Royal Institution of Great Britain held at the house of the Institution on the 17th of May, 1819,

Present
Earl Spencer, K.G. President, in the Chair.
Charles Hatchett, Esq. V.P.
James Peter Auriol, Esq. V.P.
George Duckett, Esq. V.P.
Thomas Snodgrass, Esq.
John Chamier, Esq.
James Stodart, Esq.
William Hasledine Pepys, Esq.
Lt. Col. Ansley.
Samuel Solly, Esq.
Sir S. Bernard Morland, Bart. M.P. Treasurer.
Claude Scott, Esq.
Chas. Madryll Cheere, Esq.
Louis Hayes Petit, Esq.
Daniel Moore, Esq.
Thomas Harrison, Esq. Secretary.

1. The minutes of the proceedings of the last meeting were read and confirmed.

2. Read and approved the minutes of the Committee of Accounts of this day, and signed an Order for an Advance to the Clerk of £50.–

3. (Present of Newspapers from Rev. C.P. Burney.) Read a letter from The Revd. Chas. Parr Burney, addressed to the President of the Royal Institution, stating that in case it was an offering worthy the acceptance of the Royal Institution he should be most happy in depositing there about 50 Volumes of Newspapers from the year 1767 to 1799 being Duplicates of his late Father's Collection.
Ordered.–That the special thanks of the Managers be returned to the Revd. Mr. Burney for the offer of his valuable present.

Presents received.

4. The Society for the Encouragement of Arts &c. presented,

1819.

The 36th Volume of the Society's Transactions.
Report of the Committee of the Society of Arts &c on preventing the forgery of Bank Notes.

Ordered.
That the thanks of the Managers be returned for these presents.

5. Mr Solly reported that Mr Porden would not be able to give a Course of Lectures on Architecture this Season.

6. The following persons having been duly recommended were admitted as Annual Subscribers to the Royal Institution for the year 1819.

Ann. Subsrs
Viscount Barrington 11 Hill St. –
Dr Merriman 26 Half Moon St.
T. Totton, Esqr 10 College St. Westmr
John Roberts, Esqr 60 Frith St. Soho.

To the Lectures.
Miss Stanley – Privy Gardens.

7. Adjourned to Monday the 7th of June.

Spencer

June 7.

At a Meeting of the Managers of the Royal Institution of Great Britain, held at the house of the Institution on the 7th of June, 1819.

Present
Earl Spencer K.G. President, in the Chair.
Charles Hatchett, Esqr V.P.
Claude Scott, Esqr V.P.
James Stodart, Esqr
Edmund Robt Daniell, Esqr
Sir S. Bernard Morland, Bart. M.P. Treas
Lt Col. Ansley.
Daniel Moore, Esqr
John Chamier, Esqr
Louis Hayes Petit, Esqr
James Peter Auriol, Esqr V.P.
Samuel Solly, Esqr –

1. The minutes of the proceedings of the last meeting were read and approved.

2. Read and approved the minutes of the Committee of Accounts of this day, and signed an Order for payment of the following sums, viz.

Payments.

Wm Thos Brande, Esqr 1 years Salary as Prof of Chemistry to May 1819	£200 —
John Millington, Esqr for Lectures in 1819	105 —
Ditto sundry expences for do	7.7.
Joseph Fincher on account of the Royal Institution	50 —
	£362.7.

Presents received.

3. Mr Mawe of the Strand presented
A fine Specimen of Emerald Fluor lately received from Weardale – County of Durham.
C.P. Dimond, Esqr presented Specimens of Chromate of Lead – & Emeralds in the Matrix.

1819.

The Editors of the Edinburgh Philosophical Journal, presented,
The Philosophical Journal, No 1.
Mr Charles Bell, Surgeon – presented,
An Essay on the forces which circulate the Blood.
Lt Col C. P. de Bosset, presented,
Proceedings in Parga & the Ionian Islands.
The Revd James Baker, M.R.I. presented
The Life of Sir Thos Bernard, Bart.
G. B. Greenough, Esqr President of the Geological Society presented,
A critical examination of the first principles of Geology, 8vo.
Captn Monke R.N. presented
M. Chas Dupin's Account of the Ports of England, Scotland & Ireland &c
A Philosophic & practical Inquiry into the nature & construction of Timber – presented by Mr John Lingard.
"A Guide for Gravesend" — Anonymous.

Ordered.
That the thanks of the Managers be returned for these presents.

4. Read the minutes of the last General Meeting.

5. The following person having been duly recommended was admitted as an Annual Subscriber to the Royal Institution for the year 1819.

Ann. Subsr
Arthur Aikin, Esqr of the Adelphi.

6 Adjourned to Monday the 21st of June.

Spencer

June 21.

At a Meeting of the Managers of the Royal Institution of Great Britain held at the house of the Institution on the 21st of June, 1819.

Present
Earl Spencer, K.G. President, in the Chair
Earl of Hardwicke, K.G. V.P.
Charles Hatchett, Esqr V.P.
James Peter Auriol, Esqr V.P.
Claude Scott, Esqr V.P.
James Stodart, Esqr
Samuel Solly, Esqr
John Chamier, Esqr
Thomas Snodgrass, Esqr
Edmund Robt Daniell, Esqr
Louis Hayes Petit, Esqr
Daniel Moore, Esqr
Thomas Harrison, Esqr Secretary.

1. The minutes of the proceedings of the last meeting were read and approved.

2. Read and approved the minutes of the Commee of Accounts of this day and signed an Order for payment of the following sums, viz.

Payments.

Wm Harris ¼ Salary to 31st March 1819.	£45 —
Joseph Fincher on account of the Royal Institution	55 —
	£100 —

Presents received.

3. Mr Charles M. Willich, presented,
General View of the public funded Debt of Great Britain, Jany 1819.
The Directors of the Grand Junction Water works presented
A Statement of the Question respecting the Water Companies of London.
Dr Vetch presented A Letter to Lord Palmerston on the Opthalmic Institution 3 copies.

1819.

Ordered. That the thanks of the Managers be returned for these presents.

4. Read the minutes of the last General Monthly Meeting.

5 Read the minutes of the Committee of Visitors entered in their proceedings of the 14th of June

Resolved

That the whole of the repairs be done under the direction of Mr. Brande.

Machine returned

6. The Secretary reported that R. Woodhouse, Esqr. had safely returned to the Institution with his sincere thanks, Smeaton's Machine which had been lent to him by the Board of the 5th of April last.

Spencer

June 28.

At a Meeting of the Managers of the Royal Institution of Great Britain held at the house of the Institution on the 28th of June, 1819.

Present

Earl Spencer, K.G. President, in the Chair.
Earl of Hardwicke K.G. V.P.
Charles Hatchett, Esqr. V.P.
Thomas Snodgrass, Esqr.
Samuel Solly, Esqr.
Louis Hayes Petit, Esqr.
Daniel Moore, Esqr.
Thomas Harrison, Esqr. Secretary

1. The minutes of the proceedings of the last meeting were read and approved.

2. Read and approved the minutes of the Committee of Accounts of this day, and signed an Order for an Advance to the Clerk of £50.

Presents received.

3. The Horticultural Society, presented.—
Transactions of their Society vol 3 part 3.
Mr. Professor Brande, presented—
Discourse delivered in the London Institution.
James Milligen, Esqr. presented
Medallic History of Napoleon, 4to.

Ordered.

That the thanks of the Managers be returned for these presents.

4. Read the minutes of the last General Meeting.

Coals.

5. Ordered. That 30 Chaldrons of the best Walls End Coals be laid in at the Royal Institution at 52s/ ⅌ Chaldron Metage, Shooting and all expences included.

Spencer

1819.

July 5.

At a Meeting of the Managers of the Royal Institution of Great Britain held at the house of the Institution on the 5th of July, 1819.

Present

Earl Spencer K.G. President, in the Chair
James Peter Auriol, Esqr. V.P.
Claude Scott, Esqr. V.P.
Thomas Snodgrass, Esqr.
Lt. Col. Benjn. Ansley.
Charles Hatchett, Esqr. V.P.
Sir S. Bernard Morland Bart. M.P. Treasr.
George Duckett, Esqr. V.P.
John Chamier, Esqr.
Daniel Moore, Esqr.
James Stodart, Esqr.

1. The minutes of the proceedings of the last meeting were read and approved.

2. Read and approved the minutes of the Committee of Accounts of this day, and signed an Order for an Advance to the Clerk of £70.

Lectures. Mr. Soane.

3. The President stated that he had had an interview with Mr. Soane who would be happy to deliver a Course of Lectures on Architecture gratuitously in the next Season.

— Mr. Millington

4. In consequence of the arrangement made with Mr. Millington to deliver in the next Season an additional number of Lectures,

Resolved.—That he be paid for those lectures 100 guineas.

Increase of Mr. J. Fincher's Salary.

5. Resolved.

That the Salary of Mr. Joseph Fincher the Clerk be increased from £180 to £200 to commence from the 1st of July, in consideration of the zeal diligence and constant attention which he has so long paid to the Interests of the Institution.

July 5.

Holydays to Librarians

6. Mr. Harris the Librarian & Mr. Sturt the Sub Librarian applied to the Board for leave of absence for one month each during the recess.

Resolved.—That their requests be granted, provided they be not both absent from the Library at the same time.

Adjourned to Monday the 2d. of August.

Spencer

264

Aug. 2.

1819.

At a Meeting of the Managers of the Royal Institution of Great Britain held at the house of the Institution on the 2nd of August, 1819,

Present

Earl Spencer, K.G. President, in the Chair.
Charles Hatchett, Esqr. V.P.
Thomas Snodgrass, Esqr.
James Stodart, Esqr.
Edmund Robt. Daniell, Esqr.
James Peter Auriol, Esqr. V.P.
John Chamier, Esqr.
Lt. Col. Ansley.
Daniel Moore, Esqr.
George Duckett, Esqr. V.P.
Thomas Harrison, Esqr. Secretary.

1. The minutes of the proceedings of the last meeting were read and confirmed.

2. Read and approved the minutes of the Committee of Accounts of this day, and signed an Order for an Advance to the Clerk of £120.

Life Subscr.

3. In pursuance of a nomination under the Act by the late Sir Henry Sullivan, Bart. dated the 1st of March 1814 – His Brother, Frederick Sullivan, Esqr. of Thames Ditton, was admitted to the privileges of a Life Subscriber to the Royal Institution.

Presents received.

4. Dr. James Johnson, presented the following Treatises.
"On the Influence of Tropical Climates &c. 8vo.
"On the Influence of Civic life &c. 8vo.
"On the Liver, Digestive Organs & Nervous System 8vo.
The Society of Antiquaries – presented,
Archaeologia, vol. 19 part 1.
The Geological Society – presented,
Transactions of the Society, vol. 5 part 1.

265

Aug. 2

Presented by the Editor,
The Literary Chronicle & Weekly Review – part 1.
Ordered.
That the thanks of the Managers be returned for these presents.

Annl. Subscr.

5. The following person having been duly recommended was admitted as an Annual Subscriber to the Royal Institution for the year, 1819,
Captn. Philip Carteret, C.B. R.N. 30 Bloomsbury – Square

6. Adjourned to Monday the 1st. of Novr.

Charles Hatchett V P

266

Nov. 1.

1819.

At a Meeting of the Managers of the Royal Institution of Great Britain held at the house of the Institution on the 1st. of November, 1819,

Present

Charles Hatchett, Esqr. V.P. in the Chair.
George Duckett, Esqr.
Thomas Snodgrass, Esqr.
Daniel Moore, Esqr.
Lt. Col. Benjn. Ansley.
Louis Hayes Petit, Esqr.
Edmund Robt. Daniell, Esqr.
John Chamier, Esqr.
James Stodart, Esqr.
Sir S. Bernard Morland, Bart. M.P. Treasurer

1. The minutes of the proceedings of the last meeting were read and approved.

2. Read and approved the minutes of the Committee of Accounts of this day; and signed an Order for an Advance to the Clerk of £80.

Life Subscr.

3. In pursuance of a nomination under the Act by Thomas Rowcroft, Esqr. – His Son Charles Rowcroft, Esqr. of 13 Doughty Street Bedford Row. was admitted to the privileges of a Life Subscriber to the Royal Institution.

Presents received.

4. His Grace the Duke of Bedford, presented thro' Earl Spencer, K.G. President.
Hortus Gramineus Woburnensis, folio.
Ordered.
That the special thanks of the Managers be returned for this valuable present.
Walter Fawkes, Esqr. presented
A. Catalogue of his Collection of Water Colour Drawings
John Thomson, Esqr. M.A.S. – presented,
Observations Introductory to a Work on English Etymology

267

Nov. 1.

Ordered. That the thanks of the Managers be returned for these presents.

Lectures on Theology declined.

5 Mr. Brande stated to the Board that Dr. Bathie would be willing to deliver A Course of Lectures in the next Season at this Institution, On the manners Morals & Religion of early times.
Resolved. That the thanks of the Managers be returned to Dr. Bathie for his obliging offer; but that after due deliberation they do not think the lectures proposed to be delivered by him, will come within the scope of their usual arrangements.

6. Read the minutes of the last General Monthly Meeting.

Plan of Drains

7 Read the minutes of the Committee of Visitors entered in their proceedings of the 12th. October.
Resolved. That Mr. Royce be directed to lay before the next Board, a plan of the Drains of the Royal Institn.
Resolved.
That the Hall Chair be repaired, agreeably to the Estimate of Messrs. Pringle, at an expence not exceeding £10. –

Request of Mr. Faraday.

8. Mr. Brande applied to the Board on the part of Mr. Faraday, for permission to occupy the Apartments Nos. 19 & 20 on the second Floor instead of Nos. 26 & 27 now used by Mr. Faraday.
Resolved. That Mr. Faraday's request be granted.

Annl. Subscr.

9. The following persons having been duly recommended were admitted as Annual Subscribers to the Royal Institution for the year 1820.
Arthur Mower, M.D. 6 Great Maddox St.
Revd. H. I. Hill, 12 Carburton Street Fitzroy Sqr.
Orlando Felix, Esqr.
George Mayer, Esqr. 14 Downing Street
Edward Burrard, Esqr.

10. Adjourned to Monday the 6th of December

Charles Hatchett V P

268

Dec. 6. 1819.

At a Meeting of the Managers of the Royal Institution of Great Britain held at the house of the Institution on the 6.th of December, 1819.

Present
Charles Hatchett, Esq.r V.P. in the Chair
Sir S. Bernard Morland, Bart M.P. Treasurer
John Chamier, Esq.r
Tho.s Snodgrass, Esq.r
George Duckett, Esq.r V.P.
L.t Col. Benj.n Ansley.
James Stodart, Esq.r
William Hasledine Pepys, Esq.r
Daniel Moore, Esq.r
Claude Scott, Esq.r V.P.

1. The minutes of the proceedings of the last meeting were read and approved.

2. Read and approved the minutes of the Committee of Accounts of this day; and signed an Order for payment of the following sums, viz.t

Payments.

Jos.h Fincher on account of the Royal Institution	£70 —
Reeve for 2 quarters Land & Assessed taxes to Michas 1819	79 . 2 . 4
	£149 . 2 . 4

Life Subsc.r

3. In pursuance of a nomination under the Act of Parliament by Sir John William Lubbock, Bart M.P. his Son, John William Lubbock, Esq.r of S.t James's Place, was admitted to the privileges of a Life Subscriber to the Royal Institution.

Presents received.

4. His Majesty's Commissioners of the Public Records presented — The Statutes of the Realm vol.s 4, 5. 6.
Rotuli Scotiæ, vol 2.d folio
Robert A. Slaney, Esq.r presented
An Essay on the Employment of the Poor
The Rev.d Step.n Weston — presented — Sketch of the

Dec. 6. 269

Breakwater at Plymouth Sound
D.r Schinas presented — Traité analytique de toutes les Affections Hemorroidales, par A. J. De Montegre, Paris
The Editor of the Colonial Journal — presented
The Colonial Journal N.o 4 —
Ordered. That the thanks of the Managers be returned for these presents.

Ann.l Subs.ers

5. The following persons having been duly recommended were admitted as Annual Subscribers to the Royal Institution for the year 1820.
M.r Benj.n Authur 11 Buton Street
Joseph Ashton Watson, Esq.r 54 Marchmont S.t
Major Gen.l Ogg 176 Piccadilly.

To the Lectures.
M.rs Briscoe Manchester Street

Adjourned to the 20.th Instant
J.P. Auriol

270

Dec 20.

At a Meeting of the Managers of the Royal Institution of Great Britain held at the house of the Institution on the 20.th of December, 1819.

Present
James Peter Auriol, Esq.r V.P. in the Chair.
Charles Hatchett, Esq.r V.P.
Edmund Rob.t Daniell, Esq.r
James Stodart, Esq.r
Thomas Snodgrass, Esq.r
Daniel Moore Esq.r
Sir S. Bernard Morland, Bart. M.P. Treasurer
Thomas Harrison, Esq.r Secretary.

1. Read and approved the minutes of the proceedings of the last meeting.

2. Read and approved the minutes of the Committee of Accounts of this day; and signed an Order for payment of the following sums, viz.t

Payments.

Jos.h Fincher on account of the Royal Inst.n	£50 . — —
½ years Parochial rates to Michas 1819 —	33 . 6 . 8
Dickins & C.o for Coals in 1819	78 . — —
Bulmer for Printing in d.o	57 . 5 . —
	218 . 11 . 8

And also a further advance of £50 to the Clerk on the 1.st of January 1820.

Present received.

3. The Academy of Dijon presented,
Seance Publique tenue le Jeudi 22 Avril 1819.
Ordered.
That the thanks of the Managers be returned for these presents.

4. Read the minutes of the last General Meeting.

Dec 20. 271

5. A Plan of the Drains of the Royal Institution was laid before the Board, agreeably to the Resolution of the Managers of the 1.st Nov.r last.

Lectures.
Mr Campbell.

6. In consequence of a letter from Thomas Campbell, Esq.r to M.r Fincher, wherein he stated that he should have no objections to deliver a Course of 6 Lectures at the Royal Institution in the ensuing Season, provided such a proposal were made to him by the Managers, the terms to be 100 guineas, to begin in April, and proceed punctually at the rate of 1 Lecture a week.
Resolved,
That the Secretary be directed to inform M.r Campbell, that the Board will be happy again to avail themselves of his services and to remunerate him with 100 guineas upon his completion of 6 Lectures on Poetry; the Course to commence about the middle of April, and to proceed, regularly, at the rate of one a week till finished.

Porter robbed.

7. M.r Fincher stated to the Board that the Assistant Porter Richard Hall was sent by him on Wednesday the 8.th Instant to advertize the Lectures in several of the Newspapers, and that in his way through the Strand he had been met by a Mob (who were going to a Meeting in Covent Garden) who knocked him down and robbed him of £1. 13 — the remainder of £4. 15 — given him to advertize the Lectures.
Richard Hall having been called in, and it appearing to the Board that he had met with the above loss unavoidably, and without any fault on his part,
Resolved, That the above sum of £1. 13 — be charged to the account of the Royal Institution.

Ann.l Subs.ers

8. The following persons having been duly recommended were admitted as Annual Subscribers to the Royal Institution for the year 1820 — J. H. Markland, Esq.r 12 Paper Build.gs Temple
John Bowles, Esq.r 6 G.t Maddox S.t
L.t Edw.d Hall. R.N. Jermyn S.t S.t James's

9. Adjourned to Monday 31.st Jan.y 1820
Charles Hatchett V.P.

Jan. 31. 1820

At a Meeting of the Managers of the Royal Institution of Great Britain, held at the house of the Institution on the 31st of Jany 1820.

Present

Charles Hatchett, Esqr. V.P. in the Chair.
James Peter Auriol, Esqr. V.P.
James Stodart, Esqr.
Edmund Robt. Daniell, Esqr.
George Duckett, Esqr. V.P.
John Chamier, Esqr.
Daniel Moore, Esqr.
Louis Hayes Petit, Esqr.
Sir S. Bernard Morland, Bart. M.P. Treasurer
Thomas Harrison, Esqr. Secretary.

1. The minutes of the proceedings of the last meeting were read and approved.

2. Read and approved the minutes of the Commee. of Accounts of this day; and signed an Order for an Advance to the Clerk of £350.

Presents received.

3. John Ring, Esqr. presented,
A Translation of the works of Virgil 2 vols.
By the Publishers,
The London Magazine, No. 1.
Ordered.
That the thanks of the Managers be returned for these presents.

Death of H.M. Lectures discontd.

4. Resolved. That in consequence of the lamented Death of His late Majesty the Patron of this Institution and out of respect to his Memory, the Lectures here be discontinued for the present week.

5. The following persons having been duly recommended were admitted as Subscribers to the Royal Institution.
Life Subscriber.
William Hunter Baillie, Esqr. 73 Lower Grovr. St.

Jan. 31.

Annual Subscribers.
Brigade Major Rice Jones of the Royl. Engineers Woolwich
T. Coote, Esqr. 20 Dover Street
Arthur Walpole, Esqr. 36 Half Moon Street
Lee Thornton, Esqr. 14 Queen St. May Fair.
Chas. Cazenove, Esqr. St. James's Street
J. F. Gwyn, Esqr. 9 Upper George St. Portman Sqr.
Thomas Maynard, Esqr. 21 Lower Brook St.
Warburton Davies, Esqr. 34 Queen Ann St.
George Suthie, Esqr. 28 Duke St. Piccadilly.
Revd. Edwd. R. Butcher 10 Welbeck Street.
Alexr. Nowell, Esqr. 43 Wimpole Street
Chas. Gordon, Esqr. 16 Old Burlington St.
Col. Sir Neil Campbell, 12 Charles St. St. James's Sqr.
William Lynch, Esqr. Ipswich.
Edwd. Boodle junr. Esqr. 42 Brook Street
Rear Admiral Spranger, Albany.

Charles Hatchett V.P.

Feb. 7. 1820

At a Meeting of the Managers of the Royal Institution of Great Britain held at the house of the Institution on the 7th of February, 1820

Present

Charles Hatchett, Esqr. V.P. in the Chair.
James Peter Auriol, Esqr. V.P.
Louis Hayes Petit, Esqr.
John Chamier, Esqr.
James Stodart, Esqr.
Sir S. Bernard Morland, Bart. M.P. Treasurer.

1. The minutes of the proceedings of the last meeting were read and approved.

2. Read and approved the minutes of the Committee of Accounts of this day; and signed an Order for an Advance to the Clerk of £100.

Lectures

3. Resolved,
That the Funeral of His late Majesty having been now fixed for Wednesday the 16th of February, the Lectures be further postponed till after that day; and that Mr. Millington recommence his Course of Lectures on Thursday the 17th and Mr. Brande give his Introductory Lecture on Chemistry on Saturday the 19th Instant.

4. The following persons having been duly recommended were admitted as Subscribers to the Royal Institution for the year 1820.
Annual Subscribers.
John Thurburn, Esqr. 19 Mecklenburgh Square.
Chas. E. Poulett Thomson, Esqr. 129 Park St.
Mr. Nicholl }
Mr. Jas. Stein } 9 Seymour Place.
To the Lectures,
Miss E. Blake Portland Place

5. Adjourned to the 21st. Inst. Charles Hatchett V.P.

Feb. 21.

At a Meeting of the Managers of the Royal Institution of Great Britain held at the house of the Institution on the 21st of February, 1820.

Present

Charles Hatchett, Esqr. V.P. in the Chair.
James Stodart, Esqr.
Edmund Robt. Daniell, Esqr.
Sir S. Bernard Morland, Bart. M.P. Treasurer.
John Chamier Esqr.
Daniel Moore, Esqr.
Thomas Harrison, Esqr. Secretary.

1. The minutes of the proceedings of the last meeting were read and approved.

2. Read and approved the minutes of the Committee of Accounts of this day; and signed an Order for an advance to the Clerk of £260. –

Presents received.

3. The Marquis of Buckingham, thro' Sir S. Bernard Morland, Bart. presented.
Catalogue of the Manuscripts in Stowe Library 2 vols. 4to.
Ordered. – That the special thanks of the Managers be returned for this valuable present.
The Horticultural Society – presented,
Transactions of their Society, vol. 3 part 4.
Mr. A. J. Valpy, presented,
Time, or Light & Shade, a Poem by Gompertz 4to.
Ordered. That the thanks of the Managers be returned for these presents.

4. Audited the Treasurer's account for the year 1819.

5. The following persons having been duly recommended were admitted as Subscribers to the Royal Institution.
Life Subscriber.
Warburton Davies, Esqr. 34 Queen Ann Street.

276

Annual Subscribers. 1820

William Harding, Esqr. 82 Gower Street
Richd. Digby Vicare, Esqr. Old Burlington St.
Edward Penrhyn, Esqr. – Albany
Terrick Hamilton, Esqr. 19 Orchard St.
Wm. Mussenden, Esqr. Hadengs Hotel.

To the Lectures.

Miss Elizth. Raine New Bridge Street
Mrs. Puget
Miss Puget } 35 Sackville Street
Miss Esther Puget
Miss Elizth. Puget
Miss Arnold Chandos St. Cavendish Sqr.
Miss F. Hamilton 19 Orchard Street
Miss Montgomery – 15 Manchester St.

6. Adjourned to 6th March. Charles Hatchett V.P.

277

Mar. 6.

At a Meeting of the Managers of the Royal Institution of Great Britain held at the house of the Institution on the 6th of March, 1820,

Present

Charles Hatchett, Esqr. V.P. in the Chair.
Ld. Col. Benjn. Ainsley.
John Chamier, Esqr.
Edmund Robt. Daniell, Esqr.
George Duckett, Esqr. V.P.
Daniel Moore, Esqr.
Claude Scott, Esqr. V.P.
James Stodart, Esqr. –

1. The minutes of the proceedings of the last Meeting were read and approved.

2. Read and approved the minutes of the Committee of Accounts of this day; and signed an Order for an Advance to the Clerk of £140. –

3. Presents received.

John Guillemard, Esqr. – presented 160 Volumes relating to America.

Ordered. That the special thanks of the Managers be returned for this valuable present.

The Royal Society presented, Nautical Almanac, for 1820 & 1822.

The Society of Arts – presented, Transactions of the Society, Vol. 37.

Ordered. That the thanks of the Managers be returned for these presents.

4. The following person having been duly recommended was admitted as a Subscriber to the Lectures for the year 1820.

Subscr. to Lectures — Miss Dennett 39 Guilford Street.

5. Adjourned to 20th March Geo. Duckett V.P.

278

Mar. 20. 1820

At a Meeting of the Managers of the Royal Institution of Great Britain held at the house of the Institutn. on the 20th of March, 1820,

Present

George Duckett, Esqr. V.P. in the Chair.
Charles Hatchett, Esqr. V.P.
Daniel Moore, Esqr.
Thomas Snodgrass, Esqr.
John Chamier, Esqr.
Sir S. Bernard Morland, Bart. M.P.
James Stodart, Esqr.

1. The minutes of the proceedings of the last meeting were read and approved.

2. Read and approved the minutes of the Committee of Accounts of this day, and signed an Order for an advance to the Clerk of £140 –

3. Presents received.

The British and Foreign Bible Society, presented through Earl Spencer, President.

Bibles & Testaments in 36 different Languages, uniformly bound, 44 Volumes, & Reports of the Society.

Ordered. That the special thanks of the Managers be returned for this valuable present.

Anniversary Dinner. 4. Resolved. That it be recommended by the Managers to the next General Monthly Meeting that an Anniversary Dinner of the Royal Institution be held on —

5. The following persons having been duly recommended were admitted as Subscribers to the Royal Institution

Life Subscriber. Henry Wilkinson, Esqr. of Clapham Common

To the Lectures. Mrs. James Warre Gt. George St. Hanover Sqr.

J P Auriol

279

April 3.

At a Meeting of the Managers of the Royal Institution of Great Britain held at the house of the Institution on the 3rd of April, 1820,

Present

James Peter Auriol, Esqr. V.P. in the Chair.
Charles Hatchett, Esqr. V.P.
Daniel Moore, Esqr.
James Stodart, Esqr.
Sir S. Bernard Morland, Bart. M.P. Treasurer.
Thomas Snodgrass, Esqr.
George Duckett, Esqr. V.P.
Thomas Harrison, Esqr. Secretary.

1. The minutes of the proceedings of the last Meeting were read and approved.

2. Read and approved the minutes of the Committee of Accounts of this day.

3. The following person having been duly proposed was balloted for and elected an Annual Subscriber to the Royal Institution for the year 1820.

Ann. Subsr. — John Commerell, Esqr. 5 Lower Berkeley St.

4. Adjourned to Monday the 17th Instant.

J P Auriol

1820

April 17.

At a Meeting of the Managers of the Royal Institution of Great Britain held at the house of the Institution on the 17th of April, 1820,

Present

James Peter Auriol, Esqr. V.P. in the Chair.
George Duckett, Esqr. V.P.
Claude Scott, Esqr. V.P.
Edmund Robt. Daniell, Esqr.
John Chamier, Esqr.
James Stodart, Esqr.
Daniel Moore, Esqr.
Sir S. Bernard Morland, Bart. M.P. Treasurer
Thos. Harrison, Esqr. Secretary.

1. The minutes of the proceedings of the last meeting were read and approved.

2. Read and approved the Committee of Accounts of this day, and signed an Order for an Advance to the Clerk of £50.–

Presents received.

3. The Academy of Natural Sciences at Philadelphia presented, – Journals of the Academy, vol 1 pt. 2.
Mr. Locker, presented,
The plain Englishman's Library, 3 vols.–
Ordered,
That the thanks of the Managers be returned for these presents.

Address to H.M.

4. The Address to The King, as read to the last General Monthly Meeting, condoling with His Majesty on the loss of our late Patron, and congratulating His Majesty on his accession & also soliciting the continuance of his gracious patronage of this Institution having been engrossed on Parchment; it was sealed with the Seal of the Corporation and signed by the Secretary.
Resolved. That the following Gentlemen be appointed to wait on Viscount Sidmouth at the

April 17.

Secretary of State's Office, with the Address.
James Peter Auriol, Esqr. V.P.
Charles Hatchett, Esqr. V.P.
George Duckett, Esqr. V.P.
Sir S. Bernard Morland, Bart. M.P. Treasurer
& – Thomas Harrison, Esqr. Secretary.

Dinner.

5. Read the Resolution of the last General Monthly Meeting stating that a Public Dinner of the Members of the Royal Institution be held in May under the direction of the Managers."
Resolved.
That the Dinner be fixed for Monday the 8th of May at Willis's Rooms King Street St. James' Square – The Dinner to be on Table at 6 o'clock precisely.

The Secretary stated that he had written to the Marquis of Lansdowne who had consented to take the Chair at the Dinner.
Ordered. That the thanks of the Managers be returned to the Marquis of Lansdowne for his kind acceptance of that Office.

Lectures, Dr. Crotch.

6. It having been stated to the Board that Dr. Crotch would be willing, if it were agreeable to the Managers, to commence a Course of 6 Lectures on Music, on Friday the 28th Instant – the terms 9 guineas a Lecture – the Institution to pay the hire of a Grand Piano Forte and the tuning of it, as on former occasions.
Resolved. That Dr. Crotch be engaged on the above terms.

Coals.

7. Ordered. That Messrs. Dickins & Co. lay in at the Royal Institution, 30 Chaldrons of the very best Walls-End-Coals at 54/6d. ꝑ Chaldron, Metage Shooting and all Expences included.

8. The following persons having been duly recommended

1820.

were admitted as Subscribers to the Royal Instit.n for the year 1820.
Annual Subscriber.
Frederic Alston, Esqr. 26 Gloucester Place

To the Lectures.
Mrs. Frederic Alston 26 Gloucester Place
Mrs. Richards, Bedford Square
Miss Farrer, Lincolns Inn Fields.
Miss Colebrooke }
Miss H. Colebrooke } Hampstead.
Miss E. Lee }
Mrs. Prince } Hendon
Miss Locker }
Mrs. Finch 33 Albemarle Street.

Charles Hatchett

April 24.

At a Meeting of the Managers of the Royal Institution of Great Britain held at the house of the Institution, on the 24th of April, 1820,

Present

Charles Hatchett, Esqr. V.P. in the Chair
Samuel Solly, Esqr.
Daniel Moore, Esqr.
James Stodart, Esqr.
George Duckett, Esqr. V.P.
Thomas Snodgrass, Esqr.
John Chamier, Esqr.
Thomas Harrison, Esqr. Secretary.

1. The minutes of the proceedings of the last meeting were read and approved.

2. Read and approved the minutes of the Committee of Accounts of this day, and signed an Order for payment of the following sums, viz.

Payments.

John Newman for Chemicals in 1819	£150. 13. –
1 year's Ground Rent to the City of London to Lady day 1820	150. 1. 4
	£300. 14. 4

Presents received.

3. William Thos. Brande, Esqr. – presented,
"On the Composition and Analysis of the Inflammable Gaseous Compounds."
Ordered,
That the thanks of the Managers be returned for this present.

4. Ordered. That the House Lists be filled up as follows.
President. Earl Spencer K.G. F.R.S. & F.A.S.
Treasurer. Sir S. Bernard Morland, Bt. M.P.
Secretary. Thomas Harrison, Esqr. F.R.S.

 1820.

Managers.

James Peter Auriol, Esqr. F.R.S.
Earl of Bessborough.
Rt. Hon: Reginald Pole Carew F.R.S & F.H.S.
John Chamier, Esqr.
George Duckett, Esqr. F.R.S. and F.S.A.
John Nicholas Fazakerley, Esqr. M.P.
William Haldimand, Esqr. M.P.
Charles Hatchett, Esqr. F.R.S and F.S.A.
Marquis of Lansdowne. L.L.D. F.R.S
Daniel Moore, Esqr. F.R.S. F.S.A and F.L.S
Rt. Honble Sir John Nicholl, M.P. L.L.D.
Thomas Snodgrass, Esqr.
Duke of Somerset. F.R.S
James Stodart, Esqr.
Robt. Williams, Esqr. M.P.

Visitors.
George Aust, Esqr. F.R.S. and F.S.A.
Earl of Chichester F.R.S.
Henry Thos. Colebrooke, Esqr. F.R.S.
Duke of Devonshire.
John Fuller, Esqr.
Davies Gilbert, Esqr. M.P. F.R.S. and F.L.S.
Gibbes Walker Jordan, Esqr. A.M. F.R.S.
Earl of Mountnorris, F.R.S. F.S.A.
Charles Pieschell, Esqr.
John Soane, Esqr. F.S.A.
Richard Horsman Solly, Esqr. F.R.S and F.S.A
Marquis of Stafford, K.G.
Sir Geo: Thos. Staunton, Bart. M.P. F.R.S. & L.S.
Lord Teignmouth.
Lt. Genl. Wm. Thornton.

5. The Secretary reported that himself, with the other Gentlemen appointed by the last Board, had waited on Lord Sidmouth with the Address, and that the following answer had been received from His Lordship, vizt.

April 24. — Answer to Address — from Ld. Sidmouth.

Whitehall 21st. April 1820

Gentlemen,

I have had the honour to lay before The King the very loyal and dutiful Address of the President and Members of the Royal Institution of Great Britain, established for the promotion of Science, Literature, and the Arts. — And I have the satisfaction to inform you, that His Majesty was pleased to receive the same in the most gracious manner; and that His Majesty, in compliance with your request, most graciously expressed His intention to become Patron of your distinguished and very beneficial Institution.

I have the Honour to be Gentlemen
Your most obedient humble Servant
(signed) Sidmouth.

The Managers & Secretary
of the Royal Institution of Great Britain

6. The following persons having been duly recommended were admitted as Subscribers to the Royal Institution for the year 1820.

Annual Subscriber.
Captn. George Colomb. 14 Old Cavendish St.

To the Lectures.
Mrs. & Miss Gladstone, 5 Grafton Street.
Mrs. W. Hodges. Peckham.
Mrs. Thos. Farrer, Lincolns Inn Fields.
Miss Mary Anne Pearson, 9 George St. Hanover Sqe.
The Dowager Viscountess Anson }
The Honble Miss Georgiana Anson } St. James's Square
The Honble Miss Frances Anson }
Miss Bullen

Adjourned to Monday the 8th. of May.

Bessborough Vice Prest.

May 8. At a Meeting of the Managers of the Royal Institution of Great Britain held at the house of the Institution on the 8th. of May, 1820. 1820

Present
Earl of Bessborough, in the Chair.
James Peter Auriol, Esqr.
Rt. Honble Reginald Pole Carew.
Sir S. Bernard Morland, Bart M.P. Treasurer.
Robert Williams, Esqr. M.P.
James Stodart, Esqr.
Daniel Moore, Esqr.
Thomas Harrison, Esqr. Secretary.

1. The minutes of the proceedings of the last meeting were read and approved.

2. Read and approved the minutes of the Committee of Accounts of this day, and signed an Order for payment of the following sums vizt.

Payments

Mr. Harris 3 quarters Salary to 31st. Decr. last	£135. —
Joseph Fincher on account of the Royal Instn.	50. —
	£185. —

Presents received.

3. Present of Rock Amethyst, by T. Snodgrass, Esq. — Thomas Snodgrass, Esqr. Manager R.I. presented A rare specimen of Rock Amethyst to the Mineralogical Collection of the R.I.

Ordered, That the thanks of the Managers be returned to Mr. Snodgrass for his valuable present and that a Glass be immediately prepared for it under the direction of Mr. Brande.

Mr. Dart presented in the name of the Court of Directors of the East India Company The third part of Mr. Morrison's Chinese Dictionary

May 8.

The Royal Society — presented,
Transactions of the Society for 1819.
The Society for printing and publishing the Writings of the Hon: Emanuel Swedenborg, presented,
Swedenborg's Works 11 vol.

Ordered, — That the thanks of the Managers be returned for these presents.

4. Read the minutes of the last General Annual Meeting.

5. The following persons having been duly recommended were admitted as Subscribers to the Royal Institution for the year 1820,

Annual Subscribers
John Shaw Stewart, Esqr. 5 Charles Street St. James
Richard French, Esqr. 17 Berkeley Street.
G. H. Barnett, Esqr. 2 Lower Berkeley Street.

To the Lectures.
Mrs. Thomson — Streatham
Mrs. L. Goldsmid 37 Dorset Square
Mrs. Smith }
Miss Augusta Smith } Portland Place.
Miss Emma Smith }
Miss Travis 13 Orchard St. Portman Square
Mrs. Clarke }
Miss Clarke } 20 King Street West Edgeware Road
Mrs. Richard French 17 Lower Berkeley Street
Mrs. Octavius Smith 21 Hertford Street
Miss Julia Smith — do. —
Lady Hobhouse }
Miss Hobhouse } 2 Hanover Square.
Mrs. Carter — 16 Duke St. Westmr.
Miss Cooke }
Miss C. Cooke } 21 Hertford Street

Charles Hatchett

288

May 15.

1820

At a Meeting of the Managers of the Royal Institution of Great Britain held at the house of the Institution on the 15th of May, 1820.

Present

Charles Hatchett, Esqr. in the Chair.
Rt. Hon. Reginald Pole Carew.
Thomas Snodgrass, Esqr.
Duke of Somerset
George Duckett, Esqr.
J. P. Auriol, Esqr.
Robert Williams, Esqr. M.P.
William Haldimand, Esqr. M.P.
Daniel Moore, Esqr.
Sir S. Bernard Morland, Bart. M.P. Treasurer
Thomas Harrison, Esqr. Secretary.

1\. The minutes of the proceedings of the last meeting were read and approved.

2\. Read and approved the minutes of the Committee of Accounts of this day, and signed an Order for an Advance to the Clerk of £220.

Presents received.

3\. The Geological Society, presented,
Geological Map of England – by G. B. Greenough, Esqr.
Captn. Vetch, Roy. Eng. presented,
Projection of the Globe on the Cylinder of a Meridian.
Ordered,
That the thanks of the Managers be returned for these presents.

4\. Read the minutes of the last General Monthly Meeting.

Hon. Admission. 5. J. G. Cogswell Esqr. having been recommended by the American Ambassador as a Gentleman of literary and scientific attainments was admitted to the Royal Institution under Art: 15 Chap VI of the Bye laws.

289

May 15.

6\. The following person having been duly recommended was admitted as a Subscriber to the Lectures for the year 1820

Subsr. The Honble. Mrs. Dawson 32 Clarges Street.

7\. Adjourned to Monday the 5th of June.

J. Nicholl

290

June 5.

1820

At a Meeting of the Managers of the Royal Institution of Great Britain held at the house of the Institution on the 5th of June, 1820.

Present

Rt. Honble. Sir John Nicholl, M.P. in the Chair.
Robert Williams, Esqr. M.P.
Charles Hatchett, Esqr.
Thomas Snodgrass, Esqr.
James Stodart, Esqr.
George Duckett, Esqr.
John Chamier, Esqr.
Duke of Somerset
Daniel Moore, Esqr.
Sir S. Bernard Morland, Bart. M.P. Treasr.
Thomas Harrison, Esqr. Secretary.

1\. The minutes of the proceedings of the last meeting were read and approved.

2\. Read and approved the minutes of the Committee of Accounts of this day, and signed an Order for an Advance to the Clerk of £190.–

Member Nominated. 3. Read a letter from Mrs. L. Wiggin (laid before the Board by Sir James Edward Smith) stating that her Husband Mr. John Wiggin of Craven Hill Bayswater, Middlesex, one of the original Proprietors of the Royal Institution died in February 1808 leaving her his sole Executrix and residuary Legatee; and requesting that the Managers would transfer this share in the name of her Brother Thos. Dickinson Evans Esqr. Ross Herefordshire.
Ordered.
That the name of Thos. Dickinson Evans Esqr. be added to the list of Members of this Institution.

Presents received.

4\. Dr. McCabe, presented,
Observations on the Cheltenham Waters.

291

June 5.

Dr. Jas. Man Burrows – presented
Inquiries relative to Insanity.
Henry Jeffreys, Esqr. – presented
Cases in Surgery at the St. Georges & St. James's Dispensary.
Ordered. That the thanks of the Managers be returned for these presents.

Books taken from Reading Rooms. 5. Mr. Sturt, the Assistant Librarian, reported to the Board, that notwithstanding the notice which had been put up in the Reading and Newspaper Rooms, cautioning the Subscribers against taking any Books or Papers away, the Magazines and other periodical Publications are frequently taken out of the Rooms shortly after being placed on the Tables, many of which have not been brought back.

The following are a list of those missing for the last and present year.

Journal of Science No. 12.
Medical Repository for Feby 1819.
Eclectic Review, for May 1819.
Antijacobin Review for Novr. 1819.
Monthly Review, Appendix to vol 88.
British Review, No. 29 – for Feby 1820
Repository of Arts &c. for Feby & March 1820
Massinger's Plays vol 1.

Notice. Resolved. That the following notice be put up in the Reading & Newspaper Rooms.

Whereas the undermentioned Books are missing from the Reading and Newspaper Rooms, and it being possible that they may have been taken away by some Member or Subscriber without knowing that it is contrary to the rules of the Institution to remove any book from the Room in which it is placed, the Managers particularly request that the undermentioned books may be returned in the course of one week from the date of this Notice; and if they should not be so returned, a reward of Five Guineas is hereby offered for the detection of any person who may have taken away and not returned the same within that time; and a similar reward for the detection of any such Offender in future.

292

1820.

6. The following persons having been duly recommended were admitted as Subscribers to the Royal Institution.

Life Subscriber.

Captⁿ. Thos. Herbert, R.N.

Annual Subscribers.

Sir Robt. Kennedy – 75 Wimpole Street
Henry Neave, Esqr. 30 Old Burlington S^t.

7. Adjourned to Monday the 19th. Inst^t.

Sale of Stock &c.

8. The Treasurer reported that M^r. Fuller having desired that out of the Stock standing in his and the Treasurer and Secretary's names, as much as would produce £100 should be sold and the Money paid to the Library Account, and the remaining Stock transferred to the Account of the Royal Institution, a Power was prepared and executed to that effect, and the said Sale and Transfer have since been made: vizt. £143.12.6 sold to produce £100 since paid to the Library Account and £1104.8.7 Consols transferred to the Royal Institution at the Bank.

The Treasurer also reported that the Patrons of the Library had directed £100 in addition to the usual quarterly sums should be paid to the Account of the Royal Institution in further aid of the Salaries paid by them to the Officers of the Library.

Adjourned to Monday the 19th. Inst^t.

Spencer.

293

June 19.

At a Meeting of the Managers of the Royal Institution of Great Britain held at the house of the Institution on the 19th. of June, 1820,

Present
Earl Spencer, K.G. President, in the Chair.
Charles Hatchett, Esqr.
Sir S. Bernard Morland, Bart M.P. Treasurer.
Robert Williams, Esqr. M.P.
James Stodart, Esqr.
Daniel Moore, Esqr.
Ja^s. Peter Auriol, Esqr.
John Chamier, Esqr.
Earl of Bessborough.
Thomas Snodgrass, Esqr.
Thomas Harrison, Esqr. Secretary.

1. The minutes of the proceedings of the last meeting were read and approved.

2. Read and approved the minutes of the Committee of Accounts of this day, and signed an Order for an Advance to the Clerk of £^s50.–

Presents received.

3. Captⁿ. Colby Roy: Eng^{rs}. presented
The 8th. part of the Ordnance Survey of Great Britain
Ordered, Ashhurst Majendie Esqr. pres^d.– A few Volcanic Specimens
That the thanks of the Managers be returned for these presents.

4. The following person having been duly recommended was admitted as an Annual Subscriber for the year 1820.

Ann^l. Subs^{cr}.

Philip Davies Cooke, Esqr. 22 Bruton S^t.

5. The President nominated as Vice Presidents for the present year,

Vice Presidents.

The Duke of Somerset
Marquis of Lansdowne

294

1820.

Earl of Bessborough
Ja^s. Peter Auriol, Esqr.
R^t. Hon: Reginald Pole Carew
George Duckett, Esqr.
Charles Hatchett, Esqr.
R^t. Honble Sir John Nicholl M.P.

Spencer.

295

June 26.

At a Meeting of the Managers of the Royal Institution of Great Britain held at the house of the Institution on the 26th. of June, 1820,

Present
Earl Spencer K.G. President, in the Chair.
Robert Williams, Esqr. M.P.
Daniel Moore, Esqr.
Charles Hatchett, Esqr. V.P.
James Peter Auriol, Esqr. V.P.
James Stodart, Esqr.
Sir S. Bernard Morland, Bart: M.P. Treas^r.
Thomas Snodgrass, Esqr.
George Duckett, Esqr. V.P.
Duke of Somerset, V.P.
Thomas Harrison, Esqr. Secretary.

1. The minutes of the proceedings of the last meeting were read and approved.

2. Read and approved the minutes of the Committee of Accounts of this day, and signed an Order for an Advance to the Clerk of £35.–

Lectures.

3. Read a letter from The Rev^d. T. Dudley Fosbroke, addressed to the Managers, begging to know whether he could be permitted to deliver a Course of Lectures at the Royal Institution, early in next December upon the subject of Antiquarian Science.
Resolved. That the above lectures be declined.

Presents received.

4. The Publishers of Oriental Literature, presented,
Annals of Oriental Literature.
Ordered.
That the thanks of the Managers be returned for this present.

5. Resolved.– That the thanks of the Managers be given to John Soane, Esqr. for his excellent Course of

296 1820.

Lectures on Architecture gratuitously delivered in the Theatre of the Royal Institution during the last Season at the request of this Board.

Resolved,

That this Board in acknowledging the utility of these Lectures to the Institution & and the general interest which they have so justly excited cannot but express their hopes that Mr. Soane may favour them with a similar Course during the ensuing Season.

6. Read the minutes of the last General Meeting.

Spencer

297

July 3.

At a Meeting of the Managers of the Royal Institution of Great Britain held at the house of the Institution on the 3rd of July, 1820.

Present

Earl Spencer K.G. President in the Chair.
Thomas Snodgrass, Esqr.
George Duckett, Esqr. V.P.
John Chamier, Esqr.
Charles Hatchett Esqr. V.P.
Daniel Moore, Esqr.
James Stodart, Esqr.
Earl of Bessborough V.P.
Sir J. Bernard Morland, Bart. M.P. Treasr.
Thomas Harrison, Esqr. Secretary.

1. The minutes of the proceedings of the last meeting were read and approved.

2. Read and approved the minutes of the Committee of Accounts of this day; and signed an Order for payment of the following sums, vizt.

Payments.

Wm. Thos. Brande, Esqr. — One years Salary as Professor of Chemistry to May 1820	£200. —
John Millington Esqr. for Lectures in 1820	210. —
Joseph Fincher on account of the Royal Institution	100. —
	£510. —

Presents received.

3. Mrs. Andrew Sharp, presented,
Memoirs of Granville Sharp, by Prince Hoare, Esqr. 4to.
Messrs. King Severn & Co. presented
Report of the Trial of the Action against the Imperial Insurance Company.
Lieut. Genl. William Thornton – presented,
A Dictionary of Painters, Engravers &c.
Ordered, That the thanks of the Managers be returned for these presents.

298 1820.

Increase of Mr. Sturt's Salary.

4. Resolved, That the Salary of John Sturt, the Assistant Librarian, be increased from £90 to £100 p. Ann.

Leave of Absence.

5. Mr. Sturt requested permission of the Board for leave of absence for 1 month during the recess; which was granted.

Adjourned to Monday the 6th of November next.

Charles Hatchett.

Nov. 6

At a Meeting of the Managers of the Royal Institution of Great Britain held at the house of the Institution on the 6th of November 1820.

Present

Charles Hatchett, Esqr. V.P. in the Chair.
George Duckett, Esqr. V.P.
Thomas Snodgrass, Esqr.
Daniel Moore, Esqr.
James Stodart, Esqr.
Sir J. Bernard Morland, Bart. M.P. Treasurer.
Thomas Harrison, Esqr. Secretary.

1. The minutes of the proceedings of the last meeting were read and approved.

2. Read and approved the minutes of the Committee of Accounts of this day.

Life Subscriber

3. In pursuance of a nomination under the Act of Parliament by Samuel James Arnold, Esqr., his Son Thomas James Arnold, Esqr. of Golden Square, was admitted to the privileges of a Life Subscriber to the Royal Institution.

Presents received.

4. His Grace the Duke of Wellington, K.G. presented,
The Ordnance Surveys of the Counties of Great Britain No. 2 to 8 inclusive.
No. 9 since received from Captn. Colby.
The Royal Society through Dr. Young – presented
The Nautical Almanack for 1820, 1821, 1822, 1823.
The Horticultural Society – presented
Transactions of the Horticultural Society vol. 4, part 1.
Adam Neale, M.D. presented
A Letter on the Cheltenham Waters.
John Hennen, M.D. presented
Principles of Military Surgery 8vo. 2d Edition.
Mr. George Burrell, Librarian of the Athenaeum Liverpool, pres.
A Catalogue of the Library of the Athenaeum.

Mr. Benjn. Golding – presented
An Historical Account of St. Thomas's Hospital.
Chas. Scudamore, M.D. presented
A Chemical Report of the properties of Mineral Waters
The Revd. Stephen Weston – presented
A Chinese Chronical [Chronicle] from the Persian.
The Society for the Improvement of Prison Discipline,
2 Copies of the Report of their Committee.
Monsr. J. R. Duval – presented
De l'arrangement des Secondes Dents.
The Hon: Francis Henry Egerton, presented
The Life of Lord Chancellor Egerton – part 1 - 4to.
A Second Letter to the Parisians upon Inland Navigatn.
Lt. Genl. William Thornton – presented
Army Lists from 1757 to 1808, 41 vols. 8vo.
Oliver's Scripture Lexicon,
Dunster's Synopsis of the Gospels.
Ward on the Rights of Neutral Powers.
Macdiarmid on Civil & Military Subordination.
The Age – a Poem.
Walker's Descriptive Catalogue of Pictures.
Mr. Dart presented in the name of the Court of Directors of the East India Company,
Two further Parts in continuation of Mr. Morrison's Chinese Dictionary.
Mr. S. Parker of Argyll Street presented,
A Sinumbra Lamp.
Mr. Turner, of Warren St. Fitzroy Square Engraver – presented
An Engraved Portrait of Lt. Genl. Wm. Thornton. M.R.I.

Resolved, That the thanks of the Managers be returned for these presents.

5. Read the minutes of the last General Monthly Meeting.

Adjourned to Monday the 4th. of December

J P Auriol

Dec. 4.

At a Meeting of the Managers of the Royal Institution of Great Britain held at the house of the Institution on the 4th. of December, 1820,

Present
James Peter Auriol, Esqr. V.P. in the Chair.
Charles Hatchett, Esqr. V.P.
Thomas Snodgrass, Esqr.
Daniel Moore, Esqr.
James Stodart, Esqr.

1. The minutes of the proceedings of the last meeting were read and approved.

2. Read and approved the minutes of the Committee of Accounts of this day, and signed an Order for payment of the following sums, vizt.

Payments.

Joseph Fincher on account of the Royal Institution	£250. –
Ditto — do. on the 31st. Decr.	100. –
	£350. –

Presents received.

3. Dr. Henderson presented
Mitchell's Plans and Views of Buildings, folio.
Captn. Colby – presented
The Ordnance Survey of England, No. 1.
The Royal Society – presented
Philosophical Transactions, 1820 part 1.
Major Castellani, Inspecteur des Ponts et Chaussées, Torino, presented "Influenza delle Selve sul Corso delle Aque", 2 parts, 4to.
Dr. Granville presented
"On Prussic Acid and Consumption".
Report on Midwifery,
Letter on the Plague & Contagion,
A Case on the Human Foetus found in the Ovarium,
On a Mal-conformation of the Uterine system in Women
Monsr. César Moreau presented
Memoires de l'Academie de Dijon.

John Weyland, Esqr. presented
An Engraved Portrait of himself.
Ordered, That the thanks of the Managers be returned for these presents.

4. The following persons having been duly recommended were admitted as Subscribers to the Royal Institution.
Annual Subscriber for the present year.
Sir Charles Grey.
Life Subscriber.
Francis Walpole, Esqr. Half Moon St.
Annual Subscribers for the year 1821.

Lt. George Lyon, R.N. 15 Alfred Place
Henry Spooner, Esqr. Crawford St. Portman Square.
Captn. J. Jas. Chapman } Royal Artillery Woolwich
Lt. James Colquhoun }
Captn. Dansey Royal Horse Artillery, Woolwich.
Lieut. Sanders Royal Engineers, Woolwich.
John Heriot, Esqr. Comptroller of Chelsea Hospital
John Turner, Esqr. St. James's Street
Captn. George Ackton 10 Upper Cadogan Place
Henry Bell, Esqr. 22 Holles St. Cavendish Square
John Froggatt, Esqr. 4 Hare Court, Temple.

5. Geological Drawings. Mr. Brande stated to the Board that some Drawings would be necessary to illustrate his Course of Lectures on Geology, the probable expence of which would not exceed £20.
Resolved, That the above Geological Drawings be done under the direction of Mr. Brande.

Adjourned to Monday the 22d. of January 1821.

J P Auriol V.P.

Dec. 23.

At a Special Meeting of the Managers of the Royal Institution of Great Britain, held at the house of the Institution on Saturday the 23rd. of December, 1820

Present
Charles Hatchett, Esqr. V.P. in the Chair.
Daniel Moore, Esqr.
James Stodart, Esqr.
Thomas Snodgrass, Esqr.
Sir S. Bernard Morland, Bart M.P. Treasurer
George Duckett, Esqr.
Thomas Harrison, Esqr. Secretary.

1. Books mutilated by Mr. Accum. The Secretary stated to the Board, that in consequence of the report made by Mr. Sturt, the Assistant Librarian, to the Managers on the 5th. of November last – that a number of Books, in the Reading Room, had been made imperfect by having several leaves and plates taken out of them in various parts of the Volumes, and that the Books which Mr. Sturt had discovered to be mutilated were Books that Mr. Accum, an Annual Subscriber to this Institution, was in the habit of reading, And that after Mr. Accum left the Institn. he had examined the same books which he had seen in his hand and had found leaves taken out of them. – He had directed Mr. Sturt to have holes bored in the Cupboard of the Room, adjoining the Reading Room, for the purpose of watching more closely the persons who frequented that Room; And that on the Evening of the 20th. of December, Mr. Sturt saw Mr. Accum enter the Room, where he remained nearly two hours, during which time he watched him very closely from the Cupboard, and had no doubt but that he had torn several leaves out of a Volume of Nicholson's Journal, which he had been reading, but owing to a pile of books, which was lying on the table, he could not be positive, and therefore he did not then interfere. Immediately after Mr. Accum had quitted the Institution, which he did in a hurried

304 1820.

and confused manner, Mr. Sturt examined the books which he had seen in Mr. Accum's hands, and found that Parmentier's Treatise on the Composition and use of Chocolate had been taken out of Nicholson's Journal Vol. 5, which he was positive had been entire the week before. — The Secretary further stated that Mr. Sturt was with him at Streatham Park between 8 and 9 o'clock on the next Morning, and minutely communicated to him the above; On which being prevented, by a particular engagement from coming to Town with Mr. Sturt, He wrote to Sir Robert Baker, at Bow Street, an account of the above; but it not being Sir Robert's day of attendance at the Office, Mr. Sturt requested a private hearing with Mr. R. Birnie, the sitting Magistrate, who issued a Warrant to search the house of Mr. Accum in Compton Street, and sent thither Bishop and Nicholls two Officers with Mr. Sturt. — The latter after a search of nearly two hours discovered a number of leaves which he identified as being torn out of Books belonging to the Royal Institution. The Officers brought Mr. Accum and the leaves before Mr. Birnie, when Mr. Sturt again identified sundry leaves as corresponding with books belonging to the Royal Institution which had been mutilated. — The Magistrate after hearing the whole of the Case observed, that however valuable the books might be from which the leaves found in Mr. Accum's house had been taken, yet the leaves separated from them were only waste paper. If they had weighed a pound he would have committed him for the value of a pound of waste paper, but this not being the case he discharged him.

2. The question having been put as to further proceeding against Mr. Accum,

It was resolved unanimously that Mr. Accum be further proceeded against.

Resolved. That a Bill of Indictment be preferred against Mr. Accum at the next Westminster Sessions for the above offence, and that Mr. Danl. Moore

Dec 23. 305

be requested to act therein as Solicitor to the Institution.

Resolved, That the thanks of this Meeting be given to Mr. Sturt for his zeal diligence and intelligence in this business.

J. P. Auriol V.P.

306 1821.

Jan. 22

At a Meeting of the Managers of the Royal Institution of Great Britain held at the house of the Institution on the 22d. of January, 1821,

Present

James Peter Auriol, Esqr. V.P. in the Chair.
Charles Hatchett, Esqr. V.P.
Rt. Hon: Sir John Nicholl, M.P. V.P.
James Stodart, Esqr.
Thomas Harrison, Esqr. Secretary.

1. The minutes of the proceedings of the last & Special Meeting were read and approved.

2. Read and approved the minutes of the Committee of Accounts of this day, and signed an Order for an Advance to the Clerk of £100.

Presents received.

3. Revd. Stephen Weston, presented "A Trimester in France & Swisserland"
M. Hoené Wronski, presented Address to the British Board of Longitude.
John Fearn, Esqr. presented, First lines of the Human Mind.
Sir James Edward Smith, P.L.S. &c. &c. presented A Grammar of Botany.
The Commissioners of the Public Records presented A Catalogue of the Lansdowne Manuscripts 2 vols. fol.

Ordered, — That the thanks of the Managers be returned for these presents.

4. Read the minutes of the last General Monthly Meeting.

5. Read the minutes of the Committee of Visitors entered in their proceedings of the 8th. Instant

Shelves in Library. 6. Mr. Harris requested to be allowed to have a few octavo Shelves put up in a vacant place in the Library which Mr. Royce estimated will be at an expence of about 30£.

Jan. 22 307

Resolved, That the Shelves be put up in the Library at an expence not exceeding 30£. subject to the approval of the Patrons of the Library.

Reqt. of Mr. Sturt. 7. Mr. Sturt requested permission of the Board to exchange the Rooms No. 22 & 23, which he now occupies for No. 18, and the adjoining Room at present unoccupied.

Resolved, That Mr. Sturt's request be complied with, it being understood that in case the Rooms should be wanted for any other purpose he would relinquish them.

8. The following persons having been duly recommended were admitted as Subscribers to the Royal Institution for the present Year.

Annual Subscribers.

Dr. Hawkins 37 Upper Brook Street.
Swynfen Jervis, Esqr. Clarendon Hotel.
General Wemyss 19 Cumberland Street.
Wm. Wilkinson, Esqr. 17 Bedford Place.
James W. Dunkin, Esqr. 3 Torrington Street
Edward Speer, Esqr. 40 Charles St. Westmr.
Captn. J. Harris 39 Foley Place.
Henry Entwisle, Esqr. 25 St. James's Place
Lord James Stuart, Privy Gardens.

To the Lectures.

Miss Hawkins 37 Upper Brook Street.

Adjourned to Monday the 5th. of February.

Spencer.

Feb 5. 1821.

At a Meeting of the **Managers** of the **Royal Institution of Great Britain** held at the house of the Institution on the 5th of February, 1821,

Present

Earl Spencer K.G. President, in the Chair.
James Peter Auriol, Esqr. V.P.
Charles Hatchett, Esqr. V.P.
James Stodart, Esqr.
Robert Williams, Esqr. M.P.
Thomas Snodgrass, Esqr.
Daniel Moore, Esqr.
Sir T. Bernard Morland, Bart M.P. Treasr.
Thomas Harrison, Esqr. Secretary.

1. The minutes of the proceedings of the last meeting were read and approved.

2. Read and approved the minutes of the Committee of Accounts of this day, and signed an Order for an Advance to the Clerk of £170.

3. Audited the Treasurer's Account for the year 1820.

4. In pursuance of a nomination under the Act of Parliament by Robert Bally, M.D. his Son Captn. Robert Bally of the Grenadier Guards. was admitted to the privileges of a Life Subscriber to the Royal Institution.

Presents received.

5. Lieut. Col. C. P. de Bosset, presented
Parga and the Ionian Islands.
Dr. John Thomas, presented
Practical Observations on Chronic affections of the digestive Organs.
Ordered, That the thanks of the Managers be returned for these presents.

Feb. 5.

Hon. Admission. 6. On the application of Sir H. Davy, Bart
His Excellency Count Lieven Ambassador Extray. & Minister Plenipotentiary of His Majesty the Emperor of Russia. was admitted to the Royal Institution under Art 15 Chap VI of the Bye Laws.

Request of Mrs. Greenwood. 7. Mrs. Greenwood requested permission of the Managers to be allowed, for a short time, to have with her, her Son Willm. Greenwood till he can get a situation.
Resolved. That her request be complied with; but * * she must not expect that her Son will be again employed at the Royal Institution.

8. The following persons having been duly recommended were admitted as Subscribers to the Royal Institution for the present year.

Annual Subscribers.
Hugh Mills Bunbury, Esqr. 4 Old Burlington Street
James Nugent Daniell, Esqr. of Morden Hall Surry.
Thos. Forbes Reynolds Esqr. of Wallington Surry
Revd. George Spain 24 Great Russell St.
John George Lockett, Esqr. 1 Upper Conway St.
George Dowdeswell, Esqr. 33 Montagu Square
James Hay Langham, Esqr. Langham Place
Thomas Tucker Esqr. 25 Maddox Street
Walter Mathison, Esqr. Hampstead
George Powlett Thomson Esqr. 129 Park Street
Sir Alexr. Johnston 19 Gt. Cumberland Place
Captn. Edward Stopford, R.N. 63 Welbeck Street
Rt. Hon: Sir John Newport, Bart 23 Bury Street

To the Lectures.
Miss Bunbury, 4 Old Burlington St.
Lady Dufferin, 13 Cavendish Square
Miss M. A. Hendrie & Miss Wilson 12 Titchborne Street
Miss Rickards 54 Wimpole Street.
Miss Cutler 23 Devonshire Street Portland Place.
Lady & Miss Johnston, 19 Great Cumberland Place
Mrs. & Miss Blackwood Upper Seymour Street

Adjourned to Monday 12th of February

Spencer.

Feb. 12. 1821.

At a Meeting of the **Managers** of the **Royal Institution of Great Britain** held at the house of the Institution on the 12th of February, 1821,

Present

Earl Spencer, K.G. President, in the Chair.
Charles Hatchett, Esqr. V.P.
Daniel Moore, Esqr.
Earl of Bessborough, V.P.
James Stodart, Esqr.
Thomas Harrison, Esqr. Secretary.

1. The minutes of the proceedings of the last meeting were read and approved.

2. Read and approved the minutes of the Committee of Accounts of this day, and signed an Order for an Advance to the Clerk of £170.

Presents received.

3. The Horticultural Society, presented
Transactions of the Horticultural Society vol. 4 pt. 2
The Master, Wardens and Society of Apothecaries presented Memoirs of the Botanick Garden at Chelsea.
Professor Leslie, presented
Description of Instruments designed for extending and improving Meteorological observations.
Ordered, That the thanks of the Managers be returned for these presents.

4. Read the minutes of the last General Monthly Meeting.

Hon. Admission. 5. Dr. Ekström, a Foreigner of distinguished Scientific attainmts. was admitted to the Royal Institution under Art 15 Chap VI of the Bye Laws.

Coals. 6. Ordered, That 30 Chaldrons of the best Walls' End Coals be immediately laid in the Cellars of the Royal Institution

Feb. 12

7. The following persons having been duly recommended were admitted as Subscribers to the Royal Institution for the year 1821.

Annual Subscribers.

Major Moody Royl. Engrs. 7 Alfred Place
W. H. Howard, Esqr. 3 Hyde Park Corner.
Captn. Hanway, Royl. Engrs. Woolwich
Charles Thesiger, Esqr. } 40 Bloomsbury Square
Augustus Thesiger Esqr. }
Wm. Henry Cheere, Esqr. 18 Charles St. Berkeley Square
James Moore, Esqr. 11 Conduit Street.
William Munro, Esqr. 98 Wimpole Street
Honble George Spencer Spencer House.

To the Lectures.
Mrs. Thurburn 19 Mecklenburgh Square.
Miss Magniac, Kensington
Miss Eliot, 58 Grosvenor Street.
Miss Smith }
Miss Charlotte Smith } 39 Berkeley Square
Mrs. Smith }
Miss Harman Sloane Street
Mrs. Rattray } Lower Brook Street.
Miss Waddington }
Lady Lubbock St. James's Place.

Adjourned to Monday the 26th of February.

Spencer.

312

Feb 26. | 1821.

At a Meeting of the Managers of the Royal Institution of Great Britain held at the house of the Institution on the 26th of February, 1821

Present

Earl Spencer K.G. President, in the Chair.
Charles Hatchett, Esqr. V.P.
George Duckett, Esqr. V.P.
Rt. Honble Sir John Nicholl, M.P. V.P.
Daniel Moore, Esqr.
John Chamier, Esqr.
James Stodart, Esqr.
Robert Williams, Esqr. M.P.
Thomas Snodgrass, Esqr.
Sir S. Bernard Morland, Bart M.P. Treasurer
Thomas Harrison, Esqr. Secretary.

1. The minutes of the proceedings of the last meeting were read and approved.

2. Read and approved the minutes of the Committee of Accounts of this day, and signed an Order for an Advance to the Clerk of £180.

Presents received.

3. An attempt to analyse the Automaton Chess Player of M. de Kempelen — By the Author.
The Medical Intelligencer No. 15 — By the Editor.

4. The following persons having been duly recommended were admitted as Subscribers to the Royal Institution for the present year.

Annual Subscribers.

Edward Lombe junr. Esq. of Melton Norfolk.
Captn. B. Bishop 4 Cumberland St. Portman Sq.
Captn. Hamlyn 32 Upper Grosvenor St.
Charles Mackinnon, Esqr. Camden Hill.

313

Feb 26.

To the Lectures.

Mrs. & Miss Capel. Russell Square.
Mrs. Dilkes, Upper Seymour Street.
Mrs. Charles Mills } Manchester Square
Miss Mills }
Miss L. Blois Turner
Honble Mrs. Wm. Cust — Park Street.
Miss Metcalfe 19 Upper Harley St.

Adjourned to Monday the 5th of March.

Spencer

314

Mar. 5. | 1821.

At a Meeting of the Managers of the Royal Institution of Great Britain held at the house of the Institution on the 5th of March 1821.

Present

Earl Spencer K.G. President, in the Chair.
Earl of Bessborough V.P.
Charles Hatchett, Esqr. V.P.
George Duckett, Esqr. V.P.
Daniel Moore, Esqr.
Thomas Snodgrass, Esqr.
James Stodart, Esqr.
John Chamier, Esqr.
Sir S. Bernard Morland, Bart M.P. Treasurer

1. The minutes of the proceedings of the last meeting were read and approved.

2. Read and approved the minutes of the Committee of Accounts of this day, and signed an Order for an Advance to the Clerk of £30.

Presents received.

3. Samuel Mellish, Esqr. M.R.I. presented
A Model of a double Crane for raising and lowering Goods at the same time and by the same power.
Mr. Hawkins Browne presented
An Engraved Portrait of I. H. Browne, Esqr.
Dr. Granville presented
3 Copies of a Reply to Mr. Brande
A Letter to Viscount Sidmouth — by Lachlan Macquarie
Mr. Bracy Clark presented
Description of a New Horse Shoe
Ordered,
That the thanks of the Managers be returned for these presents

4. The following persons having been duly recommended

315

Mar. 5.

were admitted as Subscribers to the Royal Institution for the present year.

Annual Subscribers.

Charles Wild, Esqr. 35 Albemarle Street
Henry B. Faulkener, Esqr. 89 Sloane Street.
Edward Parratt Esq. 12 Bridge St. Westmr.

To the Lectures.

The Honble Mrs. Douglas 67 Lower Grosvenor St.

5. Adjourned to Monday the 19th Instant.

Spencer

316

Mar 19. 1821.

At a Meeting of the Managers of the Royal Institution of Great Britain held at the house of the Institution on the 19th of March, 1821,

Present
Earl Spencer K.G. President in the Chair.
Charles Hatchett, Esqr. V.P.
George Duckett, Esqr. V.P.
Thomas Snodgrass, Esqr.
John Chamier, Esqr.
James Stodart, Esqr.
Sir S. Bernard Morland, Bart. M.P. Treasr.
Thomas Harrison, Esqr. Secretary.

1. The minutes of the proceedings of the last meeting were read and approved.

2. Read and approved the minutes of the Committee of Accounts of this day, and signed an Order for an Advance to the Clerk of £280.

Present received.

3. Daniel Hill, M.D. Surgeon presented
Practical Observations on the use of Oxygen, or Vital Air.
Ordered, That the thanks of the Managers be returned for this present

4. Read the minutes of the last General Monthly Meeting.

5. The following persons having been duly recommended were admitted as Subscribers to the Royal Institution for the present Year.

Annual Subscribers.
The Honble. Arthur Legge – Berkeley Square.
Major Handley, 61 South Molton Street.
Thos. Farquhar Chilver No. 5 New Burlington St.
George Newnham Collingwood, Esqr. Hyde Park co[rner]

Mar 19. 317

Captn. George Digby, R.N. Maddox St.
Henry Hall, Esqr. 33 Lower Brook St.
Sir Abraham Hume, Bart. Hill Street
Honble. Wm. Cust Park St. Grosvenor Square
Henry Handley, Esqr. M.P. 7 Charles St. St. James's Sq.

To the Lectures.

Dowager Lady Brownlow, Honble. Miss A. Cust, Honble. Miss L. Cust, Honble. Miss K. Cust – 30 Hill Street
Miss C. Maxwell 62 Harley St.
Mrs. Lambert Blair, Mrs. James Blair – 12 Devonshire Place
Mr. Crichlow, Miss Crichlow – 1 Grafton St. Fitzroy Square

Adjourned to Monday the 2d. of April.

Spencer

318

April 2. 1821.

At a Meeting of the Managers of the Royal Institution of Great Britain held at the house of the Institution on the 2d of April, 1821,

Present
Earl Spencer, K.G. President, in the Chair.
Charles Hatchett, Esqr. V.P.
James Peter Auriol, Esqr. V.P.
George Duckett, Esqr. V.P.
John Chamier, Esqr.
Thomas Snodgrass, Esqr.
James Stodart, Esqr.
Daniel Moore, Esqr.

1. The minutes of the proceedings of the last meeting were read and approved.

2. Read and approved the minutes of the Committee of Accounts of this day, and signed an Order for an Advance to the Clerk of £80.

Presents received

3. The Society for the Encouragement of Arts &c. presented
The 38th. Volume of their Transactions.
The Astronomical Society of London, presented
Address & Regulations of their Society.
Ordered,
That the thanks of the Managers be returned for these presents.

4. The following persons having been duly recommended were admitted as Subscribers to the Royal Institution for the present year.

Annual Subscribers.

William Whymper, Esqr.
Charles Delancey, Esqr. Cheltenham
The Revd. Edwd. Chichester 3 Sackville Street
Charles Walker, Esqr. 40 Bedford Square

April 2 319

To the Lectures.

Countess Brownlow Cavendish Square.

5. Adjourned to Monday the 16th. Instant.

Spencer

320

April 16. 1821.

At a Meeting of the Managers of the Royal Institution of Great Britain held at the house of the Institution on the 16th of April, 1821

Present
Earl Spencer, K.G. President, in the Chair.
Charles Hatchett, Esqr. V.P.
James Peter Auriol, Esqr. V.P.
George Duckett, Esqr. V.P.
Daniel Moore, Esqr.
John Chamier, Esqr.
Thomas Snodgrass, Esqr.
James Stodart, Esqr.
Sir S. Bernard Morland, Bart M.P. Treasurer
Thomas Harrison, Esqr. Secretary.

1. The minutes of the proceedings of the last meeting were read and approved.

2. Read and approved the minutes of the Committee of Accounts of this day, and signed an Order for an Advance to the Clerk of £160.

Presents received.

3. The Chevalier de Sade – presented
L'Art de faire des Lois, 8vo
M. Bergasse – presented thro' the Chevalier de Sade
Essai sur la Propriété, 8vo

Ordered,
That the thanks of the Managers be returned for these presents.

Prosecution of Mr. Accum

4. In consequence of the Resolution of the Board of Managers of the 23d of Decr. last, which directed that a Bill of Indictment should be preferred against Mr. Accum, and that Daniel Moore, Esqr. be requested to act therein as Solicitor, – Mr. Moore reported that A Bill of Indictment had been preferred at the last January Westminster Sessions against Frederick Accum for feloniously stealing and taking away 200 pieces of

April 16. 321

paper of the value of ten pence, and also for feloniously stealing and taking away four ounces weight of paper of the value of four pence, the property of the Members of the Royal Institution of Great Britain. – The Grand Jury of Westminster found a true Bill against Mr. Accum and process was issued against him. On the day after the Bill had been found, Mr. Accum's Solicitor applied to the Chairman of the Sessions for Mr. Accum to be permitted to enter into recognizances for his appearance to take his trial upon the Indictment at the ensuing April Sessions; and the application was granted. – Mr. Accum thereupon appeared in Court with his two Sureties, Randolph Ackermann of the Strand, Publisher, and John Papworth of Bath Place New Road, Architect, and entered into the usual Recognizances himself in £200, and the Sureties in £100 each.

At the April Sessions the Prosecutors were ready to proceed to the Trial, when Mr. Accum's Solicitor attended the Court, and stated that Mr. Accum was not to be found, and that Mr. Ackermann one of his Sureties had been in search of him but without success – A Motion was then made to the Court to estreat the Recognizances which was ordered accordingly, unless Mr. Accum should appear before the end of the Sessions, and as he made default in appearing, the recognizancers were estreated.

5. Read the minutes of the last General Monthly Meeting.

6. The following persons having been duly recommended were admitted as Subscribers to the Royal Institution for the present year.

Annual Subscribers.
Major Genl. Peachy 66 Baker Street
Robert Robertson, Esqr. 51 Wimpole St.
Captn. Henry Walpole 36 Half Moon St.

To the Lectures.
Mrs. Peachy, 66 Baker Street
Mrs. Leonard Currie } 17 Montagu Street
Miss F. Currie }
Miss M. Currie }

322

1821.

7. Ordered, That the House Lists be filled up as follows.

President. – Earl Spencer, K.G. F.R.S. and F.S.A.

Treasurer. Sir S. Bernard Morland, Bart M.P. LL.

Secretary. Thomas Harrison, Esqr. F.R.S.

Managers.
Samuel Boddington, Esqr.
Earl Brownlow, F.R.S. and F.S.A.
John Chamier, Esqr.
Earl of Darnley, F.R.S.
Sir H. Davy, Bart. L.L.D. Pr. R.S. &c &c &c
George Duckett, Esqr. F.R.S and F.S.A
Charles Hatchett, Esqr. F.R.S and F.S.A
Sir J. C. Hippisley, Bart. L.L.D. F.R.S and F.S.A.
Daniel Moore, Esqr. F.R.S. F.S.A and F.L.S.
Rt. Honble Sir John Nicholl, M.P. L.L.D. F.R.S
William Hasledine Pepys, Esqr. F.R.S.
Claude Scott, Esqr. F.L.S and F.H.S.
Thomas Snodgrass, Esqr.
Wm. Sotheby, Esqr. F.R.S and F.S.A.
James Stodart, Esqr.

Visitors.
George Aust, Esqr. F.R.S. and F.S.A
Lord Carrington, F.R.S.
Earl of Chichester, F.R.S
Rear Adml. Sir Edward Codrington, K.C.B.
Earl of Egremont, F.R.S. and F.S.A.
John Fuller, Esqr.
Daniel Giles, Esqr.
Sir Benjn. Hobhouse, Bart. F.R.S. and F.S.A
Lord Holland, F.R.S
Gibbes Walker Jordan, Esqr. A.M. F.R.S.
Earl of Mountnorris, F.R.S and F.S.A
Revd. Sir John Robinson, Bart.
Richd. Horsman Solly, Esqr. F.R.S. and F.S.A
Lord Teignmouth.
Giles Templeman, Esqr.

Adjourned to Monday the 30th Inst. Spencer

April 30. 323

At a Meeting of the Managers of the Royal Institution of Great Britain, held at the house of the Institution on the 30th of April 1821,

Present
Earl Spencer, K.G. President, in the Chair
James Peter Auriol, Esqr. V.P.
Charles Hatchett, Esqr. V.P.
John Chamier, Esqr.
Daniel Moore, Esqr.
Thomas Snodgrass, Esqr.
James Stodart, Esqr.
Thomas Harrison, Esqr. Secretary.

1. The minutes of the proceedings of the last meeting were read and approved.

2. Read and approved the minutes of the Committee of Accounts of this day, and signed an Order for an Advance to the Clerk of £140.

Lectures Mr. Landseer

3. Read a letter from Mr. J. Landseer, addressed to Earl Spencer dated 26th April 1821, wherein he stated that as he is now arranging his discoveries for publication respecting the Babylonian Engravings which Captn. Lockett placed in his hands, and which are in fact Astronomical Monuments of very remote Antiquity, he should like to fashion them in the shape of lectures, if it should be agreeable to the Managers of the Royal Institution for him to deliver them.

Resolved,
That the Secretary be requested to inform Mr. Landseer that the arrangements for the lectures of the present year have been completed, but that his proposed lectures appear to be of such interest that the Managers would be happy to know when, and on what terms he will be able to commence a short Course thereof.

4. Read the minutes of the last General Meeting.

324 1821.

5. The Managers having taken into their consideration the long and zealous services of Mr. Fincher the Clerk, it was unanimously resolved,

Mr. Fincher to be Asst. Secretary.

That in future he be denominated Assistant Secretary to this Institution.

6. The following persons having been duly recommended were admitted as Annual Subscribers to the Royal Institution for the present year.

Revd. George May, 41 Hertford St.
John Binny, Esqr. – 25 Upper Wimpole St.

J. C. Hippisley

325

May 7.

At a Meeting of the Managers of the Royal Institution of Great Britain, held at the house of the Institution on the 7th of May, 1821.

Present

Sir J. C. Hippisley, Bart. in the Chair.
John Chamier, Esqr.
Claude Scott, Esqr.
James Stodart, Esqr.
Charles Hatchett, Esqr.
George Duckett, Esqr.
Thomas Snodgrass, Esqr.
Daniel Moore, Esqr.
Thomas Harrison, Esqr. Secretary.

1. The minutes of the proceedings of the last meeting were read and approved.

2. Read and approved the minutes of the Commee. of Accounts of this day, and signed an Order for an Advance to the Clerk of £40.

Lectures. Mr. Landseer

3. Read a letter from Mr. Landseer, addressed to Thos. Harrison, Esqr. Secy., wherein he stated that as the present Season will close in June, he should prefer arranging with the Managers for the next Season, by which time he would engage to prepare six Lectures; the terms of remuneration to be the same as before.

Presents received.

4. T. Edward Bowdich, Esqr. presented
An Essay on the Superstitions, Customs, & Arts common to the ancient Egyptians, Abyssinians & Ashantees.
The British & French Expedition to Teembo with remarks on Civilization in Africa.
An Essay on the Geography of North Western Africa.
John Warren, Esqr. presented
The Magazine of the Fine Arts No. 1.
Henry Jeffreys, Esqr. presented
Practical Observations on the use of the Cubebs, or Java Pepper. 2 Cops.

326 1821.

Thomas Chevalier, Esqr. presented
The Hunterian Oration, delivered before the Royal College of Surgeons, Feby 14th. 1821.
Thomas Hare, Esqr. – presented
A view of the Structure and Disorders of the Stomach.
George Mackenzie, Esqr. – presented
The Royal Naval & Military Calendar.

Ordered,
That the thanks of the Managers be returned for these presents.

5. Read the minutes of the General Annual Meeting.

6. The following persons having been duly recommended were admitted as Subscribers to the Royal Institution for the present year.

Annual Subscriber.
James Sutherland, Esqr. of Margaret St. Cavendish Square

To the Lectures.
Miss Sutherland of Margaret St. Cavendish Sq.
Mrs. Townsend Grafton Street.

Adjourned to Monday the 21st. of May.

Spencer

327

May 21.

At a Meeting of the Managers of the Royal Institution of Great Britain held at the house of the Institution on the 21st of May, 1821

Present

Earl Spencer K.G. President in the Chair.
Sir J. C. Hippisley, Bart.
John Chamier, Esqr.
Sir H. Davy, Bart.
Claude Scott, Esqr.
Thos. Snodgrass, Esqr.
James Stodart, Esqr.
Charles Hatchett, Esqr.
George Duckett, Esqr.
Sir S. Bernard Morland, Bart. M.P. Treasr.
Daniel Moore, Esqr.
William Sotheby, Esqr.
Earl Brownlow.
Thomas Harrison, Esqr. Secretary.

1. The minutes of the proceedings of the last meeting were read and approved.

2. Read and approved the minutes of the Committee of Accounts of this day, and signed an Order for paymt. of the following sum, viz.

Payments

1 Years' Ground Rent to the City of London to Lady day 1821 – £150.

Life Subsr.

3. In pursuance of a nomination under the Act of Parliament by John Hornby, Esqr. – his Son William Hornby, Esqr. of Portland Place, was admitted to the privileges of a Life Subscriber to the Royal Institution.

4. Read the minutes of the last General Monthly Meeting.

5. The President nominated as Vice Presidents for the

present Year.

Vice Presidents.

Earl Brownlow.
Earl of Darnley.
Sir H Davy, Bart.
George Duckett, Esqr
Charles Hatchett, Esqr
Sir J. Nicholl, M.P.
Thomas Snodgrass, Esqr
Sir J. C. Hippisley, Bart.
Claude Scott, Esqr

Resolved,

Mr. Faraday Superintendant of the House in Mr Brande's Absence.

That Mr Faraday be allowed to occupy with the permission of Mr Brande, N.º 18 and the adjoining Room, in addition to the Rooms already occupied by him, and that he be appointed Superintendant of the House and Laboratory in the absence of Mr Brande.

7. The following persons having been duly recommended were admitted as Subscribers to the Royal Institution for the present year.

Annual Subscribers.
John Julius Wm Angerstein, Esqr Cumberland place
Wm Newnham, Esqr Jr 2 Little Stanhope S.^t
George Sydenham, Esqr 2 Regent Circus.

To the Lectures.
Mr Munn
Miss Acland 49 Albemarle S.^t

Adjourned to Monday the 4.th of June.

J. C. Hippisley

June 4.

At a Meeting of the Managers of the Royal Institution of Great Britain held at the house of the Institution on the 4.th of June 1821

Present
Earl Spencer, K.G. President, in the Chair
Sir J. C. Hippisley, Bart. V.P.
Sir H Davy, Bart. V.P.
Thomas Snodgrass Esqr V.P.
Claude Scott, Esqr V.P.
Charles Hatchett, Esqr V.P.
John Chamier, Esqr
James Stodart, Esqr
Daniel Moore, Esqr
William Sotheby, Esqr
Sir S. Bernard Morland Bart. M.P. Treas.^r
R.^t Honble Sir Jno Nicholl, M.P. V.P.
Thomas Harrison, Esqr Secretary.

1. The minutes of the proceedings of the last meeting were read and approved.

2. Read and approved the minutes of the Committee of Accounts of this day, and signed an Order for an Advance to the Assistant Secretary of £150.

3. Read a letter from Mr Stuart the Assist.^t Librarian.

Resolved,

That the subject of it be referred to Mr Brande.

Presents received.

4. Sir James Edward Smith, Pres L. S. &c. &c. presented
A Selection of the Correspondence of Linnæus & other Naturalists, 2 vol.^s 8.vo
J. Henry, Esqr presented
Report on the Criminal Law at Demerara, 8.vo

Ordered,

That the thanks of the Managers be returned for these presents.

5. Read the minutes of the last General Meeting.

6. The following persons having been duly recommended were admitted as Subscribers to the Royal Institution for the present year.

Annual Subscriber.
George Chapple Norton, Esqr of the Middle Temple

To the Lectures.
Mrs J. Weyland 5 Cork Street.
Miss Law 24 Upper Wimpole S.^t

Adjourned to Monday the 18.th of June.

Spencer

June 18.

At a Meeting of the Managers of the Royal Institution of Great Britain held at the house of the Institution on the 18.th of June, 1821

Present
Earl Spencer, K.G. President, in the Chair.
Sir J. C. Hippisley, Bart. V.P.
Thomas Snodgrass, Esqr V.P.
Charles Hatchett, Esqr V.P.
Claude Scott, Esqr V.P.
Sir H. Davy, Bart. V.P.
John Chamier, Esqr
James Stodart, Esqr
Sir S. Bernard Morland, Bart M.P. Treasurer
Daniel Moore, Esqr

1. The minutes of the proceedings of the last meeting were read and approved.

2. Read and approved the minutes of the Committee of Accounts of this day, and signed an Order for an Advance to the Assist.^t Secretary of £190.

Presents received.

3. Dr John Reid, presented
Essays on Hypochondriasis & other Nervous Affections 8.vo
The Horticultural Society, presented
Transactions of their Society vol.^e 4 part 3 – and a new List of the Society.

Ordered,

That the thanks of the Managers be returned for these presents.

4. Read the minutes of the last General Monthly Meeting.

5. Read the minutes of the Committee of Visitors entered in their proceedings of the 11.th Instant.

Repairs

Ordered,

That an Estimate of the expence of painting the outside wood work and Iron Palisades in front of the House be laid before the next Board.

John Chamier

Adjourned to the 2.^d of July.

332

July 2

1821.

At a Meeting of the Managers of the Royal Institution of Great Britain held at the house of the Institution on the 2.d of July, 1821

Present
John Chamier, Esq.r in the Chair.
James Stodart, Esq.r
Sir H. Davy, Bart: V.P.
Samuel Boddington, Esq.r
Thomas Harrison, Esq.r Secretary.

1. The minutes of the proceedings of the last meeting were read and approved.

2. Read and approved the minutes of the Committee of Accounts of this day, and signed an Order for payment of the following sums, viz.t

William Tho.s Brande, Esq.r for a year's Salary as Prof.r of Chemistry (1821)	£200 —
W.m Harris Librarian for half years Salary to 31 Dec.r 1820	90 —
Jos.h Fincher on account of the Royal Institution	197 —
	487 —

Presents received.

3. The R.t Hon: Lord de Dunstanville presented
Carew's Survey of Cornwall 4.to elegantly bound in Russia
Ordered,
That the thanks of the Managers be returned for this present.

4. Read an estimate from M.r J. C. Parker Painter & Glazier to this Institution, of the expence of painting all the outside Wood and Iron work of the Royal Institution which amounted to £37.14.0
Resolved, That it be done at an expence not exceeding the above sum.

333

July 2.

5. Read the minutes of the last General Meeting.

Leave of absence to Lib.n — 6. M.r Steat requested permission for leave of absence for a month during the recess which was granted provided he be not absent at the same time as M.r Harris.

7. Resolved, That in order to meet the Draft for £487, the Treasurer be requested to allow his account at the Bankers to be overdrawn to the amount of £400 which will be wanted for about 6 months.

Lectures. — 8. The Board having taken into consideration The Rev.d T. D Fosbroke's application to deliver a Course of Lectures on Archaeology in the ensuing Season,
Resolved. That his offer be declined.

9. Resolved, That M.r Landseer's proposal of delivering 6 Lectures, on the Engraved Hieroglyphics of Canaan Egypt & Chaldea, in the ensuing Season, be accepted.

10. M.r Stodart reported that D.r Roget would be willing to deliver a Course of 12 Lectures on Comparative Physiology in the next Season, the terms as formerly 100 guineas for the Course.
Resolved, That D.r Roget be requested to give the above lectures.

Adjourned to the 5.th of November.
Thomas Snodgrass.

334

Nov. 5.

1821

At a Meeting of the Managers of the Royal Institution of Great Britain held at the house of the Institution on the 5.th of November, 1821

Present
Thomas Snodgrass, Esq.r V.P. in the Chair.
Sir H. Davy, Bart V.P.
Charles Hatchett, Esq.r V.P.
George Druckett, Esq.r V.P.
Sir S. Bernard Morland, Bart M.P. Treas.r
Daniel Moore, Esq.r
James Stodart, Esq.r
Thomas Harrison, Esq.r Secretary.

1. The minutes of the proceedings of the last meeting were read and approved.

2. Read and approved the minutes of the Committee of Accounts of this day, and signed an Order for an advance to the Assist.t Secretary of £250.

Presents received.

3. M.r Baron Maseres, presented
Sir John Temple's History of the Irish Rebellion.
The Curse of Popery & Popish Princes
The doctrine of Permutations & Combinations
Tracts on the resolution of affected Algebraick Equat.s
Tracts on the resolution of Cubick & Biquadratick d.o
The Royal Society, presented
Philosophical Transactions for 1821 – part 1.
The Society of Antiquaries presented
Archaeologia, vol 19 part 2.
M.r Piercy Ravenstone presented
Doubts on Population & political Economy.
Sir Richard Phillips presented
The proximate causes of Material Phenomena consid.d
The Directors of the East India Company presented
Morrison's Chinese Dictionary, N.o 4 of Vol 1 quarto.
The American Academy presented
Memoirs of the American Academy of Arts & Sciences vol 4 pt 2

336

Nov. 5.

The Society for the improvement of Prison Discipline presented Third Report of the Committee.
The Hon: & Rev.d Francis Egerton presented
The Life of Lord Chancellor Egerton.
By The Author — "Advice to the Physician, the Surgeon and the Apothecary."
M.r Bracy Clark presented
An Essay on the Running Frush of Horses Feet.
Ordered,
That the thanks of the Managers be returned for these presents.

Life Subs.er — 4. M.r Samuel Acton of Wilson Street Finsbury Square was this day admitted to the privileges of a Life Subscriber to the Royal Institution under a Codicil to the Will of his late Uncle Nath.l Wright, Esq.r dated 8.th January 1818.

Hon Admissions — 5. Count de Pollon His Sardinian Majesty's Chargé d'affaires & Count de Mandelsloh Chargé d'affaires of His Majesty the King of Wurttemberg, were admitted to the Royal Institution under Art: 15 Chap VI of the Bye laws.

6. Read the minutes of the last General Monthly Meeting.

7. Read an estimate from Mess.rs Pringle of the expence of making new frames of Mahogany, light & neat, of the Ground-floor window blinds to be filled in with venetian laths painted drab colour, the cost of each Window will be about 40.s
Ordered, That it be done at an expense not exceeding the above sum for each window.

8. The following persons having been duly recommended were admitted as Subscribers to the Royal Institution for the year 1822.
Annual Subscribers.
William John Davidson, Esq.r Earls Terrace Kensington
George Lynch, Esq.r 91 Great Russell S.t

Annual Subscribers contd. 1821.

Dr. James Paterson 19 Bury St. St. James's
Giles Prodgers, Esqr. 11 Dover St. Piccadilly
Mr. Fredk. Robe – Shooters Hill.
Mr. Miles of Larkhall Lane Clapham
Dr. Harrison 7 Holles Street

9. Adjourned to Monday the 3d. of December

Charles Hatchett

Dec. 3.

At a Meeting of the Managers of the Royal Institution of Great Britain held at the House of the Institution on the 3d. of December, 1821

Present

Charles Hatchett Esqr. V.P. in the Chair.
Sir H. Davy, Bart. V.P.
George Duckett, Esqr. V.P.
Thomas Snodgrass, Esqr. V.P.
Sir S. Bernard Morland, Bart. M.P. Treas.
Daniel Moore, Esqr.

1. The minutes of the proceedings of the last meeting were read and approved.

2. Read and approved the minutes of the Committee of Accounts of this day and signed an Order for an advance to the Assist. Secretary of £100, and a further advance of £50 on the 1st. of January next.

Letters received.

Lectures. 3. Read a letter from W. M. Craig, Esqr. addressed to Thos. Harrison, Esqr. Secy. offering his services to the Managers in the next Spring to read a new Course of Lectures on Drawing & Painting as branches of elegant Education

Resolved, That Mr. Craig's obliging offer be declined as the arrangement of the Lectures for the next Season is completed.

Read a letter from Mr. J. George Wood, addressed to Charles Hatchett Esqr. offering to the consideration of the Managers, to deliver a Course of six Lectures in the ensuing Season on the Sketches of History as connected with the progress of civilization. Illustrated by numerous Drawings.

Resolved,

That Mr. Wood's offer be declined, the Lectures being complete for the ensuing Season.

Presents received.

4. The Commissioners of the Public Records presented Reports from the Commissioners of the Public Records, with an Appendix of Charters – 2 vols. folio.

1821.

Captn. Thos. Lynn presented Star Tables for 1822, 8vo.

W. M. Craig, Esqr. presented A Course of Lectures on Drawing, Painting, & Engraving 8vo.

J. S. Miller, Esqr. presented The Natural History of the Crinoidea

Apsley Pellatt, Esqr. presented Memoirs of Glass Manufactures.

Ordered,

That the thanks of the Managers be returned for these presents.

5. The following persons having been duly recommended were admitted as Annual Subscribers to the Royal Institution for the year 1822.

Annl. Subscrs.

Redmond W. Pilkington, Esqr. Whitehall Yard.
James F. Palmer, Esqr. 50 South St.
Saml. Haynes, Esqr. 71 St. James's Street.

Adjourned to Monday the 21st. January 1822.

Humphry Davy

1822
Jan 21.

At a Meeting of the Managers of the Royal Institution of Great Britain held at the house of the Institution on the 21st. of January, 1822.

Present

Sir Humphry Davy, Bart. V.P. in the Chair
Charles Hatchett, Esqr. V.P.
George Duckett, Esqr. V.P.
Sir S. Bernard Morland, Bart. M.P. Treasurer
Daniel Moore, Esqr.

1. The minutes of the proceedings of the last meeting were read and approved.

2. Read and approved the minutes of the Committee of Accounts of this day.

3. Read a letter from Messrs. Pellatt & Green of St. Pauls Church yard requesting to be allowed to place a Case of Specimens of their Crystallo Ceramie or Glass incrustation, in the Mineralogical Room of the Royal Institution.

Resolved, That this request cannot be complied with

Lectures. 4. Read a letter from Mr. John Webster addressed to W. T. Brande, Esqr. offering to the Managers to deliver a Course of three lectures on Steam Engines in the Royal Institution.

Resolved; That Mr. Webster's offer be declined.

Presents received.

The Geological Society presented Transactions of their Society vol. 5 pt. 2.

The Royal Society presented Philosophical Transactions for 1821, part 2.

Henry Goodwyn, Esqr. M.R.I. presented Introduction to a Synoptical Table of English & French Lineal measures – 2 copies.

The Revd. Stephen Weston presented A Visit to Vaucluse Nismes &c. 8vo.

340

1822

Mr. Thomas Gill presented
The Technical Repository No. 1.
Dr. John Elliotson presented
Cases of the efficacy of the Hydrocyanic Acid.
Introductory Lecture of a Course upon State-Medicine.
The Editor of the London Medical & Physical Journal presented
An Historical Essay on the progress of the Medical Sciences
The Publishers of the Royal Blue Book presented
Royal Blue Book; or Fashionable Directory.
Ordered,
That the thanks of the Managers be returned for these presents.

5. Read the minutes of the last General Monthly Meeting.

6. Read the minutes of the Committee of Visitors entered in their proceedings of the 14th. Instant.

Coals. 7. Ordered, That Messrs. Dickins & Co. do immediately supply the Institution with 20 Chaldrons of the best Walls-End Coals.

8. The following persons having been duly recommended were admitted as Annual Subscribers for the year 1822.

Annl. Subsers.
Mr. Colburn 50 Conduit Street
Robert Tierney, Esqr. Tax Office.
Alexr. J. Wilson, 4 Clarges Street
George Henry Sigel, Esqr. 91 Dean St. Soho.
J. St. Vincent Bowen, Esqr. Dover Street.
Lieut. R. H. S. Cooper, Royl. Engineers 40 Cadogan Place
Wm. Sherwood, Esqr. 5 Lower Connaught place.
Thos. Edwd. Dicey, Esqr. Kensington Gore.
Major Hardy
Josh. Sellon, Esqr. 19 Blandford Street
Adjourned to 4th. Feby.
Thomas Snodgrass V.P.

341

Feb. 4.

At a Meeting of the Managers of the Royal Institution of Great Britain, held at the house of the Institution on the 4th. of February, 1822

Present
Thomas Snodgrass, Esqr. V.P. in the Chair.
George Duckett, Esqr. V.P.
Sir H Davy, Bart V.P.
Sir S. Bernard Morland, Bart M.P. Treasurer
Daniel Moore, Esqr.
James Stodart, Esqr.
Wm. Hasledine Pepys, Esqr.
Rt. Hon: Sir John Nicholl, V.P.
Thomas Harrison, Esqr. Secretary.

1. The minutes of the proceedings of the last meeting were read and approved.

2. Read and approved the minutes of the Committee of Accounts of this day, and signed an Order for payment of the following sums, vizt.

Payments.

Reeve for 2 quarters Assessed taxes to Michas. 1821	79 . 3 . 4
Josh. Fincher on account of the Royal Institution	80 . . —
	£ 159 . 3 . 4

Presents received.

3. The Leeds Philosophical & Literary Society presented
An Introductory Discourse by Chas. Turner Thackrah, Esq.
Mr. Thomas Gill presented
The Technical Repository, No. 2.
His Majesty's Commissioners of the Public Records presented
Rymer's Foedera, vol 2 part 2.
Valor Ecclesiasticus temp. Henr VIII. vol 4.
Inquisitiones post Mortem, vol 3.
Ordered,
That the thanks of the Managers be returned for these presents.

342

1822.

4. Audited the Treasurer's Account for the year 1821.

5. R. Rush, Esqr. Envoy Extraordinary and Minister Plenipotentiary from the United States of America, was admitted to the Royal Institutn. under Art 15 Chap VI of the Bye laws:

Mr. Sturt to supply Stationery 6. Resolved, That Mr. Sturt the Assistt. Librarian be in future allowed to supply the Stationery of the Royal Institution.

7. The following persons having been duly recommended were admitted as Subscribers to the Royal Institution for the year 1822.

Annual Subscribers.

Charles Gostling, Esqr. Earls Court Brompton.
Saml. Robinson, Esqr. 81 Dean Street Soho.
Major Colebrooke 1 Clifford Street.
Thos. E. Bates Esq. Kennington.
Col. Hugh Baillie 34 Mortimer Street.
Henry Mitford Boodle, Esqr. Lower Brook St.
Mr. De Lannoy Leigh St. Burton Crescent
John Hall, Esqr. 38 St. James's Street
Mr. Wm. Keene 7 Camden Terrace.

Life Subscriber.

Everard Brande, Esqr. 10 Arlington St.

To the Lectures.
Mrs. Rice Jones Woolwich
Mr. Bidwell 8 Somerset Street
Miss Thompson, Osnaburg St. New Road.
Mrs. E. Brande, Arlington Street
Mrs. Colebrooke 1 Clifford Street
Miss Charlotte Scott 4 Fludyer St.
The Honble. Mrs. Grey.

343

Feb. 4.

To the Lectures continued.

Miss Raikes } 3 Portland Place.
Miss C. Raikes }
Miss Pocock 39 Charles St. Berkeley Square.
Mrs. Nelson }
Miss Nelson } 18 Welbeck Street.
Miss H Nelson }
Miss Bernard Morland, Pall Mall.
Mrs. Pilgrim } Kensington.
Miss Pilgrim }

Adjourned to Monday the 18th. Instant.

Spencer.

344

Feb 18.

1822.

At a Meeting of the Managers of the Royal Institution of Great Britain held at the house of the Institution on the 18th of February 1822.

Present
Earl Spencer K.G. in the Chair.
Sir H. Davy, Bart V.P.
Charles Hatchett, Esqr. V.P.
Thomas Snodgrass, Esqr. V.P.
Daniel Moore Esqr.
James Stodart Esqr.
Thomas Harrison, Esqr. Secretary.

1. The minutes of the proceedings of the last meeting were read and approved.

2. Read and approved the minutes of the Committee of Accounts of this day and signed an Order for payment of the following sums vizt.

Payments

Wm. Harris Keeper of the Library 1 years Salary to 31st. Dec: 1821	£180 —
Messrs. Dickins & Co. for Coals in 1821	81 —
J. Fincher on account of the Inst:	50 —
	£311 —

Presents received.

3. The Society for the Encouragement of Arts presented
The 39th. Volume of their Transactions.
Sir Wm. Adams presented
A Treatise on Artificial Pupil.
John Claudius Loudon, Esqr. presented
Outline for a General History of Gardening.
Mr. William Alexander presented
The Eight Beatitudes, finely engraved & neatly half-bound.
Ordered,
That the thanks of the Managers be returned for these presents.

4. Read the minutes of the last General Meeting.

345

Feb. 18.

Lectures.

5. Read a letter from Sir S. H. Clarke, Bart. M.R.I. addressed to James Peter Auriol, Esqr. recommending to the Managers Mr. Goodacre to deliver lectures on Astronomy in the Royal Institution.
Resolved, That this proposal be declined.

6. The following persons having been duly recommended were admitted, to the Royal Institution for the year 1822, as Subscribers.

Annual Subscribers.

John Blake Kirby, Esqr. 27 Great Portland Street.
Dr. Morewood 47 Welbeck Street.
Amable Berthelot, Esqr. do.
Revd. Hugh Taylor 1 Belgrave Place Pimlico.
Mr. I. Harrington 1 Grove End Road St. John's wood.
The Honble. F. Calthorpe M.P. Grosvenor Square.
Leonard Shelford, Esqr. 1 Essex Court Temple.
Lord Somerville 28 Hill Street
Lieut. Beague of the Royal Engineers Woolwich.

To the Lectures.

Mrs. Kirby 27 Great Portland St.
Mrs. & Miss Reynolds 8 Bedford row.
Mrs. Baillie } 34 Mortimer St.
Miss Smith }
Miss Louisa Dawkins, Privy Gardens.
Miss Robe, Shooters Hill.
Mrs. & Miss Sturges Bourne 15 South Audley St.
Mrs. Phillott, 6 Brill Crescent Somers Town
Miss Somerville, 28 Hill Street
Mrs. Robertson, 51 Wimpole St.
Miss Barrett, Mecklenburgh Square
Mrs. Haygarth, 2 Cork Street.

Adjourned to Monday the 4th. of March

Spencer.

346

Mar. 4.

1822.

At a Meeting of the Managers of the Royal Institution held at the house of the Institution on the 4th. of March 1822.

Present
Earl Spencer, K.G. President in the Chair.
Charles Hatchett, Esqr. V.P.
Sir Humphry Davy, Bart. V.P.
George Duckett, Esqr.
Wm. Hasledine Pepys, Esqr.
John Chamier, Esqr.
Daniel Moore, Esqr.
James Stodart, Esqr.

1. The minutes of the proceedings of the last meeting were read and approved.

2. Read and approved the minutes of the Committee of Accounts of this day, and signed an Order for an Advance to the Assist. Secretary of £150. —

Presents received.

3. Dr. Jenner presented
A Letter to Dr. Parry on the influence of artificial Eruptions, in certain Diseases incidental to the Human Body &c.
The Horticultural Society presented
Transactions of the Horticultural Society vol 4 pt. 4
Mr. I. Millard presented
The Youth's Monthly Visitor No. 2.
Ordered,
That the thanks of the Managers be returned for these presents.

on Payment of Fine to City

4. Mr. Fincher informed the Board that he had by the desire of Richd. Clarke Esqr. seen the Comptroller of London who had offered on the part of the City of London in the event of the Royal Institution immediately paying their Renewal Fine of £600,

347

Mar. 4.

which becomes due Michas 1823, to allow 5 per Cent Interest on that sum from Lady day next till the period of renewal.

5. The following persons having been duly recommd. were admitted as Subscribers to the Royal Institution for the year, 1822

Annual Subscribers.

Lieut. Genl. Long Little Marlboro House Pall Mall
S. Pearce, Esqr. 118 Gt. Russell St.

To the Lectures.

Mrs. Morse 5 Sackville Street

Adjourned to the 18th Instant

Spencer.

Mar. 18. 1822.

At a Meeting of the Managers of the Royal Institution of Great Britain held at the house of the Institution on the 18th of March 1822

Present

Earl Spencer, K.G. President, in the Chair.
Charles Hatchett, Esqr. V.P.
Sir H. Davy, Bart. V.P.
Thomas Snodgrass, Esqr. V.P.
James Stoddart, Esqr.
John Chamier, Esqr.
William Sotheby, Esqr.
Daniel Moore, Esqr.
Thomas Harrison, Esqr. Secretary.

1. The minutes of the proceedings of the last meeting were read and approved.

2. Read and approved the minutes of the Committee of Accounts of this day and signed an Order for an Advance to the Assist. Secretary of £210.

Presents received.

3. Saml. Parkes, Esqr. M.R.I. presented
The Chemical Catechism 10th Edition,
Thoughts on the Laws relating to Salt,
A Letter to Farmers & Graziers on the advantages of using Salt,
An Answer to the accusations & aspersions of Mr. Richard Phillips.
Samuel Ware, Esqr. M.R.I. presented
Tracts on Vaults and Bridges
Ordered,
That the thanks of the Managers be returned for these presents.

4. Read the minutes of the last General Monthly Meeting.

Mar. 18.

5. The Revd. Blanco White, a Spaniard, distinguished for his Literary and Scientific attainments was admitted to the Royal Institution under Art 15 Chap VI of the Bye laws.

6. Interference in Laboratory — Order thereon. Resolved, That Copies of the following Resolution be hung up in the Hall and Laboratory of the Royal Institution.

Whereas the service of the Laboratory of the Royal Institution and the preparations for Lectures have been interfered with by the presence of persons not authorized by the Bye laws, who have taken up the time of the Assistant by asking questions or making experiments; It is resolved, That no person be admitted into the Laboratory, unless he be a Member of the Royal Institution, or personally introduced by a Member or by the Honorary Professor or Professor of Chemistry; except during the Lectures therein, and that any Member of the Royal Institution or other person wishing to have experiments made in the Laboratory do submit the plan of such experiments to the Board of Managers or to the Honorary Professor or Professor of Chemistry.

7. The following persons having been duly recommended were admitted as Subscribers to the Royal Institution for the year 1822.

Annual Subscribers.

Philip Caddell Esqr. 46 York St.
Charles Vernon, Esqr. Grosvenor Square.

To the Lectures
The Hon: Mrs. N. Collingwood 5 Hyde Park corner.
Mrs. Bennett. 35 Upper Seymour Street.

Adjourned to the 1st. of April.

Spencer

April 1. 1822

At a Meeting of the Managers of the Royal Institution of Great Britain held at the house of the Institutn. on the 1st. of April, 1822

Present

Earl Spencer K.G. President, in the Chair.
Charles Hatchett, Esqr. V.P.
Sir Claude Scott, Bart. V.P.
Thomas Snodgrass, Esqr. V.P.
Daniel Moore, Esqr.
John Chamier, Esqr.

1. The minutes of the proceedings of the last meeting were read and approved.

2. Read and approved the minutes of the Committee of Accounts of this day, and signed an Order for an Advance to the Assist. Secretary of £50.

Presents received.

3. The Astronomical Society of London presented Memoirs of the Astronomical Society vol. 1.

4. Read the minutes of the Committee of Visitors entered in their proceedings of the 25th. of March.
Ordered,
That the Catalogues of the Models Mechanical and Philosophical Instruments and Apparatus of the Royal Institution be deposited in the Iron Chest of the Institution.

5. The following persons having been duly recommended were admitted as Subscribers to the Royal Institution for the year 1822.

Annual Subscribers.

The Honble. Matthew Fortescue 15 Devonshire place
Llewellyn Paxton, Esqr. Piccadilly.

April 1.

Subscribers to the Lectures.

The Honble. Mrs. Fortescue Devonshire place.
Mrs. James Martin 5 Grosvenor Gate.
Lady Gresley 22 Edward Street.

6. Adjourned to the 15th. of April.

Charles Hatchett V.P.

352

April 15 — 1822

At a Meeting of the Managers of the Royal Institution of Great Britain held at the house of the Institution on the 15th of April, 1822

Present
Charles Hatchett, Esq.r V.P. <u>in the Chair</u>.
Sir H. Davy, Bart. V.P.
Daniel Moore, Esq.r
Thomas Snodgrass, Esq.r V.P.
George Duckett, Esq.r V.P.
Thomas Harrison, Esq.r <u>Secretary</u>.

1. The minutes of the proceedings of the last meeting were read and approved.

2. Read and approved the minutes of the Committee of Accounts of this day, and signed an Order for an advance to the Assist.t Secretary of £50.

Life Subs.cr

3. In pursuance of a nomination under the Act of Parliament by Henry Hoare Esq.r of Fleet St. his Daughter Lady Acland of Killerton, Devonshire, was admitted to the privileges of a Life Subscriber to the Royal Institution.

Prosecution of Mr. Accum.

4. Mr. Daniel Moore stated that he had been applied to on the 9th Instant by Mess.rs Allen & Co. Solicitors, to obtain the consent of the Managers to Mr. Fred.k Accum's bail being discharged as they had taken great pains and been at considerable expence in endeavouring to get him back into this Country to take his Trial. and that as he had not an opportunity of making this communication to the Board, the Sessions being ended on Saturday last, He had acceded to the request of the Bail on their paying the expences incurred by the Royal Institution in the prosecution and that he had received an undertaking for the payment of the Costs.
<u>Resolved</u>, That the above proceedings be approved of.

353

April 15.

5. The following persons having been duly recommended were admitted as Subscribers to the Royal Institution for the year 1822.

<u>Annual Subscribers</u>.

Gilbert Jones, Esq.r Salisbury Square
James Stewart, Esq.r Medical Department of the Ordnance Woolwich
Julian Hibbert, Esq.r 2 Langham Place
Tho.s Barton Bowen, Esq.r 19 Portman Street.

<u>To the Lectures</u>.

Mrs Thomas Hibbert }
Miss Saccharissa Hibbert } 2 Langham Place

6. <u>Ordered</u>,
That the House Lists be filled up as follows

President. Earl Spencer K.G.

Treasurer. Sir S. Bernard Morland Bart. M.P.

Secretary. Thomas Harrison, Esq.r

<u>Visitors</u>.
George Aust, Esq.r
John Frederick Daniell, Esq.r
Duke of Devonshire.
John Fuller, Esq.r
Earl Grosvenor.
Earl of Hardwicke K.G.
Sir Benj.n Hobhouse Bart.
Gibbes Walker Jordan, Esq.r
William Manning, Esq.r M.P.
Rev.d Sir John Robinson, Bart.
Rich.d Horsman Solly, Esq.r
Sir George Tho.s Staunton, Bart.
Lord Stowell.
Giles Templeman, Esq.r
John Walker, Esq.r

354 — 1822.

<u>Managers</u>.
James Peter Auriol, Esq.r
Sir Frederick Baker, Bart.
Sir Gilbert Blane, Bart.
John Chamier, Esq.r
John George Children, Esq.r
Sir Humphry Davy Bart.
George Duckett, Esq.r
Henry Hallam, Esq.r
Charles Hatchett, Esq.r
Marquis of Lansdowne.
Daniel Moore, Esq.r
Jesse Watts Russell, Esq.r
Thomas Snodgrass, Esq.r
John Soane, Esq.r
James Stodart, Esq.r

<u>Adjourned</u> to the 6th of May. —

Thomas Snodgrass V.P.

355

May 6.

At a Meeting of the Managers of the Royal Institution of Great Britain, held at the house of the Institution on the 6th of May 1822,

Present
Thomas Snodgrass, Esq.r <u>in the Chair</u>.
John Soane, Esq.r
John Chamier, Esq.r
Jesse Watts Russell, Esq.r
James Peter Auriol, Esq.r
Sir Gilbert Blane, Bart.
Thomas Harrison, Esq.r <u>Secretary</u>.
Daniel Moore, Esq.r

1. The minutes of the proceedings of the last meeting were read and approved.

2. Read and approved the minutes of the Committee of Accounts of this day, and signed an Order for an Advance to the Assist.t Secretary of £200.

Present of Bones, from W. Salmond, Esq.

3. Read a letter from W.m Salmond, Esq.r M.R.I. addressed to the Secretary, dated York 28th April 1822, "begging to announce for the information of the Managers that he has sent them a barrel containing a considerable number of bones which are chiefly of the large-sized animals found by him at Kirkdale in this County, which has been recently described in a paper read before the Royal Society in February last by The Rev.d W. Buckland, Prof. of Geology at the University of Oxford, and he trusts these remarkable remnants of former Quadrupeds in this Island may be found worthy of a place in the Museum of the Royal Institution to which he shall always be happy to contribute whatever be desirable to render it complete."
<u>Resolved</u>, That the special thanks of the Managers be returned to William Salmond, Esq.r and that the above remnants of former Quadrupeds be deposited in the Museum of the Royal Institution.

356 1822

4. Mr. Faraday stated to the Board that the barrel containing the bones sent by Mr. Salmond had arrived and that he was then engaged in arranging them.

Lectures 5. Read a letter from Mr. James Parry, addressed to the Board of Managers; offering to deliver gratuitously Two Lectures on Education.
Resolved,
That his obliging offer be declined.

Presents received.
6. Revd. James Havilands presented
A Print of the Vase presented to Captn. Parry.
Revd. H. F. Egerton presented
Continuation of his Notes on Euripides.
Mr. Valpy presented
2 numbers of the London Museum.
The Royal College of Surgeons presented
2 Lists of the Members of the Royal College of Surgeons.
Mr. Dowson presented
Dictionary of the New Testament.
Ordered,
That the thanks of the Managers be returned for these presents.

7. In pursuance of a nomination by the President the following Vice Presidents were appointed for the present year.
James Peter Auriol, Esqr.
Sir Humphry Davy, Bart
George Duckett, Esqr.
Charles Hatchett, Esqr.
Marquis of Lansdowne.
Daniel Moore, Esqr.
Thomas Snodgrass, Esqr.

Hon. Admission 8. Le Vicomte de Chateaubriand, the French Ambassador was admitted to the Royal Institution under Art: 15 Chap. VI of the Bye laws.

May 6. 357

9. The following persons having been duly recommended were admitted as Subscribers to the Royal Institution.
Life Subscriber.
Captn. M. Montague, R.N. 90 Wimpole Street.
Annual Subscribers.
Richard Oswald, Esqr. 10 Upper Berkeley St.
The Revd. J. Blanco White 7 Paradise row Chelsea.
Revd. Deacon Morrell, 35 Albemarle St.

To the Lectures.
Mr. Robert Phillips
Mrs. Phillips
Miss Jane Phillips
Miss Sotheby 35 Albemarle Street
Miss Corbett 68 Grosvenor St.
Mrs. Richard Dawkins Privy Gardens.

10. Adjourned to Monday the 3d. of June.
Charles Hatchett V.P.

358 1822

June 3.

At a Meeting of the Managers of the Royal Institution of Great Britain held at the house of the Institution on the 3rd of June, 1822

Present
Charles Hatchett, Esqr. V.P. in the Chair.
John George Children, Esqr.
Thomas Snodgrass, Esqr. V.P.
Daniel Moore, Esqr. V.P.
John Chamier, Esqr.
James Stodart, Esqr.
George Duckett, Esqr. V.P.
Sir S. Bernard Morland, Bart. M.P. Treasr.

1. The minutes of the proceedings of the last meeting were read and approved.

2. Read and approved the minutes of the Committee of Accounts of this day.

Advance upon Treasurer. 3. Resolved, That in order to meet the current expences of the Royal Institution, the Treasurer be requested to cause an advance to be made on his account at the Bankers to the amount of £1000; – of which sum, £500 will be immediately wanted, and the remainder on the 1st. of August next.

Payments. 4. Signed an Order for payment of the following sums vizt.

Wm. Thos. Brande, Esqr.	
1 years Salary as Prof. of Chemistry £	200. –
Mr. Millington for Lectures in 1822	105. –
Dr. Roget — do. — do.	105. –
Mr. Landseer — do. — do.	52. 10. –
Mr. Newman for Chemicals in 1821	114. 4. 11
Mr. Fincher — on account	588. –
	1164. 14. 11

5. Presents received.
Earl Spencer, K.G. Pres. R.I. &c. &c. presented

June 3 359

Aedes Althorpianae; or an account of the Mansion, Books and Pictures at Althorp; by The Revd. Thos. Frognall Dibdin 2 vols. royal 8vo.
Samuel Turner, Esqr. F.R.S. presented
A Letter to the Rt. Hon: Robert Peel, on the state of the Bank of England and the resumption of Cash payments.
—— Considerations upon the Agriculture, Commerce & Manufactures of the British Empire.
The Horticultural Society presented
Transactions of the Society – supplement to vol. 4.
Ordered,
That the thanks of the Managers be returned for these presents.

6. John Randolph, Esqr. of Virginia was admitted to the Royal Institution under Art: 15 Chap. VI of the Bye laws.

7. Read the minutes of the last General Meeting.

Life Subscr. 8. In pursuance of a nomination under the Act of Parliament by the late Francis Drummond Esqr. of Sloane Street dated 21st. August 1811 – his Son
William Drummond, Esqr. Ensign in the 3d. Regiment of Foot Guards
was admitted to the privileges of a Life Subscriber to the Royal Institution.

9. The following persons having been duly recommended were admitted as Subscribers to the Royal Institution for the year 1822.
Annual Subscribers.
Revd. Henry Harvey 38 Bruton Street
Mr. Crowe 31 Bedford St. Covent Garden.
To the Lectures.
Miss Balfour Burlington Street
&
Life Subscriber.
Henry Talbot, Esqr. 31 Sackville St.

Adjourned to the 1st. of July.
Chas. Hatchett

360

July 1. | 1823.

At a Meeting of the Managers of the Royal Institution of Great Britain held at the house of the Institution on the 1st of July, 1822,

Present

Charles Hatchett, Esqr. V.P. in the Chair.
George Duckett, Esqr. V.P.
John Chamier, Esqr.
James Peter Auriol, Esqr. V.P.
Thomas Snodgrass, Esqr. V.P.
Sir S. Bernard Morland, Bart. M.P. Treasr.
Thomas Harrison, Esqr. Secretary.

1. The minutes of the proceedings of the last meeting were read and approved.

2. Read and approved the minutes of the Committee of Accounts of this day, and signed an Order for payment of the following sum on the 1st. of August

Payment — 1 Year's Ground rent to the City of London to 25th. March 1822 ———— £150. 1. 4

Presents received.

3. James South, Esqr. M.R.I. presented
Practical Observations on the Nautical Almanack,
J. P. Thomas, Esqr. presented
"A legal and constitutional argument against the alleged Judicial right of restraining the publication of reports of Judicial proceedings" 2 Copies.
The Commissioners of the Public Records presented
The Statutes of the Realm vol. 7 and 8,
Inquisitiones post Mortem, vol 2.
Mr. Hoëné Wronski presented
Introduction to a Course of Mathematics.
The Revd. Blanco White presented
Don Leucadio Doblado's Letters from Spain,
The Revd. Francis Thackeray, A.M. presented
A Defence of the Clergy of the Church of England,

361

July 1.

Saml. Solly, Esqr. M.R.I presented
Letters on the New Political Economy.
The Editor of the Council of Ten presented No. 1.
Ordered,
That the thanks of the Managers be returned for these presents.

4. Read the minutes of the last General Meeting.

5. Read the minutes of the Committee of Visitors entered in their proceedings of the 10th. of June.
Ordered, That a ladder be placed from the roof of the Laboratory to the roof of the Mineral Room, and that the Gutters be repaired.

Coals. — 6 Ordered, That 30 Chaldrons of the best Walls-end Coals be immediately laid in the Cellars of the Royal Institution at 52s. ⅌ Chaldron including all expenses.

Life Subscr. — 7. In pursuance of a nomination under the Act of Parliament by James Peter Auriol, Esqr. his Son
Charles James Auriol, Esqr. of Park lane,
was admitted to the privileges of a Life Subscriber to the Royal Institution.

Book-Case, — 8. Resolved, That a Book-case be fitted up in the Library of Reference at an expence not exceeding £7.

Leave of absence to asst. Librn — 9. Mr. Sturt, the Assist. Librarian, requested permission of the Board for leave of absence for a Month during the recess which was granted.

Adjourned to Monday the 4th. of November

J P Auriol

362

Nov. 4. | 1822.

At a Meeting of the Managers of the Royal Institution of Great Britain, held at the house of the Institution on the 4th. of November, 1822

Present

James Peter Auriol, Esqr. V.P. in the Chair.
Thomas Snodgrass, Esqr. V.P.
Daniel Moore, Esqr. V.P.
Charles Hatchett, Esqr. V.P.
George Duckett, Esqr. V.P.
Sir Gilbert Blane, Bart
James Stodart, Esqr.
John George Children, Esqr.
John Soane, Esqr.
Thomas Harrison, Esqr. Secretary.
Sir S. Bernard Morland, Bart. M.P. Treasurer.

1. The minutes of the proceedings of the last meeting were read and approved.

2. Read and approved the minutes of the Committee of Accounts of this day.

Presents received.

3. Mr. Peter Schmidtmeyer presented
Travels to Chile over the Andes, part 1,
William Spence, Esqr. F.L.S. presented
Tracts on Political Economy,
The Royal Society presented
Philosophical Transactions, 1822, part 1,
Nautical Almanack, 1825,
Mr. John Allen, junr. Hereford, presented
Bibliotheca Herefordiensis,
Mr. I. C. Hull Library, presented
Catalogue of the Hull Subscription Library,
Anthony Highmore, Esqr. presented
Philanthropia Metropolitana: a View of the Charitable Institutions established in London during the last 12 years,
Mr. J. Netherclift presented
A Fac Simile Copy of The Magna Charta, Engraved on St[one]

363

Nov. 4.

Lady Bernard presented
Three Prints of West's Picture of "Our Saviour healing the Sick in the Temple," — a Proof, — a First Impress.n in Aquafortis, — and a second more finished in Aquafortis.
Mr. Parker Cleveland of Boston in America, presented
An Elementary Treatise of Mineralogy and Geology 2 vol 8vo
The Directors of the East India Company, presented
Morrison's Chinese Dictionary, part 3d.
The Marchioness Cathe. Solari presented, thro' Earl Spencer,
Wellington, Poemetto, del Marchese Ant. Solari, &c.
Major Colby presented thro' Earl Spencer,
Survey of Middlesex &c. by the Board of Ordnance
Lieut. Col. Macdonald. F.R.S presented
A Treatise on the Harmonic System,
Mr. Hoëné Wronski presented
Trois Lettres à Sir Humphry Davy, sur l'Imposture Publique, — Petition à Parlement Britannique.
The Commissioners of the Public Records presented
The Statutes of the Realm vol 9.
The Directors of the African Institution presented
The 16th. Report, 2 Copies.
Mr. Chas. Turner Thackrah, presented
An Introductory Discourse delivered to the Leeds Philosophical and Literary Society.
Ordered,
That the thanks of the Managers be returned for these presents.

Under-Lease. — 4. Resolved, That the affixing of the Seal of the Corporation to a Lease from this Institution to Richd. Baker, Esqr. of which notice had been sent to the Managers agreeably to the Bye laws that the Seal would be affixed this day, be deferred till the next Meeting of the Managers and that the Secretary be requested to peruse the Deed in the mean time.

Binding Books — 5. The Secretary stated to the Board that he had been informed by Mr. Sturt, the Assist. Librarian, that several Volumes of Books ~~were in great want~~ of [illegible]

~~binding~~. It was resolved
That the following Gentlemen, vizt.
Charles Hatchett, Esqr.
Thomas Snodgrass, Esqr.
George Duckett, Esqr.
& Thomas Harrison, Esqr.
be requested to examine the same, and to take measures if necessary to have them immediately bound

Moniteur discontinued. 6 Resolved, That the Moniteur be discontinued from this day.

7. Read the minutes of the last General Monthly Meeting.

Adjourned to Monday the 2^d of Decem^r.
H. Davy. V.P.

At a Meeting of the Managers of the Royal Institution of Great Britain held at the house of the Institution on the 2^d of December, 1822

Present
Sir Humphry Davy, Bart. V.P. in the Chair.
Thomas Snodgrass, Esqr. V.P.
George Duckett, Esqr. V.P.
Daniel Moore, Esqr. V.P.
James Stodart, Esqr.
John Chamier, Esqr.
Sir Gilbert Blane, Bart

1. The minutes of the proceedings of the last meeting were read and approved.

2. Read and approved the minutes of the Committee of Accounts of this day, and signed an Order for an Advance to the Assist^t. Secretary of £200 on the 31st Instant, and a further Advance of £60 on the 1st of January next.

Lectures. 3. Read a letter from J. J. Halls Esqr., addressed to the Secretary, proposing to the Managers to deliver 6 Lectures on Painting as soon after Easter as may be consistent with the other arrangements of the Royal Institution.
Resolved,
That the consideration of M^r. Halls proposal be deferred till the next Meeting.

Life subs^n. 4. In pursuance of a nomination under the Act of Parliament by the late Alex^r. Marcet, M.D. dated 24th. Novr. 1821, his Son Francis Marcet, Esqr. was admitted to the privileges of a Life Subscriber to the Royal Institution.

Presents received.

5 Peter Schmidtmeyer, Esqr. presented
A series of Geological Specimens from the Andes.

M^r. Thorold presented
A Model and Drawing of a Water wheel with vertical paddles.
Col^l. Silvertop presented
Specimens of Minerals from Spain.
Charles Babbage, Esqr. F.R.S. presented
A letter to Sir Humphry Davy, Bart. on the application of Machinery to the purpose of calculating & printing Mathematical tables.
The Royal Society presented
Philosophical Transactions for 1822, part 2.
Ordered,
That the thanks of the Managers be returned for these presents.

6. Read the minutes of the last General Meeting.

Binding Books. 7. Read the following report from the Committee appointed by the last Board.
The Committee appointed by the last meeting to examine the state of the unbound Books, reported that they found a number of Volumes greatly out of repair, they had therefore directed M^r. Knapp the Bookbinder to prepare an estimate of the expence of binding them, specifying the charge of each work, which amounted to £19. 10. 7, and the Committee had ordered the same to be immediately bound at that sum.

Lease of 23 Albemarle Street. 8 Notice having been sent to the Managers agreeably to the Bye laws – The Seal of the Corporation was affixed to a Lease dated 1st July 1822 from the Members of the Royal Institution to Richard Baker Esqr. of Premises N°. 23 Albemarle Street for the term of 48 years from Midsummer last at the Rent of £70 ℔ annum. – The Deed was also countersigned by H. Davy, V.P. Chairman – George Duckett – Dan Moore Gil. Blane. Ja^s. Stodart – John Chamier – Thos. Snodgrass & Thomas Harrison Secretary.

Resolved, That the Salary of W^m. Bollaert the Assistant

in the Laboratory be increased from 10 to 12^s ℔ week.

9. Resolved unanimously, That Daniel Moore Esqr. be appointed the Solicitor of the Royal Institution.

10. Resolved, That the Assist^t. Secretary be directed to ascertain what Deeds or Papers belonging to the Royal Institution are in the possession of William Seymour Esqr. late Solicitor to the Royal Institution.

11. The following persons having been duly recommended were admitted as Annual Subscribers to the Royal Institution for the year 1823.

Ann^l. Subs^rs.
D^r. Veitch of Argyll Street.
John Braham, Esqr. Finchley, Midd^x.
Sir John Scott Lillie, North End Fulham.
Captn. Kelsall, 67 Great Portland Street
William Lukin, Esqr. 1 Orchard S^t. Portman Sq^e.
M. J. Garcia, Esqr. 21 Grosvenor Street
Alexander Trotter, Esqr. Soho Square.

Life Subscriber

Henry Madden, Esqr. 9 Cleveland row.

Adjourned to Monday the 20th of January 1823.
Charles Hatchett V.P.

368

Jan. 20.

1823.

At a Meeting of the Managers of the Royal Institution of Great Britain held at the house of the Institution on the 20th of January, 1823

Present

Charles Hatchett, Esqr. V.P. in the Chair.
~~Thomas Snodgrass, Esqr. V.P.~~
Daniel Moore, Esqr. V.P.
Thomas Snodgrass, Esqr. V.P.
Sir Gilbert Blane, Bart

1. The minutes of the proceedings of the last meeting were read and approved.

2. Read and approved the minutes of the Committee of Accounts of this day, and signed an Order for an advance to the Assistt. Secretary of £60.

Letters received.

Lectures. J. Walker, Esq.

3. Read a letter from The Revd. Blanco White, M.R.I. addressed to the Secretary, offering on the part of John Walker Esqr. his services to the Royal Institution to deliver gratuitously in the present Season, a Course of Lectures on the Scientific principles of Arithmetic (considered as a branch of Mathematics) and the Elements of Algebra.

Resolved,
That Mr. Walker's obliging offer be accepted, and that Mr. Hatchett be requested to communicate this resolution to The Revd. Mr. White.

Read a letter from Mr. John Bird, addressed to the Managers proposing to give a Course of Lectures on Astronomy at the Royal Institution, to be illustrated by his Grand Transparent Orrery, for the sum of 20 guineas.
Resolved, That Mr. Bird's offer be declined.

369

Jan. 20.

Compensation to Proprietor claimed – (beyond the time).

Read a letter from Mr. C. W. Charlton, addressed to the Secretary, requesting to be informed, on the part of Mrs. Haskins the Relict and Executrix of the late Joseph Haskins, Esqr. an original Proprietor, who died in September 1809, whether she can now avail herself of the privilege of receiving the £42 . 0 . 10 which was to be paid to those anxious to dispose of their shares three Months after the passing of the Act.

Resolved,
That the Assistt. Secretary inform Mr. Charlton that the pecuniary compensation has been in no instance paid unless demanded within the 3 months specified by the Act of Parliament.

Presents received.

4. I. D'Israeli presented
A second Series of Curiosities of Literature 3 vols.
Captn. George Powell presented
The Chart of South Shetland with various Geological Specimens,
The Revd. Thomas Jervis presented
Catalogue of Dr. Williams's Library Red Cross Street,
Mr. Peter Schmidtmeyer presented
Travels to Chile, parts 2 & 3.
The Geological Society presented
Transactions of the Geological Society, second series vol. 1 part 1.
The Horticultural Society presented
Transactions of the Horticultural Society, vol. 5 part 1.
The Editor of the Harmonicon presented
The Harmonicon.
Mr. Bernard Barton presented
Times Telescope for 1823.
Mr. John Bellamy presented
A Reply to the Quarterly Review
A Reply to Samuel Lee, M.A.
The True age of Christ,
Society for the Improvement of Prison Discipline
The Fourth Report of the Committee.

370

1823.

Ordered, That the thanks of the Managers be returned for these presents.

Deeds.

5 Mr. Fincher reported that he had received from William Seymour, Esqr. the following Deeds belonging to the Royal Institution, which were the only Deeds he had in his possession – viz.
Counterpart of Lease from Mellish to Gray dated 2d. Novr. 1776 – and
Counterpart of a Lease from Mellish to Swinton dated 6th. July 1782.
Ordered, That they be deposited in the Iron Chest.

6. Read the minutes of the last General Monthly Meeting.

—

7. Read the minutes of the Visitors entered in their proceedings of the 13th. Instant.

Insurance

8. Mr. Fincher reported that he had left the Policy of Insurance at the Sun Fire Office to have indorsed thereon their permission for using the new Stove which had been erected to warm the Lecture room.
Ordered,
That the Policy be obtained from the Sun Fire Office as soon as possible.

Repairs,

9. Read an estimate from Messrs. Brown & Co. of the charge for whitewashing the ornamental Ceiling and renovating with their liquid Composition the whole of the Paint of the Hall & Marble Chimney Piece – the whole to be completed at 10 guineas, Scaffolding included.
Ordered, That it be done at an expence not exceeding the above sum.

10. The following persons having been duly recommended were admitted as Annual Subscribers for the year 1823.

annl. Subsrs.

James Dick, Esqr. 34 Bruton Street
Frederick Musgrave Langham, Esqr. of Langham place

371

Jan. 20.

Annual Subscribers continued.

David Green, Esqr. 4 Milbank row.
Morton Lawrence, Esqr. Champion Hill Camberwell.
George Stanley, Esqr. Old Bond St.
George Head, Esqr. 22 Henrietta St. Cavendish Sq.
Sir Jas. Bland Lamb, Bart 27 Queen Ann Street.
Rupert George Esqr. 25 Bolton St.
James Henry Earle, Esqr. 7 Harley Street
Hon: Baptiste Noel, Cloisters Middle Temple.
Captn. Edwd. Pelham Brenton, R.N. – 4 Grosvenor Gate.
Robert Fourlerton, Esqr. 43 York St. Portman Sq.

Adjourned to 3d. February.

Chas. Hatchett V.P.

372

Feb. 3. 1823.

At a Meeting of the Managers of the Royal Institution of Great Britain held at the house of the Institution on the 3d. of February, 1823

Present

Charles Hatchett, Esqr. V.P. in the Chair.
Sir George Duckett Bart. V.P.
Sir Humphry Davy, Bart. V.P.
Daniel Moore, Esqr. V.P.
Sir S. Bernard Morland, Bart. M.P. Treasr.
Thomas Harrison, Esqr. Secretary.

1. The minutes of the proceedings of the last meeting were read and approved.

2. Read and approved the minutes of the Committee of Accounts of this day, and signed an Order for payment of the following sums, vizt

Payments

Reeve for Assessed taxes to Michaelmas 1822	79 . 2 . 4
Josh. Finchee on account of the Royal Institn.	80 . – . –
	159 . 2 . 4

Presents received.

3. Lieut. Genl. Wm. Thornton presented Dugdales Ancient usage of Arms, office of Lord Chancellor of England &c.
John Walker, Esqr. presented The Philosophy of Arithmetic.
Ordered That the thanks of the Managers be returned for these presents.

4. In pursuance of a nomination under the Act of Parliament by Sir Charles Hastings, Bart. his eldest Son Charles Hastings, Esqr. was admitted to the privileges of a Life Subscriber to the Royal Institution

Feb. 3. 373

5. Audited the Treasurer's Account for the year 1822.

Insurance

6. The Secretary reported that he had deposited in the Iron Chest the Policy of Insurance and the Deeds, as directed by the last Board.

7. The following persons having been duly recommended were admitted as Subscribers to the Royal Institution

Life Subscribers.

John Braham, Esqr. of Finchley
Nathl. Goldsmid, Esqr. 41 Upper Harley St.

Annual Subscribers.

J. W. Jordan, Esqr. jun: Portland Place
Thos. Luke Flood, Esqr. 23 Cheyne walk

To the Lectures.

Mrs. Caddell 46 York Street
Mrs. Wilks 4 Langham Place
Mrs. Maynard 14 Bryanstone St.
Lady Byron } 34 Upper Brook Street
Miss Doyle }

Adjourned to the 24th. Instant

H. Davy. V.P.

374

Feb. 24. 1823.

At a Meeting of the Managers of the Royal Institution of Great Britain held at the house of the Institution on the 24th. of February, 1823

Present

Sir Humphry Davy, Bart. V.P. in the Chair
Thomas Snodgrass, Esqr. V.P.
Charles Hatchett, Esqr. V.P.
Daniel Moore, Esqr. V.P.
John Chamier, Esqr.
Sir S. Bernard Morland, Bart. Treasurer.
Thomas Harrison, Esqr. Secretary.

1. The minutes of the proceedings of the last meeting were read and approved.

2. Read and approved the minutes of the Committee of Accounts of this day; and signed an Order for an Advance to the Assist. Secretary of £500.

On Finances.

3. The Secretary having laid before the Board a letter addressed to him by Earl Spencer dated 22d. Instt. inclosing a plan for renovating the finances of the Institution recently communicated to him by the Assistant Secretary, and recommending it to the immediate and serious consideration of the Managers but adding that on account of a severe domestic Calamity he is unable to attend the Board this day.
Resolved, That the Managers deeply regretting the cause of the absence of the President, and deeming it improper to come to any conclusion on so important a subject without his personal assistance, will at the rising of the Board adjourn to such day as His Lordship may appoint for his attendance at the Board, and that the Secretary do specially summon all the Members of the Board to meet on such day for the purpose of taking into consideration the present state of the finances of the Institution.

Feb. 24. 375

4. Read the minutes of the last General Monthly Meeting.

5. The following persons having been duly recommended were admitted as Subscribers to the Royal Institution for the year 1823

Annual Subscribers.

T. A. Fructuozo, Esqr. 16 Bloomsbury Square
Dr. Lee 19 St. James's Square
Sir Thos. Dyke Acland, Bart. Warrens Hotel
Dr. Charles Mitchell, R.N. 57 George St. Portman Sq.
George Saml. Collyer, Esqr. 9 Park Place
Wm. Varlo Hellyer, Esqr. 13 Paper Buildings
Frederick Mbysey, Esqr. 23 Hart St.
Charles Rennell, Esqr. of the Middle Temple

To the Lectures.

Mrs. Barker 48 Baker Street.
Miss Jane Hendrie } 12 Tichborne Street
Miss Susan Hendrie }
Mrs. Dyson } 5 Bolton Street
Miss A Dyson }
Mrs. Fitzclarence }
Miss Fitzclarence } 6 Audley Square
Mrs. Harper }
Miss Smith 27 Stamford Street
Mrs. Carbonell } 35 Golden Square
Miss Carbonell }

Spencer.

376

1823.

Mar. 1.

At a Special Meeting of the Managers of the Royal Institution of Great Britain held on Saturday the 1st of March 1823, for the purpose of taking into consideration the present state of the Finances of the Instit[n]: a special notice having been sent to each of the Managers accordingly

Present
Earl Spencer, President, in the Chair.
Sir Humphry Davy, Bart V.P.
Sir George Duckett, Bart V.P.
Daniel Moore, Esq[r]. V.P.
Charles Hatchett, Esq[r]. V.P.
John Soane, Esq[r].
Sir Gilbert Blane, Bart:
John George Children, Esq:
Henry Hallam, Esq:
James Peter Auriol, Esq[r].
John Chamier, Esq[r].
Thomas Harrison, Esq[r]. Secretary.

Finances. 1. The letter addressed to the President by the Assistant Secretary having been laid before the Board and an abstract of the plan therein contained having been read, and considerable discussion having taken place thereon, It was resolved

That the further consideration thereof be deferred till the meeting of the Managers to be held on Monday the 10th Instant.

Also that the Secretary do forthwith transmit a Copy of such plan and abstract to each of the Managers.

Spencer

377

Mar. 3.

At a Meeting of the Managers of the Royal Institution of Great Britain held at the house of the Institution on the 3d of March, 1823

Present
Earl Spencer, K.G. President, in the Chair.
Thomas Snodgrass, Esq[r]. V.P.
Daniel Moore, Esq[r]. V.P.
John Chamier, Esq[r].

1. The minutes of the proceedings of the last meeting and of the special Meeting of the Managers, were read and approved.

2. Read and approved the minutes of the Committee of Accounts of this day, and signed an Order for payment of the following sums, viz[t].

Payments

Dickins &C[o]. for Coals in 1822	£136.—
Jos[h]. Fincher on account of the Royal Institution	90.—
	226.—

Coals 3. Ordered, That Dickins &C[o]. do immediately supply the Institution with 10 Chaldrons of the best Walls-end Coals.

4. The following persons having been duly recommended were admitted as Subscribers to the Royal Institution for the year 1823.

Annual Subscriber.

W[m]. Henry Trant, Esq[r]. 6 Chandos S[t]. Cavendish Sq[r].

To the Lectures

Lady Macgregor } Camden Hill Kensington
M[rs]. Colegrave }
Miss E & Miss C Smith } 1 Upper Harley S[t].
Miss Willett }

Spencer

378

1823.

Mar. 10.

At a Meeting of the Managers of the Royal Institution of Great Britain held at the house of the Institution on the 10th of March, 1823

Present
Earl Spencer, K.G. President, in the Chair.
Sir George Duckett, Bart. V.P.
Sir Humphry Davy, Bart. V.P.
Thomas Snodgrass, Esq[r]. V.P.
Daniel Moore, Esq[r]. V.P.
Henry Hallam, Esq[r].
Sir Gilbert Blane, Bart
Jn[o]. George Children, Esq[r].
John Chamier, Esq[r].
Sir S. Bernard Morland, Bart. M.P. Treas[r].
Thomas Harrison, Esq[r]. Secretary.

1. The minutes of the proceedings of the last meeting were read and approved.

2. Read and approved the minutes of the Committee of Accounts of this day.

Lectures Mr. Frost. 3. Read a letter from M[r]. John Frost, Director & Professor of Botany to the Medico-Botanical Society of London offering his services to deliver gratuitously a Course of Lectures on Botany after Easter.

Resolved,

That M[r]. Frost's offer be accepted.

4. Read the minutes of the last General Monthly Meeting.

5. The following persons having been duly recommended were admitted as Subscribers to the Royal Institution for the year 1823.

Annual Subscribers

John Musgrave, Esq[r]. 59 Portland Place

Mar. 10.

379

Annual Subscribers, continued.

Peter Vere, Esq[r]. Grosvenor Place,
Edward James, Esq[r]. 28 Connaught Terrace,
Edw[d]. Nicholas Hart, Esq[r]. 17 Southampton Buildings.

To the Lectures.

M[rs]. Wykeham Martin, }
Miss Wykeham Martin } 32 Lower Brook S[t].
Miss Mary Wykeham Martin }
Miss Jane Wykeham Martin. }

6. The letter addressed to the President by the Assist[t]. Secretary, and an abstract of the plan therein contained having been sent round to each of the Managers, agreeable to the Resolution of the Special Meeting of the 1st March the further consideration thereof, was resumed by the Board and considerable discussion having taken place thereon, It was resolved

That the further consideration thereof be adjourned to Monday next the 17th Instant at half past One p: m.

Spencer

380 1823.

Mar. 17.

At a Meeting of the Managers of the Royal Institution of Great Britain, held at the house of the Institution on the 17th of March, 1823

Present
Earl Spencer, K.G. President, in the Chair.
Charles Hatchett, Esqr. V.P.
Daniel Moore, Esqr. V.P.
Sir George Duckett, Bart. V.P.
Thomas Snodgrass, Esqr. V.P.
Sir Gilbert Blane, Bart.
John Chamier, Esqr.
Sir S. Bernard Morland, Bart. M.P. Treasr.
Thomas Harrison, Esqr. Secretary.

1. The minutes of the proceedings of the last meeting were read and approved.

2. Read and approved the minutes of the Comme. of Accounts of this day, and signed an Order for an Advance to the Assistant Secretary of £100.

Presents received.

3. Henry Goodwyn, Esqr. presented
Introductory Remarks respecting the Imperial Gallon and Diagonal Table – 4to pamphlet
Table of the Circles arising from the division of a Unit, or any other whole number whatever, by all Integers, from 1 to 1024, 8vo.
Tabular Series of Decimal Quotients, 8vo.
Mr. John Walker, formerly Fellow of Dublin College presented
Full & plain account of the Horatian Metres, 8vo.
An Essay on the following Prize question; whether & how far the cultivation of Science and that of polite Literature assist or obstruct each other.
Selections from Lucian with a Latin Translation and English notes
A Familiar Commentary on the Compendium of Logick used in the University of Dublin, 8vo.

Mar. 17. 381

Supplementary Annotations on Livy, 8vo
Mr. Professor Millington presented
An Epitome of the Elementary principles of Natural and Experimental Philosophy part 1, 8vo
The Society for the Encouragement of Arts &c. presented
Transactions of the Society, vol 40.
Ordered,
That the thanks of the Managers be returned for these presents.

Deeds.

4. The Secretary informed the Board that he had received from Mr. Seymour, the late Solicitor to the Royal Institution, the following Deeds relating to Premises leased by the Royal Instit. to Mr. Richard Baker – vizt.

Lease from Mellish to Gray, dated 2d. Novr. 1776.
Assignment between Gray & Tanner for securing £2500 – dated 24th December, 1776.
Indorsement on said Assignment
Assignment between Tanner & Gray dated 29 Novr. 1787
Ditto between Gray & Elwes dated 24th June 1790.
Ditto between Stracey & Richd. Baker dated 17th July 1812
Counterpart Lease from the Members of the Royal Institution to Mr. Richd. Baker.
Ordered,
That the above Deeds be deposited in the Iron Chest.

Finances.

5. The state of the Finances of the Institution having been further taken into consideration and some alterations in the Bye laws respecting the admission of Members having been agreed to,
Resolved,
That the Secretary do draw up such alterations and arrange them in form for signature by 15 Members.

6. The following persons having been duly recommended were admitted as Subscribers to the Royal Institution for the present Year.

382 1823.

Annual Subscribers.

George John Gibson, Esqr. 37 Keppel St.
Viscount Folkestone 52 Lower Grosvenor St.

To the lectures,

Lady Charlotte Butler, 54 Grosvenor St.

Adjourned to Monday the 7th. of April.

H. Davy V.P.

383

April 7.

At a Meeting of the Managers of the Royal Institution of Great Britain held at the house of the Institution on the 7th of April 1823

Present
Sir Humphry Davy, Bart. V.P. in the Chair.
Charles Hatchett, Esqr. V.P.
Thomas Snodgrass, Esqr. V.P.
Henry Hallam, Esqr.
John Soane, Esqr.
Sir Gilbert Blane, Bart.
Sir S. Bernard Morland, Bart. M.P. Treasurer
John Chamier, Esqr.
Thomas Harrison, Esqr. Secretary.

1. The minutes of the proceedings of the last meeting were read and approved.

2. Read and approved the minutes of the Committee of Accounts of this day, and signed an Order for an Advance to the Assistant Secretary of £100.

Use of Lecture room declined.

3. Read a letter from G. H. Noehden, Esqr. Secy. to the Asiatic Society, addressed to the Secretary, requesting on the part of the Asiatic Society to obtain a temporary accommodation for their meetings till they are in possession of a house of their own, and suggesting that the Managers of the Royal Institution would assist them in this object by allowing the Society the use of their Lecture room, for which they would be willing to make the Royal Institution whatever compensation might be thought proper.
Resolved,
That the Secretary be requested to express to the Asiatic Society, through their Secretary, the regret of the Managers that they have not the power to grant their request.

4. Read a letter from William Manning, Esqr. M.P. M.R.I. requesting permission for Mr. Elliot of Boston in America

384

1823.

who is residing in London for a few weeks, to be allowed access to the Royal Institution as is occasionally permitted to Foreigners.
Resolved, That Mr. Manning's request cannot be granted, as Mr. Elliot is not eligible to the above privilege.

Request of Mr. Newman.

5. Read a letter from Mr. John Newman of Lisle Street requesting of the Managers the privilege of being allowed to call himself on his Cards of business "Instrument maker to the Royal Institution," as he has had the honour of being employed by the Instit: for the last fifteen years.
Resolved, That Mr. Newman's request be granted.

6. The Secretary reported that he had deposited the Deeds, as ordered by the last Board, in the Iron Chest.

Presents received.

7 Granville Penn, Esqr. presented
An Examination of the primary Argument of the Iliad, 8vo.
A Comparative Estimate of the Mineral and Mosaical Geologies, 8vo.
The Royal College of Surgeons presented
List of the Members of the Royal College of Surgeons.

8. The Secretary laid before the meeting the notice of alterations and amendments in the Bye laws as directed by the last Board.

9. The following persons having been duly recommended were admitted as Subscribers to the Royal Institution for the present year.

Annual Subscribers.

Lord Ellenborough, of Roehampton
Rt. Hon: Lieut: Genl. Sir A. Hope 15 Charles St. Berkeley Square

385

April 7.

To the Lectures.

Mrs. Fisher
Miss Fisher } 60 Upper Seymour Street

Geo. Duckett V.P.

386

1823.

April 14.

At a Meeting of the Managers of the Royal Institution of Great Britain held at the house of the Institution on the 14th of April, 1823

Present
Sir George Duckett, Bart. V.P. in the Chair.
Thomas Snodgrass, Esqr. V.P.
Daniel Moore, Esqr. V.P.
Jesse Watts Russell, Esqr. M.P.
Sir Gilbert Blane, Bart.
Jn. George Children, Esqr.
John Chamier, Esqr.
Sir S. Bernard Morland, Bart. M.P. Treas.

1. The minutes of the proceedings of the last meeting were read and approved.

2. Read and approved the minutes of the Committee of Accounts of this day, and signed an Order for an Advance to the Assistant Secretary of £30.–

Life subscr.

3. In pursuance of a nomination under the Act of Parliament by John Symmons, Esqr. his eldest Son Charles A. J. Symmons, Esqr. of Paddington House was admitted to the privileges of a Life Subscriber to the Royal Institution.

4. The following persons having been duly recommended were admitted as Subscribers to the Royal Institution for the present year.

Annual Subscribers.
A. Bello, Esqr. 8 Solls row Hampstead
P. Creutzer, Esqr. 21 Grosvenor Street

To the Lectures.
Miss Favell Bevan
Miss Frederica Bevan
Mrs. David Bevan
Miss Bevan } 42 Upper Harley St.

Spencer

387

April 21.

At a Meeting of the Managers of the Royal Institution of Great Britain held at the house of the Institution on the 21st of April, 1823

Present
Earl Spencer, K.G. President, in the Chair.
Charles Hatchett, Esqr. V.P.
James Peter Auriol, Esqr. V.P.
Sir Humphry Davy, Bart. V.P.
Daniel Moore, Esqr. V.P.
John Chamier, Esqr.
Sir S. Bernard Morland, Bart. M.P. Treasurer
Sir George Duckett, Bart. V.P.
Thomas Harrison, Esqr. Secretary.

1. The minutes of the proceedings of the last meeting were read and approved.

2. Read and approved the minutes of the Committee of Accounts of this day, and signed an Order for an Advance to the Assistant Secretary of £130.

Presents received.

3. R. Porsoni Adversaria.
Dr. Young's Account of Discoveries in Hieroglyphical Literature & Egyptian Antiquities–
presented by Dr. T. Young F.R.S. &c.
Ordered,
That the thanks of the Managers be returned for these presents.

4. The following persons having been duly recommended were admitted as Subscribers to the Royal Institution for the present year.

Annual Subscribers.
Richard Nixon, Esqr. 14 Sackville Street
Arthur de Capell Brooke, Esqr.

To the Lectures.
Miss Sampayo
Miss Mist } Peterborough House Fulham

388 1823. April 21.

5\. Ordered, That the House Lists be prepared as follows of the Managers & Visitors for the ensuing year.

President – Earl Spencer, K.G. –

Treasurer – Sir S. Bernard Morland, Bart. M.P.

Secretary – Thomas Harrison, Esqr.

Managers

Earl of Bessborough
Sir Gilbert Blane, Bart
John Chamier, Esqr.
Henry Thos. Colebrooke, Esqr.
Sir H. Davy, Bart.
Sir George Duckett, Bart
Earl of Egremont.
Davies Gilbert, Esqr. M.P.
John Guillemard, Esqr.
Charles Hatchett, Esqr.
Thomas Hoblyn Esqr.
Daniel Moore, Esqr.
Sir Claude Scott, Bart.
Thomas Snodgrass, Esqr.
Lord Teignmouth.

Visitors.
Lieut. Col. Benjn. Ansley
William Babington, M.D.
John Frederic Daniell, Esqr.
John Fuller, Esqr.
Daniel Giles, Esqr.
Sir Christr. Hawkins, Bart
Sir Benjn. Hobhouse, Bart
Earl of Mountnorris
Louis Hayes Petit, Esqr.
George Pouney, Esqr.
Revd. Sir Jno. Robinson, Bart

389

Richd. Horsman Solly, Esqr.
Sir John Thos. Stanley, Bart
Giles Templeman Esqr.
Earl of Winchilsea, K.G.

Spencer

---

390 1823.

April 28. At a Meeting of the Managers of the Royal Institution of Great Britain held at the house of the Institution on the 28th. of April, 1823

Present
Earl Spencer, K.G. President, in the Chair.
Thomas Snodgrass, Esqr. V.P.
John Chamier, Esqr.
Sir Gilbert Blane, Bart.
Jesse Watts Russell, Esqr. M.P.
Thomas Harrison, Esqr. Secretary.
Sir H. Davy, Bart. V.P.

1\. The minutes of the proceedings of the last meeting were read and approved.

2\. Read and approved the minutes of the Committee of Accounts of this day.

Presents received.

3 The Revd. Stephen Weston, B.D. F.R.S presented
The History & Antiquities of Hengrave in Suffolk, by John Gage, Esqr.
Mr. Hawkins Browne presented
Essays on Subjects of important enquiry in Metaphysics Morals and Religion, by the late Isaac Hawkins Browne, Esqr.
Mr. Joseph Joplin presented
The Septenary System of generating Curves by continued Motion.
J.R. Planché Esqr. presented
Shere Afkun the first Husband of Nourmahal a Legend of Hindostan,
Ordered,
That the thanks of the Managers be returned for these presents.

4\. The President having stated that he had some reason

391

April 28.
Lectures.
Prof. Buckland

to hope that Professor Buckland would be disposed to give a Course of Lectures on Geology gratuitously during the present Season, provided His Lordship were empowered to request him to confer that favour on the Royal Instn.:
Resolved,
That the Managers feel themselves much obliged by the President's communication, and beg leave to desire him to make such an application to Professor Buckland.

Spencer

392 1823.

May 5.

At a Meeting of the **Managers** of the Royal Institution of Great Britain held at the house of the Institution on the 5th of May, 1823

Present
Earl Spencer, K.G. President, in the Chair.
Charles Hatchett, Esqr.
John Guillemard, Esqr.
Henry Thos. Colebrooke, Esqr.
Earl of Bessborough
Sir Humphry Davy, Bart.
John Chamier, Esqr.
Thomas Snodgrass, Esqr.
Sir Gilbert Blane, Bart.
Sir Claude Scott, Bart.
Sir George Duckett, Bart.
Earl of Egremont
Daniel Moore, Esqr.
Thomas Hoblyn, Esqr.
Sir S. Bernard Morland, Bart: M.P. Treasr.
Thomas Harrison, Esqr. Secretary.

1. Read the minutes of the proceedings of the last meeting which were approved.

2. Read and approved the minutes of the Committee of Accounts of this day, and signed an Order for an advance to the Assistant Secretary of £80.

Lectures. 3 The President reported that he had been informed that there was no prospect of Profr. Buckland's lecturing here this Season.

4. In pursuance of a nomination under the Act of Parliament by William Dawson, Esqr. his Son Frederick Dawson, Esqr. of the Middle Temple was admitted to the privileges of a Life Subscriber to the Royal Institution.

Presents received.

5 Edward Hawke Locker, Esqr. F.R.S presented

May 5. 393

Views in Spain, No. 1 –
The Proprietor of the Literary Chronicle, presented The Literary Chronicle No. 190 to 206 inclusive.

Ordered,
That the thanks of the Managers be returned for these presents.

6 The President nominated the following Vice Presidents for the ensuing year,

Vice Presidents.
Earl of Bessborough
Earl of Egremont
Sir H. Davy Bart.
Sir George Duckett, Bart.
Sir Claude Scott, Bart.
Charles Hatchett, Esqr.
Daniel Moore, Esqr.
Thomas Snodgrass, Esqr.

7 Read the minutes of the Annual Meeting of the Members.

8 The following person having been duly recommended was admitted as Annual Subscriber for the present year.

Annl. Subsr. Edward Johnson, Esqr. 1 Grosvenor Square

Bessborough V.P.

394 1823.

May 12.

At a Meeting of the **Managers** of the Royal Institution of Great Britain held at the house of the Institution on the 12th of May, 1823

Present
Earl of Bessborough V.P. in the Chair.
Charles Hatchett, Esqr. V.P.
Thomas Snodgrass, Esqr. V.P.
Sir Gilbert Blane, Bart.
John Chamier, Esqr.
Henry Thos. Colebrooke, Esqr.
John Guillemard, Esqr.
Thomas Hoblyn, Esqr.
Daniel Moore, Esqr. V.P.
Thomas Harrison, Esqr. Secretary.

1. The minutes of the proceedings of the last meeting were read and approved.

2. Read and approved the minutes of the Committee of Accounts of this day, and signed an Order for an advance to the Assist. Secretary of £200.

3 Read the minutes of the last General Monthly Meeting.

4 In pursuance of Art 1 of Chap XI of the Bye Laws

Payment of City Fine. We, the undersigned Managers, having met in Committee this day, do hereby give notice, that, as a considerable sum of money will be wanted for the payment of the renewal Fine, which will become due to the City of London at Michaelmas next, and for other purposes of the Institution, they shall move at the next General Monthly Meeting, to be held on the 2d of June for power, if they should find it necessary to sell out the Stock purchased by Mr. Fuller's Donation of £1000, and now standing in the 3 pr. Cent Consolidated Bank Annuities, or such part thereof

May 12. 395

as may be required for the purposes aforesaid.
(signed) Bessborough V.P. in the Chair
Thomas Hoblyn
Charles Hatchett
H. T. Colebrooke
Gilbert Blane
Thomas Harrison Secretary
J. Guillemard
Thomas Snodgrass
Daniel Moore

Ordered, That printed Copies of the above notice be sent round to the Members.

The following persons having been duly recommended were admitted as Subscribers to the Royal Institution.

Life Subscriber,
Honble. Wm. Francis Spencer Ponsonby 19 St. James's Square

Annual Subscriber.
Henry Seymour, Esqr. 39 Upper Grosvenor St.

Adjourned to Monday the 26th Instant

Charles Hatchett V.P.

396

1823.

May 26.

At a Meeting of the **Managers** of the Royal Institution of Great Britain held at the house of the Institution on the 26th of May 1823

Present

Charles Hatchett, Esqr. V.P. in the Chair
Sir Humphry Davy, Bart. V.P.
Daniel Moore, Esqr. V.P.
Thomas Snodgrass, Esqr. V.P.
Thomas Hoblyn, Esqr.
John Chamier, Esqr.
Sir Gilbert Blane, Bart.
John Guillemard, Esqr.
Henry Thos. Colebrooke, Esqr.

1. The minutes of the proceedings of the last meeting were read and approved.

2. Read and approved the minutes of the Committee of Accounts of this day, and signed an advance to the Assistant Secretary of £105.

Spencer

397

June 2.

At a Meeting of the **Managers** of the Royal Institution of Great Britain held at the house of the Institution on the 2d. of June 1823,

Present

Earl Spencer, K.G. President, in the Chair.
Sir Humphry Davy, Bart. V.P.
Charles Hatchett, Esqr. V.P.
Daniel Moore, Esqr. V.P.
Sir George Duckett, Bart. V.P.
Thomas Snodgrass, Esqr. V.P.
Sir Claude Scott, Bart. V.P.
John Guillemard, Esqr.
Davies Gilbert, Esqr. M.P.
John Chamier, Esqr.
Sir S. Bernard Morland, Bart. M.P. Treasr.
Thomas Harrison, Esqr. Secretary.

1. The minutes of the proceedings of the last meeting were read and approved.

2. Read and approved the minutes of the Committee of Accounts of this day.

3. (acct. of Russell Institution) Read a letter from Lt. Genl. Thornton addressed to the Secretary enclosing the last year's Account of the Russell Institution, and requesting that it may be laid before the Managers:
Ordered, That Lt. Genl. Thornton's letter be acknowledged.

4. (Coals) *Ordered*, That 30 Chaldrons of the best Walls-end Coals be laid in the Cellars of the Royal Institution.

5. *Resolved*, That this meeting at its rising do adjourn to such day as may be agreed upon at the General monthly Meeting for taking into their consideration such parts of the Report of the Committee appointed by the General Monthly Meeting as may be referred to the Managers.

Spencer

398

1823.

June 4.

At a Meeting of the **Managers** of the Royal Institution of Great Britain held at the house of the Institution on the 4th. of June, 1823 for the purpose of considering such parts of the Report of the Committee appointed to take into consideration the state of the Royal Institution as relate to the diminution of Expenditure and augmentation of Income, referred to the Managers by the last General Monthly Meeting.

Present

Earl Spencer, K.G. President, in the Chair.
Sir Humphry Davy, Bart. V.P.
Charles Hatchett, Esqr. V.P.
Thomas Snodgrass, Esqr. V.P.
Sir Claude Scott, Bart. V.P.
Davies Gilbert, Esqr. M.P.
John Chamier, Esqr.
Thomas Hoblin, Esqr.
John Guillemard, Esqr.
Henry Thos. Colebrooke, Esqr.
Thomas Harrison, Esqr. Secretary.

1. The Managers were attended by William Salmond, Esqr. and Edmund Robt. Daniell Esqr. on the part of the said Committee, and a long conference with them took place on the subject.

2. (Reading Room to be closed) *Ordered*, That the Reading Room be shut up at the same hour as the Library throughout the year.

Spencer

399

June 16.

At a Meeting of the **Managers** of the Royal Institution of Great Britain held at the house of the Institution on the 16th. of June, 1823

Present

Earl Spencer, K.G. President, in the Chair.
Charles Hatchett, Esqr. V.P.
Thomas Snodgrass, Esqr. V.P.
Daniel Moore, Esqr. V.P.
Sir George Duckett, Bart. V.P.
John Chamier, Esqr.
Thomas Hoblyn, Esqr.
Henry Thos. Colebrooke, Esqr.
John Guillemard, Esqr.
Thomas Harrison, Esqr. Secretary.

1. Read and approved the minutes of the proceedings of the Managers of the 2d. and 4th. of June.

2. Read and approved the minutes of the Committee of Accounts of this day, and signed an Order for an advance to the Assistant Secretary of £60.

3. Read a letter from Lieut. Genl. Wm. Thornton, addressed to the Secretary, relating to the resolution of the last Board for shutting up the Reading Room at the same hour as the Library.
Ordered, That Genl. Thornton's letter be acknowledged.

Presents received.

4. H. C. Whites Esqr. presented
4 Specimens of Calcedony,
2 ditto of Pudding-stone
Ordered, That the thanks of the Managers be returned for this present.

5. (Removal of Journals to News-Room) *Ordered*, That the current periodical Scientific Journals be removed from the Reading Room to the Newspaper Room.

6 Read the minutes of the last General Meeting; and the resolutions referred to the Managers having been taken into consideration,

Dismissal of Mr Harris

Mr Harris the Librarian was called in and informed by the President that on account of the low state of the Finances of the Institution, it is with regret the Managers are obliged to give him notice that he will be required to quit his situation at the end of the current year.

Publication of Journal declined

7. Resolved, That the Managers do not think that the publication of a Scientific Journal will be expedient.

8. The following person having been duly recommended was admitted as an Annual Subscriber to the Royal Institution for the present year.

Annl Subscr

Lieut. Genl. Sir Edward Howorth K.C.B.
7 Berkeley Street.

9. On the application of Sir J. C. Hippisley, Bart
Monsr. de Smidt of Stutgard.
was admitted to the Royal Institution under Art: 15 Chap VI of the Bye laws.

10. Adjourned to Monday the 30th Instant specially to consider of the proposed alterations in the Bye laws.

Spencer.

June 30.

At a Meeting of the Managers of the Royal Institution of Great Britain held at the house of the Institution on the 30th of June, 1823

Present
Earl Spencer K.G. President, in the Chair.
Lord Teignmouth, V.P.
Charles Hatchett, Esq.r V.P.
Sir Claude Scott, Bart. V.P.
Thomas Snodgrass, Esq.r V.P.
Sir George Duckett, Bart. V.P.
Thomas Hoblyn, Esq.r
Sir Gilbert Blane, Bart.
John Chamier, Esq.r
Sir S. Bernard Morland, Bart. M.P. Treasurer
Thomas Harrison, Esq.r Secretary.

1. The minutes of the proceedings of the last meeting were read and approved.

2. Read and approved the minutes of the Committee of Accounts of this day, and signed an Order for an advance to the Assistant Secretary of £50.

Presents received.

3. Arthur De Capell Brooke, Esq.r presented
Travels to the North Cape, in 1820,
Edward Hawke Locker, Esq.r presented
Views in Spain, N.o 2.
Francis Ronalds, Esq.r presented
Descriptions of an Electrical Telegraph.

Ordered,
That the thanks of the Managers be returned for these presents.

Leave of absence to assist in Lab.y

4 Resolved, That on account of the illness of Wm Bollaert, Assistant in the Laboratory, leave of absence be granted to him for 6 weeks from this day.

Spencer

July 7.

At a Meeting of the Managers of the Royal Institution of Great Britain held at the house of the Institution on the 7th of July, 1823

Present
Earl Spencer K.G. President, in the Chair
Charles Hatchett, Esq.r V.P.
Earl of Bessborough, V.P.
Thomas Snodgrass, Esq.r V.P.
Sir George Duckett, Bart. V.P.
Sir Gilbert Blane, Bart.
Henry Thos. Colebrooke, Esq.r
John Chamier, Esq.r
Thomas Hoblyn, Esq.r
Sir S. Bernard Morland, Bart. M.P. Treas.r
Thomas Harrison, Esq.r Secretary.

1 The minutes of the proceedings of the last meeting were read and approved.

2. Read and approved the minutes of the Committee of Accounts of this day.

Presents received.

3. The Society of Antiquaries, presented
Archaeologia, vol. 20 part 1.

Ordered,
That the thanks of the Managers be returned for this present.

Resolved, That the following report of the Managers be laid before the General Monthly Meeting of this day.

Finances.

The Managers report that since the last General Monthly Meeting they have admitted 2 Annual Subs.rs The Managers further state that they have taken into their consideration the report on the finances of the Institution made by the Committee appointed by the General Meeting to take the state of the Royal Instit.n into consideration, and to whom the proposed alterations

July 7.

in the Bye laws were referred, and have also held a meeting with two of the three Members of that Committee who were appointed by the last General Monthly Meeting of the Members of the Institution to confer with the Managers upon this subject; and they are led to hope that by abolishing some, and curtailing other parts of their Establishment, they may accomplish a diminution in the annual expences of the Institution to the amount of £450 or thereabouts.

Journal declined.

The Managers have also given the best consideration in their power to the proposition made to them for publishing a scientific Journal and have arrived at the conclusion that such a measure would be inexpedient.

As they collect from the above mentioned report that the hopes entertained by the Committee of obtaining pecuniary assistance are likely to be realized to a certain extent, and have since understood that some individuals are ready to advance money by way of Loan to the Institution without interest not only to provide for the liquidation of the existing debt of the Institution, but also for the temporary advance of the fine which will become due to the City of London at Michaelmas next, for the renewal of their lease of the Premises held by the Instit.n the Managers conceive it unnecessary to press upon the attention of the General Monthly Meeting the question respecting the sale of the Stock purchased by Mr Fuller's Donation which when notice of the proposition for such sale was given by the Managers they conceived to be required by the urgency of the case in order to meet such a demand on the part of the City of London.

Voluntary Subscription.

The Managers therefore propose that a voluntary Subscription be opened for a Loan to the amount of £4000 not bearing Interest for the purpose of liquidating the Debts of the Institution, and the expences of the Current year, and that the Managers be authorized to give Bonds under the Seal of the Institution for the repayment of the Loan so advanced. That the loan be made for 4 years, and at the end of that period that the mode of discharging the sums so advanced be taken into consideration, and such provisions made for the discharge of them, as the

circumstances of the Institution may then admit.

Disposal of Old Newspapers. The Managers also report that in order to further the system of retrenchment in the expences of the Institution they deem it expedient that they should be authorized by the General Meeting to dispose of the old Newspapers which at present belong to the Institution and also of those Newspapers which may be received in future.

Chas. Hatchett V.P.

July 14.

At a Meeting of the Managers of the Royal Institution of Great Britain held at the house of the Institution on the 14th of July, 1823.

Present

Charles Hatchett, Esqr. V.P. in the Chair.
Sir Gilbert Blane, Bart.
John Chamier, Esqr.
Thomas Snodgrass, Esqr. V.P.
Sir Claude Scott, Bart. V.P.
Daniel Moore, Esqr. V.P.
Henry Thos. Colebrooke, Esqr.

1. The minutes of the proceedings of the last meeting were read and approved.

2. Read and approved the minutes of the Committee of Accounts of this day, and signed an Order for payment of the following sums viz.t

Payments.

Joseph Fincher on account of the Royal Institution	£550
Wm. Thos. Brande, Esqr. Professor of Chemistry, 1 years Salary, 1823.	200. —
	750. —

Presents received.

3. The Royal Society, presented
Philosophical Transactions for 1823, part 1.
The Horticultural Society, presented
Transactions of the Horticultural Society, vol 5 part 2.
Report of the Garden Committee.
The Calcutta Journal No. 306 containing Mr Buckingham's reply to the Article on his Travels in the Quarterly Review.
Ordered,
That the thanks of the Managers be returned for these presents.

4. Read the minutes of the last General Monthly Meeting.

Housekeeper's Contract. 5. Mrs Greenwood, the Housekeeper, gave in an Estimate of her Contract by which she engages to supply the Institution with Soap and Candles, Brushes Brooms &c. Wood for all the Fires, Linen, and Flannel, and Washing, & also to supply one Porter, and Women Servants at £200 per Ann. including her own wages; which she would wish to receive by Monthly payments.
Resolved,
That Mrs Greenwood's Contract be approved; and that she be allowed £200 per Annum to be paid to her by Monthly payments.

Circular resp.g Subscription. 6. Resolved, That a lithographed circular be sent round to the Members containing the resolution of the last General Monthly Meeting respecting the Voluntary Loan, with the amount already subscribed to it.

Bonds to Lenders. 7. Resolved, That Daniel Moore, Esqr., Solicitor of the Royal Institution, be requested to prepare the Bonds which are to be given by the Institution to those Members who have subscribed to the Loan.

8. The Assist.t Secretary stated to the Board that he had just received from Sir Claude Scott, Bart. his subscription to the Loan Account of £500.

Resolved, That the Treasurer be requested to invest the sum of £500 in Exchequer Bills.

Chas. Hatchett V.P.

Aug. 4.

At a Meeting of the Managers of the Royal Institution of Great Britain held at the house of the Institution on the 4th of August, 1823.

Present

Charles Hatchett, Esqr. V.P. in the Chair.
Thomas Snodgrass, Esqr. V.P.
Daniel Moore, Esqr. V.P.
Thomas Hoblyn, Esqr.
Sir S. Bernard Morland, Bart. M.P. Treas.r
John Chamier, Esqr.

1. The minutes of the proceedings of the last meeting were read and approved.

2. Read and approved the minutes of the Committee of Accounts of this day; and signed an Order for payment of the following sums viz.t

Joseph Fincher on account of the Royal Institution	£300. —
One & 1/2 years Ground rent to the City of London to Michaelmas 1823	225. —
	£525. —

Leave of absence to Ass.t Lib.n 3. Mr. Stuart, the Assistant Librarian, requested permission of the Board for leave of absence for a month during the recess which was granted.

4. Resolved, That the Treasurer be requested to invest the sum of £500 in an Exchequer Bill.

5. The following person having been duly recommended, was admitted as an Annual Subscriber to the Royal Institution for the year 1823.
Ann.l Sub.er Capt.n John Tod, Hatchetts Hotel Piccadilly —

6. It being now 2 o'Clock, the Managers present attended the General Monthly Meeting, and the Resolution

408

1823

having been confirmed by that Meeting which empowered the Managers to affix the Seal of the Corporation to the proposed Bonds to be given to those Members who should subscribe to the Voluntary Loan; The following Managers returned to the Board viz[t].

Charles Hatchett, Esq[r].
Thomas Snodgrass, Esq[r].
Daniel Moore, Esq[r].
Thomas Hoblyn, Esq[r].

And Bonds to the undermentioned persons were read, agreed to, and sealed with the Seal of the Corporation (agreeably to Art: 3 of Chap XII of the Bye laws) bearing date the 4th of August, 1823 whereby the repayment of the respective sums, without Interest, on or before the 4th of August 1827 was secured to the several Subscribers viz[t].

Subscribers to Loan.

To	Earl Spencer, K.G. for securing	£500 –
"	Daniel Moore, Esq[r]. d°	1000 –
"	Sir Claude Scott, Bart d°	500 –
	Charles Hatchett, Esq[r].	100 –
	George Aust, Esq[r].	100 –
	Sir S. Bernard Morland, Bart	100 –
	Sir George Duckett, Bart	100 –
	James Peter Auriol Esq[r].	100 –
	Sir Gilbert Blane, Bart	100 –
	Charles Nicholas Pallmer, Esq[r].	100 –
	Louis Hayes Petit, Esq[r].	100 –
	Henry Hoare, Esq[r].	200 –
	John Soane, Esq[r].	100 –
	Davies Gilbert, Esq[r]. M.P.	100 –
	Jesse Watts Russell, Esq[r]. M.P.	100 –
	Bishop of Durham	100 –
		£3400 –

Adjourned to the 3[d] of November.

Ch[s]. Hatchett V.P.

409

Nov. 3.

At a Meeting of the Managers of the Royal Institution of Great Britain held at the house of the Institution, on the 3[d] of November, 1823

Present

Charles Hatchett, Esq[r]. V.P. in the Chair
Thomas Snodgrass, Esq[r]. V.P.
Henry Tho[s]. Colebrooke, Esq[r].
Sir S. Bernard Morland, Bart: M.P. Treas[r].
Thomas Harrison, Esq[r]. Secretary.

1. The minutes of the proceedings of the last meeting were read and approved.

2. Read and approved the minutes of the Committee of Accounts of this day.

3. Presents received.

Sir George T. Staunton, Bart: M.P. presented
Journals of the House of Lords, vol[s]. 34 to 50 inclusive with Volume of Index and Calendar,
Granville Penn, Esq[r]. presented
Supplement to the Comparative Estimate of the Mineral & Mosaical Geologies,
The Royal Society presented
The Nautical Almanack, 1826,
The Court of Directors of the East India Company presented Morison's Chinese Directory, vol[s]. 2 part 1,
Edward Hawke Locker, Esqr. F.R.S. presented
Views in Spain, N°s. 4 & 5.
M. Garcio del Rio, presented,
La Biblioteca Americana, N°s. 1, 2, 3,
Le Chevalier de Kirkhoff presented
Hygiène Militaire, 8vo.
Anti-Slavery Society,
Debate in the House of Commons on the 15th of May 1823 for the Abolition of Slavery & 4 Pamphlets on the same subject.

Ordered, That the thanks of the Managers be returned for these presents.

410

1823.

4. The undermentioned Bonds were read, agreed to and sealed with the Seal of the Corporation, agreeably to Art 3 of Chap: XII of the Bye laws, bearing date this third day of November, whereby the respective sums without Interest, were secured to the Subscribers on or before the 4th of August 1827.

To John Guillemard, Esq[r]. for £100.
" Sir Edmund Antrobus, Bart £100.

Life Subs[cr].

5. In pursuance of a nomination under the Act of Parliament by the Will of the late James Stodart, Esq[r]., his Son in Law, (being also a Relation in Blood) George Stodart, Esq[r]. of 23 Keppel Street Russell Square, was admitted a Life Subscriber to the Royal Institution.

Adjourned to the 1st of December next.

H T Colebrooke

411

Dec 1.

At a Meeting of the Managers of the Royal Institution of Great Britain, held at the house of the Institution on the 1st of December, 1823,

Present

Henry Tho[s]. Colebrooke, Esq[r]. in the Chair.
John Guillemard, Esq[r].
Thomas Hoblyn, Esq[r].
Daniel Moore, Esq[r]. V.P.
Sir S. Bernard Morland, Bart M.P. Treasurer
Sir Gilbert Blane, Bart.

1. The minutes of the proceedings of the last meeting were read and approved.

2. Read and approved the minutes of the Committee of Accounts of this day, and signed an Order for payment of the following sums, viz[t].

Payments

Daniel Moore, Esq[r]. for renewal Fine to the City of London from the Royal Institution due at Mich[s]. 1823 which was advanced by him the 2[d] of October last	£600 –
Jos[h]. Fincher on account of the Royal Institution	100 –
Ditto – d° – (30th Dec[r].)	200 –
	£900 – –

Presents received.

3. D[r]. Gilchrist presented
The Hindee Moral Preceptor; or Rudimental principles of Persian Grammar.
John Frederic Daniell, Esq[r]. F.R.S. presented
Meterological Essays & Observations,
The Horticultural Society presented
Transactions of their Society vol[s]. 5 part 3,
The Royal Society presented
Philosophical Transactions 1823, part 2.

Presented by
The Author of The Count Arrezzi, a Tragedy,
Dr. John Alderson presented
An Essay on Apparitions.
The Directors of the African Institution presented
Their Seventeenth Report.

Ordered,
That the thanks of the Managers be returned for these presents.

Lectures. 4. Read a letter from Sir H. Davy, Bart addressed to the Secretary, enclosing a letter & Prospectus from Mr. James Elmes, who would be happy to deliver at the Royal Institution a Course of Lectures on the fine Arts.

5 Read a letter from Mr. Dowren, an Annl. Subr., enclosing letters from Mr. Wallis, offering to give a Course of Lectures on Astronomy, and from Mr Banks offering to give a Course of Lectures on Polite Literature.

6 The Assist. Secretary stated that Dr Gilchrist would be happy to deliver, gratuitously, a Course of Lectures "On making the English language universal."

Resolved, That the above proposals of delivering Lectures, be referred to Mr. Hatchett, Mr Danl. Moore & Mr. Harrison, and if approved of, that they be inserted in the list of lectures.

Right of Renewal forfeited. 7. The Solicitor of the Royal Institution communicated the following extract of a letter, which he had received from Mr. Spooner's Solicitor respecting the renewal fine of Premises No. 22 Albemarle Street, in which Mr Spooner says – "I know it to be my interest to have the lease "renewed, but the fact is at present, the Rent of "that house is all I have to maintain a large family "and therefore much as it is my interest to renew as "well as inclination, I have it not in my power so "to do. All I can say is that provided the Lessors "will permit me, I will undertake to renew it the moment "I have it in my power and pay interest on the fine from "the present time"

Resolved, That Mr. Danl. Moore be requested to inform Mr. Spooner, that as his renewal fine, due at Michas last, has not been paid, his right of renewal is thereby forfeited.

Adjourned to 12th. January, 1824.

Thomas Snodgrass
V. P.

Jan 12.

At a Meeting of the Managers of the Royal Institution of Great Britain, held at the house of the Institution, on the 12th. of January, 1824.

Present
Thomas Snodgrass, Esqr. V.P. in the Chair.
Henry Thos. Colebrooke, Esqr.
Daniel Moore Esqr. V.P.
Sir Gilbert Blane Bart.
Sir George Duckett, Bart. V.P.
John Guillemard, Esqr.
Thomas Harrison, Esqr. Secretary.

1. The minutes of the proceedings of the last meeting were read and approved.

2. Read and approved the minutes of the Committee of Accounts of this day.

Lectures 3. Read a letter from Mr. John Bird of Beaconsfield Bucks offering to deliver a Course of Lectures on Astronomy.

Resolved, That Mr. Birds' offer be declined.

4. Presents received

Thomas Chevalier, Esqr. F.R.S. presented
Lectures on the General Structure of the Human Body,
The Editor of the Oriental Herald & Colonial Review presented No. 1, of the Review.
Mr. Richard Morris, presented
The Botanists' Manual 8vo.

Ordered,
That the thanks of the Managers be returned for these presents.

Lectures 5. Read the following report:–
The Sub-Committee appointed by the last Board to take into consideration the proposals made by Mr. Elmes, Mr. Wallis, Mr. Banks & Dr Gilchrist,

of giving Courses of lectures in the present Season, reported that in their opinion it was inexpedient to accept these respective offers, and the Secretary had been instructed to decline them accordingly.

Dismissal of Mr. Harris. 6 Resolved, That the sum of £180 for a years' Salary due to Mr. Harris the late Librarian, at Christmas last be forthwith paid to him, and that he be informed that his duty as Librarian ceased at that time agreeably to the tenor of the notice given to him by the President on the 16th. June last.

7. Resolved, That Mr. Brande be empowered to direct an alteration to be made in the Fire places of the Great Lecture Room.

Laboratory Fund, 8. Ordered, That the following resolution be printed and sent round to the Members.

That in order to carry into effect the recommendation of the Committee, who were appointed to take the state of the Royal Institution into consideration, of placing the Laboratory upon a foundation to insure its practical efficiency without rendering it too burthensome to the rest of the Establishment;

That Copies of Chap: XXI, beginning at Art 3 of the Bye laws, and Copies of the resolution of the General Monthly Meeting of the 7th of July last, relating to the Laboratory Fund, with a List of Donations, Compositions, and Annual Contributions already subscribed, be sent round to the Members.

And that those Members who shall consent to subscribe One guinea per Annum, or who shall otherwise contribute to the Laboratory Fund, be requested to signify, as early as convenient, such their intention.

9. The following persons having been duly recommended were admitted as Subscribers to the Royal Institution for the year 1824.

416 1824

Annual Subscribers
S. H. Sutherland, Esq.r 28 Thayer Street
George Mower, Esq.r 19 Leicester Square
Capt.n Evans, 16 Charlotte St. Portland Place,
Dr Kitchener 43 Warren St. Fitzroy Square
Alexander Fraser, Esq.r 50 Tavistock Square
Thomas Copeland, Esq.r 4 Golden Square
H. Curling, Esq.r 81 Baker Street
Rev.d Rich.d Caton 32 York St. Portman Sq.re
William Garnet, Esq.r 24 Grove Place Lisson Green
Gilbert S. Mathison, Esq.r 10 Orchard Street
Tho.s Cowper Hincks, Esq.r Bond Street
W. H. Banks, Esq.r 38 Albemarle Street
Capt.n John Grover 69 Upper Norton St.
Joseph Cecil, Esq.r 171 Regent Street
Augustus Fred.k Bayford, Esq.r 16 Cheyne walk, Chelsea
Allan Gilmore Esq.r Portland Place
D. Taylor, Esq.r Colonnade Hotel Charles St.
John Parish Robertson, Esq. Suffolk Place Pall Mall
Thomas Edwards, Esq.r L.L.D. Tulse Hill Place, Brixton

To the Lectures.

Mrs Rhodes
Miss Rhodes } Tottenham Wood.
Miss Harriet Rhodes

Adjourned to Monday the 2d of February.

Chas Hatchett V.P.

Feb. 2. 417

At a Meeting of the Managers of the Royal Institution of Great Britain held at the house of the Institution on the 2d of February, 1824.

Present
Charles Hatchett, Esq.r V.P. in the Chair.
Sir Claude Scott, Bart. V.P.
Sir Humphry Davy Bart. V.P.
Daniel Moore, Esq.r V.P.
Thomas Snodgrass, Esq.r V.P.
Henry Tho.s Colebrooke, Esq.r
Sir S. Bernard Morland, Bart. M.P. Treasurer
John Guillemard, Esq.r
Sir Gilbert Blane, Bart
Thomas Harrison, Esq.r Secretary.

1. The minutes of the proceedings of the last meeting were read and approved.

2. Read and approved the minutes of the Committee of Accounts of this day, and signed an Order for an advance to the Assist.t Secretary of £300.

Lectures. 3. Read a letter from the Marquis Spineto offering to deliver at the Royal Institution, either early in the Spring, or towards the latter end of the Season, a Course of Lectures on the literature of Modern Europe, comprising the literature of England, France, Germany Spain and Italy

Resolved,
That the Marquis Spineto's offer be accepted, and that the Secretary be directed to offer the sum of £63 for a Course of 12 Lectures.

4. In pursuance of nominations under the Act of Parliament,

Life Subs.crs Walter Weldon jun.r Esq.r
nominated by his Father Walter Weldon, Esq.r

---

418 1824

and Henry Bingley, Esq.r of the Royal Mint
nominated by his Father Robert Bingley, Esq.r
were admitted as Life Subscribers to the Royal Institution.

Presents received.

5. John Boucher, Esq.r presented
The Christian Religion made plain by The Rev.d Rich.d Boucher,
The Society for the encouragement of Arts &c. presented
Transactions of the Society for 1823,
The Commissioners of the Public Records presented
Ducatus Lancastriae, Pars Prima & 2da
Edward H. Locker, Esq.r presented
Views in Spain, No 6, & 7.

Ordered,
That the thanks of the Managers be returned for these presents.

6. Audited the Treasurer's account for the year 1823.

Treasurer of Labor.y Fund. 7. Resolved,
That Francis Bernard Morland, Esq. be appointed the Treasurer of the Laboratory Fund of the Royal Institution.

8. The following persons having been duly recommended were admitted as Subscribers to the Royal Institution for the year 1824.

Annual Subscribers.
George Miller Esq.r 67 Gower Street
W.m Henderson, Esq.r 33 Charlotte St. Fitzroy Sq.r
Robert Orme, Esq.r 0 Alsops Buildings
Patrick O'Conor, Esq.r 6 Harford St.
Dougal Christie, Esq.r 3 Montagu Square
W.m Hanmer, Esq.r 17 Upper Harley St.
Rich.d Penn, Esq.r Richmond Hill

Feb. 2. 419

To the lectures.
Mrs & Miss Evans Hertford St. May Fair
Miss Bean, Albemarle Street.

Neglect of Porters &c. 9. It appearing to the Board that Richard Hall the Assistant Porter had neglected his duty,

Resolved,
That he be dismissed to-day with a week's wages.

10. It appearing also that Tho.s Slatter had neglected to fasten the Windows in the Managers Room. He was called in and admonished, and he was informed that if he be negligent again, he will be forthwith dismissed.

Adjourned to the 1st of March.

George Duckett V.P.

420

Mar. 1. 1824.

At a Meeting of the Managers of the Royal Institution of Great Britain held at the house of the Institution on the 1st of March, 1824.

Present
Sir George Duckett Bart. V.P. in the Chair.
Sir Claude Scott, Bart. V.P.
Daniel Moore, Esqr. V.P.
Thomas Snodgrass, Esqr. V.P.
Sir Gilbert Blane, Bart.
Thomas Hoblyn, Esqr.
Sir S. Bernard Morland, Bart. M.P. Treasurer
Charles Hatchett, Esqr. V.P.
John Guillemard, Esqr.

1. The minutes of the proceedings of the last meeting were read and confirmed.

2. Read and approved the minutes of the Committee of Accounts of this day: and signed an Order for an advance to the Assist. Secretary of £.500.

Lectures 3. Read a letter from T. C. Banks, Esqr. offering to deliver 3 or more lectures on Genealogy. — Mr. Banks added that he tendered his proposition gratuitously, without reference to, or expectation of reward of any kind to be given by the Institution.
Resolved,
That Sir S. Bernard Morland, Bart and Daniel Moore be requested to make inquiries whether Mr. Banks could make lectures on the above subject sufficiently interesting for public lectures, and to report to the next Board.

4. Presents received.
The Revd. Stephen Weston presented Annotations on the Psalms.
H. T. Colebrooke, Esqr. presented Letters on Intolerance by Sir Geo: Colebrooke Bart.

421

Mar. 1.

Renewal of City Lease. 5. Notice having been sent to the Managers agreeably to the Bye laws. The Seal of the Corporation was affixed to the Counterpart of Lease from the City of London to the Members of the Royal Institution dated 15th. December 1823, of three Messuages or Tenements No. 20, 21, and 22, and part of another No. 23 on the East side of Albemarle Street, part of Conduit Mead for 61 years from Michaelmas 1823 —
It was countersigned by
George Duckett, Chairman.
Claude Scott,
Gil: Blane,
Scrope B. Morland,
Thomas Snodgrass,
Thomas Hoblyn,
Dan: Moore,
} A majority of the Managers present
Jos. Fincher, Assist. Secretary.

6. The following persons having been duly recommended were admitted as Subscribers to The Royal Institution for the year 1824

Annual Subscribers.

Captn. Franklin, R.N. 55 Devonshire Street
Thomas Boddington Esqr. 23 Upper Berkeley Street
A. Augustus Warren, Esqr. 24 Charlotte Street
Lord Lilford, Grosvenor Place.
Charles Henry Smith, Esqr. Drapers hall.
Robert Addams, Esqr. 49 Jermyn Street
Captn. Wm. P. Haigh, Royl. Engineers 13 Tavistock row.
Henry Thursby, Esqr. 76 Wimpole Street.
Wm. A. A. White, Esqr.
Francis White, Esqr. } 14 College St. Westmr.
James Dickens, Esqr. 52 Parliament Street.
P. C. Cazalet, Esqr. Kensington.
Stirling James Christie, Esqr. 2 Paradise row.
Edwith Walker, Esqr. 49 Bedford Square.
Henry Whitehead, Esqr. Little Cadogan Place.
Thomas Le Blanc, Esqr. Chelsea College.
Matthew Shakespear, Esqr. Albany.

422

1824.

Annual Subscribers continued.
Captn. Brace, R.N. Harwich Street
Captn. Sotheby 13 Grosvenor Street
Captn. The Hon: Frederic Spencer St. James's Place

To the Lectures

Mrs. Boddington
Miss Boddington } 23 Upper Berkeley St.
Miss Boddington
Mrs. Andrews 7 Arlington Street
Miss Hunter 18 Sackville St.
Hon: Miss H. M. Powys.
Hon: Miss E. Powys,
Hon: Miss E. A. Powys } Grosvenor Place
Miss Solly —
Miss Lavinia Solly
Miss Sarah Solly
Miss Kingston
Mrs. Mary Ramsay Anderson, 15 Chapel Street
Miss Anne Craufurd Paterson, 16 Queen St. May Fair
Miss White
Miss L. White } 14 College Street Westmr.
Mr. Milward
Miss E. Barnard
Miss F. Barnard
Miss Barnard } Mitcham.
Mrs. Mary Christie, 17 Sloane Street
Miss Rhodes
Miss M. Rhodes
Mrs. Rhodes } Hoxton.
The Hon: Mrs. Ryder
Miss Ryder
Miss E. Ryder }
Miss Ryder, 37 Lower Grosvenor Street
Miss Emily Coles 76 Harley Street
Mrs. Charles Mills 8 Manchester Square
Miss Croft, Guilford Street
Miss Mackintosh
Miss S. Mackintosh } 56 Montagu Square.

423

Mar. 1.

To the Lectures.
Miss Smith
Miss Willet } Upper Harley Street
Mrs. Charles Williams, 6 Upper Harley Street

Hon. Admission 7. On the application of Sir George Duckett, Bart. His Excellency Baron de Cetto, the Bavarian Minister was admitted to the Royal Institution under Art: 16 Chap VI. of the Bye laws.

Spencer

424

Mar. 8. 1824

At a Meeting of the Managers of the Royal Institution of Great Britain, held at the house of the Institution, on the 8th of March, 1824.

Present

Earl Spencer, K.G. President, in the Chair.
Sir Claude Scott, Bart. V.P.
Sir George Duckett, Bart. V.P.
Thomas Snodgrass, Esqr. V.P.
Daniel Moore, Esqr. V.P.
Sir Gilbert Blane, Bart.
Sir S. Bernard Morland, Bart. M.P. Treasr.
John Chamier, Esqr.
John Guillemard, Esqr.

1. Read and Confirmed the minutes of the proceedings of the last meeting.

2. Read and approved the minutes of the Committee of Accounts of this day; and signed an Order for payment of the following sum, vizt.

Payments

Daniel Moore, Esqr. for Stamps for Bonds given to the Members who advanced the Voluntary Loan	37. 7. 9
& for Mr. Bushnell's bill of Costs for City lease paid by Mr. Moore 29th Septr. last	32. 5. –
	£ 69. 12. 9

Lectures. Genealogy.

3. Sir S. Bernard Morland, Bart reported that he had had an interview with Mr. Banks, and he was of opinion that three lectures, as proposed to be delivered by him, on Genealogy might be sufficiently interesting to be given at this Institution.

Resolved,

That Mr. Banks's offer be accepted.

Presents received.

4. The Society for the improvement of Prison Discipline

Mar. 8. 425

presented the Fifth Report of the Society,
The Conductors of the Zoological Journal presented
The Zoological Journal, No. 1,
Dr. C. Scudamore, presented
Essay on the Blood.

Ordered, That the thanks of the Managers be returned for these presents.

5. Read the minutes of the last General Monthly Meeting.

6. The following persons having been duly proposed were admitted as Subscribers to the Royal Instit.n for the year 1824.

Annual Subscribers

Captn. Henry Thomson, R.N. 13 Great George St.
Dr. Wm. Baker, 41 Conduit Street.
Jens Wolff, Esqr. Balham, Clapham.
George Craufurd Heath, Esqr. Albany.
Mr. Edward Snell – Albemarle Street.

To the Lectures.

Miss Fulton }
Miss Ellen Fulton } 4 Upper Harley Street
Mrs. P. Skelton – King Street Portman Square
Mrs. & Miss Mundy, 16 Henrietta St. Cavendish Sq.

7. Resolved, That the sum of £204. 15. – being the Compositions and Admission fees received from new Members, be invested in the 3 per Cent Consols in the name of the Corporation.

Investment of Compositions.

8. Resolved, That the sum of £199. 10. – being the Compositions from 15 old, and four new Members to the Laboratory Fund, be invested in the 3 per Cent Consols, in the names of George John Earl Spencer K.G. and Sir S. Bernard Morland, Bart. M.P. on account of the Laboratory Fund.

Adjourned to the 22d of March.

Spencer.

426

Mar. 22. 1824.

At a Meeting of the Managers of the Royal Institution of Great Britain held at the house of the Institution on the 22d of March, 1824.

Present

Earl Spencer K.G. President, in the Chair.
Sir Humphry Davy, Bart. V.P.
Thomas Snodgrass, Esqr. V.P.
Sir S. Bernard Morland, Bart. M.P. Treasr.
John Guillemard, Esqr.
John Chamier, Esqr.
Sir Gilbert Blane, Bart.

1. The minutes of the proceedings of the last meeting were read and confirmed.

2. Read and approved the minutes of the Committee of Accounts of this day; and signed an Order for payment of the following sums, vizt.

2 quarters Ground rent to the City of London to 25th Inst.	75. –
Joseph Fincher on account of the Royal Institution	80. –
	£ 155. –

and on account of the Laboratory Fund £20.

Presents received.

3. The Committee of the London Hibernian Society presented The 17th Report of the Society – 3 Copies –
Steven's Remarks on the State of Ireland, 3 copies
Edward Hawke Locker, Esqr. – presented
Views in Spain, No. 8.
Sir James Edward Smith, presented
The English Flora, Vol. 1 & 2.

Ordered,

That the thanks of the Managers be returned for these presents.

Coals.

4. Ordered, That Messrs. Dickins & Co. supply the Institution with 10 Chaldrons of best Walls-end Coals.

Mar. 22 427

Mr. Newman's Acct.

5. Resolved,

That Mr. John Newman who supplies the Institution with Chemicals, be directed to furnish his Account to the 31st Instant; and to regularly send it in at the end of each quarter.

6. The following persons having been duly recommended were admitted as Subscribers to the Royal Institution for the year 1824

Annual Subscribers.

Captn. Mundy, R.N. 9 Cork Street
John Hicks, Esqr. 39 Upper Charlotte St.
Edward Rose Tunno, Esqr. 10 Upper Brook St.
Earl of Euston, Grosvenor Place

To the Lectures.

Miss Tunno }
Miss H. Tunno } 19 Upper Brook Street
Mrs. Droz – 15 Craufurd Street
Mrs. Barton, 12 Chesterfield Street
Miss Harriet Hendrie Tichborne Street
Miss Grant }
Miss Letitia Grant } 34 Portman Square
Lady Mary Stanley }
Miss Stanley } Audley Square
Mrs. Cazalet – 15 Craufurd Street

Adjourned to the 5th of April.

Spencer.

428

April 5. 1824.

At a Meeting of the Managers of the Royal Institution of Great Britain, held at the house of the Institution on the 5th of April, 1824

Present

Earl Spencer, K.G. President, in the Chair.
Sir George Duckett, Bart. V.P.
Thomas Snodgrass, Esqr. V.P.
Henry Thos Colebrooke, Esqr.
John Guillemard, Esqr.
John Chamier, Esqr.
Sir S. Bernard Morland, Bart. M.P. Treasr.

1. The minutes of the proceedings of the last meeting were read and approved.

2. Read and approved the minutes of the Committee of Accounts of this day and signed an Order for an advance to the Assistant Secretary of £80.

Lectures or Dr Harwood. 3. Read a letter from Dr Bostock stating that Dr Harwood would be happy to deliver a Course of 4 Lectures on Zoology after Easter, if the Managers would accept of his services.
Resolved, That Dr Harwood's obliging offer be accepted.

Presents received.

4. W. S. Lewis, Esqr. presented
Biographical Sketch of the late W. Lewis, Esqr.
The London Anti-Slavery Society presented
Stephens on Slavery in the British West India Colonies.
Ordered, That the thanks of the Managers be returned for these presents.

Coals. 5. Ordered, That Messrs Dickins & Co. lay in the Cellars of the Royal Institution, 20 Chaldrons of the best

April 5. 429

Walls-End Coals at the price specified by them viz. 54s. per Chaldron, including all expences of Metage, Shooting &c.

Or Rept. of Profr. of Chemistry 6 Read the following resolution from the Visitors entered in their proceedings of the 25th of March, viz.
"That this meeting suggest to the Managers that it would materially conduce to the interests of the Royal Institution, if the Annual Report of the Professor of Chemistry were printed and circulated with the Annual Report of the Visitors."
Resolved, That the above resolution be referred to Mr Brande, and that he be requested to attend the next meeting of the Visitors thereon.

Annl. Subrs. 7. The following persons having been duly recommended were admitted as Annual Subscribers to the Royal Institution for the year 1824.
Major Genl. Bayly 15 Dover Street
Captn. De Brett Twickenham.

Adjourned to the 19th of April.
Thomas Snodgrass V.P.

430

April 19. 1824

At a Meeting of the Managers of the Royal Institution of Great Britain held at the house of the Institution on the 19th of April, 1824

Present

Thomas Snodgrass, Esqr. V.P. in the Chair.
John Chamier, Esqr.
Thomas Hoblyn, Esqr.
Henry Thos Colebrooke, Esqr.
Sir S. Bernard Morland Bart M.P. Treasr.

1. The minutes of the proceedings of the last meeting were read and approved.

2. Read and approved the minutes of the Committee of Accounts of this day.

Presents received.

3. John Taylor, Esqr. presented
Selections from the Works of Baron de Humboldt relative to Mexico,
T. C. Banks, Esqr. presented
History of the Family of Marmyun
Jacob Jones, Esqr. presented
The Fall of Constantinople and other Poems,
Ordered, That the thanks of the Managers be returned for these presents.

Repairs. 4. Read the following resolution of the Visitors entered in their proceedings of the 12th Instant.
The Visitors examined the House and found the Premises in good repair – They recommend to the Managers that Mr Faraday's room should be painted and cleaned – The Carpet in the Library of Reference wants mending, and a piece of Oil-Cloth to be placed at the Door, where the Carpet is much worn – The Clock in the lecture Room seems to have been much neglected to the great inconvenience of the persons attending the Lectures,
Resolved, That Mr Faraday's room be cleaned and

April 19. 431

if it should then require Painting that it be done. – That the Oil Cloth, as recommended, ~~should~~ be placed at the Door where the Carpet is worn, and that the Carpet ~~should~~ be mended.
Resolved, That the recommendation of the Visitors respecting the Clock in the Lecture room be referred to John Chamier, Esqr. and that he be requested to report to the next meeting thereon.

5. The Seal of the Corporation was affixed to a Bond dated this 19th of April (agreeably to Art. 3 of Chap. XII of the Bye laws) whereby the sum of £100 was secured to The Rt. Hon: Wm Sturges Bourne, M.P. without Interest on or before the 4th of August 1827.

6. Ordered, That the House Lists be filled up as follows.

President. – Earl Spencer, K.G.

Treasurer. – Sir S. Bernard Morland, Bart, M.P.

Secretary. – John Guillemard, Esqr.

Managers

Sir Frederick Baker, Bart
John Chamier, Esqr.
Edmund Robert Daniell, Esqr.
Sir Humphry Davy, Bart.
Sir George Duckett, Bart
Henry Hallam, Esqr.
Charles Hatchett, Esqr.
Marquis of Lansdowne
Daniel Moore, Esqr.
Jesse Watts Russell, Esqr.
Sir Claude Scott, Bart
Thomas Snodgrass, Esqr.
John Soane, Esqr.
Duke of Somerset
Sir George Thos Staunton, Bart.

432 1824.

Visitors.

Lt. Col. Benjn. Ansley,
Sir Edmund Antrobus, Bart.
John Bostock, M.D.
Earl of Chichester,
John George Children, Esqr.
John Frederic Daniell, Esqr.
John Fuller, Esqr.
Lieut Col. W. M. Leake,
James Heath Leigh Esqr.
Revd. Sir John Robinson, Bart.
Richard Horsman Solly, Esqr.
William Sotheby, Esqr.
Sir John Thos. Stanley, Bart.
George Watson Taylor, Esqr.
John Weyland junr. Esqr.

Adjourned to the 3d. of May.

John Chamier

433

May 3.

At a Meeting of the Managers of the Royal Institution of Great Britain held at the house of the Institution on the 3rd of May, 1824

Present
John Chamier, Esqr. in the Chair.
Sir George Duckett, Bart.
Jesse Watts Russell, Esqr. M.P.
Edmund Robert Daniell, Esqr.
John Guillemard, Esqr. Secretary.

1. The minutes of the proceedings of the last meeting were read and approved.

2. Read and approved the minutes of the Committee of Accounts of this day, and signed an Order for an advance to the Assistant Secretary of £50.

Lectures — Marquis de Spineto.

3. Read a letter from The Marquis Spineto, addressed to the Assistant Secretary, enclosing a Correspondence which took place between the late Thomas Harrison, Esqr. and himself respecting travelling expences for his lectures, and requesting it to be laid before the Managers And stating that as his lectures will end on the 25th. of May he should be happy to deliver 2 or 3 more lectures in June, and that for these additional lectures he should be satisfied with 10 guineas per lecture, travelling expences included.
Resolved, That the consideration of the Correspondence between the late Mr. Harrison and the Marquis Spineto be deferred till the next meeting. — And that the Marquis's proposal to deliver 2 or 3 additional lectures in June be declined.

4. Mr. Chamier reported that he had given directions to Mr. Vulliamy of Pall Mall to attend to the Lecture-room Clock, who had requested him to state to the Board that he would undertake the care of all the Clocks of the Institution at the rate of 10s. 6d. per Annum for each Clock.
Resolved, That Mr. Vulliamy be appointed to attend

434 1824.

to the Clocks of the Institution on the above terms in the place of Mr. Hole.

5. The President nominated the following Vice Presidents for the ensuing year.

Sir Humphry Davy, Bart. Pres: Royl. Society
Sir George Duckett, Bart
Charles Hatchett, Esqr.
Marquis of Lansdowne,
Daniel Moore, Esqr.
Sir Claude Scott, Bart.
Thomas Snodgrass, Esqr.
Duke of Somerset.

6. The following persons having been duly recommended were admitted as Subscribers to the Royal Institution for the year 1824.

Annual Subscribers.
Lieutt. John Holford 39 York Place, Portman Sq.
Revd. Robert Butler, 80 Lisson Grove, New Road
Doctor Patrick Mackie 75 King Street Westmr.

To the Lectures.
Mrs. Macgregor, } 14 Sackville Street
Miss Macgregor, }
Miss G. Macgregor, }
Mrs. Schrieber. }

Adjourned to the 17th. Instant.

Chas. Hatchett V.P.

435

May 17.

At a Meeting of the Managers of the Royal Institution of Great Britain held at the house of the Institution on the 17th. of May, 1824

Present
Charles Hatchett, Esqr. V.P. in the Chair.
Thomas Snodgrass, Esqr. V.P.
John Chamier, Esqr.
Henry Hallam, Esqr.
Sir S. Bernard Morland, Bart. M.P. Treasurer
Sir Claude Scott, Bart. V.P.
Sir George Thomas Staunton, Bart. M.P.
John Guillemard, Esqr. Secretary.

1. The minutes of the proceedings of the last meeting were read and approved.

2. Read and approved the minutes of the Committee of Accounts of this day; and signed an Order for an Advance to the Assistant Secretary of £100.

Lectures — Marquis de Spineto.

3. The Correspondence between the late Mr. Harrison and the Marquis Spineto having been taken into consideration,
Resolved, That the Board cannot agreeably to their former arrangement, make any allowance to the Marquis for the travelling expences incurred by him in order to deliver his present Course of lectures.

4. Read a letter from The Marquis Spineto, addressed to the Secretary, offering and desiring to be permitted to read at least one additional lecture to his present Course.
Resolved, That the Marquis's proposal be acceded to, and that the sum of £10 be allowed him for this additional lecture.

Presents received.

5. Edward H. Locker Esqr. presented Views in Spain No. 9.
The Editor, — Metropolitan Literary Journal, No. 1.

Mr. T. S. Winn presented
Emancipation, or Practical Advice to British Slave Holders
Mr. George Wrigman presented the Speech of J. C. Hobhouse, Esqr. on presenting the Petition of Mr. Wrigman for Reform in Parliament
Ordered,
That the thanks of the Managers be returned for these presents.

Hon. Admission. 6. Count de Villa Real, Envoy Extraordinary & Minister Plenipotentiary from His Majesty the King of Portugal, was admitted to the Royal Institution under Art. 15 Chap. VI of the Bye laws.

7. The following person having been duly recommended was admitted as a Subscriber to the Lectures of the Royal Institution for the year 1824.

Miss Augusta Furno, 19 Upper Brook Street.

Adjourned to the 7th. of June.

Spencer

June 7.

At a Meeting of the Managers of the Royal Institution of Great Britain held at the house of the Institution, on the 7th. of June, 1824.

Present
Earl Spencer, K.G. President, in the Chair
Charles Hatchett, Esqr. V.P.
Sir Claude Scott, Bart. V.P.
Thomas Snodgrass, Esqr. V.P.
John Chamier, Esqr.
Jesse Watts Russell, Esqr. M.P.
Sir S. Bernard Morland, Bart. M.P. Treasr.
John Guillemard, Esqr. Secretary.

1. The minutes of the proceedings of the last meeting were read and approved.

2. Read and approved the minutes of the Committee of Accounts of this day, and Signed an Order for an Advance to the Assistant Secretary of £70.

Lectures Dr. Roget. 3. Read a letter from Dr. Roget proposing to deliver in the ensuing Season a Course of 12 Lectures on Comparative Physiology.
Resolved,
That Dr. Roget's proposal be accepted; and that the sum of £100 be the remuneration for the Course.

— Dr. Harwood. 4. Read a letter from Dr. Harwood proposing to deliver in the next Season, a Course of 12 Lectures on Zoology.
Resolved, That Dr. Harwood's proposal be accepted, and that the remuneration, as proposed by himself, be 50 guineas for the Course.

— M. de Spineto 5. Read a letter from The Marquis Spineto offering and proposing to deliver a Course of Lectures perfectly distinct from those given by him this year, on other branches of the Literature of Modern Europe, to begin as early in March 1825, as it may suit the convenience of the Managers.

Resolved, That The Marquis's proposal be accepted, and that the sum of £100 be the remuneration for a Course of 12 lectures.

Presents received.

6. The Commissioners of the Public Records, presented Index to the Statutes of the Realm.
Ordered,
That the thanks of the Managers be returned for this present.

Lectures. 7. Resolved, That the following arrangement for Lectures in the ensuing Season be adopted.
Wm. Thos. Brande, Esqr. — Geology.
John Millington, Esqr. — Mechanics.
Dr. Crotch — Music.
Dr. Roget — Comparative Physiology.
Dr. Harwood — Zoology.
The Marquis Spineto — On the Literature of Modern Europe.

Investments. 8. Ordered, That the sum of 10 guineas, being the admission fees received from 2 new Members, be invested in the 3 Pr.Cent Consols in the name of the Corporation; — and that the sum of £42, being Compositions from 4 Members to the Laboratory Fund, be invested in the 3 Pr.Cent Consols, in the names of the President and Treasurer of the Royal Institution, on account of the Laboratory Fund.

Number of Persons attending Lectures. 9. Ordered, That the Lecture-room door-keeper be directed to keep an account of the number of persons attending each of the Public Lectures.

10. Ordered, That Mr. Royce be directed to examine the Premises and to make an estimate of the expence of the necessary repairs.

Somerset

July 5.

At a Meeting of the Managers of the Royal Institution of Great Britain held at the house of the Institution on the 5th. of July, 1824.

The Duke of Somerset, V.P. in the Chair.
Sir S. Bernard Morland, Bart. M.P. Treasurer
Charles Hatchett, Esqr. V.P.
Thomas Snodgrass, Esqr. V.P.
Daniel Moore, Esqr. V.P.
John Chamier, Esqr.
Edmund Robt. Daniell, Esqr.
John Guillemard, Esqr. Secretary.

1. The minutes of the proceedings of the last meeting were read and approved.

2. Read and approved the minutes of the Committee of Accounts of this day; and signed an Order for payment of the following sums, vizt.

Wm. Thos. Brande, Esqr. for a years Salary as Professor of Chemistry	£150. —
John Millington, Esqr. for Lectures in 1824	105. —
Dr. Crotch for ditto	56. 14 —
Joseph Fincher on account of the Royal Institution	200. —
John Hurt for ½ years Salary as Librarian to 30th. June 1824.	60. —
	571. 14 —

Lectures. 3. Read a letter from Mr. Frost offering in the next year, to deliver gratuitously another Course of Lectures on Botany.
Resolved, That Mr. Frost's offer be declined.

Presents received.

4 Edward H. Locker, Esqr. presented Views in Spain No. 10

5\. The Editors presented – The Zoological Journal No. 2.
Ordered, That the thanks of the Managers be returned for these presents.

Lectures. M. de Spineto. 6. Mr. Guillemard stated that he had seen the Marquis Spineto who informed him that he could not make out more than 10 Lectures on the proposed subjects of his next Course without entering upon new matter which would require 7, 8 or more lectures to treat it properly; but that he considered he had sufficient new observations to make on the Art of Improvising, and proposing to deliver 11 lectures instead of 12.
Resolved, That the Marquis's plan be acceded to; and that the compensation to be assigned to him for the said Course be at the rate of £100 for 12 Lectures.

Asst. in Lab[y]. wages increased. 7. Resolved, That the Salary of William Bollaert, Assist.t in the Laboratory be increased from 12 to 15s. ₽ week.

Leave of Absence to Housekeeper. 8. Mrs. Greenwood the Housekeeper having requested leave of absence for a month.
Resolved, That her request be granted.

Repairs. 9. Read an Estimate made by Mr. Royce of the expence of the necessary Repairs of the Premises, as ordered by the resolution of the last Board, which amounted to £164. 10.
Ordered, That the repairs as therein specified be done at an expence not exceeding £170.
Ordered, That a Book-case be fitted up between the Windows in the Reading Room at an expence not exceeding £10.

Adjourned to Monday the 1st of November

Thomas Snodgrass
V.P.

At a Meeting of the Managers of the Royal Institution of Great Britain, held at the house of the Institution on the 1st of November, 1824,

Present
Thomas Snodgrass, Esq. V.P. in the Chair.
Daniel Moore, Esq. V.P.
Sir S. Bernard Morland, Bart. M.P. Treas.
John Guillemard, Esq. Secretary.

1\. The minutes of the proceedings of the last meeting were read and approved.

2\. Read and approved the minutes of the Committee of Accounts of this day, and signed an Order for an advance to the Assist.t Secretary of £200.

3\. In pursuance of a nomination under the Act of Parliament by John Penn, Esq. his Nephew Granville John Penn, Esq. of Hertford Street, May Fair, was admitted a Life Subscriber to the Royal Institution.

4\. Presents received.
Orrery presented. John Disney, Esq. M.R.I. presented
An Orrery;
Edward H. Locker, Esq. presented
Views in Spain, No. 11 & 12.
C. Moreau, Esq. M.R.I. presented
Tables of the State of the Trade of Great Britain;
George C. Gorham, Esq. presented
History & Antiquities of Eynesbury & St. Neots, 2 vol.
The Horticultural Society presented
Transactions of the Society vol. 5, part 4.
Ordered, That the special thanks of the Managers be returned to John Disney Esq. for his valuable present; and that the usual thanks of the Managers be returned for the other presents.

Danl. Moore

Dec. 6.

At a Meeting of the Managers of the Royal Institution of Great Britain held at the house of the Institution on the 6th of December, 1824

Present
Daniel Moore, Esq. V.P. in the Chair.
Sir George Thos. Staunton, Bart. M.P.
Sir S. Bernard Morland, Bart. M.P. Treas.
John Guillemard, Esq. Secretary.

1\. The minutes of the proceedings of the last meeting were read and approved.

2\. Read and approved the Minutes of the Committee of Accounts of this day, and signed an Order for an advance to the Assist.t Secretary of £150.

Life Subs.rs 3. In pursuance of a nomination under the Act of Parliament by William Morley Esq. his Son John Morley Esq. of London was admitted a Life Subscriber to the Royal Instit.n

4\. Presents received.
The Rev.d Stephen Weston presented
The Englishman Abroad, 2 parts in 1,
Sir James Edward Smith presented
Compendium Florae Britannicae. 4th Edit.
The Editors of the Zoological Journal presented
their Journal, No. 3.
Ordered,
That the thanks of the Managers be returned for these presents.

Investments 5. Ordered, That the sum of £52. 10 –, being a Composition, and an admission fee from 2 new Members be invested in the 3 ₽Cents in the name of the Corporation. And that the sum of £31. 10s. – being

Compositions from 2 Members and 1 new Member – be invested in the 3 ₽Cent. Consols on account of the Laboratory Fund.

Stoppage of Glass tubes by Excise 6. Wm. Thos. Brande Esq. stated to the Board that not having been able to obtain from the London Glass works some Glass tubes for the use of the Laboratory, he had applied to those of Newcastle by whom the proper tubes were made and sent up to London; and it appeared that they had been stopped by the Officers of Excise.
Resolved, That the Secretary be requested to make a representation of the fact to the Honorable Commiss.rs of His Majesty's Excise, and to request that they will be pleased to order the articles to be forwarded to the Laboratory of the Royal Institution.

Adjourned to Monday the 17th of Jan.y 1825.

H Davy VP.

Minutes of Managers' Meetings
1799–1900

Volume VII

With an introduction by

Morris Berman
Assistant Professor of History, Rutgers University

INTRODUCTION

In the years 1814–24, the period of the previous volume, the Royal Institution began to move in the direction of practical service to the industrial and medical communities. Brande taught chemistry to medical students three days a week and published his *Manual of Chemistry* in 1819 for his students' use. He launched *The Journal of Science and the Arts* in 1816, and used the laboratory for such technical problems as the preparation of dyes for coinage and gas for illumination. As a result Faraday's early training, with the exception of his Continental tour with Davy in 1813–15, tended to be commercial in its orientation. He found himself preparing illumination gas from fish oil, doing studies of iron casting, advising in a lawsuit involving a beer patent, and testifying in one between a sugar refinery and an insurance company. Writing to his future wife, Sarah Barnard, in 1820, he confessed that "as I ponder and think on you, chlorides, trials, oil, Davy, steel, miscellanea, mercury, and fifty other professional fancies swim before me . . ." (H. Bence Jones, *The Life and Letters of Faraday* (2 vols.; London: Longman, Green, and Co., 1870), I, 322). Nor was the situation much better in 1825. In a letter to André-Marie Ampère he described himself as "an anchorite in the Scientific world", "all day long in my Laboratory," not, however, engaged in original research, but "unfortunately occupied in very commonplace employment . . ." (L. Pearce Williams (ed.), *The Selected Correspondence of Michael Faraday* (2 vols.; Cambridge University Press, 1971), I, 153–54). Brande managed to lighten his assistant's work-load by relieving Faraday of the responsibility of helping with the morning chemistry lectures early in 1826 (Minutes, pp. 59 and 61); but by the end of the year Faraday was appointed Brande's assistant in the superintendence of the House, and was frequently saddled with tasks such as procuring estimates for paint jobs (pp. 98, 101, and 114). Pure research continued to be a marginal activity for Faraday throughout the years of this volume.

With few exceptions, the technical work of 1815–25 had a random quality to it. The Institution's "open laboratory" policy was a general one, intended to provide an income for the R.I., but not directly related to the interests of an identifiable sub-group of the governorship. The owner of an iron works in Dowlais, Josiah John Guest, needed a consulting chemist, so Faraday found himself on a trip to Wales (1819). James Stodart, one of the Managers, wanted to manufacture Indian steel (wootz), so Faraday was required to work on alloys until Stodart's death (1823). A Mr. Vivian wanted some advice on his copper works, so Faraday travelled to Swansea for a fortnight's visit, and so on. Such work was important, and undoubtedly helped to create an "industrial" reputation for the R.I., but its purposes were financially rather than ideologically motivated. Ostensibly the landed interest was still the dominant group, comprising a quarter of the governorship until *ca.* 1825; but for reasons already discussed (cf. the introduction to Vol. V of the Minutes), its influence waned after 1815, and for a time no other coherent group enjoyed a plurality.

The group that ultimately inherited the control of the Institution was the legal profession, specifically its senior branch. This development was not a sudden one. The proportion of barristers in the governorship, roughly eight per cent during the first decade, grew steadily during the ensuing years, which would suggest a smooth transfer of power. Two things, however, tended to make this change discontinuous rather than comfortably gradual. Although the number of barristers increased at an even rate, it was some time around 1825 that the quantitative difference became a qualitative one. The proportion of agriculturalists dropped off sharply—from twenty-six to sixteen per cent—leaving the legal group, by then about twenty-five per cent of the governorship, in a clearly dominant position. Secondly, the law itself had undergone important changes during these years. Until the 1820's the legal profession was largely oriented to the land, involved in the conveyancing of deeds, the arrangement of entail, strict settlement, and the like. But sometime during the 1820's the preponderance of urban problems tended to push some members of the profession in new directions, as will be discussed below, and the scientific activity of the R.I. was correspondingly bent to the interests of a

definable governing group. If, in fact, we add medical and scientific-administrative types to the twenty-five per cent already indicated, the professional members amongst the governorship amounted, after 1826, to nearly fifty per cent. The period of this volume, then, can be called the Institution's age of professional takeover, with the senior branch of the legal profession emerging as the cutting edge of the new governing group.

That members of the Bar would want to bother with the administration of a scientific institution is not at all obvious. The presence of lawyers on the governing boards of institutions is common enough today, but prior to 1825 attorneys and solicitors were rarely in positions of managerial influence, and barristers, whose social prestige was very great, were not in need of any improvement in their professional status. This interest in science on the part of the Bar is thus a matter of some concern in the Royal Institution's history and one demanding an explanation.

The 1820's and the 1830's were the years during which the social havoc created by urbanization, the factory system, and the rise of an industrial economy began to reach crisis proportions, earning the general title of the "condition-of-England" question. The era was characterized by machine-breaking, a sharp increase in the crime rate, the Peterloo massacre (1819) and the Tolpuddle martyrs. It was also characterized by the beginnings of what historians have termed a "revolution in government", i.e. the creation of an administrative and civil service apparatus for enforcing the law and running the State. The two developments were not unrelated. That the law might need a bureaucracy for its enforcement was a strange notion prior to this time, for the idea of, for example, a preventative police network had hitherto been regarded as a dangerous erosion of civil liberties. By 1829, however, when Peel's Metropolitan Police were created, the security of such a coercive apparatus seemed to be a far lesser evil than the chaos and anarchy that appeared to lie just around the corner. The legislation of the period reflected this changing attitude, and ushered in the age of workhouse and treadmill, prison and prison discipline (1823), new Poor Law (1834) and Madhouse Act (1828). Modern society, in short, was taking shape.

The ideology of this "revolution in government" had been provided by England's greatest thinker in the field of jurisprudence and criminal law, Jeremy Bentham (1748–1832). Society, the Benthamites believed, was essentially rational, and hence incompatibilities in the system could only be superficial ones, requiring further scrutiny for their amelioration. Political problems were, in this schema, really technical difficulties that arose from inefficiency and lack of sufficient information. For the Utilitarians, then, the solution to the condition-of-England question was to organise and examine it: appoint enough Select Committees and Royal Commissions, compile enough statistics, and the answers would emerge from the data themselves. Bentham's codification of the law and vision of the expertly administered society was applied by disciples such as Edwin Chadwick and Samuel Romilly to issues of public health, poverty, penal administration, and so forth. Milbank Penitentiary, for example, which was completed in 1821, received the inspiration for its design directly from Bentham's model prison, the "panopticon". Many voices were raised against this abstract social engineering, and Dickens' bitter antagonism to the "commissioners of fact", as he called them, was undoubtedly widely shared; but ultimately their influence did prevail over what they themselves regarded as the misplaced sentimentality of an *ancien régime*.

Despite the importance of the Utilitarians, it is not easy to identify them as a specific school. Opinions tended to divide on specific issues and occasionally overlap with the thinking of the classical economists, especially during the 1820's and 1830's. Yet two characterizations can be made. Many of these men tended to be barristers, largely because of Bentham's emphasis on the law as a vehicle of social co-ordination; and many were subsequently founders and early members of the Statistical Society (1834), an organisation expressly devoted to the amelioration of adverse social conditions via the compilation of factual information. Herein lies the relevance of law and Utilitarianism to science and the Royal Institution: a number of important early members of the Society had served as governors of the R.I. from *ca.* 1825, and many governors are identifiable as commissioners and reformers of the Benthamite variety. The first president of the Statistical Society was Bentham's friend, the Marquis of Lansdowne, to whom the Managers of the R.I. offered the presidency in 1826 upon the death of the Earl of Chichester (cf. p. 91 of the Minutes). Other cases of overlap include governors such as Henry Hallam, barrister, legal historian, and the Society's first treasurer; future Prime Minister Lord John Russell; Charles Shaw Lefevre, Speaker of the House; the Royal Institution's President, the Duke of Somerset (Minutes, pp. 102–3); engineer I. K. Brunel; and three of Faraday's personal friends, Edward Magrath, Richard Hollier, and J. J. Guest. There is no

interlocking directorate here, as existed in the early years with the Board of Agriculture, for the Utilitarians were much too diffuse; but a case can be made that as a result of the steady increase of lawyers at the R.I., and the changing nature of the legal profession, the Institution was being run by a very atypical group.

As already implied, this group was atypical for the Bar itself. The Inns of Court had long been defunct educationally, and the Bar little more than a respectable way of life for wealthy gentlemen. As late as 1875 *The Quarterly Review* described the Bar as "an elastic band, uniting the aristocracy with the classes below it". Yet amongst this dead wood—and there were probably less than a thousand barristers in 1820—were the Broughams, the Romillies, and, in short, the professional politicians described above. Commissioners' posts, for example, were customarily staffed by the senior branch of the legal profession, and several of the Institution's governors turn out to be commissioners of lunacy, charity, bankruptcy, and public records; Masters in Chancery, police magistrates, and the like. The Utilitarian mentality of social engineering had come to the R.I. because of what it saw in science: a tool by which to construct an ordered society. This legal ideology of science was no small factor in the Institution's success and subsequent scientific activity.

The lecture programme at the R.I. affords evidence of this shift in orientation. In 1815, for example, political economy had been rejected as a lecture topic as being outside of the Institution's scope (Minutes, VI, 70). By the mid-twenties the Managers considered themselves fortunate to secure the services of J. R. McCulloch, one of England's leading statisticians and economists, in giving a private course of twenty-five to thirty lectures on the subject. McCulloch, who assumed the Professorship of Political Economy at the University of London in 1828, gave the series of talks "in honour of the late David Ricardo" (Minutes, VII, 64). The sudden emphasis on vital statistics during this time led to a series of lectures by Dr. John Gordon Smith, a noted medical reformer, on "State Medicine or Medical Jurisprudence" (pp. 64 and 164), who also went on to a professorship at London University (1829). But the real change at the R.I. was felt not so much in the lecture theatre as in the laboratory. More and more of Faraday's work fell into the category of social engineering. If the Utilitarians worshipped anything it was expertise, and within this professional value system the Royal Institution, Faraday and science were called upon to play a new role. A fever epidemic at the Milbank Penitentiary led to Faraday being appointed to disinfect the prison in 1824. This appointment was made at Davy's suggestion, but it is probably not accidental that three of the Milbank commissioners were Institution governors. The following year the "Society for the improvement of Prison Discipline" began presenting its reports to the R.I. (Minutes, p. 15). In 1830, during the last agrarian uprising, what appeared to be an incendiary device found in Oxfordshire was referred to the R.I. for chemical analysis. In 1831 Faraday prepared a disinfecting solution for crowded public places, especially for use in police offices. In 1832, the Admiralty had him analyse oatmeal samples for the victualling of convict ships. Such service became the order of the day. In 1843 Faraday was appointed to test water for Chadwick's Health of Towns Commission, and in 1844 to probe the causes of the Haswell Colliery disaster. Examples of this sort, including work on the pollution of air and water, can easily be multiplied, and were products of a philosophy that viewed social evils as technical difficulties rather than as structural or systemic. For Faraday, who was resolutely apolitical, such work was at best a distraction, yet his reputation as a "trouble-shooter of the first rank" came to be well established (L. Pearce Williams, *Michael Faraday* (New York: Basic Books, Inc., 1965), p. 483). The Royal Institution, under his leadership and a legal-Utilitarian governorship, came to play a crucial scientific role in Victorian social engineering.

The rationalization of society envisaged by the governors was extended, during these years, to the Institution itself. How to organise the R.I. more efficiently and more profitably is a major concern of this volume, and many pages are devoted to comprehensive reviews and overhauls of the Institution's most critical functions: the lectures (pp. 52ff), the *Journal* (pp. 289ff), and finances (pp. 126ff and 182ff; discussed in the introduction to Volume VIII). The increasing "streamlining" of these functions, the division of labour and appointment of separate committees, special funds, etc., serves to illustrate a salient point: that at least at the R.I. it took a legal governorship to create the modern framework of organised science. This modernization was extended to matters of protocol as well. Ballot boxes were adopted for use at general meetings, and the duties of the porters were updated and their livery clothes discarded in favour of "a plain blue Coat & Waistcoat . . ." (pp. 120, 122, and 245ff). Some of the eccentric flavour of the old R.I. is still present, such as the presentation of a stuffed Patagonian penguin or the current issue of the "Phrenological Journal" (pp. 138 and 242), but such

curiosities of polite society were by this time distinctly rare. What comes through the pages of this volume is rationalization, regularity, and bureaucracy. Rumford's premature report of 1802—"The Royal Institution of Great Britain may . . . be considered as finished and firmly established"—had finally become a reality.

The reality was, however, intellectual as well as social, and although the Institution has perhaps been too closely identified with Faraday's purely scientific interests, he was, in his own eyes, a philosopher rather than a social engineer. In the R.I. Faraday found the materials for his own personal expression. As a youth he had seized upon the method of acquiring wisdom provided by the popular Dr. Isaac Watts, in his rather banal treatise, *The Improvement of the Mind.* Watts' emphasis on lectures, conversation, reading, study, and observation as the keys to discovery proved to be critical to Faraday's devotion to the R.I., for it alone among the scientific institutions of the early nineteenth century provided facilities for the pursuit of these activities. Its superb library was becoming increasingly scientific (cf. Minutes, pp. 208, 210, and 355–56); it offered lectures and discussion; and observation and study were provided by the best laboratory then available. Faraday contributed to the improvement of all these functions, the introduction of juvenile lectures and Friday Evening Discourses being a case in point (pp. 43, 47, 63, and 176). What for Davy had been a platform, and for Brande a place of business, was for Faraday a home. As early as 1815 he referred to it, in a letter to his friend Benjamin Abbot, as "my nest", and in a letter negotiating his appointment as Professor at the Royal Military Academy (1829) he pointed out that the unique advantages of the R.I., including "a perfect laboratory with an assistant in constant occupation", made him reluctant to attach himself to another institution (Bence Jones, I, 227 and 408). The same type of consideration had led to his declining a professorship at the University of London in 1827. As Williams has pointed out, between Faraday's apartment and his basement laboratory there was little he lacked. Relatives gathered on family occasions, and in 1826 he built himself a velocipede which he raced around the corridor surrounding the lecture theatre. After hours he and his wife Sarah would occasionally use the laboratory to manufacture soda or lozenges for home consumption, and once a year for the brewing of ginger wine. Such obvious enjoyment of the place, as well as his professional service to it, stood Faraday in good stead. The Managers responded to his devotion, and to "the zeal and ability with which Mr. Faraday has discharged the duties of his situation", by admitting him "a Member of the Royal Institution without payment of the usual admission fees, fines, or compositions" (p. 39), a privilege that had not, to date, been extended to anyone else. He began attending the meetings of the Managers in 1828 (p. 218), and his attachment to the R.I. even led to his paying for half of one of the new columns that were erected in front of the building ten years later (Minutes, VIII, 541).

If the governors had something unique in Faraday, if he served their social purposes well, it should be clear that he had something unique in them. The nature of the R.I. in fulfilling the conditions of Watts' "five-fold path to truth" yielded breathtaking results. The period of this volume witnessed the discovery of benzene and electromagnetic induction, and the ensuing years brought the demonstration of the identity of electricities, specific inductive capacity, and the rotation of the plane of polarized light in a magnetic field. "*Er riecht die Wahrheit,*" Kohlrausch once remarked of Faraday; "he smells the truth". Yet it was the nature of the R.I. as a scientific institution that provided the perhaps unusual conditions for such sensitivity, and the relationship between institutional organization and intellectual achievement is one which a study of the R.I. might help to explore.

© Morris Berman

Rutgers University
June, 1973

Minutes

of the

Meetings of Managers.

Vol. VII.

(Jan. 1825 to April 1832.)

1825. Jan. 17. 1

At a Meeting of the Managers of the Royal Institution of Great Britain held at the house of the Institution on the 17th. of January, 1825

Present

Sir H. Davy, Bart: V.P. in the Chair.
Daniel Moore, Esqr. V.P.
Thomas Snodgrass, Esqr. V.P.
Sir George Duckett, Bart V.P.
Sir George Thos. Staunton, Bart: M.P.
Sir S. Bernard Morland, Bart M.P. Treasurer
John Chamier, Esqr.
John Guillemard, Esqr. Secretary.

1. The minutes of the proceedings of the last meeting were read and approved.

2. Read and approved of the Committee of Accounts of this day.

Lectures — Dr. Gordon Smith

3 Read a letter from John Gordon Smith, M.D, addressed to Dr. Paris, M.R.I. in which he stated that he should be happy to have permission of the Managers to deliver gratuitously in the present Season at the Royal Institution, a Course of 4 lectures on the application of Medicine and the Medical Sciences to the purposes of National Economy.

Resolved, That Dr. Smith's offer be accepted.

Lectures declined.

4 Read a letter from Mr. J. E. Du Val addressed to the Managers offering to deliver gratuitously a Course

of six lectures on Belles Lettres and Poetry.
Resolved, That Mr Duvals offer be declined.

Presents received.

5. Mr T. Gardiner presented
The Royal Blue Book, for 1825.
The Astronomical Society presented
Memoirs of the Society, vol 1 pt. 2.
Captn Sabine, R. A. presented
Barometrical & Trigonometrical measurements at Spitzbergen,
The Geological Society presented
Transactions of the Society vol 1. part 2
Sir George Thos. Staunton, Bart M.P. presented
Miscellaneous Notices relating to China, –
Notes of the proceedings of the Embassy to Peken.
Mr R. Ackermann presented
The Repository of Arts, 12 nos for 1824.
Mr Thos Gill presented
The Technical Repository, 12 nos for 1824.
Sir Everard Home, Bart presented
An Engraving of the Proteosaurus. –
Ordered, That the thanks of the Managers be returned for these presents.

Coals — 6 Ordered, That Messrs Dickins & Co. do supply the Institution with 20 Chaldrons of the best Walls-end Coals.

Journals presented to Sir H. Davy — 7 The Managers having been informed that Sir H Davy was without a Copy of the original Journals

of the Royal Institution which were edited by himself and Dr Young.
Resolved, That Sir H. Davy be requested to accept from the Managers a Copy of the same.

8. The following persons having been duly recommended were admitted as Subscribers to the Royal Institution for the year, 1825

Annual Subscribers.

T. Rutherford, Esqr. 21 Maddox Street.
Captn Dallas 25 Nottingham Place.
W. B. L. Hawkins, Esqr. Great Marlborough St
J. H. Peile, Esqr. 129 Mount Street,
Richard Perry Esqr. of Battersea,
Sir Thos Croft, Bart 27 Welbeck Street.
Sir Chas Joshua Smith, Bart Portland Place

To the Lectures.

Mrs & Miss Dawson 2 Great Cumberland St.
Mrs Anderson
Miss Susan Anderson } 15 Chapel Street

Adjourned to the 7th of February.
H Davy

Feb. 7.

At a Meeting of the Managers of the Royal Institution of Great Britain, held at the house of the Institution on the 7th of February, 1825

Present
Sir H. Davy Bart V.P. in the Chair.
Sir Claude Scott, Bart: V.P.
Sir George Thos. Staunton, Bart: M.P.
Daniel Moore, Esqr. V.P.
Thomas Snodgrass, Esqr. V.P.
Sir S. Bernard Morland, Bart: M.P. Trearr.
John Guillemard, Esqr. Secretary.

1. The minutes of the proceedings of the last meeting were read and approved. –

Payments — 2. Read and approved the minutes of the Committee of Accounts of this day; and signed an Order for payment of the following sums viz.t

Joseph Fincher on account of the Royal Institution	£300. –
Messrs Dickins & Co. for Coals in 1824	81. –
	£381. –

and signed an Order for payment to Jos.h Fincher of £150 on account of the Laboratory Fund.
& to J. Sturt for ½ years Salary to Xmas last £60

Lecture J. Dangerfield, Esq. — 3. J. Dangerfield, Esqr. having offered to the Board to deliver gratuitously one lecture on Stenography in the course of the present Season.
Resolved, That his offer be accepted.

Presents received.

4. The Directors of the East India Company presented Dictionary & Grammar of the Persian Language, by the King of Oude, 7 parts folio.
Morrison's Chinese Dictionary, Vol 3 pt. 1.
The American Philosophical Society presented Catalogue of the Society Library.
The Zoological Journal, No. 4 presented by the Editors – and
A Short Account of the late Thomas Harrison, Esqr. – From the Author –
Ordered,
That the thanks of the Managers be returned for these presents.

Illness of Mr Sturt; Assist.t engaged. — 5. Sir H. Davy informed the Board that he had found Mr Sturt labouring under severe illness and incapable for the present of performing his duties as Librarian – And that as one of the Honorary Librarians he had directed him to engage an Assistant, which Mr Sturt had done for the last fortnight at the rate of 30s. ℔ week.
Resolved. That Mr Sturt's Assistant be continued for one Month from this period at the expence of the Institution – and that the above two weeks be also paid by the Institution.

Extra bill of Mrs Greenwood. — 6 Mrs Greenwood the Housekeeper having brought in a bill of £2.16.6 – for extra expences.
Ordered, That the said Bill be paid; – and that Mrs Greenwood be informed that no such bill of extraordinaries will be allowed in future.

6 1825.

7. Audited the Treasurer's account for the year 1824.

8 Sir Humphry Davy having stated that he considered the talents and services of M^r. Faraday Assistant in the Laboratory entitled to some mark of approbation from the Managers, and these sentiments having met the cordial concurrence of the Board,

M^r. Faraday appointed Director of Laboratory

Resolved, That M^r. Faraday be appointed Director of the Laboratory under the superintendance of the Professor of Chemistry.

9. The following persons having been duly recommended were admitted as Subscribers to the Royal Institution

Annual Subscribers.

Dr. Fitton, 53 Upper Harley Street.
Alex^r. Mackenzie, Esqr. 1 Chapel Street Lambs Conduit St
Peter Warren Locke, Esqr 34 Bruton Street
Henry Parker, Esqr. 63 Margaret St. Cavendish Sq^r.
Wriothesly Digby, Esqr. 62 Park Street
Sir Stamford Raffles 23 Lower Grosvenor St.
Rev^d. J. Philips Potter Kensington Square.
Joseph Hunt Esqr. 11 Park Place.
M^r. Treuba 3 Hadlow St. Burton Crescent

To the Lectures.

M^{rs}. Fitton – 53 Upper Harley Street
Mrs C. Holford, Hampstead.
Miss Dyson 33 Somerset St. Portman Square
Mrs George Johnstone, 53 Tavistock Square
Miss Bidwell 99 Gloucester Place.

Feb. 7. 7.

To the Lectures continued.

Mrs Hyndman
Miss Hyndman
Miss Henry
Miss Mary Henry } of Norland house Notting-hill.
Miss Lynch, 81 Great Russell St.
Mrs Rogers, 40 Upper Bedford Place.
Miss Croft 27 Guildford Street
Mrs Roberts 30 Weymouth Street.
Lady Raffles
Miss Hull } Lower Grosvenor Street

Adjourned to the 21st. Instant

Spencer

8 1825.

1825 Feb 21.

At a Meeting of the Managers of the Royal Institution of Great Britain held at the house of the Institution on the 21st. of February, 1825

Present

Earl Spencer, K.G. President, in the Chair.
Sir H. Davy, Bart: V.P.
Daniel Moore, Esqr. V.P.
Sir George Thos. Staunton, Bart: M.P.
Charles Hatchett, Esqr. V.P.
John Guillemard, Esqr. Secretary.

1. The minutes of the proceedings of the last meeting were read and approved.

2. Read and approved the minutes of the Committee of Accounts of this day, and signed an Order for an Advance to the Assist^t. Secretary of £300.

3. The following persons having been duly recommended, were admitted as Subscribers to the Royal Institution for the year 1825.

Annual Subscribers

R. Tuckfield, Esqr. 4 Chapel St.
Chas. Augustus Fulk, Esqr. 19 Duke St. Westm^r
John Frederick Schröder, Esqr. Dulwich,
John Field, Esqr. Paddington House,
Gabriel Wood, Esqr. 168 Regent Street

To the Lectures.

M^{rs}. Tuckfield. 4 Chapel Street

Feb. 21. 9

To the Lectures, continued.

Miss Warren 151 Piccadilly.
Miss Challen, 27 Portman St.
M^{rs} & Miss Trotter, 13 Connaught Place.
Miss Harriet Croft, Guildford Street
Mrs Trevor Plowden
Miss Plowden } 27 Margaret St. Cavendish Square
Mrs Godbold, 32 Bernard St.
Mrs Brown, 2 Portland Place
Mrs Kinder 66 Portland Place.

Adjourned to the 7th of March

Chs Hatchett

10

Mar. 7. 1825.

At a Meeting of the Managers of the Royal Institution of Great Britain held at the house of the Institution on the 7th of March, 1825

Present

Charles Hatchett, Esq.r V.P. in the Chair.
Sir H. Davy, Bart. V.P.
Thomas Snodgrass, Esq.r V.P.
Edmund Rob.t Daniell, Esq.r
John Guillemard, Esq.r Secretary.

1. The minutes of the proceedings of the last meeting were read and approved.

2. Read and approved the minutes of the Committee of Accounts of this day, and signed an Order for an Advance to the Assist.t Secretary of £100.

3. Lectures declined. Read a letter from James Elmes Esq.r addressed to Sir H. Davy, Bart. in which he requested to be permitted to deliver at the Royal Institution an Introductory Lecture on the reciprocal influence on the fine Arts and Literature.
Resolved, That the offer of M.r Elmes be declined and that he be informed that the days allotted to lectures during the ensuing Season are filled up.

4. Life Subs.rs In pursuance of a nomination under the Act of Parliament by John Trotter, Esq.r his son John Trotter of Durham Park was admitted to the privileges of a Life Subscriber to the Royal Institution.

Mar. 7. 11

5. Manufacturing Glass-tubes permitted. The Secretary reported that he had received an answer from the Secretary of the Board of Excise, informing him that orders had been given to their Collector at Newcastle to proceed to manufacture the Glass tubes which had been ordered for the use of the Laboratory of the Royal Institution.

6. Resign.n of W.m Bollaert – Ass.t in Labor.y – appointment of T. Griffiths. M.r Brande having stated to the Board that William Bollaert, Assistant in the Laboratory, had given up his situation in consequence of an appointment to go out to Peru, and having recommended Thomas Griffiths in his place.
Resolved, That Tho.s Griffiths be appointed in the room of William Bollaert, and at the same Salary.

7. The following persons having been duly recommended were admitted as Subscribers to the Royal Institution for the year, 1825.

Annual Subscribers.

Capt.n Francis Vernon, Grenad.r Guards, Grosvenor Sq.
Randal Macdonnell Gibbons, Esq.r 61 Pall Mall
George Lister Esq.r 14 Southampton S.t Covent Garden
John Prior, Esq.r – Denmark-hill –

To the Lectures.

The Hon: Miss Addington
The Hon: Miss C. Addington } 16 Dover S.t
The Hon: Miss H. Addington
Miss Cartwright, Miss Chetwynd & Miss Clerke 14 Half-Moon S.t
M.rs Somerville, Hanover Square,
Miss Murray Macgregor 1 Montagu Place.

Adjourned to the 21st Instant

Spencer

12

Mar. 21. 1825.

At a Meeting of the Managers of the Royal Institution of Great Britain held on the 21st of March, 1825

Present

Earl Spencer, K.G. President, in the Chair.
Charles Hatchett, Esq.r V.P.
Thomas Snodgrass, Esq.r V.P.
Sir H. Davy, Bart. V.P.
Edmund Rob.t Daniell, Esq.r
Sir S. Bernard Morland, Bart. M.P. Treas.
John Guillemard, Esq.r Secretary.

1. The minutes of the proceedings of the last meeting were read and approved.

2. Read and approved the minutes of the Committee of Accounts of this day, and signed an Order for an advance to the Assist.t Secretary of £100. –

Presents received.

3. The Royal Society presented Philosophical Transactions, 1824 part 2.
The Society of Arts presented Transactions of the Society, vol. 42.
Ordered, That the thanks of the Managers be returned for these presents.

4. Read the minutes of the last General Meeting.

5. On closing Lib.y and Reading-rooms. The Resolution of the General Meeting requesting the Managers to reconsider the Order made by them

Mar. 21. 13

on the 4th of June, 1823 for shutting up the Reading Room at the same hour as the Library throughout the year, having been taken into consideration.
Resolved, That the above Order be rescinded.

6. Assistant in Library. M.r Sturt not being sufficiently recovered from his late illness to attend to the duties of the Library, Resolved, That his Assistant be further continued for one month.

7. Investment of Compositions. Ordered, That £57.15 – being a Composition and two admission fees from new Members, be invested in the 3 per Cent Consols in the name of the Corporation –
And that £10.10 –, being the subscription out of the Composition of a new Member, be invested in the 3 P.Cent Consols on account of the Laboratory Fund.

8. The following persons having been duly recommended were admitted as Subscribers to the Royal Institution for the year 1825,

Annual Subscribers.

William Beaumont, Esq.r Albany
Lord Calthorpe – Grosvenor Square.

To the Lectures.

Countess of Charlemont – 45 Grosvenor Place
The Hon: M.rs Caulfield 5 Cleveland row
Lady Maria Caulfield Grosvenor Place.

Adjourned to 4th April.

Spencer

14

April 18 — 1825.

At a Meeting of the Managers of the Royal Institution of Great Britain, held on the 18th of April, 1825

Present

Earl Spencer, K.G. President, in the Chair

Charles Hatchett, Esqr. V.P.

Sir Humphry Davy, Bart. V.P.

Thomas Snodgrass, Esqr. V.P.

Sir George Duckett, Bart. V.P.

Sir George Thos. Staunton, Bart. M.P.

Edmund Robert Daniell, Esqr.

John Guillemard, Esqr. Secretary.

1. The minutes of the proceedings of the last meeting were read and approved.

2. Read and approved the minutes of the Committee of Accounts of this day; and signed an Order for payment of the following sums - vizt.

Payments.

A years Ground rent to the City of London to 25th March 1825	£150. –
Joseph Fincher on account of the Royal Institution	100. –
	£250. –

Letters received.

Request of Royal Academy of Music, declined. 3. Read a letter from the Committee of the Royal Academy of Music requesting the Managers of the Royal Institution to permit a limited number of the Pupils of their Academy to attend Dr. Crotch's lectures on Musical Composition.

Resolved, That the Secretary be requested to inform

April 18 — 15

the Committee of the Academy of Music that their request cannot be complied with, as it is incompatible with the established regulations of the Institution.

Lectures – Dr. G. Smith. 4. Read a letter from Dr. Gordon Smith, addressed to the Secretary, submitting the following proposal to the Managers, that if the sketch of Public Medicine, or of the applicability of Medical Science to National Economy; which he had the honour to deliver this Season at the Royal Institution, shall have met their approbation, he would offer his services to enter minutely into some of the particular topics that belong to the study, in a set of gratuitous lectures, next Season.

Resolved, That Dr. Gordon Smith's offer be accepted.

Presents received.

5. The Committee of the Society for the improvement of Prison Discipline, presented

The 6th. Report of the Committee.

Presented by the Editors

The Dublin Philosophical Journal,

The Committee of the Cymmrodorion presented

Transactions of the Society vol 1 - 2 parts,

The Royal Asiatic Society presented,

The first part of the first Volume of the Society's Transactions.

Ordered, That the thanks of the Managers be returned for these presents.

16 — 1825.

6. Read the minutes of the proceedings of the Visitors of the 25th of March.

Repairs. Ordered, That Mr. Faraday's Passage & Dressing Room be painted.

7. On the application of Sir George Duckett Bart.

Hon. Admissions. The Baron de Montemart was admitted to the Royal Institution under Art: 15 Chap: VI of the Bye laws – and

On the application of Sir H. Davy Bart.

Dr. Theriano was admitted to the Royal Institution under the above Bye law. –

8. The following persons having been duly recommended were admitted as Subscribers to the Royal Institution for the year 1825.

Annual Subscriber.

Hans Sotheby, Esqr. 16 Hanover Street

To the Lectures.

Misses Tollet 8 Queen St. May Fair.

Mrs. H. Sotheby 16 Hanover Street

Miss Molesworth 43 Clarges Street

Lady St. Germans

9. Ordered, That the House Lists be filled up as follows.

President. – Earl Spencer, K.G.

Treasurer. – Sir S. Bernard Morland, Bart. M.P.

Secretary. – John Guillemard, Esqr.

April 18 — 17

Managers

Sir Thos. Dyke Acland, Bart. M.P.

William Babington, M.D.

George Basevi, Esqr.

Edmund Robert Daniell, Esqr.

Sir Humphry Davy, Bart. Pres. R.S.

Viscount Dudley and Ward, F.R.S.

The Marquis of Exeter,

Charles Hatchett, Esqr. F.R.S. & F.S.A.

Sir Christr. Hawkins, Bart. M.P.

Daniel Moore, Esqr. F.R.S. &: &:

Earl of Morley, F.R.S.

Hon: W. F. S. Ponsonby,

Sir Claude Scott, Bart. F.L.S.

Thomas Snodgrass, Esqr. F.R.S.

Sir George Thos. Staunton, Bart. M.P. F.R.S.

Visitors

James Alexander, Esqr.

– Col. Benjn. Ansley,

Sir Edmund Antrobus, Bart.

John Ivatt Briscoe, Esqr.

John Frederic Daniell, Esqr. F.R.S.

Earl of Egremont F.R.S. & F.S.A.

John Fuller, Esqr.

Davies Gilbert, Esqr. M.P. F.R.S. & F.L.S.

Thomas Hoblyn, Esqr. F.R.S.

James Heath Leigh Esqr.

Rt. Hon: Sir John Nicholl, M.P.

Captn. Edward Sabine, F.R.S.

Richd. Horsman Solly, Esqr. F.R.S. and F.S.A.

William Sotheby, Esqr. F.R.S. and F.S.A.

Lord Teignmouth,

18

1825

May 9.

At a Meeting of the Managers of the Royal Institution of Great Britain held on the 9th of May, 1825

Present

Sir George Thos. Staunton, Bart. M.P. in the Chair
Edmund Robt. Daniell, Esqr.
Wm. Henry Blaauw, Esqr.
C. P. Meyer, Esqr.
Benjamin Smith, Esqr.
Captn. Edward Sabine,
Sir S. Bernard Morland, Bart. M.P. Treasr.
John Guillemard, Esqr. Secretary.

1. The minutes of the proceedings of the last meeting were read and approved.

2. Read and approved the minutes of the Committee of Accounts of this day; and signed an Order for an Advance to the Assist. Secretary of £100.

3. Resignations of Visitor. Read a letter from Lt. Genl. William Thornton containing his resignation of the office of Visitor of the Royal Institution to which he had been elected on the 2d of May.

of Manager. Read a letter from Chas. Hatchett, Esqr. containing his resignation of the office of Manager to which he had been elected on the 2d of May.

of Manager. Read a letter from Thomas Snodgrass Esqr. containing his resignation of Manager to which he had been elected on the 2d of May.

of President. Read a letter from Earl Spencer containing his resignation of the Office of President.

19

May 9.

3. Resolved, That the Managers lament with deep regret that any occurrence should have induced their President to resign at this period, an office which he has so long filled to the honour and advantage of the Institution, and that while they cherish the hope he may be induced to remain at least for this season at the head of the Establishment, ** they lose no time to return their most respectful and grateful thanks for his past patronage.

4. Life Subscribers. In pursuance of nominations under the Act of Parliament the following persons were admitted as Life Subscribers to the Royal Institution viz:
Robert I. Wilberforce, Esqr. of Brompton Grove nominated by his Father William Wilberforce Esqr.
& Charles Dance, Esqr. of 57 Manchester Street, nominated by his Brother George Dance, Esqr.

5. Presents received.
Presented by the Editor
The first Volume of the American Monitor.
Presented by Walter Weldon, Esqr.
Weldon's General Laws and Elements of Chemistry.
Ordered, That the thanks of the Managers be returned for these presents.

6. Library opened for Meeting of Members. Read the following resolution of the General Monthly Meeting of the Members of this day
"That the Managers be requested to open the Library this Evening at 8 o'Clock for the purpose of a meeting of the Members."
Ordered, That the Library be opened accordingly.

20

1825

7. The following persons having been duly recommended were admitted as Subscribers to the Royal Institution for the year 1825
Annual Subscriber.
James Tucker, Esqr. 7 Lower Seymour St. Portman Sq.
To the Lectures.
Mrs. & Miss Jolly Layton house Essex
Dowager Lady Sitwell

Adjourned to Saturday the 14th Instant at One o'Clock.

Danl Moore

21

May 14.

At an adjourned Meeting of the Managers of the Royal Institution of Great Britain, held on Saturday the 14th of May, 1825

Present

Edmund Robt. Daniell, Esqr. in the Chair
Captn. Edward Sabine,
Wm. Henry Blaauw, Esqr.
Samuel Ware, Esqr.
C. P. Meyer, Esqr.
John Guillemard, Esqr. Secretary.

1. Petition of Royal Socy. of Literature. Read a letter from William Tooke, Esqr. of Grays Inn transmitting a Copy of a Petition presented to His Majesty on behalf of the Royal Society of Literature soliciting a Charter of Incorporation for the information of the Royal Institution.

2. Honry. Admission. Read a letter from Col. Ansley requesting permission under Art: 15 of Chap: VI of the Bye laws for Count Sufy a French Nobleman to have his Entree of the Royal Institution during his temporary residence in this Country.
Ordered, That Col. Ansley's request be granted.

Presents received.

3. Revd. Dr. Falconer, of Bath presented his work on
The Voyage of Hanno translated & accompanied with the Greek
& Dissertations on St. Paul's Voyage,
Arrians Voyage round the Euxine Sea.
Miscellaneous Tracts & Collections relating to Natural History.
by his Father the late Dr. Falconer, M.D. F.R.S. &c.

22 1825

Presented by the Baron de Montemart, through Sir George Duckett, Bart.

Notice sur le Troupeau de Moutons Anglois importés en 1774, par François Delporte.

Recherches sur les differentes Races de Bêtes a Laine de la Grande-Bretagne.

__Ordered__, That the thanks of the Managers be returned for these presents.

Lectures R. Addams, Esq. 4. Robt. Addams, Esqr. having offered to deliver gratuitously a short Course of lectures in the present Season on Magnetism and Electro Magnetism

__Resolved__, That his offer be accepted.

Adjourned to Saturday the 21st. Instant at One o'Clock.

Danl. Moore

23

May 21. At an adjourned Meeting of the Managers of the Royal Institution held on Saturday the 21st. of May, 1825.

Present

Edmund Robt. Daniell, Esqr. __in the Chair.__

C. P. Meyer, Esqr.

Benjamin Smith, Esqr.

Samuel Ware, Esqr.

Captn. Edward Sabine,

Daniel Moore, Esqr.

John Guillemard, Esqr. __Secretary__

Resign.n of Earl Spencer. 1 Read the following letter from Earl Spencer K.G. addressed to the Secretary.

Wimbledon Park 15th. May, 1825

Sir,

I received this morning your letter of yesterday inclosing Copies of the resolution of the Board of Managers of the Royal Institution held on the 9th. Instant and of the General Meeting of the 14th., and I take the earliest opportunity of requesting you to convey to the Managers my grateful sense of the very flattering expressions contained in their resolution; but at the same time I feel obliged to add that my resignation of the office of President having been a measure adopted by me on mature deliberation, I shall be under the necessity of informing the Gentlemen whom the General Meeting have done me the honour to depute to me, that it is impossible for me to accede to the proposal they are instructed to make of again accepting that office. I remain Sir, &c. &c.

John Guillemard, Esqr. &c. Secy. R. Instn. (signed) Spencer

24 1825

__Presents received.__

2. The Horticultural Society presented

Transactions of the Society, vol. 5 part 5.

Barclay Mounteney, Esqr. presented

Selections of various Authors on Brazil and its Gold Mines.

Edward Buckton, Esqr. presented

Collection of Opinions on the Corn Question.

__Ordered__, That the thanks of the Managers be returned for these presents.

3. Signed an Order for payment of the following sums, viz.

Payments.

Wm. Thos. Brande, Esqr. Profr. of Chemistry for a year's Salary	£150 —
The Marquis Spineto for lectures in 1825	100 —
Dr. Harwood for lectures in ditto	50 —
Joseph Fincher on account of the Royal Institution	300 —
	£600 —

Adjourned to Wednesday the 1st. of June at 3 o'Clock.

Danl. Moore

25

June 1. At an adjourned Meeting of the Managers of the Royal Institution of Great Britain, held on Wednesday the 1st. of June, 1825

Present

Edmund Robt. Daniell, Esqr. __in the Chair__

Captn. Edward Sabine,

W. H. Blaauw, Esqr.

Benjamin Smith, Esqr.

C. P. Meyer, Esqr.

Samuel Ware, Esqr.

Sir S. Bernard Morland, Bart. M.P. __Treasr.__

1 Read a letter from John Pepys, Esqr. containing his resignation of the office of Manager to which he had been elected on the 2d. of May.

Resign.n of Visitor 2. Read a letter from William Sotheby, Esqr. containing his resignation of the office of Visitor to which he had been elected on the 2d. of May.

Hony. Admission 3. On the application of Saml. Solly, Esqr.

The Baron de Maltzahn, Envoy Extraordy. and Minister Plenipotentiary from the Court of Prussia, was admitted to the Royal Institution under Art: 15 Chap: VI of the Bye Laws.

On extendg. Professorships. 4. __Resolved__, That the Committee of Managers will take into their consideration as soon as the present pressing business before them is disposed of, the expediency of extending the Professorships of the Royal Institution to other subjects than those at present established; — particularly

26 1825

Natural History –
Political Economy,
History,
Law,
Classical and English Literature either separately or combined.
Architecture,
Poetry,
Painting,
Music,
Elocution.

5 The following persons having been duly recommended were admitted as Annual Subscribers to the Royal Institution for the year, 1825.

Ann^l. Subs^rs. Hon: James Hope of the Coldstream Guards
John Davidson, Esqr. 12 Cork Street.
Rev^d. Edward Fielde of Harlington

Dan Moore

27

June 6. At a Meeting of the Managers of the Royal Institution of Great Britain held at the house of the Institution on the 6th of June, 1825,

Present
Daniel Moore, Esqr. V.P. in the Chair.
Benjamin Smith, Esqr.
Captn. Edward Sabine,
W^m. Henry Blaauw, Esqr.
C. P. Meyer, Esqr.
Edmund Robt. Daniell, Esqr.
Saml. Ware, Esqr.
Sir H Davy, Bart. V.P.
John Guillemard, Esqr. Secretary.

1. The minutes of the proceedings of the Meeting of the 9th of May, and of the adjourned Meetings of the 14th 21st May, and 1st of June were read and approved.

2. Read and approved the minutes of the Committee of Accounts of this day.

3. Read a letter from Saml. Solly, Esqr. relating to the affairs of the Institution.

Lectures. 4. Resolved, That the Secretary be requested to write to the Professor of Chemistry, to request he will lecture again in the Theatre of the Royal Institution in the course of the present Season.

5. Resolved, That the Secretary be requested to apply

28 1825

Loan of Models from Soc^y of Arts. to the Society of Arts for the loan of such Models as M^r. Addams may require to illustrate his lectures on Electro-Magnetism.

Adjourned to Thursday the 9th of June at Three o'Clock.

Dan Moore

29

June 9. At an adjourned Meeting of the Managers of the Royal Institution held on Thursday the 9th of June, 1825

Present
Daniel Moore, Esqr. V.P. in the Chair.
D^r. Somerville.
Benjamin Smith, Esqr.
Captn. Edward Sabine,
Edmund Robt. Daniell, Esqr.
Samuel Ware, Esqr.
C. P. Meyer, Esqr.
W. H. Blaauw, Esqr.

Lectures – M^r. Brande 1 Read a letter from W^m. Thos. Brande, Esqr. addressed to the Secretary, in answer to the request of the Managers at their last Board, – stating that the only lectures which he could deliver this Season are a Course on the Theory of definite proportionals which to do justice to the subject, could not be comprised in less than 8 lectures.

on Election of President. – Earl of Chichester 2. The vacancy of the office of President having been taken into consideration

Resolved,

That M^r. Daniel Moore be requested to apply to the Earl of Chichester, to ascertain whether His Lordship would allow his name to be proposed as President of the Royal Institution.

Adjourned to Monday the 13th. Instant at Three o'Clock.

30 1825

June 13.

At an adjourned Meeting of the Managers of the Royal Institution, held on Monday the 13th of June, 1825.

Present
Daniel Moore, Esqr. V.P. in the Chair.
Benjamin Smith, Esqr.
Edmund Robt. Daniell, Esqr.
Captn. Edward Sabine,
W^{m}. H^{y} Blaauw, Esqr.
Charles Pilgrim jun^{r}. Esqr.
Samuel Ware, Esqr.

On Election of President – (Earl of Chichester) 1. The Chairman reported, that agreeably to the wishes of the last Board he had waited on the Earl of Chichester, and he was happy to state that His Lordship had consented to allow his name to be proposed as President; though he regretted that his health would not allow him to give a regular attendance at the Meetings.

Resolved unanimously,

That the thanks of this Meeting be given to Daniel Moore, Esqr. for this additional proof of the great interest he takes in the prosperity of the Royal Institution.

Friday Ev^{g}. – New arrangements. 2. The Board of Managers taking into consideration the advantages likely to result to Science from the Meetings held in the Library of the Royal Institut^{n}. on Friday Evenings, and being desirous of affording them all the support in their power, have

Resolved, That one of the Board of Managers do attend at each of the meetings, in a rotation to be

June 13. 31

fixed amongst themselves, for the purpose of receiving the Visiters, and superintending the arrangements of the Evening.

Resolved, That every Member introducing Visiters to the Evening Meetings of the Royal Institution, do write the names of the persons so introduced, in a book to be provided for the purpose.

Tea & Coffee. Resolved, That Mrs Greenwood be ordered to provide Tea and Coffee &c. at the Friday Evening Meetings.

Ordered, That the above resolutions be hung up in the Hall, and Library, and Reading Rooms, of the Institution.

Comm^{r} of Acc^{ts} 3. Resolved, That M^{r}. Daniel Moore, M^{r}. Daniell and M^{r}. Pilgrim be appointed a Committee of Accounts, and that two be a quorum.

4. The following person having been duly recommended was admitted an Annual Subscriber to the Royal Institution for the year 1825.

Ann^{l}. Subs^{cr}. Charles Ogilvy, Esqr. of Sunbury Middlesex

5 Ordered, That the ballotting lists for the election of a President, Managers, and Visitors, in the room of those Officers who have lately resigned, be filled up as follows.

President – Earl of Chichester, F.R.S.

Managers. Sir J. T. Stanley Bart. F.R.S & F.S.A.
Sir Oswald Mosley Bart. D.C.L.
George Moore, Esqr. F.S.A & F.L.S.

32 1825

Visitors –
Sir Thos. Dyke Acland Bart. M.P.
Lieut. Col. W^{m}. M. Leake, F.R.S.

Adjourned to Monday the 4th. of July at One o'Clock

Danl. Moore

July 4. 33

At a Meeting of the Managers of the Royal Institution of Great Britain, held on the 4th of July, 1825.

Present
Daniel Moore, Esqr. V.P. in the Chair
Benjamin Smith, Esqr.
C. P. Meyer, Esqr.
Samuel Ware, Esqr.
Edmund Robt. Daniell, Esqr.
Sir S. Bernard Morland, Bart. M.P. Treas^{r}.
John Guillemard, Esqr. Secretary

1. The minutes of the proceedings of the Meetings of the 6th. 9th. and 13th. of June were read and approved.

2. Signed an Order for payment of the following sums, vizt.

Payments.

M^{r}. Sturt for ½ year's Salary as Librarian to 30th. June 1825.	£ 60 . –
Joseph Fincher on account of the Royal Institution	100 –
Ditto – on account of the Laboratory Fund	50 . –
	£ 210 . –

Presents received.

3. Earl Spencer, K.G. presented "A reprint of the two rare Italian Poems on the Battles of Terrouane and of Flodden Field" –

T. S. Winn presented "Speedy End to Slavery in our West India Colonies" – 2 Copies – & presented by the Editor The American Monitor N°. IV. ——

34 1825.

Ordered, That the thanks of the Managers be returned for these presents.

Letter from Mr. Harris — resp. Catalogue of Library.

4. Read a letter from Mr. William Harris, the late Librarian, addressed to John Guillemard, Esqr. Secretary, stating that he had remaining on hand about 250 Copies of the Catalogue of the Library of the Royal Institution which had been reprinted by him at his own expence, which he offered to the Managers at half price—namely 10s. a copy, and reminding them of the services which he had rendered to the Royal Institution in the purchase of Mr. Astle's Library, and in buying a variety of books at Auction on his judgment, considerably under price.

Resolved, That Mr. Harris be requested to inform the Managers what the expence of publication of the 2d. Edition was,— How many were printed,— How many were disposed of, and at what prices.

Accts. — deficiency.

5. Mr. Danl. Moore reported that he had with Mr. Daniell examined into the Accounts of the Royal Institution for the present year, and it appeared that the Expenditure would exceed the Income by about £600— But in making the Calculation of the Income, the Compositions and Admission fees of new Members that may be invested in the year, were not included.

Committee on Lectures.

6. Resolved, That the following Managers be appointed a Committee with power to add to their number, to consider what lectures shall be given at the Royal

July 4. 35

Institution in the next Season, and by what Lecturers; and any two of them shall be competent to transact the business — viz.

Mr. Daniel Moore,
Mr. Daniell,
Mr. Hare,
Mr. Smith,
Captn. Sabine.

7. The following person having been duly recommended was admitted an Annual Subscriber to the Royal Institution for the present year.

Annl. Subsrs.

Robert Staples, Esqr. of 13 Old Burlington St.

Adjourned to Monday the 7th. of November.

36 1825.

Nov 7.

At a Meeting of the Managers of the Royal Institution of Great Britain, held on the 7th. of November, 1825

Present

Sir George Thos. Staunton Bart. M.P. V.P. in the Chair
Daniel Moore, Esqr. V.P.
Edmund Robt. Daniell, Esqr.
Charles Pilgrim, junr. Esqr.
Sir S. Bernard Morland, Bart. M.P. Treas.
Captn. Sabine,
George Moore, Esqr.
John Guillemard, Esqr. Secretary.

1. The minutes of the proceedings of the last meeting were read and approved.

2. Read and approved the minutes of the Committee of Accounts of this day, and signed an Order for an advance to the Assistant Secretary of £130.

Lectures. M. de Spineto.

3. Read a letter from The Marquis Spineto, addressed to the Secretary, proposing to continue in the next Spring his lectures on the Literature of Modern Europe.

Ordered, That this letter be referred to the Committee appointed to arrange the lectures for the ensuing Season.

On Liby. Catalogue.

Read the statement of the expenses incurred by Mr. Harris for printing the 2d. Edition of the Catalogue of the Library of the Royal Institution

Nov. 7. 37

by which it appeared that the total expense of printing 500 Copies, amounted to £240 . 7 — And that he had received for Copies sold £174 . 14 - leaving a deficit of £65 . 13 —

Ordered, That the consideration of this subject be taken up at the beginning of the ensuing Season.

Presents received.

4. C. P. Meyer, Esqr. presented
Buckland's Reliquiæ Diluvianæ
The Board of Longitude presented
Captn. Sabine's Pendulum Experiments,
Nautical Almanac for 1828,—
Presented by the Editors,
The Zoological Journal Nos. 5 and 6.
P. Murphy, Esqr. presented
Inquiry into the nature and Cause of Miasmata
Lieut. Genl. Wm. Thornton presented
Vincents' Sermons,
César Moreau, Esqr. presented
State of the Revenue of the East India Company
John Penn, Esqr. presented
Records of the Outinian Society, vol. 1, part 2.
Sir James Edward Smith presented
English Flora, vol. 3 —
Dr. Granville presented
Essay on Egyptian Mummies.
Mr. Richd. Faraday presented his Paper from the Philosophical Transactions—On New Compounds of Carbon & Hydrogen —
Presented anonymously — The Anti-Lancet 3 nos.

Ordered, That the thanks of the Managers be returned for these presents

On the Apartment for Mr Faraday. — 5 Read a letter from Mr Faraday requesting the use of the Apartment No 21 as a kitchen
Ordered, That Mr Faraday's request be granted And that an estimate of the expense of fitting it up be laid before the next meeting.

Coals. — 6 Ordered, That Messrs Dikins &Co. supply the Institution with 10 Chaldrons of the best Walls-end Coals. —

Comm. for Lectures. — 7 Resolved, That the following Managers be added to the Committee appointed to arrange the lectures for the ensuing season:—
Sir George Thos. Staunton, Bart.
Sir S. Bernard Morland, Bart
John Guillemard, Esqr.
Charles Pilgrim jun. Esqr.

Lease of 23 Albemarle Street — 8. The Seal of the Corporation was affixed to a Lease of Premises No 23 Albemarle Street from Michaelmas 1823 for 60 years and 3/4, to Robt. Colver & John Dendy Esqrs. It was Countersigned by
Geo: Thos. Staunton, Chairman.
Scrope B. Morland, Treasurer
Dan: Moore,
George Moore,
C. H. Pilgrim,
Edmd. R. Daniell,
Edward Sabine,
John Guillemard Secretary.

9 Read the resolution of the General Monthly Meetg

on electing Mr Faraday Member of R.I. — of the 4th of July, requesting the Committee of Managers to consider the expediency of conferring some mark of approbation upon Mr Faraday.
Resolved, That the following resolution of the Managers be laid before the General Meeting of this day.

The Managers recommend that the best means which occur to them, under existing circumstances, of marking the approbation with which the Royal Institution have viewed the zeal and ability with which Mr Faraday has discharged the duties of his situation, and the valuable and important Chemical discoveries, by which he has exalted not only his own reputation, but that of the Institution, will be the admission of Mr Faraday as a Member of the Royal Institution without the payment of the usual admission fees, fines or compositions, and for that purpose to suspend the operation of the Bye law Chap: II Art: 6, 7, 8.

Adjourned to Monday the 21st Instant at Two o'clock.

H. Davy V.P.

Nov 21. — At a Meeting of the Managers of the Royal Institution of Great Britain, held at the house of the Institution, on the 21st of November, 1825,

Present
Sir H. Davy, Bart V.P. in the Chair.
Benjamin Smith, Esqr.
Sir Oswald Mosley, Bart.
C. P. Meyer, Esqr.

1 The minutes of the proceedings of the last meeting were read and approved.

Presents received.

2 The Royal Irish Academy presented Transactions of the Society vol 14, pt 4
Presented by the Editor, The American Monitor No 5.
Ordered, That the thanks of the Managers be returned for these presents.

Fitting up Kitchen for Mr Faraday — 3. Read an estimate from Mr Royce of the expense of fitting up No 21 as a kitchen for the use of Mr Faraday which amounted to £30.
Ordered, That it be done at an expense not exceeding the above sum.

Lectures. Mr Faraday — 4 Resolved, That it be recommended to the Sub-Committee appointed to arrange the lectures in the ensuing Season, to consider of the propriety of proposing to Mr Faraday to give a Course of 40 lectures on Practical Chemistry in the Lecture room of the Institution.

Appt of Superintendent of House — 5. Resolved, That the appointment of Superintendent of the House be taken into consideration at the next meeting.

C P Meyer

42 1825.

Dec. 5.

At a Meeting of the Managers of the Royal Institution of Great Britain, held at the house of the Institution, on the 5th of December, 1825,

Present

C. P. Meyer, Esqr in the Chair.
George Moore, Esqr
Sir S. Bernard Morland, Bart: M.P. Treas.^r
Charles Pilgrim jun.^r Esqr
John Guillemard, Esqr Secretary.

1 The minutes of the proceedings of the last meeting were read and approved.

2. Read and approved the minutes of the Committee of Accounts of this day; and signed an Order for an advance to the Assist.^t Secretary of £50.

3 Presents received.

The Royal Society, presented
Philosophical Transactions – 3 parts 1824 p.^t 3. 1825 p.^ts 1 & 2.
Presented by the Editor
Zoological Journal N.º 7.
Ordered, That the thanks of the Managers be returned for these presents.

4 Resolved, That the resolution relating to the Superintendent of the House be deferred till Monday the 19th Instant.

Dec. 5. 43

Lectures. (Juvenile) by Mr Millington.

5. The Sub-Committee appointed to consider what lectures shall be delivered in the next Season, and by what lecturers, reported that Mr Millington would be ready to deliver a set of Twenty two Lectures on Natural Philosophy suited to a Juvenile Auditory, during the Christmas, Easter, and Whitsuntide recesses.
Resolved, That Mr Millington be engaged to deliver the above lectures – And that the following be the terms of admission:– vizt – for non-subscribers to the Institution Two guineas for the whole, or one guinea and a half for the Christmas portion of 15 lectures. – The Wives and Families of Members will be admitted on the payment of One guinea and a half for the whole Course, or one guinea for the Christmas portion.

Juv. Lect.^s Proposal resp.^t Mr. Faraday.

6. Read the resolution of the General Meeting of this day recommending to the Managers to engage Mr Faraday to take a part in the Juvenile lectures proposed to be given during the Christmas and Easter recesses.
Resolved, That this resolution be referred to the Sub-Committee appointed to consider what lectures shall be delivered in the next Season.

Lectures in French – declined.

7. C. Moreau, Esqr M.R.I. having personally applied on behalf of a Gentleman desirous to deliver some lectures in this Institution in the French language.
Resolved, That it appears improper that any lecture here delivered should be in a language not understood by all the Members and Subscribers – And that the substance of this resolution be communicated to Mr Moreau.

44 1825.

8. The following persons having been duly recommended were admitted as Subscribers to the Royal Institution for the year, 1826.

Annual Subscribers.
Major John Smith of Ibbotson's Hotel,
Mr Chambers 17 Upper Montagu Street,
Dr Wm Scott 6 Sackville Street,

To the Lectures
The Marchioness Palmela.
M.^rs & Miss Le Mann, 16 Montagu Street.

Adjourned to the 12th Instant.

Dan Moore

45

Dec. 12.

At a Meeting of the Managers of the Royal Institution of Great Britain, held at the house of the Institution on the 12th of December, 1825

Present

Daniel Moore, Esqr V.P. in the Chair.
C. P. Meyer, Esqr
George Moore, Esqr
Chas Pilgrim jun.^r Esqr
Captn Edwd Sabine,
William Somerville, M.D.
Samuel Ware, Esqr
John Guillemard, Esqr Secretary.

1. The minutes of the proceedings of the last meeting were read and approved.

2. Read and approved the minutes of the Committee of Accounts of this day.

Life Subs.^r

3. In pursuance of a nomination under the Act of Parliament by Mrs Gaulier, her Son Guy Grant Aristides Gaulier was admitted to the privileges of a Life Subscriber to the Royal Institution.

4. Presents received

Messrs T. Gardner & Son presented
The Royal Blue Book for 1826,
John Frost, Esqr M.R.I. presented
An Oration delivered before the Medico-Botanical Soc.^y
C. P. Meyer, Esqr presented The system of Infants' Schools by Wm Wilson A.M.

Ordered, That the thanks of the Managers be returned for these presents.

Lectures. declined — Mr de Spineto. 5 The Committee appointed to consider what lectures shall be delivered in the Institution in the next Season, and by what lecturers, reported that they had received an application from The Marquis Spineto to continue his lectures On the Literature of Modern Europe, and that they considered it unadvisable under the present circumstances that those lectures should be continued.

G. B. Jermyn. That they had likewise received a letter from Mr George B. Jermyn offering to deliver a Course of lectures On Chivalry – and that he should be perfectly satisfied with the usual emolument – And that under the present circumstances they do not think it advisable to accept the said offer.

Mr Frost. That they had likewise received a letter from Mr Frost proposing to deliver lectures on the Materia Medica in the Laboratory of the Royal Institution And that they do not think it advisable to accept the said offer.

Mr McCulloch. That they had likewise received a letter from Mr Mc Culloch in answer to an application made by them requesting to know on what terms he would deliver a Course of lectures On Political Economy And that the terms proposed by Mr Mc Culloch appear to them incompatible with the existing regulations of the Institution.

engaged. Mr Addams. That they had received an offer from Mr Addams offering to deliver a gratuitous Course of 8 lectures on Hydrostatics, Hydraulics and Pneumatics in the next Spring, And that they think it advisable to accept the said offer.

Dec 12.

They likewise reported that in consequence of the resolution of the Managers of the 21st of November they had consulted Mr Brande the Professor of Chemy and Mr Faraday, on the practicability and propriety to engage Mr Faraday to deliver 40 lectures on Chemy And that in consequence of the representations of both those Gentlemen they were of opinion that it would not be expedient to engage Mr Faraday to give

Mr Brande & Faraday. such lectures in the ensuing Season – That Mr Brande proposes to deliver Two Courses in the ensuing Season – The first on experimental Chemistry in illustration of the Theory of definite proportionals, and of the applications of that Theory to practical purposes – & the second on the Chemistry of the Arts – And that Mr Brande further communicated to the Committee his intention of delivering in conjunction with Mr Faraday a very extended Course of lectures on Chemy to commence at the end of the ensuing year – And that it was his wish and would be his endeavour to render the said Course such as would do honour to the Institution and be highly useful to the Community.

They likewise reported that they had consulted Mr Faraday on the subject of engaging him to take a part in the Juvenile Lectures proposed to be given during the Christmas and Easter recesses, And that they found his avocations were such that it would be exceedingly inconvenient for him to engage in such lectures.

Resolved, That the above report be approved and adopted;

Ordered, That the Marquis Spineto, Mr Jermyn, Mr Frost, and Mr Mc Culloch be respectfully informed that their offers to deliver lectures at the Royal Institution

in the ensuing Season have not been accepted.

Request of Mr Brande. 6 Mr Brande stated that he felt extremely hurt on finding that any person should have been proposed by one of the Managers to give lectures in the Theatre of the Royal Institution without any assent on his part or any communication held with him; and that conceiving such a proposal included an insinuation of his inadequacy to fulfil the duties of the office which he has the honour to hold of Professor of Chemistry, he requested permission to defend his conduct in that office before the next Board of Managers.

Ordered, That Mr Brande be heard at the next meeting

7 Resolved, That Mr Millington be entitled to receive 3/4ths of whatever sums are paid from non-Subscribers to the Institution for admission to his Course of lectures suited to the capacity of a Juvenile Auditory.

Lectures. 8 Resolved, That Mr Meyer be added to the Committee appointed to consider what lectures shall be delivered at the Royal Institution in the next Season.

9 The following person having been duly recommended was admitted an Annual Subscriber to the Royal Institution for the year 1826.

Ann. Subsrs. James Taylor Esqr. 4 Montagu Square.

Adjourned to the 19th Instant.

Dec 19.

At a Meeting of the Managers of the Royal Institution of Great Britain, held at the house of the Institution, on the 19th of December, 1825,

Present

Sir H. Davy, Bart. V.P. in the Chair.
Daniel Moore, Esqr. V.P.
Charles Pilgrim, Esqr.
George Moore, Esqr.
Edmund Robt. Daniell, Esqr.
C. P. Meyer, Esqr.
Benjamin Smith Esqr.
John Guillemard, Esqr. Secretary.

1. The minutes of the proceedings of the last meeting were read and approved.

Mr Brande. 2. The Managers having heard Mr Brande, in consequence of his request made at the last meeting,

Resolved, That in their opinion no reflection was intended to be cast upon Mr Brande's conduct as Professor of Chemistry, by the resolution No. 4 of the 21st of November last.

On Superintendence of House. 3. It was moved to resolve that the Managers do not see any reason to alter the existing arrangements respecting the superintendence of the House:— On which it was moved as an amendment —"that the Managers think it essential to the interests of the Royal Institution that the Superintendent of the House should be

a resident in it" — which was negatived; and the original Motion carried.

Committee on Lectures.

4 Resolved, That a Sub-Committee of the Managers be appointed to consider if any amendments in the present mode of recompensing lectures, and of admitting persons, who are not Members or Subscribers, to the lectures, be expedient — And that

Daniel Moore, Esqr.
C. P. Meyer, Esqr.
Chas. Pilgrim, Esqr.
Edmund Robt. Daniell, Esqr.

be on the said Committee, with liberty to add to their number any other Manager of the Instit. And that the said Sub-Committee report their opinion to the Board.

5 The following persons having been duly recommended were admitted as Annual Subscribers to the Royal Institution for the year 1826.

Annl. Subscrs.

Charles Kelsall, Esqr. 19 Clifford St.
Thos. Hugh Davies, Esqr. Manchester Sqr.
Josiah Marshall Heath, Esqr. 12 Grove End St. John's Wood

Coals.

6. Ordered, That Messrs. Dickins & Co. supply the Institution with 20 Chaldrons of the best Wallsend Coals.

Adjourned to Monday the 9th of January, 1826.

Danl. Moore

1826 Jan 9.

At a Meeting of the Managers of the Royal Institution of Great Britain held at the house of the Institution on the 9th of January, 1826

Present

Daniel Moore, Esqr. V.P. in the Chair.
Edmund Robt. Daniell, Esqr.
George Moore, Esqr.
Charles Pilgrim, Esqr.
C. P. Meyer, Esqr.
Captn. Sabine
John Guillemard, Esqr. Secretary.

1. The minutes of the proceedings of the last meeting were read and approved.

2. Read and approved the minutes of the Committee of Accounts of this day; and signed an Order for an advance to the Assist. Secretary of £100.

Presents received.

3. R. B. Pitman, Esqr. presented View of the practicability of uniting the Atlantic and Pacific Oceans.
Mr. Ackermann presented Repository of the Arts, 1825 – 12 nos.
Mr. Gill presented Technical Repository for 1825 – 12 nos.
Ordered, That the thanks of the Managers be returned for these presents.

Lectures. Mr. Wallis.

4. Read the following report from the Committee appointed to consider what lectures shall be delivered at the Royal Institution in the ensuing Season. —

The Committee reported that they had proposed to Mr. Wallis to deliver a Course of six lectures on Astronomy, and to remunerate him with the sum of Thirty guineas, to which Mr. Wallis had acceded.

Resolved, That Mr. Wallis be engaged according to the above arrangement.

Report of Committee on Lectures &c.

5. Read the following report from the Committee of Managers appointed to consider if any amendments in the present mode of recompensing Lecturers and of admitting persons who are not Members or Subscribers to the Lectures be expedient. The Committee — having added Mr. George Moore to their number, report

That in taking into consideration the subject which has been referred to them they have been induced to go rather more extensively into the question of the Lectures at the Royal Institution than may appear to be warranted by the terms of the reference, but they trust that in so doing they will not be considered as having very materially departed from the line of their duty, and that the observations they have to make will be productive of advantage to the interests of the Institution.

Two kinds of Lectures — Public

There are two descriptions of lectures now delivered at the Royal Institution — namely, the public lectures which are given at 2 o'Clock to the Members, and Subscribers in the Theatre, and the more extended and

Jan 9.

Private.

scientific lectures which are given in the Laboratory early in the Morning by Mr. Brande and Mr. Faraday to the Private Pupils and Members of the Institution, and which may be termed Private Lectures. — To both these descriptions of Lectures, your Committee have thought it proper to direct their attention with the view of ascertaining whether any arrangement could be made for rendering them more extensively useful and more productive to the funds of the Institution.

Private Lectures.

With respect to the Private Lectures your Committee are of opinion that much may be done towards their improvement from which very considerable benefits may be derived both to the Institution and to the interests of Science in general. — They are also of opinion that great advantages will result from the introduction of lectures upon other scientific subjects upon similar terms.

Introduction by Mr. Brande

These lectures first commenced in the year 1813, and your Committee find from reference to the Minute Book of the Board of Managers that they originated with Professor Brande who at the time of his election to the Office of Professor was engaged in the delivery of lectures on Chemistry to the Medical Students in Windmill Street, and who suggested to the Managers the possibility of transferring the Chemical School from Windmill Street to the Laboratory of the Royal Institution — And your Committee find that at a Special Meeting of the Managers held on the 4th October 1813, the ideas suggested by Mr. Brande was recommended for consideration, and at the same time the Managers present expressed themselves to be unanimously of opinion that the measure if it

could be accomplished would be highly beneficial to the interest of Science, and of the Institution. – In consequence of this recommendation the lectures were transferred to the Laboratory of the Royal Instit^n. and have ever since continued to be delivered there and lately Mr. Faraday at the suggestion of Prof^r. Brande has taken part with Mr. Brande in the delivery of them.

Assist^ce of Mr. Faraday.

Their Success. Proposed Increase.

Your Committee think that these lectures have been productive of great advantage to the Royal Instit^n. and have greatly raised its character as a school of Chemistry, and they are of opinion that they might be made of still greater benefit if they were to be more generally known and recognized as forming part of the system of the Institution, and as adding materially to the privileges of the Members – They conceive that this might be in a great measure effected by removing them into the great Theatre which at the same time that it would contribute much to render them more generally known would afford space for an increased number of auditors, the Laboratory where they are now delivered being scarcely sufficient to contain all the persons who attend. If this were done the Laboratory might be used for the purpose of giving more extended demonstrations of the Lectures delivered up-stairs to the Members and Pupils, and to those subscribers who might be willing to pay an additional subscription.

Your Committee are also of opinion that it would greatly extend the general utility of the Royal Institution, if Lectures upon other subjects were to be delivered upon the same terms as those now delivered

in the Laboratory, that is, if Professors of eminence of different Sciences were to be admitted to give lectures to their own Pupils or Classes, upon the terms of admitting the Members and Life Subscribers to attend them gratis; by these means a source of profit might be opened to the Institution, and additional inducement would be held out to persons to become Members instead of Annual Subscribers, which your Committee consider will be a very desirable object. Your Committee are of opinion that such an arrangement as this might be carried into effect without at all interfering with the rights of the annual Subscribers, whose privilege extends to attending public lectures only – the lectures now proposed to be given being for the benefit of the Lecturers as well as for the accommodation of the Members would only be considered as Private lectures; Nor would they in any manner interfere with the advantages actually enjoyed by the Annual Subscribers since it is proposed that these lectures should be given in the Morning only, and at times when no Public lectures have ever been delivered in the Theatre. Should Lectures of the description pointed out be established in the Institution an opportunity would be afforded to the Members and Life Subscribers of attending extensive Courses of lectures upon many branches of Science without exposing the Institution to any additional expense, and your Committee think that the funds of the Institution would be thereby greatly benefited since they cannot but suppose that the additional privileges there held out will induce a great number of persons to become Members.

Public Lectures.

With respect to the Public Lectures delivered in the Theatre your Committee find considerable difficulty in suggesting any arrangement by which such lectures could be made to pay their own expenses without considerably endangering the revenues of the Institution a very large proportion of which is derived from the Annual Subscribers to whom the lectures afford the principal attraction. It appears from references to the Accounts for the last three years that the average income derived from the annual subscriptions has amounted to £1200 – forming more than one half of the Income of the Institution, and that the annual expense of providing lecturers for these years has averaged £300, exclusive of the Salary of the Professor of Chemistry; so that the income derived from the Subscribers has greatly exceeded the expense of the lectures, and as your Committee consider that the lectures form the chief inducement to individuals to become Subscribers, it appears to them that the lectures do in fact greatly add to the resources of the Institution They are of opinion however that an arrangement by which persons might be admitted to subscribe to particular Courses would be very desirable.

Subscrip^ns to Courses of Lectures recommended.

According to the present arrangements of the Institution, the lecture Season usually commences in the beginning of February, and continues till the Month of June including a period of about four months. Of course all those who subscribe to the lectures generally have the opportunity of attending the whole of the lectures during the Season, but the fashionable season in London seldom commences till after Easter, so that the persons who come to town at that time although they pay the same subscription

have only half the advantage of them who are in London during the whole period. The effect of this must be to prevent a great number of persons from subscribing who if they could be admitted upon lower terms than those who have the whole benefit would be induced to subscribe. Your Committee therefore think that by allowing such persons to subscribe to particular Courses considerable advantages would result. Of course in fixing the amount of such subscriptions care must be taken to give the general subscriber a very decided advantage in order to prevent a falling off in the general subscrip^n. Your Committee also think it will be very desirable with the view to accommodate those who come to Town after Easter that as many new Courses should be made to commence after the recess as possible, and where it is practicable they recommend that in the longer Courses of lectures some new division should be entered upon at that period.

Your Committee beg to add that in pursuance of Art: 5 of Chap: XIII of the Bye laws the above arrangement as to admitting Subscribers to particular Courses of lectures may be carried into effect without any alteration being made in the Bye laws.

Orders thereon.

Resolved, That the report of the Committee be adopted; and that it be referred to a Committee consisting of the same Gentlemen to carry into execution that part of the report which relates to the Establishment of Private Lectures, upon the same principle as those now delivered by Mr. Brande and Mr. Faraday – That the Committee have power to add to their number.

6. Resolved, That the execution of that part of the

58 1826.

report which relates to the admission of Subscribers to particular Courses of lectures in the Theatre, be referred to the Committee appointed to consider what lectures shall be delivered in the ensuing season and by what lecturers.

7. Resolved, That it be referred to the said Committee of lectures to consider of the propriety of admitting Gentlemen under a certain age to become Subscribers at a reduced payment.

Morning Lectures in Theatre. 8. Resolved, That Mr Brande and Mr Faraday be requested to deliver their Morning Lectures in the Great Theatre of the Institution.

Library to be opened on Lecture nights 9. Resolved, That the Library be opened on the Evenings of Mr Wallis's lectures on Astronomy for the accommodation of the Members & Subscribers

On lightg the House. 10. Resolved, That the mode of illuminating the House at the times of the weekly meetings of the Members in the Library, and of the Evening Lectures be referred to the Superintendent of the House.

Hour of lectures. 11. Resolved, That the Public Lectures commence at Three o'Clock in the Afternoon.

On Friday Evg. – 12. Resolved, That the weekly meetings of the Members and their Friends in the Library be resumed on Friday the 3d of February at ½ past Eight o'Clock in the Evening; and that they be continued on each succeeding Friday Evening, at the same hour till further notice

Jan. 9 59

Mr Faraday released from attendg on lectures. 13. Mr Professor Brande having suggested that in consequence of the rank which Mr Faraday holds in Science, and his many important avocations, it would be proper to relieve him from his attendance as Assistant upon the Professor of Chemistry at his lectures.

Resolved, That the suggestion of Mr Brande be adopted – And that Mr Faraday's attendance upon him be no longer required.

The following persons having been duly recommended were admitted as Subscribers to the Lectures for the year 1826

Subsrs Mrs and Miss Richards 26 Bryanstone Square

Adjourned to Monday the 6th of February.

Danl Moore

60 1826.

Feb 6. At a Meeting of the Managers of the Royal Institution of Great Britain, held at the house of the Institution on the 6th of February, 1826

Present

Daniel Moore, Esqr V.P. in the Chair.
Charles Pilgrim, Esqr
George Moore, Esqr
Captn Sabine,
Sir George Thos Staunton, Bart M.P.
William Somerville, M.D.
C. P. Meyer, Esqr
Benjamin Smith, Esqr
Edmund Robt Daniell, Esqr
Sir S. Bernard Morland, Bart M.P. Treas

1. The minutes of the proceedings of the last meeting were read and approved.

2. Read and approved the minutes of the Committee of Accounts of this day; and signed an Order for an advance to the Assist Secretary on account of the Institution of £200. —

Payments. Mr Hurt for half a years Salary as Librarian to 31st Decr last } 60 —
Reeve for ½ Yrs Assessed taxes to Michaelmas 1825 } 61 . 4 . 9
321 . 4 . 9

3. Read a letter from John Guillemard, Esqr Secy announcing that his present state of health will prevent him continuing in the regular discharge of

Feb 6. 61

Resignn of Mr Guillemard as Secy. his official duties – And that he regrets he shall be under the necessity of resigning his office of Secretary.

Resolved, That the meeting learn with extreme regret Mr Guillemard's intention. – And request he will allow his name to be continued till the next Annual Meeting.

Mr Daniell proposed as Secy. Resolved, That E. R. Daniell, Esqr be requested to perform the duties of Secretary for Mr Guillemard.

Commn from Mr Faraday 4. Read a letter from Mr Faraday in which he requested the Secretary to express to the Managers the strong sense he entertains of their kindness and consideration towards him in their late resolution made upon the suggestion of Mr Brande.

Presents received.

5. The Horticultural Society presented
Transactions of the Society, vol 6 – parts 1 & 2.
Revd Stephen Weston, B.D. presented
Historic Notices of Towns in Greece, and in other Countries that have struck Coins,
Charles Kelsall Esqr presented
Classical Excursion from Rome to Arpino,
Dr John Gordon Smith – presented
Principles of Forensic Medicine,
Analysis of Medical Evidence,
Ordered, That the thanks of the Managers be returned for these presents.

6. Read the following report from the Committee

appointed to consider what lectures shall be delivered in the ensuing Season and by what

Lectures. lecturers.—

S. Wesley. The Committee report that they had proposed to Saml. Wesley Esqr. to deliver a Course of Six Lectures on Music – To be illustrated by instrumental and Vocal examples and to remunerate him with 36 guineas.—An additional remuneration would be required by the Vocal Performers.

Dr. Harwood. That they had proposed to Dr. Harwood to deliver a Course of Lectures on Natural Philosophy, and to remunerate him with 50 guineas.

Dr. Mayo. They had also proposed to The Revd. Dr. Mayo to deliver, gratuitously, a few lectures on Education embracing the systems of Pestalozzi, and particularly of Infants' Schools to which he had acceded.

Dr. Wurm. They likewise report that they had received a proposal from Dr. Wurm to deliver a Course of lectures On German Literature which they had accepted.

Reduced payments declined. They also report that they had taken into consideration the reference made to them by the Board of Managers to consider of the propriety of admitting Gentlemen under a certain age to become Subscribers at a reduced payment – and it did not appear to them expedient to recommend any alteration in the terms of admission of the annual Subscribers.

They further report that in obedience to the reference made to them by the Board of Managers which

Proposed Admission to Courses of Lectures. empowered them to carry into execution that part of the report of the Committee, appointed to consider if any Amendments in the present mode of recompensing lecturers and of admitting persons who are not Members or Subscribers to the Lectures, be expedient, which relates to the admission of Subscribers to particular Courses of Lectures in the Theatre – they had directed that Non-Subscribers to the Institution should be admitted to Mr. Wallis and Dr. Harwoods' lectures (under the same regulations as the Annual Subscribers) on the following terms:— Gentlemen Two guineas.

Ladies – One guinea –

for each Course.

Demonstrations in Chemistry by Mr. Faraday. They further report that they had proposed to Mr. Faraday to deliver a Course of Lectures in the Theatre of the Institution in the present season, which he stated to the Managers he was under the necessity of declining as his time would be fully occupied in the present year – But that he should be happy to give to the Members, at their weekly Meetings, Demonstrations on subjects in Chemistry, similar to those given by him at the Evening Meetings in the last year.

Resolved,

That the above report be approved and adopted.

7. Read the following report from the Committee of the Managers appointed to carry into execution that part of the report, of the Committee appointed on the 19th. of December last, as relates to the establishment of Private Lectures in the Theatre of the

Private Lectures. Royal Institution on the same principle as those delivered by Mr. Professor Brande & Mr. Faraday.

J. R. McCulloch. The Committee report that they have arranged with J. R. McCulloch, Esqr. for the delivery in the Theatre of the Royal Institution, of the lectures on Political Economy instituted in honour of the late David Ricardo, Esqr.— The Course is to consist of from 25 to 30 Lectures, commencing after Easter to be delivered on three days in each week.

Dr. G. Smith. The Committee have also agreed with Dr. John Gordon Smith for the delivery of a Course of Lectures on State Medicine or Medical Jurisprudence to commence on Monday the 13th. Instant, and to be continued on each successive Monday, Wednesday and Friday till the completion of the Course.

Resolved,

That the above report be approved and adopted.

On Assignment of Lease of No. 23 Albe. St. 8. Read a letter from Messrs. Moore & Lake Solicitors to the Royal Institution, informing the Board that they had received an application from Messrs. Dendy & Morphet to procure from the Royal Institution a licence from Robt. Colmer & Saml. Dendy, Esqrs. the Executors of the late Mr. Baker to assign to Francis James Nugee of St. James's Pl. the Lease of the 9th. of May 1825 of Premises No. 23 Albemarle Street.

Resolved, That the request of Messrs. Dendy & Morphet be complied with.—

9. The following persons having been duly recommd.

were admitted as Subscribers to the Royal Instit: for the year 1826.

Annual Subscribers

Lt. Col. Farquhar, 32 Cadogan Place,
Captn. Charles Fairfield, 7 Gloucester Place.
F. Fladgate, Esqr. Queen Square Westmr.
Edward Ord Warren, Esqr. 12 Guilford St.
Wm. Doveton, Esqr. 38 Maddox Street.
Pascoe Grenfell, Esqr. M.P. 30 Charles St.
Alexr. Erskine Esqr. 1 Devonshire Place
F. C. Brant, Esqr. 21 Park Crescent Portland Place
Edward Jarman, Esqr. Albany.
Hon: Fredk. Calthorpe M.P. Grosvenor Sqr.
John Smirnove, Esqr. 46 Wigmore Street
Edward Lyon, Esqr. } 41 Jermyn Street
Lyon Winder, Esqr. }
Henry Walker, Esqr. } 49 Bedford Square
Francis Walker, Esqr. }

To the Lectures.

Mrs. Snell – 27 Albemarle Street
Miss Grenfell }
Miss H. Grenfell } Charles Street St. James's Square
Miss Emily Grenfell }
Mrs. Andrews – 7 Arlington Street
Mrs. Cutler – 3 Clifford St.
Mrs. Brodie Saville row.

To the Astronomical Lectures.

Miss Charlotte & Miss Fanny Grenfell,
Viscountess Bernard & The Honble. Miss Bernard.

Adjourned to the 20th. Instant.

[illegible]

66

1826.

Feb 20. **At a Meeting of the Managers of the Royal Institution of Great Britain held at the house of the Institution on the 20th of February, 1826.**

Present

Sir George Thos. Staunton Bt. M.P. in the Chair.
Daniel Moore, Esqr.
George Moore, Esqr.
Charles Pilgrim, Esqr.
Captn. Sabine,
Wm. Somerville, M.D.
Edmund Robt. Daniell, Esqr.

1. The minutes of the proceedings of the last meeting were read and approved.

2. Read and approved the minutes of the Committee of Accounts of this day, and signed an Order for an advance to the Assist. Secretary of £250.

Lectures. 3. Read a letter from The Marquis Spineto to Mr. Fincher on the subject of lecturing in the present Season.
Resolved, That it be referred to the Committee of Lectures.

4. Audited the Treasurer's Account for the year 1825.

Tea & Coffee. 5. Ordered, That the Housekeeper be directed to provide Tea & Coffee to the Members & Subscribers

67

Feb. 20.

who may wish to be accommodated with it in the Libraries or Reading Rooms.

Present received.

Present from Mr. Fuller. 6 John Fuller, Esqr. having presented to the Newspaper Room – four Patent Globe Lamps.
Ordered,
That the special thanks of the Managers be returned for this present.

7. The following persons having been duly recommended were admitted as Subscribers to the Royal Institution for the year 1826.

Annual Subscribers.
Frederick Buller, Esqr. Whitehall Place
Ebenezer Sheldon, Esqr. 17 Mount Street
John Leckie, Esqr. Duke St. Manchester Square
Major Marrie 1 Great Cumberland Place.

To the Lectures.
Miss Jeremie } 10 York St. Portman Square
Miss G. Jeremie Esqr. }
Miss S. M. Walker } 49 Bedford Square
Miss L. R. Walker }
Mrs. Harvey 37 Albemarle Street
Mrs. Cook } 10 Lancaster Place
Miss Cook }
Miss Caroline Fry – 21 Berners Street.

To the Astronomical Lectures
Miss Hester Rickards – Wimpole Street

68

1826.

Mr. & Miss Elliott 34 Hertford Street
James Morrison, Esqr. } 105 Fore Street
Mrs. Morrison – }
Miss Morse St. James's Square.
Mr. Holder }
Miss Holder } 33 Montagu Square
Miss H. Holder }
Mrs. Bostock } Upper Bedford Place
Mrs. Reed }
Miss Dent } Wimpole Street
Miss Wilkinson }

Adjourned to the 6th of March.

Danl. Moore

69

Mar. 6 **At a Meeting of the Managers of the Royal Institution of Great Britain held at the house of the Institution on the 6th of March, 1826**

Present

Daniel Moore, Esqr. V.P. in the Chair.
Captn. Sabine,
Charles Pilgrim, Esqr.
George Moore, Esqr.
Benjamin Smith, Esqr.
Sir S. Bernard Morland, Bart. M.P. Treasr.
Edmund Robt. Daniell, Esqr.

1. The minutes of the proceedings of the last meeting were read and approved.

2. Read and approved the minutes of the Committee of Accounts of this day, and signed an Order for an advance to the Assist. Secretary of £90.–
Payment. And signed an Order for payment of £100 – on account of the Laboratory Fund.–

3. Read a letter from John Frost, Esqr. M.R.I. proposing to deliver a Course of Lectures on Materia Medica.
Lectures. Resolved, That it be referred to the Committee appointed to arrange the Private lectures.

4. In pursuance of a nomination under the Act of Parliament by Langham Christie, Esqr.; his Brother
Life Subsr. Charles William Christie, Esqr. of Lincolns Inn was admitted to the privileges of a Life Subscriber to the Royal Institution.

70 1826.

Lectures. M. de Spineto.

5. The Committee appointed to arrange the lectures report that they have engaged The Marquis Spineto to deliver a Course of Lectures on the Literature of Modern Europe – To commence after Easter, and to remunerate him with 25 guineas; and to allow him the advantage of such subscriptions as may be paid for attending his Course separately.– They also report that they have likewise engaged The Revd. Mr. Jermyn to deliver a Course of 6 lectures on Chivalry – to be delivered in the Evening, and to remunerate him with 13 guineas, and to allow him the advantage of such subscriptions as may be paid for attending his Course separately.

Mr. Jermyn.

T. Griffiths salary raised.

6. Mr. Brande having stated that in consequence of the numerous lectures, the duties of Thomas Griffiths, Assistant in the Laboratory, had greatly increased – Resolved

That his Salary be advanced from 15s. to £1..1 ℔ week

admission of Life Subrs.

7. Resolved, That it be referred to a Committee consisting of the Managers to consider whether any alterations should be made in the terms of admission of the Life Subscribers on their becoming Members.

8. The following persons having been duly recommended were admitted as Subscribers to the Lectures for the year 1826

Subsrs. to Lectures.

Miss B. Russell } 51 Baker Street
Miss L. Russell }

Mar. 6. 71

To Dr. Harwood's Lectures.
Master Alexr. Johnston 19 Gt. Cumberland Street

Danl. Moore

72 1826.

April 3.

At a Meeting of the Managers

of the Royal Institution of Great Britain, held at the house of the Institution on the 3d. of April, 1826

Present

Daniel Moore, Esqr. V.P. in the Chair
Benjamin Smith, Esqr.
Captn. Sabine
Charles Pilgrim junr. Esqr.
George Moore, Esqr.

1. The minutes of the proceedings of the last meeting were read and approved.

2. Read and approved the minutes of the Committee of Accounts of this day.

Presents received.

3 I. F. Gyles, Esqr. M.R.I. presented
Principles of Hebrew punctuation explained.
On the meaning of a passage in St. Luke
C. Moreau, Esqr. M.R.I. presented
Tables of British and Irish Exports
The Society of Arts presented
Transactions of the Society for 1825
G. Poulett Scrope, Esqr. presented
Considerations on Volcanos,
J. Thelwall, Esqr. presented
The Panoramic Miscellany, No. 1 and 2.
The Astronomical Society presented
Transactions of the Society vol 2 pt. 1.
Royal Asiatic Society presented
Transactions of the Society vol 1 part 2.

April 3. 73

Charles Lane, Esqr. presented
The Coronation Oath considered.
Ordered, That the thanks of the Managers be returned for these presents.

4 The following persons having been duly recommended were admitted as Subscribers to the Royal Instit. for the year, 1826.

Annual Subscribers
James Trecothick junr. Esqr. 10 Bury Street
Henry Lytton Bulwer, Esqr. 6 Albany.

To the Lectures.
Mrs. Rogers 40 Upper Bedford Place
Lady Charlotte Calthorpe Grosvenor Square

Subscribers to the Marquis Spineto's Lectures.
Wm. John Bankes, Esqr. M.P.
The Countess of Falmouth
The Count & Countess Ludolf
The Earl & Countess of Hardwicke
The Earl & Countess of Caledon
Lady Elizth. Stuart
Revd. Wm. Cunningham.

Danl. Moore

April 17.

At a Meeting of the Managers of the Royal Institution of Great Britain held on the 17th of April, 1826.

Present

Daniel Moore, Esqr. V.P. in the Chair.
Charles Pilgrim, junr. Esqr.
Captn. Sabine
George Moore, Esqr.
Samuel Ware, Esqr.
Benjamin Smith, Esqr.
Sr. Bernard Morland, Bt. M.P. Treasr.
Edmund Robt. Daniell, Esqr.

1. The minutes of the proceedings of the last meeting were read and approved.

2. Read and approved the minutes of the Commee. of Accounts of this day, and signed an Order for an advance to the Assist. Secretary of £114.13.1 on account of the Laboratory Fund.

3. *Room in House for Asst. in Liby.* Read a letter from Mr. Stut requesting the Managers would allow James Simons, the present Assist. in the Library, to occupy the Room No. 23 in the Attic, which is now unoccupied, while he continues in that situation.
Resolved, That Mr. Stut's request be granted during the pleasure of the Managers, should the Superintendent of the House see no inconvenience arising from it.

Presents received.

John Smirnove, Esqr. F.R.S. & M.R.I. presented
Aperçu de la situation interieure des Etats unis d'Amerique.

Michl. Faraday Esqr. F.R.S. & M.R.I. presented a paper — On the mutual action of Sulphuric Acid and Napthaline, and on a new acid produced.
George W. Cartwright, Esqr. of Mount Pleasant State of New York — presented
Specimens of Iridescent Anthracite from the New York & Schuylkill Company's Mines Pensylvania
Copper Ore from Mount Pleasant — State of New York
Iron Ore Mountain Lodge — Highlands. —
Ordered, That the thanks of the Managers be returned for these presents.

4. *Resignn. of Mr. Guillemard.* The Assist. Secretary stated that Mr. Guillemard, the Secretary, had requested him to inform the Managers that in consequence of his present ill state of health, he was under the necessity of retiring from his official situation.
Resolved,
That the Managers learn with extreme regret Mr. Guillemard's intention of relinquishing the situation he has filled for many years with so much advantage to the Royal Institution.
Nomination of Mr. Daniell. Resolved,
That the name of Edmund Robt. Daniell, Esqr. be inserted in the House list as Secretary.

5. The following persons having been duly recommended were admitted as Subscribers to the Lectures for the year, 1826.

Mrs. Portman
Miss Portman } 38 Bryanstone Square
Miss H. Portman

Miss Williams
Miss Gerard } 36 Grosvenor Square
Mrs. Hurnall 4 Old Burlington St.

To the Marquis Spineto's Lectures.
B. Gaskell, Esqr.
Mrs. Gaskell } 30 Upper Brook St.
Countess of Westmoreland
Mr. Orby Hunter & Miss Hunter
The Marchioness of Lansdowne

To Mr. Jermyn's lectures.
Mr. & Mrs. Purcell

To Mr. Wesley's lectures on Music.
George Wallis, Esqr.

6. Ordered, That the House lists be filled up as follows.

President — Earl of Chichester, F.R.S.

Treasurer — Sir S. Bernard Morland, Bt. M.P.

Secretary — Edmund Robt. Daniell, Esqr.

Managers.
Benjamin Bond Cabbell, Esqr.
Captn. John James Chapman, R.A.
Thomas Hope, Esqr. F.R.S. and F.S.A.
Marquis of Lansdowne, L.L.D. F.R.S.
Daniel Moore, Esqr. F.R.S. F.S.A. & F.L.S.

George Moore, Esqr. F.S.A. & F.L.S.
Viscount Palmerston, M.P.
Charles Pilgrim junr. Esqr.
The Hon: Wm. Francis Spencer Ponsonby,
John Soane, Esqr. F.R.S. & F.S.A.
Captn. Edward Sabine F.R.S.
Sir Claude Scott, Bart. F.L.S. & F.H.S.
Benjamin Smith, Esqr.
Sir George Thos. Staunton, Bart. M.P. F.R.S. & F.L.S.
John Weyland, Esqr. F.R.S.

Visitors.
Henry Jas. Brooke, Esqr.
John Bostock, M.D. F.R.S. & F.L.S.
John Ivatt Briscoe, Esqr.
John George Children, Esqr. F.R.S & F.L.S.
John Frederic Daniell, Esqr. F.R.S.
Earl of Darnley, F.R.S.
Duke of Devonshire
Earl of Egremont F.R.S. and F.S.A.
John Fuller, Esqr.
Davies Gilbert, Esqr. M.P. F.R.S. & F.L.S.
John Guillemard, Esqr. F.R.S.
Lord Prudhoe, F.R.S.
Richd. Horsman Solly, Esqr. F.R.S. & F.S.A.
Chas. Hampden Turner, Esqr. F.R.S & F.L.S.
Lord Teignmouth.

G Staunton

78

1826.

May 8.

At a Meeting of the Managers of the Royal Institution of Great Britain, held on the 8th of May, 1826

Present
Sir George Thos. Staunton, Bt. M.P. in the Chair.
Daniel Moore, Esqr.
John Weyland, Esqr.
Charles Pilgrim, Esqr. junr.
Captn. Chapman,
Edmund Robt. Daniell, Esqr. Secretary.
Captn. Sabine

1. The minutes of the proceedings of the last meeting were read and approved.

2. Read and approved the minutes of the Committee of Accounts of this day; and signed an Order for an Advance to the Asist. Secretary of £150 on account of the Royal Institution, and on account of the Laboratory Fund £30.–

3. In pursuance of a nomination under the Act of Parliament by Neill Malcolm, Esqr. his Son Neill Malcolm, Esqr. of Hanover Square was admitted to the privileges of a Life Subr. to the Royal Institution.

Committee on Lectures. 4. Resolved, That the following Committee be appointed for superintending the arrangement of the lectures.
Daniel Moore, Esqr.
George Moore, Esqr.

May 8.

79

Benjamin Smith, Esqr.
Captn. Sabine
Captn. Chapman,
Benjn. Bond Cabbell, Esqr.
Sir George Thos. Staunton, Bart
Charles Pilgrim junr. Esqr.
Edmund Robt. Daniell, Esqr.
with power to add to their number.

5. Resolved,
That the following Gentlemen be appointed a Committee to arrange the weekly Evening Meetings of the Members.
Committee on Friday Evgs.
Daniel Moore, Esqr.
Captn. Sabine,
Captn. Chapman,
Chas. Pilgrim jun. Esqr.
The Secretary,
Mr. Brande,
Mr. Faraday.

6. Ordered, That the following Newspapers be discontinued – viz.
Newspapers discontinued.
The New Times,
The Representative,
Morning Advertizer,
The Star,
The Globe.

Coals. 7. Ordered, That Messrs. Dickins & Co. supply the Institution with 20 Chaldrons of the best Walls-end Coals – on the terms mentioned in their note 51s. ₽ Chaldron, including all expences.

80

1826

Apparatus lent to Dr. Paris. 8 Read an application from Dr. Paris, in the name of the Royal College of Physicians, for the loan of the following Apparatus for the illustration of lectures to be delivered before them in the next week – viz.
An Apparatus to show the decomposition of Water, & Neutral Salts, by Electricity.
Three Galvanic Troughs.
Resolved, That the above request be granted.

9 The following persons having been duly recommended were admitted as Subscribers to the Royal Institution for the year 1826.
To the Lectures
Mrs. Langford Brooke 19 Upper Brook St.

To Mr. Wesley's Lectures.
Miss Field of Christs Hospital
Mrs. Tindall } of Chelsea
Miss Howe }
Mr. & Mrs. Stokes } of Blackheath
Miss Dolling }
Miss Tunno – Upper Brook St.

Dan Moore

81

May 22.

At a Meeting of the Managers of the Royal Institution of Great Britain, held on the 22d of May, 1826,

Present
Daniel Moore, Esqr. V.P. in the Chair.
The Hon: Wm. Francis Spencer Ponsonby,
George Moore, Esqr.
Charles Pilgrim jun. Esqr.
Captn. Sabine,
Benjamin Bond Cabbell, Esqr.
Edmund Robt. Daniell, Esqr. Secretary.

1. The minutes of the proceedings of the last meeting were read and approved.

2. Read and approved the minutes of the Comme. of Accounts of this day; and signed an Order for an advance to the Asist. Secretary of £100.

3 In pursuance of a nomination under the Act of Parliament by William Blake, Esqr. his Son William John Blake, Esqr. of Christ Church, Oxford, was admitted to the privileges of a Life Subscriber to the Royal Institution.
Life Subscr.

Recommendn. to Patrons of Liby. 4. Resolved, That it is expedient that the Accounts of the Receipts and Disbursements of the Library & Mineralogical Fund be kept distinct from the General Account of the Institution.
That instead of making an allowance towards the Salaries of the Librarian and his Asistant, as heretofore, the Patrons of the Library be recommended

82 1826

to pay the whole of such Salaries, and other expenses of the Library out of the Funds appropriated to its support.– And from time to time to apply to the Managers to advance such sum of money out of the general Funds of the Institution as they may require to enable them to make such payment.

Resolved, That Mr Daniel Moore be requested to communicate the above resolutions to the Patrons of the Library at their next Meeting.

On Life Subsers becoming Members.

5. Resolved, That it would be greatly advantageous to the Institution if the Life Subscribers could be induced to become Members.

That the present payments required to be made by Life Subscribers on becoming Members of the Institution are too high, and that it will be advisable to reduce them.

That it will be expedient to procure an alteration to be made in the Bye laws for reducing the admissn fee and annual payments payable by life Subscribers who have compounded on their admission, to Two guineas for the admission fee, and two guineas for the annual payment, and to allow such Life Subsrs when elected to compound for their annual payments by paying Ten guineas in lieu thereof.

That previously to such alteration being made it will be desirable to ascertain as nearly as possible the number of Life Subscribers who will be disposed to avail themselves of the proposed alterations to become Members.

That the Secretary be requested to write to the

May 22. 83

Life Subscribers individually and to request them to signify to him their assent or dissent to the proposal.

In consequence of its having been suggested by some of the Members, at a General Meeting, that it would be desirable to extend the privilege of personally admitting, one person to the lectures &c, on payment of the sum of one guinea per annum to the Laboratory Fund, to Members elected subsequently to the 4th of Augt 1823; and a recommendation having been made to the Managers to take the subject into consideration, the Managers present proceeded to comply with such recommendation, and upon mature deliberation, Resolved, That they do not see any reason to recommend that the proposed alteration should be made.

New Times restored

6. Ordered, That the New Times be restored.

Request of Dr Harwood.

7. Dr Harwood having deposited in the Museum of the Royal Institution his Collection of Natural History, is desirous of having the Locks of the Cases so altered, as to prevent any person having access to the specimens, except himself.

Resolved, That the locks of the Cases be altered under the direction of Dr Harwood.

8 The following persons having been duly recommended were admitted as Subscribers to the Royal Institn for the year 1826.

Annual Subscribers.

Sir Henry Parnell, Bt Jermyn St

George Evelyn, Esqr 28 Gloucester Place

Revd Richard Darke of Honderton Worcestershire

84 1826.

To Dr Mayo's lectures on Education.

Miss Currie, 38 Wimpole Street

Mr Devey Dorset St

Miss Raikes 43 York Terrace Regents Park

Dan Moore

June 5. 85

At a Meeting of the Managers of the Royal Institution of Great Britain, held on the 5th of June, 1826,

Daniel Moore Esqr V.P. in the Chair.
George Moore, Esqr
Charles Pilgrim, Esqr junr
Captn Sabine,
Benjamin Bond Cabbell, Esqr
Edmund Robt Daniell, Esqr Secretary

1. The minutes of the proceedings of the last meeting were read and approved.

2. Read and approved the minutes of the Committee of Accounts of this day, and signed an Order for payment of the following sums, vizt

Payments.

Wm Thos Brande, Esqr Professor of Chemy for a year's Salary, 1826,	£150. —
Dr Harwood for lectures in ditto	52.. 10 —
The Marquis Spineto – for do	26.. 5 –
Joseph Fincher on account of the Royal Institn	100. —
	328.. 15. —

Presents received.

3. John Ranking, Esqr M.R.I. presented Historical researches on the Wars & Sports of the Mongols & Romans,

Dr Sigmond presented The unnoticed Theories of Servetus,

The London Hibernian Society presented their 20th Ann. Report
Dr. Scudamore presented
Observations on the Stethoscope &c.
An Account of the Geological Specimens from the Coasts of Australia, by Dr. Fitton,
Benjamin Travers, Esqr. F.R.S. presented
An enquiry concerning constitutional irritation,
Sir Charles Long presented
History of the College of Bon-hommes at Ashridge in the County of Buckingham, founded in 1276,— by Edmund Earl of Cornwall.
Dr. Bostocks Elementary system of Physiology, vol. 2d. presented by the Author.
Wirgmans Political & Moral contemplations 2 Copt.

<u>Ordered</u>,
That the thanks of the Managers be returned for these presents.

Comn. from Patrons of Library.

4. The Chairman reported that he had communicated to the Patrons of the Library, the resolutions of the last Board of Managers which they had adopted; and they had requested the Managers to transfer to the account of the Library the sum of £15, which had been paid by them towards the Salary of the Librarian to the 25th of March last.

Dan. Moore

June 19.

At a Meeting of the Managers of the Royal Institution of Great Britain held at the house of the Institution on the 19th of June, 1826

Present
Daniel Moore, Esqr. V.P. <u>in the Chair</u>.
Benjamin Bond Cabbell, Esqr.
Charles Pilgrim junr. Esqr.
Edmund Robt. Daniell, Esqr. <u>Secretary</u>.

1. The minutes of the proceedings of the last meeting were read and approved.

2. Read and approved the minutes of the Committee of Accounts of this day; and signed an Order for payment of the followg. sums, viz.

Josh. Fincher on account of the Royal Institution	£100 —
Ditto on account of the Laboratory Fund	100 —
To Sir Claude Scott Bart. & Co. on account of the Laboratory Fund	15 —

Mr. Harris's request respt. Catalogue referred to Patrons.

3. Read the resolution of the Managers of the 7th of November last, respecting Mr. Harris's request that the Managers would take off his hands 250 Copies of the Catalogue of the Library, which remained unsold – which he offered to them at half-price, namely, 10s. a copy.
<u>Resolved</u>, That as the funds of the Institution

will not at present, allow the Managers to make the proposed purchase, ~~that~~ Mr. Harris's proposal be submitted to the Patrons of the Library at their next meeting.

Annl. Subsr.

4. The following person having been duly proposed was ballotted for and elected an Annual Subscriber for the year, 1826
Wm. H. Tollner Esqr. 43 Cadogan Place.

Dan. Moore

July 3.

At a Meeting of the Managers of the Royal Institution of Great Britain, held on the 3d of July, 1826

Present
Daniel Moore, Esqr. V.P. <u>in the Chair</u>.
George Moore, Esqr.
Charles Pilgrim junr. Esqr.
Captn. Sabine,
Sir S. Bernard Morland, Bt. M.P. <u>Treasurer</u>.
Benjamin Bond Cabbell, Esqr.
Edmund Robt. Daniell, Esqr. <u>Secretary</u>.

1. The minutes of the proceedings of the last meeting were read and approved.

2. Read and approved the minutes of the Committee of Accounts, and signed an Order for an advance to the Assist. Secretary of £300.

<u>Presents received</u>.

3. Saml. Solly, Esqr. M.R.I. presented
The Century of Inventions of the Marquis of Worcester – by Charles Partington,
His Majesty's Commissioners of Public Records presented
Valor Ecclesiasticus, vol. 5,
Foedera – vol. 3 part 1,
The Horticultural Society presented
Transactions of the Society, vol. 6, part 3 –
Report of the Garden Committee.
Society of Antiquaries presented
Archaeologia vol. 20 pt. 2 – and vol. 21 pt. 1.

90 1826.

J. P. Yoy, Esqr. presented
Triple Chess invented by Marinelli

Ordered,
That the thanks of the Managers be returned for these presents.

Adjourned to the 6th of November

Dan Moore

91

July 10.

At a Special Meeting of the Managers of the Royal Institution of Great Britain held on Monday the 10th of July, 1826, under the authority of a Vice President.

Daniel Moore Esqr. V.P. in the Chair.
Sir Claude Scott, Bart.
Captn. Sabine.
Benjamin Bond Cabbell, Esqr.
Charles Pilgrim jun.r Esqr.
Edmund Robt. Daniell, Esqr. Secretary.

Death of Earl of Chichester —

The Secretary stated that in consequence of the death of the Earl of Chichester, the late President of this Institution, he had under the authority of a Vice President, called the present Meeting to take into consideration the necessary steps for the appointment of a successor.

Proposed ... Marq. of Lansd.ne

It was unanimously Resolved, That the Secretary be desired to write to The Marquis of Lansdowne requesting His Lordship to allow himself to be nominated as President of the Royal Institution at the ensuing Election.

Resolved, That the House of the Institution be shut up on the day of the Funeral.

Adjourned to the 22d. July

Dan Moore

92 1826.

July 22.

At an adjourned Special Meeting of the Managers of the Royal Institution of Great Britain, held on the 22d of July, 1826;

Daniel Moore, Esqr. V.P. in the Chair.
George Moore, Esqr.
Benjamin Bond Cabbell, Esqr.
Charles Pilgrim jun.r Esqr.
Benjamin Smith, Esqr.
Edmund Robt. Daniell, Esqr. Secretary.

The Secretary reported that agreeably to the resolution of the last Meeting of Managers, he had written to The Marquis of Lansdowne, and had received from His Lordship the following reply.

Letter from M. of Lansdowne ... declining Presidentship

Eaton Hall, July 17th 1826.

Sir,

I feel exceedingly flattered by the desire which the Managers of the Royal Institution have done me the honour to express through you, that I should fill the situation of President, vacant by the death of the late Earl of Chichester, and I trust they will not attribute my declining to be put in nomination on the present occasion, either to any want of zeal for the objects of the Instit. or any doubt as to its efficacy for their attainment, which the experience of so many years, and the utility of so many acquisitions to Science has indisputably proved; but to a conviction solely that the person who undertakes to preside over its

July 22 93

interests ought to bestow a degree of attention upon their administration which would be in a great measure incompatible with other duties and engagements during the limited period of my residence in London.

I request the favour of you to lay this letter before the Managers, and have the honour to be

Your obedient Servant (signed) Lansdowne

To Edmd. Robt. Daniell Esqr.
Secy. Royl. Institution.

Death of Mr. Stuart Libn.

In consequence of the decease of Mr. Stuart, the Librarian,

Ordered, That Mr. Simons who now acts as Assist. in the Library, be allowed to continue in that situatn. till the Meeting of Managers in November.

Adjourned.

Dan Moore

94

Nov. 6

1826

At a Meeting of the Managers of the Royal Institution of Great Britain, held on the 6th of November 1826,

Daniel Moore, Esqr. V.P. in the Chair
George Moore, Esqr.
Captn. Sabine,
Captn. Chapman,
Charles Pilgrim, jun^r. Esqr.
Benjamin Bond Cabbell, Esqr.
Sir S. Bernard Morland, B^t. M.P. Treas^r.
Edmund Robt. Daniell, Esqr. Secretary

1 The minutes of the proceedings of the last meeting were read and approved.

2 Read and approved the minutes of the Committee of Accounts of this day.

Life Subs^cr. 3 In pursuance of a nomination under the Act of Parliament by William Stodart, Esqr. his Son William Stodart, Esqr. of Golden Square was admitted to the privileges of a Life Subsc^r. to the Royal Institution.

Committee resp^g Lib^n 4 Resolved, That a Committee of the following Gentlemen be appointed to take into consideration the vacancy occasioned in the office of Librarian with power to add to their number, two of whom are to be a quorum – and that a meeting of the Committee be appointed for Monday next at half past Two o'clock –

Daniel Moore, Esqr.

Nov. 6 95

George Moore Esqr.
Captn. Sabine
Captn. Chapman
Charles Pilgrim jun^r. Esqr.
Benjamin Bond Cabbell Esqr.
Edmund Robt. Daniell, Esqr.

5 Resolved, That the Committee of Lectures be summoned to meet on the same day at Three o'clock.

6. The following persons having been duly recommended were admitted as Annual Subscribers for the year, 1827,

Ann^l. Subs^crs.
George Perry Esqr. Terrace House Battersea
Thomas Walker, Esqr. 1 Sackville Street
George Bradley, Esqr 12 John S^t. Berkeley Sq^e.

Adjourned to the 4th. of December

Dan Moore

96

Dec. 4.

1826.

At a Meeting of the Managers of the Royal Institution, held on the 4th. of December, 1826

Present
Daniel Moore, Esqr. V.P. in the Chair.
George Moore, Esqr.
Benjamin Bond Cabbell, Esqr.
Charles Pilgrim jun^r. Esqr.
Edmund Robt. Daniell, Esqr. Secretary.

1. The minutes of the proceedings of the last meeting were read and approved.

2. Read and approved the minutes of the Committee of Accounts of this day, and signed Orders for payment of the following sums; – vizt.

Payments.
To
Joseph Fincher on account of the Royal Institution —— £200 —
Ditto – on account of the Laboratory Fund — } 100 ——

Lamps. 3. Read a letter from M^r. Quarrell in answer to an application made to him by the Secretary, stating that he will undertake to supply and trim any number of Lamps at nine pence each burner; Also to trim, with Oil and Cotton the lamps belonging to the Royal Institution used nightly at 6^d. each burner. When a Man is required to light and attend them, the charge will be 6^d. – extra for each Night.

Dec 4 97

Presents received.

4. J. P. Yosy, Esqr. presented
Triple Chess invented by Merinelli;
John Disney, Esqr. presented
Outlines of a Penal Code,
The Board of Longitude presented
Nautical Almanac for 1829,
The Royal Society presented
Philosophical Transactions parts 1 & 2,
Michael Faraday, Esqr. presented a paper
On the existence of a limit to Vaporization,
James South, Esqr. presented a Paper
On the discordances between the Sun's observed and computed right ascensions,
John Frost, Esqr. presented
An Oration delivered before the Medico Botan^l. Society of London,
His Grace the Duke of Buckingham presented
Rerum Hibernicarum Scriptores,
W^m. Ward, Esqr. presented
A Treatise on Legacies & Bequests of Personal Property.

Ordered,
That the thanks of the Managers be returned for these presents.

Rep^t. of Com^e. on Librarian 5 Read the following report from the Committee appointed to take into consideration the vacancy in the office of Librarian.

The Committee report that having taken into consideration the proposals made by the several Candidates who had applied for the

98 1826.

situation of Librarian, they recommend Mr. S. W. Singer as a very proper person to be appointed to fill that situation, at a Salary of £120 ₩ Annum with Apartments in the House. – And they further recommend that Mr. Simons, who has for some time past assisted in the Library, be continued in the Library at a salary of One guinea per week.

Resolved, That the above report be approved and adopted; – And that the Salary of Mr. Singer, and of Mr. Simons commence on the 1st. of January next.

Library Hours. 6. Resolved, That the Library of the Institutn. be open for the use of the Members and Subscribers, every Evening on which the house is open, till Ten o'Clock from the 1st. of January to the 1st of August, 1827, except on Friday Evenings, when the Members and their Friends are alone entitled to admission.

7. Resolved,

That Mr. Daniel Moore be requested to lay before the Meeting of the Patrons of the Library of this day, the resolutions relating to the appointment of Librarian & Assistant Librarian, together with the above resolution.

Mr Faraday to assist Supt. of House. 8. Resolved, That Mr. Faraday be appointed Assistant to Mr. Brande in the superintendency of the House.

Coals. 9. Ordered, That Messrs Dickins & Co. supply the

Dec 4. 99

Royal Institution with 20 Chaldrons of the best Walls end Coals, agreeably to their estimate, at 49s. per Chaldron, including expences of Metage Shooting &c.

Porter's Salary. 10. Ordered, That the Salary of the Porter be deducted from Mr. Greenwood's Contract, and paid by the Institution after the 31st. Instant.

11. Resolved, That this Meeting do adjourn till after the Meeting of the Patrons

12. The business of the Meeting having been resumed, Mr. Danl. Moore reported that he had communicated to the Meeting of Patrons the resolution relating to the appointment of Librarian, with the Order relating to the opening of the Library in the Evening – which the Patrons had approved of.

Mr Sturt's salary contd. to Sept. 30. Mr. Danl. Moore further reported that the Patrons of the Library had ordered the Salary of the late Mr. Sturt, who died on the 22d. July last, to be paid up to the 30th. of September, – Mrs Sturt being left in very destitute circumstances.

Resolved, That the above communication be approved of.

Addl. Eveng Papers. 13. Ordered, That two additional Evening Newspapers be provided – vizt. the Star, and the Globe.

Adjourned to Monday the 22d. of January 1827.

Danl. Moore

100 1826.

Dec 11

At a Special Meeting of the Managers of the Royal Institution held, under the authority of a Vice President, on Monday the 11th. of December, 1826

Present

Daniel Moore, Esqr. V. P. in the Chair
The Hon: W. F. Spencer Ponsonby M.P.
George Moore, Esqr.
Charles Pilgrim junr. Esqr.
Benjamin Bond Cabbell, Esqr.
Edmund Robt. Daniell, Esqr. Secretary

On Election of President. 1 The Secretary stated that he had called the present meeting, under the authority of a Vice President, for the purpose of taking into consideration the vacancy in the office of President; and for considering of the propriety of shutting up the Library to the end of this Month, for the purpose of making repairs and alterations in the House and Library.

Duke of Somerset proposed. 2. Resolved, That it appears to this Meeting that it would be very desirable that The Duke of Somerset should become the President of the Royal Institution.

3. Resolved therefore, That The Honble. Mr. Ponsonby and Mr. Benjamin Bond Cabbell, be deputed to wait on His Grace to request he would allow his name to be put in nomination to fill up the vacant office of President. – –

Dec. 11. 101

Repairs 4. Resolved, That it is expedient that the Library, and the Hall and Stair-case, should be painted and whitewashed.

5 Ordered, That Mr. Faraday be directed to procure estimates of the expense of the above repairs and alterations.

6 Resolved, That a communication be opened between the Managers Room and the Reading Room.

Closing of Library 7. Resolved, That in order to carry into effect the proposed alterations and repairs, the Library and Reading Room be closed from Saturday the 16th Instant till Monday the 1st. of January next.

Adjourned to Monday the 18th. Instt. at Three o'Clock.

Danl. Moore

102 1826

Dec 18.

At an adjourned Special Meeting of the Managers of the Royal Institution of Great Britain, held on the 18th. of December, 1826

Present

Daniel Moore, Esqr. V.P. in the Chair.
Benjamin Bond Cabbell, Esqr.
Captn. Chapman.
Chas. Pilgrim junr. Esqr.
George Moore, Esqr.
Sir S. Bernard Morland Bart M.P. Treas.
Edmund Robt. Daniell, Esqr. Secretary.

1 *Nomination of Duke of Somerset as President.* Mr. Cabbell reported that he had with the Hon. Mr. Ponsonby waited on the Duke of Somerset, who had most willingly acceded to the proposal made by the Managers, informing the deputation that He should have much pleasure in allowing his name to be put in nomination as President of the Royal Institution.

2. Resolved, That the thanks of this Meeting be presented to
The Hon: Wm. Francis Spencer Ponsonby,
& Benjamin Bond Cabbell, Esqr.
for the very obliging and successful manner in which they have carried into effect the wishes of the Managers.

Ordered,
That the name of His Grace the

103

Dec 18.

Duke of Somerset be inserted in the House List as President of the Royal Institution.

3. *Repairs &c.* Mr. Faraday laid on the table Estimates from Mr. Royce, Mr. Proctor, and Mr. Parker, as to the expense of painting &c. the Hall Staircase and Library.

4. Resolved, That the Painting and Repairs be immediately commenced by Mr. Parker agreeably to his estimate at an expense not exceeding £55.

5. Resolved, That the Newspaper Room and Water Closets be painted; and that Mr. Parker be directed to furnish the Estimate.

6 Resolved, That the Curb-stone and pavement in front of the House be repaired; and that Mr. Faraday be directed to procure Estimates.

7. *Committee on Repairs.* Resolved, That Mr. George Moore and Mr. Daniell be a Committee to superintend and carry into effect the proposed repairs and alterations.

Dan Moore

104 1827

Jan 22.

At an adjourned Meeting of the Managers of the Royal Institution of Great Britain, held on the 22d. of January, 1827.

Daniel Moore, Esqr. V.P. in the Chair.
Captn. Chapman,
George Moore, Esqr.
Charles Pilgrim, junr. Esqr.
Edmund Robt. Daniell, Esqr. Secretary.

1. Read and approved the minutes of the proceedings of the Meeting of Managers of the 4th. of December last, and also the Minutes of the Special Meeting of the 11th. and 18th. December.

2. *Repairs.* Mr. Daniell and Mr. George Moore reported that the repairs and alterations resolved upon at the meetings of the Managers on the 11th and 18th. of December, had been completed.

3 *Gas Lamp.* Mr. Daniell stated that he had ordered a Gas Lamp to be put up at the Street door, and lighted from Sun-set to the hour of closing the house, and that he had ordered the tables in the Library to be covered with green baise, and had purchased Carpets for the Libraries and Managers' Room.
Resolved, That the orders made by Mr. Daniell of the above Articles be approved.

4. *Book-Shelves.* Resolved, That book-shelves be put up at the end of the Newspaper room.

105

Jan 22.

5 Mr. Daniell having presented estimates from Mr. Royce and Mr. Wade of the expense of putting up the Shelves
Resolved, That Mr. Wade be employed to put them up agreeably to his estimate (£13_) And that Mr. Parker be requested to furnish an estimate of the expense of painting such Shelves.

6. *Baize doors for Managers' room.* Ordered, That folding baise doors be placed in the Managers' room; and that estimates be procured by Mr. Faraday.

7. *Regulations Friday Evg.* Resolved, That the following regulations be adopted for the Evening Weekly Meetings of the Members.

The Library is to be opened at half past Eight o'Clock.
All the Members of the Institution have the privilege of introducing their Friends either personally or by proper tickets signed by themselves.
No persons, except Members of the Royal Institutn. or those specially invited by the Board of Managers, can be admitted, unless personally introduced by a Member, or the production of a ticket signed by a Member.
Gentlemen introduced by tickets are requested to deliver them to the Porter in attendance.
Members are requested to insert their names, and those of any Friends introduced by them in a book which will be kept open for that purpose.
Members may have on application to the Assistant Secretary an unlimited supply of tickets

8. Resolved, That M^r. Daniel Moore, M^r. Benjn. Bond Cabbell, and Captn. Chapman, be the presiding Managers at the next three Evening Meetings, in a rotation.

Lectures. 9. Read the following report from the Committee of Lectures.

Mr Reinagle. The Committee report that they had engaged R. R. Reinagle, Esqr. R. A. to deliver a Course of Lectures on Painting at the rate of Ten guineas a lectures.

Payment of M^r F's lectures. 10. Resolved, That M^r. Faraday be paid for the lectures to be delivered by him in the Theatre of the Royal Institution this Season, at the rate of six guineas for each lecture.

Payment of Mr Fincher as Collector. 11. M^r. Fincher having performed the duties of Collector during the last year.
Resolved, That taking into consideration his services to the Royal Institution, he be allowed the same sum as the late Collector – namely Thirty guineas per annum.

T J Pearsall's Sal^y raised. 12. Resolved, That the salary of Thos. J. Pearsall Assistant in the Laboratory be increased from 12 to 15^s. per week – to commence on the 1st of Jan^y.

13. The following persons having been duly recommended were admitted as Subscribers to the Royal Institution for the year 1827.

Jan 22

Annual Subscribers.
James Evans, Esqr. of Kilburn
Robert J. Pilgrim, Esqr. of Kensington
Bazett D. Colvin, Esqr. Esqr. 55 Manchester S^t.
Robert Cutlar Fergusson, Esqr. M.P. 17 Great Cumberland S^t.

To the Lectures.
Mrs Fructuozo } 16 Bloomsbury Square
Miss Ince }

Danl Moore

Feb 5.

At a Meeting of the Managers of the Royal Institution of Great Britain held on the 5th of February, 1827

Present
Daniel Moore, Esqr. V. P. in the Chair.
George Moore, Esqr.
Captn. Chapman,
Benjamin Bond Cabbell, Esqr.
Charles Pilgrim jun^r. Esqr.
Sir S. Bernard Morland, B^t. M.P. Treas^r.
Edmund Robt. Daniell, Esqr. Secretary

1. The minutes of the proceedings of the last meeting were read and approved.

2. Read and approved the minutes of the Committee of Accounts of this day, and signed an Order for an advance to the Assistant Secretary of £250. —

3 Audited the Treasurer's Account for the year 1826.
and
Audited the Treasurer's Account of the Laboratory Fund for the year 1826.

Presents received.

4 Francis Bailey, Esqr. presented
Astronomical Tables &^c.
The Astronomical Society presented
Memoirs of the Society vol 2, part 2.

Feb 5.

D^r. Roget presented
A Lecture on Human and Comparative Physiology.
The Geological Society presented
Transactions of the Society part 2 of vol^e 2.
Mr Ackerman presented
Repository of Arts for 1826. 12 n^os.
Mr Gill presented
Technical Repository for 1826 – 12 n^os.
George Stavely Smith, Esqr. presented
Encyclopedia Metropolitana 19 parts
D^r. Nicholl presented
Champollion système Hieroglyphique des anciens Egyptiens
M^r. Singer presented
Cavendish's Life of Wolsey – 2 vol^s.
M^r. Upcott presented
English Topography – 3 vol^s.
M^r. Faraday presented
Sir Thomas More's Utopia, 1551, 12°
Branca Le Machine. 4to
The Society of Arts presented
Transactions of the Society, vol^e 44
Louis Hayes Petit, Esqr. presented
Ottley's Dictionary of Chemistry.
Mr C. Moreau presented
Records of the British Navy
Thomas Alcock, Esqr. presented
An Essay on the use of the Chlorurets as disinfecting agents.
Ordered, That the thanks of the Managers be returned for these presents.

5 *Folding doors to Managers' room* — Read estimates from Mr Royce & Mr Wade of the expense of making folding baise doors in the Managers' room.

<u>Ordered</u>, That Mr Royce be employed to make them agreeably to his estimate, £5 —

6. *Repairs* — Read an estimate from Mr Weston Slater of the expense of making good the slates that have been broken in clearing away the Snow from the roof and gutters of the Institution.

<u>Ordered</u>, That Mr Faraday be directed to procure other estimates.

7. *Regulations respg Tradesmen's Accts* — The Secretary laid before the Meeting, the following regulations respecting the mode of keeping the Tradesmen's Accounts.

<u>Ordered</u>, That no Order be given for the performance of any work in the Royal Institution, and that no Goods be purchased for the use of the Royal Institution, but by the proper Officer of the department in which such Work is to be performed, or for which such Goods are to be purchased – viz. In the Secretary's Department, by the Secretary, or Assistant Secretary. In the Library, by the Librarian. In the Laboratory, by the Professor of Chemistry, or the Director of the Laboratory. In the House, by the Superintendt of the House or his Assistant.

That every Workman employed about the Royal Institution be required within a week after the completion of every job to give in an account containing the particulars of the Work done, and

of his charges for the same. And that all Tradesmen furnishing any Goods to the Institution be required to send in a Bill of Parcels with all such Goods furnished by him with the prices thereof.

That no Workman or Tradesman be employed in or by the Royal Institution who shall omit or refuse to comply with the above requisition.

That all Workmen's Accounts and Tradesmen's bills be looked over and examined by the Officer of the Department, and if correct signed by him and delivered to the Board of Managers at their next Meeting.

<u>Resolved</u>, That the above regulations be adopted.

Presents. 8 Mr Singer stated that George Staveley Smith, Esqr had presented to the Library two Library Chairs, and two Library tables:

<u>Ordered</u>, That the special thanks of the Managers be returned for this present.

9. Mr Magrath presented a Cast of the Athenaeum seal.

<u>Ordered</u>, That the thanks of the Managers be returned for this present.

10. The following persons having been duly recommended were admitted as Subscribers to the Royal Institution for the year 1827.

<u>Annual Subscribers</u>

Henry H. Penny jun: Esqr Phillimore Place Kensington

Hugh Rose, Esqr 71 Jermyn Street
Captn Mathew Taylor of the Bengal Cavalry
Lt Col Thos Broughton Park Crescent, Portland Place
Revd I. Smirnove 32 Welbeck St
Chas Montagu, Esqr
Wm Fredk Montagu, Esqr 25 Bedford Sqr
David Taylor, Esqr Colonnade Hotel Charles St

<u>To the Lectures.</u>

Mrs & Miss Clark } Savile Row
Miss Lucy & Miss C. Clark }
Miss Oakes & Miss S. Oakes Mitcham
Mr & Miss Barretto 2 Upper Portland Place
Mrs Walter Campbell 20 Grosvenor Place
Mrs Grey

Adjourned to Monday next at Three o'clock.

Confirmed this twelfth day of February 1827 Somerset Pres.

At a Meeting of the Managers of the Royal Institution of Great Britain held on the 12th of February, 1827

Present

The Duke of Somerset, President, <u>in the Chair</u>
Daniel Moore, Esqr V.P.
Sir S. Bernard Morland, Bart M.P. <u>Treasr</u>
Benjamin Bond Cabbell, Esqr
George Moore, Esqr
Charles Pilgrim jun: Esqr
Edmund Robt Daniell, Esqr <u>Secretary</u>

1. The minutes of the proceedings of the last meeting were read and approved.

2. The President signified to the Meeting that he had appointed the following Managers to be Vice Presidents for the present year, viz.

Vice Presidents
The Marquis of Lansdowne,
Viscount Palmerston,
Sir Claude Scott, Bart.
Sir George Thos Staunton, Bart.
Daniel Moore, Esqr

3 *Life Subs &c* — In pursuance of a nomination under the Act of Parliament by John Jacob, Esqr M.R.I. his Son Mr Charles Jacob of Greenwich was admitted to the privileges of a Life Subr to the Royal Institution.

4 Mr Brande having stated to the Managers that he was about to enter into an agreement

114 1827.

On Mr Brande's agreement resp. Journal

with Mr Colburn to publish a new series of the Journal of Science and the Arts, edited at the Royal Institution.

Resolved, That Mr Brande should after he has made his agreement with Mr Colburn, lay a copy of such agreement before the Managers at their next Meeting.

Newsr room, &c.

5. *Resolved*, That the Newspaper Room be in future used as a Reading Room; and that the Reading Room and Committee Room adjoining it, be used as a Newspaper Room, except when the Committee Room shall be required for the Meetings of the Managers or Visitors.

Repairs.

6. Mr Faraday laid before the meeting an estimate from Mr Bird, Bricklayer, of the expense of repairing the Slating that had been broken in the Roofs and Gutters, amounting to to £2.

Ordered, That Mr Bird be directed to repair the Slates at an expense not exceeding £2.

7. The following persons having been duly recommended were admitted as Subscribers to the Royal Institution for the present year

Annual Subscribers.

Walter Campbell, Esqr. 20 Grosvenor Place
John Irving junr. Esqr. Richmond Terrace, Whitehall.
John Story Penleaze, Esqr. 104 Piccadilly.

Feb. 12. 115

To the Lectures.

Miss Andrews, 7 Arlington Street
Mrs Paterson } 1 Upper Portland Place
Miss Barker }
Miss Bartram } 27 Princes St. Leicester Sq.
Miss E. Bartram }
Miss Sturch York Terrace, Regents Park

To Mr Brande's Lectures.

Mr George Murray of Kilburn
The Misses Rhodes of Muswell hill
Mr & Mrs Drinkwater Fitzroy Square.

Adjourned to the 5th of March.

Confirmed this 5th of March 1827
Somerset, Pres.

116 1827.

Mar. 5.

At a Meeting of the Managers of the Royal Institution of Great Britain held on the 5th of March, 1827

Present

His Grace the Duke of Somerset *President in the Chair*
Daniel Moore, Esqr. V.P.
George Moore, Esqr.
Captn. Chapman,
Benjamin Bond Cabbell, Esqr.
Charles Pilgrim, junr. Esqr.
Sir S. Bernard Morland, Esqr. M.P. *Treasr.*
Edmund Robt. Daniell, Esqr. *Secretary*.

1 The minutes of the proceedings of the last meeting were read and approved.

2. Read and approved the minutes of the Committee of Accounts of this day, and signed an Advance to the Assistant Secretary of £150.

Life Subsr.

3. In pursuance of a nomination under the Act of Parliament by William Simons, Esqr. his Grandson – Thomas Simons Watson, Esqr. of Berners Street, was admitted to the privileges of a Life Subscriber to the Royal Institution.

Presents received.

4. The Royal Society of Literature presented
Transactions of the Society, vol. 1 pt. 1.
Louis Hayes Petit, Esqr. presented
Essay on the Phonetic system of Hieroglyphics,

Mar. 5 117

Hieroglyphics collected by the Egyptian Society
60 Lithographic plates in Atlas folio.
Thomas James Mathias, Esqr. presented
Spencers Legend of the Red Cross Knight in Italian Verse
Il Cavalier della Croce Rossa tradotto dal Inglese di Spenser.
His Italian Lyric Poetry.
Le Sue Poesie Liriche.
John Ranking, Esqr. presented
Ayeen Akbary, or the institutes of the Emperor Akbar 2 vols.
Haym Tesoro Britannico, 2 vols; 4to.
Avril's Travels into China,
The Board of Longitude presented
Mayers' Tables, folio,
Edward Magrath, Esqr. presented
Memoirs of Antonio Canova.
Arthur de Capell Brooke, Esqr. F.R.S. presented
Winter Sketches in Lapland, folio.
W. T. Brande, Esqr. presented
Manual of Chemistry, 3 vols. 8vo. –

Ordered,
That the thanks of the Managers be returned for these presents.

Journal.

5 Mr Brande reported that he had not yet made his agreement with Mr Colburn, but that he expected he should be able to do so by the next Meeting of the Managers.

on Comm. for Weekly Evgs.

6. *Resolved*, That Mr George Moore's name be added to the Committee for arranging the weekly Evening Meetings of the Members; – And that Meetings of the Committee be held on Thursdays at 4 o'Clock.

118 1827.

[Request of Mr Faraday respt dedication.] 7. Mr Faraday having requested permission to dedicate his work on Chemistry which he was about publishing, to His Grace the Duke of Somerset the President – the Vice Presidents – Managers and Members of the Royal Instit.n Resolved, That Mr Faraday's request be acceded to – And as a mark of the approbation of the Managers they further resolve that he be permitted to use the arms of the Institution in the title page of his Work.

8. The following persons having been duly recommended were admitted as Subscribers to the Royal Institution for the present year.

Annual Subscribers.
Hon: Captn Abbott, R.N. 18 Spring Gardens
Captn Wm Rooke 77 Gloucester Place
To the Lectures.
Mrs Nicholl 12 Old Burlington Street

Subscribers to Mr Webster's Lectures.
Mr Herbert Langham 36 Lincolns Inn Fields.
Major Mitchell Thistle Grove
Subscribers to Mr Millington's Lectures.
Mr Dowdeswell Park Place St James's
Mr R. C. Nicholl 1 New Court Temple.

Adjourned to Monday the 19th Instant at Three o'Clock.

Danl Moore

119

Mar 19. At a Meeting of the Managers of the Royal Institution of Great Britain held on Monday the 19th of March, 1827,

Present
Daniel Moore, Esqr V.P. in the Chair.
Benjamin Bond Cabbell, Esqr
Charles Pilgrim junr Esqr
George Moore, Esqr
Captn Chapman
Edmund Robt Daniell, Esqr Secretary.

1. The minutes of the proceedings of the last meeting were read and approved.

[Under Lease sealed] 2. Notice having been sent to the Managers agreeably to the Bye laws, the Seal of the Corporation was affixed to a Lease from the Members of the Royal Institution to Mr Paynter of a piece of Ground and part of a Messuage No 16 in New Grafton Street for 60 years & 3/4 from Michas 1823; renewable every 14 years.
It was Countersigned by
Danl Moore – Chairman
Chas Pilgrim junr
George Moore, Esqr
Benjamin Bond Cabbell
Edmund R. Daniell Esq

[on journal.] 3. Mr Brande stated that the agreement had been prepared, but had not been signed by Mr Colburn, but that he hoped to be able to lay it before the next meeting.

120 1827.

[London Gazette supplied gratis.] 4. Read a letter from G. R. Clarke, Esqr of the Gazette Office stating that he is authorized to supply the Royal Institution with the London Gazettes, free of expense, from the beginning of next month.
Resolved, That the thanks of the Managers be returned to Mr Clarke for his kindness in procuring the London Gazettes for the Royal Institution.

[Petition for Macadamizing Albe Street] 5 Read a Petition signed by several of the Inhabitants of Albemarle Street requesting the Paving Committee of this Parish to have the Street Macadamized.
Resolved, That the Secretary be requested to sign the Petition on the part of the Managers.

[Porters' Clothes.] 6 The Secretary having stated that the Porters wished not to wear Livery Clothes.
Resolved, That a plain blue Coat & Waistcoat be provided for each of them.

[Loan of Organ.] 7 Mr Chas Holford, M.R.I. having sent an Organ to be placed in the Lecture room for the use of Mr Wesley, during the time of his Course of lectures on Music.
Resolved, That the thanks of the Managers be returned to Mr Holford for his obliging loan of the Organ.

[Payment for Periodicals by Patrons.] 8 Resolved, That the Patrons of the Library be requested in future to take the superintendence

121

Mar 19. of the Periodical publications, and to defray the expenses out of the Funds appropriated to the Library.

[Price of Catalogues.] 9. Mr Daniel Moore Secretary to the Patrons of the Library reported that the Patrons had agreed to dispose of the Catalogues of the Library at 12s each copy in boards.
Ordered, That notice of the same be put up in the Library and Reading Rooms.

10. The following persons having been duly recommended were admitted as Subscribers to the Royal Institution for the year 1827

Annual Subscribers.
John Hardwicke, Esqr Albany
J. F. Sampayo, Esqr 37 Albemarle Street
John Francis Gil, Esqr } 52 Bruton St.
Eustace Joseph Torres Esqr }
Wm Fraser, Esqr of the Parliament St Chambers

To the Lectures
Miss Browne No 2 Portland Place.

Confirmed this 26th day of March 1827.
Somerset
Pres.

Mar. 26

At a Meeting of the Managers of the Royal Institution of Great Britain held on the 26th of March, 1827

1827.

Present

His Grace The Duke of Somerset, Presidt. in the Chair.
Daniel Moore, Esqr. V.P.
George Moore, Esqr.
Benjamin Bond Cabbell, Esqr.
Charles Pilgrim, Esqr. junr.
Sir S. Bernard Morland, Bt. M.P. Treasurer
Edmund Robt. Daniell, Esqr. Secretary

1. The minutes of the proceedings of the last meeting were read and approved.

Contract between Messrs. Brande & Colburn resp. Journal.

2. Mr. Professor Brande produced the Contract entered into between himself and Mr. Colburn dated respecting the publication of the Journal.
Resolved, That the Contract be approved of.

3. Mr. Brande reported that Mr. Colburn had agreed to furnish the Members of the Royal Institution with the numbers of the Journal at the publication price on their application to Mr. Colburn.
Resolved,
That notice of such agreement be placed in the Libraries and Reading Rooms.

Ballotting Boxes.

4. Resolved, That in future two ballotting boxes be used at the General Meetings.

Confirmed this 2nd day of April 1827
Somerset Pres.

April 2.

At a Meeting of the Managers of the Royal Institution of Great Britain held on the 2nd of April, 1827,

Present

His Grace The Duke of Somerset Pres.t in the Chair
Daniel Moore, Esqr. V.P.
George Moore, Esqr.
Benjamin Bond Cabbell, Esqr.
Charles Pilgrim junr. Esqr.
Sir Claude Scott, Bart. V.P.
Sir S. Bernard Morland, Bt. M.P. Treasr.
Captn. Chapman,
Edmd. Robt. Daniell, Esqr. Secretary

1. The minutes of the proceedings of the last meeting were read and approved.

Presents received.

2. Arthur de Capell Brooke Esqr. presented
Winter Sketches in Lapland,
Wm. Brockendon, Esqr. presented
The Passes of the Alps, No. 1.
Lt. Genl. Wm. Thornton presented
Prose e Rime de due Buonaccorse de Montemagno.
Ordered, That the thanks of the Managers be returned for these presents.

3. Read and approved the minutes of the Commt. of Accounts of this day.

Investments.

4. Ordered, That the sum of One hundred and fifty guineas be invested on account of the Laboratory Fund, being the Contribution from

1827.

Daniel Moore, Esqr. as Patron of the Laboratory and the sum of Fifty guineas paid out of the Composition of five new Members.

Resoln. of Visitors.

5 Read the following resolution of the Visitors entered in their proceedings of the 26th of March; – viz.t

Repairs &c.

That the house is generally in good order, though a few repairs are wanting to the roof and Chimnies, and to Mr. Greenwood's room, & one of Mr. Faraday's; but the Visitors call the attention of the Managers particularly to the state of the flue which crosses the Model room which appears to be such as threatens danger to the building.
Resolved,
That the Surveyor of the Sun Fire Office be requested to examine into the Flue passing over the Model room.

Come. of Accts.

6. Resolved, That a Committee be appointed of the following Gentlemen – namely
Captn. Chapman
George Moore, Esqr.
& the Secretary.
to investigate the Accounts of the Institution with the view of ascertaining the amount of the Debt of the Royal Institution – And from what period it has existed.

7. The following persons having been duly recommended were admitted as Subscribers to the Royal Institution

April 2

for the year, 1827.

Annual Subscribers
Donald Maclean, Esqr. Cadogan Place
Henry Edwd. Thornton, Esqr. Collen St. Hanover Sq.

To Mr. Wesley's lectures.
Miss Field
Miss Poynder
Miss Montague of Bedford Square
Miss Reece – Regent Street
Miss Bedingfeld
Mrs. P. Milford 64 Upper Berkeley Street
Miss Swinnove
Revd. Thos. Fuller 59 Montagu Square
Miss Davies, of Blackheath
Mr. & Miss Farrer of Bedford Square

Adjourned to Saturday next at 4 oClock.

Confirmed this 7th day of April 1827
Somerset
Pres

April 7. 1827.

At a Meeting of the Managers of the Royal Institution of Great Britain, held on the 7th of April, 1827

Present

His Grace The Duke of Somerset Presid.t in the Chair
Daniel Moore, Esq.r V.P.
Hon: W.m F. Spencer Ponsonby, M.P.
Benjamin Bond Cabbell, Esq.r
Sir S. Bernard Morland, Bart. M.P. Treas.r
Capt.n Chapman
George Moore, Esq.r
Charles Pilgrim, jun.r Esq.r
Edmund Robert Daniell, Esq.r Secretary.

1. The minutes of the proceedings of the last meeting were read and approved.

2. Read a letter from A. B. Vallé, Esq.r requesting an admission to the Royal Institution, for Doctor Volpas – Physician to the King of Naples. *Hon. Admission.*

Resolved,

That the Secretary be requested to send D.r Volpas the usual letter of admission.

3. The Committee appointed at the last meeting to investigate the Accounts with the view of ascertaining the amount of the Debt of the Royal Institution, and from what period it has existed, presented the following report, which was read by the Secretary. *Rep.t of Com.e resp.g Debt.*

In performing the duty entrusted to them the Committee have thought it right to confine their investigation to the period embraced by the Annual

reports of the Visitors appointed under the authority of the Act of Parliament. These reports commence from the year 1809, and the Committee have ascertained that at the beginning of that year, the Debt due by the Royal Institution amounted to £3490. 8. 9 *Debt.. 1809.* The annexed Account shews the Annual Receipts and Expenditure of the Institution, and the consequent variations which have taken place in the amount of the Debt down to the end of the year 1823, from which time a new pecuniary arrangement took place, in consequence of the suggestions of a Committee of Members, appointed to take the state of the Royal Institution into consideration. From this account it appears that at no period between the years 1809 and 1823 inclusive was the Royal Institution free from debt, and that although the amount of it varied from time to time, yet that at the end of the year 1823, it was very nearly the same as it was in the year 1809. *— in 1823.* – In all these years the sums received from the compositions of Members were invariably considered as part of the Annual Income of the Royal Institution, and applied accordingly, consequently no permanent Fund was created which could be applied to the liquidation of the Debt, or to increase the Capital of the Establishment. *Acc.ts from 1809–23 satisfactory.* The result however of the Account up to the period above mentioned is so far satisfactory as it shews that although the receipts of the Royal Institution were subject to fluctuations (as those of every Establishment which depends for its support chiefly upon popular favour must be) they were upon the average of fourteen years more than sufficient to meet its' Expenditure.

1827.

At the close of the year 1823, the debt of the Royal Institution amounted to the sum of £3259. 7. 2. *Debt in 1823.* – In the course of that year several Members of the Institution most liberally combined to advance a very large sum of money by way of loan without Interest to be applied in liquidation of the debt. The sum so advanced amounted to £3600, which exceeded the debt by the sum of £340. 12. 10. *Loan without interest.* In the same year (1823,) a material alteration took place in the Financial situation of the Institution by an alteration in the Bye laws, made at the suggestion of the Committee before referred to, by which the admission fees and compositions of Members were directed to be funded, and a separate Fund was formed for the support of the Laboratory. *Investment of Compositions & Adm.n Fees.* In pursuance of these Bye laws the Admission fees and Compositions of Members instead of being applied in payment of the Current expenses, as had been before invariably done, were during the years 1824 and 1825, invested in the purchase of Stock. The consequence of this measure was that during the years, 1824 and 1825, an apparent increase of the Debt took place, as will appear from the following statement viz.t – *Consequence of Investments.*

1823	– Amount of the Debt at the close of the year	£3259. 7. 2
	Voluntary Loan	3600. –
	Balance in favour of the Institution	340. 12. 10
1824	Income	2031. 12. 3
		2372. 5. 1
	Expenditure	2543. 11
	Deficiency	171. 5. 11
1825	Expenditure	2272. 19. 3
		2444. 5. 4
	Income	1702. 17. 2
		741. 8. 2

So that at the end of the year 1825 there was an apparent debt against the Institution of £741. 8. 2, beyond the amount of the voluntary loan. The effect however of the alterations in the Bye Laws above alluded to, having been to withdraw from the Income of the Institution, the whole amount of the Compositions and admission fees, which had before always been considered as Income, and to invest them in the purchase of Stock, it is necessary that in taking an account of the debt of the Institution, these sums should be included – They are as follow.

1824	– Admission fees and Compositions invested	£267. 15 –
1825.	Ditto	231. –
		498. 15. –

This sum being deducted from the debt of £741. 8. 2 above stated, reduces it to the sum of £242. 13. 2, thereby making the whole amount of the debt due to the Royal Institution at the end of the year, 1825, including the voluntary loan, £3842. 13. 2 *Debt in 1825.* In order however to arrive at the exact situation of the Finances of the Royal Institution, it is necessary to deduct from the debt as above stated, the value of a renewal fine of £240 which was paid by the Institut.n to the City of London on the renewal of their Lease, and which ought to have been repaid by the Tenant of one of the Houses held under the Institution but which was omitted to be done. *Non-payment of Renewal Fine by Tenant.* The effect of this omission was an apparent increase of the debt of the Institution, but in reality it was a very advantageous investment of the sum in question, and it cannot now be estimated at less than £300, which being deducted from

from the debt will reduce it to £3542 . 13 . 2. It also appears by the Visitors' report of last year that there was a very considerable sum of money due to the Institution for arrears from Members, a very small part only of these arrears have since been received, and, although the Committee cannot flatter themselves that any sum nearly approaching the whole amount of the remainder will ever be recovered, they think they may with confidence estimate those which are available at not less than £100, which ought also to be set off against the debt which will be thereby still further reduced to £3442 . 13 . 2.

Arrears.

In the statement which they have made the Committee have not included the amount of Mr. Fuller's donation of £1000, which has been invested in the purchase of £1250 . 3 per Cent Consols, nor the amount of the Laboratory Fund which amounted at the end of the year 1825 to £335 . 19 . 10 – three per Cent Consols, Of course the value of these sums if set off against the debt would very materially diminish its amount.

The Committee cannot but think that the conclusion to be drawn from the statement which they have made is highly satisfactory since it shews that at the close of the year, 1825, the Debt due by the Royal Institution was not greater than it was at the time when its existence as a public Body may fairly be said to have commenced, and that for a period of sixteen years its receipts and payments have nearly balanced each other. It certainly would have been desirable that in the course of that time something should have been

Debt about the same after 16 years.

done towards the discharge of the Incumbrance – But when it is considered how much has been done by the Institution during the period referred to, what the Establishment is which it is required to maintain, and how much its income depends upon sources which must always be fluctuating and uncertain they think the investigation which has been made must in its results prove most cheering to the friends of the Establishment since it shews that notwithstanding these disadvantages the Royal Institution has been enabled to maintain itself for so long a period without any increase of the debt which was incurred at its first formation.

The Account before the Board of Visitors will shew that during the last year the equilibrium which has been maintained between the Receipts and Expenditure of the Institution has not been disturbed, and the Committee cannot but express the hope they feel themselves justified in entertaining from the great success which has during this season attended the exertions of the Managers that when the Accounts of the present year shall be laid before the Visitors it will appear that the balance has very considerably inclined in favour of the Institution.

An Account shewing the increase and diminution of the debt of the Royal Institution from the year 1809 being the year from which the Accounts of the Visitors appointed under the Act of Parliament commence to the year, 1823, both inclusive.

Account of Debt

Debt due at the commencement	3490 . 8 . 9	Debt 1809
Expenditure	6652 . 14 . 4	
	10143 . 3 . 1	
Receipts	6736 . 18 . 10	
	3406 . 4 . 3	Debt 1810
Expenditure	4181 . 19 . 2	
	7588 . 3 . 5	
Receipts	4728 . 18 . 1	
	2859 . 5 . 4	Debt 1811
Expenditure	3248 . 2 . 9	
	6107 . 8 . 1	
Receipts	4321 . 12 . 8	
	1785 . 15 . 5	Debt 1812
Expenditure	3277 . 13 . –	
	5063 . 8 . 5	
Receipts	3330 . 15 . –	
	1732 . 13 . 5	Debt 1813
Expenditure	3200 . 12 . 8	
	4933 . 6 . 1	
Receipts	3105 . 16 . –	
	1827 . 10 . 1	Debt 1814
Expenditure	3926 . 8 . 2	
	5753 . 18 . 3	
Receipts	4413 . – . –	
	1340 . 18 . 3	Debt 1815
Expenditure	3357 . 8 . 1	
	4698 . 6 . 4	
Receipts	2480 . 12 . –	
	2217 . 14 . 4	Debt 1816
Expenditure	2905 . 5 . 11	
	5123 . 0 . 3	
Receipts	2858 . 0 . 4	
	2264 . 19 . 11	Debt 1817

Bro forward	2264 . 19 . 11	
Expenditure	2641 . 19 . 8	
	4906 . 19 . 7	
Receipts	2775 . 8 . 4	
	2131 . 11 . 3	Debt 1818
Expenditure	2802 . 11 . 5	
	4934 . 2 . 8	
Receipts	3446 . 0 . 4	
	1488 . 2 . 4	Debt 1819
Expenditure	2867 . 10 . 5	
	4355 . 12 . 9	
Receipts	3194 . 4 . 8	
	1161 . 8 . 1	Debt 1820
Expenditure	3179 . 8 . 4	
	4340 . 16 . 5	
Receipts	2659 . 1 . 7	
	1681 . 14 . 10	Debt 1821
Expenditure	2924 . 2 . 4	
	4605 . 17 . 2	
Receipts	2650 . 18 . –	
	1954 . 19 . 2	Debt 1822
Expenditure	2808 . 7 . 8	
	4763 . 6 . 10	
Receipts	2367 . 16 . 8	
	2395 . 10 . 2	Debt 1823
Expenditure	3421 . 11 . –	
	5817 . 1 . 2	
Receipts	2557 . 14 . –	
	3259 . 7 . 2	Debt at the end of the year 1823.

Resolved, That the above report be approved of, and that it be laid before the Board

of Visitors at their Meeting on Monday next, accompanied by the following statement.

Infringement of Bye Laws in not investing — Reasons. The Managers of the Royal Institution in submitting the Accounts of the Institution to the inspection of the Visitors feel it their duty to state that in order to meet the current expenses of the year, they have ventured to depart from the Article in the Bye laws, by which it is directed that the Compositions of Members shall be invested in the purchase of Stock in the Public Funds.

In taking this step the Managers have been actuated by the conviction that it is really more to the interest of the Institution that this money should have been thus applied rather than in conformity with the Bye laws in question, because money funded during the last year would have produced an interest very little exceeding 3 and ½ ₽ Cent, whilst that which must have been borrowed would have borne an interest of 5 ₽ Cent. When therefore they call the attention of the Visitors to this fact, and when they farther call their attention to the present state of the House, the Laboratory and the Museum and beg them to compare it with what it was last year, they hope they do not expect too much in trusting that the Visitors will not only approve of the measure but recommend to the General Meeting to give this departure from the Bye laws their sanction.

Enquiry into Accounts. The Managers have been induced in consequence of the reports which at the commencement of the last year were widely and industriously circulated, that the Institution was in an embarrassed state, and that

it would shortly be necessary to break up the Establishment, and to close its doors, to cause an enquiry to be made into the Accounts of the Royal Institution, with the view of satisfying themselves whether such injurious reports rested on any substantial foundation, they accordingly appointed a select Committee of their own Members to investigate the Accounts of the Royal Instn for the purpose of ascertaining the present amount of the debt, and from what period it has existed.

The report of that Committee the Managers have now the honour of laying before the Visitors, in the full confidence that it will prove as satisfactory to them as it has been to the Board of Managers, since it shews that the Debt of the Institution is not greater than it was when its existence as a Public Body commenced — And that notwithstanding the fluctuations which have taken place in its concerns, its receipts have for a period of sixteen years equalled its expenditure.

On Subsns to Labry Fund. The Managers beg to add for the information of the Board of Visitors that in pursuance of recommendations contained in their last report, they caused a circular letter to be sent to all the Members of the Institution elected previously to the 4th of August 1823, who had not already expressed their intention of contributing the annual sum of one guinea to the support of the Laboratory requesting them to signify to the Secretary whether they would consent to be charged with that sum or not. The Managers regret to state that to nearly 300 letters which were so sent, only nineteen answers have been received - The names of those who have become

Contributors to the Laboratory Fund since the last Annual Meeting will be laid before the Board of Visitors.

Keys. 6. The box containing the Seal of the Corporation with four keys, was laid on the table, and Counterpart Keys were delivered to the President and the Treasurer, and to Daniel Moore Esqre and the Secretary.

7. The Secretary reported that he had seen the Surveyor of the Sun Fire Office who was perfectly satisfied with the Flue as it now stood.

Resolved, That Mr Feetham be directed to examine the Flue.

Adjourned to Monday the 23d Instant at Three o'Clock.

Confirmed this 23rd day of April 1827
Somerset
Pres.

April 23.

At a Meeting of the Managers of the Royal Institution of Great Britain, held on the 23d of April, 1827

Present

His Grace The Duke of Somerset Presᵗ in the Chair
Daniel Moore, Esqr V.P.
George Moore, Esqr
Benjamin Bond Cabbell, Esqr
Captn Chapman
Charles Pilgrim junr Esqr
Sir S. Bernard Morland, Bart. M.P. Treasr
Edmund R. Daniell, Esqr Secretary.

1. The minutes of the proceedings of the last meeting were read and approved.

Life Subscriber. 2. In pursuance of a nomination under the Act of Parliament by His Grace The Duke of Somerset, his Son Edward Adolphus Lord Seymour was admitted to the privileges of a Life Subsr to the Royal Institution.

Mr Faraday allowed to take Pupil 3. Mr Faraday having requested permission of the Managers to be allowed to take a Pupil in the Laboratory.

Resolved, That Mr Faraday's request be complied with. —

4. Presents received

Major Genl Hardwick presented to the Museum, numerous and various Specimens in Natural History.

138 1827.

Michael Shepley, Esq.r presented to the Museum – a Patagonian Penguin stuffed.

The Zoological Society – presented a Porcupine Fish.

Lt. Col. Combs – presented to the Museum – a Poisoned Arrow and Insects.

Ordered, That the thanks of the Managers be returned for these presents.

Mummy of Cat. T. L. Donaldson, Esq.r presented The mummy of a Cat.

Resolved, That the thanks of the Managers be returned to Mr. Donaldson for the above present; and that Dr. Granville be requested to examine it.

5. *Ordered*, That the House lists be filled up as follows.

President – The Duke of Somerset –

Treasurer – Sir T. Bernard Morland. Bt. M.P.

Secretary – Edmund Robt. Daniell, Esq.r

Managers

Sir Thos. Dyke Acland, Bart. M.P.

Decimus Burton, Esq.r

Benjamin Bond Cabbell, Esq.r

Captn. John James Chapman.

Henry Thos. Colebrooke, Esq.r

Hudson Gurney, Esq.r M.P.

April 23. 139

James Heath Leigh, Esq.r

Daniel Moore, Esq.r

George Moore, Esq.r

Sir Oswald Mosley, Bart.

Whitlock Nicholl, M.D.

Charles Pilgrim junr. Esq.r

The Hon: Wm. Francis Spencer Ponsonby M.P.

Sir Claude Scott, Bart.

Samuel Ware, Esq.r

Visitors

Henry James Barchard, Esq.r

John Ivatt Briscoe, Esq.r

James Broadwood, Esq.r

Lord George Cavendish

John Frederic Daniell, Esq.r

John Fuller, Esq.r

Davies Gilbert, Esq.r M.P.

John Guillemard, Esq.r

Earl Grosvenor

Lt. Col. Wm. Martin Leake.

William Pole, Esq.r

Wm. Noble Rule, Esq.r

Peter Mark Roget, M.D.

George Stavely Smith, Esq.r

Richd. Horsman Solly, Esq.r

Confirmed the 7th day of May 1827

Somerset, Pres.

140 1827.

May 7.

At a Meeting of the Managers of the Royal Institution of Great Britain held on the 7th of May, 1827

Present

His Grace The Duke of Somerset, in the Chair

Daniel Moore, Esq.r

Henry Thos. Colebrooke, Esq.r

George Moore, Esq.r

Dr. Nicholl

James Heath Leigh, Esq.r

Sir Oswald Mosley, Bart.

Charles Pilgrim junr. Esq.r

Decimus Burton, Esq.r

Edmund Robert Daniell, Esq.r Secretary

1. The minutes of the proceedings of the last meeting were read and approved.

2. Read and approved the minutes of the Committee of Accounts of this day, and signed Orders for an advance to the Assistant Secretary of the following sums, vizt.

On account of the Royal Institution £300 – & on account of the Laboratory Fund £100.

3. The President reported that he had nominated the following Vice Presidents for the ensuing year. – vizt.

Vice Presidents. Sir Thos. Dyke Acland, Bart. M.P.

Hudson Gurney, Esq.r M.P.

Daniel Moore, Esq.r

Sir Oswald Mosley, Bart.

May 7. 141

The Hon: Wm. Francis Spencer Ponsonby, M.P.

Sir Claude Scott, Bart.

commn. from Mr. Rolandi resp. of MSS. of J.J. Rousseau. 4. Read a letter from Mr. Rolandi acquainting the President and Members of the Royal Institution, that he has received the Manuscript Herbal of Jean Jacques Rousseau, in eight Volumes; the illustrations being entirely in the hand writing of that celebrated Man, with some additional notes by Bernardin de St Pierre to whom this book belonged; and if it will give the President and Members of the Royal Institution any satisfaction to see this curious relict, Mr. Rolandi will feel highly honoured by a visit when he will have the pleasure of shewing it.

Resolved, That the thanks of the Managers be returned to Mr. Rolandi for his obliging communication.

Presents received.

5. Revd. Alexr. Dyer, B.A. presented Specimens of British Poetesses

John Ranking, Esq.r presented The Voyage and Travaile of Sir John Maundevile & Historical Researches on the Conquest of Peru

Sir Thomas Phillipps, Bart. presented Wiltshire Institutions 3 parts folio. & Deed relating to the Abbey of Shaftesbury

G. Poulett Scrope, Esq.r Geology of Central France – 2 vols.

Dr. Bevan – presented "Natural History, Physiology and Management of the Honey Bee, 8vo.

T. S. Winn, Esq.r presented
A Speedy End to Slavery.
Michael Faraday, Esq.r presented
Chemical Manipulation,
Royal Humane Society presented
Fifty third Annual Report of the Society.
Resolved,
That the thanks of the Managers be returned for these presents.

Presents. 6 Read a letter from Godfrey Higgins, Esq.r accompanying a present of a Book entitled the Celtic Druids published by him, in which it was stated that he had received an anonymous letter requesting him to present to the Royal Institution a small tract published by him called the Horae Sabbaticae.
Resolved, That the thanks of the Managers be returned for this present; and that the Secretary be requested to communicate to him the regret of the Managers that an application professing to come from the Royal Institution should have been communicated in such a manner.

Coals. 7. Read an estimate from Mr E. Chater offering to supply the Royal Institution with Coals for a twelvemonth on being allowed to send them in at any time they may think most expedient at 46s. ℔ Chald.n exclusive of Shooting and Metage.
Read an estimate from Mess.rs Pickins & Co offering to supply in the same manner the best Walls-end Coals at 48s. ℔ Chaldron including all expenses of

Metage Shooting &c.
Resolved, That Mess.rs Pickins & Co's Contract be approved of.—

Meetings of Patrons in Managers' Room. 8. Ordered, That the Patrons of the Library be allowed in future to hold their Meetings in the Room appropriated for the Meetings of the Committees of Managers and Visitors.

Comm.ee for Friday Ev.gs 9. Resolved, That the following be the Committee for regulating the Friday Evening Meetings of the Members—viz.t The Duke of Somerset
Daniel Moore, Esq.r
Capt.n Chapman
George Moore, Esq.r
Charles Pilgrim jun.r Esq.r
Whitlock Nicholl, M.D.
James Heath Leigh, Esq.r
The Secretary
Mr Brande
Mr Faraday.

Com. for Lectures. 10. Resolved, That the following be the Committee of Lectures for the ensuing Season.
The Duke of Somerset
Daniel Moore, Esq.r
George Moore, Esq.r
Capt.n Chapman
Benjamin Bond Cabbell, Esq.r
Charles Pilgrim jun.r Esq.r
James Heath Leigh, Esq.r
Dr Nicholl
Edmund Rob.t Daniell, Esq.r

11. The following persons having been duly recommended were admitted as Subscribers to the Royal Institution for the year, 1827
Annual Subscriber
J. Thornton, Esq.r of the Colonial Office

To the Lectures
Mrs Gray
Miss Gray } British Museum
Miss S. Gray

Mrs Alexander } Devonshire Street
Miss Bruce } Portland Place.

Adjourned to Monday the 21st Instant at Three o'Clock.
Confirmed this 4th day of June 1827
Somerset Pres.

Meeting resp.g erection of Furnace for Glass Manuf.e

At a Special Meeting of the Managers of the Royal Institution of Great Britain, held on the 14th of May, 1827 (under the authority of the President) to take into consideration the request made by the Committee of the Royal Society and the Board of Longitude for the improvement of Glass for optical purposes, to erect at the Royal Institution a Furnace for making experiments on a small scale.—
Present.
His Grace the Duke of Somerset Pres.t in the Chair
Daniel Moore, Esq.r V.P.
Sir Oswald Mosley, Bart. V.P.
Henry Thos. Colebrooke, Esq.r
Benjamin Bond Cabbell, Esq.r
Whitlock Nicholl, M.D.
James Heath Leigh, Esq.r
Charles Pilgrim, jun.r Esq.r
Capt.n Chapman,
Edmund Rob.t Daniell, Esq.r Secretary.

1 Read the following letter from J. F. Herschell Esq.r Sec. R.S. addressed to the Secretary.

Letter from J. F. Herschell, Sec. R.S.

I have the honour to inform you that it has been resolved by the Committee of the Royal Society and the Board of Longitude for the improvement of Glass for optical purposes, to request permission from the Managers of the Royal Institution to erect on their premises (but at their own expense) a furnace proper for making experiments on a small scale, and to make a series of such experiments therewith

146

1827.

in case there should be found on the Premises of the Royal Institution, a situation convenient for such a purpose, and I am directed by the Committee to request you will inform the Managers, of this their wish, and at your earliest convenience, communicate their determination on the subject"

2 **Resolved**, That the Secretary be directed to acknowledge the receipt of Mr. Herschell's letter, and to acquaint him of the wish of the Managers to accede to the request of the Committee of the Royal Society and Board of Longitude, provided the accommodation afforded by the house of the Royal Institution will allow of it, and to request the Committee to appoint a time to meet a Committee of the Board of Managers to confer with them on the subject.

3. That it appears to the Managers from the report of Mr. Professor Brande that the Battery Room may be applied to the purpose required by the Committee of the Royal Society and Board of Longitude, and that it be shewn to them accordingly.

4. **Resolved**, That a Committee of this Board be appointed to confer with the Committee of the Royal Society and Board of Longitude on the subject and to report to this Board.

Resolved,

That the Committee consist of

May 14 & 21.

147

the following Managers

The President

Captn. Chapman

George Moore, Esqr.

James Heath Leigh, Esqr.

Dr. Nicholl

The Secretary

and that Mr. Brande be requested to give them his assistance.

Adjourned to Monday the 21st. Instant at 3 o'clock.

Confirmed the 4th day of June 1827

Somerset Pres.

21st. May 1821.

Three Managers not being in attendance at Three o'Clock this day, no meeting took place.

148

1827.

June 4.

At a Meeting of the Managers of the Royal Institution of Great Britain held on the 4th. of June, 1827

Present

His Grace The Duke of Somerset, in the Chair.

~~Daniel Moore, Esq. V.P.~~

Henry Thos. Colebrooke, Esqr.

~~George Moore, Esqr.~~

Dr. Nicholl,

James Heath Leigh, Esqr.

Sir Oswald Mosley, Bart.

Hudson Gurney, Esqr. M.P.

Sir S. Bernard Morland Bart. Treasr.

Edmund Robert Daniell, Esqr. Secy.

1 The minutes of the proceedings of the meeting of the 7th. of May, and the minutes of the Special Meeting of Managers of the 14th. of May were read and approved.

2 Read and approved the minutes of the Committee of Accounts of this day; and signed an Order for an advance to the Assistant Secretary of £300.

Life Subscrs. 3. In pursuance of nominations under the Act of Parliament, the following persons were admitted as Life Subscribers to the Royal Institution

Hugh Percy Davison, Esqr.

nominated by his Father Alexander Davison Esqr.

and The Hon. Charles J. Shore of Portman Sq.

nominated by his Father Lord Teignmouth

June 4.

149

Presents received.

4. His Royal Highness the Duke of Sussex &c. presented Bibliotheca Sussexiana vol 1. parts 1 & 2.

The Commissioners of Public Records presented Calendar of the proceedings in Chancery in the Reign of Elizabeth, vol 1.

Calendar to pleadings &c. Dutchy of Lancaster vol 2.

William Brockedon, Esqr. presented Passes of the Alps, No. 2.

John Frederic Daniell, Esqr. F.R.S. presented Meterological Essays & Observations, part 2,

The Horticultural Society presented Transactions of the Society, vol 6 part 4.

Rept. respt. Glass Manuf. 5. Read the following report from the Committee of Managers appointed to take into consideration the request made by the Committee of the Royal Society and the Board of Longitude for the improvement of Glass for optical purposes.

The Committee report that they met together on Monday the 21st. of May – and that their meeting was attended by Davies Gilbert, Esqr. John George Children, Esqr. and J. F. Herschell, Esqr. on behalf of the Committee of the Royal Society and the Board of Longitude, who proceeded to inspect the Battery room and Yard adjoining, with the view of ascertaining whether those places would afford the accommodation required for the purposes of the Committee of the Royal Society and Board of Longitude, and that the Secretary of the Royal Institution has since received the following letter from Mr. Herschell.

Letter from Mr Herschell. dated 25th May, 1827 —

"I have the honour to inform you that the Committee appointed to confer with the Managers of the Royal Institution respecting the erection of a furnace on their Premises, having laid before the General Committee of the Royal Society and the Board of Longitude the result of that conference I am directed to return through you to the President and Managers the thanks of the Committee for the very handsome and liberal manner in which they have acceded to their request by placing at their disposal a spot every way so well adapted to the purposes of any experiments on a small scale which their objects may render it necessary to institute."

on Partition in Model room. 6. Resolved, That a partition be made across the Model Room, at the third pillar from the windows; and that Mr Faraday be requested to procure a specification of the same with estimates of the cost.

7. Resolved, That the Weekly Evening Meetings be concluded on the 15th of June.

Subscrs to Lectures. 8. The following subscribers were admitted to Mr Reinagle's Lectures on Painting.

Madame de Polier Vernand of 21 Holles St
Mr & Miss Jennings of East Acton.
The three Misses Willimott of Little Ryder St
Mrs & Miss Dodsworth of 13 Dorset Place Dorset Sq.
Mr & Miss Panter of North End Lodge Fulham

June 13. At a Special Meeting of the Managers of the Royal Institution, held on the 13th of June, 1827, by order of the President, to take into consideration the state of the debt of the Royal Instit:

Present
Daniel Moore, Esqr V.P. in the Chair.
Sir Claude Scott, Bart. V.P.
Sir Oswald Mosley, Bart.
James Heath Leigh, Esqr
George Moore, Esqr
Henry Thos Colebrooke, Esqr
Dr Nicholl,
Charles Pilgrim junr Esqr
Samuel Hare, Esqr
Edmund Robert Daniell, Esqr Secretary.
Captn Chapman.

on Repayment of Volr Loan. 1. Read the resolution of the General Monthly Meeting of the 7th of July 1823, respecting the repayment of the Voluntary Loan advanced without Interest to the Royal Institution.

Liberality of Mr D. Moore. 2. Mr Daniel Moore having informed the Board that as long as the loan will be of use to the Institution he has no intention of calling upon them for the repayment of his proportion of it; and that he was even disposed to go greater lengths for the benefit of the Institution.

Resolved, That the thanks of the Managers be given to Mr Danl Moore for his liberality to the Royal Institution.

on Accounts. 3 Read the report of the Committee appointed to investigate the Accounts with the view of ascertaining the amount of debt and from what period it has existed, as entered in the proceedings of the Managers of the 7th of April last.

Extension of time for repayment of Loan. 4 Resolved, That the Secretary be directed to write to the Contributors to the Voluntary Loan, to inform them that the period for the repayment of it will arrive on the 1st of August next, and to request a further extension of the time of payment.

Resolved, That Mr D. Moore and Sir Claude Scott be requested to call a Meeting of the Contributors to the loan for the purpose of laying before them the state of the Finances of the Institution, and of making such arrangements with them as the circumstances may require.

Adjourned to Friday next the 22d Instant at Three o'Clock.

June 22. At an adjourned Special Meeting of the Managers of the Royal Institutn held on the 22d of June to take into consideration the state of the debt of the Royal Institution.

Present
Daniel Moore, Esqr V.P. in the Chair.
Captn Chapman,
James Heath Leigh, Esqr
Dr Nicholl,
Sir Oswald Mosley, Bart
Henry Thos Colebrooke, Esqr
Benjamin Bond Cabbell, Esqr
George Moore, Esqr
Edmund Robt Daniell, Esqr Secretary.

On repayment of Loan. The Secretary reported that he had written to the Contributors to the loan, agreeably to the resolution of the Managers at their last meeting, and that he had received the following answers.

Time of repayment extended by Earl Spencer. 1. Read a letter from Earl Spencer K.G. stating that His Lordship very willingly agrees to the proposal of the Managers to extend the time of repayment of the loan of £500 advanced by him, till the Funds of the Institution shall be more equal to supply means for that purpose, without any intention of ever calling for any interest upon it.

by Sir S. Morland. 2. Read a letter from Sir S. Bernard Morland, Bart. stating that he shall readily concur in any plan for extending the period of the loan

to the Royal Institution which may be approved of by those who formerly took the lead in that measure, and contributed most largely to it.

J. Soane, Esq. 3. Read a letter from John Soane, Esq.r requesting the Managers to retain his Voluntary Loan of £100 as long as it will be of use to the purposes of the Royal Institution.

L.H. Pettit. 4. Read a letter from Louis Hayes Petit, Esq.re M.P. stating that he cannot use better language than that of his friend Mr Daniel Moore – not to call for the repayment of the £100 advanced by him as long as it can be of any service to the Royal Institution.

Sir E. Antrobus 5 Read a letter from Sir Edmund Antrobus, Bart. the Executor of the late Sir Edmund Antrobus, stating that he is willing to allow the repayment of the bond for £100 to stand over on the same footing that Mr Moore has, – namely, – till it is quite convenient for the Institution to repay it.

C.N. Pallmer, Esq 6. Read a letter from Chas. N. Pallmer Esq.r M.P. stating that he shall cordially accede to any desire of the Managers with respect to his little loan to the Institution which has his sincere desire for its prosperity.

Resolved, That the thanks of the Managers be given to Earl Spencer, K.G.

Sir S. Bernard Morland, Bart. M.P.
John Soane, Esq.r
Louis Hayes Petit, Esq.r
Sir Edmund Antrobus Bart.
Charles Nicholas Pallmer, Esq.r M.P.

for their liberal acquiescence in the wishes of the Managers to extend the period of repayment of the loan.

amount of Loan ~~resum~~ presented by Sir G. Duckett. 7. Read a letter from Sir George Duckett, Bart. stating his wish that his contribution of £100 should be considered as a gift.

Resolved,

That the thanks of the Managers be given to Sir George Duckett, Bart. for his very liberal donation to the Royal Institution.

Sir C. Scott, C. Hatchett, ~~G. Aust~~ extend Loan for two years. G. Aust 8. Read a report of the Meeting of the Contributors to the loan held on the 20th. of June in pursuance of the resolution of the last Meeting by which it appeared that in addition to the sums mentioned in the letters above referred to – Sir Claude Scott, Bart. Charles Hatchett, Esq.r and George Aust, Esq.r had agreed to extend the period of their contributions to the Voluntary loan for two years from the 1st. of August next.

Resolved,

That the thanks of the Managers be given to Sir Claude Scott, Bt. Charles Hatchett Esq.r and George Aust, Esq.r for their obliging proposal.

Resolved, That it appears to this meeting

Constitution of Sinking Fund. that the Compositions invested in the Funds pursuant to Chap. XI Art. 2 of the Bye laws since the year 1822 have produced the sum of £539.14.8. 3 p Cent Consols.

Resolved, That the above sum be carried to a separate account to be entitled – The Sinking Fund.

That the surplus of the Compositions at the end of each year which shall remain after deducting all the current expenses of the Institution be invested in the purchase of like Stock to be carried to the same account.

That the Dividends of the Sinking Fund be from time to time invested in the purchase of Stock to be added to the Capital of the Fund in the nature of compound Interest.

Adjourned to Monday the 2d of July at half past One o'clock p.m.

Danl. Moore

At a Meeting of the Managers of the Royal Institution of Great Britain held on the 2d. of July, 1827

Present

Daniel Moore, Esq.r V.P. in the Chair.
Sir Oswald Mosley, Bart.
Sir S. Bernard Morland, Bart. M.P. Treas.r
James Heath Leigh, Esq.r
Capt.n Chapman,
Decimus Burton, Esq.r
Henry Thos. Colebrooke, Esq.r
George Moore, Esq.r
Edmund Robt. Daniell, Esq.r Secretary.

1. Read and approved of the minutes of the 4th. of June, – and also the minutes of the Special Meeting of the Managers of the 13th. and the adjourned Special Meeting of the 22d. of June.

2. Read and approved the minutes of the Committee of Accounts of this day; and signed an Order for payment of the following sums, viz.t – To

Payments

William Thos. Brande, Esq.r for a year's Salary as Professor of Chemistry in 1827	£150 —
Joseph Fincher on account of the Royal Institution	300 —
	£450 —

3. Letters received.

Payment of Loan postponed – J. Guillemard. Read a letter from John Guillemard Esq.r addressed to the Secretary, stating that he does not ask, or

158

1827.

intend to ask for present payment of any sum which may be due to him from the Institution, but to follow the liberal example of Mr. Moore and other contributors.

4 Sir G. Blane. Read a letter from Sir Gilbert Blane, Bart stating that he shall be most happy to accede to the wishes of the Managers in allowing the repayment of the bond for £100, advanced by him, to stand over for 4 years from the period at which it will become due; – And that he shall not require any interest whatever from the Royal Institution.

5 ~~Amt of~~ Loan presented by G. Aust. Esq. Read a letter from George Aust, Esqr. stating that as he finds Sir George Duckett has presented the loan advanced by him to the Royal Institutn. he shall be most happy to follow his example, and begs to present the £100 advanced by him to the Royal Institution, on their acquiescing with his wish of being in future exonerated from all payments as Member.

Resolved,

That the thanks of the Managers be presented to George Aust, Esqr. for his very liberal donation to the Royal Institution; and that Mr. Aust's wish of being in future exonerated from all payments as Member, be reported to the General Meeting of this day.

Resolved, That the thanks of this Meeting be presented to John Guillemard, Esqr. and Sir Gilbert

July 2

159

Blane, Bart for their liberal acquiescence in the wishes of the Managers to extend the period of repayment of the loan.

6. Resolved, That a report of the state of the Voluntary Loan be laid before the General Meeting of the Members of this day.

7. Life Subscrs. In pursuance of nominations under the Act of Parliament, the following persons were admitted as Life Subscribers to the Royal Institution.

William Terry jun. Esqr. of High Wycombe nominated by his Father William Terry, Esqr.

& Frederick Manning, Esqr. of Combe Bank Kent nominated by his Father William Manning, Esqr. M.P.

8 Presents received for the Library.

The Horticultural Society presented Transactions of the Society vol 6 part 5 and vol 7. part 1

William Marsden, Esqr. presented Bibliotheca Marsdeniana – 4to.

The Editors of the Zoological Journal presented No. 9 of their Journal.

The Geological Society presented Transactions of the Society, second series vol 2 pt. 2

Thomas James Mathias, Esqr. presented Il Castello dell' Ozio.

Presents received for the Mineralogical Colln.

Geological Specimens collected by Captn. Parry and his Companions – presented by Mr. Marshall.

Different Specimens of Rock Salt from the Mines in the neighbourhood of Northwick in the County of

160

1827.

Chester – presented by James Heath Leigh, Esqr.

Ordered,

That the thanks of the Managers be returned for these presents.

9. On erection of Glass Furnaces &c. Read the following report from the Committee of Managers appointed to take into consideration the request made by the Committee of the Royal Society and the Board of Longitude for the improvement of Glass.

The Committee report that they had received a communication from Mr. Herschell intimating that the Committee of the Royal Society and the Board of Longitude had obtained plans and estimates of the furnace and building proposed to be erected on the Premises of the Royal Institution. And that they wished to commence the erection as soon as possible. –

And that the Committee had informed them that they were at liberty to commence their erection on the Premises of the Royal Institution whenever it might suit their convenience.

10 On Closing Library Resolved, That the Library of Reference after this day be closed at six o'clock in the Evening, for the Summer months; but Gentlemen desirous of being accommodated with books from it, after that hour may have the use of them in the Reading Room by previous intimation of their wish to the Librarians. – And that from the 1st of Novr. next until the 1st of Feby 1828 the Library be open from ten o'clock till four in the day, and

July 2

161

from seven o'clock till Ten in the Evening, and from the 1st of February till the 1st of June from Ten o'Clock till five in the day, and from seven till Ten in the Evening.

11. The following persons having been duly recommended were admitted as Subscribers to the Royal Institution for the year, 1827 –

Annual Subscriber.

Dr. Charles Holland 32 Canonbury Square

To Dr. Harwood's lectures

Miss Malcolm }
Miss Mary Malcolm } 16 Hanover Square

Dan Moore

Adjourned to Monday the 6th of August.

162

1827.

Aug. 6.

At a Meeting of the Managers of the Royal Institution of Great Britain, held on the 6th of August, 1827

Present
Daniel Moore, Esq.r V.P. in the Chair.
George Moore, Esq.r
Benjamin Bond Cabbell, Esq.r
Capt.n Chapman,
Hon.ble Tho.s Colebrooke, Esq.r
Sir S. Bernard Morland, Esq.r M.P. Treas.r
Edmund Robert Daniell, Esq.r Sec.y

1. The minutes of the proceedings of the last meeting were read and approved.

2. Read and approved the minutes of the Committee of Accounts of this day, and signed an Order for payment of the following sums, viz.t

Payments.

Joseph Fincher on account of the Royal Institution	£ 200 —
D.r Harwood for a Course of lectures in 1827	52.10 —

Repayment of Loan deferred – M.r Auriol

3. Read a letter from M.r Auriol stating that if it will be of any use to the Institution he has no objection to allow the loan of £100 advanced by his late Father, to stand over for two years without interest upon the understanding that payment will be required

163

Aug. 6.

at the expiration of that time.

R.t Hon. Sturges Bourne

4. The Assistant Secretary stated that The R.t Hon: W.m Sturges Bourne had authorized him to inform the Managers that he should be most happy to extend the period of the repayment of the loan of £100 advanced by him till convenient for the Royal Institution to repay it.
Resolved, That the thanks of the Managers be returned to the above Gentlemen for their acquiescence in the wishes of the Managers.

5. Presents received.

Vicomte de Santarem presented
Memorias Para a Historia, e Theoria das Cortes Geraes
Memorias dos Alcaides Mores da Villa de Santarem.
Analyze de Huma Medalha de Oiro do Imperador Honorio.
The Prince of Lieven presented two rare and beautiful specimens of Emerald Copper from Siberia
John Lingard, Esq.r presented – Inquiry into the nature of Timber.
John Frost Esq.r F.R.S. presented Introductory Lecture on Botany.
Ordered, That the special thanks of the Managers be returned to Vicomte de Santarem & the Prince of Lieven for the above presents and the usual thanks to M.r Lingard & M.r Frost.

164

1827.

Lectures.

6. Read the following report from the Committee of Lectures.
The Committee report that they had given permission to D.r Gordon Smith to deliver a Course of lectures in the Theatre of the Royal Institution on Medical Jurisprudence to alternate with M.r Brande and M.r Faraday's Morning lectures – To commence on Monday the 1st of October at nine in the Morning.

Repairs.

7. Read the minutes of the Visitors as entered in their proceedings of the 26th of March last
Resolved,
That the repairs as therein stated be done under the direction of M.r George Moore and the Secretary
Resolved, That M.r Faraday's rooms

Close of Lib.y

8. That the Library of Reference be closed for the purpose of cleaning and arranging the books from the 3.d to the 15th of September inclusive.

Payment for Lectures.

9. Resolved, That D.r Harwood be paid 2/3.ds of the amount received from his Course of Juvenile lectures delivered during the last Easter recess.

Adjourned to the 5th of Nov.r

165

Nov. 5.

At a Meeting of the Managers of the Royal Institution of Great Britain held on Monday the 5th of November, 1827

Present
Charles Pilgrim jun.r Esq.r in the Chair.
George Moore, Esq.r
Benjamin Bond Cabbell, Esq.r
Hon.ble Tho.s Colebrooke, Esq.r
D.r Nicholl
Sir S. Bernard Morland, Bart. M.P. Treas.r
Edmund Rob.t Daniell, Esq.r Secretary

1. The minutes of the proceedings of the last meeting were read and approved.

2. Read and approved the minutes of the Committee of Accounts of this day, and signed Orders for payment of the following sums.

Payments

Joseph Fincher on account of the Royal Institution	£ 200 —
Ditto on account of the Labor.y Fund	£ 100 —

Presents received.

3. The Society for the improvement of Prison discipline presented
The 7th Report of the Committee of their Society.
The Board of Longitude presented
Nautical Almanac for 1830.
Cesar Moreau Esq.r F.R.S. presented
Statistical state of Ireland, in a series of tables.

166

1827.

John Frost, Esq.r presented
Remarks on the Mustard Tree of the new Testament.
Bingley's practical introduction to Botany
53.d Annual Report of the Royal Humane Society.
Septimus Burton, Esq.r presented
Three prints of Egyptian Antiquities,
The Royal Asiatic Society presented
Transactions of the Society, vol 1, part 2.
The Editor of The Phrenological Journal presen.d
vol 4, No. 16 of the Journal.
Signor D. Paoli – presented through Mr. Mawe,
Ricerche sui Moti Molecolari de Solidi
H. Sharpe, Esq.r presented
Genealogical History of the family of Brabazon
Joseph Fincher presented
The Achievements of Prayer; selected exclusively from The Holy Scriptures.

Ordered,

That the thanks of the Managers be returned for these presents.

4. Resolved, — *Committee for lighting R.I.*

That the following Gentlemen be a Committee for lighting the Institution for the ensuing year – viz.t
Mr. Pilgrim,
Mr. Cabbell,
Mr. Brande,
Mr. Faraday,
The Secretary.

167

Nov. 5.

5. Resolved, That the Committee of lectures do meet on Monday next the 12th Instant, at Three o'Clock.

6. *Ass.t in Lab.y – Wages raised*

Resolved, That the wages of Thomas Pearsall, Assistant in the Laboratory, be raised from 15s. to One guinea per week – To commence from this day.

7. The following persons having been duly recommended were admitted as Subscribers to the Royal Institution for the year, 1828

Subs.rs

George Edward Wood, Esq.r 35 Regency Square Brighton
Alex.r Mc.Donnell, Esq.r West India Dock-house Billiter Square.

8. Ordered, — *Rep.t up.g accounts*

That the following report of the Managers be presented to the General Meeting of this day.

The Managers report that they have investigated the probable Receipts and Expenditure of the Royal Institution for the present year, And they find that the total receipts on the General Account of the Institution will probably exceed the sum of £2750, including the Compositions of Members, the amount of which will be about £800; which being deducted from the above sum of £2750, will leave the sum of £1950 which the Managers conceive will be the amount of the Annual Income of the Institution up to the end of the year. That the expenditure of the Institution as far as they have been able to estimate the same will amount to about £2460, including the expenses of repairs and all outgoings up to Christmas, which will

168

1827.

exceed the receipts on account of the annual income by the sum of £510 – The Managers therefore recommend that they should be authorized to make up that sum out of the Compositions of Members which will thereby leave a surplus of receipts beyond the expenditure of the Institution for the present year amounting to nearly £300

9 Adjourned to Monday the 19th Instant at Three o'Clock p.m.

Confirmed the 19th November 1827

C H Pilgrim

169

Nov. 19.

At a Meeting of the Managers of the Royal Institution held on the 19th of November 1827

Present

Charles H. Pilgrim, Esq.r in the Chair.
George Moore Esq.r
Benjamin Bond Cabbell, Esq.r
Dr. Nicholl
Samuel Ware, Esq.r
Edmund Robt. Daniell, Esq.r Secretary.

1. The minutes of the proceedings of the last meeting were read and approved.

Adjourned to Monday the 3rd of December, 1827 at One o'Clock, p.m.

Confirmed the 3rd of December 1827

Somerset
Pres.

170

1827.

Dec 3.

At a Meeting of the Managers of the Royal Institution of Greatt Britain held on the 3d of Decemr 1827

Present

The Duke of Somerset, Presidt in the Chair
Henry Thos Colebrooke, Esqr
Benjamin Bond Cabbell, Esqr
George Moore, Esqr
Chas H. Pilgrim, Esqr
James Heath Leigh, Esqr
Dr Nicholl
Sir S. Bernard Morland, Bt M.P. Treasr
Samuel Ware, Esqr
Edmund Robt Daniell, Esqr Secretary.

1. The minutes of the proceedings of the last meeting were read and approved.

2. Read and approved the minutes of the Committee of Accounts of this day.

Presents received.

3. Wm Brockedon, Esqr presented
Passes of the Alps – pt 3 & 4
By the Editors
The Zoological Journal No 10
The Commissioners of Public Records presented
Parliamentary Writs vol 1.
Wm Pitts, Esqr presented
Series of Outlines from Æneis of Virgil
Saml Weller Singer, Esqr presented
The Correspondence of Henry Hyde, Earl of Clarendon

171

Dec 3.

and of his Brother Laurence Hyde, Earl of Rochester with the Diary of Lord Clarendon, 2 vols 4to
Dr Bostock M.R.I. presented
Elements of Physiology vol 3.
H. Sharpe, Esqr presented
Genealogical History of the Family of Brabazon
Ordered,
That the thanks of the Managers be returned for these presents.

4 Painting, &c Resolved, That the Passage from the Staircase to the Lecture-room be coloured in distemper, and the dado and the cornice be painted.

5. The following persons having been duly proposed were admitted as Annual Subscribers for the year 1828. –

annl subsrs
Charles Moss, Esqr 13 Henrietta St Cavendish Sqe
Robt Hugh Wilson Ingram, Esqr 6 Pump Court Temple
Peter F. Aikin, Esqr 17 Henrietta Street Brunswick Sqe

6. Ordered,

Rept respt [er]ections for Glass Manufe

That the following report of the Managers be presented to the General Meeting of this day.

The Managers report that in the Month of May last a letter was addressed to them by J. F. Herschell, Esqr one of the Secretaries of the Royal Society, on behalf of a Committee appointed by the Royal Society and Board of Longitude for the improvement of Glass for optical purposes requesting

172

1827.

permission to erect at their own expense on the Premises of the Royal Institution a furnace proper for making experiments on a small scale and to make a series of such experiments therewith. That the Managers considering that by acceding to such request they should be forwarding the objects of the Royal Institution and contributing to the interests of Science in general, agreed to grant the permission required. And that the said Committee have accordingly proceeded to erect and have completed at their own expense in the yard of the Royal Institution adjoining the Laboratory, a Furnace and other Buildings in which they are conducting experiments upon the Manufacture of Glass according to the proposition contained in Mr Herschell's letter.

Adjourned to the 17th December at Three o'Clock.

Confirmed the 17th Day of December 1827
Somerset Pres.

173

Dec 17.

At a Meeting of the Managers of the Royal Institution of Greatt Britain held on the 17th December, 1827,

The Duke of Somerset Prest in the Chair
George Moore, Esqr
Henry Thos Colebrooke, Esqr
Dr Nicholl,
Sir S. Bernard Morland, Bt M.P. Treasr

1. The minutes of the proceedings of the last meeting were read and approved.

Request from Mr T. Griffiths.

Read a letter from Mr Thomas Griffith late Assistant in the Laboratory, requesting the Managers would allow him some remuneration for the expenses incurred by him in painting & papering, and fitting up some Bookcases which cannot be removed without injuring the Rooms which he occupied when he resided in the Royal Institution.

Resolved, That the sum of £5 be allowed to Mr Griffiths.

Adjourned to Monday the 21st of January 1828. –

Confirmed the 21st day of January 1828
Somerset Pres.

174

Jan. 21. 1828.

At a Meeting of the Managers of the Royal Institution held on the 21st of January, 1828,

The Duke of Somerset Presid.t in the Chair,
Henry Thos. Colebrooke, Esqr.
George Moore, Esqr.
Charles H. Pilgrim, Esqr.
Edmund Robt. Daniell, Esqr. Secretary.

1. The minutes of the proceedings of the last meeting were read and approved.

Decease of D. Moore, Esq. &c.

2. The Secretary communicated the lamented decease of Daniel Moore, Esqr V.P. And also the resignation of Captn. Chapman, Manager. Resolved,

That Sir George Thos. Staunton Bt. and Captn. Sabine be requested to allow their names to be put in nomination to fill up the Vacancies.

3. Read the following letter from W. H. Booth Esqr. one of the Executors of the late Danl. Moore, Esqr. containing the following extract from the Will

Loan by D. Moore, Esq. presented

"I remit unto the Royal Institution of Great Britain the sum of One thousand Pounds, which I some years since advanced to the Institution on bond without Interest."

Resolved That the above letter be reported to the next General Meeting.

Jan 21. 175

Painting, &c.

4. Read the resolution of the Managers of the 3d. of December, directing that the Passage from the Stair-case to the Lecture room should be coloured in distemper.

The Secretary reported that finding the difference of expense would be very trifling he had directed the whole to be painted in Oil which had been done. Approved of.

on Lighting with Portable Gas.

5. The Committee appointed for lighting the Institution in the ensuing Season report "that they recommend the use of Portable Gas for lighting the Library and the Theatre on the Friday Evening Meetings. And that they had entered into a Contract with Mr. Drew to fit up the Theatre and Rooms in the manner in the Contract stated, at an expense not exceeding £93. with a condition that if it should be thought adviseable to discontinue the use of Portable Gas at the end of the Season, he would allow £45 for the whole of the fittings.

Approved of.

Repairs.

6. Read the resolution of the Committee of Visitors as entered in their proceedings of the 14th. Instant Resolved,

That Mr. Parker, the Painter, be directed to examine and report on the repairs therein pointed out.

7. The Committee of Lectures report that they have made arrangements for the delivery of the following

176 1828.

Course of Lectures in the ensuing Season.

Lectures.

Chemistry – By Mr. Brande & Mr. Faraday.

Music – By Mr. Saml. Wesley – The Course six lectures – To be remunerated with 5 guineas each lecture; and to be allowed for Vocal & instrumental examples.

Architecture – By Mr. Alfred Ainger – The Course six lectures. To be remunerated with 5 guineas each lecture.

Painting and the Fine Arts – By Mr. R. R. Reinagle at 10 guineas each lecture

Hieroglyphics – By The Marquis Spineto at 30 guineas for the Course and the subscriptions of non-Subscribers who may attend his lectures.

Natural History By Dr. Harwood – at 50 guineas for the Course.

Mechanics – By Mr. Millington at 5 guineas a lecture.

8. Resolved, That the sum of Fifty guineas be paid to Mr. Faraday for his Course of Juvenile lectures.

Adjourned to 4th. Feby

Confirmed the 4th day of February 1828
Somerset, Presd.

Feb. 4. 177

At a Meeting of the Managers of the Royal Institution of Great Britain held on the 4th. of February, 1828

Present

The Duke of Somerset, President, in the Chair
Benjamin Bond Cabbell, Esqr.
George Moore, Esqr.
Sir S. Bernard Morland, Bart. M.P. Treasr.
Henry Thos. Colebrooke, Esqr.
Edmund Robt. Daniell, Esqr. Secretary.

1. The minutes of the proceedings of the last meeting were read and approved.

2. Read and approved the Minutes of the Committee of Accounts of this day, and signed an Order for an Advance to the Assist. Secretary of £400.

3. Audited the Treasurer's Account of the Royal Institution for the year 1827 – And also the Treasurer's Account of the Laboratory Fund for the same year.

on Donation by D. Moore, Esq

4. Read the following letter from W. H. Booth, Esqr. dated 16th. January 1828.

"As an Executor with The Revd. Dr. Maddy and Mr. Tomkyns of the Will of the late Daniel Moore Esqr. which is dated the 16th. of December last; I send you an extract from the Will as regards the Royal Institution of Great Britain

"I remit unto the Royal Institution of Great Britain the sum of £1000, which I some years

178

1828.

since advanced to the Royal Institution on bond without Interest."

The Will contains a general direction that all legacy duty should be defrayed out of the Estate.

I have the honour to be &c. &c.
signed / W. H. Booth

To Edmd. R. Daniell Esqr.
Sec: Royal Institution.

Marble Bust of D. Moore proposed.

5. Resolved, That the Managers of the Royal Institution deeply impressed with the unremitting attention and liberal support afforded by the late Daniel Moore Esqr. from the commencement of the Royal Institution to the termination of his life, recommend that a Marble Bust should be erected in the Library of the Royal Institution as a small tribute of sincere respect to his memory; and that the expense thereof be defrayed by a private subscriptn. amongst the Members.

Request from Mr. Lake

6. Read a letter from William Lake, Esqr. containing a request, in consequence of the decease of his late partner Mr. Daniel Moore, to be allowed to transact any Professional Concerns in which the Royal Institution may be engaged.

Resolved, That Mr. Lakes request be complied with; And that he be appointed Solicitor to the Royal Institution.

7. Presents received.

The Astronomical Society presented
Memoirs of the Society vol. 3 pt. 1.

Feb. 4. 179

J. C. Beltrami, Esqr. presented
A Pilgrimage in Europe & America 2 vols. 8vo.
Presented by the Editor
The Harmonicon new series No. 1.
The Society of Arts presented
Transactions of the Society vol. 45
W. V. Hellyer, Esqr. presented
Tables of the Representation of the Commons in Parliament
Charles T. Lane Esqr. presented
The Coronation Oath considered, 8vo.
The Linnean Society presented Transactions of the Society vol. 15, pt. 2
Captn. J. Shaw, R.N. presented
T. S. Pockston's Chronological Chart of the Patriarchs. 2 copies.
Presented by the Editors,
Zoological Journal No. 11.

Ordered,
That the thanks of the Managers be returned for these presents

on Election of two Managers

8. The Secretary reported that he had received answers from Sir George Thos. Staunton, Bart and Captn. Sabine, who had allowed their names to be put in nomination as Managers in the room of the late Daniel Moore, Esqr. and Captn. Chapman.

Ordered, That ballotting lists be filled up in the names of Sir George T. Staunton & Captn. Sabine.

9. Read the resolution of the General Meeting

180

1828.

of the 2d. of July last ordering that the Compositions invested in the Funds pursuant to Art. 2 Chap XI of the Bye laws since the year 1822, amounting to the sum of £539.14.8 three per Cent Consols, be carried to a separate Account to be entitled "The Sinking Fund."

Resolved,
That an account be opened in the Books of the Institution entitled "The Sinking Fund Account"; and that the sum of £539.14.8 three per Cent Consols, part of the sum of £1789.14.8 standing to the General Account of the Institution, be carried to that account together with the Interest accrued due thereon in January last.

Resolved,
That the Treasurer be directed to invest such Dividends and all future Dividends hereafter to accrue due on the said £539.14.8 three per Cent Consols, and future Investments in the purchase of like Stock to be carried to the same Account.

10. The following persons having been duly recommended were admitted as Subscribers to the Royal Institution for the year 1828.

Subscribers.

Annual Subscribers.
George Stratton 56 Montagu Square
George Smith junr. Esqr. 85 Baker Street
T. Valentine, Esqr. 49 Frith St. Soho.
Richard Onslow, Esqr. 2 New St. Spring Gardns.
John Frederic Hahn, Esqr. of Wandsworth

Feb. 4. 181

To the Lectures.
Lady Cunynghame } Savile row
Miss Cunynghame }
Mrs. Tulloh 5 Portugal St. Grosvenor Sqr.
Mrs. Brodie Savile row.
Lady Raffles Grosvenor Street
Miss Holland 35 Montagu Square
Mrs. Hare 32 Argyll Street
Mrs. Tyringham Bernard Pall Mall
Miss Potticary Upper Baker St.
Lady Louisa Fitzmaurice } Lansdowne House
Miss Little }
Miss Sturch York Terrace Regents Park
Mrs. Hume 6 Bryanstone Square
Mrs. B. Droz 23 Park Crescent

To Mr. Brandes' Lectures.
Charles Goding Esqr. Knightsbridge.
Mrs. & Miss King 73 Gower St.
Lady Dashwood Montagu Place
Miss Curtis Grosvenor St. West.

Adjourned to Monday the 18th. Instant at Three o'Clock. –

Confirmed the 18th day of February 1828
Somerset Pres.

182

Feb. 18 — 1828.

At a Meeting of the Managers of the Royal Institution of Great Britain held on the 18th of Feby, 1828

Present
The Duke of Somerset, Presid.t in the Chair
Benjamin Bond Cabbell, Esq.r V.P.
Henry Tho.s Colebrooke, Esq.r
George Moore, Esq.r
Sir S. Bernard Morland, B.t M.P. Treas.r
Samuel Ware, Esq.r
Edmund R. Daniell Esq.r Secretary

1. The minutes of the proceedings of the last meeting were read and approved.

2. Resolved, — *Permanent Com. of Acc.ts – Regulations.*

That it is expedient to form a permanent Committee of Accounts, for the purpose of superintending the General Accounts of the Institution.

That such a Committee be formed consisting of five Members of the Board of Management besides the President, of whom two are to be a quorum.

That the Treasurer and Secretary of the Institution for the time being be ex-officio two of the Members of such Committee; that M.r George Moore, M.r Benj.n Bond Cabbell and D.r Nicholl be the other Members of the Committee for the current year.

That the Committee of Accounts do meet on the Wednesday before each General Monthly

183

Feb. 18

Meeting And that they have power to adjourn to any day which they may consider expedient; And that in case of necessity any two Members of the Committee may call a special Meeting of the Comm.e by letter under their hands addressed to the Assistant Secretary.

That at every Monthly Meeting of the Committee of Accounts there be laid before them all the Tradesmens' Bills and Workmens Accounts for Goods or Work delivered or performed to or in the Institution during the preceding Month signed by the proper Officer pursuant to the regulations of the 5th of February, 1827.

That no Bill or Account be paid or discharged, unless the same shall have been previously laid before the Committee of Accounts pursuant to the last resolution, and signed by the Chairman of such Committee with his initials.

That the Committee of Accounts do cause regular minutes of their proceedings to be entered in a Book to be provided for that purpose;— And that they do also make a report of the state of the Accounts of the Institution to the Board of Managers at each Monthly Meeting of such Board.

3. *Mourning for Porters &c.* The Secretary reported that as a mark of respect to the memory of the late M.r Dan.l Moore he had taken upon himself to order that each of the Porters should have a suit of Mourning; and that M.r Simons the Assistant in the Library should have the sum of 5 guineas, and M.rs

184

1828.

Greenwood 3 guineas, and the Housemaid 2 guineas to provide themselves with Mourning.
Approved of.

4. Presents received.
H. M. Grover, Esq.r presented Socrates – a Dramatic Poem
W.m Tho.s Brande, Esq.r presented Tables of definite Proportionals.
M.r La Beaume presented Observations on Galvanism &c.
Ordered, That the thanks of the Managers be returned for these presents.

Adjourned to Monday the 3.d of March.

Confirmed the 3.rd day of March 1828
Somerset Pres

185

Mar. 3.

At a Meeting of the Managers of the Royal Institution of Great Britain held on the 3.d of March, 1828.

Present
The Duke of Somerset, Presid.t in the Chair
Benjamin Bond Cabbell, Esq.r V.P.
George Moore, Esq.r
Sir S.t Bernard Morland, Bart. M.P. Treas.r
C. H. Pilgrim, Esq.r
D.r Nicholl,
James Heath Leigh, Esq.r
Edmund Rob.t Daniell, Esq.r Secretary.

1. The minutes of the proceedings of the last meeting were read and approved.

2. *Rep.t from Com. of Acc.ts* Read the following report from the Comm.e of Accounts.

The Committee of Accounts report that they have examined the General Accounts of the Royal Institution from the 1.st of January last to the 27th of February, and they found a balance in the hands of the Treasurer of £773 . 8 . 4; And in the hands of the Assist.t Secretary of £196 . 3 . 10.

And that they have examined and approved the following bills which they recommend to be immediately paid.

Payments

M.r Parker for Painting &c.	155 . 5 . 7
M.r Royce for Repairs &c.	47 . 10 . 10
M.r Bird for Bricklayers work	26 . 16 . 6
Norris for Carpeting for the Lib.y	42 . 17 —
Carr.d forward	272 . 9 . 11

186

1828

Bro^t forw^d.	272 . 9 . 11
Jones & Yarrell for Newsp.^s	84 . 9 . 2
Houghton — ditto	29 . 2 . 10
Ditto for Stationery in 1827	4 . 16 . 8
Parochial rates to Christmas	57 . 10 —
Water rate to Mich^s 1827	10 . 10 —
Vulliamy for repair of Clocks for the year 1827	2 . 17 . 6
Lighting the lamp in front of the Institution with Gas for the year 1827	5 . 2 —
Brown for matting in ditto	1 . 16 . 6
Dickins & Co. for Coals in 1827	85 — —
	553 . 14 . 7

Approved of; And signed an Order for an Advance to the Assistant Secretary of £550.

Presents received.

3. Charles Fellows, Esqr. presented
Narrative of an Ascent to Mont Blanc.
Henry Sass, Esqr. presented
A Journey to Rome & Naples.
H. H. Wilson, Esqr. presented
Select specimens of the Theatre of the Hindus, 3 vol^s 8vo
The Commissioners of Public Records presented
Reports respecting the Public Records of Ireland.
Samuel Solly jun^r Esqr. presented to the Museum
numerous Skeletons of British Birds
M^r. Lingard presented, A polished Specimen
of Cotham Marble.
Ordered, That the thanks of the Managers be

Mar 3. 187

returned for these presents.

Librarian permitted to accept another situation.

4. Read the minute of the Patrons of the Library respecting M^r. Singer's application to them to be allowed to accept the situation of Secretary to the Travellers Club which he stated would in no wise interfere with his duties as their Librarian; And the resolution of the Patrons approving of his acceptance of that office provided the Managers saw no objection to it.

Resolved,
That the Managers concur with the Patrons of the Library in approving of Mr Singer's acceptance of the Secretaryship of the Travellers' Club, And grant him their permission accordingly.

Repairs &c.

5. In pursuance of a resolution of the Managers of the 21st of January last; Read the following report from M^r. Parker the Painter, "that he had carefully examined the Lantern light belonging to the Lecture room, and it does not appear to be necessary to do any thing to it at present: — In the Autumn it would be desirable to paint it, and take off the loose pully. The wood work in general is in good condition.

Loan of Piano.

6. James Broadwood, Esqr. M.R.I. having offered to lend a Piano-forte to the Royal Institution during M^r. Wesley's lectures
Resolved, That Mr Broadwood's obliging offer be accepted with the thanks of the Managers.

7. Read the recommendation of the Patrons, (as

188

1828

entered in their proceedings of the 27th. Feby)
to increase the Salary of M^r. Simons the Assistant Librarian to 30^s. ℘ week,
Resolved, That the above recommendation be approved of; And that the Salary of M^r. Simons be increased to 30^s. a week. — To commence from the 1st. of January last

8 The following persons having been duly recommended were admitted as Subscribers to the Royal Institution for the year, 1828.

Annual Subscribers.
John Green, Esqr. 67 Baker Street
Charles Wray, Esqr. 7 Sackville Street
John Kenyon, Esqr. 39 Devonshire Place
Meriono Balcarce, Esqr 76 Grosvenor Square
Charles Wilding, Esqr. 4 Wimpole Street

To the Lectures.
Mrs Kenyon 39 Devonshire Place.
Miss Duncan 5 Hanover Street
Miss Anthony John St Bedford row
Mrs Smith & 4 Daughters – 6 Portland Place.

To M^r. Wesley's Lectures.
M^r. R. W. S. Lutwidge 21 Old Square Lincolns Inn
Miss S. Jones Salisbury Square
M^r. W^m. Cahusac of Maida-hill
Lady Charlotte Calthorpe's Grosvenor Square
Mrs Clark 3 Hyde Park Place West.

Mar 3. 189

To M^r. Brande's Lectures.
Thomas Baring, Esqr. } 41 Devonshire Place
Miss Baring — }
~~Mrs Clark, 3 Hyde Park West~~

Adjourned to Monday the 17th. Instant at Three o'Clock.

Confirmed the 17th day of March 1828
Somerset Pres.

190

1828

Mar 17.

At a Meeting of the Managers of the Royal Institution of Great Britain held on the 17th. of March, 1827

Present
The Duke of Somerset, Presid.t in the Chair
Benjamin Bond Cabbell, Esq.r V.P.
George Moore, Esq.r
Henry Tho.s Colebrooke, Esq.r
Dr. Nicholl,
C. H. Pilgrim, Esq.r
Edmund Rob.t Daniell, Esq.r Secretary.

1. The minutes of the proceedings of the last meeting were read and approved.

2. Presents received.
Chevalier Max: V. Speck presented
Speck'sche Gemälde-Sammlung. Leipsig
Sir James Edward Smith Pres Linn: Soc.
presented English Flora vol 4.
Ordered, That the thanks of the Managers be returned for these presents.

Adjourned to Monday the 7th. of April

Confirmed the 7th day of April 1828
Somerset. Pres.

191

April 7.

At a Meeting of the Managers of the Royal Institution of Great Britain held on the 7th. of April, 1828.

Present
His Grace The Duke of Somerset, Pres.t in the Chair
C. H. Pilgrim, Esq.r
Capt.n Sabine.
Dr. Nicholl,
George Moore, Esq.r
Benjamin Bond Cabbell, Esq.r V.P.
Sir S. Bernard Morland, Bart. M.P. Treas.
Edmund Rob.t Daniell, Esq.r Secretary.

1. The minutes of the proceedings of the last meeting were read and approved.

2. Read the following report from the Comm.e of Accounts.

Rep.t of Com. of Acc.ts

The Committee of Accounts report, that they have examined the General Accounts of the Royal Institution of the last month, and they found a balance in the Treasurer's hands of £430.3.4. And in the hands of the Assistant Secretary of £101.11.7

They likewise report that they have examined the Laboratory Account of the Royal Institution from the 1st. of January last, and they found a balance in the hands of the Treasurer of £262.0.10 And in the hands of the Assistant Secretary of £25.14.7. They recommend that the sum of £136.10 – being the amount of Compositions of 13 new Members, which were paid in the last year

192

1828

but were not invested pursuant to the Bye Law; And the sum of £42 being the amount of four Compositions paid in the present year, amounting together to the sum of £178.10 should be invested in the purchase of Stock to be carried to the Account of the Laboratory Fund.

They recommend the following sums to be paid out of the General Fund;

Payments

Mr. Wesley on account of his lectures	£31.10 –
Mr. Quarrell the balance of his Account	41.4.8
Mr. Drew for Gas fittings, on account	50. —
	122.14.8

They also recommend that the following bills should be paid on the Laboratory Account.

Fisher for Chemicals in 1827	£15.15 –
Newman for ditto in d.o	42.5 –
Dickins for Coals in d.o	7.10 –
Richards for Spirits of Wine in d.o	5.10 –
	£71.1.10

Approved of; and signed Orders for payment to the Assistant Secretary of the following sums.

On account of the Royal Inst.n £130. –
and on account of the Labor.y £80 –
and an Order that the Treasurer of the Laboratory Fund do invest the sum of £178.10. on account of that Fund.

The Committee of Accounts further report

193

April 7.

Arrears

that they have examined into the list of Arrears due from Members with the view of distinguishing those which may be available from those which are not likely to be recovered. And they find that the total amount of Arrears due to the Institution from Members on the 31st. of December last, amounted to the sum of £485.2 – of which the sum of £108.3 – only can be considered as available. And they recommend that a Statement to this effect should be made to the Board of Visitors.

Approved of.—

Life Subs.r

3. In pursuance of a nomination under the Act of Parliament by Thomas Day, Esq.r – his Son John Day, Esq.r of 28 Montagu St. Russell Square, was admitted to the privileges of a Life Subscriber to the Royal Institution.

Presents received.

4. Sir Thomas Phillipps, Bart. presented
Catalogue Sommaire des M.s.s. de la Bibl: de St. Omer
B. L. Vulliamy Esq.r presented
Considerations on Public Clocks.
Rev.d Richard Burgess presented
Description of the Circus on the Via Appia,
George Gladwin, Esq.r presented
Print of the North Elevation of St. Pauls Cathedral
Louis Hayes Petit, Esq.r M.P. presented
Hieroglyphics continued by the Royal Society of Literature.
Hugh Barclay, Esq.r presented
Select Views of the Roman Antiquities.

194

1828

M. Cesar Moreau, Esqr. presented
Chronological Records of British Finance
A. A. Linde, Esqr. presented
A pamphlet on the injurious tendency of
modifying the Navigation laws,
The Astronomical Society presented
Monthly Notices of their proceedings Nos. 8 & 9.
George Bennett, Esqr. presented
A part of the leaf of the Talipot Palm
from the Island of Ceylon to be placed in the
Museum.
Dr. Boott presented
Specimen of the Wax from the berries of the
Myrica Cerifera.
Michael Shepley, Esqr. presented
Specimen of Octoëdral Iron ore.
Resolved,
That the thanks of the Managers
be returned for these presents.

Resn. of Visitors — 5. Read the following resolution of the Visitors
of the 25th of March last.
The Visitors recommend the following to the
Repairs — consideration of the Managers. The cornice at the
back of the House is in a bad state. The guide
Gas light in the Model room is not sufficiently
protected, and the Chimney in the Stove of the
Committee room is defective. The attention of the
Managers is requested to the inconvenient state of
the Water Closet.
Resolved, That the above recommendations of
the Visitors be referred to the Superintendant of the House

195

April 7.

6. Read a letter from Dr. Edmund Clark offering to
Lectures — give a Course of six lectures on Botany in the
present Season at the Royal Institution.
Resolved, That Dr. Clark's proposal be
referred to the Committee of lectures.

7. The following persons having been duly proposed
were admitted as Subscribers to the Royal Institution
for the year, 1828.

Annual Subscribers.
Howard Elphinstone, Esqr. 21 Green St. Grosvr. Square
Captn. Phillott of Kensington Gore
Thomas Little, Esqr. 9 Upp. Gloucester St. Dorset Sq.
Francis Wm. Staines, Esqr. 72 Charlotte St. Fitzroy Sq.
The Hon: J. S. Wortley 77 South Audley Street
To the Lectures,
Miss Hope 37 Upper Seymour St.
Mrs. Reeve 11 Gt. Cumberland Place
To Mr. Wesley's lectures.
Miss Hill
Miss Callcott } 6 Portman Square

Adjourned to Monday the 21st.
Instant at Three o'Clock.
Confirmed the 21st day of April 1828
— Somerset Pres.

196

1828.

April 21. At a Meeting of the Managers
of the Royal Institution of Great Britain,
held at the house of the Institution on
the 21st. of April, 1828.
Present
His Grace The Duke of Somerset Presdt. in the Chair.
Sir Claude Scott, Bart. V.P.
Sir S. Bernard Morland, Bart. M.P. Treasr.
Benjamin Bond Cabbell, Esqr. V.P.
Henry Thos. Colebrooke, Esqr.
Chas. H. Pilgrim, Esqr.
George Moore, Esq.
Edmund Robt. Daniell, Esqr. Secretary.
Captn. Sabine.

1. The minutes of the proceedings of the
last meeting were read and approved.

2. The Secretary communicated the following
Gold Medals presented by J. Fuller, Esq. — proposal from John Fuller, Esqr.
That in order to give an additional
stimulus to the progress of Chemistry, Mr.
Fuller intends to appropriate a Gold Medal
as a reward for Chemical discoveries, and he
proposes that the Managers of the Royal Institutn.
of Great Britain together with the Professor of
Chemistry for the time being, should nominate
every two years on his Birth Day, (the 20th of
February,) any person being a Member of the
Institution, whom they may consider as having
merited the distinction), or if no Member of the
Institution should in the opinion of the Managers

197

April 21.

and Professor of Chemistry be entitled to it, then they
may name such other person as they may judge worthy
of such an honour, provided however that if any
material difference of opinion should occur amongst
the Members of the Board of Managers (which Mr.
Fuller can hardly contemplate,) or if an equality of
Votes should take place, Mr. Fuller is to be at liberty
to suspend the delivery of the Medals till such differ-
ences are reconciled; or, in the latter case, (that of an
equality of Votes) Mr. Fuller himself shall have the
casting or decisive vote.
The Dies are meant for the Royal Institution
in case of Mr. Fullers' decease.
Mr. Fuller has also presented Gold Medals
to the following Members, vizt.
Gold Medals. — Sir Humphry Davy, Bart.
Dr. Wollaston
Chas. Hatchett, Esqr.
William Thos. Brande, Esqr.
John George Children, Esqr.
John Frederic Daniell, Esqr.
Michael Faraday, Esqr.
Silver Medal. — and a Silver Medal to Mr. John Newman,
occasional Assistant in the Laboratory, and
Philosophical Instrument Maker to the Roy. Instn.
Resolved, That the thanks of the Managers be returned to
Mr. Fuller for his liberal offer. —
3. On the application of Wm. Manning, Esqr. M.P.
Hon. Admn. — The Baron Wetterstedd
was admitted to the Lectures &c. under Art: 15 Chap:
VI of the Bye laws.

4. The Secretary reported that he had arranged

1828

Lectures. Dr Clarke. with the assent of the President, that Dr Clarke should deliver a Course of 4 Lectures on Botany to commence on Wednesday the 7th of May at Three o'clock.

Foreign Correspondt. 5 Read a letter from M. Frederick de Bibra informing the Managers that Dr. I. Theophilus Schinko wished to correspond with the Royal Institution of Great Britain.
Resolved, That the Secretary be requested to reply to the letter.

Lectures 6. Read letters from Mr. Gilbert Burnett proposing to deliver lectures on Botany.
Resolved, That Mr. Burnett's proposal be referred to the Committee of Lectures.

Presents received.

7. C. Moreau, Esqr. presented
Chronological Records of British Finance,
A. A. Lindo, Esqr. presented a pamphlet on the injurious tendency of modifying the Navigation Laws.
Dr. Whitelaw Ainslie presented
Clemenza; or the Tuscan Orphan, a Tragedy.
_ Burke, Esqr. presented
Print of Richard Burke Recorder of Bristol,
The Revd. Stephen Weston presented
Italian Classics edited by Mathias, 18 vols.
Davies Gilbert, Esqr. M.P. Pres: R.S. presented
The Creation and Noah's Flood written in Cornish with an English Translation.
Mount Calvary a Mystery written in Cornish

with an English Translation by Keigwin.
Ordered, That the thanks of the Managers be returned for these presents.

8. Read the Visitors Report.
Resolved,
Special Meeting of Visitors called resp. Investment. That the President be requested to call a Meeting of the Visitors for the purpose of reconsidering that part of their Report which relates to the rescinding so much of the resolution of last year as directs the surplus of the Compositions at the end of each year to be invested in the purchase of 3 Pr Ct. Consols, and which recommends the Members to direct the £188. 6. 8 in the Report mentioned, to be applied towards discharging the floating debt; and to confer with a deputation from the Board of Managers upon the subject.

Whereupon the President in pursuance of the above request, appointed a Special Meeting of the Visitors to be called on Thursday next the 24th Instant at half past Two o'Clock, p.m.

Resolved, That the following Gentlemen be a deputation to confer with the Board of Visitors on the subject of their Report _ viz.
Sir S. Bernard Morland, Bart
Captn. Sabine,
Mr George Moore,
Mr. B. B. Cabbell,
& The Secretary.

1828

Ordered, That the House lists of Officers be filled up as follows.

President _ The Duke of Somerset

Treasurer _ Sir S. Bernard Morland, Bt. M.P.

Secretary _ Edmund Robt. Daniell, Esqr.

Managers.
Charles Barclay, Esqr. M.P.
Benjamin Bond Cabbell, Esqr.
Henry Thos. Colebrooke, Esqr.
Sir George Duckett, Bart
Isaac Lyon Goldsmid, Esqr.
Joseph Jekyll, junr. Esqr.
George Moore, Esqr.
R. I. Murchison, Esqr.
Whitlock Nicholl, M.D.
Wm. Hasledine Pepys, Esqr.
Charles Pilgrim, junr. Esqr.
Captn. Edward Sabine
Sir Claude Scott, Bart
Wm. Somerville, M.D.
Edward Sterling, Esqr.

Visitors.
Thos. Tyringham Bernard
James Broadwood, Esqr.
Arthur de Capel Brooke Esqr.
Lord Carrington
John Frederic Daniell, Esqr.

John Disney Esqr.
Earl of Darnley
John Fuller, Esqr.
Davies Gilbert, Esqr. M.P. Pres. R.S.
Thomas Hoblyn, Esqr.
Henry Gally Knight, Esqr.
George Long, Esqr.
William Noble Rule, Esqr.
George Stavely Smith, Esqr.
Richard Horsman Solly, Esqr.

Confirmed the 5th day of May 1828
Somerset Pres.

1828.

May 5. **At a Meeting of the Managers of the Royal Institution of Great Britain held on the 5th of May 1828**

Present

His Grace The Duke of Somerset, Prest in the Chair.
Benjamin Bond Cabbell, Esqr.
Sir George Duckett, Bart.
Joseph Jekyll junr. Esqr.
George Moore, Esqr.
Dr. Nicholl,
William Hasledine Pepys, Esqr.
Charles Pelgrim, junr.
Captn. Sabine,
Dr. Somerville,
Edward Sterling, Esqr.
Sir S. Bernard Morland, Bart. M.P. Treasr.

1 The minutes of the proceedings of the last meeting were read and approved.

Rept. of Com. of Accts. 2. Read the following report from the Commee of Accounts.

The Committee of Accounts report that they have examined the General Accounts of the Royal Institution of the last month, and they found a balance in the hands of the Treasurer of £576.12.10, and in the hands of the Assistant Secretary of £90.12.11.

They recommend the following Accounts to be paid.

Payments.
Mr. Wesley the balance of his acct. £51.14.–
Gardner for Oil in 1827 — 28.6.9
Pringle for Upholstery in ditto — 36.7.–

May 5.

Approved of; and signed an Order for an advance to the Assistant Secretary of £200.—

Donation from G. Garrow, Esq. 3. Read a letter from George Garrow Esqr. M.R.I. addressed to the Assistant Secretary dated Madras 20th August 1827, regretting that he had not the satisfaction of seeing and taking leave of his friends at the Royal Institution previous to his leaving England and begging to enclose a small pledge of his attachment to the Royal Institution (£10) which he wished to be carried to the Laboratory Account in lieu of his annual subscription.

Resolved,
That the thanks of the Managers be returned to Mr. Garrow for his kind communication.

4. Presents received.

The Directors of the Maclurian Museum presented Nos. 1 and 2 of the Contributions of the Museum
Dr. Whitlock Nicholl presented 180 Volumes of Medical Books,
John Carne, Esqr. presented Tales of the West – 2 vols. 12mo
William Yarrell, Esqr. F.L.S. presented The Osteology of the Chlamyphorus Truncatus
John Disney, Esqr. M.R.I. presented Specimens of green Serpentine Marble, red Porphory and the gallo Antico of the Ancients and a Lithographic Drawing of his suspension Bridge.

Ordered, That the thanks of the Managers be returned for these presents –

1828.

5 Resolved, That the following Committees be appointed for the present year.

Committees. Committee for regulating the Friday Evening Meetings of the Members.
The Duke of Somerset Prest.
George Moore, Esqr.
Dr. Nicholl
Captn. Sabine
Wm. Hasledine Pepys, Esqr.
Edmund Robt. Daniell, Esqr. Secy. R.I.
William Thos. Brande, Esqr.
Michael Faraday, Esqr.

Committee for arranging the Lectures for the ensuing Season
The Duke of Somerset
Benjamin Bond Cabbell, Esqr.
Chas. H. Pelgrim, Esqr.
Sir George Duckett, Bart.
Dr. Somerville
George Moore, Esqr.
Dr. Nicholl,
Wm. Hasledine Pepys, Esqr.
Edmund R. Daniell, Esqr. Secy.

Committee of Accounts. —
The President
Treasurer
Secretary
George Moore, Esqr.
Benjamin Bond Cabbell Esqr.
Dr. Nicholl.

May 5.

6 The following persons having been duly recommended were admitted as Subscribers for the year, 1828.

Annual Subscriber.
Revd. Edward Williams 4 Hanover St. Hanover Square

To the Lectures.
Mrs. Lathom 112 Mount Street

To Mr. Reinagle's Lectures
Sir Thomas Baring, Bart. }
Thomas Baring, Esqr. }
Miss Baring } 21 Devonshire Place
Miss E. Baring }
Miss Sealy }
Miss Maitland
Mr. & Miss Domville } of Highgate
Miss S. Domville }
Miss Paton – Upper Gower Street

To Mr. Faraday's Lectures.
Mr. Charles Palairet
Thomas Baring, Esqr.
Miss Baring, Miss E. Baring
Miss Sealy.

Adjourned to Wednesday the 21st Instant at Three o'Clock.

Claude Scott

1828.

May 21. At a Meeting of the Managers of the Royal Institution of Great Britain, held on the 21st of May, 1828

Present

Sir Claude Scott, Bart. V.P. in the Chair
Sir George Duckett, Bart. V.P.
George Moore, Esq.r V.P.
Benjamin Bond Cabbell, Esq.r V.P.
Dr. Somerville, V.P.
Edward Sterling Esq.r
Chas. H. Pilgrim, Esq.r
Wm. Hasledine Pepys, Esq.r
Captn. Sabine,
Isaac Lyon Goldsmid, Esq.r
Sir S. Bernard Morland, Bart M.P. Treas.r
Edmund Robt. Daniell, Esq.r Secretary

1. The minutes of the proceedings of the last meeting were read and approved.

2. (Life Subs.er) In pursuance of a nomination under the Act of Parliament by William Bridgman, Esq.r of Frogmore, Weston, Herefordshire, his Son William Bridgman jun. Esq.r of Weston Herefordshire was admitted to the privileges of a Life Subscriber to the Royal Institution.

Presents received.

3. Robert Bakewell, Esq.r presented Introduction to Geology, 2d Edn. 8vo.
The Editors of the Zoological Journal presented No. 12 of their Journal,

May 21.

Granville Penn, Esq.r presented "Comparative Estimate of the Mineral & Mosaical Geologies, 2d Edn. 2 vols. 8vo.
William Terry, Esq.r presented Græca Grammatica, 8vo.
Edwin Atherstone, Esq.r presented The fall of Nineveh. 8vo.
Isaac Lyon Goldsmid, Esq.r presented A Letter to the Duke of Wellington.

Charles B. Calmady Esq.r M.R.I. presented A Specimen of Schorl rock from Dartmoor Devonshire.
Charles R. Bigge, Esq.r M.R.I. presented Models of some very extraordinary Diamonds now in this Country.
Mr. Bennett presented "Ornamental Printing"
Ordered, That the thanks of the Managers be returned for these presents.

4. Read the minutes of the last General Meeting.

(Bond — continuance of thought necessary) Resolved, That the Managers are of opinion that under the provisions of the Act of Parliament the taking of a security from Members of the Royal Institution who do not compound for their annual payments, cannot be dispensed with; and that a bond is the most eligible species of security that can be adopted.

5. The Secretary having communicated to the Board

1828.

(Formation of Med. & Scientific Library) that Dr. Nicholl who has presented a large collection of Medical Books to the Library had suggested that it would be desirable to appropriate a Room to the formation of a Medical and Scientific Library, and that he had made the donation in question with the view of forwarding such an arrangement.
Resolved, That this meeting do approve of the suggestion of Dr. Nicholl, and request the Secretary to communicate the same to the Patrons at their next meeting.
(Newspapers removed to Lower Library) Resolved, That for the purpose of affording facilities to the adoption of Dr. Nicholl's suggestion the Lower Library be in future appropriated to the purposes of a Newspaper Room, and that the Newspapers &c. be placed therein.

6 (Carpet) Resolved, That the Secretary be authorized to purchase the Carpet now in the Lower Library for the use of that Room.

7. The following persons having been duly recommended were admitted as Subscribers for the year 1828.

To the Lectures.
Mrs. Hanham 29 Park St. Grovr. Square
To Mr. Faraday's Lectures.
Mr. C. Herries.

Adjourned to Monday the 2d. of June.
Confirmed the second day of June 1828
Somerset Pres.

June 2. At a Meeting of the Managers of the Royal Institution held on the 2d. of June, 1828

Present

His Grace The Duke of Somerset Pres.t in the Chair
Captn. Sabine,
George Moore, Esq.r V.P.
Benjamin Bond Cabbell, Esq.r V.P.
Chas. H. Pilgrim, Esq.r
Edmund Robt. Daniell, Esq.r Secretary.

1. The minutes of the proceedings of the last meeting were read and approved.

2. (Rept. of Com. of Accts.) Read the following report from the Committee of Accounts.
The Committee of Accounts report that they have examined the General Accounts of the Royal Institution of the last Month, and they found a balance in the hands of the Treasurer of £564. 11. 10 and in the hands of the Assistant Secretary of £139. 19. 11.
They recommend that the following Accounts be paid.

(Payments)

Weston Slater for repairs in 1826 & 1827	£5. 3. 4
R. R. Reinagle for a Course of Lectures in 1828	63. —
Dr. Harwood for ditto	52. 10
Mr. Alfred Ainger, Esq.r do.	31. 10 —
Wm. Thos. Brande, Esq.r for a years Salary as Professor of Chem.y	150. —
	302. 3. 4

Approved of; And signed an Order for an Advance to the Assistant Secretary of £300.

3. On Med. & Scientific Library. The Secretary reported that he communicated to the Patrons of the Library at their last meeting, Dr Nicholl's proposition to appropriate a Room to the formation of a Medical and Scientific Library, who had given directions to the Librarian to take measures for carrying it into execution at the earliest convenient Season.

4. Present received.

George Philip Reinagle, Esqr presented Illustrations of the Battle of Navarin.

The Horticultural Society presented Transactions of the Society vol 7 part 2.

John Whitchurch Bennett, Esqr presented "Fishes of Ceylon".

Ordered, That the thanks of the Managers be returned for these presents.

5. Foreign Correspondence. Read a letter from Messrs G. C. Johnson and I. T. Baldwin accompanied with the Transactions of the Maclurian Lyceum of Philadelphia, expressing a desire to cultivate a Correspondence with the Royal Institution.

Resolved, That the Secretary be desired to convey the thanks of the Managers to the above Gentlemen for the Transactions of the Maclurian Lyceum, and to express their readiness on the part of the Royal Institution to meet the wishes of the Maclurian Lyceum by cultivating any Correspondence

which may tend to promote the diffusion and extension of useful knowledge throughout the World.

6. Lectures. Read a letter from Mr Brayley proposing to deliver a Course of Lectures on the Phenomena and History of Igneous Meteors and Meteorites.

Resolved, That Mr Brayley's letter be referred to the Committee of Lectures.

7. Alterations in House. Ordered, That the Superintendent of the House be directed to procure estimates of the expense of taking down the partition between the Apparatus Room and the passage to the Lecture Room, and of making good the different parts, to be laid before the next Meeting of Managers.

Confirmed the 7th day of July 1828
Somerset Pres.

July 7.

At a Meeting of the Managers of the Royal Institution of Great Britain held on the 7th of July, 1828,

Present

His Grace The Duke of Somerset, Pres.t in the Chair.
George Moore, Esqr V.P.
Benjamin Bond Cabbell, Esqr V.P.
Sir S. Bernard Morland, Bart. M.P. Treas.r
Isaac Lyon Goldsmid, Esqr
Dr Nicholl.
Wm Hasledine Pepys, Esqr
Edmund Robt Daniell, Esqr Secretary.

1. The minutes of the proceedings of the last meeting were read and approved.

2. Rept of Com. of Acc.ts Read the following report from the Committee of Accounts.

The Committee of Accounts report that they have examined the General Accounts of the Royal Institution of the last month, and they found a balance in the hands of the Treasurer of £601.12.10; And in the hands of the Assistant Secretary of £18.0.8.

They recommend that the following Accounts be paid.

Mr Faraday for a Course of Lectures On the operations of the Laby delivered in 1828	£52.10.–
Mr Millington for lectures in do	52.10.–
The Marquis Spineto for a Course of lectures on Hieroglyphics delivered in ditto	31.10.–
	136.10.–

Brot forwd	136.10.0
2 quarters Assessed taxes & Easter Offering to Lady day 1828	63.6.8
2 quarters Parochial rates to do	38.6.8
Ridgway for Newspapers in 1826	2.14.10
Allnutt for Matting in do	6.1.–
Mortlock for China for the Weekly Evening Meetings	13.18.0
Gas Company for Gas in 1828	20.15.4
Dr E. J. Clark for his expenses in delivering Lectures on Botany	9.10.–
Mrs Stutt for Stationery in 1828	28.15.4
Joseph Fincher on account of the Royal Institution	240.–
	£559.17.10

And signed an Order for an Advance of £560 to Treas.

3. Lectures. M. de Spineto. Read a letter from The Marquis Spineto, accompanied with various letters of recommendation offering to deliver a Course of lectures on the nature and grammar of the most popular Foreign languages, and of the Italian in particular.

Resolved, That The Marquis Spineto's proposal be referred to the Committee of lectures.

4. Dr Clarke. Read a letter from Dr Edmund J. Clark proposing to deliver a Course of 6 Lectures on Botany in the next season.

Resolved, That Dr Clark's proposal be referred to the Committee of lectures.

5. Alterations & repairs. Resolved, That a Committee be appointed to superintend the alterations and repairs necessary to be done in the house during the ensuing Vacation,

1828.

And that they be instructed to limit the expense to the sum of £100.

That the Committee consist of

Mr. G. Moore

Mr. Pepys

and The Secretary

to be assisted by Professor Brande and Mr. Faraday.

6. Alterations. Read estimates of the expense of taking down the partition between the Apparatus room and the passage to the Lecture room, pursuant to resolution (No. 7) of the last meeting.

Resolved, That Mr. Ramsay's estimate be approved of; and that he be directed to proceed with the work at an expense not exceeding £17 . 10 –

7. Blinds. Resolved, That blinds be put up in the Windows of Mr. Faraday's rooms.

8. Old Newspapers. Ordered, That Mr. Singer do look over the old Newspapers in the Model room and report to the Managers what they are and which of them will be proper to retain for the use of the Institution.

Confirmed the 3rd Day of November 1828
Somerset Pres.

Nov. 3.

At a Meeting of the Managers of the Royal Institution of Great Britain, held on the 3d. of November, 1828

Present

His Grace The Duke of Somerset, President in the Chair

Isaac Lyon Goldsmid, Esqr.

George Moore Esqr.

Joseph Jekyll junr. Esqr.

Chas. H. Pilgrim, Esqr.

Dr. Nicholl,

Wm. Hasledine Pepys, Esqr.

Sir S. Bernard Morland, Bt. M.P. Treasr.

Edmund Robert Daniell, Esqr. Secy.

1. The minutes of the proceedings of the last meeting were read and approved.

2 Life Subsr. Read the following nomination of a Life Subscriber.

In pursuance of a nomination under the Act of Parliament by The Revd. Basil Woodd, his Son William James Tilson Woodd, Esqr. of No. 10 Paddington Green, was admitted to the privileges of a Life Subscriber to the Royal Institution.

3. Lectures Mr. Philipps. Read a letter from G. S. Pattison Esqr. M.R.I. accompanied with a prospectus of a Course of lectures on Singing and Vocal Music proposed to be delivered by Mr. Philipps at the Royal Institution in the ensuing Season.

Resolved, That it be referred to the Committee of lectures.

1828.

4. Dr. G. Smith. Read a letter from Dr. Gordon Smith proposing to resume his Course of lectures on State Medicine, and to deliver in the ensuing season A Course of Lectures "On the Philosophy of ordinary Life".

Resolved, That it be referred to the Committee of lectures.

5. Mr. Thelwall. Read a letter from Mr. J. Thelwall proposing to deliver a Course of lectures on Elocution.

Resolved, That Mr. Thelwall's proposal be referred to the Committee of lectures.

6. Mr. T. Stackhouse. Read a letter from Mr. T. Stackhouse offering to deliver gratuitously a Course of lectures on the still existing Monuments of primeval Britain.

Resolved, That Mr. Stackhouse's proposal be likewise referred to the Committee of lectures

Presents received.

7. Wm. Shaw Mason, Esqr. presented

Statistical account; or Parochial Survey of Ireland, 2 vols. 8vo.

James Burton junr. Esqr. presented

Excerpta Hieroglyphica, No. 2

R. Duppa Esqr. M.R.I. presented

Travels in Italy, Sicily & the Lipari Islands 12mo.

Board of Longitude presented

Nautical Almanac for 1831.

The Royal Society presented

Transactions of the Society pts. 3 & 4, 1826, pt. 1 & 2 1827, & part 1 & 2. 1828.

Nov. 3.

The Editors presented The Zoological Journal No. 14.

His Excelly. The Visconde de Santarem presented

Memorias para a Historia e Thoria das Cortes Geraes.

Sir George Duckett, Bart. presented

A Brief Commentary on the Revelation of St. John & The Burial and Resurrection of Jesus Christ, from the German of Michaelis.

Captn. Brant M.R.I. presented to the Museum: Three Shells.

John Hodgson, Esqr. M.R.I. presented to the Museum, two Land Crabs.

Ordered, That the thanks of the Managers be returned for these presents.

8. Mr. Singer agreeably to resolution No. 8 of the last meeting, laid before the Board the following list of Newspapers.

Newspapers.

List of Newspapers.

The Times from 1800 to 1827 half bound

The Courier from 1800 to 1827 ditto

Dublin Evening Post from 1810 to 1827

Edinburgh Courant from 1801 to 1827

Irish Farmers Journal from 1817 to 1826.

Moniteur – 11 Vols. half bound.

Public Advertiser 14 Vols. half bound.

Morning Post from 1800 to 1810 half bound

Morning Chronicle – ditto ditto.

London Gazette from 1800 to 1827, half bound.

Literary Gazette from 1820 to 1827 ditto.

218

1828.

Journal de Manheim, Journal des Debats & others
19 Vol^s. 4to. half bound.
30 Volumes Miscellaneous half bound.
Large quantity of Newspapers in Model-room unbound.

Mr Faraday to attend at Managers' Meetings

9. Resolved, That Mr. Faraday be requested to attend at the Sittings of the Board of Managers.

Adjourned to Monday the 17th. Instant at Three o'Clock.

Confirmed the 17 Nov 1828
Geo Duckett

219

Nov. 17. At a Meeting of the Managers of the Royal Institution of Great Britain, held on the 17th. of November, 1828

Present
Sir George Duckett, Bart. V.P. in the Chair.
Chas. H. Pilgrim, Esqr.
Edward Sterling, Esqr.
Joseph Jekyll jun^r. Esqr.
George Moore, Esqr.
R. I. Murchison, Esqr.
Benjamin Bond Cabbell, Esqr.
Dr. Nicholl.
Dr. Somerville V.P.
Edmund Robt. Daniell, Esqr. Secretary.
William Thos. Brande, Esqr. Prof^r. Chem^y.
Michael Faraday, Esqr. Director of the Lab^y.

1. The minutes of the proceedings of the last meeting were read and approved.

Presents received.

2. John Hatherley, Esqr. of Teignmouth, presented Specimens of Bovey Coal and of Iron Ores from Devonshire.
John Frederic Daniell, Esqr. presented Specimens of Birds.
Mr. Pitts presented the second number of a series of outlines from the Æneis of Virgil.
Ordered, That the thanks of the Managers be returned for these presents.

3. Resolved, That a Committee be appointed to

220

1828.

take into consideration the manner in which the Royal Institution is to be lighted during the ensuing year, consisting of the following Gentlemen, vizt.

Comm^e. resp^g Lighting.

Mr. Moore,
Mr. Jekyll,
The Secretary,
Mr. Brande,
Mr. Faraday.

4. Read letters from Mr. Drew relating to the fitting up of Gasometers in the Royal Institution.
Resolved, That they be referred to the above Committee.

Repairs.

5. Read the following report.
The Committee appointed to superintend the alterations and repairs directed to be made in the Royal Institution during the Vacation, report
That the Laboratory has been completely whitewashed.
That the lower Library has been painted all over, except the interior of the Book cases.
That the partition between the Apparatus room and the passage to the Lecture room has been taken down, and the different parts have been made good and the whole painted agreeably to the estimate of Mr. Joshua Ramsay.
That the passage from the Hall to the Reading room has been cleansed and whitewashed.
The Committee likewise report that the

221

Nov. 17.

repairs recommended by the Visitors have been completed, vizt.
The lantern roof and the Cornice in the back part of the House has been repaired. –
The Chimney in the Stove of the Committee room and the Hearth have been repaired.
And the Window broken in the Museum has been replaced.
The Committee also report that the estimated amount of the above mentioned repairs and alterations will not exceed the sum of forty guineas.

Mr. Singer exonerated from payment as Member.

6. Resolved, That as a mark of their approbation of the conduct of Mr. Singer in his office as Librarian of this Institution, the Managers recommend that the payments made by him as a Member of the Royal Institution be returned to him; And that he be exonerated from all future payments, during such time as he shall continue to fill the Office of Librarian. And they submit the above resolution to the next General Meeting of the Members.

7. Resolved, That the Weekly Evening Meetings of the Members do commence for the ensuing season on Friday the 23^d of January next.

8. Resolved, That Mr. Jekyll be added to the Committee for conducting the Friday Evening Meetings.

9. Resolved, That the following Gentlemen be

222 1828

appointed on the Committee for arranging the lectures viz[t].

Mr. Sterling
Mr. Murchison
Mr. Jekyll

10 Resolved, That Mr. Faraday be at liberty to order a Carpet for his Room.

11. Life-preserving Waistcoat. Read a letter from the Commissioners of His Majesty's Customs transmitting for the inspection of the Managers a Life Preserving Waistcoat invented by Lt. Andrew Kennedy R.N.

Resolved, That the Secretary be requested to acknowledge this communication, And that it be referred to the Committee for conducting the Friday Evening Meetings.

12. Regulations for Hall & Porters. The following regulations for the Hall of this Institution were presented by Mr. Faraday. And the Porter having been called in, they were read to him by the Chairman.

1. – The Hall is to be in order and the doors to be opened by 9 o'clock in the Morning.
2. – The Porter in charge of the Hall must not leave the house for any purpose until his companion be ready to take his place.
3. The Hall is never to be left without a Porter during the hours the house is open, except for the brief purposes of answering a question or delivering a message to some one in the house, or stirring a fire in the public rooms.

Nov. 17. 223

4. The Porter in charge of the Hall is to be so evidently in the exercise of his duty, that he may see and be seen of every person who enters the house. It is a strict rule that the Hall never be left unwatched except for the necessary and short intervals above mentioned.
5. No manufacturing employment is to be carried on in the Hall of the Institution. The Porters are to be clean and to be disengaged from any other duty whilst in the Hall, than that of attending to applications and to the convenience of the Members and Subscribers.
6. The Porters are to preserve a clean copy of these regulations affixed to a Board; and to produce them when required.

Confirmed the first day of December 1828
Somerset Pres.

224 1828

Dec. 1. At a Meeting of the Managers of the Royal Institution of Great Britain, held on the 1st of December, 1828,

Present

His Grace The Duke of Somerset Pres.t in the Chair
Sir George Duckett, Bart. V.P.
Wm. Hasledine Pepys, Esqr.
Charles H Pilgrim, Esqr.
Edward Sterling, Esqr.
Benjamin Bond Cabbell, Esqr. V.P.
George Moore, Esqr. V.P.
Joseph Jekyll jun.r Esqr.
Sir S. Bernard Morland Bt. M.P. Treas.r
Edmund Robt. Daniell, Esqr. Secy.
Captn. Sabine,
Michael Faraday, Esqr. Director of the Laby.

1. The minutes of the proceedings of the last meeting were read and approved.

2. Rept. of Com.e of Acc.ts Read the following report from the Committee of Accounts.

The Committee of Accounts report that they have examined the General Accounts of the Institution during the recess. And they found a balance in the hands of the Treasurer of £ 319 . 14 . 10; And in the hands of the Assistant Secretary of £ 78 . 5 . 1.

They recommend the following Accounts to be paid.

2 quarters Assessed taxes to Michas	£ 61 . 4 . 9
2 ditto Parochial rates to ditto	40 . 16 . 8
	102 . 1 . 5

Dec. 1. 225

Brot. forwd	102 . 1 . 5
4 quarters Water rate to Michas	10 . 10 . –
Mr. Ramsay for making good & painting the room adjoining the lecture room	17 . 10 . –
Mr. Wade for repairs in Mr. Singer's Apartments	2 . 3 . 7
H Greenwood Sundries for the house	2 . 2 . 2
Joseph Fincher on account of the Royal Institution	165 . 12 . 10
	300

And signed an Order for an advance to the Assistant Secretary of £ 300.–

3. Life Subs.crs Read the following nomination of a Life Subscriber.

In pursuance of a nomination under the Act of Parliament by Thomas William Carr, Esqr. his Son William Ogle Carr, Esqr. of Stone Buildings, Lincolns Inn, was admitted to the privileges of a Life Subscriber to the Royal Institution.

4. Mr. Singer. Read a letter from Mr. Singer the Librarian, returning his sincere thanks to the Managers for their kind intentions towards him as expressed in the minute of a resolution passed by them at their last meeting.

Presents received.

5. Messrs. Kennedy & Grainger presented State of the Tenancy of Land in Great Britain
Sir Gilbert Blane, Bart. presented Select Dissertations on Medical Sciences.

1828.

Earl Spencer presented
"Havelok the Dane" 4to.
Ordered, That the thanks of the Managers be returned for these presents.

Rept. of Accts. 6. Ordered, That the following report of the Managers be presented to the General Monthly Meeting of this day.

The Managers report that they have investigated the probable Receipts and Expenditure of the Royal Institution for the present year; And they find that the total receipts on the General Account of the Institution will probably amount to £2400, including the Composition of Members, the amount of which will be about £700, which being deducted from the above sum of £2400, will leave the sum of £1700.

That the expenditure of the Institution as far as they have been able to estimate the same will amount to £2300 including the expenses of Repairs and all Outgoings to Christmas.

Adjourned to Monday the 15th Instant at Three o'Clock.

Confirmed the 15th day of December 1828
Somerset Pres.

Dec 19. At a Meeting of the Managers of the Royal Institution of Great Britain held on Monday the 19th of December 1828,

Present

His Grace The Duke of Somerset Pres: in the Chair
Sir George Duckett, Bart. V.P.
George Moore, Esqr. V.P.
Benjamin Bond Cabbell, Esqr. V.P.
Chas. H. Pilgrim, Esqr.
Dr. Somerville V.P.
Wm. Hasledine Pepys, Esqr.
Joseph Jekyll junr. Esqr.
Isaac Lyon Goldsmid, Esqr.
Sir S. Bernard Morland, Bart. M.P.
Dr. Nicholl,
Captn. Sabine,
Edmund Robt. Daniell, Esqr. Secy.
Michael Faraday, Esqr.

1. The minutes of the proceedings of the last meeting were read and approved.

on Lighting R.I. 2. Read the following report.

The Committee appointed to take into consideration the manner in which the Royal Institution is to be lighted during the ensuing year, report, that it does not appear to them advisable to make any alteration in lighting the House from that adopted last year.

3. Read an estimate of the expense of making a Curtain for the Ante lecture room which amounted to £10. –

1828.

Resolved, That the estimate be approved of.

"Standard" ordered. 4. Ordered, That the Standard Newspaper be taken in commencing on the 1st. of Jany. next.

5. Resolved, That the sum of £2 . 7 . 6 be allowed to Mrs. Greenwood the Housekeeper, for Candles supplied to Mr. Singer. –

Confirmed the 2d. Day of Feb. 1829
George Moore

1829.
Feb. 2. At a Meeting of the Managers of the Royal Institution of Great Britain held on the 2d. of February, 1829

Present

George Moore, Esqr. V.P. in the Chair.
Sir S. Bernard Morland, Bart. M.P. Treas.
His Grace The Duke of Somerset, Pres.
Sir George Duckett, Bart. V.P.
Benjamin Bond Cabbell, Esqr.
Dr. Nicholl,
Chas. Pilgrim, Esqr.
Joseph Jekyll junr. Esqr.
Captn. Sabine,
Edmund Robert Daniell, Esqr. Secy.
Michael Faraday, Esqr. Director of the Laby.

1. The minutes of the proceedings of the last meeting were read and approved.

Rept. of Comm. of Accts. 2. Read the following report from the Committee of Accounts.

The Committee of Accounts report that they have examined the General Accounts of the Royal Institution, and they found a balance in the hands of the Treasurer on the 31st. of December last of £42 . 18 . 6; And in the hands of the Assistant Secretary of £76 . 4 . 3.

They also report that they have examined the Laboratory Account, and found a balance in the hands of the Treasurer of £172 . 8 . 11, And a balance due to the Assistant Secretary on the 31st. December last of £120 . 19 . 3.

They likewise report that they have audited the Treasurers' General Account for the year 1828, and also the Treasurers Account for the Laboratory Fund for the same period.

3. In pursuance of a nomination under the Act of Parliament by Thomas Lewis, Esqr. his Daughter Arabella Lewis of Sussex Place Regents Park, was admitted to the privileges of a Life Subscriber.

Presents received.

4 George Stavely Smith, Esqr. M.R.I. presented Lord Byrons Works complete in 1 vol. 8vo.
Revd. S. Weston presented Supplement to German Grammar,
Thos. I. Mathias, Esqr. presented La Mutabilita Poema in due Canti
Sir George Thos. Staunton, Bart presented Miscellaneous Notices relating to China,
The Commissioners of Public Records presented Calendarium Inquisitionem post mortem vol. 4.
Dr. Granville, M.R.I. presented Travels to St. Petersburgh, 2 vol. 8vo.
The Royal Asiatic Society presented Transactions of the Society vol. 2 pt. 1st.
Geological Society presented Transactions of the Society vol. 2 pt. 3.
Wm. Brockedon, Esqr. presented Passes of the Alps Nos. 7 and 8.
Ordered, That the thanks of the Managers be returned for these presents —

Porters' Wages. 5 Read a letter from the Porters respectfully requesting the Managers to allow them an increase of Wages.

Mr. Pearsall's Salary. 6. Mr. Faraday in the absence of Mr. Brande applied to the Board on the part of Mr. Thos. Pearsall, Assistant in the Laboratory for an increase of Salary.
Resolved, That the above requests be taken into consideration after the Annual Meeting in May.

House keeper's Estimate ~~Salary~~ 7. Read an estimate from Mrs. Greenwood the Housekeeper of the Articles to be supplied by her agreeably to her Contract.
Resolved, That the Secretary, Mr. Brande, and Mr. Faraday be a Committee to make an agreement with Mrs. Greenwood.

8 The following persons having been duly recommended were admitted as Subscribers to the Royal Institution for the year, 1829

Annual Subscribers.
The Revd. James Dunn 2 North Bank Regents Park
John Malleson, Esqr. 12 Queen St. May Fair.
Francis Douce, Esqr. 15 Upper Gower St.
Wm. Crichton, Esqr. 44 Upper Gower St.
To the Lectures.
Mrs. & Miss Hume 6 Bryanstone Square
Mrs. Balmanno 7 Craven St. Strand
Mrs. Jas. Thompson George St. Hanover Sq.
Mrs. Roy 283 Regent St.
Mrs. C. Richardson, Miss Richardson & Miss Wilcott & Golburn(?)
Confirmed the 2nd day of March 1829
Somerset Pres.

Feb. 16. At a Special Meeting of the Managers of the Royal Institution held on Monday the 16th. of February, 1829

Present
The Duke of Somerset Prest. in the Chair.
Sir George Duckett, Bart. V.P.
Dr. Nicholl,
Chas. Pilgrim,
Captn. Sabine,
Joseph Jekyll, junr. Esqr.
Sir S. Bernard Morland, Bt. M.P. Tres.
Michl. Faraday, Esqr. Director of the Laby.

1. The object of the meeting having been stated,
Robbery in Hall. &c. Resolved, That Wm. Basil Wake be prosecuted forthwith for stealing Mr. Wade's Cloak and also two books belonging to the Royal Institution.
That the Costs of the prosecution be defrayed by this Institution.
That Mr. Wm. Ballard, Police Officer of Marlborough Street be employed to conduct the prosecution on the part of the Institution.
That the books belonging to this Institution and now in pawn, be redeemed.

2. That a Committee be appointed to take into consideration the state of the Hall, and to report to the Managers at their next meeting, what measures it would recommend for insuring a regular attendance and securing the property of the House.

That the Committee do consist of
Sir George Duckett, Bart.
Captn. Sabine
Mr. Pilgrim,
Dr. Nicholl
The Secretary
and Mr. Faraday.
of whom any three to be a quorum.
Confirmed the 2nd day of March 1829
Somerset Pres.

234

Feb 18 — Feb. 18.

At a Special Meeting of the Managers of the Royal Institution held on the 18th of February 1829

Present

The Duke of Somerset, Pres.t in the Chair.
George Moore, Esq.r V.P.
Sir George Duckett Bt. V.P.
Cha.s Barclay, Esq.r M.P.
Cha.s H Pilgrim, Esq.r
Edward Sterling, Esq.r
Dr. Somerville
Sir S. Bernard Morland Bt. M.P.
Dr. Nicholl,
Mich.l Faraday, Esq.r Director of the Lab.y

Prosecution of Wm. B. Wake.

The Meeting having been appointed in consequence of an application of a Solicitor, on the part of Wm. Basil Wake

Resolved, That this Meeting does not think it expedient to see the Solicitor who attends on the part of Wm. Basil Wake, but that the determination of the Board to proceed with the prosecution be communicated to him in writing.

Resolved, That Mr. Wade be subpœnaed as a witness in the prosecution of W. B. Wake.

Resolved, That this Board do prefer another Bill of Indictment against Mr. W. B. Wake for stealing Books found at Mr. Harrison's the Pawnbroker.

Resolved, That the Librarian of the Institut.n do go with the Police Officer, and examine

235

1829.

the books that have been found at the Pawnbrokers, to ascertain which of them belong to this Institution.

Confirmed the 2nd Day of March 1829
Somerset Pres.

236

Mar. 2 — 1829.

At a Meeting of the Managers of the Royal Institution of Great Britain held on the 2d of March, 1829,

Present

His Grace The Duke of Somerset Pres.t in the Chair
Sir George Duckett, Bart. V.P.
George Moore, Esq.r V.P.
William Hasledine Pepys, Esq.r
Joseph Jekyll, jun.r Esq.r
Charles H. Pilgrim, Esq.r
Capt.n Sabine,
R. I. Murchison, Esq.r
Isaac Lyon Goldsmid, Esq.r
Dr. Nicholl,
Benjamin Bond Cabbell, Esq.r
Sir S. Bernard Morland, Bart. M.P. Treas.r
Michael Faraday, Esq.r Director of the Lab.y
Wm. Tho.s Brande, Esq.r

1. The minutes of the proceedings of the 2d of February, and also the minutes of the proceedings of the Special Meetings of the 16th and 18th of February were read and confirmed.

2. Read the following report from the Committee of Accounts.

Rep.t of Com.e of Accounts

The Committee of Accounts report that they have examined the General Accounts from the 1st of January last and they found a balance in the hands of the Treasurer of £755. 18 – And in the hands of the Assistant Secretary of £0. 15. 9

They recommend the following accounts to be paid.

237

Mar. 2.

	£	s	d
Nicol for Printing in 1827	101	19	–
Feetham Sundries for the house in ditto	22	18	7
Bird for Bricklayers work in 1828	12	1	–
Royce for Repairs in d.o	20	8	7
Pringle for Upholstery in d.o	54	17	9
Houghton for Newspapers & Stationery in d.o	39	18	10
Vulliamy for winding Clocks in d.o	2	12	6
Gas Light Comp.y for a years lighting a lamp at the Hall door	5	12	–
	260	8	3
Joseph Fincher on account of the Royal Institution	239	11	9
	500	–	–

And also the sum of £120. 19. 2 to the Assist.t Secretary being the balance due to him on the Laboratory Account on the 31st Dec.r last.

Signed an Order for Advances to the above amounts.

Presents received.

3. Rev.d Mr. Caton presented
A Specimen of Basalt with Calcidony
A ditto – of Metallo-Chrone by Sign. Nobili
Specimens of Barbadoes Tar by Mr. Caldicot
Mr. Wheatstone presented African Bow, Quivers and Arrows, and Sciagraphia Lithologica &c.
P. S. Robinson, Esq.r presented
The Vitruvius Britannicus, Woburn Abbey.
Francis Bayley, Esq.r presented Appendix to Astronomical tables.

1829.

William Alex.r Provis, Esq.r presented An Historical and Descriptive Account of Menai Bridge.

Benjamin Green, Esq.r presented Lecture on the Study of Ancient Coins.

Ordered,

That the thanks of the Managers be returned for these presents.

Prosecution of W. B. Wake. 4 Resolved, That the sum of £6 be paid to W.m Ballard for conducting the prosecution against Will.m Basil Wake.

5. The following persons having been duly recommended were admitted as Subscribers to the Royal Institution for the year 1829

Annual Subscribers

George Horatio Smith, Esq.r S.t James's Palace

M.r Blackwell 5 Maddox Street

To the Lectures.

M.rs Chater & Miss Wilson of Kilburn

Miss M. Dew 59 Guilford Street

Miss Joanna Sturch 21 York Terrace

Miss Malcolm } 11 Hanover Square
Miss M. Malcolm }

Confirmed the sixth day of April 1829

Somerset Pres.

April 6. At a Meeting of the Managers of the Royal Institution of Great Britain held on the 6th of April 1829

Present

His Grace The Duke of Somerset in the Chair

George Moore, Esq.r V.P.

Benjamin Bond Cabbell, Esq.r V.P.

Charles Pilgrim jun.r Esq.r

Sir George Duckett, Bart. V.P.

D.r Nicholl,

R. I. Murchison, Esq.r

W.m Hasledine Pepys, Esq.r

D.r Somerville V.P.

Joseph Jekyll jun.r Esq.r

Capt.n Sabine,

Edmund Rob.t Daniell, Esq.r Sec.y

W.m Tho.s Brande, Esq.r Superint.t of the House

Michael Faraday, Esq.r Director of the Lab.y

1. The minutes of the proceedings of the last meeting were read and approved.

Rep.t of Com. of Acc.ts 2. Read the following report from the Committee of Accounts.

The Committee of Accounts report that they have examined the General Accounts of the Royal Institution of the last month, and they found a balance in the hands of the Treasurer of £387. 17. 6. And in the hands of the Assistant Secretary of £140. 16. 6.

They likewise report that they have examined

1829.

the Laboratory Account of the Royal Institution from the 1st of January last. And they found a balance in the hands of the Treasurer of £240. 1. 2

They recommend the following sums to be paid out of the General Fund.

Parker for Painting in 1828	£ 52. 5. 2
Feetham Ironmongery in d.o	25. 12. 8
Drew for Gas fittings	53. 16. 6
	131. 14. 4

They likewise recommend the following sums to be paid on the Laboratory Account

Newman for Chemicals in 1828	£ 99. 19. –
Fisher for ditto in d.o	11. 13. –
Dickins for Coals in d.o	7. 10. –
Rickards for Spirits of Wine in d.o	2. 12. –
Wedgwood Sundries for the Lab.y	–. 7. 6
	122. 1. 6
and Joseph Fincher on account of the Laboratory	60 – –
	182. 1. 6

Signed Orders for an advance to the Assist.t Secretary of £130 on the General Account, & £100 on account of the Laboratory Fund.

Lectures. 3. Read a letter from Gilbert T. Burnett, Esq.r stating that he was desirous should it meet with the approbation of the Managers, to be proposed as a Candidate for the vacant Botanical Professorship.

Resolved, That the Secretary be requested to inform M.r Burnett that it is not at present the

April 6.

intention of the Managers to propose any person to fill up the vacant Professorship.

Opening of Library in Evening 4. Read a letter from Capt.n Cook, M.R.I. complaining that the Library of Reference had not been opened in the Evenings as was the case last year.

Resolved,

That the Secretary be requested to reply to Capt.n Cook and to express the Managers' regret that he should have been put to any inconvenience from the circumstance of the Library of Reference not having been open; And to inform him that as it is his wish as well as that of other Members of the Institution that access should be had to the Library of Reference in the Evening they have directed that it shall be reopened after Monday next, from 7 till 11 o'Clock in the Evening.

Presents received.

5 J. Brookes, Esq.r presented Observations On a new Genus of the order of Rodentia

The Magazine of Natural History N.o 6 presented by The Proprietor.

Rev.d Robert Vaughan presented Life and Opinions of John de Wycliff D.D.

The Rev.d M.r Kinsey presented Illustrations of Portugal.

George Wigman, Esq.r presented Political and Moral Contemplations.

M.r Wheatstone presented Sciagraphia Lithologica Curiosa,

1829.

L. Kennedy, Esq.r presented
a Tract on the Cultivation of Waste lands.
The Phrenological Journal No. 20
presented by the Proprietor.
Ordered,
That the thanks of the Managers
be returned for these presents.

State of Hall

6. Read the report of the Committee
appointed to take into consideration
the state of the Hall.
Resolved, That it be taken into
consideration at the next Meeting of the
Managers.

7. The following persons having been duly
recommended were admitted as Subscribers
to the Royal Institution for the year
1829.

Annual Subscribers
William Martins, Esq.r of St. James's Palace
John Kennaway Esq.r of Escot nr. Exeter.

To the Lectures.
Lady Frances Shirley No. 2 Harley St.

Adjourned to Monday next the
13th Instant at Two o'Clock.
Confirmed 13 April 1829
Benjn. Bond Cabbell V.P.

April 13.

At a Meeting of the Managers of
the Royal Institution of Great Britain
held on the 13th of April, 1829,

Present
Benjamin Bond Cabbell Esq. V.P. in the Chair
George Moore, Esq.r V.P.
Sir George Duckett, Bart. V.P.
Dr. Somerville, V.P.
Edward Sterling, Esq.r
Wm. Hasledine Pepys, Esq.r
Captn. Sabine
Charles Pilgrim, Esq.r
Joseph Jekyll jun.r Esq.r
Dr. Nicholl
Isaac Lyon Goldsmid, Esq.r
Sir S. Bernard Morland Bart. M.P.
Edmund Robt. Daniell, Esq.r Secy. Treas.r

Michael Faraday, Esq.r Director of the Laby.

1. The minutes of the proceedings of the
last meeting were read and approved.

Visitors.
Repairs.

2. Read the following resolution of the
Visitors as entered in their proceedings of the
25th Ult.o
The Visitors examined the House and
Premises and remark that the Window frames
in the Attics are very much in want of paint.
They are informed that no portion of that
part of the House had been painted for 24
years; And they are of opinion that the painting

1829.

the Window frames can no longer be delayed. —
They likewise observe a slight repair is wanting
in the roof of the Laboratory which has been
injured by the bursting of a Water pipe, and
the whole of the basement floor is greatly
in want of paint and whitewashing. They
likewise recommend, whenever an opportunity
should offer, that the Stove to warm the
lecture room be removed to a more suitable
situation as they are of opinion it is not
altogether safe in its present place.
Resolved,
That the above recommendation of
the Visitors be taken into consideration after
the Annual Meeting.

3. Octavius Morgan, Esq.r presented
A Specimen of Metallic Titanium
& A Specimen of Carburet of Iron from the
Iron Furnaces of S. Wales.
Ordered,
That the thanks of the Managers
be returned for these presents.

4. Proceeded to take into consideration the
following report of the Hall Committee.

Regulation of Hall.

5. The Committee appointed to take into
consideration the state of the Hall of the
Royal Institution report that they find after
examination the following to be the present state
of the attendance in the Hall, viz.t

April 13.

Porters.

Two Men are employed, of whom the first is
Thomas Hatter Door-keeper who has now been in
that situation 24 years. He is married has a
family and lives out of the House. He comes every
morning at Nine o'Clock, and stays until Eleven
o'Clock at Night with the exception of two
hours for Meals. His duties are to attend the
Hall, clean the lamps, attend the Fires and
Post boxes; take out Cards and Notices of
Meetings to Members and Managers, also for
the Friday Evening lectures, on which Evenings
he also assists in taking care of the Coats &c.
Thomas Hatter is also Door keeper during the
Lectures in the Theatre for which he receives
Ten guineas per Annum in addition to one
guinea per week making together £62.10.—
per Annum.
Morris Kitton, Porter, the second Servant,
has lived in the House and service of the
Royal Institution five years. Previous to Nine
o'Clock in the Morning, he cleans the Tables
Chairs &c. of all the Public Rooms of the
Institution (except the Library.) At 9 o'Clock,
he is ready and in attendance at the Hall
where he remains, except when called away to
carry Coals to any of the Fires, take messages
to persons living in the House, or notices &c.
of Meetings out of doors, till Eleven at Night
dining in or out of the Hall as it may
happen. His duties also extend to dusting
Books, assisting in carrying up Apparatus &c.
brought in or required for the Lectures, attending

246

1829.

also upon the Friday Evening Meetings in taking Cloaks and serving at the Tea Tables; for these various employments he receives 20.s ℔ week, but cannot support a respectable appearance with that salary.

That both Servants agree in stating their inability to conform with the regulations more strictly than at present, and that much additional duties have accrued from the Friday Evening Meetings.

That the Committee have received great assistance and advantage during this investigation from Mr. Faraday who has stated to them the difficulties and embarrassment felt by the different Lecturers in procuring the arrangement of the Apparatus, there being no person at present to whom they can apply; as well as by himself in the Correspondence and other business imposed upon him unavoidably, by his exertions for the benefit of the Royal Institution in arranging the lectures of Friday Evenings to whom likewise the Committee acknowledge themselves indebted for the definition for the duties which follows in detail, the manner in which most of the Departments especially those of the Laboratory and Lecture-room are cramped by the present insufficient attendance.

Upon the above evidence the Committee cannot avoid drawing the attention of the Board of Managers to the mixed and irregularly defined nature of the Servants' duties, and

to April 13.

247

feeling convinced of the inadequacy of the existing system to meet what is due to a proper regard to the interests and property of the Institution, they beg to recommend the following resolutions viz.

Engagement of 3rd Porter.

That it is expedient to have a third Servant to be called the Hall Porter, and who shall wear a distinguishing dress or livery, and whose duties shall be as follow

Duties of Porters.

Hall Porter. — Dust the Hall sweep the Doorway and crossing before the Doors are opened to Members.

On Laboratory-Lecture Mornings to be ready by half past Eight o'Clock to attend to the Members and Subscribers who come to them.

To take down the names of all Members and Subscribers who come to the House either for the purposes of reading or attending the Lectures as well as those who come on the Friday Evenings. To receive letters for Members who give their address at the Royal Institution.

To answer the inquiries of all persons who come into the Hall and to ring for the Servants whose attention is wanted in the Rooms.

To take in, stamp and iron all Newspapers and To take in letters parcels and messages left for those in the house and immediately forward them by the proper Servant. To take charge of the Post boxes and expedite the letters to the office by the proper Servant 10 minutes before the appointed hours. To behave respectfully to Members and Subscribers in discharge of his duties by standing in their presence, opening the

248

1829.

door for them and such other services as are performed by a Servant for his Master.

The hours in the Hall to be from 9 o'Clock a.m. to 11 o'Clock p.m. with exception of one hour at   for dinner and one hour at   for Tea; but he is not to leave the Hall on those or any other occasions until he has given it in charge for the time to the first or second Servant and seen him in the performance of his duty, for the neglect of which he shall be fully responsible.

The Committee further suggest that the remaining duties be divided between the two other Servants as follows, viz.

1st. Servant. To look after all Fires in the public parts of the House, keep them in order, and especially to see that they are out when the house closes:

To attend carefully to the Stove attached to the lecture Room.

To take messages parcels and letters from the Post and deliver them immediately to the persons in the house for whom they are intended.

To trim light and extinguish the lamps at the proper hour.

To take the letters to the Post Office at the proper times.

To carry Coals to all parts of the house,

To take charge of and file the Newspapers.

To assist in moving Apparatus of Lectures.

To dust and keep in order the Apparatus Room

April 13.

249

To assist the Lecturers and give attendance during lectures if required.

To dust books in Library in the Summer. —

When in the House his place to be in and about the Hall and Public Rooms; part of his duty to be to attend at all times to the inquiries and convenience of the Members and to behave with constant respect.

His hours to be from   a.m. to 11 p.m. except one hour for dinner at   and half an hour   , during which time he is to be clean and engaged in the service of the Institution.

Second Servant — will be responsible to the first Servant. Must assist him at all times. He will have especially to deliver all messages &c. to Tradesmen; All Notices of Summonses to Members of Committees, Managers &c. — All notices of elections, thanks &c. All Correspondence relating to the Friday Evening Meetings.

To go for and take home the objects required for these and other occasions.

When not employed directly in or out of the house, his place will be in the Hall and public Rooms attending to the convenience of the Members &c. and watching the property of the Institution.

When one Servant is engaged out of Doors either in his duties, or in getting his Meals, the other must remain in the house unless it be for a very short period, as going to the Post with letters; so that there shall always remain

1829.

two in the House, either the Porter and one Servant or the two Servants.

It is absolutely necessary that one of three sleep in the house. First the Porter, or in his place one of the others to be arranged by the Superintendent.

Any neglect of duty to be followed by a reprimand and being repeated the Superintendent will suspend the party and report to the next Managers' Meeting.

6. Resolved,
That this report be received and adopted; And that the thanks of the Managers be presented to the Committee for their valuable exertions.

7 Resolved, That the Report be referred back to the same Committee to complete the details and to devise the best means of carrying the same into effect and to report thereon.

Adjourned to Wednesday the 22d Instant at half past Two o'clock.
Geo Duckett

April 22

At a Meeting of the Managers of the Royal Institution of Great Britain held on the 22d of April, 1829.

Present

Sir George Duckett, Bart. V.P. in the Chair.
George Moore, Esqr. V.P.
Dr. Somerville, V.P.
Benjamin Bond Cabbell, Esqr. V.P.
Dr. Nicholl
Captn. Sabine
Charles Pilgrim, Esqr.
Sir S. Bernard Morland, Bart. M.P. Treasr.
Edmund Robt. Daniell, Esqr. Secy.

Michael Faraday, Esqr. Director of the Laby.
Willm. Thos. Brande, Supt. of the House

1. The minutes of the proceedings of the last meeting were read and approved.

2. Ordered, That the House lists of Officers be filled up as follows.

President. The Duke of Somerset

Treasurer. Sir S. Bernard Morland, Bart.

Secretary. Edmund R. Daniell, Esqr.

Managers.
Benjamin Bond Cabbell, Esqr.
Lord John Campbell
Sir George Duckett, Bart.

1829.

Col. Sir William Gomme K.C.B.
Edmund S. Halswell, Esqr.
The Marquis of Lansdowne,
Charles Shaw Lefevre, Esqr.
James Heath Leigh, Esqr.
George Moore, Esqr.
Whitlock Nicholl, M.D.
Charles H. Pilgrim, Esqr.
William Pole, Esqr.
Captn. Edward Sabine,
Sir Claude Scott, Bart.
Sir Giffin Wilson.

Visitors.

Thomas Tyringham Bernard, Esqr.
Henry James Barchard, Esqr.
John Ivatt Briscoe, Esqr.
Lord George Augustus Henry Cavendish M.P.
John Disney, Esqr.
James Wm. Farrer, Esqr.
John Fuller, Esqr.
Davies Gilbert, Esqr. M.P. Pres. R.S.
Lt. Col. Wm. Martin Leake
George Long, Esqr.
The Hon: Wm. F. S. Ponsonby, M.P.
Edward Adolphus Lord Seymour,
Sir Oswald Mosley, Bart.
George Stavely Smith, Esqr.
Richard Horsman Solly, Esqr.

Confirmed 4th day of May 1829
Somerset Pres.

May 4.

At a Meeting of the Managers of the Royal Institution of Great Britain held on the 4th of May, 1829,

Present

His Grace The Duke of Somerset Pres: in the Chair
Charles Shaw Lefevre, Esqr.
William Pole, Esqr.
Benjamin Bond Cabbell, Esqr.
George Moore, Esqr.
Dr. Nicholl
Chas. H. Pilgrim, Esqr.
Captn. Sabine
Edmund S. Halswell, Esqr.
~~Captn. [illegible]~~

Michael Faraday, Esqr. Director of the Laby.

1. The minutes of the proceedings of the last meeting were read and approved.

Reprt. of Com. of Accts. 2. The Committee of Accounts report that they have examined the General Accounts of the last Month, And they found a balance in the hands of the Treasurer of £344.10.6; And in the hands of the Assistant Secretary of £128.0.2.

They recommend the following sums to be paid out of the Funds of the General Account.

Sewer rate	12.10 –
2 quarters Parochial rates to Lady Day last	40.16.8
ditto Assessed taxes to do.	61.4.8
ditto Water rate to ditto	5.5 –
	£119.16.4

1829.

And signed an Order for £119 . 16 . 4 accordingly.

3. Read the following letter from Sir Claude Scott, Bart addressed to the Secretary.

Donation by Sir Cl. Scott

Bruton Street 4th May 1829

Dear Sir,

I beg you will signify to the Board of Managers that I wish the loan of £500 from me which will become payable the 4th of next August without Interest, may be considered as a Donation from this day for the benefit of the Institution. I herewith send the bond to be cancelled and am Dear Sir, Yours truly

(signed) Claude Scott

To
Edmund Robt. Daniell, Esqr. F.R.S.
Secy. Royal Institution.

Resolved, That the grateful thanks of the Managers be returned to Sir Claude Scott Bart. for his munificent Donation and that it be communicated to the General Monthly Meeting of this day.

4. Resolved, That the following Committees be appointed for the present Year.

Committee for regulating the Friday Eveng. Meetgs.
The President
George Moore, Esqr.
Dr. Nicholl,
Captn. Sabine,

May 4.

Benjamin Bond Cabbell, Esqr.
Edmund Robt. Daniell, Esqr. Secy.
William Thos. Brande, Esqr. Supert. of the House
Michael Faraday, Esqr. (Director of the Laby.

Committee for arranging the Lectures.
The President
Benjamin Bond Cabbell, Esqr.
Chas. H. Pilgrim, Esqr.
Sir George Duckett, Bart.
George Moore, Esqr.
Dr. Nicholl,
Edward S. Halswell, Esqr.
William Pole, Esqr.
Edmund Robt. Daniell, Esqr. Secy.

Committee of Accounts
The President
Treasurer
Secretary
George Moore, Esqr.
Benjamin Bond Cabbell, Esqr.
Dr. Nicholl.

Committee for regulating the Hall.
was reappointed consisting of the following Gentlemen.
Sir George Duckett, Bart.
Captn. Sabine
Chas. H Pilgrim Esqr.
Dr. Nicholl
George Moore, Esqr.
The Secretary
& Mr. Faraday, Director of the Laby.

1829.

5. Read a letter from H. W. Dewhurst Esqr. offering to deliver a Course of Lectures on Human and Comparative Anatomy.

Lectures.

Resolved, That Mr. Dewhurst's offer be declined.

6. Resolved, That Mr. Pilgrim be authorized to procure an Organ for Dr. Crotch's Lectures.

7. Resolved, That the Superintendents of the Library and Laboratory be ordered to lay upon the Managers' table at every meeting, correct lists of all presents made to the Royal Institution in their respective Departments.

8. The following persons having been duly recommended were admitted as Subscribers to the Royal Institution for the year, 1829

Annual Subscribers.
Henry Wm. Beavan, Esqr. 26 Sackville St.
Lord Selsey 33 Grosvenor Square

To Mr. Faraday's Lectures.
Mr. Fisher }
Mrs. Maierhouse } 60 Upper Seymour St.

Adjourned to Monday the 18th Instant at Three o'clock.

Confirmed the 18th day of May 1829
Somerset Pres.

May 18.

At a Meeting of the Managers of the Royal Institution of Great Britain held on the 18th of May, 1829.

Present
His Grace The Duke of Somerset Prest. in the Chair
Sir Claude Scott, Bart. V.P.
Benjamin Bond Cabbell, Esqr. V.P.
Dr. Nicholl, V.P.
Charles H. Pilgrim, Esqr.
Captn. Sabine,
George Moore, Esqr. V.P.
Edmund S. Halswell, Esqr.
James Heath Leigh, Esqr.
Sir Giffin Wilson
Sir T. Bernard Morland Bt. M.P. Treasr.
William Pole, Esqr.
Edmund Robt. Daniell, Esqr. Secy.

William Thos. Brande, Esqr. Supt. of the House
Michael Faraday, Esqr. Director of the Laby.

1. The minutes of the proceedings of the last meeting were read and approved.

2. Read a letter from Messrs. White Blake & Houseman applying on the part of the Executors of the late Henry Hoare, Esqr. for repayment of the sum of £200 advanced by him on loan to the Royal Institution.

Repayment of loan.

Resolved, That the consideration of the above letter be postponed till the next Meeting of the Managers.

258

1829.

Lectures. 3 Read a letter from Mr J. Buckingham proposing to deliver a Course of lectures on Asiatic Literature.
Resolved, That Mr Buckingham's proposal be declined.

Presents received.

4 L. Kennedy, Esqr. presented
Tract on the cultivation of Waste Lands in the united Kingdom.
Reverend Mr Hawkins presented
Elements of Medical Statistics.
Alexander Trotter, Esqr.
presented Travels in Phensloaasta,
Royal Society of Literature presented
Transactions of the Society vol 1 part 2.
The Asiatic Oriental Translation Committee presented The Travels of Ibn Batuta
John Soane, Esqr. presented
Designs for completing some of the Public Buildings in Westminster.
J. P. Meyer, Esqr. presented
Geometrical Landscape containing the altitudes of the principal Buildings in London &c.
Wm Brockedon, Esqr. presented
Passes of the Alps No. 10.
The Horticultural Society presented
Transactions of the Society, vol. 7 pt. 3,
J. Vincent Thompson, Esqr. presented
Storia Pittorica della Itallia 6 tom 12mo
Idea del Tempio della Pittura di Gio. Carlo Lomazzo
Wollaston's Syllabus of Chemical Lectures.
Farish's ditto — of lectures on Chemical Arts & Manufact.

259

May 18

Presented to the Chemical & Mineral Department Specimens of finely crystallized Carbonate of Soda and Sulphate of Copper. — Also a specimen of crystallized Glass — By —— Cookson, Esqr. of Newcastle.
Miscellaneous Specimens of Minerals — By J. Vincent Thompson, Esqr. M.R.I.
Specimens of Sulphuret Sulphate and Carbonate of Lead from Spain. By John Mac Donnell, Esqr.
Ordered,
That the thanks of the Managers be returned for these presents.

Organ. 5. Mr Pilgrim reported that he had succeeded in obtaining an Organ for Dr. Crotch's lectures. at an expense of fifteen guineas.

Rept. on State of Hall. 6. Read the following report from the Committee appointed to take into consideration the state of the Hall of the Royal Institution.

That in execution of the trust confided to them, they have engaged as Hall Porter

Engagement of J. Roberts — John Roberts of whom they have received a good character at a Salary of £67 pr Annm including Clothes and Board wages.

Resignation of T. Statter &c. — That they made an offer of the situation of first or indoor Servant to Thomas Statter with a Salary of £71, who has declined continuing in the service of the Institution, giving as his reasons his dissatisfaction at the proposed salary, and also his dislike to wear a livery or button, the Committee have therefore engaged in his room Thomas

260

1829.

Th. Williams engaged. Williams whose recommendation and character have proved quite satisfactory, at a Salary of £67 including Clothes and Board wages. That the other Servant Morris Kirtlan at first declined the situation of second or out-door Servant offered to him, but having since expressed his regret at having so done and his wish to be re-engaged the Committee have admitted his appeal and have appointed him second or out-door Servant at a Salary also of £67 to include as before.

That in all their arrangements the Committee have reserved to themselves the power to substitute at any time Liveries in lieu of £10. 10 — pr annum now forming part of the Salaries as above mentioned.

Badge — That the Committee recommend to the Board of Managers the adoption of a Badge, similar to one procured for their inspection by Mr Faraday to be worn by the Hall Porter while in the Hall and by the Servant taking his place, while he is not there, the expense of which is estimated at Five guineas.

Button. — They are indebted to the active exertions of Mr Faraday in providing the Badge above alluded, to as also a drawing of a button bearing the insignia of the Royal Institution which they also submit to the Managers with a recommendation for its adoption, which button if adopted, will, with a blue Coat form a sufficient distinction for the servants of the

261

May 18

Institution. It has been considered expedient by the Committee to take away the distinction of first and second Servants, and to make the duties of both equal and the responsibility confined to Mr Faraday.

It is matter of regret to the Committee to have been obliged to discharge a Servant who has been so many years in the service of the Royal Institution, but no alternative remained to the Committee on this point.

Since drawing up the foregoing report, Mr Faraday has reported that Morris Kirtlan has also declined the new situation allotted him, the Committee have applied themselves to the selection of a Servant in his room,

Engagement of Th. Webster. — and have engaged Thomas Webster who is highly recommended by his last Master, to supply his place, on the same terms as were offered to Morris Kirtlan.

Resolved,
That the above report be adopted; and that Mr Faraday be directed to procure a Badge of the size of the Model shewn, but without the Crown, for the Porter, and order the die for the Buttons according to the pattern agreed upon.

Resolved, That the thanks of the Managers be presented to the Committee, and that they be requested to continue their services till the arrangements have been completed.

Lectures.

7. It having been stated to the Managers that Dr. Harwood was willing to postpone his lectures till the next Season, And it appearing from the great number of lectures which are already in course of delivery, that it will be difficult to fix such a time for the commencement of Dr. Harwoods as will enable him to conclude them within the usual season,

It was resolved, that Dr. Harwood's lectures be postponed till the next Season.

Mr Pearsall's Salary raised.

8. Resolved, That the proposal to raise Mr. Pearsall's Salary be taken into consideration.

Ordered,

That the rules for the regulation of his duties as Assistant in the Laboratory be laid upon the table.

Resolved, That Thomas Pearsall's Salary be increased to £80 ℔ Ann. to commence from the 1st. of January last; And that he be informed that the Managers expect that he will strictly conform to the rules of the Laboratory.

9. Resolved, That Mr. Pole be added to the Committee of Accounts.

10. Resolved, That Mr. Pilgrim, Mr. Heath Leigh and Mr. Pole be added to the Committee for conducting the Friday Evening Meetings.

Confirmed June 1st.

Whitlock Nicholl.

June 1. At a Meeting of the Managers of the Royal Institution of Great Britain held on the 1st. of June, 1829,

Present

Whitlock Nicholl, M.D. Vice Prest. in the Chair
George Moore, Esqr. V.P.
Col. Gomm, K.C.B.
Benjamin Bond Cabbell, Esqr. V.P.
Chas. H. Pilgrim, Esqr.
William Pole, Esqr.
Edmund S. Halswell, Esqr.
James Heath Leigh, Esqr.
Sir S. Bernard Morland, Esq. Bt. Treasr.
Edmund Robt. Daniell, Esqr. Secy.

Michael Faraday, Esqr. Director of the Laby.

1. The minutes of the proceedings of the last meeting were read and approved.

Rept. from Com. of Accts.

2. Read the following report from the Commee. of Accounts.

The Committee of Accounts report that they have examined the General Accounts of the last month And that they have found a balance in the hands of the Bankers of £607. 19. 2 – and in the hands of the Assistant Secretary of £85. 1. 2.

They further report that they have examined the General Accounts and have found that the sum of £188. 6. 8, being the surplus over the current expenses of the year 1827, and

the sum of £99. 14. 11, being the surplus over the current expenses of the year, 1828, making together the sum of £288. 1. 7 have not been invested pursuant to the resolution of the General Meeting of the 2d. July, 1828.

The Committee of Accounts therefore recommend that the same be forthwith invested and carried to the Account of the Sinking Fund.

Resolved,

That the suggestion of the Commee. be adopted; And that the Treasurer be directed to invest the above sum of £288. 1. 7 and place the same to the Account of the Sinking Fund.

On Repayment of Loan.

3. Mr. Pilgrim stated that he had seen Mr. H. Hoare and had communicated with him, upon the subject of Messrs. White Blake & Houseman's letter, and proposed that the consideration of that letter be further postponed till the next monthly Meeting.

Lectures declined.

4. Read a letter from Mr. Buckingham proposing to deliver at the Royal Institution one lecture gratuitously on either of the Countries that might be selected by the Managers either at the next Friday Evening Meeting, or on any vacant day in the present week.

Resolved, That Mr. Buckingham's proposal be declined.

5. An application was made on the part of Mr.

June 1.

Lectures in French declined.

Victor proposing to read in the french language a Course of lectures On French Literature.

Resolved,

That Mr. Victor's proposal be declined.

5. Presents received.

Messrs. Parke & Scoles presented
Map of Nubia,
J. Vincent Thompson, Esqr. presented
Le Parfait Ambassadeur 2 vol. 12mo.
Kellæi de Lapide Philosophorum 12mo.
Necker de l'Administration des Finances de la France 3 tomes 8vo.
Ditto – Compte rendu au Roi 4to. and Regulations for the Prussian Infantry
Fableaux et Contes des Poëtes François 4 tomes 8vo.
Chalmers on the Christian Revelation.
Sherlock's (Dr. W.) Scripture Proofs of our Saviours Divinity
——— Practical Discourse concerning a future Judgment.
Tuckers, (Dean) four Tracts on Political and Commercial subjects.
Ditto four letters on important National subjects
Sheffield's (Lord John) Observations on the Corn Bill,
Howlett's (Revd. John) Examination of Dr. Prices' Essay.
De Pinto's Essay on Circulation and Credit
Antiques du Musée Royal
The Royal Humane Society presented
Transactions of the Society for 1829, part 1.

266

1829.

J. C. Loudon, Esqr. presented
Magazines of Natural History, No. 7,
Alexr. Trotter, Esq. presented
Plan of Improvements between the old and new Towns of Edinburgh
J. H Curtis, Esqr. presented
Essay on the Deaf and Dumb.
Presented to the Mineralogical & Geological Collection.
Various Mineralogical and Geological Specimens from India - by Captn. Blake.
Ordered,
That the thanks of the Managers be returned for these presents.

6. The following persons having been duly recommended were admitted as Subscribers to the Royal Institution for the present year.

To Mr. Webster's Lectures.
John Wm. Bowden, Esqr. 9 Chesterfield Street
Mr Grant of the Athenaeum.
Mr. Wilson, Chapel St. Grosvenor Place
Count de Sambury
Captn. Swinburne
Earl of Harrowby
Sir John Johnstone, Bart
Mr. Holbech 6 Grosvenor Place.
To Mr. Phillips's Lectures
Mrs Murray
Miss Murray } Albemarle Street
Miss E. Murray
Mrs Partridge – ditto
Mrs Sturch Ulster Terrace

June 1. 267

Lady Sitwell Manchester Square
Mrs C. P. Montagu,
Mrs Roget & Miss Danton,
Mrs Shaw Lefevre
Smith Wright, Esqr.
Miss Smith Wright } Manchester Square
Mrs Reid of Granville Street Brunswick Square
To Dr Crotch's Lectures.
Mrs Erskine
Miss E. Yates
Mrs George Freere
Miss Freere
Mrs Coltman 34 Gloucester Place
To Mr Faraday's Lectures
Sir John Johnstone, Bart.
Miss Cartwright
Miss Clerke } 15 Half Moon Street

Annual Subscriber,
J. W. Braine, Esqr. George St. Hanover Square

Adjourned to Monday the 22d Instant at Three o'Clock.

Confirmed June 22d. 1829,
Giffin Wilson

268

June 22 June 22

At a Meeting of the Managers of the Royal Institution held on the 22d of June, 1829,

Present
Sir Giffin Wilson, in the Chair.
Sir George Duckett, Bart. V.P.
William Pole, Esqr.
Benjamin Bond Cabbell, Esqr. V.P
Dr. Nicholl, V.P
Edmund Robt. Daniell, Esqr. Secy.

Michael Faraday, Esqr. Director of the Laby

1. The minutes of the proceedings of the last meeting were read and approved.

2. Presents received.
Giovanni Berchet, Esqr. presented
Le Fantasie, Romanza,
Cesar Moreau, Esqr. presented
Jomard, Description de la Ville et des Environs du Caire
Academie de Dijon, Compte Rendu de 1826 – 1827
Academie Royale des Sciences &c. de Bordeaux
Séance Publique, 1828,
Extrait du Bulletin de la Société de Geographie No. 63
Astronomical Society presented
Transactions of the Society vol 3 part 2
The Linnean Society presented
Transactions of the Society, vol 16 part 1
J. M. Rendel, Esqr. presented
Particulars of the Construction of Lary Bridge
W. Thos. Brande, Esqr. presented
Outlines of Geology.

1829. 269

F. C. Balfour, Esqr. presented The Travels of Macarius, p. 1
Mrs Jones presented to the Museum.
A fine Coral Venice Fan.
Ordered, That the thanks of the Managers be returned for these presents.

3. Repairs. Resolved, That Mr Faraday be requested to procure Estimates of the expense of painting the outside of the Windows &c. back and front, white-washing and painting Mrs Greenwood's Room, and of white-washing the outside of the House

Confirmed the 6th day of July 1829
Somerset Pres.

270

1829.

July 6. At a Meeting of the Managers of the Royal Institution of Great Britain held on the 6th of July, 1829.

Present

His Grace The Duke of Somerset Pres: in the Chair.
Sir George Duckett, Bart. V.P.
Sir S. Bernard Morland, Bart. M.P. Treas.
Benjamin Bond Cabbell, Esq. V.P.
William Pole, Esq.
Capt. Sabine
James H. Leigh, Esq.
Wm Thos. Brande, Esq. Superint. of the House
Michael Faraday, Esq. Director of the Laby.

1. The minutes of the proceedings of the last meeting were read and approved.

2. Read the following report from the Committee of Accounts.

Rept. from Com. of Accts. 3. The Committee of Accounts report that they have examined the General Accounts of the last month, And that they have found a balance in the hands of the Bankers of £659.0.7; and a balance due to the Assistant Secretary of £32.0.6

They recommend that the following sums be paid

Mr Millington for lectures	£52.10 –
Ditto Sundries for ditto	5.5 –
Dr Crotch for lectures	56.14 –
Sundries for ditto	18. –
Carrd. forwd.	132.9.

July 6 — 271

Brot. forward	132.9 –
Mr Webster for lectures	42.0.0
Mr Faraday for ditto	52.10 –
Mr Brande for a year's Salary as Professor of Chemistry, 1829	150 –
Mr Phillips for a Course of lectures	73.10 –
	450.9 –
The balance due to J Fincher on the 1st of July	32.0.6
Joseph Fincher on account of the Royal Institution	150 –
	632.9.6

And signed an Order for an advance to the Assistant Secretary of £632.9.6.

On Lectures by Dr G. Smith 4. Mr Fincher stated that in consequence of an application having been made to him by Dr Gordon Smith for remuneration for two Courses of lectures delivered by him at the Royal Instit. in the years 1824 and 1826, he had waited upon him on the subject, and that he had received a letter from Dr Gordon Smith since that interview stating, "that with reference to the subject of his application he begged to state that he waived his claims upon the Royal Institution as far as any pecuniary object is concerned; and that he was confident the Managers would, on the score of injured feelings, not hesitate to announce through the medium of a respectable Journal, that Dr Gordon Smith's lectures on the Philosophy of Ordinary life are necessarily reserved till next Season".

272

1829.

Resolved, That Dr Gordon Smith's letter be referred to the Committee of lectures; And that a letter be written by the Secretary informing him of this arrangement and to add that no injury to his feelings was ever intended by the Managers of this Institution.

Lectures. 5 Read a letter from Mr Brayley proposing to deliver in the next Season, a Course of lectures on Igneous Meteors and Meteorites.

Read a letter from Mons. d'Emden proposing to deliver two lectures on French Literature.

Read a letter from Mr T. Phillipps proposing to deliver a Course of four or six lectures in the next winter on Singing and Vocal Music.

Resolved, That the above letters be referred to the Committee of lectures.

Presents received.

6. The Society of Arts presented
Transactions of the Society vol. 46,
The Editors presented
The Zoological Journal, No. 16,
Dr Gordon Smith presented
Hints for the examination of Medical evidence
J. C. Loudon, Esq. presented
The Magazine of Natural History,
The Honble. East India Compy. presented
Anwari Sohili, by Hussein Vaez Kashify.

July 6. — 273

Lt. Genl. Willm. Thornton presented
Gil Blas, 4 vols.
Henry Coddington, Esq. presented
Treatise on Reflexion and Refraction of Light
The Society of Antiquaries presented
Archaeologia, vol. 22.
The Editor presented
Law Magazine first five numbers.

Ordered, That the thanks of the Managers be returned for these presents.

Repairs 7 Mr Faraday laid before the Meeting Mr Parker's estimate for painting agreeably to resolution No. 3, at the last Meeting, which was agreed to, the amount being £39.13.6

8 Mr Faraday also laid before the Meeting Mr Bird's estimate for whitewashing the whole of the front of the House which amounted to £33.

Resolved, That the whitewashing be deferred till the next year.

Leave of Absence granted. 9. Mr Pearsall, the Assistant in the Laboratory, and Mrs Greenwood the Housekeeper having applied for leave of absence for two or three weeks each during the recess, Leave was granted reserving it to Mr Brande and Mr Faraday to arrange the time of their going.

10 Mr Fincher requested leave of absence for four weeks during the recess, which was granted.

11. The following person having been duly recommended was admitted an Annual Subscriber to the Royal Institution for the present year.

Ann^l. Subs^cr. Charles Gilmour, Esqr. 1 York Gate Regents Park

Adjourned to Monday the 2^d of November

Confirmed the 2nd day of November 1829
Somerset Pres.

Oct. 2.

At a Special Meeting of the Managers of the Royal Institution of Great Britain held under the appointment of a Vice President on the 2^d of October, 1829.

Present

George Moore, Esqr. Vice Pres^t. in the Chair.
Captn. Sabine
D^r. Nicholl
Edmund S. Halswell, Esqr.
Benjamin Bond Cabbell, Esqr. V.P.

William Thos. Brande, Esqr. Superintend^t. of the House
Michael Faraday, Esqr. Director of the Lab^y.

The Chairman stated that he had called the present meeting for the purpose of taking into consideration certain propositions relating to the Lectures.

Lectures M^r. Ritchie. Captn. Sabine informed the Board that he had received a proposal from M^r. Ritchie to deliver a Course of lectures at the Royal Institution on Natural Philosophy.

Resolved,

That M^r. Ritchie be received as a Lecturer, And that he be requested to send a prospectus of his intended Course early in the ensuing Month, And that Captn. Sabine be requested to make the above communication to M^r. Ritchie.

M^r. Brande. M^r. Brande communicated to the Meeting his intention, if it met with the approbation of the

Managers, to deliver a Course of lectures on the History of the late Sir H. Davy's Chemical Discoveries.

Note-books of Sir H. Davy. Resolved, That M^r. Brande's proposal be approved; And that he be requested to apply to the Executors of Sir H. Davy for the note Books which were used by him in the Laboratory of the Royal Institution while prosecuting his discoveries which it appeared had several years ago been taken away by Sir H Davy.

Confirmed the 2nd day of November 1829
Somerset Pres.

Nov. 2.

At a Meeting of the Managers of the Royal Institution of Great Britain held on the 2^d of November, 1829.

Present

His Grace The Duke of Somerset, Pres^t. in the Chair
George Moore, Esqr. V.P.
Sir George Duckett, Bar^t. V.P.
Edmund S. Halswell, Esqr.
William Pole, Esqr.
Benjamin Bond Cabbell, Esqr. V.P.
Sir S. Bernard Morland, Bar^t. M.P. Treas^r.
Edmund Robt. Daniell, Esqr. Secretary.

1. The minutes of the proceedings of the Meeting of the 6th of July, and also the minutes of the proceedings of the 2^d of October were read and approved.

2. Read the following report from the Committee of Accounts.

Rep^t. from Com^e. of Acc^ts. The Committee of Accounts report that they have examined the General Accounts since the last meeting in July, and that they have found a balance in the hands of the Bankers of £486 . 11 . 1; and a balance due to the Assistant Secretary of £84 . 2 . 4.

They recommend that the following sums be paid.

2 quarters Assessed taxes to Mich^as last	61 . 4 . 8
2 ditto Parochial rates to d^o.	36 . 13 . 4
Carr^d. up	97 . 18 . –

278

1829.

Bro^t forw^d.	97 . 18 –
Edinburgh Courant from Jany 1828 to Jany 1829	4 . 17 –
Mrs Sturt for Stationery to June last	25 . 19 . 5
Gardner for Oil in 1828	66 . 17 . 9
Joseph Fincher on account of the Royal Institution	204 . 7 . 10
	£ 400 –

and signed an Order accordingly.

3. Letter from Dr G Smith. The Secretary stated that he had written to Dr. Gordon Smith pursuant to the resolution of the 6th of July last; And that he had received in answer the following communication.

Dear Sir, University July 16. 1829

Your letter of the 13th has reached me, and given me great satisfaction. If you think it worth while, have the goodness to inform the Managers of the Royal Institution that I am quite satisfied as to the integrity of their intentions towards me, and also as to their unimpeachable conduct: I have traced the slander (alluded to in my letter to the Duke of Somerset) and in my conversations with Mr. Fincher, to its source, and acquit all parties officially connected with the Royal Institution of any participation in or approbation of it whatever. It was, however, a most intolerable stigma; and I do not know that I have even now shaken myself quite free of its consequences.

To re-appear in the Theatre of the Royal Institution will give me great pleasure, and

Nov. 2. 279

when I have the opportunity of doing so, I think you will find that, (though a Man of comparatively obscure name, and slender pretensions,) I shall acquit myself there, much better than I have done upon any former occasion.

I remain &c. (signed) J Gordon Smith

To E. R. Daniell, Esqr.

4. Repayment of Loan requested. Read a letter from Charles James Auriol, Esqr. addressed to the Assistant Secretary requesting him to state to the Managers that the prolonged period for which he consented to continue to them the loan of £100 without Interest, has expired, and that he should be obliged for the early repayment of the amount.

Resolved,

That the Secretary be directed to inform Mr. Auriol that the Royal Institution will be prepared to discharge the amount of his bond in the course of the month of January next.

5. On Withdrawal. Read a letter from Granville Sharp Pattison, Esqr. addressed to the Managers stating that he addressed a letter to the Secretary of the Royal Institution informing him that it was his wish to retire as a Member of the Royal Institution, but it appeared that the letter had not been received, and whether under these circumstances he should be required to pay the annual subscription for this year.

Resolved, That Mr. Granville Pattison be informed that as the letter announcing his intention to

280

1829.

withdraw from the Royal Institution has not been received by the Secretary or any of the Officers of the Royal Institution, the Managers cannot feel themselves justified in recommending to the General Meeting of Members that the payment of Mr. Granville Pattison's annual contribution for the current year should be remitted.

6. Presents received.

Wm Brockedon, Esqr. presented
Passes of the Alps, N°. 11.
Rev^d. Mr. Edwards presented
The Journal of a Naturalist
William Salmond, Esqr. M.R.I. presented
Illustrations of the Geology of Yorkshire.
William Watson, Esqr. presented
Letter to Sir James Mackintosh on the affairs of Portugal,
James Burton jun^r. Esqr. presented
Excerpta Hieroglyphica, N°. 3,
The Editors presented
The Edinburgh Journal of Natural Science,
– Madden, Esqr. presented
Examination of Remarks on the Glossary of Havelok,
Robert Gouger, Esqr. presented
A Letter from Sydney.

Presented to the Museum

Two cakes of Bombay Wootz, by Captn. Bagnold
A specimen of Chalk in Iron stone from Hastings by Miss Lewis

Nov. 2. 281

A Specimen of Jet from the North Coast of Yorkshire By Mrs Myddleton of Grinkle Park
Two elastic speaking tubes – By Mr Hancock.

Ordered, That the thanks of the Managers be returned for these presents.

Adjourned to Monday the 16th Instant at One o'clock.

Confirmed the 16th day of November 1829
Somerset Pres.

282

1829.

Nov 16.

At a Meeting of the Managers of the Royal Institution, held on the 16th of November, 1829.

Present

His Grace The Duke of Somerset Pres.t in the Chair
Sir George Duckett, Bart. V.P.
George Moore, Esq.r V.P.
Dr. Nicholl, V.P.
Edmund S. Halswell, Esq.r
Benjamin Bond Cabbell, Esq.r V.P.
Edmund Rob.t Daniell, Esq.r Sec.y

William Tho.s Brande, Esq.r Superint.t of the house

1\. The minutes of the proceedings of the last meeting were read and approved.

Adjourned to Monday the 7th of December at half-past One o'Clock.

Confirmed the 7 Day of Dec 1829
George Moore V.P.

283

Dec 7.

At a Meeting of the Managers of the Royal Institution of Great Britain, held on the 7th of December, 1829

Present

George Moore, Esq.r V.P. in the Chair.
Sir S. Bernard Morland Bart. M.P. Treas.
Benjamin Bond Cabbell, Esq.r V.P.
Sir George Duckett, Bart. V.P.
Whitlock Nicholl, M.D. V.P.
Edmund S. Halswell, Esq.r
William Pole, Esq.r
Capt.n Sabine
Edmund Rob.t Daniell, Esq.r Sec.y

Michael Faraday, Esq.r Director of the Laby

1\. The minutes of the proceedings of the last meeting were read and approved.

2\. Read the following report from the Comm.e of Accounts. *(margin: Rep.t from Com. of Acc.ts)*

The Committee of Accounts report that they have examined the General Accounts since the last meeting, And that they have found a balance due to the Bankers of £3. 12. 3 – And a balance in the hands of the Assistant Secretary of £71. 18. 5

They likewise report that they have exam.d the Laboratory account of the Royal Institution And they found a balance in the hands of the Bankers, after repaying the sum of £91. 16. 4 due to the Assistant Secretary on the 2.d Instant, of £146. 10. 11.

284

1829.

They recommend that the following bills be paid on account of the Royal Institution if the receipts of the year should be adequate for that purpose

Jones & Yarrell Newspapers	}	70. 10. 9
Dickins & C.o, Coals in ditto in 1828		85. —

and signed an Order for an Advance to the Assistant Secretary of £200.

They likewise recommend that the sum of £91. 16. 4 the balance due to the Assist.t Secretary on the Laboratory Account be repaid, And also that the further sum of £100 be advanced to him on account of the Laboratory;
And signed an Order accordingly.

3\. Presents received.

John Auldjo, Esq.r presented
Narrative of an Ascent to Mont Blanc
Gilbert Burnett, Esq.r M.R.I. presented
Eidodendron, N.o 5
J. G. Wilkinson, Esq.r presented
Materia Hieroglyphica part 1 & 2
Richard Evans, Esq.r
Paradise Lost translated into Welsh

Ordered,
That the thanks of the Managers be returned for these presents.

4\. Capt.n Sabine gave notice of his intention to bring under the consideration of the Managers at their next meeting, the subject of the

285

Dec 7.

publication of the Journal of the Royal Institution (Chap XVII of the Bye laws.)

(margin: Notebooks 5 of Sir H Davy produced) Mr Faraday produced the Note books belonging to the Laboratory of the Royal Institution, which had been used by the late Sir H. Davy while prosecuting his Chemical Discoveries from October 1805 to October 1812.

Confirmed the 21st day of December 1829
Somerset Pres.

286

Dec 21. 1829

At a Meeting of the Managers of the Royal Institution of Great Britain held on the 21st of December, 1829,

Present

His Grace The Duke of Somerset, President in the Chair

Sir George Duckett, Bart. V.P.

Dr. Nicholl, V.P.

George Moore, Esqr. V.P.

Benjamin Bond Cabbell, Esqr. V.P.

Sir S. Bernard Morland, Bt. M.P. Treasr.

Edmund S. Halswell, Esqr.

Charles Pilgrim, Esqr.

Captn. Sabine,

Edmund Robt. Daniell, Esqr. Secretary

Michael Faraday, Esqr. Director of the Laby

1. The minutes of the proceedings of the last meeting were read and approved.

Notebooks of Sir H. Davy 2. Mr. Faraday having paged the Note books of Sir H Davy, and prefixed a Note to each of them giving an account of their present state,

It was ordered, That they be deposited in the Library of Reference with particular charge respecting their safety.

Requests of Dr. Paris 3. Read an application from Dr. Paris addressed to the Secretary requesting permission of the Managers to inspect the Manuscript Notes

287

Dec 21.

of the late Sir H. Davy, and to make from them such extracts, as he may consider essential for the illustration of the Biography he has undertaken to write.

Resolved, That Dr. Paris's request be granted

4. Mr. Faraday stated that Dr. Paris had applied through him, for permission from the Managers to send an Artist to the house of the Institution to make such drawings of the original Safety Lamps in their possession as might best illustrate the account of that discovery which he would have to give in the Life of Sir H. Davy.

Resolved, That the above request be granted

on Journal of R. I. 5. Resolved, That a Sub-Committee be appointed to take into consideration Chap: XVII of the Bye laws relating to the publication of the Journal of the Royal Institution And that the following Gentlemen be appointed the Committee, viz.

The Duke of Somerset,

Captn. Sabine,

George Moore, Esqr.

Benjamin Bond Cabbell, Esqr.

Edmund Robt. Daniell, Esqr. Secy

And that Mr. Professor Brande and Mr. Faraday,

be requested to assist in their deliberation.

6 Captn. Sabine informed the Board that Davies

288

1829.

Donation from D. Gilbert, Esq. Gilbert, Esqr. M.P. the President of the Royal Society had desired that he might be considered as relinquishing the sum of £100 due to him on Bond from the Royal Institution in favour of the Library Fund.

7 Presents received.

Isaac Cookson, Esqr. of Newcastle presented Various Specimens of Glass in different states of C

Ordered, That the thanks of the Managers be returned for this present.

8. Resolved, That Edmund S. Halswell, Esqr. be added to the Committee for conducting the Friday Evening Meetings.

Adjourned to the third Monday in January 1830 at Three o'Clock

Confirmed the 18th day of January 1830

Somerset Pres.

289

1830.
Jan. 18.

At a Meeting of the Managers of the Royal Institution of Great Britain held on the 18th of January, 1830

Present

His Grace The Duke of Somerset, Pres. in the Chair.

Sir S. Bernard Morland Bt. M.P. Treasr.

George Moore, Esqr. V.P.

Whitlock Nicholl, M.D. V.P.

Edmund S. Halswell, Esqr.

Benjamin Bond Cabbell, Esqr. V.P.

Charles H. Pilgrim, Esqr.

William Pole, Esqr.

Sir George Duckett, Bart. V.P.

Edmund Robt. Daniell, Esqr. Secy.

Michael Faraday Esqr. Director of the Laby

1. The minutes of the proceedings of the last meeting were read and approved.

Report respg Journal 2. Mr. Daniell in the absence of Captn. Sabine presented the report from the Sub-Committee appointed to take into consideration Chap: XVII of the Bye laws relating to the publication of the Journal of the Royal Institution, viz.

The Committee appointed to take into consideration the seventeenth Chapter of the Bye laws report,

Bye Laws That by the terms of the Chapter of the Bye laws referred to them the Managers of the Institution are directed to cause Journals

of the Royal Institution to be published from time to time and if possible at regular times: The Journals so published are to contain such reports of the Committees and such original communications from the Members and others as the Managers may think worthy of public attention and also such Abstracts from the Transactions of learned Societies at home and abroad and notices of discoveries as may be thought likely to diffuse science and to render it of practical utility.

It is also directed that the selection of Papers, the form of publication and the manner of distribution or of sale shall be at the discretion of the Managers.

It appears upon referring to the Minutes of the Committee of Managers in the year

1800. 1800, that on the 31st of March in that year resolutions were passed relative to the publication of a Journal by which it was resolved, "That Count Rumford should be requested to take such measures as should be necessary in order to facilitate the speedy publication of the Journals of the Institution And that he should be requested to superintend that publication employing such persons as his co-operators and Assistants as he should think proper."

"That the Journals of the Royal Institution of Great Britain should be exclusively devoted to the diffusion of the knowledge of new and

interesting scientific discoveries, and of useful improvements in Mechanics, Arts, and Manufactures and particularly in making known all such new inventions and contrivances as tend to facilitate labour; and render it more productive to promote domestic economy, and increase the conveniences comforts and enjoyments of life. That neither political discussions nor religious disputes nor the common News of the day should ever find a place in those Journals: Nor were the common Advertizements of individuals ever to be published in them on any pretext whatever and "that in determining the price at which the Journals of the Institution should be sold care should be taken to put such a price upon the work as should be at least sufficient to cover all the expenses of its publication and pay those who should be employed on wages in its compilation.

1800. First Journal.

In pursuance of the above resolutions a Journal was commenced and published periodically, in numbers, containing about one or two sheets each; the whole of which were printed in the house of the Royal Instit: at a press established there under the direction of Mr William Savage who appears to have been engaged as Printer to the Institution. There is a book now in the Library containing all the Journals which have been published under the above resolutions bound up together. This book comprizes in all about 22 sheets of which the first 21 seem to have formed

a complete Volume comprising title page Index &c the remaining sheets appear to have constituted the first two numbers of the second Volume from which point the publication has been suddenly discontinued. Your Committee however have been unable to discover from any entry in the Minute book what was the immediate reason of its abandonment. From the preface prefixed to the first Volume and from sundry resolutions entered in the Minute books it appears that during the publication of this Journal the superintendance was occasionally varied and that the first three sheets were published under Count Rumford's direction; that Dr Young was the Editor of the next four, and that the subsequent parts were conducted jointly by Dr Young and Sir H Davy.

Count Rumford & Dr Young

Dr Young & H. Davy

It does not appear that during the period of the publication of the Journal above alluded to that there was any Bye law in existence by which its publication was regulated or authorized. But it seems to have been carried into execution by the Managers upon their own authority and according to regulations made at their own Board; they appear however to have been at that time careful not to part with the control over the contents since by an entry in the minutes of their proceedings on the 31st of April 1802, the Managers reserve to themselves the right of inspecting and controling the contents of the Journals whenever they think

Controlled by Managers

proper so to do".

Bye Law of 1811.

The present Bye-law, by which the publication of Journals of the Royal Institution is sanctioned and directed, was not made till the 6th of May 1811, when certain previously existing regulations were repealed and the Chapter referred to the consideration of your Committee was substituted in their place. No steps however were taken towards the publication of the Journals till the 8th of January 1816, when the following entries occur in the Minutes - viz!

1816. Extract Minutes.

"Read the resolution of the Committee appointed "to revise the Bye laws of the 6th Inst. recommending "Mr Brande's prospectus of a Journal of the "Royal Institution to be laid before the Managers".

Mr Brande's Journal

"Mr Brande laid on the table a Prospectus "of a Journal of the Royal Institution to be "published four times in each year.

"The Bye law respecting publication having "been read, resolved, That in execution of "the said Bye law a Journal to the effect "above mentioned be regularly published to "commence as soon as convenient."

"That the said Prospectus be forthwith "printed, and a Copy sent to each Manager, "and also to each Member of the Committee "appointed to revise the Bye laws in the "course of the present week, and that the further "consideration thereof be adjourned to Monday "next."

"That Mr Brande be requested to undertake "the office of Editor, and to employ such Assistants "as he may find requisite".

Brande's Journal.

In pursuance of these resolutions the publication of a Journal was commenced under the superintendance of Mr Brande and has ever since been regularly published once in every three months. The last number was published on the 1st of January last. The first 44 numbers appear to have been published by Mr Murray, and the title of the first 6 volumes is, "a Journal of Science and the Arts edited at the Royal Institution of Great Britain," the usual crest &c of the Royal Institution being stamped on the title page. From the 6th Volume, the words, "edited at the Royal Institution," appear to have been omitted from the title, but the Crest &c. has always been prefixed except on one occasion, when as your Committee have been informed it was omitted. After the publication of the 44th number, a new series of the Journal appears to have commenced of which Mr Colburn is the publisher. This new series is entitled, "a Journal of Science Literature and the Arts." The words "edited at the Royal Institution", have been omitted in all the numbers hitherto published; but the stamp containing the Crest &c of the Royal Institution has been made use of.

44 Nos published by Murray.

New Series — by Mr Colburn.

Not under control of Managers.

The Committee regret that in consequence of Mr Brande having declined to attend their Meetings they are unable to state, as fully as

they could wish, the principles upon which this Journal has hitherto been conducted; but from what they have been able to collect from other sources, it seems, that the whole management of the Journal has been confided to Mr Brande and that the Board of Managers have not interfered in the conduct of it in any manner whatsoever. That no reference has ever been made to them as to the selection of any of the Papers for publication; and that in the first series as published by Mr Murray, little or no notice has been taken of the proceedings of the Royal Institution. In fact the Managers appear to have concluded themselves to be entirely divested by the appointment of an Editor of all control over the publication and not to be in any way responsible for its contents; a conclusion which your Committee cannot think they were warranted by the construction of the Bye law, in drawing.

Remarks.

In remarking upon this Journal, the Committee beg to be understood as not expressing any opinion as to the mode in which it has hitherto been conducted by Mr Brande. The acquiescence of the Board of Managers appears to have left Mr Brande entirely to the exercise of his own discretion with regard to the publication, and the Committee believe that Mr Brande in the exercise of that discretion has exerted himself to the best of his skill and ability for the benefit of the parties interested in the Journal. The Committee

feel themselves also bound to mention; that from statements which have been made to them, they are led to believe, that Mr Brande has received very little (if any) remuneration for his trouble and exertions in the performance of the task intrusted to him. It was, as far as the Committee have been enabled to ascertain the fact, originally intended by the Board of Managers who appointed Mr Brande to the Editorship of the Journal that his remuneration should arise from a participation in the profits of the publication with the Publishers, but they believe that the profits have hitherto been too trifling to afford any adequate compensation for the exercise of the talents care and industry necessary to conduct such a Concern.

Not profitable to Mr Brande.

The Committee cannot however avoid expressing their feeling that if the Journal had been made more substantially a publication emanating from the Managers of the Royal Institution than it appears to have been considered, and that if more advantage had been taken of the resources of the Royal Institution in the preparation of its different parts, it would have been more productive of advantage to the interests of Science and more in conformity with the spirit of the Bye law. And they are of opinion that the Board of Managers in so completely divesting themselves of the control over the publication by delegating of their

authority to an Editor have notwithstanding the acknowledged high character and attainments of the Editor to whom they have committed it, in some measure neglected the interests of the Institution and exceeded the power vested in them by the Bye laws.

Examination of Agreement between Messrs Brande & Colburn.

As it appears by a Minute of the Board of Managers dated the 26th of March, 1827 that the Agreement entered into by Mr Brande with Mr Colburn relative to the publication of the new series of the Journal, was laid before the Board and approved of, your Committee have thought it right to make themselves acquainted with its contents; and they find it to be an agreement under the hands and seals of Mr Brande and Mr Colburn dated the 26th of March 1827, by which Mr Colburn agrees with Mr Brande to undertake the future publication and management of the periodical work entitled the Quarterly Journal of Science Literature and Art (theretofore published by Mr Murray,) and to conduct and carry on the same upon the conditions and subject to the regulations thereinafter contained.

Conditions.

The principal of these conditions appear to be, that the Journal should in future be conducted upon a more extended and popular plan than before and that instead of being chiefly confined to the abstract Sciences and principally addressed to the proficient, its contents should be rendered more suitable to the general reader. That it

should be published quarterly. That the Copyright of the Journal except as after mentioned should be held and enjoyed by Mr Brande and Mr Colburn in equal Moieties and that each of the parties should be at liberty to dispose of his share as he should think fit. That Mr Brande should be the Editor and Conductor of Publication and should edit and conduct the same to the best of his skill and ability so long as he and Mr Colburn should mutually agree on the conduct thereof with a provision that in case any difference should thereafter arise between the parties as to the manner in which the same is to be conducted and edited, the same should be referred to arbitration in the manner mentioned in the Agreement. That Mr Brande should duly deliver the complete copy of each number to the Printer on the days therein specified and in case he should neglect to do so then Mr Brande should pay the charges and expenses occasioned by such default in which Mr Colburn should be put to in payment to other persons in order to bring out the numbers of the work in regular Succession at the proper periods of publication. That Mr Brande in consideration of his receiving one Moiety of the profits of the work, should prepare, compose, obtain contributions from writers, edit and conduct the publication &c to the best of his ability. And that he should himself compose or furnish three sheets of each

number and obtain from other writers such contributions and articles as may be fit and necessary for the other parts of each number and pay for all such contributions and articles out of the sum of £63 per number to be paid to him by Mr Colburn provided that in case the average sale of the work on the day of publication of any number should amount to 1500 then that Mr Colburn should pay to Mr Brande the sum of £84 for the purpose of paying such contributions with a further provision for the payment on the part of Mr Colburn of the sum of £105 per number in case the average sale should amount to 2000 Copies Mr Brande undertaking to expend the whole of such sums in paying for such contributions and Articles. —

The agreement also contains a clause by which it is provided that in case Mr Brande should at any time by ill health, absence from London or any other cause become unable or should neglect well and sufficiently to attend to fulfil or perform his said duties as Editor and Conductor of the said Work or from contributing to each number one fifth part of the various matter and Articles therein contained then Mr Brande with the consent and approbation of Mr Colburn should during the time Mr Brande should be incapacitated engage some other person to conduct and manage the same And in case Mr Brande should not do so then Mr Colburn is thereby authorized by Mr Brande to engage and appoint such

person and to pay him such sum or sums of money out of Mr Brande's share of the profits as he shall find necessary and deem requisite. It is also agreed that Mr Colburn should act as Treasurer and receive and pay all monies relating to the publication and should generally leave the whole management of the printing and publication of the work except as by the third Article Mr Brande is to act as Editor and Literary Conductor of the same And that in case any dispute should arise then Arbitrators are to be appointed, Mr Colburn is to be at liberty to insert Advertisements in or on the work, and to receive the profits thereof. Mr Colburn is also to be permitted to make the usual deductions for Commission &c before any division of the profits and he engages to render Annual Accounts to Mr Brande if required so to do. He also undertakes in case the undertaking should be attended with any loss to bear the whole of such loss himself but Mr Colburn is to be at liberty at any time to withdraw from carrying on the work and from being at any further expense therein in case he should find the same to be a losing Concern. The Committee have thought it right to draw the attention of the Board of Managers thus particularly to the terms of the above Contract in order that the Managers may be the more fully competent to judge of the propriety of the conclusion at which the Committee have arrived namely, that the Agreement in question

On Powers of Managers

is not such an agreement as the Board of Managers were authorized in sanctioning. It is the opinion of the Committee that the Managers of the Royal Institution have no power under the 17th Chapter of the Bye laws to sanction the publication of any Journal in which the insertion of original communications, the selection of Papers, the form of publication and the manner of distribution or of sale is not subject to their discretion. The Agreement above stated does not provide for the exercise of this discretion on the part of the Managers in any of the points mentioned and is moreover as the Committee are of opinion extremely objectionable inasmuch as it contains a Clause by virtue of which under certain circumstances Mr Colburn is authorised to perform the duties of Editor without any reference whatever to the discretion of the Managers in so important a point. This being the opinion of the Committee with respect to the Agreement under which the publication of the Journal is now conducted and feeling as the Committee do most strongly that the original course adopted by the Managers of appointing an Editor and of delegating all their authority and control over the publication to him is not sanctioned by the Bye law referred to their consideration, the Committee think that they cannot do otherwise than recommend to the Board of Managers to rescind the resolution of the 8th of January 1816,

Order to be rescinded

and take upon themselves the conducting of a Journal to be published in conformity with the provisions of

Proposed appointment of Publication Committee for New Journal

the 17th Chapter of the Bye laws.

Your Committee are of opinion that such publications would be best afforded by vesting the superintendance of it to a Sub-Committee of Managers to be called The Publication Committee And that such Committee should consist of the President Treasurer and Secretary who should be Members of it Ex-officio and of five other Managers who should be annually nominated by the Board of Managers at their first Meeting after the 1st of May, and that three Members of the Publication Committee should constitute a quorum.

That the duty of the Publication Committee should be to select the Papers and Articles and to prepare the Journal for publication in order to assist them in which they should be at liberty to call in the aid of the several Professors of the Institution, the ~~Superintendent~~ Director of the Laboratory and of such other Scientific persons whether Members of the Institution or otherwise as they may think proper and to form Sub-Committees consisting either of original or associated Members or of both. It should however be an invariable rule that all the contents of each number should be approved of by the majority of the Members of the Publication Committee present either at an Ordinary or Extraordinary Meeting and that all payments on account of the Journal should be authorized in the same manner.

The Publication Committee ought also to have

power to frame Bye laws for their own regulation and to appoint a Secretary to conduct the Correspondence and to superintend the details of the publication who should be paid such sums of money by way of Salary as the Publication Committee may think reasonable.

Your Committee are also of opinion that the contents of the Journal should be comprised conformably to the Bye laws under the following heads.

Proposed Contents

1 – Original communications on subjects connected with Science and the Arts.

2. The proceedings of the Royal Institution.

3. The proceedings of Scientific and Literary Institutions in Great Britain to be procured if possible from the proper Officers of such Societies or Institutions.

4. The proceedings of Foreign Scientific and Literary Institutions.

5. Notices or Analyses of original Scientific and Literary works both British & Foreign &

6. Miscellanea comprising Scientific and Literary intelligence collected from authentic sources and from periodical publications of acknowledged respectability.

In pointing out the subjects to be embraced by such a ~~Committee~~ Journal, your Committee cannot do more than give an outline of the objects to which the Publication Committee ought to direct their attention. The filling up the details

must necessarily be left to the discretion of the Committee when appointed. Your Committee however think that their best endeavours should be exerted to keep the British Public acquainted with the progress of Science and Literature in this and other Countries as well by procuring Analyses and Translations to be made by competent persons of the most important Papers in the Transactions of the British and Foreign Scientific and Literary Institutions as early as conveniently can be done after their publication as by full analyses and notices of such other Publications and Papers connected with Science Literature and the Arts as may appear to them deserving of such attention.

Your Committee also think that it would add greatly to the interest and utility of the Journal if in those cases where the publication to be noticed describe any novel or interesting experiments admitting of verification in the Royal Institution such experiments were (as far as it might in other respects be convenient) to be repeated by the Professor to whose department they may appertain and an account of their repetition with any remarks that may have occurred in its progress made to accompany the notice to be given in the Journal of the original Experiments. It would also tend to increase the confidence of the Public in the Journal if it were understood to be a rule of the Publicat. Committee that no original communication should be published unless the name of the Author be

known to the Committee or to one at least of its Members, the Committee however being at liberty to publish the Paper without affixing the Authors name if it should be his wish to withhold it.

Your Committee have not directed their attention to the mode in which the Publication of the Journal should take place as they are of opinion that the subject would be more fit for the discussion of the Publication Committee, if appointed; As however it will be impossible in first embarking in such an undertaking to foresee all the points to which it may become necessary for the Committee to direct their attention and many circumstances may arise which may render it necessary that modifications should take place in the Arrangements made for the publication, your Committee recommend that provision should be made for reconsidering the mode of conducting the Journal at the expiration of a twelve-month from its commencement and for introducing such alterations as experience shall have shewn to be expedient.

The Committee further suggest that in case the Board of Managers should think fit to adopt the recommendation contained in this report, and to rescind the resolution of the 8th of January, 1816, as suggested, immediate notice of such determination should be given to Mr. Brande.

(Signed) Somerset
Chairman.

Resolved,

Consideration of Report deferred.

That the report of the Committee be entered on the minutes; but that the consideration

1830

Communication to Mr Brande

of it be postponed till Monday next at 3 o'clock; and that in the mean time a Copy of the Report and of this resolution be communicated to Mr Brande.

That Mr Brande be requested to confer with the Board of Managers upon the subject of the Report on the above mentioned day.

Giving Testimonials declined.

Read a letter from Mr Griffiths late Assistant in the Laboratory requesting the honour of an audience, to solicit the favour of the testimonials of the Managers of his having for several years attended the Laboratory and Lectures of the Royal Institution.

Resolved, That Mr Griffiths be informed that it is not usual for the Managers to grant testimonials of the nature required by him.

Present received.

William Ogle Carr, Esqr. M.R.I. presented Specimens of true and false Pepper.

Ordered, That the thanks of the Managers be returned for this present.

Repayment of Loan ordered.

Resolved, That the sum of £100 be paid to Mr Auriol; And that an application be made to the General Meeting as early as can be done in conformity with the Bye Law for authority to sell out so much of the Stock standing to the account of the Sinking Fund as shall be sufficient to meet the demand.

Jan 18.

Annl Subscrs

The following persons having been duly recommended were admitted as Annual Subscribers for the year 1830.

Robert Gairdner, Esqr. 12 Bolton St. Piccadilly
Lt. Chas. Wm. Webber R.N. 17 Bolton Street
J. C. Ireland, Esqr. Albany
Robt. J. Lundie Esqr. 2 Charles St. Manchester Sq.
Thos. Noble Elwyn, Esqr. 36 Albemarle Street
William Gordon Pearse, Esqr. 17 Duke St. Grosvr. Sq.

Adjourned to Monday next at 3 o'clock.

Confirmed the 25th day of January 1830

Somerset Pres.

Jan 25. 1830

At a Meeting of the Managers of the Royal Institution of Great Britain held on the 25th of January, 1830,

Present

His Grace The Duke of Somerset, Pres. in the Chair
Sir George Duckett, Bt. V.P.
Sir Giffin Wilson,
William Pole, Esqr.
Charles Pilgrim, Esqr.
Dr. Nicholl, V.P.
George Moore Esqr. V.P.
Benjamin Bond Cabbell, Esqr. V.P.
Edmund S. Halswell, Esqr.
Edmund Robert Daniell, Esqr. Secy
Captain Sabine,

Wm. Thos. Brande, Esqr. Superintt. of the House
Michael Faraday, Esqr. Director of the Lab.

1. The minutes of the proceedings of the last meeting were read and approved.

Agreement of Mr Brande to Report on Journal

2. Mr Brande having attended in pursuance of the resolution of the last meeting, and expressed his readiness to give the Board of Managers every assistance in his power towards carrying the 17th Chapter of the Bye laws into effect.

Resolved, That the principle of the report of the Committee appointed to consider the 17th Chapter of the Bye laws, be adopted; and

Jan 25.

that it be referred back to a Committee consisting of the same Managers to consider and point out by what means it may be most conveniently carried into execution; And that Mr Brande and Mr Faraday be requested to give the Committee their assistance.

List of Persons attending Lectures

3 Resolved, That the person appointed to attend the Lecture room door be directed to keep a list of the number of persons who on each day attend each Course of lectures which shall be delivered in the Theatre of this Institution, distinguishing every separate Course; And that the Book be laid on the table at the Meetings of the Managers.

4. The following persons having been duly recommended were admitted as Subscribers to the Royal Institution for the year 1830

Annual Subscriber;

Col. James Veney, 88 Pall Mall

To the Lectures

Lady Duckett 28 Norfolk St. Park Lane

Confirmed the 1st Feb. 1830

Somerset Pres.

310

Feb. 1. 1830.

At a Meeting of the Managers of the Royal Institution of Great Britain held on the 1st of February, 1830,

Present

His Grace The Duke of Somerset Pres.t in the Chair
Sir George Duckett, Bart. V.P.
Benjamin Bond Cabbell, Esq.r V.P.
George Moore, Esq.r V.P.
Dr Whitlock Nicholl, V.P.
Captn Sabine,
C. H. Pilgrim, Esq.r
William Pole, Esq.r
Edmund Robt Daniell, Esq.r Secy

William Thos Brande, Esq.r

1\. The minutes of the proceedings of the last meeting were read and approved.

Rept from Com of Accts 2. Read the following report from the Committee of Accounts.

The Committee of Accounts report that they have examined the General Accounts since the last Meeting on the 7th of December 1829; And that they have found a balance in favour of the Institution of £312.3.4

They recommend that the following bills be paid on account of the Royal Institutn

Mr Nicol for Printing in 1828 £91.5 –
Mr Hurt for Stationery in 1829 14.15.6
Joseph Fincher for Collecting in do 31.10 –
Knapp for Bookbinding – in do 3.16.5
Carrd forwd 141 – 6.11

311

Feb. 1.

Brot forward 141.6.11
Gas Company for Gas in 1829 21.6.5
Joseph Fincher on account of the Royal Institution 100 —
262.13.4

Member Absent abroad 3. Read a letter from Dr Quin addressed to the Secretary, informing him that he unexpectedly left England after his election as Member of the Royal Institution and that as he is now established in Paris and does not expect to visit England for some years, he wishes to be informed how he can conform to the regulations of the Royal Institution with respect to the annual contributn and also if there any Bye laws regarding Members not resident in England.

Resolved, That the above letter be taken into consideration at the next Meeting of the Managers

4 Presents received.

R. Watson, Esq.r presented
Plan for preventing Ships foundering at Sea,
R. Duppa, Esq.r presented
Travels on the Continent, Sicily &c. 2d Edn
G. W. Johnson, Esq.r presented
The History of English Gardening,
B. L. Vulliamy, Esq.r presented
Supplement to his Treatise on Public Clocks,
R. F. Newnham, Esq.r presented
The Pleasures of Anarchy,
T. Clissold, Esq.r presented
Practical hints for the developement of the Human Mind.

312

1830.

J. H. Curtis, Esq.r presented
A Synoptical Chart of the diseases of the Ear.
T. Gill, Esq.r presented
12 nos of Gill's Technological Repository
J. Britton, Esq.r presented
Britton's Beauties of Wiltshire

Ordered, That the thanks of the Managers be returned for these presents.

Payment of Deficiency in Substns for Mr Moore's Bust 5. Read the minute of the General Meeting of the 1st of June last, relating to Mr Sievier's bill of £94.10 for Mr Moore's Bust.

Resolved, That as the total subscription has only amounted to £66.3 – out of which £57.15 – has yet been received, that the deficiency amounting to £36.15 – be paid out of the general Funds of the Institution agreeably to the resolution of the General Meeting of the 1st of June, 1829,

6\. The Assistant Secretary had the pleasure to inform the Board that John Pepys, Esq.r M.R.I. had presented to the Royal Institutn the sum of £100 to be applied to the general purposes of the Institution.

Resolved, That the above communication be reported to the General Meeting.

7\. The Secretary in the absence of Mr Faraday

313

Feb. 1.

Admissions to Friday Evg limited to two for each Member informed the Board that the Committee appointed to superintend the arrangements of the Weekly Evening Meetings considering that it will be expedient to limit the privilege of Members with respect to the admission of Strangers on the Friday Evenings to two persons each Member had resolved that after Friday the 22d of January, no Member of the Royal Institution be permitted to introduce more than two persons (either personally or by ticket) to each of the Evening Meetings; And that they had directed a Notice to that effect to be suspended in the Hall, Library and Newspaper Room announcing such regulation.

Resolved, That the above resolution of the Committee for superintending the arrangements of the Friday Evenings be approved of.

Payment to Mr F for Juvenile Lectures 8. Ordered, That the sum of 50 guineas be paid to Mr Faraday for his Course of Juvenile Lectures delivered during the Christmas Vacation.

9\. Ordered, That Mr Moore's Bust be placed over the Door leading from the Ante-Room into the Lecture Room.

Mr Fuller's Gold Medal 10. The Secretary having communicated to the meeting that John Fuller, Esq.r had announced to him his intention of giving this year his Gold Medal to such person as the Managers of

314

1830.

the Royal Institution together with the Professor of Chemistry should consider worthy of that distinction.

Resolved, That the thanks of the Managers be returned to Mr. Fuller for the above communication.

Resolved, That in order to carry Mr. Fuller's intention into effect, that a Committee be appointed consisting of His Grace the President and two other Members of this Board, the President of the Royal Society, the Professor of Chemistry and the Director of the Laboratory and such Members of the Royal Institution as have already received the Medal, to report to this Board such person as they may think to have merited the distinction.

Resolved,

That Captn. Sabine and Dr. Nicholl be upon the Committee.

11. The following person having been duly recommended was admitted as an Annual Subscriber for the year 1830

A Hamilton, Esqr. 26 Cumberland St.

Adjourned to Monday the 15th. Instant at Three o'Clock.

Confirmed the 15th day of February 1830

Somerset Pres.

315

Feb. 15.

At a Meeting of the Managers of the Royal Institution of Great Britain, held on the 15th. of February, 1830

Present

His Grace The Duke of Somerset, Pres: in the Chair

Sir George Duckett, Bart. V.P.

George Moore, Esqr. V.P.

Benjamin Bond Cabbell, Esqr. V.P.

William Pole, Esqr.

C. H. Pilgrim, Esqr.

Captn. Sabine,

Edmund S. Halswell, Esqr.

Edmund Robt. Daniell, Esqr. Secy.

William Thos. Brande, Esqr. Supt. of the House

Michael Faraday, Esqr. Director of the Laby.

1. The minutes of the proceedings of the last meeting were read and approved.

Rept. of Com. on Mr. Fuller's Medal.

2. Read the following report of the Committee on Mr. Fuller's Medal.

The Committee appointed by the Managers to report upon the proper person to be nominated by Mr. Fuller to receive his Gold Medal for discoveries in Chemistry, report,

Presented to W. H. Pepys. Esq.

That it is the opinion of the present Committee that Mr. Fuller's Gold Medal should be presented to William Hasledine Pepys, Esqr. M.R.I. for his various researches in Chemical Science, and for his inventions and improvemts.

316

1830

Researches of Mr. Pepys.

of Chemical Apparatus. The Committee especially refer to Mr. Pepys' invention and employment of a new Eudiometer, to his researches connected with the results of the combustion of the Diamond and of the varieties of Charcoal and Plumbago; and to his investigation upon the subject of Respiration, in conjunction with Mr. W. Allen. Among his contributions to Chemical Apparatus, the contrivances of the Air-holders in common use, and of Mercurial Gasometers, are especially alluded to by the Committee.

Resolved,

That the report of the Committee be approved; and that William Hasledine Pepys, Esqr. be nominated to Mr. Fuller as the proper person to receive his Medal. And That the above resolution be communicated to Mr. Fuller and to Mr. Pepys; and that Mr. Pepys be requested to attend the General Meeting on the 1st. of March for the purpose of receiving the Medal from Mr. Fuller.

Relinquishment of Loan by H. Hoare Esq. by compounding for Payments as Member.

3. The Secretary stated that Mr. Pilgrim having informed him of Mr. Hoare's intention of relinquishing his claim on the Royal Institn. of £200 advanced by his late Father Henry Hoare, Esqr. And of appropriating that sum in becoming a Member of the Royal Institution and Patron of the Library and Mineralogical Collection; he had written to Mr. Hoare to acquaint him with the terms required for his

317

Feb. 15.

qualification; And he had received in reply the following letter.

Fleet Street 5th. Feby 1830

Sir,

In answer to your letter of the 29th. Jany I beg to say that I should wish to become a Member of the Royal Institution by paying over Eighty guineas; and also Patron of the Library by allowing the remnant of the Two hundred Pounds due to the late Mr. Hoare to be carried to the account of the Library.

I have the honour to be &c.

(signed) Henry Hoare

Ordered,

That the necessary steps be taken for Mr. Hoare's election as Member of the Royal Institution and Patron of the Library.

4 Resolved, That the consideration of Dr. Quin's letter be postponed till the next Meeting.

5 Resolved, That the resolution of the Managers at their last Meeting (No. 9.) be reconsidered.

Commee. for placing Mr. Moore's Bust.

6. Resolved, That a Committee be appointed to consider the placing of Mr. Moore's Bust.

That the Committee consist of

His Grace The Duke of Somerset.

Mr. Moore,

Mr. Halswell,

and

Mr. Brande,

Mr. Daniell.

318

1830.

That the Committee be authorized to make the necessary arrangements for carrying their report when made into execution.

7. Engraved Portrait of M. Faraday presented.

Mr. Colnaghi presented An Engraved Portrait of Michael Faraday Esqr.

Ordered, That the thanks of the Managers be returned for this present.

8 Resolved, That it be referred to the Committee appointed to consider the placing of Mr. Moore's Bust to consider the proper place for the Print

9 Letter from Mr. Ritchie resp.g Apparatus.

Read the following letter from Mr. Ritchie addressed to Captn. Sabine.

Royal Institution 15th. Feby 1830

Dear Sir,

Having examined the Apparatus in the Institution I find it very imperfect for illustrating various parts of my Course particularly Electro-Dynamics and Light. I therefore feel that unless this part of the Apparatus be rendered more complete, I shall be unable to do justice to these interesting subjects. Should the Managers allow me to get constructed under my own inspection the various pieces of Apparatus absolutely necessary. I think that fifteen or twenty pounds would accomplish all I should want in this Course. I also take the liberty of stating that for want of a separate Room, I am

Feb 15

319

unable to carry on any original investigation which may occur, or even to prepare the experiments for my Course in the way I could wish. Trusting you will pardon this liberty. I am, Dear Sir,

With much respect Your most obedient Servant

To Captn. Sabine. (signed) William Ritchie

Resolved,

That Mr. Ritchie be requested to furnish a list of the Apparatus required to Mr. Brande and Mr. Faraday and that they be at liberty to order such of them as they may think necessary provided the amount do not exceed the sum of £20.

Dr. Harwood.

Resolved, That Mr. Ritchie be allowed the use of the Room lately occupied by Dr. Harwood to prepare for his lectures.

Resolved, That a letter be written to Dr. Harwood to request him to deliver up the key of the Room lately occupied by him and also to send up the keys of the several cases in the Museum.

Complaints resp.g Talking in Reading rooms. 10

The Superintendent of the House having stated that great complaints have been made of the practice indulged in by many of the Members and Subscribers of talking very loud in the Reading Rooms of the Institution, which statement was confirmed by several Members of this Board from their own experience.

Resolved, That a notice of the above complaint be communicated to the next

320

1830.

General Monthly Meeting.

Adjourned to Monday the 1st. of March.

Confirmed the 1st. of March 1830

Somerset Pres.

Mar. 1.

321

At a Meeting of the Managers of the Royal Institution of Great Britain, held on the 1st. of March, 1830,

Present

His Grace The Duke of Somerset Pres. in the Chair
Sir George Duckett, Bart. V.P.
Benjamin Bond Cabbell, Esqr. V.P.
George Moore, Esqr. V.P.
Edmund S. Halswell, Esqr.
Dr. Whitlock Nicholl, V.P.
Charles Pilgrim, Esqr.
William Pole, Esqr.
Captn. Sabine,
Sir S. Bernard Morland, Bart. M.P. Treas.
Edmund Robt. Daniell, Esqr. Secy.

William Thos. Brande, Esqr.
Michael Faraday, Esqr.

1. The minutes of the proceedings of the last meeting were read and approved.

2. Rept. of Comm.e of Acct.s

Read the following report from the Comm.e of Accounts.

The Committee of Accounts report that they have examined the General Accounts since the last meeting, and that they have found a balance in the hands of the Bankers of £446. 6.– and a balance in the hands of the Assistant Secretary of £33. 17.

They likewise report that they have examined the Laboratory Account, and that they have

found a balance in favour of that account of £220 - 9 . –

They recommend that the following bills be paid on the General Account.

Feetham for Ironmongery in 1829 £ 16 - 9 - 3
Pringle for Upholstery in do. - 12 - 10 - 10
Norris for Carpeting in do. — 29 - 6 - 9

and

On the Laboratory Account

Fisher for Chemicals in 1829. £ 20 - 9 - 11

They also report that they have audited the General and Laboratory Accounts for the year, 1829.

Report of Commee. respt. Journal.

3. Read the following report from the Committee for considering the mode in which the report of the Committee of the 18th January should be carried into effect.

In preparing the arrangements for the proposed Journal, the subject which appeared to require consideration in the first instance, was the price per sheet, for which it would be necessary that funds should be placed at the disposal of the publication Committee. After maturely considering this point the Committee came to the opinion that a Journal, which should thoroughly fulfil the purposes named in the Bye law, and should be worthy of the responsibility and trouble proposed to be incurred in its management, would require funds for its support at the rate of ten guineas a sheet; which

sum would include every expense connected with the preparation of the Manuscript.

Offer to Mr Colburn.

At the request of the Committee Mr. Brande notified to Mr. Colburn the Publisher of the present Journal, that the agreement under which it is conducted should terminate at the conclusion of the next Volume; the last number of which will appear on the 1st. of July next. The Committee also empowered Mr. Brande to offer Mr. Colburn the publication of the new Journal on the plan and terms proposed by the Committee: which offer Mr. Colburn declined.

Declined.

The Committee beg leave to observe that hitherto the communication with Mr. Colburn has been merely verbal and that they have not considered themselves authorized in giving formal notice of their resolution to Mr. Colburn without the sanction of the Board of Managers. They recommend however that immediate notice in writing should be given to Mr Colburn on the part of the Managers and of Mr. Brande of their intention to put an end to the agreement referred to in the former report of the 18th. of January; they also think that Mr. Colburn should be directed to discontinue the use of the Stamp containing the Crest of the Royal Institution.

Close of Agreement between Messrs. Brande & Colburn. recommended

Agreement with Mr. Murray

The Committee next authorized Mr Brande, Mr. Daniell and Captn. Sabine to make the same offer to Mr. Murray; who, in an interview with those gentlemen, stated his readiness to concur fully with the Managers of the Royal Institution in making a fair experiment, whether a Scientific

Journal of the superior character contemplated by the Managers would meet with a sufficient sale for its support: that, for that purpose, he would undertake to print and publish at his own expense and risk, the two first numbers of such a Journal, on being furnished with the Manuscript of them without payment: and should the experiment succeed, to continue the Journal, paying the expenses of Manuscript as well as of publication, and repaying the sums, not exceeding 10 guineas a sheet, which should have been advanced on the proportion of the Manuscript of the two first numbers.

The arrangements have thus arrived at a stage when it becomes proper that the Committee should again bring the subject under the consideration of the Board of Managers. It appears that for the establishment of the Journal on the footing which gives the best promise of its usefulness and permanency an advance is necessary of a sum not exceeding £150 for a first number to be published in October; and of the same sum for a second number to be published in January; the repayment of these sums and the continuance of the Journal beyond those two numbers depending upon the public support which they may receive.

The Meeting was adjourned till after the General Meeting.

Meeting resumed.

Resolved, That the report of the Committee be approved of and adopted.

Close of Agreemt. with Mr Colburn.

That a formal notice be given to Mr. Colburn on the part of the Managers of the Royal Institution and of Mr. Brande of their intention to put an end to the agreement entered into by him with Mr Brande respecting the publication of the Journal after the conclusion of the next Volume.

That the Secretary be directed also to inform Mr. Colburn and his Partner that after the conclusion of the next volume of the present Journal they will not be at liberty to make use of the Crest or Arms of the Royal Institution in or upon any future publication to be made by them without the leave of the Board of Managers first had and obtained.

Appointment of Committee

Resolved, That the resolution of the 8th. of January 1816, & referred to in the report of the Committee of the 18th. of January last, be rescinded.

That a Committee be appointed to superintend the future publication of the Journals of the Institution to be called "The Publication Committee".

That the Committee do consist of eight Members of the Board of Managers, and that
The President
The Treasurer
& The Secretary
} be Members of the Committee Ex Officio, and that three Members shall constitute a quorum

1 That the following Managers be upon the Committee vizt.
George Moore, Esqr.
Benjamin Bond Cabbell, Esqr.
D^r. Whitlock Nicholl
Captn. Sabine
Charles Pilgrim Esqr

Duties of Committee. That the duty of the publication Comm^e. be to select the Papers and Articles for the Journal and to prepare each number for publication and that in the performance of their duty the Committee do carry into execution as far as it may be practicable the suggestions of the Committee of the 18th. of January.

That the Committee be at liberty to call in the assistance of the several Professors of the Institution; the Director of the Laboratory and of such other scientific persons as they may think proper.

That the publication Committee be at liberty to form Sub-Committees consisting either of original or associated Members, or of both.

That the Publication Committee be authorized to enter into an agreement with M^r. Murray for the publication of a Journal upon the bases stated in the Report, and that they do report the same to this Board.

That the Publication Committee do report to the Board of Managers the best method of providing the Funds that will be requisite to

defray the expense of providing contributions for the first two numbers.

Bye-Laws That the Publication Committee do frame proper Bye laws for the regulation of their proceedings and that they do alter and amend such Bye laws and frame new ones as often as they shall deem it expedient, provided that no such Bye law be contrary or inconsistent with the Charter and Bye laws of this Institution and the resolutions of the Board of Managers.

That the Publication Committee be at liberty to appoint a Secretary to assist them in conducting their Correspondence and in superintending the details of the Publication, and to pay him out of the funds at their disposal such sum of money by way of Salary as to y^e. Committee shall seem expedient.

That the duty of the Publication Committee be to select the Papers and Articles of the Journal and to prepare the Manuscript of each number previous to its being sent to the Publisher and to generally superintend the Printing ~~and Publication~~.

That it be an invariable rule that all the contents of each number of the Journal should be approved of by a majority of the Members of the Publication Committee present either at an Ordinary or extraordinary Meeting.

That the Publication Committee shall have power of disposing of all the Funds placed at their disposal by the Board of Managers provided that no payment thereout be made on account

of the Journal unless authorized by a majority of Members of the Publication Committee present either at an Ordinary or Extraordinary Meeting.

That no original communication be published in the Journal unless the name of the Author be known to the Publication Committee or to one at least of its Members. The Committee however are to be at liberty to publish ~~every~~ any communication without affixing the name of the Author, if it be the Authors wish that it should be withheld.

Resignation of D^r. Harwood. 4. Read a letter from D^r. Harwood announcing that in consequence of his residence out of London being incompatible with the performance of his duties as Professor of Natural History in this Institution he was desirous of resigning that honour and informing the Secretary that M^r. Greenwood had the key of the Room appropriated to his use but that it would be inconvenient to him to send up the other keys; and requesting that the delivery of them might be postponed till after his return to Town from Liverpool.

Resolved,
That D^r. Harwood's resignation be accepted; and that it be reported to the General Monthly Meeting in April.

Resolved, That D^r. Harwood's request to postpone the delivery of the keys of the Cases in the Museum be complied with.—

5 Ordered, That the key of the Room lately occupied by D^r. Harwood be delivered to M^r. Ritchie.

Presents received.

6 The Geological Society presented
Transactions of the Society Vol 3 part 1.
Louis Hayes Petit, Esqr. presented
Tattam & Young's Egyptian Grammar.
John Marshall, Esqr. presented
A Popular Summary of Versification
Thomas Alcock, Esqr. presented
Lectures on Practical and Medical Surgery
William Brockedon, Esqr. presented
Passes of the Alps, N°. 12
E. Atherstone, Esqr. presented
The fall of Nineveh, Vol 2^d.
Sir George Duckett, Bart. presented
Stones from the foundation of London Bridge

Ordered,
That the thanks of the Managers be returned for these presents.

7. Signed an Order for an Advance to the Assistant Secretary of £200 on account of the Royal Institution, and £50 on account of the Laboratory Fund.

Adjourned to Monday the 15th. Inst^t.

Confirmed the 15th day of March 1830
Somerset Pres.

330

1830.

Mar 15.

At a Meeting of the Managers of the Royal Institution of Great Britain held on the 15th of March, 1830,

Present

His Grace The Duke of Somerset Pres: in the Ch[r]

Sir S. Bernard Morland, B[t] M.P. Treas[r]

George Moore, Esq[r] V.P.

D[r] Whitlock Nicholl, V.P.

Benjamin Bond Cabbell, Esq[r] V.P.

Capt[n] Sabine,

William Pole, Esq[r]

Cha[s] H. Pilgrim, Esq[r]

Edmund S[t] Halswell, Esq[r]

Edmund Rob[t] Daniell, Esq[r] Sec[y]

Michael Faraday, Esq[r] Director of the Lab[y]

1\. The minutes of the proceedings of the last meeting were read and approved.

Report from Publication Committee.

2\. Read the following report from the Publication Committee.

The Committee report that in pursuance of the direction of the last meeting of the Board of Managers they had proceeded to enquire into the best method of defraying the expense of providing contributions for the two first numbers of the Journal, and that they are of opinion that the same will best be effected by a voluntary Contribution amongst the Members of the Institution; to be repaid

331

Mar 15.

out of the first profits of the Journal to be received from the Publisher. The Committee also report that the following Members had signified their willingness to become Contributors to the amount set opposite to their respective names.

Subscribers to publication of Journal.

His Grace The Duke of Somerset	£ 50
Benjamin Bond Cabbell, Esq[r]	10
George Moore, Esq[r]	10
Edmund Rob[t] Daniell Esq[r]	10
Charles H Pilgrim, Esq[r]	10
Michael Faraday, Esq[r]	10
R. G Clarke, Esq[r]	10
George Meredith, Esq[r]	10
W[m] Hasledine Pepys, Esq[r]	10
R. H. Solly, Esq[r]	10
Sir George Tho[s] Staunton B[t]	10
Sir George Duckett, Bart.	10
William Pole, Esq[r]	10
D[r] Whitlock Nicholl	10

Lectures declined

3\. Read a letter from Robert Owen, Esq[r] proposing to deliver one or more lectures on the Friday Evenings in the Royal Institution on the Science of Society.

Resolved, That M[r] Owen's proposal be declined as not coming within the objects of the Institution.

Presents received.

4\. Rich[d] Davenport Esq[r] M.R.I presented The Amateurs' Perspective with Supplem[t]

Miles Marley, Esq[r] presented "Treatise on the diseases of Children".

332

1830

Ordered, That the thanks of the Managers be returned for these presents.

Letters from M[r] Brande & M[r] Daniell resp[g] Journal.

5\. The Secretary reported that in pursuance of the resolution of the last Meeting M[r] Professor Brande had written to M[r] Colburn to inform him that it is his intention to put an end to the Agreement entered into by him with M[r] Colburn respecting the publication of the Journal, after the conclusion of the next Volume.

The Secretary also reported that in pursuance of the same resolution he had written to Mess[rs] Colburn and Bentley to request them after the conclusion of the present Journal not to use the Crest or Arms of the Royal Institution on any future publication to be made by them unless sanctioned by the Board of Managers.

Resolved, That the above letters be entered on the Minutes.

Copy of a letter from W[m] Tho[s] Brande Esq[r] to Henry Colburn, Esq[r] dated 4th March 1830

Dear Sir,

In order to prevent any future misunderstanding between us, I think it right to give you formal notice that I shall withdraw from editing and conducting the Journal now published by you under the title of a Quarterly Journal of Science Literature and the Arts, under

333

Mar 15

and by virtue of the Agreement entered into between us dated the 21st of March 1827, after the publication of the next Volume, the last number of which is to appear in July next, shall have been completed. And that from and after the publication of the above mentioned Volume, I shall consider the said Agreement to be at an end.

I am, Dear Sir, yours truly

(signed) W[m] Tho[s] Brande

Copy of a letter from Edmund Rob[t] Daniell Esq[r] Sec[y] R.I to Mess[rs] Henry Colburn and Richard Bentley dated 4th March, 1830.

Royal Institution 4th March 1830

Gentlemen,

I am desired by the Managers of the Royal Institution to request that after the conclusion of the next Volume of the Quarterly Journal published by you under the title of a Quarterly Journal of Science Literature and the Arts, the last number of which will be published in July next, you will discontinue the use of the stamp containing the Crest and name of the Royal Institution of Great Britain. And that you will not after that period make use of the Arms or Crest of the Royal Institution of Great Britain, or of the name of the Royal Institution in or upon any Journal or Work thereafter to be published by you without the leave of the Managers of the Royal Institution first had and obtained.

I am Gentlemen,

Your most obedient Servant

(signed) Edmund R. Daniell

1830.

Sinking of Book cases in Library. 7. Mr. Faraday reported that one of the Bookcases in the Library of Reference appeared to have sunk.

Resolved, That Mr. George Moore, Mr. Brande, Mr. Faraday and the Secretary be requested to examine into the state of the Building in the part referred to; and to report their opinion upon the same to this Board.

Confirmed 5 April 1830
Benjn. Bond Cabbell
V.P.

Mar 15

At a Meeting of the Managers of the Royal Institution of Great Britain, held on the 5th of April, 1830.

Present
Benjamin Bond Cabbell, Esqr. V.P. in the Chair
Sir George Duckett, Bart. V.P.
Edmund S. Halswell, Esqr.
George Moore, Esqr. V.P.
Charles Pilgrim jun. Esqr.
William Pole, Esqr.
Captn. Edwd. Sabine,
Edmund Robt. Daniell, Esqr. Secretary.

1. The minutes of the proceedings of the last meeting were read and approved.

Repr. from Com. of Accts 2. Read the following report from the Committee of Accounts.

The Committee of Accounts report that they have examined the General Accounts since the last meeting, and that they have found a balance in favour of the Royal Institution of £380.16.5

They recommend that the following sums be paid on account of the Royal Institution.

H. Greenwood on account of Tea & Coffee for Friday Evening Meetings	30. 0. 0
Jones & Yarrell for Newspapers in 1829	82.14.7
Mitchell for Votes House of Commons in ditto	11.12.—
Joseph Fincher on account of the Royal Institution	100. 0. 0
	224. 6. 7

1830.

and on account of the Laboratory Fund
Newman for Chemicals in 1829. £60-16.6

And signed Orders accordingly.

Repr. from Comme. resp. Liby. Floor 3. Read the following report from the Committee appointed to examine the Floor of the Library.

The Committee report that they have carefully surveyed the same, and find that it has gradually sunk in certain parts, although the appearances are not very evident on the Ceiling of the Newspaper Room beneath.

It is impossible to form an accurate opinion of the strength or soundness of the Timbers without taking up a part of the Floor boards in the Library so as to obtain a general and complete view of them, which would of course require Workmen to be introduced into the Room, and occasion it to be closed for a certain period.

Under these circumstances the Committee recommend that for the remainder of the Season during which the Friday Evening Meetings are held, the parts which appear to have been acted upon by the additional weight introduced on those occasions, should be carefully watched, and if the depression is found to increase, that immediate steps should be taken for a full investigation of the failure, by a competent Committee empowered to make such arrangements as may render the Floor perfectly secure.

The Committee are also of opinion that at

Mar 15.

the conclusion of the season, even if the appearances of depression should not increase, that it would be desirable a Committee should be appointed to examine and report upon the subject for the purpose of allaying any alarms or apprehensions that may already have been excited.

Resolved,

That the above Report be approved of; And that the same Committee be requested to watch the parts of the Floor which appear to have been acted upon by the weight of the Friday Evening Meetings, and to report to this Board in case they shall see necessary so to do.

Repaymt. of loan to Mr. Hatchett 4. Read a letter from Charles Hatchett, Esqr. dated Belle Vue House Chelsea 20th March, 1830 applying for the repayment of the sum of £100 advanced by way of loan to the Royal Institution.

Resolved, That Mr. Hatchett be informed that the Managers of the Royal Institution are ready to pay the amount of his bond on the production of the bond and of a proper receipt.

5 Resolved, That an application be made to the General Meeting pursuant to Art 1, Chap: XI of the Bye laws for their sanction to sell out as much Stock standing to the Account of the Royal Institution Sinking Fund as will be sufficient to pay the amount of Mr. Auriol and Mr. Hatchett's Bonds.

338

1830

6\. On Superseding Climbing-boys in Cleaning Chimneys.

Read a letter from Robert Steven, Esqr. Hon^y Sec^y to the Society for superseding the necessity of Climbing Boys in sweeping Chimnies enclosing testimonials in favour of the Machine and requesting in the name of the Committee that one of the Agents of the Society may be employed in the Royal Institution.

Resolved, That it be referred to the Assistant Superintendent of the House to ~~[illegible]~~ adopt such ~~[illegible]~~ method of sweeping the Chimnies as he shall think best.

7\. On Prof^r of Botany.

Read a letter from M^r. Gilbert T. Burnett respectfully soliciting the honourable office now vacant of Professor of Botany in the Royal Institution.

Resolved, That M^r. Burnett be informed that it is not the intention of the Managers of the Royal Institution to propose any person to fill the office of Professor of Botany on the present occasion.

8\. Lectures. Capt. Manby.

Read a letter from Captn. G. W. Manby, addressed to the Secretary, requesting him to state to the Managers that if they think it would be attended with any beneficial consequences, he should be happy to repeat the three lectures that he has delivered before the Scientific Society at Yarmouth upon the most effectual means of saving the lives of Shipwrecked Mariners.

Resolved, That Captn. Manby's letter be referred to the Committee for regulating the Friday Evening Meetings.

339

Mar 15

9\. **Presents received.**

The Honble East India Company presented
Catalogue of Oriental Manuscripts 2 vol^s 8vo
Madras Observatory Papers,
Robert Hoblyn, Esqr. presented
Translation of the first Book of the Georgics,
Michael Faraday, Esqr. presented
The Bakerian Lecture on the Manufacture of Glass:
The Editors of the Zoological Journal presented N°. 17 of their Journal.

Ordered, That the thanks of the Managers be returned for these presents.

10\. Rep^t from Visitors.

Read the following report from the Visitors as entered in their proceedings of the 25th of March.

Repairs. The Visitors found the Premises generally in good repair. They recommend that the Kitchen and Passage leading thereto be white-washed. They are also of opinion that the interior of the upper part of the House will shortly require painting.

11\. Prospectus of Journal.

The Publication Committee reported, That the Prospectus of the Journal of the Royal Institution was now printed.

Ordered, That the Prospectus of the Journal be laid before the General Meeting.

The Publication Committee also reported

340

1830

Subscriptions to Journal.

that Thomas Snodgrass, Esqr.
and Captn. Sabine
had each become Contributors of £10 towards providing contributions for the Journal.

12\. The following persons having been duly recommended were admitted as Subscribers to the Royal Institution for the year 1830

Annual Subscribers

Alexander Colvin, Esqr. 15 Lower Seymour S^t
Henry F. Broadwood, Esqr. Bryanstone Sq^r
Major Genl. Sir Colquhoun Grant, K.C.B.
George Hobson, Esqr. 5 Harley Place
Christian Paul Meyer jun^r Esqr. Cavendish Sq^r.

To the Lectures

Miss Dyneley Manchester Square
Miss Howard 5 Harley Place

Adjourned to the 22^d Instant at Three o'clock.

Confirmed April 22^d 1830
Whitlock Nicholl

341

April 22.

At a Meeting of the Managers of the Royal Institution held on the 22^d of April, 1830.

Present

Dr Whitlock Nicholl V. P. in the Chair.
Benjn. Bond Cabbell, Esqr.
George Moore, Esqr.
Captn. Sabine,
Chas. H. Pilgrim, Esqr.
Edmund Robt. Daniell, Esqr. Sec^y.

1\. The minutes of the proceedings of the last meeting were read and approved.

2\. Proceeded to prepare the House list of the Officers of the Royal Institution for the ensuing year.

3 Decease of Sir S. Morland. Treasurer — Sir G. Duckett proposed.

The Assistant Secretary stated that a Vacancy had occurred in the office of Treasurer by the decease of Sir S. Bernard Morland, Bart.

Resolved, That Sir George Duckett Bart. be proposed as Treasurer of the Royal Instit^n.

4\. Comm^n from C. Hatchett resp^g Vol^y Loan.

Read a letter from Charles Hatchett, Esqr. addressed to the Secretary informing him that he had left his bond and a receipt with his Bankers Messrs Coutts.

Resolved, That the sum of £100 be paid to Messrs. Coutts agreeably to Messrs. Hatchett's request; And signed an Order accordingly.

342 1830.

5\. Re Repayment of Loan to Sir G. Blane. Read a letter from Sir Gilbert Blane Bart requesting that the £100 which he granted first as a loan without Interest for four years and afterwards extended to four years more, be paid to himself at the expiration of the second period, or to his Executor should he die before that time; And that it should be paid to the house of Herries & Co. Bankers St. James's Street.

Resolved, That Sir Gilbert Blane be informed that the Managers will take care that the loan shall be repaid at the expiration of the extended term of four years and paid to Messrs. Herries as requested.

6\. Lectures declined. Read a letter from Major Bell proposing to deliver before the Members a gratuitous Lecture on Universal History.

Resolved,

That Major Bell's proposal be declined

Presents received.

7\. Dr. Davy presented Mineralogical Specimens from Ceylon and various parts of the East Indies

Mr. Smily M.R.I. presented

A Specimen of an Alligator brought by him from Prince of Wales's Island.

Mr. Reveley presented

An Essay on the Mosaic Dispensation not intended to be perpetual.

Ordered, That the thanks of the Managers be returned for these presents.

April 22. 343

8\. Ordered, That the House lists be filled up as follows.

President. The Duke of Somerset. F.R.S.

Treasurer – Sir George Duckett Bt. F.A.S.

Secretary. Edmund Robt. Daniell, Esq. F.R.S.

Managers.

Benjamin Bond Cabbell, Esq. F.S.A.
John George Children, Esq. F.R.S. & F.L.S.
Earl Grosvenor,
I. L. Goldsmid, Esq. F.R.S.
Edmund S. Halswell, Esq. A.M. F.R.S.
Thomas Hoblyn, Esq. F.R.S.
Joseph Jekyll junr. Esq.
George Moore, Esq. F.S.A. & F.L.S.
R. I. Murchison, Esq. F.R.S.
Whitlock Nicholl, M.D. F.R.S. &c. &c.
William Hasledine Pepys, Esq. F.R.S.
Charles Pilgrim junr. Esq.
William Pole, Esq. F.R.S.
Lord Prudhoe, F.R.S.
Captn. Edward Sabine Royal Arty. Sec. R.S.

Visitors.

Sir Thos. Dyke Acland, Bart M.P.
Henry James Barchard, Esq.
Thomas Tyringham Bernard, Esq.
John Frederic Daniell, Esq. F.R.S.
The Duke of Devonshire, K.G.

344 1830.

James William Farrer, Esq.
John Fuller, Esq.
Augustus B. Granville M.D. F.R.S.
George Long, Esq.
Peter M. Roget M.D. Sec: R.S.
Edward Adolphus Lord Seymour
Earl Stanhope, F.R.S.
Richd. Horsman Solly, Esq. F.R.S.
N. A. Vigors, Esq. F.R.S. & F.L.S.
George Warry, Esq.

Confirmed 3 May 1830
Geo. Duckett

May 3. 345

At a Meeting of the Managers of the Royal Institution of Great Britain held on the 3d. of May, 1830

Present

Sir George Duckett, Bart: Treas. in the Chair
Benjamin Bond Cabbell, Esq.
George Moore, Esq.
Wm. Hasledine Pepys, Esq.
John George Children, Esq.
Charles H Pilgrim, Esq.
William Pole, Esq.
Captn. Edwd. Sabine,
Edmund Robt. Daniell, Esq. Sec.

Michael Faraday, Esqre. Director of the Laby.

1\. The minutes of the proceedings of the last meeting were read and approved.

2\. Rept from Com. of Accts. Read the following report from the Committee of Accounts.

The Committee of Accounts report that they have examined the General Accounts since the last meeting, And that they have found a balance in favour of the Royal Institution of £336. 8. 4

They recommend that the following sums be paid on account of the Royal Institution

Mr. Adcock for a Course of lectures on Mechanics	47. 5. –
2 quarters Parochial rates to Lady day last	36. 13. 4
	83. 18. 4

346

1830.

And signed an Order accordingly.

Presents received.

3. C. P. Cooper, Esqr. presented Proceedings in Parliament relating to the Court of Chancery.
Lettres sur la Cour de la Chancelerie.
Louis Hayes Petit, Esqr. M.P. presented
Egyptian Hieroglyphics, part 5th.
Henry Coddington, Esqr. presented
Treatise On Optics part 2d.
Dr. Christie presented
Observations on Cholera
W. T. Brande, Esqr. presented
Manual of Chemistry 3d. Edn.
The Astronomical Society presented
Transactions of the Society Vol 4 pt. 1.

Ordered, That the thanks of the Managers be returned for these presents.

4. Resolved, That the following Committees be appointed for the ensuing year.

Committee for regulating the Friday Evening Meetings.

His Grace The President,
George Moore, Esqr.
Benjamin Bond Cabbell, Esqr.

347

May 3.

Dr. Whitlock Nicholl
Captn. Sabine
William Pole, Esqr.
I. L. Goldsmid, Esqr.
John George Children, Esqr.
Chas. H. Pilgrim, Esqr.
Edmund Robt. Daniell, Esqr. Secy R.I.
William Thos. Brande Esqr. Prof. of Chem.
Michael Faraday, Esqr. Director of Laby

Committee for arranging the Lectures.

His Grace The President
Sir George Duckett, Bart.
George Moore, Esqr.
Benjamin Bond Cabbell, Esqr.
Dr. Whitlock Nicholl
C. H. Pilgrim, Esqr.
Edmund S. Halswell, Esqr.
William Pole, Esqr.
I. L. Goldsmid Esqr.
John George Children, Esqr.
Edmund Robt. Daniell, Esqr. Secy R.I.

Committee of Accounts.

The President
The Treasurer
The Secretary
George Moore, Esqr.
Benjamin Bond Cabbell Esqr.
Dr. Whitlock Nicholl

348

1830.

William Pole, Esqr.
Wm. Hasledine Pepys, Esqr.

Adjourned to Tuesday the 18th Instant at Three o'Clock.

Confirmed May 18 Whitlock Nicholl

349

May 18.

At a Meeting of the Managers of the Royal Institution of Great Britain held on the 18th of May, 1830.

Present

His Grace The Duke of Somerset Presdt. in the Chair
Dr. Whitlock Nicholl,
Benjamin Bond Cabbell, Esqr.
Joseph Jekyll junr. Esqr.
William Hasledine Pepys, Esqr.
Charles H. Pilgrim junr. Esqr.
Edmund S. Halswell, Esqr.
Thomas Hoblyn, Esqr.
William Pole, Esqr.
Edmund Robt. Daniell, Esqr. Secy.

Wm. Thomas Brande, Esqr. Supt. of the House
Michael Faraday, Esqr. Director of the Laby

1. The minutes of the proceedings of the last meeting were read and approved.

Mr. Hatchett's Donation to Laby Fund

2. Read a letter from Charles Hatchett, Esq dated Belle Vue House Chelsea, 12th May, 1830, addressed to Sir George Duckett, Bart. Treasurer of the Royal Institution, enclosing a Donation of One hundred guineas to the Laboratory Fund.
Resolved, That the thanks of the Managers be returned to Mr. Hatchett for his liberal donation.

3. Resolved, That it be proposed to Mr. Ritchie

Lectures of Mr Ritchie – – Public – to give a Course of Public Lectures on Natural Philosophy in the ensuing year on the same terms as during the present Season.

– – – Private. That Mr Ritchie be allowed the use of the Theatre for the purpose of giving lectures upon Natural Philosophy to a private Class on the usual terms of admitting the Members of the Institution, and of paying one third of the receipts to the use of the Institution. The arrangements of these lectures to be subject to the approval of the Lecture Committee

That the care of the Apparatus connected with the Department of Natural Philosophy be confided to Mr Ritchie during the ensuing year.

Confirmed 7 June 1830
Benjn. Bond Cabbell

June 7.

At a Meeting of the Managers of the Royal Institution of Great Britain, held on the 7th of June, 1830,

Present

Benjamin Bond Cabbell Esqr. in the Chair.
George Moore Esqr.
R. I. Murchison, Esqr.
John George Children, Esqr.
Wm. Hasledine Pepys, Esqr.
Joseph Jekyll junr. Esqr.
Sir George Duckett, Bart. Treasurer.
Charles H. Pilgrim, Esqr.
Edmund Robt. Daniell, Esqr. Secretary.

Michael Faraday, Esqr. Director of the Laby

1. The minutes of the proceedings of the last meeting were read and approved.

2. Read the following report from the Committee of Accounts.

Rept. of Com. of Accts. 3. The Committee of Accounts report that they have examined the General Accounts since the last Meeting, and that they have found a balance in favour of the Royal Institution of £297.13.4.

They recommend that the following sums be paid on account of the Royal Institution.

Joseph Fincher on account of the Royal Institution	£100 – –
2 quarters Assessed taxes to Lady day 1830	63. 6. 9
Royce for Carpenters work in 1829	22. 6. 7
	185. 13. 4

Resolved, That the report be approved and signed an Order for payment of £185.13.4, on the General Account.

Payment for Lectures. 4 Mr Ritchie having applied for the payment of his Course of lectures on Natural Philosophy which he had just completed.
Ordered, That the sum of £84 be paid to Mr Ritchie the amount due to him for sixteen lectures at 5 guineas ⅌ lecture.
And signed an Order accordingly.

5. Presents received.

John Burke, Esqr. presented
Official Kalender for 1830
James Burton, Esqr.
Excerpta Hieroglyphica, No 4,
Monsr. Bidone presented
Expériences sur les Courans d'Eau
Memoire Théorique des Veines liquides,
Earl Spencer, K. G. presented
Mirror of Parliament for the years 1828 & 1829

Ordered,
That the thanks of the Managers be returned for these presents.

On Donation by Mr Pepys. 6. Mr Samuel Solly having attended the meeting for the purpose of suggesting that the donation of £100 given by Mr I. Pepys to the general purposes of the Institution should be transferred to the Library & Mineralogical Fund in order that Mr Pepys should thereby

June 7.

be constituted a Patron of those Funds, It was Resolved, That it is the opinion of this Meeting that the Managers have no power to interfere with the appropriation of the sum in question made by Mr Pepys.

Foreign Journals recommended. 7. Mr Faraday communicated to the Meeting a List of Foreign Scientific Journals which it would be desirable to have in addition to those already taken in for the purpose of affording greater facilities to the preparation of the Journal.
Resolved, That the list delivered in by Mr Faraday be referred to Mr Singer in order that he may prepare an estimate of the expense of taking in each of the Journals specified in it and also of any other Publicatn. of the same nature which may suggest itself to him as proper to be taken in for the purposes of the Journal.

8. The following persons having been duly recommended were admitted as Subscribers to the Royal Institution for the present year.

To the Lectures.
Mrs & Miss Jervoise Connaught Square

To Mr Montgomery's Lectures.
The Countess of Morton
Miss Mosley
Miss Emily Mosley
Miss Mary Ann Mosley

1830.

Mr. & Mrs. Kinnerley 127 Regent Street
Miss Hope
Miss H Hope
Viscountess Bernard
Miss Sankey
Mrs. Vaughan
Miss Vaughan
Miss Cartwright
Miss Donkin
Miss Elphinstone
Miss Joanna Baillie
Mrs. Hoare
Miss Hoare

Adjourned to Monday the 21st Instant at Three o'Clock.

Confirmed the 21st day of June 1830.
Somerset Pres.

June 21. **At a Meeting of the Managers of the Royal Institution of Great Britain, held on the 21st of June, 1830,**

Present
His Grace The Duke of Somerset Pres.t in the Chair
Benjamin Bond Cabbell, Esq.r
Dr. Whitlock Nicholl
Charles H Pilgrim, Esq.r
William Pole, Esq.r
Sir George Duckett, Bart. Treasurer.
Edmund Robt. Daniell, Esq.r Secy.

Michael Faraday, Esq.r Director of the Laby.

1. The minutes of the proceedings of the last meeting were read and approved.

2. Read the following list of Foreign Scientific Journals with the price affixed to each publication as estimated by Mr. Singer agreeably to the resolution of the last Meeting - viz.t

Foreign Journals ordered.

Annales des Sciences Naturelles, 30s. pr. Ann:
Journal de Pharmacie — 15s. ₽ Ann:
Annales de L'Industrie — 30s. ₽ Ann:
Schweiger's Journal £2. 8. ₽ Ann: 25 ₽ Ct. off
Annallen der Physick £2. 8. ₽ Ann: 25 ₽ Ct. off.
Italian Journals of Science - Brugnatelli is discont.d
Omodeis Medical Journal £2. 8. 0 ₽ Ann: 25 ₽ Ct. off
Leonhardts Journal Heidelberg 18s. ₽ Ann:
Karstens Journal, Berlin — £2. 2. ₽ Ann: 25 ₽ Ct. off
Hufflands Journal 34s. ₽ Ann: 25 ₽ Ct. off
The whole amount of the above Journals will be about £13. 13 — ₽ Ann:

1830.

Ordered, That the periodicals mentioned in the foregoing list be taken in; and that the question as to the Fund, out of which the expense is to be defrayed, be postponed till after the publication of the two first numbers of the Journal.

3. *Repairs in Theatre.* Resolved, That it be referred to Mr. Faraday to procure an estimate of the expense of putting the Matting and Cushions in the Theatre into proper condition.

4 *On Number of Lectures.* Resolved, That it be recommended to the Committee of Lectures to consider of the propriety of limiting the lectures to be given between Christmas and Easter to two each week, and of fixing the hour for such lectures to Two o'Clock.

5 His Grace The President nominated the following Vice Presidents for the present year: - viz.t

Vice President.

Benjamin Bond Cabbell, Esq.r
Earl Grosvenor,
George Moore, Esq.r
Dr. Whitlock Nicholl
William Pole, Esq.r
Lord Prudhoe

Adjourned to Monday the 5th of July at half past One o'Clock.

Confirmed the 5th day of July 1830
Somerset Pres.

July 5 **At a Meeting of the Managers of the Royal Institution of Great Britain, held on the 5th of July 1830,**

Present
His Grace The Duke of Somerset Pres.t in the Chair
Sir George Duckett Bart. Treasurer.
Benjamin Bond Cabbell, Esq.r V. P.
George Moore, Esq.r V. P.
Wm. Hasledine Pepys, Esq.r
Isaac Lyon Goldsmid, Esq.r
Charles H. Pilgrim, Esq.r
Edmund Robt. Daniell, Esq.r Secy.

William Thos. Brande Esq.r Supt. of the House
Michael Faraday, Esq.r Director of the Laby.

1. The minutes of the proceedings of the last meeting were read and approved.

2. *Report from Com. of Acct.s* Read the following report from the Committee of Accounts.

The Committee of Accounts report that they have examined the General Accounts since the last meeting, and that they have found a balance in favour of the Royal Institution of £301. 5. 6.

They likewise report that they have examined the Laboratory Account, and that they have found a balance in favour of that Account of £333. 0. 2.

They recommend that the following sums be paid on account of the Royal Institution.

358

1830.

William Thos. Brande Esqr. for a year's Salary as Professor of Chemistry	150. —
James Montgomery, Esqr. for a Course of lectures in 1830	52. 10 —
Thomas Phillips Esqr. for do.	73. 10 —
Mr. Maudsley for Sundries for the Weekly Eveng. Meetings	13. 12. 7
H. Greenwood Sundries for the Friday Eveng. Meetings (the balance of her Account for Tea & Coffee)	51. 3. 7
H. Greenwood Sundries for the house	2. 0. 6
Mr. Faraday Sundries for the Friday Eveng. Meetgs. &c.	11. 19. 4
Dr. Clark expenses for ditto	4. 0. 0
Bolton for attending as Waiter at ditto	2. 17 —
Joseph Fincher on account of the Royal Institution	100. 0. 0
	461. 13. 0

Sale of Stock recommended, &c. They also recommend to the Managers to sell out sufficient of Stock from the Sinking Fund to raise £161. 10. 1 which together with the sum of £38. 9. 11 the surplus of last year, will amount to the sum of £200 to replace the sums paid on account of Mr. Hatchett and Mr. Auriol's bonds.

They further recommend that the sum of

359

July 5. £105 given by Mr. Hatchett to the Laboratory Fund be invested in the purchase of £3 Ⓟ Cent Consols to the Laboratory Account.

And that the sum of £100 be advanced to Mr. Fincher on the Laboratory Account.

3 Resolved, That the above report be approved of; And that the Treasurer be empowered to take the necessary steps for selling out sufficient Stock from the Sinking Fund to raise the sum of £161. 10. 1, to be carried to the General Accot. of the Royal Institution.

4. Investment. Resolved, That the Treasurer of the Laboratory Fund be requested to invest the sum of £105 presented to that Fund by Mr. Hatchett in the purchase of Stock to be carried to the account of the Laboratory Fund.

5. Advance of £50 to Patrons. Read an application from the Patrons of the Library requesting that the Managers would grant them a Loan of £70, on account of the Library and Mineralogical Fund.

Resolved, That the above request of the Patrons of the Library be complied with And signed an Order on the Treasurer for payment of the above sum of £70 accordingly.

6 Presents received

The Linnean Society presented Transactions of the Society vol. 16 pt. 2.

360

1830.

F. C. Belfour, Esqr. presented The Life of Sheikh Mohammed.

J. Audall, Esqr. presented History of Armenia 2 vols. 8vo.

Royal College of Surgeons Catalogue of the Hunterian Collection pt. 1.

Ordered,

That the thanks of the Managers be returned for these presents.

7. An address to the King (W. IV.) Read the minutes of the General Meeting of the 6th. of March 1820, with reference to the necessary proceedings on the demise of the Patron of the Royal Institution.

Resolved,

That an application be made to the General Meeting of this day to appoint a Committee to prepare the Address to be presented to His Majesty soliciting His patronage to this Institution.

8. Repairs. Mr. Faraday laid before the Meeting the estimate of the expense of repairing the Lecture room Seats.

Resolved, That the Cushions be repaired by Messrs. Pringle agreeably to their estimate at 1s. per foot.

Adjourned to Monday the 2d. of August at half past One o'Clock.

Aug. 2 — Confirmed

Whitlock Nicholl.

361

Aug. 2 At a Meeting of the Managers of the Royal Institution of Great Britain, held on the 2d. of August, 1830,

Present

Dr. Whitlock Nicholl, V.P. in the Chair.

George Moore, Esqr. V.P.

Charles H Pilgrim, Esqr.

Wm. Hasledine Pepys, Esqr.

John George Children, Esqr.

Edmund S. Halswell, Esqr.

Joseph Jekyll junr. Esqr.

Sir George Duckett, Bart. Treasurer

Edmund Robt. Daniell, Esqr. Secretary.

1. The minutes of the proceedings of the last meeting were read and approved.

2. Lectures — Mr. Buckingham. Read a letter from Benjamin Bond Cabbell, Esqr. enclosing a letter from Mr. Buckingham containing a proposition to deliver a regular lecture at the Royal Institution more fully explaining the objects of his late Address delivered at the Royal Institution, interspersed with descriptive anecdote and peculiarities of the Countries lying East of China &c.

Resolved, That Mr. Buckingham's proposal be declined; And that he be desired in future not to prefix the name of the Royal Institution at the head of his Advertizements & Prospectuses.

3. Read a letter from Captn. Frist addressed to the Secretary stating that he has a small Case of

1830

Bath waters from St. Thomas's in the East of the Island of Jamaica to his address and wishing to be informed whether the Managers would wish it for the Royal Institution for analyzation.

Resolved, That Capt.n Frost be informed that the Managers beg leave to decline receiving the Case of Bath Waters for the Royal Institution.

Power for sale of Stock declined.

4. Notice having been sent to the Managers agreeably to the Bye laws, the Seal of the Corporation was affixed to a Power for Sale of £200 out of the Stock entitled "The Sinking Fund of the Royal Institution".

The Power was countersigned by

Whitlock Nicholl, Chairman
George Moore V.P.
Wm. Hasledine Pepys
Edmund R. Daniell Sec.y

and witnessed by

Jos.h Fincher Assist. Secretary
Tho.s Williams, Porter R.I.

5. The key of the box containing the Seal of the Corporation having been mislaid, the box was broken open in the presence of the Board, and the Seal was deposited in the Secretary's tin box.

6. Presents received.

The Royal Asiatic Society presented Transactions of the Society Vol. 2.

The Royal Society presented Transactions

Aug. 2.

of the Society for 1829 parts 2 & 3, and for 1830 part 1.

Ordered, That the thanks of the Managers be returned for these presents.

Address to H.M. Will. IV.

7. Ordered, That the Secretary be requested to affix the Seal of the Corporation and to sign the Address to be presented to the King soliciting His Majesty's patronage to this Institution.

1830.

Nov. 1.

At a Meeting of the Managers of the Royal Institution of Great Britain, held on the 1st of Nov.r 1830

Present

Dr. Whitlock Nicholl, V.P. in the Chair
Isaac Lyon Goldsmid, Esq.r
Benjamin Bond Cabbell, Esq.r V.P.
George Moore, Esq.r V.P.
Edmund S. Halswell, Esq.r
William Hasledine Pepys, Esq.r
Edmund Rob.t Daniell, Esq.r Sec.y

Michael Faraday, Esq.r Director of the Lab.y

1. The minutes of the proceedings of the last meeting were read and approved.

Rep.t of Committee on Lib.y Floor.

2. Read the following report from the Committee appointed to examine and report upon the Library Floor.

The Committee beg to state that having directed a few of the boards to be taken up, they have been enabled to ascertain the dimension and arrangement of the principal Timbers, and they find that the separation of the joints between the tops and standards of the Bookcases which led to the reference to the Committee, is in no way connected with the sinking of the Floor; but occasioned by a shrinking of the materials, without injury to any of the other parts. In their examination however upon this subject, their attention was called to a piece of

Nov. 1.

Timber of considerable scantling and about twenty five feet in length, without those means having been adopted for its support which are generally applied to Girders and Beams of long bearings. Upon a close inspection of this Timber, they also discovered, that it had several rents or shakes, likely to reduce its strength, and if liable at times to great weights, ultimately to produce considerable depression and perhaps fracture. Although the Committee were under no apprehension of any immediate danger from the state of the Floor, which having already remained many years without occasioning either accident or alarm, might probably have continued unaltered for a great length of time, yet when they took into their consideration the responsibility which is attached to every opinion passed upon subjects relating to public safety, they were induced to recommend that some steps should be taken to strengthen that part of the Floor, which for a short time during the Friday Evening Meetings, is subjected to considerable pressure.

Mr. Ainger

Mr. Ainger a Member of the Royal Institution, who kindly assisted the Committee during their investigation, suggested a plan for the purpose, which seemed calculated to give the security required, at the least possible expense.

Measure recommended.

Mr. Ainger proposed to fix a cast iron Column of small diameter about twelve feet from the South end of the Newspaper room, and in the centre of its width upon a foundation of brickwork in cement, carried up as a spandrel

366

1830

between the springings of the Arches in the Basement beneath, and supporting two pieces of Oak bolted together; and carefully secured at each end to the present Girders, by strong wrought iron bolts with Nuts and Screws.

Mr. Royce the Carpenter of the Royal Institution made an estimate of the expense of the works which amounted to £27.5 — and stated, that he would ✱✱ undertake to complete it in the course of a fortnight

The Committee being of opinion that it was desirable the business should be performed in the course of the present month, during which the closing of the Library and Newspaper rooms for a short time would be less inconvenient than at any other period, have ventured to take upon themselves the responsibility of giving the requisite orders; and they are now enabled to report that the works are completed.

Resolved, That the above report be approved of,

Resolved, That the thanks of the Managers be given to Mr. Aiger for the assistance rendered by him to the Committee.

Hon. Admission. Read a letter from The Chevalier de Barboza a Foreigner attached to the Portuguese Legation requesting to be allowed to frequent the Reading Rooms of the Royal Institution.

Resolved, That The Chevalier de Barboza be admitted to the Royal Institution agreeably to Art: 13 Chap: VI of the Bye laws.

Nov 1. 367

Presents received

Dr. Hancock presented Fruit of the Locust Tree and a Section of a Twining Plant from the Onowoks Country - the pith at one edge the accretions not concentric

Captn. Wild R.A. Woolwich presented Poison from the North American Indians as prepared by them

Crystals of Salt from Nantwich presented by Mr. Clarke of the Naval School Chelsea.

Richard Pering, Esqr. presented A treatise on the Anchor

J Bennet, Esqr. presented A Treatise on a new metallic Alloy.

John Fearn, Esqr. presented A Treatise on Cerebral Vision

The London Institution presented Catalogue of their Library.

W. H. Potter, Esqr. presented A Treatise on the use of Chlorine in Consumption

The Editors presented Zoological Journal No. 18.

Sir Thomas Phillipps, Bart presented Glamorganshire Antiquities,

The Natural History Society of Northumberland presented Transactions of the Society Vol 1 part 1

Dr. Turner presented Elements of Chemistry

Sir Charles Knowles Bart presented Observations on Naval Tactics.

The Royal Humane Society presented Report for 1830

368

1830

Gilbert Burnett, Esqr. presented Eidodendron No. 6

Ordered, That the thanks of the Managers be returned for these presents.

The Secretary laid before the Board the box containing the Seal of the Corporation; four keys having been provided agreeably to Chap XII Art 4 of the Bye laws.

Disposal of Keys. Resolved, That one of the keys (A) be kept by His Grace the President one (B) by Sir George Duckett, Bart the Treasurer one (B) by George Moore, Esqr. V.P. and the fourth (A) by Edmund Robert Daniell, Esqr. the Secretary.

Carpets. Ordered, That the Carpets be put down in the Libraries; and that the Libraries be opened in the Evening as usual.

Resolved, That the name of W. H. Pepys Esqr. be added to the Committee of Lectures.

Geo Duckett

Nov. 20. 369

At a Special Meeting of the Managers of the Royal Institution of Great Britain, held on Saturday the 20th of November 1830

George Moore Esqr. V.P. in the Chair
Wm. Hasledine Pepys, Esqr.
Sir George Duckett, Bart Treasurer
Benjamin Bond Cabbell, Esqr.
Edmund S. Halswell, Esqr.
Dr. Whitlock Nicholl V.P.
Edmund Robt. Daniell, Esqr. Secy.

The Secretary stated that this Meeting had been called under the authority of a Vice President, in the absence from town of His Grace the President, for the purpose of taking into consideration the following communication which he had received from the Royal Asiatic Society.

Letter from Royal Asiatic Soc. requesting Loan of Theatre — granted

Royal Asiatic Society's house
November 15th. 1830.

Sir,

I am directed by the Council of the Royal Asiatic Society to request that the Managers of the Royal Institution will obligingly accommodate them with the use of their Theatre for the dissection by Dr Granville of a Mummy that belongs to the Society.

The Mummy would be dissected at the end of this month or the beginning of December if either of those times would be convenient to the Members of the Royal Institution who would

370

1830.

of course be entitled to be present at the operation.

I presume to mention that it is probable from 100 to 200 of the Members of the Royal Asiatic Society would attend on the occasion.

I have the honour to be &c. &c.

signed) Wm. Huttmann

Assistant Secretary

To E. R. Daniell, Esqr.
Secretary to the Royal Institution,

Resolved,

That the request of the Royal Asiatic Society be complied with; And that Mr. Huttman be informed that the Managers will be happy to accommodate them with the use of the Theatre of the Royal Institution on any day that they may appoint except Monday the 6th. of December being the General Monthly Meeting. &c.

Geo Duckett

371

Dec. 6.

At a Meeting of the Managers of the Royal Institution of Great Britain, held on the 6th. of December, 1830,

Present

Sir George Duckett Bt. Treasurer, in the Chair
George Moore, Esqr. V. P.
William Hasledine Pepys, Esqr.
Joseph Jekyll, junr. Esqr.
Edmund S. Halswell, Esqr.

Michael Faraday, Esqr. Director of the Lab.

1. The minutes of the proceedings of the last meeting were read and approved, and also the minutes of the Special Meeting of the 20th. of November.

Rept. from Com. of Accts.

2. Read the following report from the Commee. of Accounts.

The Committee of Accounts report that they have examined the General Accounts since the last meeting on the 30th. of June, and that they have found a balance in the hands of the Bankers of £222 . 0 . 6 and a balance due to the Assistant Secretary of £93 . 14 . 11.

They likewise report that they have examined the Laboratory Account from the same period, and that they have found a balance in the hands of the Bankers of £179 . 18 . 7; and in the hands of the Assistant Secretary of £22 . 9 . 3.

372

1830

They recommend that the following sums be paid on account of the Royal Institution amounting to £411 . 13 . 7. which will make an overdraw on the Bankers' Account of £189 . 13 . 1.

½ yrs Assessed taxes to Mich.s 1830	£61 . 4 . 6
ditto - Parochial do. to ditto	44 . 3 . 4
A year's Water rate to ditto	10 . 10 . –
Mrs. Sturt for Stationery in 1830	32 . 18 . 2
H Greenwood Sundries for the house	3 . 1 . 8
Duggan for watering the front of the Royal Institution	3 . 0 . 0
Mr. Burnett Sundries for his lectures on Botany in 1830	8 . 16 . –
Knapp for Bookbinding in ditto	4 . 5 . –
Balance due to J. Fincher on the 1st Decr.	93 . 14 . 11
They also recommend that the sum of be advanced to Mr. Fincher on the General Account; and	150 –
	411 . 13 . 7

the sum of £60 on account of the Laboratory Fund. &c.

Resolved, That the above report be approved of; And signed Orders accordingly.

Present received.

3. Mr. Bollaert (formerly Chemical Assistant in the Laboratory,) presented

Minerals from South America

Ordered, That the thanks of the Managers be returned for this present

373

Dec. 6.

Loan of Theatre

4. Read a letter from Mr. Wm. Huttmann Assistant Secretary of the Royal Asiatic Society appointing Saturday the 11th. of December at 12 o'Clock for the dissection of the Mummy in the Theatre of the Royal Institution.

Confirmed the 7. of Feb. 1831

George Moore (V.P.)

374

Feb. 7. 1831.

At a Meeting of the Managers of the Royal Institution of Great Britain held on the 7th of February, 1831,

Present

George Moore, Esqr. V.P. in the Chair
Sir George Duckett, Bart. Treasurer.
Edmund S. Halswell, Esqr.
Wm. Hasledine Pepys, Esqr.
Edmund Robt. Daniell, Esqr. Secretary.

Wm. Thos. Brande Esqr. Profr. of Chemistry
Michael Faraday, Esqr. Director of the Lab.y

1. The minutes of the proceedings of the last meeting were read and approved.

Repr. from Com. of Accts. 2. Read the following report from the Commt. of Accounts.

The Committee of Accounts report that they have examined the General Accounts since the last meeting on the 6th of December, And that they have found a balance in the hands of the Bankers of £618. 9. 1; and in the hands of the Assistant Secretary of £11. 19. 10

They recommend that the following bills be paid viz.

Parker for Painting Plumbing & Glazing in 1829	62. 12. 9
Dickins &c. for Coals in do.	92. 0. 0
Houghton for Newspapers &c. in do.	31. 3. 2
Carrd. forwd.	185. 15. 11

375

Feb. 7.

Bro.t forward	185. 15. 11
Gardner for Oil in 1829	75. 8. 3
Nicol for Printing in do.	67. 8. 6
Bird for Bricklayers work in do.	1. 16. –
Vulliamy for repairing Dials in do.	2. 12. 6
	333. 1. 2
And also an Advance to the Assistant Secretary of	100. –. –
	433. 1. 2

Resolved, That the above report be approved of; and signed an Order on the Treasurer for payment of the above sum.

Lectures. 3. The Secretary read the following report from the Committee of lectures.

The Committee report that they have made the following arrangements for Lectures during the present season.

Mr. Professor Brande – A Course of lectures On the Chemical History and applications of the Metals.

Mr. Vigors – A Course of lectures On Ornithology. To be delivered gratuitously.

Mr. Ritchie – A Course of lectures On Natural and Experimental Philosophy. To be remunerated with 5 guineas a lecture.

Courses of Lectures to be delivered after Easter.

Mr. James Montgomery – six lectures on Poetry. To be remunerated with 15 guineas a lecture.

Mr. John Lindley – On Vegetable Physiology

376

1831.

and Botany. To be remunerated with 5 guineas a lecture.

Mr. Robert Willis – A Course of six lectures on Sound. To be remunerated with 10 guineas a lecture.

Mr. Thos. Webster – On some important points in Geology. To be remunerated with 40 guineas for the Course.

and

Mr. Faraday – A Course of lectures on a subject which will be announced on a future occasion.

Resolved, That the above report be approved of.

Payment for Mr. Webster's Juvenile Lectures. 4. The Secretary stated that the Committee of Lectures had proposed to remunerate Mr. Webster for his lectures on Geology adapted to a Juvenile Auditory, with £30; but that as he had been formerly allowed £40 for a similar Course, and as these lectures had given great satisfaction, he submitted to the Board that the amount should be increased to that sum.

Resolved, That Mr. Webster be allowed £40 for his lectures delivered during the Christmas Vacation; and signed an Order on the Treasurer accordingly.

5. The Secretary also reported from the Comm. of Lectures ~~that~~ ~~in~~ the resolutions of the Managers of the 18th of May, 1830, by which it was resolved

377

Feb. 7.

Lectures by Mr. Ritchie. That it be proposed to Mr. Ritchie to give a Course of public lectures on Natural Philosophy in the ensuing year on the same terms as during the present season; and that Mr. Ritchie should be allowed the use of the Theatre for the purpose of giving lectures upon Natural Philosophy to a private Class on the terms therein mentioned.

which resolution had been communicated to Mr. Ritchie; in answer to which Mr. Ritchie had written to the Secretary a letter dated the 7th of December, 1830 in which he expressed himself to the following effect – viz.

Letter from Mr. R. "I feel highly gratified with the proposal of the Managers of the Royal Institution of resuming my lectures next season; but however anxious I am for the honour of being connected with that Establishment, my friends are decidedly against my resigning a permanent situation for what may be only a single Course of lectures. As I have however resolved to devote myself to Science and as the Royal Institution possesses such facilities for Experimental research, should the Managers agree to propose me as Professor of Natural & Experimental Philosophy, at the first General Meeting for that purpose, I shall immediately resign my present situation and repair to London as soon as possible. Were my name and interest thus closely connected with that noble Establishment, a large portion of my time would be devoted to original researches and to

verifying and extending novel and interesting experiments for the Journal. I beg you will lay this communication before the Board of Management and let me know the result as soon as possible."

Professorship of Nat. Phil recommended. 610

That the Secretary had laid the above letter of Mr. Ritchie before the Committee upon which the Committee were of opinion that it would be very desirable that Mr. Ritchie should be attached to this Institution and they had resolved to recommend to the Board of Managers to take the necessary steps towards the establishment of a Professorship of Natural and Experimental Philosophy in order that Mr. Ritchie might be proposed to the General Meeting to fill that office as soon as convenient. In the mean time the Committee had agreed with Mr. Ritchie for the delivery of a Course of lectures upon Natural & Experimental Philosophy upon the terms which have been specified in the former part of their Report.

Resolved,

That the report of the Committee of lectures be approved of and adopted; and that the necessary notice be prepared and read at the General Meeting this day, for the Establishment of a Professorship of Natural Philosophy with a Salary of £50 per Annum

Lectures by Mr Vigors gratuitous.

The Secretary further reported that Mr. Vigors had declined to receive any remuneration for the Course of lectures proposed to be delivered by

him; but that he had expressed a wish that the Zoological Society might in some way have the benefit of the Lectures And that in consequence thereof, had agreed that such Members of the Zoological Society as wished to attend Mr. Vigors' lectures should be admitted on production of a ticket signed by the proper Officer of the Society.

Resolved,

That this meeting do approve of the arrangement entered into with Mr. Vigors.

Resolved,

That Mr. Vigors be thanked for his proposal to deliver Lectures in the Royal Instit.n and to be informed that the Managers have given directions for the admission of the Members of the Zoological Society to his lectures on the production of a proper ticket.

Report from Publication Committee resp.g Journal 6. Read the following report from the Publication Committee.

The Committee to whose superintendence the publication of the Journal of the Royal Institution has been confided report that two numbers of the Journal have already been published, The first number having appeared on the 1st of October 1830, and the second number on the 1st of the present Month. Of the contents of those two numbers as they are before the Members, it will not be necessary for the Committee to speak; but they have the satisfaction of knowing that they have been well received, and that the greatest portion of the impression has been sold. – The

Committee have found that most of the distinguished Men of Science to whom they have addressed themselves have promised their aid and many of them have been prompt in furnishing them with papers.

Subscriptions to Journal.

Subscription towards the expense of Contributions to the Journal of the Royal Institution.

His Grace The President	£50
Benj.n Bond Cabbell Esq.r V.P.	10
George Moore, Esq.r V.P.	10
Edmund Robert Daniell, Esq. Sec.	10
Charles H. Pilgrim, Esq.r	10
Michael Faraday, Esq.r	10
R. G. Clarke, Esq.r	10
George Meredith, Esq.r	10
W. H. Pepys, Esq.r	10
R. H. Solly, Esq.r	10
Sir George Tho.s Staunton Bt.	10
Sir George Duckett, Bart V.P.	10
William Pole, Esq.r	10
Dr. Whitlock Nicholl	10
Thomas Snodgrass, Esq.r	10
Edward Sabine, Esq.r	10
Sir Oswald Mosley Bt.	10
J. F. Daniell, Esq.r	7 . 7 .
W. Heathfield, Esq.r	3 . 3
	220 . 10

Disbursed for Contributions to the first two numbers of the Journal of the Royal Institution. –

To the first number	£108 .
To the second ditto	111 . 11
	219 . 11

Resolved, That the above report be approved of.

Thanks vot.d for loan of Theatre to Roy. As.c Soc. 7. Read a letter from T. D. Broughton, Esq.r Sec.y of the Royal Asiatic Society, addressed to the Sec.y of the Royal Institution, transmitting the following resolution:

At a Meeting of the Council of the Royal Asiatic Society held on the 15th of January, 1831,

It was unanimously resolved,

That the cordial thanks of the Council be returned to the Managers of the Royal Institution for their obligingly allowing the use of the Theatre for the dissection of the Society's Mummy; and for their presenting admissions to the Friday Evening Meetings of the Institution to the Members of that Council.

Hon.y Admissions. 8. Read a letter from The Chevalier de Barboza addressed to the Secretary, requesting permission of the Managers on behalf of the Chevaliers D. Souza and Rebeiro, attached to the Portuguese legation and also for the Chevalier Sarmento a Portuguese Magistrate, to attend not only the Friday Evening Meetings of the Members of the Royal Institution, but likewise the Reading Rooms in the Morning.

Resolved, That permission be granted to the above Gentlemen agreeably to Chap: VI Art: 15 of the Bye laws.

Presents received.

9. Wm Pole, Esq.r V.P. presented a magnificent

382

1831.

Collection of Volcanic specimens from Vesuvius.
Mr. Marshall presented Two Shells
from the Eastern Coast of Africa brought
by Captn. Kendall.
Dr. Hancock presented
Beaks of the Toucan
Seed of the Tulip Tree
Bone of a Fish's tongue

The Royal College of Surgeons presented
Catalogue of the Contents of the Museum
& Catalogue of the Hunterian Collection part 4.
Michl. Faraday, Esqr. presented part 2.
Chemical Manipulations, 2d. Edn.
Experiments in Electro-Magnetism by Sturgeon.
Presented by the Proprietors. The Life of
The Duke of Wellington.
Wm. Walton, Esqr. presented various Foreign
Political Tracts.
P. Borthwick, Esqr. presented
Observations on the Prophecies of Daniel.
Wm. Crum Esqr. presented
Experiments on the primary Colours,
Dr. Paris presented
The Life of Sir Humphry Davy,
Monsr. le Comte de Penhoent
Esquisses sur la Bretagne,
Monsr. de Fuss presented
Memoires de l'Academie des Sciences,
H. T. De la Beche, Esqr. presented
Sections & Views illustrative of Geological Phenomena
John Gould, Esqr. presented the first number of a Century of
Birds from the Himalaya Mountains.

Feb. 7. 383

The Society of Arts presented Transactions of
the Society Vol. 48 pt. 1.
Wm. Hamilton, Esqr. presented
Aegyptica, with a folio Volume of plates
Dr. Moseley presented
A Dictionary of Latin quantities.
Ordered,
That the thanks of the Managers be
returned for these presents.

10. The following persons having been duly
recommended were admitted as Subscribers
to the Royal Institution for the year, 1831.

Annual Subscribers.
Charles Neate Esqr. 3 Garden Place Lincolns Inn fields
The Revd. Thos. Smith Turnbull 68 Baker St.
Henry Gilbert, Esqr. Piccadilly.
Thomas Beavan, Esqr. 20 Sackville Street
The Honble. Major Fane, M.P. 31 Grosvenor Square
Robert Palmer, Esqr. M.P.

To the Lectures
Mrs. & Miss Arbuthnot Wimpole Street
Mrs. Rogers 56 Manchester Street
Miss Edmonstone 23 Park Crescent
Mrs. Grover 12 Egremont Place New Road.

Adjourned to Monday the 28th Inst.
at Three o'Clock.

Confirmed this 28th day of February 1831
Somerset Pres.

384

1831.

Feb 28. At a Meeting of the Managers
of the Royal Institution of Great
Britain held on the 28th of February
1831.

Present
His Grace The Duke of Somerset Presidt. in the Chair
George Moore, Esqr. V.P.
R. I. Murchison, Esqr.
Chas. H. Pilgrim, Esqr.

William Thos. Brande, Esqr. Prof. of Chem.
Michael Faraday, Esqr. Superintt. of the house

1. The minutes of the proceedings of the
last meeting were read and approved.

2. In pursuance of a nomination under
Life Subsr. the Act of Parliament by the late John
Chamier, Esqr. his Son
Frederick Chamier, Esqr. of Park Crescent
was admitted to the privileges of a Life
Subscriber to the Royal Institution.

3. Resolved, That the names of His
Trustees of Labr. Fund. Grace The Duke of Somerset President,
Sir George Duckett, Bart. Treasurer, and
Edmund Robert Daniell, Esqr. Secretary be
united with that of Earl Spencer, K.G. as
Trustees of the Stock invested on account
of the Laboratory Fund of the Royal
Institution.

Feb. 28. 385

Presents received.
4. Dr. Paterson Ann Subr. R.I. presented
Remarks on the present state of Affairs
Monsr. Beral presented
Nomenclature Pharmaceutiques
Ordered,
That the thanks of the Managers be
returned for these presents

Confirmed 7 March 1831
Whitlock Nicholl

386

Mar. 7. 1831.

At a Meeting of the Managers of the Royal Institution of Great Britain, held on the 7th of March, 1831

Present

Dr. Whitlock Nicholl V.P. in the Chair.
C. H. Pilgrim, Esqr.
W. H. Pepys Esqr.
Edmund S. Halswell, Esqr.
George Moore, Esqr. V.P.
Benjamin Bond Cabbell, Esqr. V.P.
Edmund Robt. Daniell, Esqr. Secretary

William Thos. Brande, Esqr. Prof. of Chem.
Michael Faraday, Esqr. Superintendent of the house

1. The minutes of the proceedings of the last meeting were read and approved.

Rept. from Com. of Accts. 2. Read the following report from the Committee of Accounts.

The Committee of Accounts report that they have examined the General Accounts since the last Meeting and that they have found a balance in the hands of the Bankers of £334. 19 – And in the hands of the Assistant Secretary of £118. 12. 4

They recommend that the following bills be paid.

Jones & Yarrell for Newspapers in 1830 } 93 - 6 -
Houghton ditto. & Stationery in do. 28. 6. 5
Carrd. up – 121. 12. 5

387

Mar. 7.

Brt. forward 121 - 12 - 5
Bentley for Prospectuses, & Instr. 9 - 4 -
Portable Gas Compy. for Gas in 1830 } 29 - 6 - 7
Drew for attendance for ditto 2 - 10 -
Royce for repairs and alterations of the House in 1830 } 88 - 13 - 5
251 - 6 - 5

Approved of, and signed an Order accordingly.

On Request that Apparatus may come free from Duty. 3. The Secretary stated that he had applied officially to the Lords Commissioners of His Majesty's Treasury requesting their Lordships would allow a Case of Chemical Apparatus and Chemicals which have been shipped from Hamburgh for the use of the Royal Institutn. to be delivered to this Institution free of duty.

Resolved, That the above application made by the Secretary be approved of.

4. The following persons having been duly recommended were admitted to the Royal Institution as Subscribers for the year 1831

Annual Subscribers.
Viscount Bernard, M.P. of Connaught Place
John Gordon, Esqr. 3 William St. Golden Sq.
William Hodges, Esqr. 20 Devonshire Place

To the Lectures.
Lady Midleton
The Hon: Miss C. Brodrick

388

1831.

Honble Miss Fitzwilliam
— Miss E. Fitzwilliam
— Miss F. Fitzwilliam

To Mr. Ritchie's Lectures.

Mrs Spitta
Miss Porter
Miss Harrop
Miss Warner
Miss Hankey
Miss C. Hankey
Miss McLeod
Miss Mowat
Mrs Roberts
Miss Roberts
Miss E Roberts

Portrait of Mr Faraday presented by Mr Pickersgill 5. Read a letter from Mr. Pickersgill accompanied with a Portrait of Mr. Faraday in a gilt frame which he requested to be presented in his name to the Managers of the Royal Institution of Great Britain as a testimony of respect for Mr Faraday's great talents and for an Institut. which does so much to promote and encourage the advancement of Science.

Resolved, That the thanks of the Managers be returned to Mr. Pickersgill for his very valuable and acceptable present And that Mr Pickersgill's present be reported to the General Meeting this day.

Adjourned to Monday the 14th Instt at Half past Two o'Clock
Geo Duckett

389

Mar. 14.

At a Meeting of the Managers of the Royal Institution of Great Britain held on the 14th of March, 1831.

Sir George Duckett, Bart V.P. in the Chair.
Benjamin Bond Cabbell, Esq. V.P.
George Moore, Esqr. V.P.
William Hasledine Pepys, Esqr.
Edmund S. Halswell, Esqr.
Dr. Whitlock Nicholl, V.P.
Charles H Pilgrim, Esqr.
Joseph Jekyll junr. Esqr.
Edmund Robert Daniell, Esqr. Secretary.

Michael Faraday, Esqr. Director of the Laby.

1. The minutes of the proceedings of the last meeting were read and approved.

Copy of Portrait of Sir H. Davy ordered 2. Read the following resolution of the General Monthly Meeting of the Members of the 7th Instant.

"That it appears desirable that a Copy should be obtained of the Picture of Sir Humphry Davy, by Sir Thomas Lawrence, in the possession of the Royal Society, for the purpose of being put up in the Royal Institution.

Resolved, That the Managers be directed to make an application to the Council of the Royal Society for leave to have such copy made, And in case such permission be granted, to have it taken by a competent Artist."

Ordered, That the Secretary be requested to

390

1831.

apply to the Royal Society for such permission.

3. Engraved Portt of Sir H. D. presented. Mr E. Magrath presented An Engraved Portrait of Sir H. Davy Bt.
Ordered, That the thanks of the Managers be returned for this present.

4. Resolved, That the following notice be hung up in the Newspaper Room Reading Room and Libraries.

On talking in Reading rooms. Much dissatisfaction having been expressed to the Board of Managers on the part of many of the Members who daily resort to the Reading Rooms of the Royal Institution for the purpose of Reading, that their attention is distracted by the loud conversations of certain Individuals, The Managers earnestly request that the practice complained of, be discontinued.

Adjourned to Monday the 28th Instant at 3 o'clock.

Confirmed March 28 1831
Whitlock Nicholl

391

Mar 28.

At a Meeting of the Managers of the Royal Institution of Great Britain held on the 28th of March, 1831.

Present

Dr Whitlock Nicholl, V.P. in the Chair.
George Moore, Esqr. V.P.
Wm Hasledine Pepys, Esqr.
Edmund S. Halswell, Esqr.
Chas. H. Pilgrim, Esqr.
Edmund Robt. Daniell, Esqr. Secy.

Michael Faraday, Esqr. Director of the Laby.

1. The minutes of the proceedings of the last meeting were read and approved.

2. Apparatus delivered free from duty. Read a letter from T. Spring Rice, Esqr. dated Treasury Chambers 14th March 1831 stating that in answer to the application made by the Managers, the delivery of the Case of Chemicals, which has been shipped from Hamburgh, had been authorized by the Lords Commissioners of His Majesty's Treasury to be delivered to the Royal Institution free of duty.

3. On copy of Sir H. Davy's portrait. Read a letter from Dr. Roget Secy Royal Socy dated 18th March 1831 stating that in answer to the application on behalf of the Managers for permission to have a Copy taken of the Portrait of Sir H. Davy in order to be put up in the Royal Institution, he has to acquaint

392

1831.

the Managers that His Royal Highness the President and Council will willingly grant such permission, on condition that their Picture be not removed from the Apartments of the Royal Society.

4. Hon. Admissions. On the application of The Chevalier de Barboza, Messrs Jozé de Silva Carvalho and Manoel Goncalves de Miranda, attached to the Portuguese Embassy were admitted to the Libraries and other public parts of the Institution under Art 15 Chap: VI of the Bye laws.

Presents received.

5. M. Bourdon presented A Hydrostatic Circulation in Glass.
C. Turner Esqr. presented A Proof Impression of an Engraving of Dr. T. Young,
E. Magrath, Esqr. presented A Proof Impression of an Engraving of Count Rumford
J. Britton, Esqr. presented Busts of Ben Johnson & Shakespear
Dr. Marsh of Woolwich presented Specimens of New Zealand Flax in various states of preparation
Col. Pasley presented Specimens of Artificial Cement.
Ordered, That the thanks of the Managers be returned for these presents.

6 Resolved, That C.H. Pilgrim, Esqr. be added to the Committee of Accounts.

393

Mar 28.

7. Persons under 15 not to be admitted on Friday Evgs. Objections having been made to the introduction of Children at the Friday Evening Meetings of the Members.
Resolved, That the Members be informed that no young persons under the age of 15 be admitted to the Weekly Evening Meetings of the Members.

8. On admission of Children to Lectures. Read the following proposal from the minutes of the Visitors entered in their proceedings of this day.
The Visitors are of opinion that if the Children under fifteen years of age of the Members of the Royal Institution were admitted to the Lectures upon an annual subscription of One guinea, a considerable increase of Income might be obtained and a taste for Science diffused among the rising generation.
Resolved, That the above proposal be taken into consideration at the next Meeting of the Managers.

Confirmed 4 April 1831
Benjn Bond Cabbell V.P.

394

April 4.

1831.

At a Meeting of the Managers of the Royal Institution of Great Britain held on the 4th of April, 1831

Present

Benjamin Bond Cabbell, Esq.re V.P. in the Chair
George Moore, Esq.r V.P.
Joseph Jekyll jun.r Esq.re
Edmund Robt. Daniell, Esq.re Secretary

William Thos. Brande, Esq.re Prof. of Chem.y

1. The minutes of the proceedings of the last meeting were read and approved.

Rept. from Com. of Accts.

2. Read the following report from the Committee of Accounts.

The Committee of Accounts report that they have examined the General Accounts of the Royal Institution since the last Meeting And that they have found a balance in favour of the Royal Institution of £ 359 . 10 . 5

They also report that they have examined the Laboratory Account since the Meeting of the 6th December last And that they have found a balance in favour of that account of £ 345 . 17 . 1

They recommend that the following sums be paid on account of the Royal Institution.

2 quarters Parochial rates to 25th March 1831	44 . 3 . 4
2 quarters Assessed taxes to d.o	61 . 4 . 6
Carr.d forw.d	105 . 7 . 10

395

April 4.

Bro.t forward	105 . 7 . 10
Sewer rate	12 . 10 –
Mr. Hurt for Stationery in 1830	15 . 14 . 4
Mitchell for Session Paper House of Commons	12 . 3 . 0
Joseph Fincher on account of the Royal Institution	100 . 0 . 0
	245 . 15 . 2

And on account of the Laboratory Fund

Newman for Chemicals in 1830	89 . 1 . 0
Fisher for ditto — d.o	15 . 13 . 10
Coals in 1830	12 . 0 . 0
Rickards for Spirits of Wine d.o	5 . 0 . 0
Joseph Fincher on account of the Laboratory of the Royal Institution	80 . 0 . 0
	201 . 14 . 10

On Payment for Model of Steam Engine — Steam Eng.e returned.

The Committee further report that the sum of £60 had been charged the Institut.n for a Model of a Steam Engine which had been supplied by Mr. Newman in pursuance of an Order given many years ago but of which no minute could be found in the Books – And that they had not thought proper to recommend the payment of the sum so charged without the sanction of the Board of Managers.

The Committee also report that Mr. Newman had attended their Meeting and had expressed his readiness to take back the Model in case the same was not approved of by the Managers.

396

1831.

Mr. Brande having stated to the Board that the Model of the Steam Engine would not be immediately required, And Mr. Newman having expressed his willingness to take back the same

Resolved, That the Steam Engine be returned to Mr. Newman.

Resolved, That the report of the Committee of Accounts be approved of; And signed Orders for payment of the sums before specified.

Coals.

3. Resolved, That the Committee of Accounts be requested to make the necessary arrangements for supplying the Royal Institution with Coals, the Contract with Mess.rs Dickens & Co having now ceased.

Presents received.

4. The Board of Admiralty presented
Nautical Almanac for 1833
The Company of Clockmakers presented
Catalogue of their Library.
John Gorton, Esq.re presented
Topographical Dictionary of Great Britain and Ireland vol. 1.

Ordered, That the thanks of the Managers be returned for these presents.

Ordered, That the following Professors be

397

April 4.

put in nomination from the Chair at the General Meeting of this day.

William Thos. Brande Esq.re Prof. of Chem.y
Reverend William Ritchie Prof. of Nat.l Natural Philosophy

On Alt.n in Bye Laws

5. Resolved, That a notice be prepared to be laid before the General Meeting of this day for the alteration of Art 3 Chap. XIII of the Bye laws.

Adjourned to Monday the 11th Instant at Three o'Clock.

Confirmed George Moore
April 11 – 1831

April 11.

1831.

At a Meeting of the Managers of the Royal Institution of Great Britain held on the 11th of April, 1831

Present

George Moore, Esqr. V.P. in the Chair.
Wm Hasledine Pepys, Esqr.
William Pole, Esqr. V.P.
Chas. H. Pilgrim, Esqr.
Benjamin Bond Cabbell Esqr. V.P.
Edmund Robt. Daniell, Esqr. Secretary

1. The minutes of the proceedings of the last meeting were read and approved.

Children (under 15) of Members to be admitted for reduced subscriptions.

2. The Secretary stated that there having been no Meeting of the Managers when the Card announcing the lectures to be delivered after Easter, was required to be circulated, he had made the following addition to the terms on which non-Subscribers are admitted to subscribe to separate Courses of lectures.

"The Sons and Daughters of the Members of the Royal Institution, under fifteen years of age, may be admitted on payment of half the sum required of Nonsubscribers for each Course, viz: One guinea for the Sons; and 10/6 for the Daughters.

Resolved, That the above arrangement be approved of.

3. Read the following resolution of the Committee

April 11.

of Visitors entered in their proceedings of this date.

On Payment of Salary of Profr. of Chemy.

The Visitors recommend to the Committee of Managers that in future the Salary paid to the Professor of Chemistry shall be charged to the account of the Laboratory Fund, as the Salaries of the Director of the Laboratory and his Assistant are; It appearing to the Visitors that the Laboratory Fund is now in a condition to pay its own expenses, whilst the general Funds of the Institution call for every retrenchment that can be made in the different items of Expenditure.

Resolved, That the consideration of the above recommendation be deferred for the present.

Lectures — Mr Vigors

4. The Secretary reported that Mr. Vigors had completed his Course of gratuitous lectures on Ornithology.

Resolved,

That the thanks of the Managers be returned to N. A. Vigors Esqr. for the above Course of Lectures.

5 ~~Proceeded to prepare the House lists of Managers and Visitors for the ensuing year.~~

Confirmed April 18. 1831.
George Moore V.P.

April 18.

1831.

At a Meeting of the Managers of the Royal Institution of Great Britain, held on the 18th of April, 1831

Present

George Moore, Esqr. V.P. in the Chair
Chas. H. Pilgrim, Esqr.
Edmund S. Halswell, Esqr.
Sir George Duckett, Bart. Treasurer
Edmund Robt. Daniell, Esqr. Secy.

1. The minutes of the proceedings of the last meeting were read and approved.

Theatre lent to Zool. Socy.

2. The Secretary stated that he had received a letter from N. A. Vigors, Esqr. dated the 13th Instant applying on behalf of the Council of the Zoological Society for permission from the Managers to make use of the Lecture Room of the Royal Institution for the purpose of holding the Anniversary Meeting of the Zoological Society on Friday the 29th Instant. And that as under their Bye law they are obliged to advertize the time and place of the Meeting at least fourteen days before the Meeting it was necessary to obtain an answer before Friday.

The Secretary informed the Meeting that having ascertained from Mr. Faraday that the proposed Meeting would not interfere with any of the arrangements of the Royal Institution on that Friday Evening he had

April 18

authorized Mr Vigors to announce that the Meeting would be held in the Royal Institution.

Resolved, That the above communication be approved of.

Ordered, That the House Lists be filled up as follows.

President — The Duke of Somerset

Treasurer — Sir George Duckett Bart.

Secretary — Edmund Robt. Daniell Esqr.

Managers

Benjamin Bond Cabbell, Esqr.
Capt. John James Chapman,
Davies Gilbert, Esqr.
Sir Henry Halford, Bart.
Henry Hallam, Esqr.
Wm. Hamilton, Esqr.
The Marquis of Lansdowne,
The Hon: W. F. S. Ponsonby,
Lord John Russell, M.P.
Edmund S. Halswell, Esqr.
George Moore Esqr.
Dr. Whitlock Nicholl,
Wm. Hasledine Pepys, Esqr.
Charles Pilgrim, Esqr.
William Pole, Esqr.

402

1831.

Visitors
Thomas Tyringham Bernard, Esqr.
Sir Gilbert Blane Bart
Robert George Clarke Esqr.
Rt Hon: Reginald Pole Carew
John Frederic Daniell, Esqr
John Fuller, Esqr.
Dr Granville
Henry Gally Knight, Esqr
Lt Col Wm Martin Leake
Viscount Palmerston
Edward Adolphus Lord Seymour
Richard Horsman Solly, Esqr
Wm Sotheby Esqr.
Sir John Thos Stanley Bart
N. A Vigors, Esqr

Confirmed the 9th day of May 1831
Somerset Pres.

403

May 9.

At a Meeting of the Managers of the Royal Institution of Great Britain, held on the 9th of May, 1831

Present
His Grace The Duke of Somerset Prest in the Chair
George Moore, Esqr. —
William Pole, Esqr.
Captn Chapman,
Dr Whitlock Nicholl,
William Hamilton, Esqr.
Edmund S. Halswell, Esqr.
Charles H. Pilgrim, Esqr.
Edmund Robert Daniell Esqr. Secy.

Michael Faraday, Esqr. Director of the Laby

1 The minutes of the proceedings of the last Meeting were read and approved.

2. The Secretary reported that he had received from N. A Vigors, Esqr. Secretary of the Zoological Society, the following resolution which had been unanimously agreed to at the Anniversary Meeting of the Zoological Socy.

Theatre lent to Zool. Soc.

Resolved, That the thanks of this Meeting be given to the President and Managers of the Royal Institution for the use of the Theatre of their Establishment upon this occasion.
Zoological Society Bruton Street
30 April, 1831.

3. Presents received for the Library

404

1831.

E. W. Brayley, Esqr. presented a Treatise on "The utility of the knowledge of Nature."
I. H Curtis, Esqr. presented "Treatise on the Ear."
Dr Donati presented
Pomona Italiana,
Theoria della reproduzione Vegetale.

Ordered,
That the thanks of the Managers be returned for these presents.

3. Resolved, That the following Committees be appointed for the ensuing year.

Committee for regulating the Weekly Evening Meetings.

His Grace The President
Benjamin Bond Cabbell, Esqr.
Captn Chapman, R.A.
Edmund Robert Daniell, Esqr.
Edmund S. Halswell, Esqr.
George Moore, Esqr.
Whitlock Nicholl, M.D.
Charles H Pilgrim, Esqr.
William Pole, Esqr.
William Thomas Brande, Esqr.
Michael Faraday, Esqr.

Committee for arranging the Lectures
His Grace The President
Sir George Duckett Bart

405

May 9.

Benjamin Bond Cabbell, Esqr.
Captn Chapman, R. A
~~John George Children Esqr.~~
Edmund Robert Daniell, Esqr.
Edmund S. Halswell Esqr.
Whitlock Nicholl, M.D.
Charles H Pilgrim, Esqr.
William Pole Esqr.
William Hasledine Pepys
George Moore, Esqr.

Committee of Accounts.
His Grace the President
The Treasurer
The Secretary
Benjamin Bond Cabbell, Esqr.
George Moore, Esqr.
Dr Whitlock Nicholl
William Hasledine Pepys, Esqr.
Chas H Pilgrim, Esqr.
William Pole, Esqr.

Publication Committee.
His Grace The President.
Sir George Duckett, Bart.
Benjamin Bond Cabbell Esqr.
Edmund Robert Daniell, Esqr.
John Frederic Daniell, Esqr.
Captn Chapman R. A.
Edmund S. Halswell, Esqr.
William R. Hamilton, Esqr.
George Moore, Esqr.
Dr Whitlock Nicholl

Charles Pilgrim, Esq.
Dr. Roget
William Thos. Brande, Esq.
Michael Faraday, Esq.

Presents received for the Museum &c.

4. Portion of a Sponge stave dissected by the Larvæ of Insects – presented by Mr. Marsh of Woolwich.
Proof Engraving of Dr. Wollaston presented by William Skelton, Esq.
Engraved Portrait of Dr. Wollaston presented by G. H. Wollaston, Esq.
Portrait of D. Hervey and four other prints presented by John Marshall, Esq.
Ordered, That the thanks of the Managers be returned for these presents.

5. His Grace The President announced that he had nominated the following Vice Presidents for the ensuing year.

Vice Presidents. The Marquis of Lansdowne
Benjamin Bond Cabbell, Esq.
Captn. John James Chapman R.A.
Davies Gilbert, Esq. M.P.
George Moore, Esq.
Charles Pilgrim junr. Esq.

Confirmed June 6th 1831
George Moore V.P.

June 6.

At a Meeting of the Managers of the Royal Institution of Great Britain held on the 6th. of June, 1831

Present

George Moore, Esqr. V.P. in the Chair
Sir Henry Halford, Bart.
Davies Gilbert, Esq. M.P.
Sir George Duckett, Bart. Treasurer.
Henry Hallam, Esq.
William Hamilton, Esq.
Edmund S. Halswell, Esq.
William Pole, Esq.
William Hasledine Pepys, Esq.
Captn. John James Chapman, R.A.
Charles H. Pilgrim, Esq.
Dr. Whitlock Nicholl
Benjamin Bond Cabbell, Esqr.
Edmund Robt. Daniell, Esq. Secy.

William Thos. Brande, Esqr. Prof of Chemy.
Michael Faraday Esq. Director of the Laby.

1. The minutes of the proceedings of the last meeting were read and approved.

Repts. from Com. of Accts. 2. Read the following report from the Commee. of Accounts.

The Committee of Accounts report that they have examined the General Accounts since the last Meeting on the 4th. of April, and that they have found a balance in the hands of the Bankers of £493 . 15 . 11, and in

the hands of the Assistant Secretary of £59 . 15 . 10

They also report that they have examined the Laboratory Account and that they have found a balance in the hands of the Bankers of £232 . 12 . 8 and in the hands of the Assistant Secretary of £50 . 0 . 6.

They recommend that the following sums be paid on account of the Royal Institution

H. Greenwood on account of Tea & Coffee for the Weekly Evening Meetings	40 . 0 . 0
Messrs. Pringle for Upholstery in 1830	50 . 0 . 0
Bird for Bricklayers work in do.	4 . 19 . 6
Vulliamy for repairing & winding Clocks in ditto	2 . 12 . 6
Gas Light & Coke Company for a years lighting in ditto	5 . 12 . 0
Mortlock for China in ditto	5 . 1 . 4
Brown for Brushes in do.	3 . 2 . 0
Allnutt for Matting in do.	19 . 2
James Montgomery, Esq. for a Course of Lectures on Poetry	94 . 10 –
Mr. Webster for ditto on Geology	42 . 0 . 0
Joseph Fincher on account of the Royal Institution	150 . 0 . 0
	398 . 16 . 6

The Committee of Accounts also report that agreeably to the resolution of the Managers of the 4th. of April last requesting them to make the

Coals. necessary arrangements for supplying the Royal Institution with Coals, the Contract with Messrs. Dickins & Co. having ceased; they had applied to the following persons from whom they had received tenders to supply the Royal Institution with Coals.

H. Golding, Duke Street, at 38s. per Chaldron including Metage & Shooting and every expense, also the usual Ingrain

W. Edwards of Middle Wharf Scotland Yard ⅌ Chald. at 39s. including Metage & Shooting

Mr. Wyburd at Bryanstone St. at 38s. ⅌ Chaldron of 36 Imperial Bushels

James & Felton Belvidere Wharf at 37s. ⅌ Chald. delivered free of every expense.

Mr. Hill of the firm of Duke & Hill stating that their business is confined to the disposal of Coals to Coal Merchants, but they should have much pleasure in recommending a highly respectable House who would furnish the Institution with the best Coals on the lowest possible terms.

And that they had received a letter from Messrs. Dickins & Co. hoping to have a continuance of their favours as they had had the honour of serving the Royal Institution with Coals upwards of 25 years and pledging themselves to supply the Institution as low as any other house in the Trade.

The Committee further report that they had directed Mr. Fincher to ascertain from Messrs. Dickins at what price they would supply the Royal Institution with the best Coals.

Resolved, That the above report be approved

of; And signed an Order for payment of the foregoing sums.

3. Mr. Ritchie elected Prof. of Nat. Phil. 378

The Revd. William Ritchie having been elected to the Professorship of Natural Philosophy in the Royal Institution with a Salary of £50 ℔ Ann.

Resolved, That Mr. Ritchie be paid his Salary of £50 for the current year, including the remuneration to be paid for the Course of lectures on Natural & Experimental Philosophy delivered by him before Easter. And signed an Order on the Treasurer accordingly.

4. Resoln. of Visitors ~~re~~ Payment of Salary of Profr. of Chemistry from Lab.y Fund ordered.

Read the resolution of the Visitors as entered in the proceedings of the Managers of the 11th of April last recommending that the Salary of the Professor of Chemistry should in future be charged to the account of the Laboratory Fund.

Resolved, That the above recommendation of the Visitors be approved of; And that the Salary of the Professor of Chemistry for the present year be paid out of the surplus Income of the Laboratory Fund after discharging the usual demands payable out of that Fund.

5. Lectures – Mr. Willis.

Mr. Faraday having stated to the Board that Mr. Willis would be glad to extend his Course of Lectures on Sound to two or three additional Lectures.

Resolved, That Mr. Halswell be requested to communicate with Mr. Willis upon the subject, and to ascertain what remuneration he would require for the additional Lectures.

7. No. of Lectures.

A suggestion having been made by the Secretary that it might be desirable that some of the Courses of Lectures should be given at the rate of two or three lectures in each week.

Resolved, That this suggestion be referred to the consideration of the Lecture Committee

8. Mr. Atherstone

Read a letter from Mr. Atherstone proposing to give a Course of Lectures on Poetry.

Referred to the consideration of the Committee of Lectures.

Presents received.

9. George Bennett, Esqr. presented to the Museum

A Model of a Malay Prow.

A Specimen of Fragments of Shells cohering together found on the beach at the Island of Ascension

Native Cloth of the Sandwich Islands manufactured from Bark.

Native Cloth of the Island of Tahiti &c &c.

Specimen of the Bark from which the Cloth is prepared. &c &c.

Mr. Bollaert presented various Articles from Terra del Fuego.

An Engraving of Mr. Winsor presented by his Son J. A. Winsor, Esqr.

Sir Gilbert Blane, Bart. presented

Reflections on the present Crisis.

Hercules Sharpe, Esqr. presented

Memoir of Brass Crosby, Esqr.

The Royal College of Surgeons presented Catalogue of their Library

Catalogue of the contents of the Museum part 3.

Ordered, That the thanks of the Managers be returned for these presents.

Resolved, That the following Committee be appointed to ascertain the amount of expense likely to be incurred in painting the Lecture Room; and also to consider whether it will be advisable to make any and what alterations in the method of warming and lighting the same.

Resolved, That such Committee do consist of

His Grace the President
Mr. George Moore
Mr. Pole
Mr. Pepys
The Secretary
Mr. Professor Brande
& Mr. Faraday

Donation by D. Gilbert, Esq.

Mr. Davies Gilbert having conditionally relinquished his claim on the Royal Institution of the sum of £100 advanced by way of loan to the Royal Institution.

Resolved, That the thanks of the Managers be returned to Mr. Davies Gilbert for this mark of his liberality towards the Royal Institution.

Confirmed the 4th day of July 1831

Somerset Pres

July 4. **At a Meeting of the Managers of the Royal Institution of Great Britain, held on the 4th of July, 1831.** 1831.

Present

His Grace The Duke of Somerset. Presᵗ in the Chair
Sir George Duckett, Bart. Treasurer.
George Moore, Esq. V.P.
Captⁿ. J. James Chapman, R.A.
Davies Gilbert, Esq. M.P. V.Pres.
Chas. H. Pilgrim, Esq.
William Pole, Esq.
William Hamilton, Esq.
Edmund Robt. Daniell, Esq. Secy.
Michael Faraday, Esq. Director of the Laby.

1. The minutes of the proceedings of the last meeting were read and approved.

Repᵗ. from Com. of Accᵗˢ. 2. Read the following report from the Committee of Accounts.

The Committee of Accounts report that they have examined the General Accounts since the last meeting and that they have found a balance in favour of the Royal Institution of £191.10.7.

They recommend that the following sums be paid on account of the Royal Institution.

J. Lindley, Esq. for a Course of lectures on Botany	21.0.0
Michael Faraday, Esq. for a Course of lectures	52.10.–
Dr. Edmund Clark Sundries for the Friday Evening Meetgs.	5.0.0
	78.10.–

July 4.

Brᵗ forward	78.10.–
C. Bolton for attending as Waiter at the Friday Evening Meetings	2.17.–
Joseph Fincher on account of the Royal Institution	100.0.0
	181.7.0

Resolved, That the Report be approved of; and signed an Order on the Treasurer accordingly.

Coals. The Committee of Accounts further report that they had received ~~the following~~ a letter from Messrs. Dickins & Co. in answer to their enquiries made by the Assistant Secretary, proposing to supply the Royal Institution with any quantity of the best Coals they may require, within two months with the usual Ingrain; and also free from all charges of Metage, Shooting, &c at 39s. per Chaldron; the account to be paid at the end of the quarter; or to supply any number of Tons of the same sort of Coals at 35s. per Ton, clear of all expenses. And that the Committee had ascertained that in the tender made by Messrs. James and Feltoe of supplying the Royal Institution with Coals at 37s per Chaldron, including the usual expenses, the Ingrain was not included; and that in Mr. Wyburd's Contract for 38s per Chaldron, the Ingrain was not included; but that in Mr. Edwards tender of 39s per Chaldron the Ingrain was included.

Resolved, That the above report be approved of

1831.

Resolved, That Messrs. Dickins & Co's Contract to supply the Royal Institution with Coals at 39s. per Chaldron be approved of; And that they be allowed to supply the Institution with Coals on the terms before stated; subject to any alteration that may be made in the new Act of Parliament in favour of the price of Coals.

Presents received.

3. The Revd. G. Peacock presented A Treatise on Algebra
A Hayward, Esq. presented "The Vocation of our age for Legislation."
Thomas Alcock, Esq. presented Travels in Russia &c.
J. G. Wilkinson Esq. presented Extracts from several Hieroglyphical subjects

Ordered, That the thanks of the Managers be returned for these presents.

Repairs. 4. Read the following report from the Committee appointed to ascertain the expense likely to be incurred in painting the Lecture Room.

The Committee report that they had directed Estimates to be made by Mr. Parker and Mr. Royce of the expense of painting the Lecture Room, including a wood moulding to the Pediment, which amounted together to the sum of £58-2-5.

Resolved, That the above report be adopted; And that the Painting & Repairs be proceeded with forthwith.

July 4.

Copy (by W. Pickersgill, jun) of Portt. of Sir H. Davy. —Frame presented by Mr. Faraday. 5. The Secretary reported that agreeably to the resolution of the General Meeting of the 7th of March, directing the Managers to have a copy made by a competent Artist of the Picture of Sir Humphry Davy, by Sir Thomas Lawrence, in the possession of the Royal Society for the purpose of being put up in the Royal Instit.; Mr. W. Pickersgill junr. had completed the Picture the remuneration of which amounted to 25 guineas. And that Mr. Faraday had had the Picture placed in an elegant gilt frame which he had presented to the Royal Institution.

Resolved, That the thanks of the Managers be returned to Mr. Faraday for his liberal present; And that the sum of 25 guineas be paid to Mr. W. Pickersgill junr. for the Picture.

6. Ordered, That Mr. Faraday's sitting room be painted and whitewashed, agreeably to the resolution of the Visitors of the 2d of April last.

Life Subscriber. 7. In pursuance of a nomination under the Act of Parliament by the late Wm. Luxmoore, Esqr. his Grand Nephew Coryndon Henry Luxmoore of Albany, was admitted a Life Subscriber to the Royal Institution.

Committee for arranging Portraits, &c. 8. Resolved, That a Committee be appointed to take into consideration the hanging of the

418 1831.

Pictures and placing the Busts in the Royal Institution consisting of the following Managers.

His Grace The President
Mr. Pole
Mr. Moore
Mr. Halswell
Mr. Davies Gilbert
Mr. Pilgrim
Sir George Duckett
Mr. Hamilton
The Secretary.

And that a Meeting of the Committee be appointed for Wednesday next at Three o'Clock.

Loan of Piano-fortes 9. Mr. Faraday stated that Messrs. Broadwood had supplied the Royal Institution with two Piano-fortes during the whole of Mr. Willis' Lectures.

Resolved, That the thanks of the Managers be returned to Messrs. Broadwood for the Loan of the Piano-fortes.

Holyday to Housekeeper. 10 Mrs. Greenwood having requested leave of absence for a short time during the recess. Leave was granted.

11. The following persons having been duly recommended were admitted as Annual Subscribers to the Royal Institution for the year 1831.

July 4 419

John F. Foxhall, Esqr. 71 Baker Street
Edwin Hill Handley, Esqr. 6 Grays Inn Square

Confirmed 1st day of August 1831
Somerset Pres.

420 1831.

Aug. 1.

At a Meeting of the Managers of the Royal Institution of Great Britain held on the 1st of August, 1831

Present

His Grace The Duke of Somerset Pres. in the Chair
George Moore Esqr. Vice President
Dr. Whitlock Nicholl
Edmund Robt. Daniell, Esqr. Secy.

Michael Faraday, Esqr. Director of the Laby.

1. The minutes of the proceedings of the last meeting were read and approved.

Rept. from Com. of Accts. 2 Read the following report of the Committee appointed to take into consideration the hanging of the Pictures and placing the Busts in the Royal Institution.

The Committee report that they consider the best place for hanging the Pictures, is the front wall of the Staircase, one in each pannel. And they recommend the Busts to be placed on Pedestals on the Landings of the Staircase.

Resolved, That this report be approved of, and adopted.

3. Presents received.

The Society of Antiquaries presented Archaeologia Vol. 23.

James Yates, Esqr. presented Remarks on the formation of Alluvial deposits

Aug. 1. 421

The Royal Humane Society presented The 57th. Report of the Society.

The Royal College of Surgeons presented Catalogue of the Contents of the Museum pt. 5.

An engraved Portrait of Davies Gilbert, Esqr. M.P. presented by himself.

Ordered, That the thanks of the Managers be returned for these presents.

4 Resolved, That Mr. Willis be allowed the sum of Ten guineas for his two additional lectures on Sound.

5. Signed an Order for an advance to the Assistant Secretary of £350.

Adjourned to Monday the 7th of November

Confirmed 7th Nov/31 HHamilton

422

1831.

Nov 7. At a Meeting of the Managers of the Royal Institution of Great Britain, held on the 7th of Nov.r 1831

Present

William Hamilton, Esq.r in the Chair.
Henry Hallam, Esq.r
George Moore, Esq.r V.P.
Dr Whitlock Nicholl,
Wm Hasledine Pepys, Esq.r
Benjamin Bond Cabbell, Esq.r V.P.
Michael Faraday Esq.r Direct.r of the Lab.y

1. The minutes of the proceedings of the last meeting were read and approved.

Repayment of Loan requested. 2. Read a letter from Sir Gilbert Blane Bart. requesting the repayment of the £100 advanced by him to the Royal Institution by way of loan, as the extended term of four years had now expired.

Resolved, That the consideration of the above letter be postponed till the next meeting.

Loan of Lamps. 3. Read a letter from Mr Parker of Argyll Street, addressed to Mr Faraday, requesting the Managers would allow a couple of his fountain lamps to be placed on the table of the Institution in exchange for two which he had for sometime past lent to the Royal Institution; and proposing to charge two guineas each for the exchange, the Lamps to be the property of the Institution.

Nov. 7. 423

Resolved, That Mr Parker be thanked for the loan of his two lamps; and that his liberal proposal be acceded to.

Presents received.

4. John Ranking, Esq.r presented
The Conquest of Mexico and Peru with an Appendix.
Frederick Madden, Esq.r presented
Privy Purse expenses of Princess Mary.
Dr George Johnston presented
The Flora of Berwick upon Tweed
The Cambridge Philosophical Society presented
Transactions of the Society 3 vol.s and first part of vol 4.
The Rev.d H Barry presented
Cæsar and the Britons,
Mr Brande presented
Elements of Chemistry,
Mr Faraday presented
"On a peculiar class of Acoustical figures.
Mr Dale, Butler to Lady Warburton, presented
Wasps' Nests.

Ordered, That the thanks of the Managers be returned for these presents.

Arrangem.t of Friday Ev.s 5. Mr Faraday Sec.y to the Committee for regulating the Weekly Evening Meetings, stated that he had filled up the list of the Friday Evening Meetings for the ensuing Season as follows.

424

1831.

January 20th – 27th
February 3d. 10th. 17th. 24th
March 2. 9. 16. 23. 30
April 6. 13 –
May 4. 11. 18. 25th
June 1. 8.

Resolved, That the above arrangement be approved of.

Adjourned to Monday the 21st Instant at Three o'Clock.

Confirmed Nov 21. 1831
George Moore V.P.

425

Nov. 21. At a Meeting of the Managers of the Royal Institution of Great Britain, held on the 21st. of Novem.r 1831

Present

George Moore, Esq.r V.P. in the Chair.
Benjamin Bond Cabbell, Esq.r V.P.
William Pole, Esq.r
Wm Hasledine Pepys, Esq.r
Edmund R. Daniell, Esq.r

Michael Faraday, Esq.r Director of the Lab.y

1. The minutes of the proceedings of the last meeting were read and approved.

On Repayment of Loan. 2. The application from Sir Gilbert Blane Bart. postponed from the last meeting, for the repayment of the £100 advanced by him to the Royal Institution having been taken into consideration, the usual notice to the General Meeting to empower the Managers to sell out so much of the Stock invested on account of the Sinking Fund as will be sufficient to pay off the above sum of £100, was prepared and signed by the Managers present.

3. Read the following report from the Comm.e of Lectures.

The Committee beg to recommend to

On Subscriptions to Lectures.

the Managers that the terms of subscription to the Lectures in the next Season be as follows. vizt

Gentlemen as well as Ladies to be admitted to the Lectures & Museum on payment of Two guineas each; or One guinea to each Course. ✕

And that the Sons and Daughters of Members under the age of 21, be admitted to all the Lectures and Museum on an annual payment of One guinea each.

The Committee further report that they have made arrangements with Mr J. Rennie Prof^r of Zoology Kings College London to deliver a Course of Juvenile lectures on Zoology during the Christmas Vacation.

Resolved, That the above Report be approved of, and adopted.

Position of Pictures.

4. It having been represented to the Board that the situation proposed for the Pictures of Sir H. Davy, Mr Brande & Mr Faraday by the resolution of the 1st of August, might be injurious to the Pictures,

It was resolved, That the resolution of the 1st of August be reconsidered; And that for the present the Pictures be suspended in the Ante-room

Ordered, That the plates of the tickets of admission to the Institution be repaired.

Confirmed Dec 5. 1831
George Moore V.P.

Dec. 5.

At a Meeting of the Managers of the Royal Institution of Great Britain, held on the 5th of December, 1831,

Present

George Moore, Esqr. V.P. in the Chair
Benjamin Bond Cabbell, Esqr. V.P.
William Hasledine Pepys, Esqr
William Pole, Esqr
Henry Hallam Esqr
Edmund S. Halswell Esqr
Dr Whitlock Nicholl
Edmund Robt Daniell Esqr Sec^y

1. The minutes of the proceedings of the last Meeting were read and approved.

Rep^t from Comm. of Acc^ts

2 Read the following report from the Committee of Accounts.

The Committee of Accounts report that they have examined the General Accounts since the last Meeting on the 29th of June; And that they have found a balance in the hands of the Bankers of £132. 11. 9; And in the hands of the Assistant Secretary of £47. 4 –

They also report that they have examined the Laboratory Account since the Meeting on the 1st of June and that they have found a balance in favour of that account of £166. 18. 10.

They recommend that the following sums be paid.

H Greenwood the balance of her Account for Tea & Coffee for the Weekly Evening Meetings	47. 13. 7
2 quarters Assessed taxes to Mich^s.	62. 4. 6
ditto Parochial rates to d^o:	58. 6. 8
Mrs Stert for Stationery	32. 16. 7
Joseph Fincher on account of the Royal Institution	150. 0. 0
Hedges for Candles for Mr Singer	3. 11 –
	354. 12. 4

And an advance to the Assistant Secretary of £100 on account of the Laboratory Fund.

Resolved, That the above report be approved of; and signed Orders accordingly

Portrait of Count Rumford presented.

3. Read a letter from Mr Royce stating that the Countess of Rumford had requested him to inform the Managers that she has a Portrait of her Father, half length in oil, which had been executed on the Continent and esteemed a great likeness, and if the Managers would accept of it and place it in the Royal Institution, the Countess

Dec. 5.

would feel great pleasure in presenting the Portrait to the Managers.

Resolved, That the thanks of the Managers be returned to the Countess of Rumford for her obliging proposal which they shall feel most happy in accepting.

Ordered, That the above communication be reported to the General Meeting.

Presents received.

4. Dr Marshall Hall presented a paper On the circulation of the Blood.
Reginald Orton, Esqr presented Essay on the Cholera of India,
C. P. Cooper, Esqr presented Lettres sur la Cour de la Chancellerie
Michael Faraday, Esqr presented a paper On Acoustical figures,

Ordered, That the thanks of the Managers be returned for these presents.

5 Prepared a Notice for the alteration of Chap. II Art. 18 of the Bye laws, by introducing in line 3, the following words, "or any Codicil thereto attested by two or more witnesses."

6. Ordered, That an interleaved copy of the Bye laws with the alterations which have been made since they were last printed be prepared for the use of the Managers.

480

1831.

Adjourned to 6th Jany 1832

Confirmed 6 Jan 1832
Benjn. Bond Cabbell V.P.

431

1832
Jan. 6.

At a Meeting of the Managers of the Royal Institution of Great Britain, held on the 6th of January, 1832;

Present

Benjamin Bond Cabbell Esqr. V.P. in the Chair.
William Pole, Esqr.
George Moore, Esqr. V.P.
William Hamilton, Esqr.
Dr. Whitlock Nicholl,
Edmund Robt. Daniell, Esqr. Secretary.
Michael Faraday, Esqr. Director of the Laby.

1. The minutes of the proceedings of the last meeting were read and approved.

Presents received.

2. Mr. Sadler presented a series of samples of the strata passed through in boring a well at South Lambeth

Mr. Bollaert presented a series of Geological Specimens from South America.

Ordered,

That the thanks of the Managers be returned for these presents.

on Mr. Fuller's Medals

3. Read the minute of the proceedings of the Managers of the 21st of April 1828 relating to Mr. Fuller's Gold Medal, which he had appropriated as a reward for Chemical discoveries, and which he had directed should be awarded every two years on his birth-day (the 20th. of February,) to such person,

432

1832

~~being a Member of the Royal Institution~~, as ~~whom~~ the Managers together with the Professor of Chemistry should consider to have merited the distinction.

Read also the minute of the proceedings of the Managers of the 1st. of February, 1830, relating to the means which had been adopted for carrying Mr. Fuller's intention into effect on his Birth-Day in February 1830.—

Resolved, That as the present is the year appointed for nominating a suitable person to receive the honour of Mr. Fuller's Gold Medal, that a Committee be appointed consisting of

His Grace The President.
Mr. Davies Gilbert.
Dr. Whitlock Nicholl.
Mr. Wm. R. Hamilton.
Mr. W. H. Pepys.
Mr. Professor Brande
& Mr. Faraday.

to confer with

Mr. Charles Hatchett
Mr. Children
& Mr. J. Frederic Daniell

to point out such person as they may think ~~[illegible]~~ to merit the distinction.

confirmed, February 6th
J. J. Chapman V.P.

433

Feb 6

At a Meeting of the Managers of the Royal Institution of Great Britain, held on the 6th. of February, 1832,

Present

Capt. J. J. Chapman V.P. in the Chair.
Sir George Duckett, Bart. Treasurer.
Wm. R. Hamilton, Esqr.
George Moore, Esqr. V.P.
Edmund S. Halswell, Esqr.
Edmund Robt. Daniell, Esqr. Secy.

Michael Faraday, Esqr. Director of the Laby

1. The minutes of the proceedings of the last meeting were read and approved.

Repts. from Com. of Accts.

2. Read the following report of the Committee of Accounts.

The Committee of Accounts report that they have examined the General Accts. since the last meeting on the 5th. of Decr. and that they have found a balance in the hands of the Bankers of £546..19..8 and in the hands of the Assistant Secretary of £23..11..1.

They also report that they have examd. the Laboratory Account since the last meeting and that they have found a balance in favour of that Account of £188..16..2.

They recommend that the following

1832

outstanding bills of the year, 1830, be paid, viz

Dickins & Co for Coals	92.0.0
Gardner for Oil	89.14.6
Parker for Painting &c	22.4.4
Feetham for Ironmongery	14.16.9
Nicol for Printing	79.0.0
Snell for Furniture	10.16.9
Mr Colburn for Copies of Mr Brande's Journal presented to Scientific and Literary Instit.n	29.8.0
	338.0.4
& to the Assistant Secretary on account of the Institution	100.0.0
	438.0.4

And an advance to the Assistant Secretary on account of the Laboratory of £100.

Resolved, That the above report be approved of; and signed Orders accordingly.

£100 advance to Patrons. 3. Read an application from the Patrons of the Library for an advance of £100 on account of the Library as entered in their proceedings of the 1st Instant.

Resolved, That an advance be made to the Patrons of the Library of the sum of £100 and signed an Order accordingly.

4. The Assistant Secretary reported that

Feb 6.

Donation by J Pepys, Esq. John Pepys Esq. M.R.I. had presented a donation of One hundred Pounds to be applied to the general purposes of the Royal Institution.

Ordered, That the above communication be reported to the General Meeting this day.

Presents received

5 H. Lloyd, Esq. presented Englische Sprachlehre für Deutsche

John Gorton, Esq. presented Topographical Dictionary of Great Britain and Ireland vol. 2.

Robert Hannay, Esq. presented History of the representation of England

Revd W. D. Conybeare presented Inaugural Address.

Mr M. A. Schless presented Ehrenberg Beitrage zur Kentniss der Infusionsthierchen 4to 1830.

William Pole, Esq. presented Eight Etchings from Medals by a new process.

Mr Francis Giles presented A Fossil Shell, and a Crystal of Sulphate of Barytes.

Ordered, That the thanks of the Managers be returned for these presents.

On Talking in Reading rooms 6 The Secretary stated that it had been represented to him by several Members of the Institution that the practice of loud

1832.

conversation in the Reading Rooms of the Royal Institution alluded to in the Report of the Managers of the 14th of March 1831 was still persevered in by certain Individuals frequenting those Rooms, particularly the Newspaper Room (the names of which Individuals had been mentioned to him) and that several Members had declared that it is their intention to withdraw from the Institution unless some steps are taken to put a stop to the interruption which they experience.

Resolved, That the following report be laid before the General Meeting this day.

The Managers beg again to call the attention of the Members of the Royal Institution to the practice alluded to in their Report of the 14th March, 1831, and to express their regret at having to state to the Meeting that notwithstanding the discussion which took place upon the subject at the General Monthly Meeting of the 7th of March and the Notices which were in consequence put up in the Reading Rooms, the practice referred to still continues to a very great extent; and that they have received several communications from Members expressing their intention of withdrawing from the Institution unless some steps were put a stop to the annoyance complained of.

The Managers are unwilling to refer personally to the Individuals who have been

Feb 6.

pointed out to them as most in the habit of indulging in the practice referred to, as they hope those individuals will be induced whilst in the Reading Rooms of the Institution to hold their private conversations in a manner less obnoxious to the other Members and Subscribers. Should however their hopes in this respect not be realized, the Managers will feel themselves under the painful necessity of reporting the names of the persons alluded to, to the General Meeting. And they entertain no doubt but that the Members of the Institution will empower them to adopt such measures as may enable them to preserve that order and decorum which ought to distinguish the Reading Rooms of a Scientific and Literary Institution.

7. Read the following report from the Comm.e of Lectures

Lectures. Mr Webster. The Committee report that they have made arrangement with Mr Webster to deliver a Course of eight Lectures on Geology and to remunerate him with five guineas a lecture.

Dr Clark. And that they have accepted a proposal made by Dr Edmund Clark to deliver a Course of eight lectures On the Evidence of Design in Creation, and to remunerate him with five guineas a lecture.

Mr B Smart. And that they have made arrangements with Mr B. H. Smart to deliver a Course of

438

1832

Twelve Lectures illustrative of the English Drama at and before the time of Shakspeare and to remunerate him with 50 guineas for the Course.

And that they had made the following arrangements for the Lectures to be delivered after Easter.

Mr Willis. Mr. Robert Willis, M.A. to deliver a Course of Lectures on Sound on the same terms as during the last Season

Mr G. Burnett. Mr Gilbert Burnett to deliver a Course of Lectures on Vegetable Utilities, at five guineas a lecture.

Mr J. Britton. Mr. John Britton to deliver a course of six Lectures on the Architectural Antiquities of the middle ages: the terms of remuneration to be five guineas a lecture.

Mr Faraday. Mr. Faraday to deliver a Course of lectures, the subject of which would be announced on a future occasion.

8. The following persons having been duly recommended were admitted as Annual Subscribers for the year 1832

Robert Ainslie, Esqr. 8 Park St. Westmr.
George Cornewell Lewis, Esqr. 17 Henrietta St. Cavendish Sqr.
Captn. Orme, Edith Weston Rutlandshire.
Luke Trapp Flood, Esqr. Belle Vue lodge Chelsea
George Vernon Colebrooke, Esqr. 30 Argyll St.

Feb. 6 439

Robert Mark Walt Ormsby, Esqr. 33 Dean St. Soho.
Thomas Nettleshipp, Esqr Lower Seymour St.
Archibald Smith, Esqr. 27 Cumberland Terrace

Adjourned to the 20th Instant at Three o'Clock.

confirmed J. J. Chapman
Vice President

440

1832

Feb 20. At a Meeting of the Managers of the Royal Institution of Great Britain held on the 20th of February, 1832

Present
Capt J. J. Chapman V.P. in the Chair
William Pole, Esqr.
Davies Gilbert, Esqr. M.P.
Benjamin Bond Cabbell, Esqr. V.P.

Michael Faraday, Esqr. Director of the Laby

1. The minutes of the proceedings of the last meeting were read and approved.

Presents received.

2 Dr. Granville presented
Catechism of Health
John Braddock, Esqr. presented
Memoir on Gunpowder.
Ordered, That the thanks of the Managers be returned for these presents.

3. In pursuance of the power which the Managers have under Chap VI Art 6, of the Bye Laws, of admitting to the Royal Institution, Ambassadors, Ministers of foreign Powers and States resident at the Court of Great Britain, and also Foreigners of elevated rank and of distinguished Literary attainments who may occasionally visit this Country.

Feb 20. 441

Hon. Admissn. Resolved, That the Rajah Mohun Roy, and Mr. Julian Niemcewitz, President of the Royal Society at Warsaw, have the above privilege conferred upon them; and that Cards of invitation from the Managers to the Weekly Evening Meeting be transmitted to them.

4 Hon. Admissions. The Chevalier de Barboza having requested an invitation from the Managers to the Weekly Evening Meetings and to the Library and Newspaper Rooms in favour of His Excellency the Chevalier de Lima, Portuguese Minister at this Court.

Resolved, That His Excellency the Chevr. de Lima be admitted under the above Article in Chap VI of the Bye laws. And that a Card of invitation from the Managers to the Weekly Evening Meetings be transmitted to him.

[illegible]
Chairman

442

1832

Mar 4

At a Meeting of the Managers of the Royal Institution of Great Britain, held on the 5th. of March, 1832,

Present

Sir George Duckett, Bart Treas^r. in the Chair
William R. Hamilton, Esqr.
William Pole, Esqr.
Chas. H. Pilgrim, Esqr.
W^m. Hasledine Pepys, Esqr.
Benjamin Bond Cabbell, Esqr. V. P.
Captn. J. J. Chapman
Edmund S. Halswell, Esqr.
D^r. Whitlock Nicholl,
Edmund Robt. Daniell, Esqr. Sec^y

Michael Faraday, Esqr. Director of the Lab^y

1. The minutes of the proceedings of the last meeting were read and approved.

Rep^t. of Comm. of Acc^ts. 2. Read the following report of the Comm^e. of Accounts

The Committee of Accounts report that they have examined the General Accounts since the last Meeting, And that they have found a balance in the hands of the Bankers of £209 . 16 . 4, and in the hands of the Assistant Secretary of £104 . 15 . 7.

They recommend that an Advance be made to the Assistant Secretary of £150.

443

Mar 5

Resolved, That the above report be approved of; And signed an Order on the Treasurer accordingly.

Presents received.

3. The Royal Society presented Transactions of the Society, 1830 p^t. 2 and 1831 p^t. 1 & 2.
Sir Thos. C. Banks B^t. presented
The Case of Alexander Earl of Selkirk
William Nicol, Esqr. M. R. I. presented
Critical Catalogue of rare and valuable Italian prints
& Tragedy of Hamlet
H^y. T. de la Beche, Esqr. presented
Geological Manual 2^d Ed^n.

Ordered,
That the thanks of the Managers be returned for these presents.

Lectures 4. Read a letter from The Marquis Moscati proposing to deliver a Course of lectures on Italian Poetry.

Resolved,
That it be referred to the Committee of lectures.

Indisposition of Mr. Brande. 5. Mr. Faraday reported that Mr. Brande had communicated to him, that in consequence of severe indisposition it would be impossible for him to deliver his Course of lectures before Easter.

444

1832

Resolved, That it be referred to the Committee of lectures to make the necessary alteration in the arrangement for the Lectures.

6. Ordered, That the Morning Advertizer and Ledger be discontinued.

Adjourned till after the General Meeting.

The Managers having reassembled.

Mr. Pole, in the Chair.

Resolved, That the report of the Managers ordered by the last General Meeting to be put up in the Newspaper Room, be now taken down.

Sale of Stock. Resolved, That the Treasurer be requested to sell out so much of the Stock invested on account of the Sinking Fund as will be sufficient to pay off the sum of £100 due to Sir Gilbert Blane Bart on bond from the Royal Institution And that he also be requested to pay the same to Messrs. Herries & Co. on receiving from them the bond and a receipt from Sir Gilbert Blane, which he has deposited with them.

Adjourned to the 19th. Instant

Somerset
Chairman

445

Mar. 19.

At a Meeting of the Managers of the Royal Institution of Great Britain, held on the 19th. of March, 1832

Present

His Grace The Duke of Somerset Pres^t. in the Chair
D^r. Whitlock Nicholl
Benjamin Bond Cabbell, Esqr. V. P.
Charles Pilgrim jun^r. Esqr.
Henry Hallam, Esqr.
William R. Hamilton, Esqr.
George Moore, Esqr. V. P.
Edmund Robt. Daniell, Esqr. Sec^y

Michael Faraday, Esqr. Director of the Lab^y.

1. The minutes of the last meeting were read and approved.

Power of Sale. 2. Notice having been sent to the Managers agreeably to the Bye laws ~~agreeably to the Bye laws~~, the Seal of the Corporation was affixed to a Power for Sale of sufficient Stock to pay off the sum of £100 due to Sir Gilbert Blane, Bart.

The Power was countersigned by

Witnessed by	Somerset
Jos. Fincher Assist Sec^y	Benjn. Bond Cabbell
Thos. Williams Porter	Edmund R. Daniell

Portrait of Mr. Fuller. 3. The Secretary stated that Mr. Fuller had presented his Portrait in a handsome

gilt frame to the Royal Institution.

Resolved, That the thanks of the Managers be returned to Mr. Fuller for his valuable present.

Comm. for Lighting Theatre.

4. Resolved, That a Committee be appointed to examine and report upon the lighting of the Lecture room table and Ante-lecture room consisting of the following persons

Mr. George Moore
Captn. Chapman
Mr. Daniell
& Mr. Faraday

Lectures.... Mr. Gillt. Burnett.

5. Read the following report from the Committee of Lectures.

The Committee report that they had arranged with Mr. Gilbert Burnett to commence his lectures on Tuesday the 20th. Instant And that they had engaged with The Marquis Moscati to deliver a Course of Lectures on Italian Literature and to remunerate him with the sum of 30 guineas for the Course

Adjourned to Monday the 26th. Instant at Three o'Clock.

Confirmed 26th March 1832 Somerset Pres.

Mar. 26.

At a Meeting of the Managers of the Royal Institution of Great Britain, held on the 26th. of March, 1832

Present

His Grace The Duke of Somerset Presdt. in the Chair
George Moore, Esqr. V.P.
Benjamin Bond Cabbell, Esqr. V.P.
Captn. J. J. Chapman
Edmund S. Halswell, Esqr.
Charles H. Pilgrim, Esqr.
Wm. R. Hamilton, Esqr.
William Pole, Esqr.
Dr. Whitlock Nicholl
Edmund Robt. Daniell, Esqr. Secy.

1. The minutes of the proceedings of the last meeting were read and approved.

Award of Fuller Medal to M. Gay Lusac.

2. Read the following report from the Committee appointed by the resolution of the last meeting to nominate a proper person to receive Mr. Fuller's Gold Medal.

The Committee report that it is their unanimous opinion that Mr. Fuller's Gold Medal should be presented to Mons. Gay Lusac, an honorary Member of this Institution, for his various Chemical discoveries

Resolved, That the report of the Committee be approved of and adopted.

3. The Secretary laid before the Meeting

the following statement of the amount due to the Royal Institution from the Estate of Sir George Duckett & Co.

Acct. with Sir G. Duckett's Estate.

	£
On the General Account	59.15.10
Laboratory Account	69.9.0
Sinking Fund	11.11.1
And the sum of £63, being a Check from The Rt. Honble Sir R. Gordon, and a Check from Mr. Saml. Paynter of £9 which had been paid into the Assistant Secretary's private Account at Sir George Duckett's	72.0.0
	212.15.11
Jany. Dividends of the Sinking Fund omitted	11.11.1
	224.7.0

Messrs. Drummond appointed Bankers.

Resolved, That Messrs. Drummond be appointed the Bankers of the Royal Institution And that they be requested to open an account in the name of the Royal Institution of Great Britain.

Hon. Admissions.

On the application of The Chevalier de Barboza,

Mr. Rodrigo Magathaens
& Mr. Antonio Gomes

were admitted to the Royal Institution under Chap: VI Art 6 of the Bye laws, and to the Weekly Evening Meetings of the Members during the present year.

Resolved, That the letter of admission be

also sent to The Chevalier Augustus Willm. Schleigal.

His Grace the President reported that he had received a proposal from The Chevalier A.W. Schleigal to deliver a Course of Lectures in the Institution during the present Season.

Ordered, That the proposal of the Chevalier Schleigal be referred to the Committee of Lectures.

Bankruptcy of Sir G. Duckett & Co.

Resolved, That Mr. Lake the Solicitor of the Royal Institution be requested to take the necessary steps towards proving the debt due to the Royal Institution from the Estate of Sir George Duckett & Co. under the Commission of Bankrupt issued against that Firm.

Confirmed the 2nd day of April 1832
Somerset Pres.

1832

April 2

At a Meeting of the Managers of the Royal Institution of Great Britain, held on the 2.d of April, 1832

Present

His Grace The Duke of Somerset Pres.t in the Chair
George Moore, Esq.r V.P.
William Hamilton, Esq.
William Pole, Esq.
W.m Hasledine Pepys, Esq.
Benjamin Bond Cabbell, Esq.r V.P.
Capt.n John James Chapman V.P.
D.r Whitlock Nicholl
Cha.s H Pilgrim, Esq.
Michael Faraday Esq.r Director of the Laboratory

1\. The minutes of the proceedings of the last meeting were read and approved.

Rep.t of Com. of Acc.ts — 2 Read the following report from the Committee of Accounts.

The Committee of Accounts report that they have examined the General Accounts since the last Meeting and that they have found a balance in favour of the Royal Institution of £314 . 8 . 2 – viz.t £266 . 19 . 0 Receipts since the last Meeting, and the sum of £47 . 9 . 2, the balance in the Assistant Secretary's hands on the account of Payments In which sum of £266 . 19 . – is included £72 paid into the hands of the Bankers to

April 2

Private Acc.t of M.r Fincher — the private Account of M.r Fincher, which account had been kept by him, under the direction of the late Sir S. Bernard Morland, Bart when Treasurer.

The Committee recommend that M.r Fincher be ordered to retain the balance of £266 . 19 – in his hands on accounts of Receipts since the last Meeting and all future Receipts until further orders.

The Committee also recommend that the following sums be paid

To	M.r Greenwood on account of Tea & Coffee	£40 —
	M.r Webster, Lectures.	42
	A years Insurance	63 . 16 . 6

to be discontinued. — The Committee further recommend that in future all sums of money received by the Assistant Secretary & Collector on account of the Institution as well such as are received on account of y.e Library & Mineralog.l Fund and of the Laboratory as such as are received on the General Account of the Institution be paid into the Bankers of the Royal Institution to the Credit of the Treasurer And that the Treasurer of the Institution do at each Meeting of the Committee of Accounts pay over the sums due to the Library & Laboratory Funds to the Treasurer of such Funds respectively.

Resolved, That the above report be approved of.

1832

Directions to Bankers. — 3. Resolved, That Mess.rs Drummond be directed to honour the Drafts which may be drawn upon them on this account by the Treasurer for the time being, whose appointment will be duly notified to Mess.rs Drummond by the Secretary of the Royal Institution.

On Prof.r of Botany. — 4. Read a letter from M.r Gilbert Burnett respecting the Professorship of Botany and enquiring whether the Managers design to propose any Candidate, or have any objection that the office should be filled at the approaching Annual election, and ~~venturing to~~ requests them to favour him by their nomination.

Confirmed 21 April 1832
Benj.n Bond Cabbell V.P.

April 21.

At a Meeting of the Managers of the Royal Institution of Great Britain held on the 21.st of April, 1832

Present

Benjamin Bond Cabbell, Esq.r V.P in the Chair.
Capt.n J. J. Chapman.
George Moore, Esq.r V.P.
Charles H Pilgrim, Esq.
William R. Hamilton, Esq.
Edmund Rob.t Daniell, Esq.r Secy

1\. The minutes of the proceedings of the last meeting were read and approved.

On Prof.r of Botany — 2. Resolved, That no resolution having been made respecting M.r Burnett's application made at the last meeting, that the Assistant Secretary be directed to acquaint M.r Burnett that in consequence of his letter not having been submitted to them till within half an hour before the General Meeting when it was necessary to nominate the Professors, they were prevented from taking his proposal into consideration.

Lectures. — 3. Read a letter from The Marquis Moscati requesting either an hour in the day or Evening to fulfil the promise made to his Auditors, of delivering an extra Lecture On the Modern Arcadia and its Constitution.

Resolved, That the Marquis Moscati's proposal be accepted And that it be referred to the Secretary to make the arrangement.

454 1832 April 21. 455

Presents received.

4. The Committee of the Prison Discipline Society presented The eighth Report of the Committee.
Alex.r Gordon Esq.r presented Treatise on Locomotion.
The Geological Society presented Transactions of the Society vol 3 part 2.
Michael Faraday Esq.r presented Experimental Researches on Electricity.
Society of Arts presented Transactions of the Society p.t 2 of Vol 48.
John Fearn, Esq.r presented Tract on the human Sensorium.
David Maddock Esq.r presented "The Immortality of the Soul.
Ordered, That the thanks of the Managers be returned for these presents.

Resolved, That the House Lists be filled up as follows.

President. The Duke of Somerset F.R.S

Treasurer. William R Hamilton, Esq.r F.R.S.

Secretary Edmund Rob.t Daniell, Esq.r F.R.S

Managers

Richard Berens, Esq.r
Benjamin Bond Cabbell, Esq.r
Capt.n John James Chapman, R.A.
Hudson Gurney, Esq.r M.P. F.R.S
Edmund S. Halswell, Esq.r A.M.
William Varlo Hellyer, Esq.r
George Long Esq.r
George Moore Esq.r F.S.A. & F.L.S.
Whitlock Nicholl M.D. F.R.S &c &c
John Ayrton Paris, M.D. F.R.S
Charles Pelgrim jun.r Esq.r
William Pole, Esq.r F.R.S
Lord Prudhoe, F.R.S
Peter Mark Roget M.D. Sec.R.S.
George Watson Taylor, Esq.r M.P.

Visitors

Alfred Ainger, Esq.r
Sir Gilbert Blane Bart M.D. F.R.S
John George Children, Esq.r F.R.S & F.L.S.
Robert George Clarke, Esq.r
James William Farrer, Esq.r
John Fuller, Esq.r
Augustus B. Granville, M.D. F.R.S.
Richard Hollier, Esq.r
Octavius Morgan, Esq.r
Joseph Neeld, Esq.r M.P. F.S.A.
Lord Seymour.
Richard Horsman Solly, Esq.r F.R.S. & F.S.A.
William Somerville, M.D. F.R.S. Lond: & Edin:
William Sotheby, Esq.r F.R.S & F.S.A.
Rev.d George Thursby, A.M. F.R.S & F.L.S

Confirmed 7th May 1832

Somerset Pres.

INDEX

ILLUSTRANS
COMMODA
VITÆ